T0202170

COMPUTER SIMULATION OF LIQUIDS

Computer Simulation of Liquids

Second Edition

Michael P. Allen

Department of Physics, University of Warwick, UK
H. H. Wills Physics Laboratory, University of Bristol, UK

Dominic J. Tildesley

Centre Européen de Calcul Atomique et Moléculaire (CECAM),
EPFL, Switzerland

OXFORD
UNIVERSITY PRESS

OXFORD
UNIVERSITY PRESS

Great Clarendon Street, Oxford, OX2 6DP,
United Kingdom

Oxford University Press is a department of the University of Oxford.
It furthers the University's objective of excellence in research, scholarship,
and education by publishing worldwide. Oxford is a registered trade mark of
Oxford University Press in the UK and in certain other countries

© M. P. Allen and D. J. Tildesley 2017

The moral rights of the authors have been asserted

First Edition published in 1987
Second Edition published in 2017

Published in the United States of America by Oxford University Press
198 Madison Avenue, New York, NY 10016, United States of America

British Library Cataloguing in Publication Data
Data available

Library of Congress Control Number: 2017936745

ISBN 978–0–19–880319–5 (hbk.)
ISBN 978–0–19–880320–1 (pbk.)

Printed and bound by
CPI Group (UK) Ltd, Croydon, CR0 4YY

To Diane, Eleanor, Pauline, and Charles

Preface

In the years following the publication of the first edition, we have frequently discussed producing an updated version, and indeed have been nagged on many occasions by colleagues to do so. Despite its increasingly dated content, with quaint references to microfiche, magnetic tapes, and Fortran-77 language examples, the first edition has continued to sell well for three decades. This is quite surprising, bearing in mind the tremendous development of the field and the computer technologies on which it is based. To an extent, the material in our book has been complemented by the publication of other books and online resources which help to understand the underlying principles. Also, it is much easier than it used to be to find technical details in the primary literature, in papers, appendices, and supplementary information. New and improved techniques appear all the time, and the problem is almost that there is too much information, and too much rediscovery of existing methods. The widespread use of simulation packages has provided enormous leverage in this research field. There is much to gain by carefully reading the manual for your chosen package, and we strongly recommend it!

Nonetheless, it remains true that 'getting started' can be a significant barrier, and there is always the need to understand properly what is going on 'under the hood', so as not to use a packaged technique beyond its range of validity. Many colleagues have reaffirmed to us that there is still a need for a general guide book, concentrating on the strengths of the first edition: providing practical advice and examples rather than too much theory. So, we agreed that an updated version of our book would be of value. We intended to produce this many years ago, and it is a sad fact that the demands of academia and industry left too little time to make good on these aspirations. We wish to acknowledge the patience of our editor at Oxford University Press, Sönke Adlung, who has stuck with us over this long period.

Although the field has grown enormously, we resisted the temptation to change the title of the book. It was always focused on the liquid state, and this encompasses what are now known as complex fluids, such as liquid crystals, polymers, some colloidal suspensions, gels, soft matter in general, some biological systems such as fluid membranes, and glasses. The techniques will also be of interest outside the aforementioned fields, and there is no well-defined dividing line, but we try not to stray too far outside our expertise. Rather than give a long list in the title, we hope that 'Computer Simulation of Liquids', interpreted with some latitude, is still sufficiently descriptive.

The content of the book, although structured in the same way as the first edition, has changed to reflect the above expansion in the field, as well as technical advances. The first few chapters cover basic material. Molecular dynamics in various ensembles is now regarded as basic, rather than advanced, and we devote whole chapters to the handling of long-range forces and simulating on parallel computers, both of which are now mainstream topics. There are a few more chapters covering advanced simulation methods, especially those for studying rare events, mesoscale simulations (including coarse graining), and the study of inhomogeneous systems. Instead of concentrating some scientific examples in a single chapter, we have scattered them through the text, to illustrate still further what can be done with the techniques we describe. These examples very much reflect our personal preferences, and we have tried to resist the temptation to turn our book into a collection of scientific or technical reviews, so many otherwise suitable 'highlights' have been omitted. To give a balanced overview of such a huge field would probably be impossible and would certainly have resulted in a very different, and much larger, book. We have dropped material, when methods have been superceded (such as predictor–corrector algorithms), or when they were really of limited or specialized interest (such as the use of integral equations to extend correlation functions to longer distance).

The examples of program code which accompanied the first edition were first provided on microfiche, and later online, courtesy of Cornell University and CCP5. We continue to use such code examples to illustrate ideas in the text, and provide them online. We give the individual filenames, the first few lines of each example, and some guidance on usage, in the book. The full set of codes is available online at

http://www.oup.co.uk/companion/allen_tildesley

Although we stick to Fortran 2008 in the main, some online files are also provided in Python, to widen the accessibility. Some relevant programming considerations may be found in Appendix A.

We wish to reiterate our thanks to those who supported us at the start of our careers (see below) and we have many more people to thank now. J. Anwar, P. Carbone, J. H. Harding, P. A. Madden, S. C. Parker, M. Parrinello, D. Quigley, P. M. Rodger, M. B. Sweatman, A. Troisi, and M. R. Wilson all provided advice and/or encouragement during the early stages of writing. S. Bonella, P. J. Daivis, S. Khalid, P. Malfreyt, B. D. Todd, and R. Vuilleumier advised us on specific topics. G. Ciccotti, A. Humpert, and G. Jackson read and commented on a complete first draft. Any mistakes or misconceptions, naturally, remain our own responsibilities. Our colleagues over the years, at Bristol, Warwick, Southampton, Imperial College London, Unilever plc, and CECAM Lausanne, have also provided a stimulating working environment and a challenging intellectual atmosphere. MPA also wishes to acknowledge helpful study leave periods spent in Germany, at the Universities of Mainz and Bielefeld, and in Australia, at the Universities of Swinburne, Monash, and Deakin. DJT acknowledges an important and stimulating collaboration with the chemistry department at the Université Blaise Pascal, Clermont-Ferrand, France.

Our families have remained an ongoing source of support and inspiration. DJT thanks Eleanor for her unwavering encouragement, while MPA particularly wishes to thank Pauline and Charles, whose holidays frequently had to coincide with conferences and summer schools over the years!

Bristol   MPA
Lausanne   DJT
August 2016

From the Preface to the First Edition

This is a 'how-to-do-it' book for people who want to use computers to simulate the behaviour of atomic and molecular liquids. We hope that it will be useful to first-year graduate students, research workers in industry and academia, and to teachers and lecturers who want to use the computer to illustrate the way liquids behave.

Getting started is the main barrier to writing a simulation program. Few people begin their research into liquids by sitting down and composing a program from scratch. Yet these programs are not inherently complicated: there are just a few pitfalls to be avoided. In the past, many simulation programs have been handed down from one research group to another and from one generation of students to the next. Indeed, with a trained eye, it is possible to trace many programs back to one of the handful of groups working in the field 20 years ago. Technical details such as methods for improving the speed of the progam or for avoiding common mistakes are often buried in the appendices of publications or passed on by word of mouth. In the first six chapters of this book, we have tried to gather together these details and to present a clear account of the techniques, namely Monte Carlo and molecular dynamics. The hope is that a graduate student could use these chapters to write his own program.

Both of us were fortunate in that we had expert guidance when starting work in the field, and we would like to take this opportunity to thank P. Schofield (Harwell) and W. B. Streett (Cornell), who set us on the right road some years ago. This book was largely written and created at the Physical Chemistry Laboratory, Oxford, where both of us have spent a large part of our research careers. We owe a great debt of gratitude to the head of department, J. S. Rowlinson, who has provided us with continuous encouragement and support in this venture, as well as a meticulous criticism of early versions of the manuscript. We would also like to thank our friends and colleagues in the physics department at Bristol and the chemistry department at Southampton for their help and encouragement, and we are indebted to many colleagues, who in discussions at conferences and workshops, particularly those organized by CCP5 and CECAM, have helped to form our ideas. We cannot mention all by name but should say that conversations with D. Frenkel and P. A. Madden have been especially helpful. We would also like to thank M. Gillan and J. P. Ryckaert, who made useful comments on certain chapters, and I. R. McDonald who read and commented on the completed manuscript.

Books are not written without a lot of family support. One of us (DJT) wants to thank the Oaks and the Sibleys of Bicester for their hospitality during many weekends over the last three years. Our wives, Diane and Pauline, have suffered in silence during our frequent disappearances, and given us their unflagging support during the whole project. We owe them a great deal.

Bristol	MPA
Southampton	DJT
May 1986	

Contents

1

Introduction

1.1 A short history of computer simulation

What is a liquid? As you read this book, you may be mixing up, drinking down, sailing on, or swimming in, a liquid. Liquids flow, although they may be very viscous. They may be transparent or they may scatter light strongly. Liquids may be found in bulk, or in the form of tiny droplets. They may be vaporized or frozen. Life as we know it probably evolved in the liquid phase, and our bodies are kept alive by chemical reactions occurring in liquids. There are many fascinating details of liquid-like behaviour, covering thermodynamics, structure, and motion. Why do liquids behave like this?

The study of the liquid state of matter has a long and rich history, from both the theoretical and experimental standpoints. From early observations of Brownian motion to recent neutron-scattering experiments, experimentalists have worked to improve the understanding of the structure and particle dynamics that characterize liquids. At the same time, theoreticians have tried to construct simple models which explain how liquids behave. In this book, we concentrate exclusively on atomic and molecular models of liquids, and their analysis by computer simulation. For excellent accounts of the current status of liquid science, the reader should consult the standard references (Barker and Henderson, 1976; Rowlinson and Widom, 1982; Barrat and Hansen, 2003; Hansen and McDonald, 2013).

Early models of liquids (Morrell and Hildebrand, 1936) involved the physical manipulation and analysis of the packing of a large number of gelatine balls, representing the molecules; this resulted in a surprisingly good three-dimensional picture of the structure of a liquid, or perhaps a random glass, and later applications of the technique have been described (Bernal and King, 1968). Assemblies of metal ball bearings, kept in motion by mechanical vibration (Pieranski et al., 1978), have been used as models of granular materials and show some analogies with molecular systems (Olafsen and Urbach, 2005). Clearly, the use of large numbers of macroscopic physical objects to represent molecules can be very time-consuming; there are obvious limitations on the types of interactions between them, and the effects of gravity are difficult to eliminate. However, modern research on colloidal suspensions, where the typical particle size lies in the range 1 nm–1000 nm, with the ability to manipulate individual particles and study large-scale collective behaviour, has greatly revitalized the field (Pusey and van Megen, 1986; Ebert et al., 2009; Lekkerkerker and Tuinier, 2011; Bechinger et al., 2013).

The natural extension of this approach is to use a mathematical, rather than a physical, model, and to perform the analysis by computer. It is now over 60 years since the first computer simulation of a liquid was carried out at the Los Alamos National Laboratories in the United States (Metropolis et al., 1953). The Los Alamos computer, called MANIAC, was at that time one of the most powerful available; it is a measure of the continuing rapid advance in computer technology that handheld devices of comparable power are now available to all at modest cost.

Rapid development of computer hardware means that computing power continues to increase at an astonishing rate. Using modern parallel computer architectures, we can expect to enjoy exaflop computing by 2020 (an exaflop is 10^{18} floating-point operations per second). This is matched by the enormous increases in data storage available to researchers and the general public. Computer simulations, of the type we describe in this book, are possible on most machines from laptops to continental supercomputers, and we provide an overview of some opportunities with respect to architecture and computing languages, as they relate to the field, in Appendix A.

The very earliest work (Metropolis et al., 1953) laid the foundations of modern Monte Carlo simulation (so-called because of the role that random numbers play in the method). The precise technique employed in this study is still widely used, and is referred to simply as 'Metropolis Monte Carlo'. The original models were highly idealized representations of molecules, such as hard spheres and disks, but, within a few years, Monte Carlo (MC) simulations were carried out on the Lennard-Jones interaction potential (Wood and Parker, 1957) (see Section 1.3). This made it possible to compare data obtained from experiments on, for example, liquid argon, with the computer-generated thermodynamic data derived from a model.

A different technique is required to obtain the dynamic properties of many-particle systems. Molecular dynamics (MD) is the term used to describe the solution of the classical equations of motion (Newton's equations) for a set of molecules. This was first accomplished, for a system of hard spheres, by Alder and Wainwright (1957; 1959). In this case, the particles move at constant velocity between perfectly elastic collisions, and it is possible to solve the dynamic problem without making any approximations, within the limits imposed by machine accuracy. It was several years before a successful attempt was made to solve the equations of motion for a set of Lennard-Jones particles (Rahman, 1964). Here, an approximate, step-by-step procedure is needed, since the forces change continuously as the particles move. Since that time, the properties of the Lennard-Jones model have been thoroughly investigated (Verlet, 1967; 1968; Johnson et al., 1993).

After this initial groundwork on atomic systems, computer simulation developed rapidly. An early attempt to model a diatomic molecular liquid (Harp and Berne, 1968; Berne and Harp, 1970) using molecular dynamics was quickly followed by two ambitious attempts to model liquid water, first by MC (Barker and Watts, 1969), and then by MD (Rahman and Stillinger, 1971). Water remains one of the most interesting and difficult liquids to study by simulation (Morse and Rice, 1982; McCoustra et al., 2009; Lynden-Bell, 2010; Lin et al., 2012). From early studies of small rigid molecules (Barojas et al., 1973) and flexible hydrocarbons (Ryckaert and Bellemans, 1975), simulations have developed to model more complicated systems such as polymers (Binder, 1995), proteins, lipids, nucleic acids, and carbohydrates (Monticelli and Salonen, 2013). Simulations containing half a

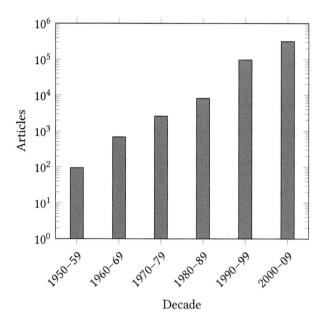

Fig. 1.1 The approximate number of articles concerning the computer simulation of condensed phases published in each complete decade. The search was carried out using the Web of Science® by searching on Monte Carlo, molecular dynamics, Brownian dynamics, lattice Boltzmann, dynamical density functional theory, Car–Parrinello, QM/MM in both the TITLE and TOPIC search fields.

million atoms have been conducted for 50 million timesteps to study the surface tension of a small liquid droplet (van Giessen and Blokhuis, 2009) and the massive parallel molecular dynamics code, ls1 mardyn, has been used to simulate a trillion Lennard-Jones atoms (Niethammer et al., 2014). It is now possible to follow the folding of a solvated protein using simulations in the microsecond-to-millisecond range (ca. 10^9–10^{12} timesteps) on a special purpose computer (Piana et al., 2014).

The growth of the field of computer simulation over the last 60 years, as evidenced by the number of publications in refereed journals, has been dramatic. In Fig. 1.1, we have attempted to calculate the number of papers published in this field during each complete decade. While bibliometric exercises of this kind will fail to capture some important papers and will often include some unwanted papers in related disciplines, the overall trend in the number of articles is clear.

This is, in part, due to the continuing and substantial increase in computing power, which follows the celebrated Moore's law curve over this period (see Appendix A). It is also due to the application of these methods to a wide range of previously intractable problems in the materials and life sciences. However, it is also, in no small part, due to the ingenuity of its practitioners in extending the early methods to areas such as: the calculation of free energies and phase diagrams (Chapter 9); the simulation of rare events (Chapter 10); the development of nonequilibrium methods for calculating transport coefficients (Chapter 11); the development of coarse-grained methods to extend the length and timescales that can be simulated (Chapter 12); and in the extension to include quantum mechanical effects (Chapter 13). This level of activity points to the proposition that computer simulation now sits alongside experiment and theory as a third and equally important tool in modern science. We start by asking: what is a computer simulation? How does it work? What can it tell us?

1.2   Computer simulation: motivation and applications

Some problems in statistical mechanics are exactly soluble. By this, we mean that a complete specification of the microscopic properties of a system (such as the Hamiltonian of an idealized model like the perfect gas or the Einstein crystal) leads directly, and perhaps easily, to a set of useful results or macroscopic properties (such as an equation of state like $PV = Nk_BT$). There are only a handful of non-trivial, exactly soluble problems in statistical mechanics (Baxter, 1982); the two-dimensional Ising model is a famous example.

Some problems in statistical mechanics, while not being exactly soluble, succumb readily to an analysis based on a straightforward approximation scheme. Computers may have an incidental, calculational, part to play in such work; for example, in the evaluation of cluster integrals in the virial expansion for dilute, imperfect gases (Rosenbluth and Rosenbluth, 1954; Wheatley, 2013). The problem is that, like the virial expansion, many 'straightforward' approximation schemes simply do not work when applied to liquids. For some liquid properties, it may not even be clear how to begin constructing an approximate theory in a reasonable way. The more difficult and interesting the problem, the more desirable it becomes to have exact results available, both to test existing approximate methods and to point the way towards new approaches. It is also important to be able to do this without necessarily introducing the additional question of how closely a particular model (which may be very idealized) mimics a real liquid, although this may also be a matter of interest. Computer simulations have a valuable role to play in providing essentially exact results for problems in statistical mechanics which would otherwise only be soluble by approximate methods, or might be quite intractable. In this sense, computer simulation is a test of theories and, historically, simulations have indeed discriminated between well-founded approaches, such as integral equation theories (Hansen and McDonald, 2013), and ideas that are plausible but, in the event, less successful, such as the old cell theories of liquids (Lennard-Jones and Devonshire, 1939a,b). The results of computer simulations may also be compared with those of real experiments. In the first place, this is a test of the underlying model used in a computer simulation. Eventually, if the model is a good one, the simulator hopes to offer insights to the experimentalist, and assist in the interpretation of new results. This dual role of simulation, as a bridge between models and theoretical predictions on the one hand, and between models and experimental results on the other, is illustrated in Fig. 1.2. Because of this connection role, and the way in which simulations are conducted and analysed, these techniques are often termed 'computer experiments'.

Computer simulation provides a direct route from the microscopic details of a system (the masses of the atoms, the interactions between them, molecular geometry, etc.) to macroscopic properties of experimental interest (the equation of state, transport coefficients, structural order parameters, and so on). As well as being of academic interest, this type of information is technologically useful. It may be difficult or impossible to carry out experiments under extremes of temperature and pressure, while a computer simulation

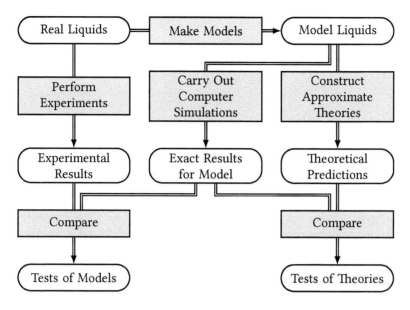

Fig. 1.2 The connection between experiment, theory, and computer simulation.

of the material in, say, a shock wave, a high-temperature plasma, a nuclear reactor, or a planetary core, would be perfectly feasible. Quite subtle details of molecular motion and structure, for example in heterogeneous catalysis, fast ion conduction, or enzyme action, are difficult to probe experimentally but can be extracted readily from a computer simulation. Finally, while the speed of molecular events is itself an experimental difficulty it represents no hindrance to the simulator. A wide range of physical phenomena, from the molecular scale to the galactic (Hockney and Eastwood, 1988), may be studied using some form of computer simulation.

In most of this book, we will be concerned with the details of carrying out simulations (the central box in Fig. 1.2). In the rest of this chapter, however, we deal with the general question of how to put information in (i.e. how to define a model of a liquid) while in Chapter 2 we examine how to get information out (using statistical mechanics).

1.3   Model systems and interaction potentials

1.3.1   Introduction

In most of this book, the microscopic state of a system may be specified in terms of the positions and momenta of a constituent set of particles: the atoms and molecules. Within the Born–Oppenheimer (BO) approximation (see also Chapter 13), it is possible to express the Hamiltonian of a system as a function of the nuclear variables, the (rapid) motion of the electrons having been averaged out. Making the additional approximation that a classical description is adequate, we may write the Hamiltonian $\mathcal{H}$ of a system of N molecules as a sum of kinetic- and potential-energy functions of the set of coordinates $\mathbf{q}_i$

and momenta $\mathbf{p}_i$ of each molecule i. Adopting a condensed notation

$$\mathbf{q} = (\mathbf{q}_1, \mathbf{q}_2, \cdots, \mathbf{q}_N) \tag{1.1a}$$

$$\mathbf{p} = (\mathbf{p}_1, \mathbf{p}_2, \cdots, \mathbf{p}_N) \tag{1.1b}$$

we have

$$\mathcal{H}(\mathbf{q}, \mathbf{p}) = \mathcal{K}(\mathbf{p}) + \mathcal{V}(\mathbf{q}). \tag{1.2}$$

Usually, the Hamiltonian will be equal to the total internal energy E of the system. The generalized coordinates $\mathbf{q}_i$ may simply be the set of Cartesian coordinates $\mathbf{r}_i$ of each atom (or nucleus) in the system, but, as we shall see, it is sometimes useful to treat molecules as rigid bodies, in which case $\mathbf{q}$ will consist of the Cartesian coordinates of each molecular centre of mass together with a set of variables Ω_i that specify molecular orientation. In any case, $\mathbf{p}$ stands for the appropriate set of conjugate momenta. For a simple atomic system, the kinetic energy $\mathcal{K}$ takes the form

$$\mathcal{K} = \sum_{i=1}^{N} \sum_{\alpha} p_{i\alpha}^2 / 2m_i \tag{1.3}$$

where m_i is the molecular mass, and the index α runs over the different (x, y, z) components of the momentum of atom i. The potential energy $\mathcal{V}$ contains the interesting information regarding intermolecular interactions: assuming that $\mathcal{V}$ is fairly sensibly behaved, it will be possible to construct, from $\mathcal{H}$, an equation of motion (in Hamiltonian, Lagrangian, or Newtonian form) which governs the entire time-evolution of the system and all its mechanical properties (Goldstein, 1980). Solution of this equation will generally involve calculating, from $\mathcal{V}$, the forces $\mathbf{f}_i$ and torques $\mathbf{\tau}_i$ acting on the molecules (see Chapter 3). The Hamiltonian also dictates the equilibrium distribution function for molecular positions and momenta (see Chapter 2). Thus, generally, it is $\mathcal{H}$ (or $\mathcal{V}$) which is the basic input to a computer simulation program. The approach used almost universally in computer simulation is to separate the potential energy into terms involving pairs, triplets, etc. of molecules. In the following sections we shall consider this in detail.

Recently, there has been a spectacular growth in the number of simulation studies which avoid the use of effective potentials by considering the electrons explicitly using density functional theory (Martin, 2008). In an early approach, the electron density was represented by an extension of the electron gas theory (LeSar and Gordon, 1982; 1983; LeSar, 1984). In most of the current work, the electronic degrees of freedom are explicitly included in the description. The electrons, influenced by the external field of the nuclei, are allowed to evolve during the course of the simulation by an auxiliary set of dynamical equations (Car and Parrinello, 1985). This method, known as *ab initio* molecular dynamics (Marx and Hutter, 2012), is now sufficiently well developed that it may become the method of choice for simulations in materials and the life sciences as the speed of computers increases. We will consider this approach in more detail in Chapter 13.

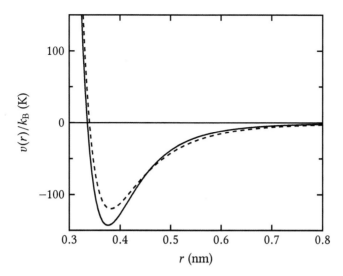

Fig. 1.3   Argon pair potentials. We illustrate (solid line) a recent pair potential for argon calculated by *ab initio* methods (see Patkowski and Szalewicz, 2010). Also shown is the Lennard-Jones 12–6 potential (dashed line) used in computer simulations of liquid argon.

1.3.2   Atomic systems

Consider first the case of a system containing N atoms. The potential energy may be divided into terms depending on the coordinates of individual atoms, pairs, triplets, etc.:

$$\mathcal{V} = \sum_i v_1(\mathbf{r}_i) + \sum_i \sum_{j>i} v_2(\mathbf{r}_i, \mathbf{r}_j) + \sum_i \sum_{j>i} \sum_{k>j} v_3(\mathbf{r}_i, \mathbf{r}_j, \mathbf{r}_k) + \dots . \qquad (1.4)$$

The $\sum_i \sum_{j>i}$ notation indicates a summation over all distinct pairs i and j without counting any pair twice (i.e. as ij and ji); the same care must be taken for triplets. The first term in eqn (1.4), $v_1(\mathbf{r}_i)$, represents the effect of an external field (including, e.g. the container walls) on the system. The remaining terms represent particle interactions. The second term, v_2, the pair potential, is the most important. The pair potential depends only on the magnitude of the pair separation $r_{ij} = |\mathbf{r}_{ij}| = |\mathbf{r}_i - \mathbf{r}_j|$, so it may be written $v_2(r_{ij})$. Figure 1.3 shows one of the more recent estimates for the pair potential between two argon atoms, as a function of separation (Patkowski and Szalewicz, 2010). This potential was determined by fitting to very accurate *ab initio* calculations for the argon dimer. The potential provides a position for the minimum and a well-depth that are very close to the experimental values. It can be used to calculate the spectrum of the isolated argon dimer and it produces a rotational constant and dissociation energy that are in excellent agreement with experiment (Patkowski et al., 2005). In fact, the computed potential is accurate enough to cast some doubt on the recommended, experimental, values of the second virial coefficient of argon at high temperatures (Dymond and Smith, 1980).

   The potential shows the typical features of intermolecular interactions. There is an attractive tail at large separations, essentially due to correlation between the electron clouds surrounding the atoms ('van der Waals' or 'London' dispersion). In addition, for charged species, Coulombic terms would be present. There is a negative well, responsible for cohesion in condensed phases. Finally, there is a steeply rising repulsive wall at short distances, due to non-bonded overlap between the electron clouds.

The v_3 term in eqn (1.4), involving triplets of molecules, is undoubtedly significant at liquid densities. Estimates of the magnitudes of the leading, triple-dipole, three-body contribution (Axilrod and Teller, 1943) have been made for inert gases in their solid-state face centred cubic (FCC) lattices (Doran and Zucker, 1971; Barker and Henderson, 1976). It is found that up to 10 % of the lattice energy of argon (and more in the case of more polarizable species) may be due to these non-additive terms in the potential; we may expect the same order of magnitude to hold in the liquid phase. Four-body (and higher) terms in eqn (1.4) are expected to be small in comparison with v_2 and v_3.

Despite the size of three-body terms in the potential, they are only rarely included in computer simulations (Barker et al., 1971; Attard, 1992; Marcelli and Sadus, 2012). This is because, as we shall see shortly, the calculation of any quantity involving a sum over triplets of molecules will be very time-consuming on a computer. In most cases, the pairwise approximation gives a remarkably good description of liquid properties because the average three-body effects can be partially included by defining an 'effective' pair potential. To do this, we rewrite eqn (1.4) in the form

$$\mathcal{V} \approx \sum_i v_1(\mathbf{r}_i) + \sum_i \sum_{j>i} v_2^{\text{eff}}(r_{ij}). \tag{1.5}$$

The pair potentials appearing in computer simulations are generally to be regarded as effective pair potentials of this kind, representing all the many-body effects; for simplicity, we will just use the notation $v(r_{ij})$, or $v(r)$. A consequence of this approximation is that the effective pair potential needed to reproduce experimental data may turn out to depend on the density, temperature, etc., while the true two-body potential $v_2(r_{ij})$, of course, does not.

Now we turn to the simpler, more idealized, pair potentials commonly used in computer simulations. These reflect the salient features of real interactions in a general, often empirical, way. Illustrated, with the accurate argon pair potential, in Fig. 1.3 is a simple Lennard-Jones 12–6 potential

$$v^{\text{LJ}}(r) = 4\epsilon \left[(\sigma/r)^{12} - (\sigma/r)^6 \right] \tag{1.6}$$

which provides a reasonable description of the properties of argon, via computer simulation, if the parameters ϵ and σ are chosen appropriately. The potential has a long-range attractive tail of the form $-1/r^6$, a negative well of depth ϵ, and a steeply rising repulsive wall at distances less than $r \sim \sigma$. The well-depth is often quoted in units of temperature as ϵ/k_{B}, where k_{B} is Boltzmann's constant; values of $\epsilon/k_{\text{B}} = 120$ K and $\sigma = 0.34$ nm provide reasonable agreement with the experimental properties of liquid argon. Once again, we must emphasize that these are not the values which would apply to an isolated pair of argon atoms, as is clear from Fig. 1.3.

For the purposes of investigating general properties of liquids, and for comparison with theory, highly idealized pair potentials may be of value. In Fig. 1.4, we illustrate three

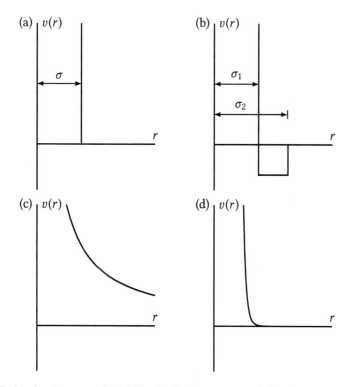

Fig. 1.4 Idealized pair potentials. (a) The hard-sphere potential; (b) the square-well potential; (c) The soft-sphere potential with repulsion parameter $v = 1$; (d) The soft-sphere potential with repulsion parameter $v = 12$. Vertical and horizontal scales are arbitrary.

forms which, although unrealistic, are very simple and convenient to use in computer simulation and in liquid-state theory. These are: the hard-sphere potential

$$v^{\mathrm{HS}}(r) = \begin{cases} \infty & \text{if } r < \sigma \\ 0 & \text{if } \sigma \leq r; \end{cases} \tag{1.7}$$

the square-well potential

$$v^{\mathrm{SW}}(r) = \begin{cases} \infty & \text{if } r < \sigma_1 \\ -\epsilon, & \text{if } \sigma_1 \leq r < \sigma_2 \\ 0, & \text{if } \sigma_2 \leq r; \end{cases} \tag{1.8}$$

and the soft-sphere potential

$$v^{\mathrm{SS}}(r) = \epsilon(\sigma/r)^v = ar^{-v}, \tag{1.9}$$

where v is a parameter, often chosen to be an integer. The soft-sphere potential becomes progressively 'harder' as v is increased. Soft-sphere potentials contain no attractive part. It is often useful to divide more realistic potentials into separate attractive and repulsive

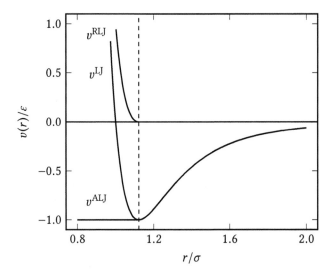

Fig. 1.5 The separation of the Lennard-Jones potential v^{LJ} into attractive and repulsive components, v^{ALJ} and v^{RLJ}, respectively. The vertical dashed line shows the position of r_{min}.

components, and the separation proposed by Weeks et al. (1971) involves splitting the potential at the minimum. For the Lennard-Jones potential, the repulsive and attractive parts are, as illustrated in Fig. 1.5,

$$v^{RLJ}(r) = \begin{cases} v^{LJ}(r) + \epsilon & \text{if } r < r_{min} \\ 0 & \text{if } r_{min} \leq r \end{cases} \tag{1.10a}$$

$$v^{ALJ}(r) = \begin{cases} -\epsilon & \text{if } r < r_{min} \\ v^{LJ}(r) & \text{if } r_{min} \leq r, \end{cases} \tag{1.10b}$$

where $r_{min} = 2^{1/6}\sigma \approx 1.12\sigma$. In perturbation theory (Weeks et al., 1971), a hypothetical fluid of molecules interacting via the repulsive potential v^{RLJ} is treated as a reference system and the attractive part v^{ALJ} is the perturbation. It should be noted that the potential v^{RLJ} is significantly harder than the inverse twelfth power soft-sphere potential, which is also sometimes thought of as the 'repulsive' part of $v^{LJ}(r)$.

For ions, of course, these potentials are not sufficient to represent the long-range interactions. A simple approach is to supplement one of these pair potentials with the Coulomb charge–charge interaction

$$v^{qq}(r_{ij}) = \frac{q_i q_j}{4\pi\epsilon_0 r_{ij}} \tag{1.11}$$

where q_i, q_j are the charges on ions i and j and ϵ_0 is the permittivity of free space (not to be confused with ϵ in eqns (1.6)–(1.10)). For ionic systems, induction interactions are important: the ionic charge induces a dipole on a neighbouring ion. This term is not pairwise additive and hence is difficult to include in a simulation. The shell model is a crude attempt to account for this polarizability (Dixon and Sangster, 1976; Lindan, 1995).

Each ion is represented as a core surrounded by a shell. Part of the ionic charge is located on the shell and the rest in the core. This division is always arranged so that the shell charge is negative (it represents the electronic cloud). The interactions between ions are just sums of the Coulombic shell–shell, core–core, and shell–core contributions. The shell and core of a given ion are coupled by a harmonic spring potential. The shells are taken to have zero mass. During a simulation, their positions are adjusted iteratively to zero the net force acting on each shell: this process makes the simulations expensive. We shall return to the simulation of polarizable systems in Section 1.3.3.

When a potential depends upon just a few parameters, such as ϵ and σ, it may be possible to choose an appropriate set of units in which these parameters take values of unity. This results in a simpler description of the properties of the model, and there may also be technical advantages within a simulation program. For Coulomb systems, the factor $4\pi\epsilon_0$ in eqn (1.11) is often omitted, and this corresponds to choosing a non-standard unit of charge. We discuss such reduced units in Appendix B. Reduced densities, temperatures, etc. are often denoted by an asterisk, that is, ρ^*, T^* etc.

1.3.3   Molecular systems

In principle, there is no reason to abandon the atomic approach when dealing with molecular systems: chemical bonds are simply interatomic potential-energy terms (Chandler, 1982). Ideally, we would like to treat all aspects of chemical bonding, including the reactions which form and break bonds, in a proper quantum mechanical fashion. This difficult task has not yet been accomplished but there are two common simplifying approaches. We might treat the bonds as classical harmonic springs (or Morse oscillators) or we could treat the molecule as a rigid or semi-rigid unit, with fixed bond lengths and, sometimes, fixed bond angles and torsion angles.

Bond vibrations are of very high frequency (and hence difficult to handle, certainly in a classical simulation). It quite possible that a high-frequency vibration will not be in thermal equilibrium with the fluid that surrounds it. These vibrations are also of low amplitude (and are therefore unimportant for many liquid properties). For these reasons, we prefer the approach of constraining the bond lengths to their equilibrium values. Thus, a diatomic molecule with a strongly binding interatomic potential-energy surface might be replaced by a dumb-bell with a rigid interatomic bond.

The interaction between the nuclei and electronic charge clouds of a pair of molecules i and j is clearly a complicated function of relative positions $\mathbf{r}_i$, $\mathbf{r}_j$ and orientations Ω_i, Ω_j (Gray and Gubbins, 1984). One way of modelling a molecule is to concentrate on the positions and sizes of the constituent atoms (Eyring, 1932). The much simplified 'atom–atom' or 'site–site' approximation for diatomic molecules is illustrated in Fig. 1.6. The total interaction is a sum of pairwise contributions from distinct sites a in molecule i, at position $\mathbf{r}_{ia}$, and b in molecule j, at position $\mathbf{r}_{jb}$:

$$v(\mathbf{r}_{ij}, \Omega_i, \Omega_j) = \sum_a \sum_b v_{ab}(r_{ab}). \tag{1.12}$$

Here a, b take the values 1, 2, v_{ab} is the pair potential acting between sites a and b, and r_{ab} is shorthand for the inter-site separation $r_{ab} = |\mathbf{r}_{ab}| = |\mathbf{r}_{ia} - \mathbf{r}_{jb}|$. The interaction sites are usually centred, more or less, on the positions of the nuclei in the real molecule, so as to

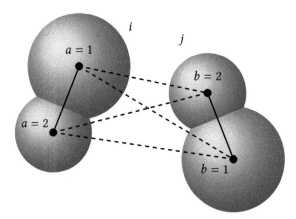

Fig. 1.6 The atom–atom model of a diatomic molecule. The total interaction is a sum of terms involving the distances $|\mathbf{r}_{ia} - \mathbf{r}_{jb}|$, indicated by dashed lines.

represent the basic effects of molecular 'shape'. A very simple extension of the hard-sphere model is to consider a diatomic composed of two hard spheres fused together (Streett and Tildesley, 1976), but more realistic models involve continuous potentials. Thus, nitrogen, fluorine, chlorine, etc. have been depicted as two 'Lennard-Jones atoms' separated by a fixed bond length (Barojas et al., 1973; Cheung and Powles, 1975; Singer et al., 1977).

The description of the molecular charge distribution may be improved somewhat by incorporating point multipole moments at the centre of charge (Streett and Tildesley, 1977). These multipoles may be equal to the known (isolated molecule) values, or may be 'effective' values chosen simply to yield a better description of the liquid structure and thermodynamic properties. A useful collection of the values of multipole moments is given in Gray and Gubbins (1984). Price et al. (1984) have developed an efficient way of calculating the multipolar energy, forces and torques between molecules of arbitrary symmetry up to terms of $O(r_{ij}^{-5})$. However, it is now generally accepted that such a multipole expansion of the electrostatic potential based around the centre of mass of a molecule is not rapidly convergent.

A pragmatic alternative approach, for ionic and polar systems, is to use a set of fictitious 'partial charges' distributed 'in a physically reasonable way' around the molecule so as to reproduce the known multipole moments (Murthy et al., 1983). For example, the electrostatic part of the interaction between nitrogen molecules may be modelled using five partial charges placed along the axis, while for methane, a tetrahedral arrangement of partial charges is appropriate. These are illustrated in Fig. 1.7. For the case of N_2, taking the molecular axis to lie along z, the quadrupole moment Q is given by (Gray and Gubbins, 1984)

$$Q = \sum_{a=1}^{5} q_a z_a^2 \qquad (1.13)$$

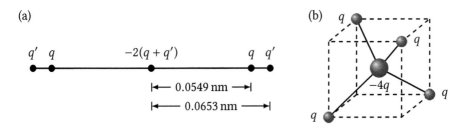

Fig. 1.7 Partial charge models: (a) A five-charge model for N_2. There is one charge at the bond centre, two at the positions of the nuclei, and two more displaced beyond the nuclei. Typical values (with $e = 1.602 \times 10^{-19}$ C): $q = +5.2366\, e$, $q' = -4.0469\, e$, giving $Q = -4.67 \times 10^{-40}$ C m^2 (Murthy et al., 1983). (b) A five-charge model for CH_4. There is one charge at the centre and four others at the positions of the hydrogen nuclei. Typical values are CH bond length 0.1094 nm, $q = 0.143\, e$ giving $O = 5.77 \times 10^{-50}$ C m^3 (Righini et al., 1981).

with similar expressions for the higher multipoles (all the odd ones vanish for N_2). The first non-vanishing moment for methane is the octopole O

$$O = \frac{5}{2} \sum_{a=1}^{5} q_a x_a y_a z_a \qquad (1.14)$$

in a coordinate system aligned with the cube shown in Fig. 1.7. The aim of all these approaches is to approximate the complete charge distribution in the molecule. In a calculation of the potential energy, the interaction between partial charges on different molecules would be summed in the same way as the other site–site interactions.

The use of higher-order multipoles has enjoyed a renaissance in recent years. This is because we can obtain an accurate representation of the electrostatic potential by placing multipoles at various sites within the molecule. These sites could be at the atom positions, or at the centres of bonds or within lone pairs, and it is normally sufficient to place a charge, dipole and quadrupole at any particular site. This approach, known as a distributed multipole analysis (Stone, 1981; 2013, Chapter 7), is illustrated for N_2 and CO in Fig. 1.8. In the case of N_2 the multipoles are placed at the centre of the bond and on the two nitrogen atoms, with their z-axis along the bond. Each site has a charge and a quadrupole and, in addition, the two atoms have equal and opposite dipoles. These are calculated using an accurate density functional theory B3LYP (Martin, 2008). In atomic units (see Appendix B), the overall quadrupole of the molecule calculated from this distribution is $-1.170\, ea_0{}^2$ corresponding to the experimental estimate of $(-1.09 \pm 0.07)\, ea_0{}^2$. A similar calculation for CO produces charges, dipoles and quadrupoles on all three sites (the C and O atoms and the centre of the bond). The overall dipole and quadrupole moments from this distribution are $0.036\, ea_0$ and $-1.515\, ea_0{}^2$ respectively, compared with the experimental estimates of $0.043\, ea_0$ and $-1.4\, ea_0{}^2$. The electrostatic energy between two molecules is now the sum of the multipole interactions between the atoms or sites in different molecules. The energy of interaction between two sets of distributed multipoles $\{q_a, \mu_a, Q_a\}$ and $\{q_b, \mu_b, Q_b\}$,

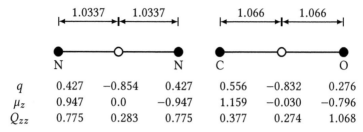

	N		N	C		O
q	0.427	−0.854	0.427	0.556	−0.832	0.276
μ_z	0.947	0.0	−0.947	1.159	−0.030	−0.796
Q_{zz}	0.775	0.283	0.775	0.377	0.274	1.068

Fig. 1.8 The distributed multipoles required to represent the electrostatic potential of a N_2 and CO molecule calculated using a cc-p-VQZ basis set. Multipoles are placed at the positions of the atoms (black circles) and at the midpoint of the bond (white circles). The distances are in atomic units, $a_0 = 0.529$ Å; charges, q, are in units of $e = 1.602 \times 10^{-19}$ C; dipoles, μ, are in units of $ea_0 = 8.478 \times 10^{-30}$ C m; and quadrupoles, Q, are in units of $ea_0^2 = 4.487 \times 10^{-40}$ C m^2 (see Appendix B). Data from Stone (2013).

on atoms a and b at $\mathbf{r}_a$ and $\mathbf{r}_b$, is given by

$$v_{ab}^{elec} = T q_a q_b + T_\alpha (\mu_{a\alpha} q_b - q_a \mu_{b\alpha})$$
$$+ T_{\alpha\beta} \left(\tfrac{1}{3} q_a Q_{b\alpha\beta} - \mu_{a\alpha}\mu_{b\beta} + \tfrac{1}{3} Q_{a\alpha\beta} q_b \right)$$
$$+ \tfrac{1}{3} T_{\alpha\beta\gamma} \left(\mu_{a\alpha} Q_{b\beta\gamma} - Q_{a\alpha\beta}\mu_{b\gamma} \right) + \tfrac{1}{9} T_{\alpha\beta\gamma\delta} Q_{a\alpha\beta} Q_{b\gamma\delta} \quad (1.15)$$

where we take the sum over repeated Cartesian indices α, β etc. The interaction or 'T' tensors are given by

$$T_{\alpha,\beta...\gamma} = (-1)^n \nabla_\alpha \nabla_\beta ... \nabla_\gamma \frac{1}{r_{ab}}, \quad (1.16)$$

where n is the order of the tensor. Thus

$$T = \frac{1}{r_{ab}}, \quad T_\alpha = \frac{(r_{ab})_\alpha}{r_{ab}^3}, \quad T_{\alpha\beta} = \frac{3(r_{ab})_\alpha (r_{ab})_\beta - r_{ab}^2 \delta_{\alpha\beta}}{r_{ab}^5}, \quad (1.17)$$

and so on. Note that the T tensors are defined for $\mathbf{r}_{ab} = \mathbf{r}_a - \mathbf{r}_b$. This is a useful formulation of the electrostatic energy for a computer simulation where the T tensors are readily expressed in terms of the Cartesian coordinates of the atoms. In addition, it is also straightforward to evaluate the derivative of the potential to obtain the force (the field) or the field gradient. The electrostatic potential, ϕ, at a distance r from a charge q is $\phi(r) = q/r$ and the corresponding electric field is $\mathcal{E} = -\nabla\phi(r)$. The field is simply the force per unit charge. For example the field ($\mathcal{E}$) and field gradient ($\mathcal{E}'$) arising from a charge q_b at b are

$$\mathcal{E}_\alpha = -\nabla_\alpha q_b T = q_b T_\alpha$$
$$\mathcal{E}'_{\alpha\beta} = -\nabla_\alpha \nabla_\beta q_b T = -q_b T_{\alpha\beta}. \quad (1.18)$$

The quadrupole tensor used in eqn (1.15) is defined to be traceless,

$$Q_{\alpha\beta} = \sum_a q_a \left(\tfrac{3}{2}(r_a)_\alpha (r_b)_\beta - \tfrac{1}{2} r_a^2 \delta_{\alpha\beta} \right). \quad (1.19)$$

Code 1.1 Calculation of T tensors

This file is provided online. For a pair of linear molecules, electrostatic energies and forces are calculated using both the angles between the various vectors, and the T tensors.

```
! t_tensor.f90
! Electrostatic interactions: T-tensors compared with angles
PROGRAM t_tensor
```

The components of the dipole and quadrupole will initially be defined in an atom-fixed axis frame centred on an atom (or site) and at any given point in a simulation it will be necessary to transform these properties to the space-fixed axis system for use in eqn (1.15) (Dykstra, 1988; Ponder et al., 2010). This can be simply achieved with a rotation matrix which we discuss in Section 3.3.1. An example of the calculation of T tensors is given in Code 1.1.

Electronic polarization refers to the distortion of the electronic charge cloud by the electrostatic field from the other molecules. In a molecular fluid it can be an important contribution to the energy. It is inherently a many-body potential and unlike many of the interactions already discussed in this chapter, it cannot be broken down to a sum over pair interactions. For this reason, it is expensive to calculate and was often omitted from earlier simulations. In these cases, some compensation was obtained by enhancing the permanent electrostatic interactions in the model. For example, in early simulations of water, the overall permanent dipole of the molecule was set to ca. 2.2 D rather than the gas-phase value 1.85 D (where $1\,\text{D} = 0.299\,79 \times 10^{-30}\,\text{C m}$) in order to fit to the condensed phase properties in the absence of polarization (Watanabe and Klein, 1989). Nevertheless, polarization can be included explicitly in a model and there are three common approaches: the induced point multipole model; the fluctuating charge model; and the Drude oscillator model (Antila and Salonen, 2013; Rick and Stuart, 2003).

The induced multipole approach (Applequist et al., 1972) is based on a knowledge of the atomic dipole polarizability, $\alpha^a_{\alpha\beta}$, on a particular atom a. Consider a molecule containing a set of charges, q_a, on each atom. The induced dipole at a contains two terms

$$\Delta\mu^a_\gamma = \alpha^a_{\alpha\gamma}\left(\mathcal{E}^a_\alpha + \sum_{b\neq a}T_{\alpha\beta}\Delta\mu^b_\beta\right),\tag{1.20}$$

where we sum over repeated indices α, β. The first term in the field, $\mathcal{E}$, comes from the permanent charges at the other atoms and the second term comes from the dipoles that have been induced at these atoms. We ignore contributions from the field gradient at atom a by setting the higher-order polarizabilities to zero. Eqn (1.20) can be formally solved for the induced dipoles

$$\Delta\mu^a_\alpha = \sum_{b\beta}A_{\alpha\beta}\mathcal{E}^b_\beta\tag{1.21}$$

Example 1.1 Water, water everywhere

The earliest simulations of molecular liquids focused on water (Barker and Watts, 1969; Rahman and Stillinger, 1971) and, since then, there have been over 80 000 published simulations of the liquid. Considerable effort and ingenuity have gone into developing models of the intermolecular potential between water molecules. There are three types of classical potential models in use: rigid, flexible, and polarizable.

The simplest rigid models use a single Lennard-Jones site to represent the oxygen atom and three partial charges: at the centre of the oxygen and the position of the hydrogen atoms. There are no specific dispersion interactions involving the H atoms and the charges are set to model the effective condensed-phase dipole moment of water, 2.2 D–2.35 D. Examples include the the SPC and the SPC/E models (Berendsen et al., 1981; 1987) used in the GROMOS force field, and the TIP3P model (Jorgensen et al., 1983) implemented in AMBER and CHARMM. The precise geometry and the size of the charges are different in each of these models. They predict the experimental liquid densities at a fixed pressure but tend to overestimate the diffusivity. The addition of a fourth negative charge along the bisector of the H–O–H bond creates the TIP4P model (Jorgensen et al., 1983) and its generalization TIP4P/2005 (Abascal and Vega, 2005). These models are capable of producing many of the qualitative features of the complicated water phase diagram. The TIP5P potential model (Mahoney and Jorgensen, 2000) supplements the three charges on the atoms with two negative charges at the position of the lone pairs. This model correctly predicts the density maximum near 4 °C at 1 bar, and the liquid structure obtained from diffraction experiments.

Flexibility can be included in models such as SPC/E using the intramolecular potential of Toukan and Rahman (1985), in which anharmonic oscillators are used to represent the O–H and H–H stretches. These flexible models predict many of the features of the vibrational spectrum of the liquid (Praprotnik et al., 2005).

A recent study by Shvab and Sadus (2013) indicates that rigid models underestimate the water structure and H-bond network at temperatures higher than 400 K and that none of the models so far discussed can predict the heat capacities or thermal expansion coefficients of the liquid. To improve on this position it is necessary to include polarization in the potential. Li et al. (2007a) show that the Matsuoka–Clementi–Yoshimine potential fitted from quantum calculations can be adapted to include three-body dispersion interactions for O atoms and fluctuating charges to create the more accurate MCYna model (Shvab and Sadus, 2013). These enhancements produce good agreement with experimental data over the entire liquid range of temperatures. Jones et al. (2013) have taken a different approach by embedding a quantum Drude oscillator (QDO) and using adiabatic path-integral molecular dynamics to simulate 4000 water molecules. Sokhan et al. (2015) show that this approach can produce accurate densities, surface tensions, and structure over a range of temperatures. Models of water in terms of pseudo-potentials to describe the nuclei and core electrons, and a model of the exchange correlation function to describe the non-classical electron repulsion between the valence electrons will be described in Chapter 13.

where the relay matrix $\mathbf{A} = \mathbf{B}^{-1}$ and

$$B_{\alpha\beta} = \begin{cases} (\alpha^a)^{-1} & \text{if } a = b \\ -T_{\alpha\beta} & \text{if } a \neq b. \end{cases} \tag{1.22}$$

Here $\mathbf{A}$ and $\mathbf{B}$ have dimensions of the number of sites involved in the polarization; this can be a large matrix, so practically eqn (1.20) is solved in a simulation by iterating the induced dipoles until convergence is achieved (Warshel and Levitt, 1976). This method can also be used with the distributed multipole analysis where the field at a polarizable atom might contain terms from the charge, dipole and quadrupole at a neighbouring atom, while the induction still occurs through the dipole polarizability (Ponder et al., 2010).

There is a well-known problem with these point polarizability models in which the elements of $\mathbf{A}$ diverge at short separations: the so-called polarization catastrophe. This is caused by the normal breakdown in the multipole expansion at these distances. It can be mitigated by smearing the charges on a particular site (Thole, 1981). The effect of this modification is to change the interaction tensor to

$$\tilde{T}_{\alpha\beta} = \frac{3f_t(r_{ab})_\alpha(r_{ab})_\beta - f_e r_{ab}^2 \delta_{\alpha\beta}}{4\pi\epsilon_0 r_{ab}^5} \tag{1.23}$$

where f_e and f_t are two, simple, damping functions. A useful discussion of the various possible choices for these damping functions is given by Stone (2013). The modified tensor $\tilde{T}_{\alpha\beta}$ can now be used in eqn (1.20) to calculate the induced moments. Once the induced dipole at atom a has been consistently determined then the induction energy associated with that atom is

$$v_a^{\text{ind}} = -\tfrac{1}{2}\mathcal{E}_\alpha^a \Delta\mu_\alpha^a. \tag{1.24}$$

The second method of including polarization in a model is the fluctuating charge model, sometimes referred to as the electronegativity equalization model. The partial charges are allowed to fluctuate as dynamical quantities. We can illustrate this approach by considering a model for water (Sprik, 1991). In addition to the three permanent charges normally used to represent the electrostatic moments, four additional fluctuating charges are disposed in a tetrahedron around the central oxygen atom (see Fig. 1.9). The magnitudes of the charges $q_i(t)$ fluctuate in time, but they preserve overall charge neutrality

$$\sum_{i=1}^{4} q_i(t) = 0 \tag{1.25}$$

and they produce an induced dipole

$$\Delta\mu = \sum_{i=1}^{4} q_i(t)\mathbf{r}_i \tag{1.26}$$

where $\mathbf{r}_i$ are the vectors describing the positions of the tetrahedral charges with respect to the O atom. If $|\mathbf{r}_i| \ll r_{\text{OH}}$ then the higher moments of the fluctuating charge distribution can be neglected. The potential energy from the four charges is the sum of the electrostatic

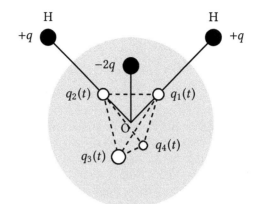

Fig. 1.9 A polarizable model for water (Sprik, 1991). The oxygen nucleus, O, is at the centre of the small tetrahedron. The three permanent charges $+q$, $+q$ and $-2q$, at the filled black spheres, are arranged to model the permanent electrostatic potential of water. The four fluctuating charges, $q_i(t)$, located at the white spheres, respond to the surrounding field and can be used to model the polarization of the molecule.

energy $(-\Delta\boldsymbol{\mu}\cdot\boldsymbol{\mathcal{E}})$ and a self energy term $(\Delta\mu^2/2\alpha_O)$ where α_O is the dipole polarizability associated with the oxygen atom. The fluctuating charges in this model can be determined by minimizing the potential energy in a given configuration subject to the constraint of charge neutrality, eqn (1.25),

$$\frac{\partial}{\partial q_i}\left(\frac{\left|\sum_{i=1}^{4}q_i\mathbf{r}_i\right|^2}{2\alpha_O} - \sum_{i=1}^{4}q_i\mathbf{r}_i\cdot\boldsymbol{\mathcal{E}}\right) = 0. \tag{1.27}$$

This approach can be extended to more complicated molecules by adding the appropriate number of fluctuating charges; and simplified to study spherical ions by including just two fluctuating charges within the spherical core. In these models, the fluctuating charges, q_i, are a crude representation of the electronic charge density and these can be usefully replaced by more realistic Gaussian charge distributions of width σ

$$\rho_i(\mathbf{r}) = q_i\left(\frac{1}{2\pi\sigma^2}\right)^{3/2}\exp\left(-\frac{|\mathbf{r}-\mathbf{r}_i|^2}{2\sigma^2}\right). \tag{1.28}$$

This improves the description of the polarization, particularly at short intermolecular separations (Sprik and Klein, 1988). We note that these models can be readily included in a molecular dynamics simulation by setting up separate equations of motions for the fluctuating charges (Sprik and Klein, 1988; Rick et al., 1994), and we shall consider this approach in Section 3.11.

The third approach is the Drude oscillator model or shell model. A polarizable site is represented as a heavy core particle of charge q_d and a massless or light shell particle of charge $-q_d$. These two particles are connected by a harmonic spring with a spring constant k. The minimum in the spring potential is obtained when the core and shell are coincident. The small charge q_d is in addition to the permanent charge at a particular site. The shell and core can separate to produce an induced dipole moment

$$\Delta\boldsymbol{\mu} = -q_d\Delta\mathbf{r} \tag{1.29}$$

where $\Delta\mathbf{r}$ is the vector from the core to the shell. The repulsion–dispersion interactions associated with a particular site are normally centred on the shell part of the site.

In the adiabatic implementation the shell is massless and at each step of a simulation the positions of the shells are adjusted iteratively to achieve the minimum energy configuration. In the dynamic model the shells are given a low mass (0.5 u) and an extended Lagrangian approach is used to solve the dynamics for short timesteps. In this case the shell particles are coupled to a heat bath at a low temperature (see Section 3.11). For these models, the atomic polarizability is isotropic and given by $\alpha^a = q_d^2/k$. Procedures are available for parameterizing the shell models to produce the correct molecular polarizabilities and electrostatic moments (Anisimov et al., 2005).

The model for water, shown in Fig. 1.9, begs the question as to whether we need to use a separate intermolecular potential to represent the hydrogen bond between two molecules. The hydrogen bond, between an H atom in one molecule and a strongly electronegative atom in another, is part permanent electrostatic interaction, part induced interaction, and some charge transfer. The evidence as reviewed by Stone (2013) indicates that the attractive electrostatic interaction is the most important term in determining the structure of the hydrogen-bonded dimer but that induced interactions will make an important contribution in condensed phases. It should be possible to avoid a separate hydrogen-bond potential by including an accurate representation of the electrostatic interactions (by using, for example, the distributed multipole approach) and by including polarization.

For larger molecules it may not be reasonable to 'fix' all the internal degrees of freedom. In particular, torsional motion about bonds, which gives rise to conformational interconversion in, for example, alkanes, cannot in general be neglected (since these motions involve energy changes comparable with normal thermal energies). An early simulation of n-butane, $CH_3CH_2CH_2CH_3$ (Ryckaert and Bellemans, 1975; Maréchal and Ryckaert, 1983), provides a good example of the way in which these features are incorporated in a simple model. Butane can be represented as a four-centre molecule, with fixed bond lengths and bond-bending angles, derived from known experimental (structural) data (see Fig. 1.10). A very common simplifying feature is built into this model: whole groups of atoms, such as CH_3 and CH_2, are condensed into spherically symmetric effective 'united atoms'. In fact, for butane, the interactions between such groups may be represented quite well by the ubiquitous Lennard-Jones potential, with empirically chosen parameters. In a simulation, the C_1–C_2, C_2–C_3 and C_3–C_4 bond lengths are held fixed by a method of constraints, which will be described in detail in Chapter 3. The angles θ and θ' may be fixed by additionally constraining the C_1–C_3 and C_2–C_4 distances; that is, by introducing 'phantom bonds'. If this is done, just one internal degree of freedom, namely the rotation about the C_2–C_3 bond, measured by the angle ϕ, is left unconstrained; for each molecule, an extra term in the potential energy, $v^{torsion}(\phi)$, appears in the Hamiltonian. This potential would have a minimum at a value of ϕ corresponding to the *trans* conformer of butane, and secondary minima at the *gauche* conformations. It is easy to see how this approach

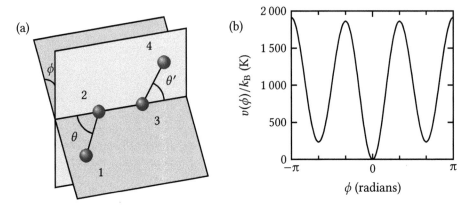

Fig. 1.10 (a) Geometry of a model of butane defining bending angles θ, θ' and the torsional angle ϕ (Ryckaert and Bellemans, 1975). (b) The torsional potential, in the AUA(2) model of Padilla and Toxvaerd (1991) as reviewed in Dysthe et al. (2000).

may be extended to much larger flexible molecules. The consequences of constraining bond lengths and angles will be treated in more detail in Chapters 2 and 4.

As the molecular model becomes more complicated, so too do the expressions for the potential energy, forces, and torques, due to molecular interactions. In Appendix C, we give some examples of these formulae, for rigid and flexible molecules, interacting via site–site pairwise potentials, including multipolar terms. We also show how to derive the forces from a simple three-body potential.

1.3.4   Coarse-grained potential models

Coarse graining a potential involves avoiding the full atomic representation of the molecules to find a description of the interaction at a longer or coarser length scale. We have already seen one simple example of this in the use of a united-atom potential for the methylene and methyl groups in butane. Coarse graining will reduce the number of explicit pairs that are needed for the calculation of the energy and force for a particular system and will reduce the computer time or, alternatively, allow us to study a much larger system. Normally an increase in the characteristic length scale in the model goes hand in hand with an increase in the timestep that we can use in a dynamical simulation of the problem. Coarse graining will allow us to use a longer timestep and to cover more 'real' time in our simulation.

One flavour of coarse-grained model has been widely used to study liquid crystalline systems, exhibiting some long-range orientational order. For example, for the nematogen quinquaphenyl, a large rigid molecule that forms a nematic phase, a substantial number of sites would be required to model the repulsive core. A crude model, which represented each of the five benzene rings as a single Lennard-Jones site, would necessitate 25 site–site interactions between each pair of molecules; sites based on each carbon atom would be more realistic but require 900 site–site interactions per pair. An alternative coarse-grained representation of intermolecular potential, introduced by Corner (1948), involves a single site–site interaction between a pair of molecules, characterized by energy and length parameters that depend on the relative orientation of the molecules.

A version of this family of molecular potentials that has been used in computer simulation studies is the Gay–Berne potential (Gay and Berne, 1981). This is an extension of the Gaussian overlap model generalized to a Lennard-Jones form (Berne and Pechukas, 1972). The basic potential acting between two linear molecules is

$$v^{GB}(r_{ij}, \hat{\mathbf{e}}_i, \hat{\mathbf{e}}_j) = 4\epsilon(\hat{\mathbf{r}}, \hat{\mathbf{e}}_i, \hat{\mathbf{e}}_j)\left[(\sigma_s/\rho_{ij})^{12} - (\sigma_s/\rho_{ij})^6\right], \tag{1.30a}$$

$$\text{where} \qquad \rho_{ij} = r_{ij} - \sigma(\hat{\mathbf{r}}, \hat{\mathbf{e}}_i, \hat{\mathbf{e}}_j) + \sigma_s. \tag{1.30b}$$

Here, r_{ij} is the distance between the centres of i and j, and $\hat{\mathbf{r}} = \mathbf{r}_{ij}/r_{ij}$ is the unit vector along $\mathbf{r}_{ij}$, while $\hat{\mathbf{e}}_i$ and $\hat{\mathbf{e}}_j$ are unit vectors along the axis of the molecules. The molecule can be considered (approximately) as an ellipsoid characterized by two diameters σ_s and σ_e, the separations at which the side-by-side potential, and the end-to-end potential, respectively, become zero. Thus

$$\sigma(\hat{\mathbf{r}}, \hat{\mathbf{e}}_i, \hat{\mathbf{e}}_j) = \sigma_s \left[1 - \frac{\chi}{2}\left(\frac{(\hat{\mathbf{e}}_i \cdot \hat{\mathbf{r}} + \hat{\mathbf{e}}_j \cdot \hat{\mathbf{r}})^2}{1 + \chi(\hat{\mathbf{e}}_i \cdot \hat{\mathbf{e}}_j)} + \frac{(\hat{\mathbf{e}}_i \cdot \hat{\mathbf{r}} - \hat{\mathbf{e}}_j \cdot \hat{\mathbf{r}})^2}{1 - \chi(\hat{\mathbf{e}}_i \cdot \hat{\mathbf{e}}_j)}\right)\right]^{-1/2} \tag{1.31a}$$

$$\text{where} \qquad \chi = \frac{\kappa^2 - 1}{\kappa^2 + 1}, \quad \text{and} \quad \kappa = \sigma_e/\sigma_s. \tag{1.31b}$$

κ is the elongation and χ is the shape anisotropy parameter ($\kappa = 1$, $\chi = 0$ for spherical particles, $\kappa \to \infty$, $\chi \to 1$ for very long rods, and $\kappa \to 0$, $\chi \to -1$ for very thin disks).

The energy term is the product of two functions

$$\epsilon(\hat{\mathbf{r}}, \hat{\mathbf{e}}_i, \hat{\mathbf{e}}_j) = \epsilon_0\, \epsilon_1^\nu(\hat{\mathbf{e}}_i, \hat{\mathbf{e}}_j)\, \epsilon_2^\mu(\hat{\mathbf{r}}, \hat{\mathbf{e}}_i, \hat{\mathbf{e}}_j) \tag{1.32a}$$

where

$$\epsilon_1(\hat{\mathbf{e}}_i, \hat{\mathbf{e}}_j) = \left(1 - \chi^2(\hat{\mathbf{e}}_i \cdot \hat{\mathbf{e}}_j)^2\right)^{-1/2} \tag{1.32b}$$

$$\epsilon_2(\hat{\mathbf{r}}, \hat{\mathbf{e}}_i, \hat{\mathbf{e}}_j) = 1 - \frac{\chi'}{2}\left(\frac{(\hat{\mathbf{e}}_i \cdot \hat{\mathbf{r}} + \hat{\mathbf{e}}_j \cdot \hat{\mathbf{r}})^2}{1 + \chi'(\hat{\mathbf{e}}_i \cdot \hat{\mathbf{e}}_j)} + \frac{(\hat{\mathbf{e}}_i \cdot \hat{\mathbf{r}} - \hat{\mathbf{e}}_j \cdot \hat{\mathbf{r}})^2}{1 - \chi'(\hat{\mathbf{e}}_i \cdot \hat{\mathbf{e}}_j)}\right) \tag{1.32c}$$

and the energy anisotropy parameter is

$$\chi' = \frac{\kappa'^{1/\mu} - 1}{\kappa'^{1/\mu} + 1}, \quad \text{where} \quad \kappa' = \epsilon_{ss}/\epsilon_{ee}. \tag{1.32d}$$

ϵ_{ss} and ϵ_{ee} are the well depths of the potentials in the side-by-side and end-to-end configurations respectively. The potential is illustrated for these arrangements, as well as for T-shaped and crossed configurations, in Fig. 1.11. The original model, with exponents $\mu = 2$, $\nu = 1$, and parameters $\kappa = 3$, $\kappa' = 5$, was used to mimic four collinear Lennard-Jones sites (Gay and Berne, 1981). The potential and corresponding force and torque can be readily evaluated and the functional form is rich enough to create mesogens of different shapes and energy anisotropies that will form the full range of nematic, smectic, and discotic liquid crystalline phases (Luckhurst et al., 1990; Berardi et al., 1993; Allen, 2006a;

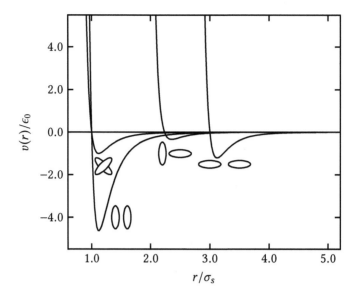

Fig. 1.11 The Gay-Berne potential, with parameters $\mu = 1$, $\nu = 3$, $\kappa = 3$, $\kappa' = 5$ (Berardi et al., 1993), as a function of centre–centre separation, for various molecular orientations.

Luckhurst, 2006). It is discussed further in Appendix C. Extensions of the potential, and its use in modelling liquid crystals, are discussed by Zannoni (2001).

The MARTINI approach is a coarse-grained potential developed for modelling lipid bilayers (Marrink et al., 2004; 2007) and proteins (Monticelli et al., 2008). In this model the bonded hydrogen atoms are included with their heavier partners, such as C, N, or O. These united atoms are then further combined using a 4:1 mapping to create larger beads (except in the case of rings where the mapping is normally 3:1). For these larger beads, there are four different bead types: charged (Q), polar (P), nonpolar (N), and apolar (C). Each of these types is further subdivided depending on the bead's hydrogen-bond forming propensities or its polarity. Overall there are 18 bead-types and each pair of beads interacts through a Lennard-Jones potential where the σ and ϵ parameters are specific to the atom types involved. Charged beads also interact through Coulombic potentials. The intramolecular interactions (bonds, angles, and torsions) are derived from atomistic simulations of crystal structures. This kind of moderate coarse graining has been successfully applied to simulations of the clustering behaviour of the membrane bound protein syntaxin-1A (van den Bogaart et al., 2011) and the simulation of the domain partitioning of membrane peptides (Schäfer et al., 2011).

It is possible to coarse grain potentials in a way that results in larger beads, that might contain 1–3 Kuhn chain-segments of a polymer or perhaps ten solvent molecules. We will consider this approach more fully in Chapter 12. However, at this point, we mention a very simple coarse-grained model of polymer chains due to Kremer and Grest (1990) and termed the finitely extensible nonlinear elastic (FENE) model. The bonds between beads

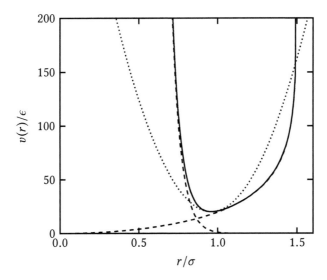

Fig. 1.12 The potential between bonded atoms in a coarse-grained polymer (solid line) together with its component parts (dashed lines): the attractive FENE potential, eqn (1.33) with $R_0 = 1.5\sigma$ and $k = 30\epsilon/\sigma^2$, and the repulsive Lennard-Jones potential, eqn (1.10a). Also shown (dotted line) is a harmonic potential, fitted to the curvature at the minimum. See Kremer and Grest (1990) for details.

within the chain are represented by the potential energy

$$v^{\text{FENE}}(r) = \begin{cases} -\frac{1}{2}kR_0^2 \, \ln\left(1 - (r/R_0)^2\right) & r < R_0 \\ \infty & r \geq R_0. \end{cases} \tag{1.33}$$

This is combined with the potential $v^{\text{RLJ}}(r)$ of eqn (1.10a), representing the effects of excluded volume between every pair of beads (including those that are bonded together). The key feature of this potential is that it cannot be extended beyond $r = R_0$. This is important when studying entanglement effects: the simpler harmonic potential could, in principle, extend enough to let chains pass through one another, in some circumstances.

Finally, there has been considerable effort to develop a simple, single-site coarse-grained potential for water. One approach (Molinero and Moore, 2009; Moore and Molinero, 2011) has been to abandon the long-range electrostatics conventionally associated with hydrogen bonds, and use instead short-range directional interactions, of the kind previously used to model silicon (Stillinger and Weber, 1985). The resulting monatomic water (mW) model is very cheap to simulate but surprisingly successful in reproducing experimental structural and thermodynamic properties. Can one go further? It is difficult to imagine that a spherical, isotropic potential will be able to capture the strong association interactions in the fluid. Nevertheless, Lobanova et al. (2015) have used a Mie potential, a versatile form of the standard Lennard-Jones potential, where

$$v_{\text{Mie}}(r) = C\epsilon\left[\left(\frac{\sigma}{r}\right)^n - \left(\frac{\sigma}{r}\right)^m\right], \quad \text{with} \quad C = \left(\frac{n}{n-m}\right)\left(\frac{n}{m}\right)^{m/(n-m)}. \tag{1.34}$$

A potential with $n = 8$ and $m = 6$ can be used with temperature-dependent energy and length parameters to represent the thermophysical properties of water over a broad range of conditions. However, a simpler form where ϵ and σ are independent of temperature can be used to represent water in the calculation of mixture phase diagrams such as CO_2/H_2O (Müller and Jackson, 2014). We briefly discuss this approach to coarse graining in Section 12.7.3. The examples just given are two amongst many attempts to model water in a coarse-grained way (for a review see Hadley and McCabe, 2012).

1.3.5   Calculating the potential

This is an appropriate point to introduce a piece of computer code, which illustrates the calculation of the potential energy in a system of Lennard-Jones atoms. Simulation programs are written in a range of languages: Fortran, C, and C++ are the most common, sometimes with a wrapper written in Python or Java. Here we shall use Fortran, which has a compact notation for arrays and array operations, and is simple enough to be read as a 'pseudo-code'. Appendix A contains some discussion of different programming approaches, and a summary of some of the issues affecting efficiency. We suppose that the coordinate vectors of our atoms are stored in an array r of rank two, with dimensions (3,n), where the first index covers the x, y, and z components, and the second varies from 1 to n (equal to N, the number of particles). The potential energy will be stored in a variable pot, which is zeroed initially, and is then accumulated in a double loop over all distinct pairs of atoms, taking care to count each pair only once. This is shown in Code 1.2. The Lennard-Jones parameters ϵ and σ are assumed to be stored in the variables epslj and sigma respectively. The colon ':' is short for an implied loop over the corresponding index, so the statement rij(:) = r(:,i) - r(:,j) stands for the vector assignment $\mathbf{r}_{ij} = \mathbf{r}_i - \mathbf{r}_j$. The SUM function simply adds the components of its (array) argument, which in this case gives $r_{ij}^2 = x_{ij}^2 + y_{ij}^2 + z_{ij}^2$. Code 1.2 takes no account of periodic boundary conditions (we return to this in Section 1.6.2). Some measures have been taken here to avoid unnecessary use of computer time. The value of σ^2 is computed once beforehand, and stored in the variable sigma_sq; the factor 4ϵ, which appears in every pair potential term, is multiplied in once, at the very end. The aim is to avoid many unnecessary operations within the crucial 'inner loop' over index j. The more general questions of time-saving tricks in this part of the program are addressed in Chapter 5. The extension of this type of double loop to deal with other forms of the pair potential, and to compute forces in addition to potential terms, is straightforward, and examples will be given in later chapters. For molecular systems, the same general principles apply, but additional loops over the different sites or atoms in a molecule may be needed. For example, consider the site–site diatomic model of eqn (1.12) and Fig. 1.6. Then the intermolecular interactions might be computed as in Code 1.3. Note that, apart from the dependence of the range of the j loop on the index i, the order of nesting of loops is a matter of choice. Here, we have placed a loop over molecular indices innermost; assuming that n is relatively large, and depending on the machine architecture, this may improve the efficiency of fetching the relevant coordinates from memory (in Fortran, the arrays are stored so that the first indices vary rapidly, and the last indices vary slowly, so there is usually an advantage in accessing contiguous blocks of memory, or cache, in sequence). Simulations of molecular systems may also

Code 1.2   Double loop for Lennard-Jones potential

This code snippet illustrates the calculation of the potential energy for a system of Lennard-Jones atoms, using a double loop over the atomic indices. The declarations at the start are given just to remind us of the types and sizes of variables and arrays (some notes on precision of variables appear in Appendix A).

```
INTEGER                    :: n, i, j
REAL , DIMENSION(3,n)  ::  r
REAL , DIMENSION(3)    ::  rij
REAL                       ::  epslj, sigma, sigma_sq
REAL                       ::  pot, rij_sq, sr2, sr6, sr12
sigma_sq = sigma ** 2
pot = 0.0
DO i = 1, n-1
   DO j = i+1, n
      rij(:) = r(:,i) - r(:,j)
      rij_sq = SUM ( rij ** 2 )
      sr2    = sigma_sq / rij_sq
      sr6    = sr2 ** 3
      sr12   = sr6 ** 2
      pot    = pot + sr12 - sr6
   END DO
END DO
pot = 4.0 * epslj * pot
```

involve the calculation of intramolecular energies, which, for site–site potentials, will necessitate a triple summation (over i, a, and b).

These examples are essentially summations over pairs of interaction sites in the system. Any calculation of three-body interactions will, of course, entail triple summations over distinct triplets of indices i, j, and k; these will be much more time consuming than the double summations described here. Even for pairwise-additive potentials, the energy or force calculation is the most expensive part of a computer simulation. We will return to this crucial section of the program in Chapter 5.

1.4   Constructing an intermolecular potential from first principles

1.4.1   Introduction

There are two approaches to constructing an intermolecular potential for use in a simulation. For small, simple molecules and their mixtures, it is possible to customize a model, with considerable freedom in choosing the functional form of the potentials and in adjusting the parameters for the problem at hand. For larger molecules such as polymers, proteins, or DNA, either in solution or at a surface, or for multi-component mixtures

Code 1.3   Site–site potential energy calculation

The coordinates r_{ia} of site a in molecule i are stored in the elements $r(:,i,a)$ of a rank-3 array; for a system of diatomic molecules na=2.

```
INTEGER                  :: n, i, j, a, b
REAL, DIMENSION(3,n,na) :: r
REAL, DIMENSION(3)       :: rij
DO a = 1, na
  DO b = 1, na
    DO i = 1, n - 1
      DO j = i + 1, n
        rij(:) = r(:,i,a) - r(:,j,b)
        ... calculate the i-j interaction ...
      END DO
    END DO
  END DO
END DO
```

containing many different types of molecule, then it will be more usual to employ one of the standard force fields (consisting of fixed functional forms for the potentials combined with parameters corresponding to the many different atom types in the simulation). We will cover the first aspect of model building in this section and consider force fields in Section 1.5.

There are essentially two stages in setting up a model for a realistic simulation of a given system. The first is 'getting started' by constructing a first guess at a potential model. This will allow some preliminary simulations to be carried out. The second is to use the simulation results, in comparison with experiment, to refine the potential model in a systematic way, repeating the process several times if necessary. We consider the two phases in turn.

1.4.2   Building the model potential

To illustrate the process of building up an intermolecular potential from first principles, we consider a small molecule, such as N_2, OCS, or CH_4, which can be modelled using the interaction site potentials discussed in Section 1.3. The essential features of this model will be an anisotropic repulsive core, to represent the shape, an anisotropic dispersion interaction, and some partial charges or distributed multipoles to model the permanent electrostatic effects. This crude effective pair potential can then be refined by using it to calculate properties of the gas, liquid, and solid, and comparing with experiment. Each short-range site–site interaction can be modelled using a Lennard-Jones potential. Suitable energy and length parameters for interactions between pairs of identical atoms in different molecules are available from a number of simulation studies. Some of these are given in Table 1.1. The energy parameter ϵ increases with atomic number as the polarizability goes up; σ also increases down a group of the Periodic Table, but decreases

Table 1.1 Atom–atom interaction parameters

Atom	Source	ϵ/k_B (K)	σ (nm)
H	Murad and Gubbins (1978)	8.6	0.281
He	Maitland et al. (1981)	10.2	0.228
C	Tildesley and Madden (1981)	51.2	0.335
N	Cheung and Powles (1975)	37.3	0.331
O	English and Venables (1974)	61.6	0.295
F	Singer et al. (1977)	52.8	0.283
Ne	Maitland et al. (1981)	47.0	0.272
S	Tildesley and Madden (1981)	183.0	0.352
Cl	Singer et al. (1977)	173.5	0.335
Ar	Maitland et al. (1981)	119.8	0.341
Br	Singer et al. (1977)	257.2	0.354
Kr	Maitland et al. (1981)	164.0	0.383

from left to right across a period with the increasing nuclear charge. For elements which do not appear in Table 1.1, a guide to ϵ and σ might be provided by the polarizability and van der Waals radius respectively. These values are only intended as a reasonable first guess: they take no regard of chemical environment and are not designed to be transferable. For example, the carbon atom parameters in CS_2 given in the table are quite different from the values appropriate to a carbon atom in graphite (Crowell, 1958).

Interactions between unlike atoms in different molecules can be approximated using the venerable Lorentz–Berthelot combining rules. For example, in CS_2 the cross-terms are

$$\sigma_{CS} = \tfrac{1}{2}\left(\sigma_{CC} + \sigma_{SS}\right), \quad \epsilon_{CS} = \left(\epsilon_{CC}\epsilon_{SS}\right)^{1/2}. \tag{1.35}$$

These rules are approximate; the ϵ cross-term expression, especially, is not expected to be appropriate in the majority of cases (Delhommelle and Millié, 2001; Haslam et al., 2008).

In tackling larger molecules, it may be necessary to model several atoms as a unified site. We have seen this for butane in Section 1.3, and a similar approach has been used in a model of benzene (Evans and Watts, 1976). The specification of an interaction site model is made complete by defining the positions of the sites within the molecule. Normally, these are located at the positions of the nuclei, with the bond lengths obtained from a standard source (CRC, 1984).

Rapid progress has been made in fitting the parameters for many classical pair potentials using *ab initio* quantum mechanical calculations. For example, symmetry-adapted perturbation theory, based on a density-functional approach, can be used to calculate separable and transferable parameters for the dispersion and electrostatic interactions (McDaniel and Schmidt, 2013). Calculations on monomers are used to estimate asymptotic properties such as charge and polarizability, while dimer calculations are used to estimate the parameters depending on charge density overlaps. The resulting parameters can be used with simple functional forms in simulations and the technique has recently been applied to the parameterization and simulation of an ionic liquid (Son et al., 2016).

The site–site Lennard-Jones potentials include an anisotropic dispersion which has the correct r^{-6} radial dependence at long range. However, this is not the exact result for the anisotropic dispersion from second-order perturbation theory. The correct formula, in an appropriate functional form for use in a simulation, is given by Burgos et al. (1982). Its implementation requires an estimate of the polarizability and polarizability anisotropy of the molecule.

It is also possible to improve the accuracy of the overall repulsion–dispersion interaction by considering an anisotropic site–site potential in place of $v_{ab}(r_{ab})$ in eqn (1.12). In other words, in a diatomic model of a chlorine molecule, the interatomic potential between chlorine atoms in different molecules would depend on r_{ab} and the angles between r_{ab} and intramolecular bonds. This type of model has been used to rationalize the liquid and solid structures of liquid Cl_2, Br_2, and I_2 (Rodger et al., 1988a,b).

The most straightforward way of representing electrostatic interactions is through partial charges as discussed in Section 1.3. To minimize the calculation of site–site distances they can be made to coincide with the Lennard-Jones sites, but this is not always desirable or possible; the only physical constraint on partial charge positions is that they should not lie outside the repulsive core region, since the potential might then diverge if molecules came too close. The magnitudes of the charges can be chosen to duplicate the known gas-phase electrostatic moments (Gray and Gubbins, 1984, Appendix D). Alternatively, the moments may be taken as adjustable parameters. For example, in a simple three-site model of N_2 representing only the quadrupole–quadrupole interaction, the best agreement with condensed phase properties is obtained with charges giving a quadrupole 10 %–15 % lower than the gas-phase value (Murthy et al., 1980). However, a sensible strategy is to begin with the gas-phase values, and alter the repulsive core parameters ϵ and σ before changing the partial charges.

Partial charges can also be developed using theoretical calculations. Bayly et al. (1993) have developed the widely used restrained electrostatic potential (RESP) method. In this technique:

(a) a molecule is placed in a 3D grid of points;

(b) the electrostatic potential is calculated at each grid point, outside the repulsive core, using a quantum mechanical calculation;

(c) a charge at each atom of the molecules is adjusted to reproduce the electrostatic potential at the grid points as accurately as possible.

Typically, accurate enough quantum mechanical estimates of the electrostatic field can be obtained using the 6-31G* level of the Gaussian code (Frisch et al., 2009). In order to make this fitting procedure robust and to obtain charges that are transferable between different molecules, it is necessary to minimize the magnitude of the charges that will fit the field. This is achieved using a hyperbolic restraint function in the minimization that pulls the magnitude of the charges towards zero.

Distributed multipoles and polarizabilities, for molecules containing up to about 60 atoms, can be calculated from first principles using the camCASP package developed by Stone and co-workers (Misquitta and Stone, 2013).

1.4.3   Adjusting the model potential

The first-guess potential can be used to calculate a number of properties in the gas, liquid, and solid phases; comparison of these results with experiment may be used to refine the potential, and the cycle can be repeated if necessary. The second virial coefficient is given by

$$B(T) = -\frac{2\pi}{\Omega^2} \int_0^\infty r_{ij}^2 dr_{ij} \int d\Omega_i \int d\Omega_j \exp\left[-v(r_{ij}, \, \Omega_i, \, \Omega_j)/k_B T\right] - 1 \qquad (1.36)$$

where $\Omega = 4\pi$ for a linear molecule and $\Omega = 8\pi^2$ for a non-linear one. This multidimensional integral (four-dimensional for a linear molecule and six-dimensional for a non-linear one) is easily calculated using a non-product algorithm (Murad, 1978). Experimental values of $B(T)$ have been compiled by Dymond and Smith (1980). Trial and error adjustment of the Lennard-Jones ϵ and σ parameters should be carried out, with any bond lengths and partial charges held fixed, so as to produce the closest match with the experimental $B(T)$. This will produce an improved potential, but still one that is based on pair properties.

The next step is to carry out a series of computer simulations of the liquid state, as described in Chapters 3 and 4. The densities and temperatures of the simulations should be chosen to be close to the orthobaric curve of the real system, that is, the liquid–vapour coexistence line. The output from these simulations, particularly the total internal energy and the pressure, may be compared with the experimental values. The coexisting pressures are readily available (Rowlinson and Swinton, 1982), and the internal energy can be obtained approximately from the known latent heat of evaporation. The energy parameters ϵ are adjusted to give a good fit to the internal energies along the orthobaric curve, and the length parameters σ altered to fit the pressures. If no satisfactory fit is obtained at this stage, the partial charges may be adjusted. It is also possible to adjust potential parameters to reproduce structural properties of the liquid, such as the site–site pair distribution functions (see Section 2.6), which can be extracted from coherent neutron diffraction studies using isotopic substitution (Cole et al., 2006; Zeidler et al., 2012).

Although the solid state is not the province of this book it offers a sensitive test of any potential model. Using the experimentally observed crystal structure, and the refined potential model, the lattice energy at zero temperature can be compared with the experimental value (remembering to add a correction for quantum zero-point motion). In addition, the lattice parameters corresponding to the minimum energy for the model solid can be compared with the values obtained by diffraction, and also lattice dynamics calculations (Neto et al., 1978) used to obtain phonons, librational modes, and dispersion curves of the model solid. Finally, we can ask if the experimental crystal structure is indeed the minimum energy structure for our potential. These constitute severe tests of our model-building skills (Price, 2008).

1.5   Force fields

In approaching the simulation of a complicated system, there might be 30 different atom types to consider and several hundred different intra- and inter-molecular potentials to fit. One would probably not want to build the potential model from scratch. Fortunately, it is possible to draw on the considerable body of work that has gone into the development of consistent force fields over the last 50 years (Bixon and Lifson, 1967; Lifson and Warshel, 1968; Ponder and Case, 2003).

A force field, in the context of a computer simulation, refers to the functional forms used to describe the intra- and inter-molecular potential energy of a collection of atoms, and the corresponding parameters that will determine the energy of a given configuration. These functions and parameters have been derived from experimental work on single molecules and from accurate quantum mechanical calculations. They are often refined by the use of computer simulations to compare calculated condensed phase properties with experiment. This is precisely the same approach described in Section 1.4.3, but on a bigger scale, so that the transferable parameters developed can be used with many different molecules. Some examples of widely used force fields are given in Table 1.2. This list is representative and not complete. The individual force fields in the table are constantly being updated and extended. For example, the OPLS force field has been refined to allow for the modelling of carbohydrates (Kony et al., 2002) and the OPLS and AMBER force fields have been used as the basis of a new field for ionic liquids (Lopes et al., 2004). Extensions and versions are often denoted by the ffXX specification following the force field name. A short search of the websites of the major force fields will establish the latest version and the most recent developments.

Force fields are often divided into three classes. Class I force fields normally have a functional form of the type

$$\mathcal{V} = \sum_{\text{bonds}} \tfrac{1}{2} k_r (r_{ij} - r_0)^2 + \sum_{\text{angles}} \tfrac{1}{2} k_\theta (\theta_{ijk} - \theta_0)^2$$

$$+ \sum_{\text{torsions}} \sum_n k_{\phi,n} [\cos(n\phi_{ijk\ell} + \delta_n) + 1] + \sum_{\substack{\text{non-bonded} \\ \text{pairs}}} \left[\frac{q_i q_j}{4\pi\epsilon_0 r_{ij}} + \frac{A_{ij}}{r_{ij}^{12}} - \frac{B_{ij}}{r_{ij}^{6}} \right]. \quad (1.37)$$

The first term in eqn (1.37) is a sum over all bonds, with an equilibrium bond-length r_0. There is one term for every pair ij of directly connected atoms. In some force fields the harmonic potential can be replaced by a more realistic functional form, such as the Morse potential, or the bonds can be fixed at their equilibrium values. The second term is a sum over all bond angles. There is one term for each set of three connected atoms ijk and it usually has a quadratic form. The third term is the sum over all torsions involving four connected atoms $ijk\ell$. In principle, this is an expansion in trigonometric functions with different values of n, the multiplicity (i.e. the number of minima in a rotation of 2π around the j–k bond); many force fields fix $n = 3$. This term can also include improper torsions, where the four atoms defining the angle are not all connected by covalent bonds; such terms serve primarily to enforce planarity around sp^2 centres and use a variety of functional forms (Tuzun et al., 1997). The fourth term is a sum over the non-bonded interactions (between molecules and within molecules). In particular, it describes the electrostatic and repulsion–dispersion interactions. It invariably excludes 1–2 and 1–3 pairs in the same molecule. Some force fields do include a non-bonded 1–4 interaction but the parameters A'_{ij}, B'_{ij} describing this interaction can be different from the values for atoms separated by more than three bonds (a scaling factor of 0.4 is used in the param19 force field of CHARMM (Brooks et al., 1983)). In some force fields, the r_{ij}^{-12}

Table 1.2 Force fields and their domains of application. This list is not complete and simply includes representative examples of some of the force fields commonly used in liquid-state simulations.

Force field	Class	Domain of Application	Source
OPLS	I	peptides, small organics	Jorgensen et al. (1996)
CHARMM22	I	proteins with explicit water	Mackerell et al. (1998)
CHARMM27	I	DNA, RNA, and lipids	Mackerell et al. (1998)
AMBER ff99	I	peptides, small organics, RESP charges	Wang et al. (2000)
GAFF	I	small organics, drug design	Wang et al. (2004)
GROMOS ffG45a3	I	lipids, micelles	Schuler et al. (2001)
COMPASS	II	small molecules, polymers	Sun (1998)
clayFF	II	hydrated minerals	Cygan et al. (2004)
MM4	II	small organics, coordination compounds	Allinger et al. (1996)
UFF	II	full Periodic Table (including actinides)	Rappe et al. (1992)
AMBER ff02	III	polarizable atoms	Cieplak et al. (2001)
AMOEBA	III	polarizable multipoles, distributed multipoles	Ponder et al. (2010)
MARTINI	III	coarse-grained, proteins, lipids, polymers	Marrink et al. (2007)
ReaxFF	III	chemical reactions	van Duin et al. (2001)

repulsion (associated with the Lennard-Jones potential) is replaced by an r_{ij}^{-9} repulsion which can produce better agreement with direct quantum calculations of the repulsion (Hagler et al., 1979; Halgren, 1992). The exponential form of the repulsion ($A \exp(-Br_{ij})$) was used in earlier versions of the AMBER force fields (MM2 and MM3) but has now been replaced by the r_{ij}^{-12} repulsion. The cross-interactions for the parameters in the repulsion–dispersion potential are often described using the Lorentz–Berthelot combining rules or an alternative such as the Slater–Kirkwood formula (Slater and Kirkwood, 1931). If these crossed interactions are important in the model they can be determined directly by fitting to experiment. In class I force fields, a simple Coulombic term is used to describe the interaction between the partial charges, which represent the electrostatic interactions between molecules.

Different parameters are required for different atoms in different environments, and all of the atom types in the model must be specified. For example, in the GROMOS force field ffG45a3 (Schuler et al., 2001), there are 12 types of C atoms, six Os, six Ns, four Cls, three Hs, two Ss, two Cus and one type for each of the remaining common atoms. The parameters $\{k_r, k_\theta, k_{\phi,n}, \delta_n, q_i, q_j, A_{ij}, B_{ij}\}$ are then specified for combinations of the atom types. For example, in a peptide chain, which contains C, N, and C_α atom types along the backbone (where C is a carbon additionally double-bonded to an oxygen and C_α is a carbon additionally connected to a hydrogen and a side chain) we would require k_r for the C–N stretch, a different k_r for the N–C_α stretch, k_θ for the C–N–C_α bend, $k_{\phi,n}$, for the C–N–C_α–C torsion, and additional parameters for the other bends and torsion in the backbone.

All-atom force fields provide parameters for every type of atom in a system, including hydrogen, while united-atom force fields treat the hydrogen and carbon atoms in each terminal methyl and each methylene bridge as a single interaction centre.

A class II force field normally adds cubic or anharmonic terms to the stretching potentials and defines explicit off-diagonal elements in the force constant matrix. Thus, the force field will contain terms of the form

$$v^{\text{str-str}}(r_{12}, r_{23}) = k_{12,23}(r_{12} - r_{12,0})(r_{23} - r_{23,0})$$
$$v^{\text{bend-str}}(\theta_{123}, r_{12}) = k_{123,12}(\theta_{123} - \theta_{123,0})(r_{12} - r_{12,0}) \tag{1.38}$$

where r_{12} and r_{23} are two adjacent bonds in the molecule, which include the angle θ_{123}. These additional potentials represent the fact that bonds, angles and torsions are not independent in molecules. Most cross-terms involve two internal coordinates and Dinur and Hagler (1991) have used quantum mechanical calculations to show that the stretch–stretch, stretch–bend, bend–bend, stretch–torsion, and bend–bend–torsion are the important coupling terms. The cross-terms are essential to include in models when attempting to calculate accurate vibrational frequencies. Despite the additional complexity, Class II force fields, such as COMPASS and CFF, have been used to good effect in liquid-state simulations (Peng et al., 1997; Sun, 1998).

Class III force fields go beyond the basic prescription to include more accurate representations of the electrostatic interactions between molecules and the inclusion of polarizability (as discussed in Section 1.3.3). For example, the AMOEBA force field includes distributed multipoles and the atom polarizabilities with the Thole modification of the interaction tensor. This class would also include coarse-grained force fields such as MAR-TINI used to model lipids, proteins, and carbohydrates (see Section 1.3.4) and force fields specifically designed to model chemical reactions such as REAXFF. REAXFF includes a set of relationships between the bond distance and the bond order of a particular covalent bond. Once the bond order is determined, the associated bond energy can be calculated. This procedure results in proper dissociation of bonds to separated atoms at the appropriate distances.

After many decades of force field development, there are still considerable differences between the predictions from even the Class I force fields. In an excellent review of the field, Ponder and Case (2003) compare simulations of a solvated dipeptide using CHARMM27, AMBER94, and OPLS-aa force fields to map the free energy of the dipeptide as a function of the two torsional angles, ψ and ϕ. All three force fields exhibit $\psi-\phi$ maps that are different from one another and different from the results of an *ab initio* simulation of the same problem. In contrast, in considering the liquid-state properties for butane, methanol, and N-methylacetamide, Kaminski and Jorgensen (1996) demonstrated reasonable agreement between the AMBER94 and OPLS force field, both of which had been fitted to liquid-state properties. In this study the MMFF94 force field, that had been optimized for gas-phase geometries, needed to be adjusted to obtain the same level of agreement when applied to the liquids. One important point is that it is not possible to mix and match different force fields. They have been optimized as a whole and one should not attempt to use parts of one field with parts of another. This means that devising force fields to simulate very different materials interacting with each other is a particular challenge. As an illustration, the steps taken to model the adsorption of biomolecules on the surface of metallic gold, in water, are discussed in Example 1.2.

It is difficult to make blanket recommendations concerning the use of particular force fields. Individual researchers will need to understand the kind of problems for which the force field has been optimized to know if it can be applied to their particular problem. One sensible strategy would be to check the effect of using a few of the more common force fields on the problem to understand the sensitivity of the results to this choice.

An important advantage of the force-field approach is that that particular fields are often associated with large simulation programs. The acronyms CHARMM, AMBER, and GROMOS can also stand for large molecular dynamics codes which have been designed to work with the particular forms of a field and there are many examples of other codes such as LAMMPS (Plimpton, 1995) and DL_POLY (Todorov and Smith, 2011) that can take standard force fields with some adjustments. There is also a huge industry of analysis and data manipulation programmes that have grown with the major force fields and codes.

Of course, using these programmes as black-boxes is never a good idea and we plan in this book to dig into the principles behind such codes. Equally, if one can take advantage of the many years of careful development that have gone into producing these packages in an informed way, an enormous range of complicated and important applications can be tackled fairly quickly.

Example 1.2 Peptide–gold potentials

Peptides, short chains of amino acids, may be designed so as to specifically favour adsorption on certain material surfaces. This underpins a range of possible bio-nanotechnology applications (Care et al., 2015). Understanding this selectivity and specificity is a great challenge to molecular simulation: clearly the adsorption free energy depends on many factors, including changes in peptide flexibility, its solvation, and displacement of the water layer at the surface. Measurement of adsorption free energies requires advanced simulation techniques (see Chapters 4 and 9); modelling the potential energy of interaction between the surface and individual amino acids is itself challenging, involving the cross-interaction between two very different materials (Di Felice and Corni, 2011; Heinz and Ramezani-Dakhel, 2016). Here we focus on recent attempts to model peptide interactions with the surface(s) of metallic gold.

A simple Lennard-Jones force field for a range of FCC metals, including gold, has been proposed (Heinz et al., 2008): ϵ_{AuAu} and σ_{AuAu} are chosen to reproduce various experimental bulk and surface properties, under ambient conditions. Water and peptide atom–Au parameters are obtained by standard combining rules. Feng et al. (2011) have used this potential to study the adsorption of individual amino acids on gold, while Cannon et al. (2015) have used it to highlight solvent effects in peptide adsorption. A different parameterization, similar in spirit, has been derived independently (Vila Verde et al., 2009; 2011). The whole method has been generalized to cover a range of other materials (Heinz et al., 2013). Compatibility with standard force fields, such as CHARMM, is an advantage of this approach; polarization of the metal, and chemisorption, however, are neglected.

A purely dispersive potential of this kind may have limitations when one considers structure: adsorption (of water molecules or peptide atoms) onto hollow sites on the surface is strongly favoured. On metallic surfaces, however, adsorption on top of surface atoms is often preferred, as indicated by first-principles simulations. In the GolP force field (Iori et al., 2009), dynamical polarization of gold atoms is represented by a rotating dipole, and virtual interaction sites are introduced to tackle the hollow-site adsorption problem. GolP is parameterized using extensive first-principles calculations and experimental data, with special consideration given to surface interactions with sp^2-hybridized carbons. An extension, GolP–CHARMM, reparameterized for compatibility with CHARMM, also allows consideration of different gold surfaces (Wright et al., 2013b,a), opening up the study of facet selectivity (Wright et al., 2015). In GolP, the gold atoms are held fixed during the simulation.

Tang et al. (2013) have compared GolP results with experimental studies of peptide adsorption, and with the force field of Heinz et al. (2008). While both models perform reasonably well in describing the trend in amino acid adsorption energies, there are areas such as the prediction of water orientation in the surface layer where GolP–CHARMM agrees better with first-principles simulations (Nadler and Sanz, 2012). This approach may allow one to separate the enthalpic contributions to the binding free energy, and ascribe them to individual residues (Corni et al., 2013; Tang et al., 2013).

1.6   Studying small systems

1.6.1   Introduction

Simulations are usually performed on a small number of molecules, $10 \le N \le 10\,000$. The size of the system is limited by the available storage on the host computer, and, more crucially, by the speed of execution of the program. The time taken for a double loop used to evaluate the forces or potential energy is proportional to N^2. Special techniques (see Chapter 5) may reduce this dependence to $O(N)$, for very large systems, but the force/energy loop almost inevitably dictates the overall speed and, clearly, smaller systems will always be less expensive. If we are interested in the properties of a very small liquid drop, or a microcrystal, then the simulation will be straightforward. The cohesive forces between molecules may be sufficient to hold the system together unaided during the course of a simulation, otherwise our set of N molecules may be confined by a potential representing a container, which prevents them from drifting apart (see Chapter 13). These arrangements, however, are not satisfactory for the simulation of bulk liquids. A major obstacle to such a simulation is the large fraction of molecules which lie on the surface of any small sample; for 1000 molecules arranged in a $10 \times 10 \times 10$ cube, $8^3 = 512$ lie in the interior, leaving 488 (nearly half!) on the cube faces. Even for $N = 100^3 = 10^6$ molecules, 6 % of them will lie on the surface. Whether or not the cube is surrounded by a containing wall, molecules on the surface will experience quite different forces from those in bulk.

1.6.2   Periodic boundary conditions.

The problem of surface effects can be overcome by implementing periodic boundary conditions (Born and von Karman, 1912). The cubic box is replicated throughout space to form an infinite lattice. In the course of the simulation, as a molecule moves in the original box, its periodic image in each of the neighbouring boxes moves in exactly the same way. Thus, as a molecule leaves the central box, one of its images will enter through the opposite face. There are no walls at the boundary of the central box, and no surface molecules. This box simply forms a convenient axis system for measuring the coordinates of the N molecules. A two-dimensional version of such a periodic system is shown in Fig. 1.13. The duplicate boxes are labeled A, B, C, etc., in an arbitrary fashion. As particle 1 moves through a boundary, its images 1_A, 1_B, etc. (where the subscript specifies in which box the image lies) move across their corresponding boundaries. The number density in the central box (and hence in the entire system) is conserved. It is not necessary to store the coordinates of all the images in a simulation (an infinite number!), just the molecules in the central box. When a molecule leaves the box by crossing a boundary, attention may be switched to the image just entering. It is sometimes useful to picture the basic simulation box (in our two-dimensional example) as being rolled up to form the surface of a three-dimensional torus or doughnut, when there is no need to consider an infinite number of replicas of the system, nor any image particles. This correctly represents the topology of the system, if not the geometry. A similar analogy exists for a three-dimensional periodic system, but this is more difficult to visualize!

   It is important to ask if the properties of a small, infinitely periodic, system and the macroscopic system which it represents are the same. This will depend both on the range of the intermolecular potential and the phenomenon under investigation. For a fluid of

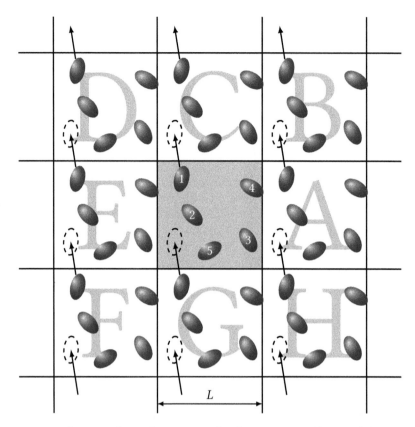

Fig. 1.13 A two-dimensional periodic system. Molecules can enter and leave each box across each of the four edges. In a three-dimensional example, molecules would be free to cross any of the six cube faces.

Lennard-Jones atoms it should be possible to perform a simulation in a cubic box of side $L \approx 6\,\sigma$ without a particle being able to 'sense' the symmetry of the periodic lattice. If the potential is long range (i.e. $v(r) \sim r^{-\nu}$ where ν is less than the dimensionality of the system) there will be a substantial interaction between a particle and its own images in neighbouring boxes, and consequently the symmetry of the cell structure is imposed on a fluid which is in reality isotropic. The methods used to cope with long-range potentials, for example in the simulation of charged ions ($v(r) \sim r^{-1}$) and dipolar molecules ($v(r) \sim r^3$), are discussed in Chapter 5. We know that even in the case of short-range potentials the periodic boundary conditions can induce anisotropies in the fluid structure (Mandell, 1976; Impey et al., 1981). These effects are pronounced for small system sizes ($N = 100$) and for properties such as the g_2 light scattering factor (see Chapter 2), which has a substantial long-range contribution. Pratt and Haan (1981) have developed theoretical methods for investigating the effects of boundary conditions on equilibrium properties.

The use of periodic boundary conditions inhibits the occurrence of long-wavelength fluctuations. For a cube of side L, the periodicity will suppress any density waves with a wavelength greater than L. Thus, it would not be possible to simulate a liquid close

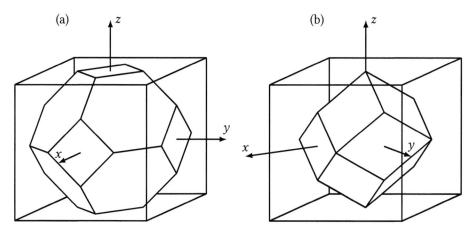

Fig. 1.14 Non-cubic, space-filling, simulation boxes. (a) The truncated octahedron and its containing cube; (b) the rhombic dodecahedron and its containing cube. The axes are those used in Code 1.4 and Code 1.5 of Section 1.6.4.

to the gas–liquid critical point, where the range of critical fluctuations is macroscopic. Furthermore, transitions which are known to be first order often exhibit the characteristics of higher-order transitions when modelled in a small box, because of the suppression of fluctuations. Examples are the nematic–isotropic transition in liquid crystals (Luckhurst and Simpson, 1982) and the solid–plastic-crystal transition for N_2 adsorbed on graphite (Mouritsen and Berlinsky, 1982). The same limitations apply to the simulation of long-wavelength phonons in model solids, where in addition, the cell periodicity picks out a discrete set of available wavevectors (i.e. $\mathbf{k} = (n_x, n_y, n_z)2\pi/L$, where n_x, n_y, n_z, are integers) in the first Brillouin zone (Klein and Weis, 1977). Periodic boundary conditions have also been shown to affect the rate at which a simulated liquid nucleates and forms a solid or glass when it is rapidly cooled (Honeycutt and Andersen, 1984).

Despite the preceding remarks, the common experience in simulation work is that periodic boundary conditions have little effect on the equilibrium thermodynamic properties and structures of fluids away from phase transitions and where the interactions are short-ranged. It is always sensible to check that this is true for each model studied. If the resources are available, it should be standard practice to increase the number of molecules (and the box size, so as to maintain constant density) and rerun the simulations. The cubic box has been used almost exclusively in computer simulation studies because of its geometrical simplicity. Of the four remaining semi-regular space-filling polyhedra, the rhombic dodecahedron (Wang and Krumhansl, 1972), and the truncated octahedron (Adams, 1979; 1980) have also been studied. These boxes are illustrated in Fig. 1.14. They are more nearly spherical than the cube, which may be useful for simulating liquids, whose structure is spatially isotropic. In addition, for a given number density, the distance between periodic images is larger than in the cube. This property is useful in calculating distribution functions and structure factors (see Chapters 2 and 8). As we shall see in Section 1.6.4, they are only slightly more complicated to implement in simulations than cubic boxes.

(a)

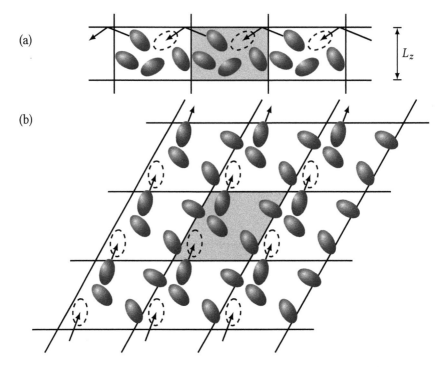

L_z

(b)

Fig. 1.15 Periodic boundary conditions used in the simulation of adsorption (see e.g. Severin and Tildesley, 1980). (a) A side view of the box. There is a reflecting boundary at height L_z. (b) A top view, showing the rhombic shape (i.e. the same geometry as the underlying graphite lattice). Periodic boundary conditions in this geometry are implemented in Code 1.6.

So far, we have tacitly assumed that there is no external potential, that is, no v_1, term in eqns (1.4) and (1.5). If such a potential is present, then either it must have the same periodicity as the simulation box, or the periodic boundaries must be abandoned. In some cases, it is not appropriate to employ periodic boundary conditions in each of the three coordinate directions. In the simulation of CH_4 on graphite (Severin and Tildesley, 1980) the simulation box, shown in Fig. 1.15, is periodic in the plane of the surface. In the z-direction, the graphite surface forms the lower boundary of the box, and the bulk of the adsorbate is in the region just above the graphite. Any molecule in the gas above the surface is confined by reversing its velocity should it cross a plane at a height L_z above the surface. If L_z is sufficiently large, this reflecting boundary will not influence the behaviour of the adsorbed monolayer. In the plane of the surface, the shape of the periodic box is a rhombus of side L. This conforms to the symmetry of the underlying graphite. Similar boxes have been used in the simulation of the electrical double layer (Torrie and Valleau, 1979), of the liquid–vapour surface (Chapela et al., 1977), and of fluids in small pores (Subramanian and Davis, 1979).

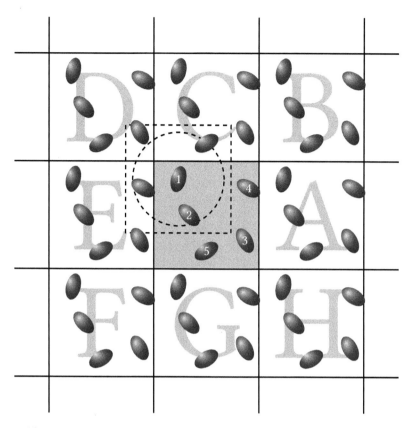

Fig. 1.16 The minimum image convention in a two-dimensional system. The central 'box' contains five molecules. The dashed 'box' constructed with molecule 1 at its centre also contains five molecules. The dashed circle represents the cutoff.

1.6.3 Potential truncation

Now we must turn to the question of calculating properties of systems subject to periodic boundary conditions. The heart of the MC and MD programs involves the calculation of the potential energy of a particular configuration, and, in the case of MD, the forces acting on all molecules. Consider how we would calculate the force on molecule 1, or those contributions to the potential energy involving molecule 1, assuming pairwise additivity. We must include interactions between molecule 1 and every other molecule i in the simulation box. There are $N - 1$ terms in this sum. However, in principle, we must also include all interactions between molecule 1 and images i_A, i_B, etc. lying in the surrounding boxes. This is an infinite number of terms, and of course is impossible to calculate in practice. For a short-range potential-energy function, we may restrict this summation by making an approximation. Consider molecule 1 to rest at the centre of a region which has the same size and shape as the basic simulation box (see Fig. 1.16). Molecule 1 interacts with all the molecules whose centres lie within this region, that is, with the closest periodic images of the other $N - 1$ molecules. This is called the 'minimum image convention': for example, in Fig. 1.16, molecule 1 could interact with molecules 2, 3_D, 4_E, and 5_C. This technique, which is a natural consequence of the periodic boundary conditions, was first used in simulation by Metropolis et al. (1953).

In the minimum image convention, then, the calculation of the potential energy due to pairwise-additive interactions involves $\frac{1}{2}N(N-1)$ terms. This may still be a very substantial calculation for a system of (say) 1000 particles. A further approximation significantly improves this situation. The largest contribution to the potential and forces comes from neighbours close to the molecule of interest, and for short-range forces we normally apply a spherical cutoff. This means setting the pair potential $v(r)$ to zero for $r \geq r_c$, where r_c is the cutoff distance. The dashed circle in Fig. 1.16 represents a cutoff, and in this case molecules 2, 4_E and 5_C contribute to the force on 1, since their centres lie inside the cutoff, whereas molecule 3_D does not contribute. In a cubic simulation box of side L, the number of neighbours explicitly considered is reduced by a factor of approximately $4\pi r_c^3/3L^3$, and this may be a substantial saving. The introduction of a spherical cutoff should be a small perturbation, and the cutoff distance should be sufficiently large to ensure this. As an example, in the simulation of Lennard-Jones atoms the value of the pair potential at the boundary of a cutoff sphere of typical radius $r_c = 2.5\,\sigma$ is just 1.6 % of the well depth. Of course, the penalty of applying a spherical cutoff is that the thermodynamic (and other) properties of the model fluid will no longer be exactly the same as for (say) the non-truncated, Lennard-Jones fluid. As we shall see in Chapter 2, it is possible to apply long-range corrections to such results so as to recover, approximately, the desired information.

The cutoff distance must be no greater than $\frac{1}{2}L$ for consistency with the minimum image convention. In the non-cubic simulation boxes of Fig. 1.14, for a given density and number of particles, r_c may take somewhat larger values than in the cubic case. Looked at another way, an advantage of non-cubic boundary conditions is that they permit simulations with a given cutoff distance and density to be conducted using fewer particles. As an example, a simulation in a cubic box, with r_c set equal to $\frac{1}{2}L$, might involve $N = 256$ molecules; taking the same density, the same cutoff could be used in a simulation of 197 molecules in a truncated octahedron, or just 181 molecules in a rhombic dodecahedron.

1.6.4   Computer code for periodic boundaries

How do we handle periodic boundaries and the minimum image convention in a simulation program? Let us assume that, initially, the N molecules in the simulation lie within a cubic box of side L, with the origin at its centre, that is, all coordinates lie in the range $(-\frac{1}{2}L, \frac{1}{2}L)$. As the simulation proceeds, these molecules move about the infinite periodic system. When a molecule leaves the box by crossing one of the boundaries, it is usual to switch attention to the image molecule entering the box by simply adding L to, or subtracting L from, the appropriate coordinate. One simple way to do this uses an IF statement to test the positions immediately after the molecules have been moved (whether by MC or MD). For example,

```
IF ( r(1,i) >  box2 ) r(1,i) = r(1,i) - box
IF ( r(1,i) < -box2 ) r(1,i) = r(1,i) + box
```

where the first index 1 selects the x coordinate. Similar statements are applied to the y and z coordinates, or a vector assignment may be applied to all components at once

```
WHERE ( r(:,i) >  box2 ) r(:,i) = r(:,i) - box
WHERE ( r(:,i) < -box2 ) r(:,i) = r(:,i) + box
```

Here, box is a variable containing the box length L, and box2 is just $\frac{1}{2}L$. An alternative to the IF statement is to use arithmetic functions to calculate the correct number of box lengths to be added or subtracted. For example,

```
r(:,i) = r(:,i) - box * ANINT ( r(:,i) / box )
```

The function ANINT(x) returns the nearest integer to x, converting the result back to type REAL; thus ANINT(-0.49) has the value 0.0, whereas ANINT(-0.51) is −1.0. In Fortran, this function returns an array-valued result, computed component by component, if given an array argument. As we shall see in Chapter 5, there are faster ways of coding this up, especially for large system sizes.

By using these methods, we always have available the coordinates of the N molecules that currently lie in the 'central' box. It is not strictly necessary to do this; we could, instead, use uncorrected coordinates, and follow the motion of the N molecules that were in the central box at the start of the simulation. Indeed, as we shall see in Chapters 2 and 8, for calculation of transport coefficients it may be most desirable to have a set of uncorrected positions on hand. If it is decided to do this, however, care must be taken that the minimum image convention is correctly applied, so as to work out the vector between the two closest images of a pair of molecules, no matter how many 'boxes' apart they may be. This means, in general, adding or subtracting an integer number of box lengths (rather than just one box length).

The minimum image convention may be coded in the same way as the periodic boundary adjustments. Of the two methods just mentioned, the arithmetic formula is usually preferable, being simpler; the use of IF statements inside the inner loop may reduce program efficiency (see Appendix A). Immediately after calculating a pair separation vector, the following statements should be applied:

```
rij(:) = rij(:) - box * ANINT ( rij(:) / box )
```

This code is guaranteed to yield the minimum image vector, no matter how many 'box lengths' apart the original images may be. For cuboidal, rather than cubic, boxes, the variable box may be an array of three elements, holding the x, y, and z box lengths, without essentially changing the code.

The calculation of minimum image distances is simplified by the use of reduced units: the length of the box is taken to define the fundamental unit of length in the simulation. By setting $L = 1$, with particle coordinates nominally in the range $(-\frac{1}{2}, +\frac{1}{2})$, the minimum image correction becomes

```
rij(:) = rij(:) - ANINT ( rij(:) )
```

which is simpler, and faster, than the code for a general box length. This approach is an alternative to the use of the pair potential to define reduced units as discussed in Appendix B, and is more generally applicable. For this reason a simulation box of unit length is adopted in most of the examples given in this book.

Code 1.4 Periodic boundaries for truncated octahedron

This code snippet applies the truncated octahedron periodic boundary correction to a position vector r_i, or equivalently the minimum image convention to a displacement vector r_{ij}, provided as the array r. The box is centred at the origin and the containing cube is of unit length (see Fig. 1.14(a)). The Fortran AINT function rounds towards zero, producing a real-valued integer result: for example AINT(-0.51) and AINT(0.51) both have the value 0.0, whereas AINT(-1.8) is −1.0. The result of the Fortran SIGN function has the absolute value of its first argument and the sign of its second.

```
REAL , DIMENSION(3) :: r
REAL               :: corr
REAL , PARAMETER   :: r75 = 4.0 / 3.0

r(:) = r(:) - ANINT ( r(:) )
corr = 0.5 * AINT ( r75 * SUM ( ABS ( r(:) ) ) )
r(:) = r(:) - SIGN ( corr, r(:) )
```

Code 1.5 Periodic boundaries for rhombic dodecahedron

This code snippet applies the rhombic dodecahedron periodic boundary correction to a position vector r_i, or equivalently the minimum image convention to a displacement vector r_{ij}, provided as the array r. The box is centred at the origin and the side of the containing cube is $\sqrt{2}$ (see Fig. 1.14(b)).

```
REAL , DIMENSION(3) :: r
REAL               :: corr
REAL , PARAMETER   :: rt2 = SQRT(2.0), rrt2 = 1.0 / rt2

r(1) = r(1) - ANINT ( r(1) )
r(2) = r(2) - ANINT ( r(2) )
r(3) = r(3) - rt2 * ANINT ( rrt2 * r(3) )
corr = 0.5 * AINT ( ABS(r(1)) + ABS(r(2)) + rt2*ABS(r(3)) )
r(1) = r(1) - SIGN ( corr, r(1) )
r(2) = r(2) - SIGN ( corr, r(2) )
r(3) = r(3) - SIGN ( corr, r(3) ) * rt2
```

There are several alternative ways of coding the minimum image corrections, some of which rely on the images being in the same, central box (i.e. on the periodic boundary correction being applied whenever the molecules move). Some of these methods, for cubic boxes, are discussed in Appendix A. We have also mentioned the possibility of conducting simulations in non-cubic periodic boundary conditions. An implementation of the minimum image correction for the truncated octahedron (Adams, 1983a) is given

Code 1.6 Periodic boundaries for rhombus

Here we apply corrections for the rhombic box in two dimensions x, y. In most applications the molecules will be confined in the z direction by real walls rather than by periodic boundaries, so we assume that this coordinate may be left unchanged. The box is centred at the origin. The x axis lies along one side of the rhombus, which is of unit length (see Fig. 1.15). The acute angle of the rhombus is 60°.

```
REAL , DIMENSION(3) :: r
REAL , PARAMETER      :: rt3 = SQRT(3.0), rrt3 = 1.0 / rt3
REAL , PARAMETER      :: rt32 = rt3 / 2.0, rrt32 = 1.0 / rt32

r(1) = r(1) - ANINT ( r(1) - rrt3 * r(2) ) &
       &     - ANINT ( rrt32 * r(2) ) * 0.5
r(2) = r(2) - ANINT ( rrt32 * r(2) ) * rt32
```

in Code 1.4. A similar correction for the rhombic dodecahedron (Smith, 1983) appears in Code 1.5. This is a little more complicated than the code for the truncated octahedron, and the gain small, so that the latter is usually preferable. We also give in Code 1.6 the code for the two-dimensional rhombic box often used in surface simulation.

Now we turn to the implementation of a spherical cutoff, that is, we wish to set the pair potential (and all forces) to zero if the pair separation lies outside some distance r_c. It is easy to compute the square of the particle separation r_{ij} and, rather than waste time taking the square root of this quantity, it is fastest to compare this with the square of r_c which might be computed earlier and stored in a variable r_cut_sq. After computing the minimum image intermolecular vector, the following statements would be employed:

```
rij_sq = SUM ( rij(:) ** 2 )
IF ( rij_sq < r_cut_sq ) THEN
   ... compute i-j interaction ...
END IF
```

In a large system, it may be worthwhile to apply separate tests for the x, y, and z directions or some similar scheme.

```
IF ( ABS ( rij(1) ) < r_cut ) THEN
   IF ( ABS ( rij(2) ) < r_cut ) THEN
      IF ( ABS ( rij(3) ) < r_cut ) THEN
         rij_sq = SUM ( rij(:) ** 2 )
         IF ( rij_sq < r_cut_sq ) THEN
            ... compute i-j interaction ...
         END IF
      END IF
   END IF
END IF
```

The time saved in dropping out of this part of the program at any early stage must be weighed against the overheads of extra calculation and testing. In Chapter 5 we discuss the more complicated time-saving tricks used in the simulations of large systems.

1.6.5   Spherical boundary conditions

As an alternative to the standard periodic boundary conditions for simulating bulk liquids, a two-dimensional system may be embedded in the surface of a sphere without introducing any physical boundaries (Hansen et al., 1979), and the idea may be extended to consider a three-dimensional system as being the surface of a hypersphere (Kratky, 1980; Kratky and Schreiner, 1982). The spherical or hyperspherical system is finite: it cannot be considered as part of an infinitely repeating periodic system. In this case, non-Euclidean geometry is an unavoidable complication, and distances between particles are typically measured along the great circle geodesics joining them. However, the effects of the curved geometry will decrease as the system size increases, and such 'spherical boundary conditions' are expected to be a valid method of simulating bulk liquids. Interesting differences from the standard periodic boundary conditions, particularly close to any solid–liquid phase transition, will result from the different topology. Periodic boundaries will be biased in favour of the formation of a solid with a lattice structure which matches the simulation box. Spherical boundaries, on the other hand, are not consistent with periodic lattices, so the liquid state will be thermodynamically favoured in most simulations using this technique, and crystalline phases will inevitably contain defects. Similar considerations may apply to liquid-crystalline phases.

1.6.6   Periodic boundary conditions for three-body potentials

Finally, we note that some care is required when using the minimum image convention with three-body potentials such as the Axilrod–Teller potential (see Appendix C). This problem is illustrated in Fig. 1.17. In Fig. 1.17(a), atom 1 is at the centre of its box, of side L, and atoms 2 and 3_E are the two minimum images used in the calculation of the pair potential. However atom 3 is the minimum image of atom 2 and a straightforward application of the minimum image algorithm will lead to the incorrect triplet 123 rather than 123_E.

Attard (1992) has shown that this problem can be solved using the following statements for the separation vector

```
REAL , DIMENSION(3) :: rij, rik, rjk, tij, tik
tij(:) = box * ANINT ( rij(:) / box )
tik(:) = box * ANINT ( rik(:) / box )
rij(:) = rij(:) - tij(:)
rik(:) = rik(:) - tik(:)
rjk(:) = rjk(:) + tij(:) - tik(:)
```

Normally the three-body potential is set to zero if one side of the triangle is greater than $L/2$.

Some workers have taken a more brute-force approach (Sadus and Prausnitz, 1996; Marcelli and Sadus, 2012). If the potential cutoff r_c is set to $L/4$, the only triplets that contribute to the potential are those where all of the three atoms are within a box of side

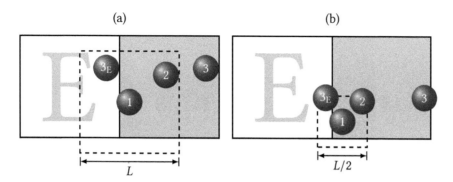

Fig. 1.17 Periodic boundary conditions and the minimum image convention for a triplet interaction: (a) an inconsistency in the triplet configuration for a cutoff of $L/2$; (b) a consistent triplet with a cutoff of $L/4$.

$L/2$ (as shown in Fig. 1.17(b)). Each of the atoms is then always the unique minimum image of the other two and the triplet is unambiguously determined with the normal minimum image calculation. This method works well. However, at a fixed density the simulation will need to include eight times as many atoms in circumstances where the additional calculation of the three-body force is particularly expensive.

2

Statistical mechanics

Computer simulation generates information at the microscopic level (atomic and molecular positions, velocities, etc.) and the conversion of this very detailed information into macroscopic terms (pressure, internal energy, etc.) is the province of statistical mechanics. It is not our aim to provide a text in this field since many excellent sources are available (Hill, 1956; McQuarrie, 1976; Landau and Lifshitz, 1980; Friedman, 1985; Chandler, 1987; Tuckerman, 2010; Swendsen, 2012; Hansen and McDonald, 2013). In this chapter, our aim is to summarize those aspects of the subject which are of most interest to the computer simulator.

2.1 Sampling from ensembles

Let us consider, for simplicity, a one-component macroscopic system; extension to a multicomponent system is straightforward. The thermodynamic state of such a system is usually defined by a small set of parameters (such as the number of particles N, the temperature T, and the pressure P). Other thermodynamic properties (density ρ, chemical potential μ, heat capacity C_V, etc.) may be derived through knowledge of the equations of state and the fundamental equations of thermodynamics. Even quantities such as the diffusion coefficient D, the shear viscosity η, and the structure factor $S(k)$ are state functions: although they clearly say something about the microscopic structure and dynamics of the system, their values are completely dictated by the few variables (e.g. NPT) characterizing the thermodynamic state, not by the very many atomic positions and momenta that define the instantaneous mechanical state. These positions and momenta can be thought of as coordinates in a multidimensional space: phase space. For a system of N atoms, this space has $6N$ dimensions. Let us use the abbreviation Γ for a particular point in phase space, and suppose that we can write the instantaneous value of some property $\mathcal{A}$ (it might be the potential energy) as a function $\mathcal{A}(\Gamma)$. The system evolves in time so that Γ, and hence $\mathcal{A}(\Gamma)$ will change. It is reasonable to assume that the experimentally observable 'macroscopic' property $\mathcal{A}_{\text{obs}}$ is really the time average of $\mathcal{A}(\Gamma)$ taken over a long time interval:

$$\mathcal{A}_{\text{obs}} = \langle \mathcal{A} \rangle_{\text{time}} = \Big\langle \mathcal{A}\big(\Gamma(t)\big) \Big\rangle_{\text{time}} = \lim_{t_{\text{obs}} \to \infty} \frac{1}{t_{\text{obs}}} \int_0^{t_{\text{obs}}} \mathcal{A}\big(\Gamma(t)\big)\, \mathrm{d}t. \tag{2.1}$$

The equations governing this time evolution, Newton's equations of motion in a simple classical system, are of course well known. They are just a system of ordinary differential

Computer Simulation of Liquids. Second Edition. M. P. Allen and D. J. Tildesley.
© M. P. Allen and D. J. Tildesley 2017. Published in 2017 by Oxford University Press.

equations: solving them on a computer, to a desired accuracy, is a practical proposition for, say, 10^5 particles, although not for a truly macroscopic number (e.g. 10^{23}). So far as the calculation of time averages is concerned, we clearly cannot hope to extend the integration of eqn (2.1) to infinite time, but might be satisfied to average over a long finite time τ_{obs}. This is exactly what we do in a molecular dynamics simulation. In fact, the equations of motion are usually solved on a step-by-step basis, that is, a large finite number τ_{obs} of timesteps, of length $\delta t = t_{obs}/\tau_{obs}$, are taken. In this case, we may rewrite eqn (2.1) in the form

$$\mathcal{A}_{obs} = \langle \mathcal{A} \rangle_{time} = \frac{1}{\tau_{obs}} \sum_{\tau=1}^{\tau_{obs}} \mathcal{A}(\Gamma(t)). \tag{2.2}$$

In the summation, τ simply stands for an index running over the succession of timesteps. This analogy between the discrete τ and the continuous t is useful, even when, as we shall see in other examples, τ does not correspond to the passage of time in any physical sense.

The practical questions regarding the method are whether or not a sufficient region of phase space is explored by the system trajectory to yield satisfactory time averages within a feasible amount of computer time, and whether thermodynamic consistency can be attained between simulations with identical macroscopic parameters (density, energy, etc.) but different initial conditions (atomic positions and velocities). The answers to these questions are that such simulation runs are indeed within the power of modern computers, and that thermodynamically consistent results for liquid state properties can indeed be obtained, provided that attention is paid to the selection of initial conditions. We will turn to the technical details of the method in Chapter 3.

The calculation of time averages by MD is not the approach to thermodynamic properties implicit in conventional statistical mechanics. Because of the complexity of the time evolution of $\mathcal{A}(\Gamma(t))$ for large numbers of molecules, Gibbs suggested replacing the time average by the ensemble average. Here, we regard an ensemble as a collection of points Γ in phase space. The points are distributed according to a probability density $\rho(\Gamma)$. This function is determined by the chosen fixed macroscopic parameters (NPT, NVT, etc.), so we use the notation ρ_{NPT}, ρ_{NVT}, or, in general, ρ_{ens}. Each point represents a typical system at any particular instant of time. Each system evolves in time, according to the usual mechanical equations of motion, quite independently of the other systems. Consequently, in general, the phase space density $\rho_{ens}(\Gamma)$ will change with time. However, no systems are destroyed or created during this evolution, and Liouville's theorem, which is essentially a conservation law for probability density, states that $d\rho/dt = 0$ where d/dt denotes the total derivative with respect to time (following a state Γ as it moves). As an example, consider a set of N atoms with Cartesian coordinates $\mathbf{r}_i$, and momenta $\mathbf{p}_i$, in the classical approximation. The total time derivative is

$$\frac{d}{dt} = \frac{\partial}{\partial t} + \sum_i \dot{\mathbf{r}}_i \cdot \nabla_{\mathbf{r}_i} + \sum_i \dot{\mathbf{p}}_i \cdot \nabla_{\mathbf{p}_i} \tag{2.3a}$$

$$= \frac{\partial}{\partial t} + \dot{\mathbf{r}} \cdot \nabla_{\mathbf{r}} + \dot{\mathbf{p}} \cdot \nabla_{\mathbf{p}}. \tag{2.3b}$$

In eqn (2.3a), $\partial/\partial t$ represents differentiation, with respect to time, of a function; $\nabla_{\mathbf{r}_i}$, and $\nabla_{\mathbf{p}_i}$, are derivatives with respect to atomic position and momentum respectively; and $\dot{\mathbf{r}}_i$,

$\dot{p}_i$, signify the time derivatives of the position and momentum. Equation (2.3b) is the same equation written in a more compact way, and the equation may be further condensed by defining the Liouville operator L

$$iL = \left(\sum_i \dot{r}_i \cdot \nabla_{r_i} + \sum_i \dot{p}_i \cdot \nabla_{p_i} \right) = \left(\dot{r} \cdot \nabla_r + \dot{p} \cdot \nabla_p \right) \tag{2.4}$$

so that $d/dt = \partial/\partial t + iL$ and, using Liouville's theorem, we may write

$$\frac{\partial \rho_{ens}(\Gamma, t)}{\partial t} = -iL\rho_{ens}(\Gamma, t). \tag{2.5}$$

This equation tells us that the rate of change of ρ_{ens} at a particular fixed point in phase space is related to the flows into and out of that point. This equation has a formal solution

$$\rho_{ens}(\Gamma, t) = \exp(-iLt) \, \rho_{ens}(\Gamma, 0) \tag{2.6}$$

where the exponential of an operator really means a series expansion

$$\exp(-iLt) = 1 - iLt - \tfrac{1}{2}L^2t^2 + \cdots . \tag{2.7}$$

The equation of motion of a function like $\mathcal{A}(\Gamma)$, which does not depend explicitly on time, takes a conjugate form (McQuarrie, 1976):

$$\dot{\mathcal{A}}\big(\Gamma(t)\big) = iL\mathcal{A}\big(\Gamma(t)\big) \tag{2.8}$$

or

$$\mathcal{A}\big(\Gamma(t)\big) = \exp(iLt)\mathcal{A}\big(\Gamma(0)\big). \tag{2.9}$$

To be quite clear: in eqns (2.5) and (2.6) we consider the time-dependence of ρ_{ens} at a fixed point Γ in phase space; in eqns (2.8) and (2.9), $\mathcal{A}\big(\Gamma(t)\big)$ is time-dependent because we are following the time evolution $\Gamma(t)$ along a trajectory. This relationship is analogous to that between the Schrödinger and Heisenberg pictures in quantum mechanics.

If $\rho_{ens}(\Gamma)$ represents an equilibrium ensemble, then its time-dependence completely vanishes, $\partial \rho_{ens}/\partial t = 0$. The system evolution then becomes quite special. As each system leaves a particular state $\Gamma(\tau)$ and moves on to the next, $\Gamma(\tau + 1)$, another system arrives from state $\Gamma(\tau - 1)$ to replace it. The motion resembles a long and convoluted conga line at a crowded party (see Fig. 2.1). There might be several such processions, each passing through different regions of phase space. However, if these are all connected into just one trajectory that passes through all the points in phase space for which ρ_{ens} is non-zero (i.e. the procession forms a single, very long, closed circuit) then each system will eventually visit all the state points. Such a system is termed 'ergodic' and the time taken to complete a cycle (the Poincaré recurrence time) is immeasurably long for a many-particle system (and for many parties as well it seems).

One way of answering the question 'was it a good party?' would be to interview one of the participants, and ask for their time-averaged impressions. This is essentially what we do in a molecular dynamics simulation, when a representative system evolves deterministically in time. However, as indicated in Fig. 2.1, this time average might not

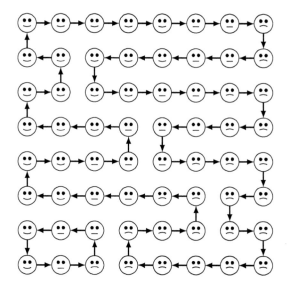

Fig. 2.1 A schematic representation of phase space. The circles represent different state points $(\mathbf{q}, \mathbf{p})$, and they are connected by a path representing the classical trajectory, analogous to a conga line at a party. Each state is characterized by some property (e.g. 'happiness' at the party). In an ergodic system, the single long trajectory would eventually pass through (or arbitrarily near) all states; in the bottom left corner of the diagram we symbolically indicate a disconnected region of six states which may or may not be practically important.

be representative of the whole trajectory: to be sure, it would have to be long enough to sample all the states. An alternative route to the average properties of our partygoers, would be to take photographs of all of them at the same time, assemble the complete collection of 'happy' and 'sad' faces, and take an average over them. This corresponds to replacing the time average in eqn (2.1) by an average taken over all the members of the ensemble, 'frozen' at a particular time:

$$\mathcal{A}_{\text{obs}} = \langle \mathcal{A} \rangle_{\text{ens}} = \left\langle \mathcal{A} \mid \rho_{\text{ens}} \right\rangle = \sum_{\Gamma} \mathcal{A}(\Gamma) \rho_{\text{ens}}(\Gamma). \tag{2.10}$$

The $\langle \mathcal{A} | \rho \rangle$ notation reminds us of the dependence of the average on both $\mathcal{A}$ and ρ: this is important when taking a thermodynamic derivative of $\mathcal{A}_{\text{obs}}$ (we must differentiate both parts) or when considering time-dependent properties (when the Schrödinger–Heisenberg analogy may be exploited). Actually, we will be concerned with the practical question of efficient and thorough sampling of phase space, which is not quite the same as the rigorous definition of ergodicity (for a fuller discussion, see Tolman, 1938). In terms of our analogy of conga lines, there should not be a preponderance of independent closed circuits ('cliques') in which individuals can become trapped and fail fully to sample the available space (this is important in parties as well as in simulations). An MD simulation which started in the disconnected six-state region of Fig. 2.1, for example, would be disastrous. On the other hand, small non-ergodic regions are less likely to be dangerous and more likely to be recognized if they are unfortunately selected as starting points for a simulation. In a similar way, regions of phase space which act as barriers and cause bottlenecks through which only a few trajectories pass can result in poor sampling by the relatively short simulation runs carried out in practice, even if the system is technically ergodic.

Finally, we might use a different kind of evolution to sample the states of the system: a random walk. This is the Monte Carlo approach: it may be more or less efficient than molecular dynamics. It also fits quite well the analogy of a party, in which the participants sample the different situations randomly, rather than systematically. Once again, trajectory averages are calculated over a finite duration, so these are not necessarily identical to full ensemble averages, and the approach might or might not alleviate some of the ergodicity issues.

It is sometimes convenient to use, in place of $\rho_{ens}(\Gamma)$, a 'weight' function $w_{ens}(\Gamma)$, which satisfies the following equations:

$$\rho_{ens}(\Gamma) = Q_{ens}^{-1} w_{ens}(\Gamma) \tag{2.11}$$

$$Q_{ens} = \sum_{\Gamma} w_{ens}(\Gamma) \tag{2.12}$$

$$\langle \mathcal{A} \rangle_{ens} = \sum_{\Gamma} w_{ens}(\Gamma) \mathcal{A}(\Gamma) \Big/ \sum_{\Gamma} w_{ens}. \tag{2.13}$$

The weight function is essentially a non-normalized form of $\rho_{ens}(\Gamma)$, with the partition function Q_{ens} (also called the sum over states) acting as the normalizing factor. Both w_{ens} and Q_{ens} contain an arbitrary multiplicative constant, whose choice corresponds to the definition of a zero of entropy. Q_{ens} is simply a function of the macroscopic properties defining the ensemble, and connection with classical thermodynamics is made by defining a thermodynamic potential Ψ_{ens} (see e.g. McQuarrie, 1976)

$$\Psi_{ens} = -\ln Q_{ens}. \tag{2.14}$$

This is the function that has a minimum value at thermodynamic equilibrium. For example, Ψ_{ens} might be the negative of the entropy S for a system at constant NVE, where V is the volume and E the total internal energy, or the Gibbs function G for a constant-NPT system, where P is the pressure and T the temperature.

Throughout the foregoing discussion, although we have occasionally used the language of classical mechanics, we have assumed that the states are discrete (e.g. a set of quantum numbers) and that we may sum over them. If the system were enclosed in a container, there would be a countably infinite set of quantum states. In the classical approximation, Γ represents the set of (continuously variable) particle positions and momenta, and we should replace the summation by a classical phase-space integral. w_{ens} and Q_{ens} are then usually defined with appropriate factors included to make them dimensionless, and to match up with the usual semiclassical 'coarse-grained' phase-space volume elements. On a computer, of course, all numbers are held to a finite precision and so, technically, positions and momenta are represented by discrete, not continuous, variables; we now have a countable and finite set of states. We assume that the distinction between this case and the classical limit is of no practical importance, and will use whichever representation is most convenient.

One conceivable approach to the computation of thermodynamic quantities, therefore, would be a direct evaluation of Q_{ens} for a particular ensemble, using eqn (2.12). This summation, over all possible states, is not feasible for many-particle systems: there are too many states, most of which have a very low weight due to non-physical overlaps between the repulsive cores of the molecules, rendering them unimportant. We would like to conduct the summation so as to exclude this large number of irrelevant states, and include only those with a high probability. Unfortunately, it is generally not possible to estimate Q_{ens} directly in this way. However, the underlying idea, that of generating (somehow) a set of states in phase space that are sampled from the complete set in accordance with the probability density $\rho_{ens}(\Gamma)$, is central to the Monte Carlo technique.

We proceed by analogy with molecular dynamics in the sense that the ensemble average of eqn (2.13) is replaced by a trajectory average like eqn (2.2). Newton's equations generate a succession of states in accordance with the distribution function ρ_{NVE} for the constant-NVE or microcanonical ensemble. Suppose we wish to investigate other ensembles; experiments in the laboratory, for example, are frequently performed under conditions of constant temperature and pressure, while it is often very convenient to consider inhomogeneous systems at constant chemical potential. For each such case, let us invent a kind of equation of motion, that is, a means of generating, from one state point $\Gamma(\tau)$, a succeeding state point $\Gamma(\tau + 1)$. This recipe need have no physical interpretation, and it could be entirely deterministic or could involve a stochastic, random, element. It might be derived by modifying the true equations of motion in some way, or it may have no relation whatever with normal dynamics.

To be useful, this prescription should satisfy some sensible conditions:

(a) the probability density $\rho_{ens}(\Gamma)$ for the ensemble of interest should not change as the system evolves;

(b) any 'reasonable' starting distribution $\rho(\Gamma)$ should tend to this stationary solution as the simulation proceeds;

(c) we should be able to argue that ergodicity holds, even though we cannot hope to prove this for realistic systems.

If these conditions are satisfied, then we should be able to generate, from an initial state, a succession of state points which, in the long term, are sampled in accordance with the desired probability density $\rho_{ens}(\Gamma)$. In these circumstances, the ensemble average will be equal to a kind of 'time average':

$$\mathcal{A}_{obs} = \langle \mathcal{A} \rangle_{ens} = \frac{1}{\tau_{obs}} \sum_{\tau=1}^{\tau_{obs}} \mathcal{A}\big(\Gamma(\tau)\big). \tag{2.15}$$

Here τ is an index running over the succession of τ_{obs} states or trials generated by our prescription; in a practical simulation, τ_{obs} would be a large finite number. This is exactly what we do in Monte Carlo simulations. The trick, of course, lies in the generation of the trajectory through phase space, and the different recipes for different ensembles will be discussed in Chapter 4. In general, because only a finite number of states can be generated in any one simulation, Monte Carlo results are subject to the same questions of initial condition effects and satisfactory phase space exploration as are molecular dynamics results.

2.2   Common statistical ensembles

Let us consider four ensembles in common use: the microcanonical, or constant-NVE, ensemble just mentioned, the canonical, or constant-NVT, ensemble, the isothermal–isobaric constant-NPT ensemble, and the grand canonical constant-μVT ensemble. For each ensemble, the aforementioned thermodynamic variables are specified, that is, fixed. Other thermodynamic quantities must be determined by ensemble averaging and, for any particular state point, the instantaneous values of the appropriate phase function will deviate from this average value, that is, fluctuations occur.

The probability density for the microcanonical ensemble is proportional to

$$\delta[\mathcal{H}(\Gamma) - E]$$

where Γ represents the set of particle positions and momenta (or quantum numbers), and $\mathcal{H}(\Gamma)$ is the Hamiltonian. The delta function selects those states of an N-particle system in a container of volume V that have the desired energy E. When the set of states is discrete, δ is just the Kronecker delta, taking values of 0 or 1; when the states are continuous, δ is the Dirac delta function. The microcanonical partition function may be written:

$$Q_{NVE} = \sum_{\Gamma} \delta[\mathcal{H}(\Gamma) - E] \tag{2.16}$$

where the summation takes due note of indistinguishability of particles. In the quasi-classical expression for Q_{NVE}, for an atomic system, the indistinguishability is handled using a factor of $1/N!$

$$Q_{NVE} = \frac{1}{N!}\frac{1}{h^{3N}} \int d\mathbf{r}\, d\mathbf{p}\, \delta[\mathcal{H}(\mathbf{r}, \mathbf{p}) - E]. \tag{2.17}$$

Here, $\int d\mathbf{r}\, d\mathbf{p}$ stands for integration over all $6N$ phase space coordinates. The appropriate thermodynamic potential is the negative of the entropy

$$-S/k_B = -\ln Q_{NVE}. \tag{2.18}$$

The factor involving Planck's constant h in eqn (2.17) corresponds to the usual zero of entropy for the ideal gas (the Sackur–Tetrode equation).

For a classical system, Newton's equations of motion conserve energy and so provide a suitable method (but not the only method (Severin et al., 1978; Creutz, 1983)) for generating a succession of state points sampled from this ensemble, as discussed in the previous section. In fact, for a system not subjected to external forces, these equations also conserve total linear momentum $\mathbf{P}$, and so molecular dynamics probes a subset of the microcanonical ensemble, namely the constant-$NVEP$ ensemble (for technical reasons, as we shall see in Chapter 3, total angular momentum is not conserved in most MD simulations). Since it is easy to transform into the centre-of-mass frame, the choice of $\mathbf{P}$ is not crucial, and zero momentum is usually chosen for convenience. Differences between the constant-NVE and constant-$NVEP$ ensembles are minor: for the latter, an additional three constraints exist in that only $(N-1)$ particle momenta are actually independent of each other.

The probability density for the canonical ensemble is proportional to

$$\exp[-\mathcal{H}(\Gamma)/k_BT]$$

and the partition function is

$$Q_{NVT} = \sum_{\Gamma} \exp[-\mathcal{H}(\Gamma)/k_BT] \qquad (2.19)$$

or, in quasi-classical form, for an atomic system

$$Q_{NVT} = \frac{1}{N!}\frac{1}{h^{3N}} \int dr\, dp\, \exp[-\mathcal{H}(r,p)/k_BT]. \qquad (2.20)$$

The appropriate thermodynamic function is the Helmholtz free energy A

$$A/k_BT = -\ln Q_{NVT}. \qquad (2.21)$$

In the canonical ensemble, all values of the energy are allowed, and energy fluctuations are non-zero. Thus, although $\rho_{NVT}(\Gamma)$ is indeed a stationary solution of the Liouville equation, the corresponding mechanical equations of motion are not a satisfactory method of sampling states in this ensemble, since they conserve energy: normal time evolution occurs on a set of independent constant-energy surfaces, each of which should be appropriately weighted, by the factor $\exp[-\mathcal{H}(\Gamma)/k_BT]$. Our prescription for generating a succession of states must make provision for transitions between the energy surfaces, so that a single trajectory can probe all the accessible phase space, and yield the correct relative weighting. We shall encounter several ways of doing this in the later chapters.

Because the energy is always expressible as a sum of kinetic (p-dependent) and potential (q-dependent) contributions, the partition function factorizes into a product of kinetic (ideal gas) and potential (excess) parts

$$Q_{NVT} = \frac{1}{N!}\frac{1}{h^{3N}} \int dp\, \exp(-\mathcal{K}/k_BT) \int dq\, \exp(-\mathcal{V}/k_BT) = Q_{NVT}^{id}\, Q_{NVT}^{ex}. \qquad (2.22)$$

Again, for an atomic system, we see (by taking $\mathcal{V} = 0$)

$$Q_{NVT}^{id} = \frac{V^N}{N!\,\Lambda^{3N}} \qquad (2.23)$$

Λ being the thermal de Broglie wavelength

$$\Lambda = (h^2/2\pi m k_BT)^{1/2}. \qquad (2.24)$$

The excess part is

$$Q_{NVT}^{ex} = V^{-N} \int dr\, \exp[-\mathcal{V}(r)/k_BT]. \qquad (2.25)$$

Instead of Q_{NVT}^{ex}, we often use the configuration integral

$$Z_{NVT} = \int dr\, \exp[-\mathcal{V}(r)/k_BT]. \qquad (2.26)$$

Some workers include a factor $N!$ in the definition of Z_{NVT}. Although Q_{NVT}^{id} and Q_{NVT}^{ex} are dimensionless, the configuration integral has dimensions of V^N. As a consequence of the separation of Q_{NVT}, all the thermodynamic properties derived from A can be expressed as a sum of ideal gas and configurational parts. In statistical mechanics, it is easy to evaluate ideal gas properties (Rowlinson, 1963), and we may expect most attention to focus on the configurational functions. In fact, it proves possible to probe just the configurational part of phase space according to the canonical distribution, using standard Monte Carlo methods. The corresponding trajectory through phase space has essentially independent projections on the coordinate and momentum sub-spaces. The ideal gas properties are added onto the results of configuration-space Monte Carlo simulations afterwards.

The probability density for the isothermal–isobaric ensemble is proportional to

$$\exp[-(\mathcal{H} + PV)/k_B T].$$

Note that the quantity appearing in the exponent, when averaged, gives the thermodynamic enthalpy $H = \langle \mathcal{H} \rangle + P\langle V \rangle$. Now the volume V has joined the list of microscopic quantities (**r** and **p**) comprising the state point. The appropriate partition function is

$$Q_{NPT} = \sum_{\Gamma} \sum_{V} \exp[-(\mathcal{H} + PV)/k_B T] = \sum_{V} \exp(-PV/k_B T) Q_{NVT}. \tag{2.27}$$

The summation over possible volumes may also be written as an integral, in which case some basic unit of volume V_0 must be chosen to render Q_{NPT} dimensionless. This choice is not practically important for our purposes, but has been discussed in detail elsewhere (Wood, 1968b; Attard, 1995; Koper and Reiss, 1996; Corti and Soto-Campos, 1998; Han and Son, 2001). In quasi-classical form, for an atomic system, we write:

$$Q_{NPT} = \frac{1}{N!} \frac{1}{h^{3N}} \frac{1}{V_0} \int dV \int d\mathbf{r} \, d\mathbf{p} \, \exp[-(\mathcal{H} + PV)/k_B T]. \tag{2.28}$$

The corresponding thermodynamic function is the Gibbs free energy G

$$G/k_B T = -\ln Q_{NPT}. \tag{2.29}$$

The prescription for generating state points in the constant-NPT ensemble must clearly provide for changes in the sample volume as well as energy. Once more, it is possible to separate configurational properties from kinetic ones, and to devise a Monte Carlo procedure to probe configuration space only. The configuration integral in this ensemble is

$$Z_{NPT} = \int dV \, \exp(-PV/k_B T) \int d\mathbf{r} \, \exp[-\mathcal{V}(\mathbf{r})/k_B T]. \tag{2.30}$$

Again some definitions include $N!$ and V_0 as normalizing factors.

The density function for the grand canonical ensemble is proportional to

$$\exp[-(\mathcal{H} - \mu N)/k_B T]$$

where μ is the specified chemical potential. Now the number of particles N is a variable, along with the coordinates and momenta of those particles. The grand canonical partition function is

$$Q_{\mu VT} = \sum_{N} \sum_{\Gamma} \exp[-(\mathcal{H} - \mu N)/k_B T] = \sum_{N} \exp(\mu N/k_B T) Q_{NVT}. \tag{2.31}$$

In quasi-classical form, for an atomic system,

$$Q_{\mu VT} = \sum_N \frac{1}{N!} \frac{1}{h^{3N}} \exp(\mu N/k_B T) \int dr\, dp\, \exp(-\mathcal{H}/k_B T). \qquad (2.32)$$

Although it is occasionally useful to pretend that N is a continuous variable, for most purposes we sum, rather than integrate, in eqns (2.31) and (2.32). The appropriate thermodynamic function is just $-PV/k_B T$:

$$-PV/k_B T = -\ln Q_{\mu VT}. \qquad (2.33)$$

Whatever scheme we employ to generate states in the grand ensemble, clearly it must allow for addition and removal of particles. Once more, it is possible to invent a Monte Carlo method to do this and, moreover, to probe just the configurational part of phase space; however, it turns out to be necessary to include the form of the kinetic partition function in the prescription used.

So far the discussion has been limited to one-component systems but each of the ensembles considered can be readily extended to multi-component mixtures. For example in the grand ensemble, the density function for a c-component mixture containing N_i particles of type i is proportional to

$$\left[\prod_{i=1}^{c} \frac{\exp\left(\mu_i N_i/k_B T\right)}{N_i!} \right] \exp(-\mathcal{H}/k_B T) \qquad (2.34)$$

where μ_i is the chemical potential of species i and $\mathcal{H}$ is the Hamiltonian of the c-component mixture. In quasi-classical form, for an atomic system, the grand partition function is

$$Q_{\mu_1, \mu_2, \ldots \mu_n VT} = \sum_{N_1, N_2 \ldots N_n} \frac{1}{h^{3N}} \left[\prod_{i=1}^{c} \frac{1}{N_i!} \exp(\mu_i N_i/k_B T) \right] \int dr\, dp\, \exp(-\mathcal{H}/k_B T).$$

$$(2.35)$$

The appropriate thermodynamic function is

$$-PV/k_B T = -\ln Q_{\mu_1, \mu_2, \ldots \mu_n VT}. \qquad (2.36)$$

It is also useful to study mixtures in the semi-grand ensemble (Kofke and Glandt, 1988). Here the total number of particles is fixed at N but the identities of the individual particles can change. The chemical potential of an arbitrary species, say 1, is defined as μ_1 and the $c-1$ chemical potential differences, $(\mu_2 - \mu_1) \ldots (\mu_n - \mu_1)$, are fixed. When V and T are also fixed, the probability density is proportional to

$$\left[\prod_{i=1}^{c} \exp[(\mu_i - \mu_1)N_i/k_B T] \right] \exp(-\mathcal{H}/k_B T)$$

where $\mathcal{H}$ is the Hamiltonian for a system of N particles ($N = \sum_i N_i$) and where each particle is defined to have a specific identity from 1 to c. In quasi-classical form, for an atomic system, the semi-grand partition function is

$$Q_{\{\mu_i|i\neq1\}NVT} =$$

$$\sum_{i_1=1}^{c}\cdots\sum_{i_N=1}^{c}\frac{1}{N!h^{3N}}\left[\prod_{i=1}^{c}\exp\big((\mu_i-\mu_1)N_i/k_BT\big)\right]\int dr\,dp\,\exp(-\mathcal{H}/k_BT) \quad (2.37)$$

where the sums are now over the particle identities (e.g. i_1 is the identity of particle 1). The corresponding thermodynamic potential is

$$-(PV-\mu_1N)/k_BT = -\ln Q_{\{\mu_i|i\neq1\}NVT}. \quad (2.38)$$

It is also possible to develop a semi-grand ensemble at constant pressure rather than constant volume. In this case, the chemical potential difference is conveniently replaced by the fugacity fraction as the independent variable, as discussed in Section 4.7 (Kofke and Glandt, 1988; Frenkel and Smit, 2002). An important advantage of the semi-grand ensemble is that the chemical potential difference can be defined as a continuous function. For example, in a polydisperse fluid of hard spheres, the distribution $\mu(\sigma)-\mu_1$ as a function of the hard-sphere diameter σ would be fixed, and through the semi-grand ensemble we could predict the distribution of particle sizes.

It is possible to construct many more ensembles, some of which are of interest in computer simulation. When comparing molecular dynamics with Monte Carlo, it may be convenient to add the constraint of constant (zero) total momentum, that is, fixed centre of mass, to the constant-NVT ensemble. It is also permissible to constrain certain degrees of freedom (e.g. the total kinetic energy (Hoover, 1983b,a), or the energy in a particular chemical bond (Freasier et al., 1979)) while allowing others to fluctuate. Also, nonequilibrium ensembles may be set up (see Chapter 11). The possibilities are endless, the general requirements being that a phase-space density $\rho_{ens}(\Gamma)$ can be written down, and that a corresponding prescription for generating state points can be devised. The remaining questions are ones of practicality.

Not all ensembles are of interest to the computer simulator. The properties of generalized ensembles, such as the constant-μPT ensemble, have been discussed (Hill, 1956). Here, only intensive parameters are specified: the corresponding extensive quantities show unbounded fluctuations, that is, the system size can grow without limit. Also, μ, P, and T are related by an equation of state, so, although this equation may be unknown, they are not independently variable. For these reasons, the simulation of the constant-μPT ensemble and related pathological examples is not a practical proposition. In all the ensembles dealt with in this book, at least one extensive parameter (usually N or V) is fixed to act as a limit on the system size.

Finally, it is by no means guaranteed that a chosen prescription for generating phase-space trajectories will correspond to any ensemble at all. It is easy to think of extreme examples of modified equations of motion for which no possible function $\rho_{ens}(\Gamma)$ is a stationary solution. In principle, some care should be taken to establish which ensemble, if any, is probed by any novel simulation technique.

Example 2.1 Entropy and disorder

At the very least, molecular simulations provide an experimental route to check the predictions of statistical mechanics, as indicated in Fig. 1.2. In addition, however, simulations have provided the impetus to revise some basic ideas, especially in connection with the concept of entropy. Some of the earliest simulations (Wood and Jacobson, 1957; Alder and Wainwright, 1957) demonstrated that the hard-sphere system exhibited a phase transition between solid and liquid phases. Because there are no energetic terms in this model, the thermodynamic driving force must be the entropy: at sufficiently high density, the ordered, solid, phase has a higher entropy than the liquid. This result was not immediately accepted, as it required a rethinking of the definition of disorder, and its connection to the entropy. Roughly speaking, the loss in entropy, associated with the localization of particles around positions on a regular lattice, is more than compensated by the entropy gain associated with the increased free volume that may be explored by each particle around its lattice site. On compressing a disordered hard-sphere system, when the volume occupied by the spheres reaches $\eta \approx 64\,\%$, the free volume becomes zero (random close packing); the freezing transition occurs before this, at $\eta \approx 49\,\%$ (for comparison, in the FCC structure, close packing occurs at $\eta \approx 74\,\%$). Recent computer simulations of polyhedral hard particles have shown that shape-related entropic effects alone can give rise to a huge variety of solid structures (Damasceno et al., 2012; van Anders et al., 2014).

A more fundamental debate concerns the factor $N!$ appearing in eqns (2.17), (2.20), and usually associated with particle indistinguishability. In quantum mechanics, identical particles are indistinguishable as a matter of principle. However, our simulations are of classical, distinguishable (i.e. labelled) particles! Also, statistical mechanics is applied successfully to colloidal systems, for which the constituent particles are of mesoscopic size and clearly distinguishable, even when nearly monodisperse. Should the factor $N!$ be included or not? The question arises whenever two systems, under identical conditions, in a classical simulation or a colloidal experiment, are brought into contact and allowed to exchange particles: is the entropy additive (i.e. extensive) or is there an entropy of mixing term? The answer is that the factor $N!$ *should* be present, for N very similar but distinguishable particles, in order to obtain extensive entropy and Helmholtz free energy functions, and hence it is not intimately connected with quantum mechanics. This has been discussed in the context of simulations by Swendsen (2002) and for colloidal systems by Warren (1998) and Swendsen (2006) (see also Frenkel, 2014, and references therein). These considerations come from first principles, not computer experiments. However, interestingly, they are crucial in practical attempts to quantify the entropy of a granular system, in terms of the number of ways of realizing a jammed structure, extending ideas of Edwards and Oakeshott (1989) and Edwards (1990). Computer simulations (Asenjo et al., 2014) involving the preparation of jammed configurations by quenching equilibrated nearly-hard-sphere liquids, have confirmed the need to include the $N!$ term in order to define an extensive granular entropy, even for systems of distinguishable particles.

2.3   Transforming between ensembles

Since the ensembles are essentially artificial constructs, it would be reassuring to know that they produce average properties which are consistent with one another. In the thermodynamic limit (for an infinite system size) and as long as we avoid the neighbourhood of phase transitions, this is believed to be true for the commonly used statistical ensembles (Fisher, 1964). Since we will be dealing with systems containing a finite number of particles, it is of some interest to see, in a general way, how this result comes about. The method of transformation between ensembles is standard (Hill, 1956; Lebowitz et al., 1967; Münster, 1969; Landau and Lifshitz, 1980) and a useful summary for several ensembles of interest has appeared (Graben and Ray, 1993). We merely outline the procedure here; nonetheless, the development is rather formal, and this section could be skipped on a first reading.

We shall be interested in transforming from an ensemble in which an extensive thermodynamic variable F is fixed to one in which the intensive conjugate variable f is constant. Typical conjugate pairs are (β, E), $(\beta P, V)$, $(-\beta \mu, N)$, where $\beta = 1/k_B T$. If the old partition function and characteristic thermodynamic potential are Q_F, and Ψ_F, respectively, then the new quantities are given by

$$Q_f = \int dF'\, \exp(-F'f) Q_{F'} \tag{2.39}$$

$$\Psi_f = \Psi_F + Ff. \tag{2.40}$$

Equations (2.19)–(2.33) provide specific examples of these relations. Equation (2.40) corresponds to the Legendre transformation of classical thermodynamics. For example, when moving from a system at constant energy to one at constant temperature (i.e. constant β), the characteristic thermodynamic potential changes from $-S/k_B$ to $-S/k_B + \beta E = \beta A$. Similarly, on going to constant temperature and pressure, the thermodynamic potential becomes $\beta A + \beta PV = \beta G$.

The average $\langle \mathcal{A} \rangle_f$ calculated in the constant-f ensemble is related to the average $\langle \mathcal{A} \rangle_F$ calculated at constant F by (Lebowitz et al., 1967)

$$\langle \mathcal{A} \rangle_f = \exp(\Psi_f) \int dF'\, \exp(-\Psi_{F'} - F'f) \langle \mathcal{A} \rangle_{F'}. \tag{2.41}$$

The equivalence of ensembles relies on the behaviour of the integrand of this equation for a large system: it becomes very sharply peaked around the mean value $F' = \langle F \rangle_f$. In the thermodynamic limit of infinite system size, we obtain simply

$$\langle \mathcal{A} \rangle_f = \langle \mathcal{A} \rangle_F \tag{2.42}$$

where it is understood that $F = \langle F \rangle_f$. Thus, the averages of any quantity calculated in, say, the constant-NVE ensemble and the constant-NVT ensemble, will be equal in the thermodynamic limit, as long as we choose E and T consistently so that $E = \langle E \rangle_{NVT}$. In fact, there are some restrictions on the kinds of functions $\mathcal{A}$ for which eqn (2.42) holds. $\mathcal{A}$ should be, essentially, a sum of single-particle functions,

$$\mathcal{A} = \sum_{i=1}^{N} \mathcal{A}_i \qquad (2.43)$$

or, at least, a sum of independent contributions from different parts of the fluid, which may be added up in a similar way. All of the thermodynamic functions are of this short-ranged nature, insofar as they are limited by the range of intermolecular interactions. For long-ranged (e.g. dielectric) properties and long-ranged (e.g. Coulombic) forces, this becomes a more subtle point.

The situation for a finite number of particles is treated by expanding the integrand of eqn (2.41) about the mean value $\langle F \rangle_f$. If we write $F' = \langle F \rangle_f + \delta F'$ then we obtain (Lebowitz et al., 1967):

$$\langle \mathcal{A} \rangle_f = \langle \mathcal{A} \rangle_{F = \langle F \rangle_f} + \frac{1}{2} \left(\frac{\partial^2}{\partial F^2} \langle \mathcal{A} \rangle_F \right)_{F = \langle F \rangle_f} \langle \delta F^2 \rangle_f + \cdots . \qquad (2.44)$$

The correction term, which is proportional to the mean-square fluctuations $\langle \delta F^2 \rangle$ of the quantity F in the constant-f ensemble, is expected to be relatively small since, as mentioned earlier, the distribution of F values should be very sharply peaked for a many-particle system. This fluctuation term may be expressed as a straightforward thermodynamic derivative. Since F and f are conjugate variables, we have

$$\langle F \rangle_f = -\partial \Psi_f / \partial f \qquad (2.45)$$

$$\langle \delta F^2 \rangle_f = \partial^2 \Psi_f / \partial f^2 = -\partial \langle F \rangle_f / \partial f . \qquad (2.46)$$

We may write this simply as $-(\partial F / \partial f)$. Equation (2.44) is most usefully rearranged by taking the last term across to the other side, and treating it as a function of f through the relation $F = \langle F \rangle_f$. Thus

$$\langle \mathcal{A} \rangle_F = \langle \mathcal{A} \rangle_f - \frac{1}{2} \langle \delta F^2 \rangle_f \frac{\partial^2}{\partial F^2} \langle \mathcal{A} \rangle_f \qquad (2.47)$$

$$= \langle \mathcal{A} \rangle_f + \frac{1}{2} \frac{\partial F}{\partial f} \frac{\partial^2}{\partial F^2} \langle \mathcal{A} \rangle_f$$

$$= \langle \mathcal{A} \rangle_f + \frac{1}{2} \frac{\partial}{\partial f} \frac{\partial}{\partial F} \langle \mathcal{A} \rangle_f$$

$$= \langle \mathcal{A} \rangle_f + \frac{1}{2} \frac{\partial}{\partial f} \left(\frac{\partial f}{\partial F} \right) \frac{\partial}{\partial f} \langle \mathcal{A} \rangle_f .$$

Bearing in mind that F is extensive and f intensive, the small relative magnitude of the correction term can be seen explicitly: it decreases as $O(N^{-1})$.

Although the fluctuations are small, they are nonetheless measurable in computer simulations. They are of interest because they are related to thermodynamic derivatives (like the specific heat or the isothermal compressibility) by equations such as eqn (2.46). In general, we define the root mean square (RMS) deviation $\sigma(\mathcal{A})$ by the equation

$$\sigma^2(\mathcal{A}) = \langle \delta\mathcal{A}^2 \rangle_{\text{ens}} = \langle \mathcal{A}^2 \rangle_{\text{ens}} - \langle \mathcal{A} \rangle_{\text{ens}}^2 \qquad (2.48)$$

where

$$\delta\mathcal{A} = \mathcal{A} - \langle \mathcal{A} \rangle_{\text{ens}}. \qquad (2.49)$$

It is quite important to realize that, despite the $\langle \delta\mathcal{A}^2 \rangle$ notation, we are not dealing here with the average of a mechanical quantity like $\mathcal{A}$; the best we can do is to write $\sigma^2(\mathcal{A})$ as a difference of two terms, as in eqn (2.48). Thus, the previous observations on equivalence of ensembles do not apply: fluctuations in different ensembles are not the same. As an obvious example, energy fluctuations in the constant-NVE ensemble are (by definition) zero, whereas in the constant-NVT ensemble, they are not. The transformation technique may be applied to obtain an equation analogous to eqn (2.47) (Lebowitz et al., 1967). In the general case of the covariance of two variables $\mathcal{A}$ and $\mathcal{B}$ the result is

$$\langle \delta\mathcal{A}\delta\mathcal{B} \rangle_F = \langle \delta\mathcal{A}\delta\mathcal{B} \rangle_f + \left(\frac{\partial f}{\partial F} \right) \left(\frac{\partial}{\partial f} \langle \mathcal{A} \rangle_f \right) \left(\frac{\partial}{\partial f} \langle \mathcal{B} \rangle_f \right). \qquad (2.50)$$

Now the correction term is of the same order as the fluctuations themselves. Consider, once more, energy fluctuations in the microcanonical and canonical ensembles, that is, let $\mathcal{A} = \mathcal{B} = F = E$ and $f = \beta = 1/k_B T$. Then on the left of eqn (2.50) we have zero, and on the right we have $\sigma^2(E)$ at constant-NVT and a combination of thermodynamic derivatives which turn out to equal $(\partial E/\partial\beta) = -k_B T^2 C_V$ where C_V is the constant-volume heat capacity.

2.4   Simple thermodynamic averages

A consequence of the equivalence of ensembles is that, provided a suitable phase function can be identified in each case, the basic thermodynamic properties of a model system may be calculated as averages in any convenient ensemble. Accordingly, we give in this section expressions for common thermodynamic quantities, omitting the subscripts which identify particular ensembles. These functions are usually derivatives of one of the characteristic thermodynamic functions Ψ_{ens}. Examples are $P = -(\partial A/\partial V)_{NT}$ and $\beta = (1/k_B T) = (1/k_B)(\partial S/\partial E)_{NV}$.

The kinetic, potential, and total internal energies may be calculated using the phase functions of eqns (1.1)–(1.3).

$$E = \langle \mathcal{H} \rangle = \langle \mathcal{K} \rangle + \langle \mathcal{V} \rangle. \qquad (2.51)$$

The kinetic energy is a sum of contributions from individual particle momenta, while evaluation of the potential contribution involves summing over all pairs, triplets, etc. of molecules, depending upon the complexity of the function as discussed in Chapter 1.

The temperature and pressure may be calculated using the virial theorem, which we write in the form of 'generalized equipartition' (Münster, 1969):

$$\langle p_k \partial \mathcal{H}/\partial p_k \rangle = k_B T \tag{2.52a}$$

$$\langle q_k \partial \mathcal{H}/\partial q_k \rangle = k_B T \tag{2.52b}$$

for any generalized coordinate q_k or momentum p_k. These expressions are valid (to $O(N^{-1})$) in any ensemble.

Equation (2.52a) is particularly simple when the momenta appear as squared terms in the Hamiltonian. For example, in the atomic case, we may sum up $3N$ terms of the form $p_{i\alpha}^2/m_i$, to obtain

$$\left\langle \sum_{i=1}^{N} |\mathbf{p}_i|^2/m_i \right\rangle = 2\langle \mathcal{K} \rangle = 3Nk_B T. \tag{2.53}$$

This is the familiar equipartition principle: an average energy of $k_B T/2$ per degree of freedom. It is convenient to define an instantaneous 'kinetic temperature' function

$$\mathcal{T} = 2\mathcal{K}/3Nk_B = \frac{1}{3Nk_B} \sum_{i=1}^{N} |\mathbf{p}_i|^2/m_i \tag{2.54}$$

whose average is equal to T. Obviously, this is not a unique definition. For a system of rigid molecules, described in terms of centre-of-mass positions and velocities together with orientational variables, the angular velocities may also appear in the definition of $\mathcal{T}$. Alternatively, it may be useful to define separate 'translational' and 'rotational' temperatures each of which, when averaged, gives T. In eqn (2.52a) it is assumed that the independent degrees of freedom have been identified and assigned generalized coordinates q_k, and momenta p_k. For a system of N atoms, subject to internal molecular constraints, the number of degrees of freedom will be $3N - N_c$ where N_c is the total number of independent internal constraints (fixed bond lengths and angles) defined in the molecular model. Then, we must replace eqn (2.54) by

$$\mathcal{T} = \frac{2\mathcal{K}}{(3N - N_c)k_B} = \frac{1}{(3N - N_c)k_B} \sum_{i=1}^{N} |\mathbf{p}_i|^2/m_i. \tag{2.55}$$

We must also include in N_c, any additional global constraints on the ensemble. For example, in the 'molecular dynamics' constant-$NVEP$ ensemble, we must include the three extra constraints on centre-of-mass motion.

Equations (2.52) are examples of the general form

$$\left\langle \mathcal{A} \frac{\partial \mathcal{H}}{\partial q_k} \right\rangle = k_B T \left\langle \frac{\partial \mathcal{A}}{\partial q_k} \right\rangle, \qquad \left\langle \mathcal{A} \frac{\partial \mathcal{H}}{\partial p_k} \right\rangle = k_B T \left\langle \frac{\partial \mathcal{A}}{\partial p_k} \right\rangle,$$

valid for any dynamical variable $\mathcal{A}$, which may be easily derived in the canonical ensemble (see e.g. Landau and Lifshitz, 1958, p100, eqn (33.14)). These are generally termed 'hyper-virial' relations (Hirschfelder, 1960). Setting $\mathcal{A} = \partial \mathcal{H}/\partial q_k = \partial \mathcal{V}/\partial q_k$ gives an alternative way of calculating the temperature from purely configurational properties, independent of the momenta. For example, for a simple atomic system

$$k_B T = \frac{\langle (\partial V / \partial r_{i\alpha})^2 \rangle}{\langle \partial^2 V / \partial r_{i\alpha}^2 \rangle} = \frac{\langle f_{i\alpha}^2 \rangle}{\langle \partial^2 V / \partial r_{i\alpha}^2 \rangle}. \tag{2.56}$$

Naturally, in a simulation it is usual to average this expression over all atoms i and all coordinate directions α, when the numerator becomes the mean-square force and the denominator becomes the average Laplacian of the potential. This 'configurational temperature' is useful in Monte Carlo simulations, in which the momenta do not appear (Rugh, 1997), and comparing it with the usual kinetic expression, (2.54), or with the prescribed temperature, is a useful check that a simulation is working properly (Butler et al., 1998). More details of how to calculate the configurational temperature appear in Appendix F.

The pressure may be calculated via eqn (2.52b). If we choose Cartesian coordinates, and use Hamilton's equations of motion (see Chapter 3), it is easy to see that each coordinate derivative in eqn (2.52b) is the negative of a component of the force $\mathbf{f}_i$ on some molecule i, and we may write, summing over N molecules,

$$-\frac{1}{3} \left\langle \sum_{i=1}^{N} \mathbf{r}_i \cdot \nabla_{\mathbf{r}_i} V \right\rangle = \frac{1}{3} \left\langle \sum_{i=1}^{N} \mathbf{r}_i \cdot \mathbf{f}_i^{\text{tot}} \right\rangle = -N k_B T. \tag{2.57}$$

We have used the symbol $\mathbf{f}_i^{\text{tot}}$ because this represents the sum of intermolecular forces and external forces. The latter are related to the external pressure, as can be seen by considering the effect of the container walls on the system:

$$\frac{1}{3} \left\langle \sum_{i=1}^{N} \mathbf{r}_i \cdot \mathbf{f}_i^{\text{ext}} \right\rangle = -PV. \tag{2.58}$$

If we define the 'internal virial' W

$$-\frac{1}{3} \sum_{i=}^{N} \mathbf{r}_i \cdot \nabla_{\mathbf{r}_i} V = \frac{1}{3} \sum_{i=1}^{N} \mathbf{r}_i \cdot \mathbf{f}_i = W \tag{2.59}$$

where now we restrict attention to intermolecular forces, then

$$PV = N k_B T + \langle W \rangle. \tag{2.60}$$

This suggests that we define an instantaneous 'pressure' function (Cheung, 1977)

$$\mathcal{P} = \rho k_B \mathcal{T} + W/V = \mathcal{P}^{\text{id}} + \mathcal{P}^{\text{ex}} \tag{2.61}$$

whose average is simply P. Again, this definition is not unique; apart from the different ways of defining W which we shall see later, it may be most convenient (say in a constant-temperature ensemble) to use

$$\mathcal{P}' = \rho k_B T + W/V = \langle \mathcal{P}^{\text{id}} \rangle + \mathcal{P}^{\text{ex}} \tag{2.62}$$

instead. Both $\mathcal{P}$ and $\mathcal{P}'$ give P when averaged, but their fluctuations in any ensemble will, in general, be different. Note that the preceding derivation is not really valid for the infinite periodic systems used in computer simulation: there are no container walls and no external forces. Nonetheless, the result is the same (Erpenbeck and Wood, 1977).

For pairwise interactions, $\mathcal{W}$ is more conveniently expressed in a form which is explicitly independent of the origin of coordinates. This is done by writing $\mathbf{f}_i$ as the sum of forces $\mathbf{f}_{ij}$ on atom i due to atom j

$$\sum_i \mathbf{r}_i \cdot \mathbf{f}_i = \sum_i \sum_{j \neq i} \mathbf{r}_i \cdot \mathbf{f}_{ij} = \tfrac{1}{2} \sum_i \sum_{j \neq i} \left(\mathbf{r}_i \cdot \mathbf{f}_{ij} + \mathbf{r}_j \cdot \mathbf{f}_{ji} \right). \tag{2.63}$$

The second equality follows because the indices i and j are equivalent. Newton's third law $\mathbf{f}_{ij} = -\mathbf{f}_{ji}$ is then used to switch the force indices

$$\sum_i \mathbf{r}_i \cdot \mathbf{f}_i = \tfrac{1}{2} \sum_i \sum_{j \neq i} \mathbf{r}_{ij} \cdot \mathbf{f}_{ij} = \sum_i \sum_{j > i} \mathbf{r}_{ij} \cdot \mathbf{f}_{ij} \tag{2.64}$$

where $\mathbf{r}_{ij} = \mathbf{r}_i - \mathbf{r}_j$ and the final form of the summation is usually more convenient. It is essential to use the $\mathbf{r}_{ij} \cdot \mathbf{f}_{ij}$ form in a simulation that employs periodic boundary conditions. So we have at last

$$\mathcal{W} = \tfrac{1}{3} \sum_i \sum_{j > i} \mathbf{r}_{ij} \cdot \mathbf{f}_{ij} = -\tfrac{1}{3} \sum_i \sum_{j > i} \mathbf{r}_{ij} \cdot \nabla_{\mathbf{r}_{ij}} v(r_{ij}) = -\tfrac{1}{3} \sum_i \sum_{j > i} w(r_{ij}) \tag{2.65}$$

where the intermolecular pair virial function $w(r)$ is

$$w(r) = r \frac{dv(r)}{dr}. \tag{2.66}$$

Like $\mathcal{V}$, $\mathcal{W}$ is limited by the range of the interactions, and hence $\langle \mathcal{W} \rangle$ should be a well-behaved, ensemble-independent function in most cases.

For molecular fluids we may write

$$\mathcal{W} = \tfrac{1}{3} \sum_i \sum_{j > i} \mathbf{r}_{ij} \cdot \mathbf{f}_{ij} = -\tfrac{1}{3} \sum_i \sum_{j > i} \mathbf{r}_{ij} \cdot \left(\nabla_{\mathbf{r}_{ij}} \mathcal{V} \right)_{\Omega_i, \Omega_j} = -\tfrac{1}{3} \sum_i \sum_{j > i} w(r_{ij}) \tag{2.67}$$

where $\mathbf{r}_{ij}$ is the vector between the molecular centres. Here we have made it clear that the pair virial is defined as a position derivative at constant orientation of the molecules

$$w(r_{ij}) = r_{ij} \left(\frac{\partial v(r_{ij}, \Omega_i, \Omega_j)}{\partial r_{ij}} \right)_{\Omega_i, \Omega_j}. \tag{2.68}$$

The pressure function $\mathcal{P}$ is defined through eqn (2.61) as before. For interaction site models, we may treat the system as a set of atoms, and use eqns (2.65), (2.66), with the summations taken over distinct pairs of sites ia and jb (compare eqn (1.12)). When doing this, however, it is important to include all intramolecular contributions (forces along the bonds for example) in the sum. Alternatively, the molecular definition, eqns (2.67), (2.68) is still valid. In this case, for computational purposes, eqn (2.68) may be rewritten in the form

$$w(r_{ij}) = \sum_a \sum_b \frac{w_{ab}(r_{ab})}{r_{ab}^2} (\mathbf{r}_{ab} \cdot \mathbf{r}_{ij}) \tag{2.69}$$

where $\mathbf{r}_{ab} = \mathbf{r}_{ia} - \mathbf{r}_{jb}$ is the vector between the sites and $w_{ab}(r_{ab})$ is the site–site pair virial function. This is equivalent to expressing $\mathbf{f}_{ij}$ in eqn (2.67) as the sum of all the site–site forces acting between the molecules. Whether the atomic or molecular definition of the virial is adopted, the ensemble average $\langle W \rangle$ and hence $\langle \mathcal{P} \rangle = P$ should be unaffected. In inhomogeneous systems, the pressure is a tensor; see Section 2.12.

In systems with discontinuous interactions, such as the hard-sphere model, the usual expressions for the pressure cannot be applied. As we shall see in Chapter 3, in MD simulations of hard particles, we solve the classical equations of motion for the motion in between discrete collisions; at the moment of collision, an impulse acts between the two colliding particles, and changes their momenta. This is responsible for the non-ideal contribution to the pressure. The virial expression (2.65) can be recast into a form involving a sum over collisions, by time-averaging it:

$$\langle W \rangle = \frac{1}{t_{obs}} \int_0^{t_{obs}} dt \left(\tfrac{1}{3} \sum_i \sum_{j>i} \mathbf{r}_{ij} \cdot \mathbf{f}_{ij} \right) = \frac{1}{3t_{obs}} \sum_{colls} \mathbf{r}_{ij} \cdot \delta \mathbf{p}_{ij} \qquad (2.70a)$$

where i and j represent a pair of molecules colliding at time t_{ij}, $\mathbf{r}_{ij}$ is the vector between the molecular centres at the time of collision, and

$$\delta \mathbf{p}_{ij} = \delta \mathbf{p}_i = -\delta \mathbf{p}_j = \int_{t_{ij}^-}^{t_{ij}^+} dt\, \mathbf{f}_{ij} \qquad (2.70b)$$

is the collisional impulse, that is, the change in momentum. The sum in eqn (2.70a) is over all collisions occurring in time t_{obs}, and the integral in eqn (2.70b) is over an infinitesimal time interval around t_{ij}. This expression may also be written in terms of the collision rate and the average of $\mathbf{r}_{ij} \cdot \delta \mathbf{p}_{ij}$ per collision. Equation (2.70a) replaces eqn (2.65) in the average pressure equation (2.60). Further details, including a discussion of the system-size dependence of these formulae may be found elsewhere (Alder and Wainwright, 1960; Hoover and Alder, 1967; Erpenbeck and Wood, 1977). In MC simulations of hard systems, a less direct approach must be used to estimate P, and this is discussed in Section 5.5.

Quantities such as $\langle N \rangle$ and $\langle V \rangle$ are easily evaluated in the simulation of ensembles in which these quantities vary, and derived functions such as the enthalpy are straightforwardly calculated. Now we turn to the question of evaluating entropy-related ('statistical') quantities such as the Gibbs and Helmholtz functions, the chemical potential μ, and the entropy itself. A direct approach is to conduct a simulation of the grand canonical ensemble, in which μ, or a related quantity, is specified. It must be said at the outset that there are some technical difficulties associated with grand canonical ensemble simulations, and we return to this in Chapter 4. There are also difficulties in obtaining these functions in the other common ensembles, since they are related directly to the partition function Q, not to its derivatives. To calculate Q would mean summing over all the states of the system. It might seem that we could use the formula

$$\exp\left(A^{ex}/k_BT\right) = Q_{NVT}^{ex}{}^{-1} = \left\langle \exp(\mathcal{V}/k_BT) \right\rangle_{NVT} \qquad (2.71)$$

to estimate the excess statistical properties, but, in practice, the distribution ρ_{NVT} will be very sharply peaked around the largest values of $\exp(-\mathcal{V}/k_BT)$, that is, where $\exp(\mathcal{V}/k_BT)$ is comparatively small. Consequently, any simulation technique that samples according to the equilibrium distribution will be bound to give a poor estimate of A by this route. Special sampling techniques have been developed to evaluate averages of this type (Valleau and Torrie, 1977) and we return to this in Chapter 9. It is comparatively easy to obtain free-energy differences for a given system at two different temperatures by integrating the internal energy along a line of constant density:

$$\left(\frac{A}{Nk_BT}\right)_2 - \left(\frac{A}{Nk_BT}\right)_1 = \int_{\beta_1}^{\beta_2}\left(\frac{E}{Nk_BT}\right)\frac{d\beta}{\beta} = -\int_{T_1}^{T_2}\left(\frac{E}{Nk_BT}\right)\frac{dT}{T}. \qquad (2.72)$$

Alternatively, integration of the pressure along an isotherm may be used:

$$\left(\frac{A}{Nk_BT}\right)_2 - \left(\frac{A}{Nk_BT}\right)_1 = \int_{\rho_1}^{\rho_2}\left(\frac{PV}{Nk_BT}\right)\frac{d\rho}{\rho} = -\int_{V_1}^{V_2}\left(\frac{PV}{Nk_BT}\right)\frac{dV}{V}. \qquad (2.73)$$

To use these expressions, it is necessary to calculate ensemble averages at state points along a reversible thermodynamic path. To calculate absolute free energies and entropies, it is necessary to extend the thermodynamic integration far enough to reach a state point whose properties can be calculated essentially exactly. In general, these calculations may be expensive, since accurate thermodynamic information is required for many closely spaced state points.

One fairly direct, and widely applicable, method for calculating μ is based on the thermodynamic identities

$$\exp(-\mu/k_BT) = Q_{N+1}/Q_N = Q_N/Q_{N-1} \qquad (2.74)$$

valid at large N for both the constant-NVT and constant-NPT ensembles. From these equations we can obtain expressions for the chemical potential in terms of a kind of ensemble average (Widom, 1963; 1982). If we define the excess chemical potential $\mu^{ex} = \mu - \mu^{id}$ then we can write

$$\mu^{ex} = -k_BT\ln\left\langle\exp(-\mathcal{V}_{test}/k_BT)\right\rangle \qquad (2.75)$$

where $\mathcal{V}_{test}$ is the potential energy which would result from the addition of a particle (at random) to the system. This is the 'test particle insertion' method of estimating μ. Eqn (2.75) also applies in the constant-μVT ensemble (Henderson, 1983). A slightly different formula applies for constant-NVE because of the kinetic temperature fluctuations (Frenkel, 1986):

$$\mu^{ex} = -k_B\langle\mathcal{T}\rangle\ln\left[\langle\mathcal{T}\rangle^{-3/2}\langle\mathcal{T}^{3/2}\exp(-\mathcal{V}_{test}/k_BT)\rangle\right] \qquad (2.76a)$$

where $\mathcal{T}$ is the instantaneous kinetic temperature. Similarly, for the constant-NPT ensemble, it is necessary to include the fluctuations in the volume V (Shing and Chung, 1987):

$$\mu^{ex} = -k_BT\ln\left[\langle V\rangle^{-1}\langle V\exp(-\mathcal{V}_{test}/k_BT)\rangle\right]. \qquad (2.76b)$$

In all these cases the 'test particle', the $(N+1)$th, is not actually inserted: it is a 'ghost', that is, the N real particles are not affected by its presence. The Widom method can also be applied to inhomogeneous systems, see Section 2.12.

There is an alternative formula which applies to the removal of a test particle (selected at random) from the system (Powles et al., 1982). This 'test particle' is not actually removed:

it is a real particle and continues to interact normally with its neighbours. In practice, this technique does not give an accurate estimate of μ^{ex}, and for hard spheres (for example) it is completely unworkable (Rowlinson and Widom, 1982). We defer a detailed discussion of the applicability of these methods and more advanced techniques until Chapter 9.

2.5 Fluctuations

We now discuss the information that can be obtained from the RMS fluctuations calculated as indicated in eqn (2.48). The quantities of most interest are the constant-volume specific heat capacity $C_V = (\partial E/\partial T)_V$ or its constant-pressure counterpart $C_P = (\partial H/\partial T)_P$, the thermal expansion coefficient $\alpha_P = V^{-1}(\partial V/\partial T)_P$, the isothermal compressibility $\beta_T = -V^{-1}(\partial V/\partial P)_T$, the thermal pressure coefficient $\gamma_V = (\partial P/\partial T)_V$, and the adiabatic (constant-S) analogues of the last three. The relationship $\alpha_P = \beta_T \gamma_V$ means that only two of these quantities are needed to define the third. In part, formulae for these quantities can be obtained from standard theory of fluctuations (Landau and Lifshitz, 1980), but in computer simulations we must be careful to distinguish between properly defined mechanical quantities such as the energy or Hamiltonian $\mathcal{H}$, the kinetic temperature $\mathcal{T}$ or the instantaneous pressure $\mathcal{P}$, and thermodynamic concepts such as T and P, which can only be described as ensemble averages or as parameters defining an ensemble. Thus, a standard formula such as $\sigma^2(E) = \langle\delta E^2\rangle = k_B T^2 C_V$ can be used to calculate the specific heat in the canonical ensemble (provided we recognize that E really means $\mathcal{H}$), whereas the analogous simple formula $\sigma^2(P) = \langle\delta P^2\rangle = k_B T/V\beta_T$ will not be so useful (since P is not the same as $\mathcal{P}$).

Fluctuations are readily computed in the canonical ensemble, and accordingly we start with this case. As just mentioned, the specific heat is given by the fluctuations in the energy:

$$\langle\delta\mathcal{H}^2\rangle_{NVT} = k_B T^2 C_V. \tag{2.77}$$

This can be divided into kinetic and potential contributions which are uncorrelated (i.e. $\langle\delta\mathcal{K}\delta\mathcal{V}\rangle_{NVT} = 0$):

$$\langle\delta\mathcal{H}^2\rangle_{NVT} = \langle\delta\mathcal{V}^2\rangle_{NVT} + \langle\delta\mathcal{K}^2\rangle_{NVT}. \tag{2.78}$$

The kinetic part can be calculated easily, for example in the case of a system of N atoms:

$$\langle\delta\mathcal{K}^2\rangle_{NVT} = \frac{3N}{2}(k_B T)^2 = 3N/2\beta^2 \tag{2.79}$$

yielding the ideal gas part of the specific heat $C_V^{id} = (3/2)Nk_B$. For this case, then, potential-energy fluctuations are simply

$$\langle\delta\mathcal{V}^2\rangle_{NVT} = k_B T^2\left(C_V - \tfrac{3}{2}Nk_B\right). \tag{2.80}$$

Consideration of the cross-correlation of the potential-energy and virial fluctuations yields an expression for the thermal pressure coefficient γ_V (Rowlinson, 1969)

$$\langle\delta\mathcal{V}\delta\mathcal{W}\rangle_{NVT} = k_B T^2\left(V\gamma_V - Nk_B\right) \tag{2.81}$$

where $\mathcal{W}$ is defined in eqns (2.65)–(2.69). In terms of the pressure function defined in eqn (2.61) this becomes

$$\langle\delta\mathcal{V}\delta\mathcal{P}\rangle_{NVT} = k_B T^2\left(\gamma_V - \rho k_B\right) \tag{2.82}$$

once more valid for a system of N atoms. Equation (2.82) also applies if $\mathcal{P}$ is replaced by $\mathcal{P}'$ or by $\mathcal{P}^{\text{ex}}$ (eqn (2.82)), which is more likely be available in a (configuration-space) constant-NVT Monte Carlo calculation. Similar formulae may be derived for molecular systems. When we come to consider fluctuations of the virial itself, we must define a further 'hypervirial' function

$$\mathcal{X} = \tfrac{1}{9} \sum_i \sum_{j>i} \sum_k \sum_{\ell>k} \left(\mathbf{r}_{ij} \cdot \boldsymbol{\nabla}_{\mathbf{r}_{ij}} \right) \left(\mathbf{r}_{k\ell} \cdot \boldsymbol{\nabla}_{\mathbf{r}_{k\ell}} \right) \mathcal{V} \tag{2.83}$$

which becomes, for a pairwise additive potential

$$\mathcal{X} = \tfrac{1}{9} \sum_i \sum_{j>i} x(r_{ij}) \tag{2.84}$$

where

$$x(r) = r\frac{\mathrm{d}w(r)}{\mathrm{d}r} \tag{2.85}$$

$w(r)$ being the intermolecular viral defined in eqn (2.66). It is then easy to show that

$$\langle \delta \mathcal{W}^2 \rangle_{NVT} = k_{\mathrm{B}} T \left(N k_{\mathrm{B}} T + \langle \mathcal{W} \rangle_{NVT} - \beta_T^{-1} V + \langle \mathcal{X} \rangle_{NVT} \right) \tag{2.86}$$

or

$$\langle \delta \mathcal{P}^2 \rangle_{NVT} = \frac{k_{\mathrm{B}} T}{V} \left(\frac{2N k_{\mathrm{B}} T}{3V} + \langle \mathcal{P} \rangle_{NVT} - \beta_T^{-1} + \frac{\langle \mathcal{X} \rangle_{NVT}}{V} \right). \tag{2.87}$$

The average $\langle \mathcal{X} \rangle$ is a non-thermodynamic quantity. Nonetheless, it can be calculated in a computer simulation, so eqns (2.86) and (2.87) provide a route to the isothermal compressibility β_T. Note that Cheung (1977) uses a different definition of the hypervirial function. In terms of the fluctuations of $\mathcal{P}'$, the analogous formula is

$$\langle \delta \mathcal{P}^{\text{ex}2} \rangle_{NVT} = \langle \delta \mathcal{P}'^2 \rangle_{NVT} = \frac{k_{\mathrm{B}} T}{V} \left(\langle \mathcal{P}' \rangle_{NVT} - \beta_T^{-1} + \frac{\langle \mathcal{X} \rangle_{NVT}}{V} \right) \tag{2.88}$$

and this would be the formula used most in constant-NVT simulations.

The desired fluctuation expressions for the microcanonical ensemble may best be derived from the preceding equations, by applying the transformation formula, eqn (2.50) (Lebowitz et al., 1967; Cheung, 1977) or directly (Ray and Graben, 1981). The equivalence of the ensembles guarantees that the values of simple averages (such as $\langle \mathcal{X} \rangle$) are unchanged by this transformation. In the microcanonical ensemble, the energy (of course) is fixed, but the specific heat may be obtained by examining fluctuations in the separate potential and kinetic components (Lebowitz et al., 1967). For N atoms,

$$\langle \delta \mathcal{V}^2 \rangle_{NVE} = \langle \delta \mathcal{K}^2 \rangle_{NVE} = \tfrac{3}{2} N k_{\mathrm{B}}^2 T^2 \left(1 - \frac{3N k_{\mathrm{B}}}{2 C_V} \right). \tag{2.89}$$

Cross-correlation of the pressure function and (say) the kinetic energy may be used to obtain the thermal pressure coefficient:

$$\langle \delta \mathcal{P} \delta \mathcal{K} \rangle_{NVE} = \langle \delta \mathcal{P} \delta \mathcal{V} \rangle_{NVE} = \frac{N k_{\mathrm{B}}^2 T^2}{V} \left(1 - \frac{3V \gamma_V}{2 C_V} \right). \tag{2.90}$$

Finally the expression for fluctuations of $\mathcal{P}$ in the microcanonical ensemble yields the isothermal compressibility, but the formula is made slightly more compact by introducing the adiabatic compressibility β_S, and using $\beta_S^{-1} = \beta_T^{-1} + TV\gamma_V^2/C_V$

$$\langle \delta \mathcal{P}^2 \rangle_{NVE} = \frac{k_B T}{V} \left(\frac{2Nk_B T}{3V} + \langle \mathcal{P} \rangle_{NVE} - \beta_S^{-1} + \frac{\langle X \rangle_{NVE}}{V} \right). \tag{2.91}$$

In eqns (2.89)–(2.91) T is short for $\langle \mathcal{T} \rangle_{NVE}$. All these expressions are easily derived using the transformation technique outlined earlier, and they are all valid for systems of N atoms. The same expressions (to leading order in N) hold in the constant-$NVEP$ ensemble probed by molecular dynamics. Analogous formulae for molecular systems may be derived in a similar way.

Conversion from the canonical to the isothermal–isobaric ensemble is easily achieved. Most of the formulae of interest are very simple since they involve well-defined mechanical quantities. At constant T and P, both volume and energy fluctuations may occur. The volume fluctuations are related to the isothermal compressibility

$$\langle \delta V^2 \rangle_{NPT} = V k_B T \beta_T. \tag{2.92}$$

The simplest specific heat formula may be obtained by calculating the 'instantaneous' enthalpy $\mathcal{H} + PV$, when we see

$$\left\langle \delta(\mathcal{H} + PV)^2 \right\rangle_{NPT} = k_B T^2 C_P. \tag{2.93}$$

This equation can be split into the separate terms involving $\langle \delta \mathcal{H}^2 \rangle$, $\langle \delta V^2 \rangle$, and $\langle \delta \mathcal{H} \delta V \rangle$. Finally the thermal expansion coefficient may be calculated from the cross-correlations of 'enthalpy' and volume:

$$\left\langle \delta V \delta(\mathcal{H} + PV) \right\rangle_{NPT} = k_B T^2 V \alpha_P. \tag{2.94}$$

Other quantities may be obtained by standard thermodynamic manipulations. Finally, to reiterate, although P is fixed in these expressions, the functions $\mathcal{P}$ and $\mathcal{P}'$ defined in eqns (2.61)–(2.62) will fluctuate around the average value P.

In the grand canonical ensemble, energy, pressure, and number fluctuations occur. The number fluctuations yield the isothermal compressibility

$$\langle \delta N^2 \rangle_{\mu VT} = k_B T (\partial N/\partial \mu)_{VT} = \frac{N^2}{V} k_B T \beta_T. \tag{2.95}$$

Expressions for the other thermodynamic derivatives are a little more complicated (Adams, 1975). The simplest formula for a specific heat is obtained by considering (by analogy with the enthalpy) a function $\mathcal{H} - \mu N$:

$$\left\langle \delta(\mathcal{H} - \mu N)^2 \right\rangle_{\mu VT} = k_B T^2 C_{\mu V} = k_B T^2 \left(\frac{\partial \langle \mathcal{H} - \mu N \rangle}{\partial T} \right)_{\mu V} \tag{2.96}$$

and the usual specific heat C_V (i.e. C_{NV}) is obtained by thermodynamic manipulations:

$$C_V = \tfrac{3}{2} N k_B + \frac{1}{k_B T^2} \left(\langle \delta \mathcal{V}^2 \rangle_{\mu VT} - \frac{\langle \delta \mathcal{V} \delta N \rangle_{\mu VT}^2}{\langle \delta N^2 \rangle_{\mu VT}} \right). \tag{2.97}$$

The thermal expansion coefficient may be derived in the same way:

$$\alpha_P = \frac{P\beta_T}{T} - \frac{\langle \delta V \delta N \rangle_{\mu VT}}{N k_B T^2} + \frac{\langle V \rangle_{\mu VT} \langle \delta N^2 \rangle_{\mu VT}}{N^2 k_B T^2}. \tag{2.98}$$

Finally, the thermal pressure coefficient is given by

$$\gamma_V = \frac{N k_B}{V} + \frac{\langle \delta V \delta N \rangle_{\mu VT}}{VT} \left(1 - \frac{N}{\langle \delta N^2 \rangle_{\mu VT}} \right) + \frac{\langle \delta V \delta W \rangle_{\mu VT}}{V k_B T^2}. \tag{2.99}$$

Except within brackets $\langle \cdots \rangle$, N in these equations is understood to mean $\langle N \rangle_{\mu VT}$ and similarly P means $\langle P \rangle_{\mu VT}$. As emphasized by Adams (1975), when these formulae are used in a computer simulation, it is advisable to cross-check them with the thermodynamic identity $\alpha_P = \beta_T \gamma_V$.

The impression may arise that particular thermodynamic derivatives (such as α_P) are best calculated by conducting a simulation in the corresponding ensemble (e.g. constant-*NPT*). This is not the case, and Lustig (2012) has provided a systematic approach to calculating such quantities in a wide variety of ensembles. Care must be taken in the application of any formulae to the zero-momentum ensemble usually employed in molecular dynamics (Çağin and Ray, 1988; Lustig, 1994a,b). Also, it is important to bear in mind that significant deviations from the thermodynamic limit will happen when the system size is small (Ray and Graben, 1991; Shirts et al., 2006; Uline et al., 2008).

2.6   Structural quantities

The structure of a simple monatomic fluid is characterized by a set of distribution functions for the atomic positions, the simplest of which is the pair distribution function $g_2(\mathbf{r}_i, \mathbf{r}_j)$, or $g_2(r_{ij})$ or simply $g(r)$. This function gives the probability of finding a pair of atoms a distance r apart, relative to the probability for a completely random distribution at the same density. To define $g(r)$, we integrate the configurational distribution function over the positions of all atoms except two, incorporating the appropriate normalization factors (McQuarrie, 1976; Hansen and McDonald, 2013). In the canonical ensemble

$$g(\mathbf{r}_1, \mathbf{r}_2) = \frac{N(N-1)}{\rho^2 Z_{NVT}} \int d\mathbf{r}_3 d\mathbf{r}_4 \cdots d\mathbf{r}_N \exp[-\beta \mathcal{V}(\mathbf{r}_1, \mathbf{r}_2, \cdots \mathbf{r}_N)]. \tag{2.100}$$

Obviously the choice $i = 1$ and $j = 2$ is arbitrary in a system of identical atoms. An equivalent definition begins with the pair density

$$\rho^{(2)}(\mathbf{r}' + \mathbf{r}, \mathbf{r}') = \left\langle \sum_i \sum_{j \neq i} \delta(\mathbf{r}' + \mathbf{r} - \mathbf{r}_i) \delta(\mathbf{r}' - \mathbf{r}_j) \right\rangle$$

for positions separated by a vector $\mathbf{r}$. This is independent of $\mathbf{r}'$ in a homogeneous system (we discuss inhomogeneous systems in Section 2.12). $g(\mathbf{r})$ is defined as the ratio

$$g(\mathbf{r}) = \frac{\rho^{(2)}(\mathbf{r}' + \mathbf{r}, \mathbf{r}')}{\rho^2} = \frac{V^2}{N^2} \frac{1}{V} \int d\mathbf{r}' \, \rho^{(2)}(\mathbf{r}' + \mathbf{r}, \mathbf{r}')$$

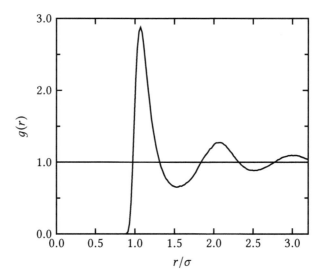

Fig. 2.2 Pair distribution function for the Lennard-Jones fluid (shifted, $r_c = 2.5\sigma$) close to its triple point ($T^* = 0.8$, $\rho^* = 0.8$).

where we average over $\mathbf{r}'$. The result of integrating over the delta functions is an ensemble average over pairs

$$g(r) = \frac{V}{N^2}\left\langle\sum_i\sum_{j\neq i}\delta(\mathbf{r}-\mathbf{r}_{ij})\right\rangle, \quad \text{where } \mathbf{r}_{ij} = \mathbf{r}_i - \mathbf{r}_j, \tag{2.101}$$

and we note that the result depends only on $r = |\mathbf{r}|$ in an isotropic liquid. The pair distribution functions of simple model systems, such as hard spheres, may be predicted accurately by integral equation theories, and are frequently used as a basis for modelling the structure of a range of fluids (Hansen and McDonald, 2013); they may be easily calculated (Smith et al., 2008).

Equation (2.101) may be used in the evaluation of $g(r)$ by computer simulation; in practice, the delta function is replaced by a function which is non-zero in a small range of separations, and a histogram is compiled of all pair separations falling within each such range (see Chapter 8). Fig. 2.2 shows a typical pair distribution function for the Lennard-Jones liquid close to its triple point.

The pair distribution function is useful, not only because it provides insight into the liquid structure, but also because the ensemble average of any pair function may be expressed in the form

$$\left\langle a(\mathbf{r}_i, \mathbf{r}_j)\right\rangle = \frac{1}{V^2}\int d\mathbf{r}_i d\mathbf{r}_j\, g(\mathbf{r}_i, \mathbf{r}_j)a(\mathbf{r}_i, \mathbf{r}_j) \tag{2.102a}$$

or, in an isotropic fluid,

$$\langle \mathcal{A}\rangle = \left\langle\sum_i\sum_{j>i}a(r_{ij})\right\rangle = \tfrac{1}{2}N\rho\int_0^\infty a(r)g(r)4\pi r^2\, dr. \tag{2.102b}$$

For example, we may write the energy (assuming pair additivity)

$$E = \tfrac{3}{2}Nk_BT + 2\pi N\rho \int_0^\infty r^2 v(r)g(r)\, dr \tag{2.103}$$

or the pressure

$$PV = Nk_BT - \tfrac{2}{3}\pi N\rho \int_0^\infty r^2 w(r)g(r)\, dr \tag{2.104}$$

although in practice, a direct evaluation of these quantities, as discussed in Section 2.4, will usually be more accurate. Even the chemical potential may be related to $g(r)$

$$\mu = k_BT \ln(\rho\Lambda^3) + 4\pi\rho \int_0^1 d\xi \int_0^\infty r^2 v(r)g(r;\xi)\, dr \tag{2.105}$$

with Λ given by eqn (2.24). As usual with the chemical potential, there is a twist: the formula involves a pair distribution function $g(r, \xi)$ which depends upon a parameter ξ coupling the two atoms, and it is necessary to integrate over ξ (McQuarrie, 1976).

The definition of the pair distribution function may be extended to the molecular case when the function $g(r_{ij}, \Omega_i, \Omega_j)$ depends upon the separation between, and orientations of the molecules. This may be evaluated in a simulation by compiling histograms, as in the atomic case, but of course there is now the problem that many more variables are involved, and a very large, multidimensional table will be needed. A number of different approaches which give partial descriptions of the orientational ordering have been developed (Gray and Gubbins, 1984):

(a) sections through $g(r_{ij}, \Omega_i, \Omega_j)$ are calculated as a function of r_{ij} for fixed relative orientations (Haile and Gray, 1980);

(b) $g(r_{ij}, \Omega_i, \Omega_j)$ can be expressed as a spherical harmonic expansion, where the coefficients are functions of r_{ij} (Streett and Tildesley, 1976; Haile and Gray, 1980);

(c) a set of site–site distribution functions $g_{ab}(r_{ab})$ can be calculated in the same way as the atomic $g(r)$ for each type of site.

The first method proceeds by compiling histograms, just as for $g(r)$, but restricting the accumulation of data to pairs of molecules which are close to a few specific relative orientations. Thus for pairs of linear molecules, parallel configurations and T-shapes might be of interest.

The spherical harmonic expansion for a pair of linear molecules would take the form

$$g(r_{ij}, \Omega_i, \Omega_j) = 4\pi \sum_{\ell=0}^\infty \sum_{\ell'=0}^\infty \sum_m g_{\ell\ell'm}(r_{ij}) Y_{\ell m}(\Omega_i) Y_{\ell'\bar{m}}(\Omega_j) \tag{2.106}$$

where the functions $Y_{\ell m}(\Omega) \equiv Y_{\ell m}(\theta, \phi)$ are spherical harmonic functions of the polar angles defining the direction of the molecular axis, and $\bar{m} = -m$. The range of the sum over m values is either $(-\ell, \ell)$ or $(-\ell', \ell')$, whichever is the smaller. Note that the orientations are measured relative to the vector r_{ij} in each case. In a simulation, the coefficients $g_{\ell\ell'm}$ would be evaluated by averaging a product of spherical harmonics over a spherical shell around each molecule, as described in Chapter 8. The function $g_{000}(r)$ is the isotropic component, that is, the pair distribution function for molecular centres averaged over all orientations. This approach is readily extended to non-linear molecules. The expansion can be carried out in a molecule-fixed frame (Streett and Tildesley, 1976) or in a space-fixed frame (Haile and Gray, 1980). The coefficients can be recombined to give the total distribution function, but this is not profitable for elongated molecules, since many terms are required for the series to converge. Certain observable properties are related to limited numbers of the harmonic coefficients. The angular correlation parameter of rank ℓ, g_ℓ, may be expressed in the molecule-fixed frame

$$g_\ell = 1 + \frac{4\pi\rho}{2\ell + 1} \sum_{m=-\ell}^{\ell} (-1)^m \int_0^\infty g_{\ell\ell m}(r) r^2 \mathrm{d}r = 1 + \frac{1}{N}\left\langle \sum_i \sum_{j\neq i} P_\ell(\cos \gamma_{ij}) \right\rangle \quad (2.107)$$

where $P_\ell(\cos \gamma)$ is a Legendre polynomial and γ_{ij} is the angle between the axis vectors of molecules i and j. g_1 is related to the dielectric properties of polar molecules, while g_2 may be investigated by depolarized light scattering. Formulae analogous to eqns (2.106) and (2.107) may be written for non-linear molecules. These would involve the Wigner rotation matrices $\mathcal{D}^\ell_{mm'}(\Omega_i)$ instead of the spherical harmonics (Gray and Gubbins, 1984, Appendix 7). In liquid crystals, where the isotropic rotational symmetry is broken, these expansions involve many more terms (Zannoni, 2000).

As an alternative, a site–site description might be more appropriate. Pair distribution functions $g_{ab}(r_{ab})$ are defined for each pair of sites on different molecules, using the same definition as in the atomic case. The number of independent $g_{ab}(r_{ab})$ functions will depend on the complexity of the molecule. For example, in a three-site model of OCS, the isotropic liquid is described by six independent $g_{ab}(r_{ab})$ functions (for OO, OC, OS, CC, CS, and SS distances), whereas for a five-site model of CH_4, the liquid is described by three functions (CC, CH, HH). While less information is contained in these distribution functions than in the components of $g(r_{ij}, \Omega_i, \Omega_j)$, they have the advantage of being directly related to the structure factor of the molecular fluid (Lowden and Chandler, 1974) and hence to experimentally observable properties (for example neutron and X-ray scattering). We return to the calculation of these quantities in Chapter 8.

Finally, we turn to the definitions of quantities that depend upon wavevector rather than on position. In a simulation with periodic boundaries, we are restricted to wavevectors that are commensurate with the periodicity of the system, that is, with the simulation box. Specifically, in a cubic box, we may examine fluctuations for which $\mathbf{k} = (2\pi/L)(n_x, n_y, n_z)$ where L is the box length and n_x, n_y, n_z are integers. This is a severe restriction, particularly at low k. One quantity of interest is the spatial Fourier transform of the number density

$$\rho(\mathbf{k}) = \sum_{i=1}^{N} \exp(-\mathrm{i}\mathbf{k} \cdot \mathbf{r}_i). \quad (2.108)$$

Fluctuations in $\rho(\mathbf{k})$ are related to the structure factor $S(k)$

$$S(k) = N^{-1}\left\langle \rho(\mathbf{k})\rho(-\mathbf{k}) \right\rangle \quad (2.109)$$

which may be measured by neutron or X-ray scattering experiments, and depends only on $k = |\mathbf{k}|$ in an isotropic system. Thus, $S(k)$ describes the Fourier components of the density fluctuations in the liquid. It is related, through a three-dimensional Fourier transform (see Appendix D) to the pair distribution function

$$S(k) = 1 + \rho\hat{h}(k) = 1 + \rho\hat{g}(k) = 1 + 4\pi\rho \int_0^\infty r^2 \frac{\sin kr}{kr} g(r)\, dr \qquad (2.110)$$

where we have introduced the Fourier transform of the total correlation function $h(r) = g(r) - 1$, and have ignored a delta function contribution at $k = 0$. In a similar way, k-dependent orientational functions may be calculated and measured routinely in computer simulations.

2.7   Time correlation functions and transport coefficients

Correlations between two different quantities $\mathcal{A}$ and $\mathcal{B}$ are measured in the usual statistical sense, via the correlation coefficient $c_{\mathcal{A}\mathcal{B}}$

$$c_{\mathcal{A}\mathcal{B}} = \langle \delta\mathcal{A}\delta\mathcal{B} \rangle / \sigma(\mathcal{A})\sigma(\mathcal{B}) \qquad (2.111)$$

with $\sigma(\mathcal{A})$ and $\sigma(\mathcal{B})$ given by eqn (2.48). Schwartz inequalities guarantee that the absolute value of $c_{\mathcal{A}\mathcal{B}}$ lies between 0 and 1, with values close to 1 indicating a high degree of correlation. The idea of the correlation coefficient may be extended in a very useful way, by considering $\mathcal{A}$ and $\mathcal{B}$ to be evaluated at two different times. For an equilibrium system, the resulting quantity is a function of the time difference t: it is a 'time correlation function' $c_{\mathcal{A}\mathcal{B}}(t)$. For identical phase functions, $c_{\mathcal{A}\mathcal{A}}(t)$ is called an autocorrelation function and its time integral (from $t = 0$ to $t = \infty$) is a correlation time $t_{\mathcal{A}}$. These functions are of great interest in computer simulation because:

(a) they give a clear picture of the dynamics in a fluid;
(b) their time integrals $t_{\mathcal{A}}$ may often be related directly to macroscopic transport coefficients;
(c) their Fourier transforms $\hat{c}_{\mathcal{A}\mathcal{A}}(\omega)$ may often be related to experimental spectra measured as a function of frequency ω.

Useful discussions of time correlation functions may be found in the references (Steele, 1969; 1980; Berne and Harp, 1970; McQuarrie, 1976; Hansen and McDonald, 2013). A few comments may be relevant here. The non-normalized correlation function is defined

$$C_{\mathcal{A}\mathcal{B}}(t) = \left\langle \delta\mathcal{A}(t)\delta\mathcal{B}(0) \right\rangle_{\text{ens}} = \left\langle \delta\mathcal{A}\big(\Gamma(t)\big)\delta\mathcal{B}\big(\Gamma(0)\big) \right\rangle_{\text{ens}} \qquad (2.112)$$

so that

$$c_{\mathcal{A}\mathcal{B}}(t) = C_{\mathcal{A}\mathcal{B}}(t)/\sigma(\mathcal{A})\sigma(\mathcal{B}) \qquad (2.113a)$$

or

$$c_{\mathcal{A}\mathcal{A}}(t) = C_{\mathcal{A}\mathcal{A}}(t)/\sigma^2(\mathcal{A}) = C_{\mathcal{A}\mathcal{A}}(t)/C_{\mathcal{A}\mathcal{A}}(0). \qquad (2.113b)$$

Just like $\langle \delta\mathcal{A}\delta\mathcal{B} \rangle$, $C_{\mathcal{A}\mathcal{B}}(t)$ is different for different ensembles, and eqn (2.50) may be used to transform from one ensemble to another. The computation of $C_{\mathcal{A}\mathcal{B}}(t)$ may be thought of as a two-stage process. First, we must select initial state points $\Gamma(0)$, according to the desired distribution $\rho_{\text{ens}}(\Gamma)$, over which we will subsequently average. This may be done using any of the prescriptions mentioned in Section 2.1. Second, we must evaluate $\Gamma(t)$. This means solving the true (Newtonian) equations of motion. By this means, time-dependent properties may be calculated in any ensemble. In practice, the mechanical equations of motion are almost always used for both purposes, that is, we use molecular dynamics to calculate the time correlation functions in the microcanonical ensemble.

Some attention must be paid to the question of ensemble equivalence, however, since the link between correlation functions and transport coefficients is made through linear response theory, which can be carried out in virtually any ensemble. This actually caused some confusion in the original derivations of the expressions for transport coefficients (Zwanzig, 1965). In the following, we make some general observations, and refer the reader elsewhere (McQuarrie, 1976; Frenkel and Smit, 2002; Tuckerman, 2010; Hansen and McDonald, 2013) for a fuller discussion.

Transport coefficients such as the diffusion coefficient, thermal conductivity, and shear and bulk viscosities appear in the equations of hydrodynamics, such as the mass and thermal diffusion equations and the Navier–Stokes equation. Accordingly, they describe the relaxation of dynamical variables on the macroscopic scale. Provided the long-time and large-length-scale limits are considered carefully, they can be expressed in terms of equilibrium time correlation functions of microscopically defined variables. Such relations are generally termed Green–Kubo formulae (Hansen and McDonald, 2013). Linear response theory can be used to provide an interpretation of these formulae in terms of the response of the system to a weak perturbation. By introducing such perturbations into the Hamiltonian, or directly into the equations of motion, their effect on the distribution function ρ_{ens} may be calculated. Generally, a time-dependent nonequilibrium distribution $\rho(t) = \rho_{ens} + \delta\rho(t)$ is produced. Hence, any nonequilibrium ensemble average (in particular, the desired response) may be calculated. By retaining the linear terms in the perturbation, and comparing the equation for the response with a macroscopic transport equation, we may identify the transport coefficient.

The Green–Kubo expression is usually written as the infinite time integral of an equilibrium time correlation function of the form

$$\gamma = \int_0^\infty dt \left\langle \dot{\mathcal{A}}(t)\dot{\mathcal{A}}(0) \right\rangle \tag{2.114}$$

where γ is the transport coefficient, and $\mathcal{A}$ is the appropriate dynamical variable. Associated with any expression of this kind, there is an 'Einstein relation'

$$\left\langle \left(\mathcal{A}(t) - \mathcal{A}(0)\right)^2 \right\rangle = 2t\gamma, \quad \text{as } t \to \infty, \text{ or } \quad \gamma = \lim_{t \to \infty} \frac{d}{dt} \frac{1}{2} \left\langle \left(\mathcal{A}(t) - \mathcal{A}(0)\right)^2 \right\rangle \tag{2.115}$$

which holds at large t compared with the correlation time of $\mathcal{A}$. The connection between eqns (2.114) and (2.115) may be easily established by integration by parts. Note that only a few genuine transport coefficients exist; that is, for only a few 'hydrodynamic' variables $\mathcal{A}$ do eqns (2.114) and (2.115) give a non-zero γ (McQuarrie, 1976).

In computer simulations, transport coefficients may be calculated from equilibrium correlation functions, using eqn (2.114), by observing Einstein relations, eqn (2.115), or indeed going back to first principles and conducting a suitable nonequilibrium simulation. The details of the calculation via eqns (2.114), (2.115) will be given in Chapter 8, and we will examine nonequilibrium methods in Chapter 11. For use in equilibrium molecular dynamics, we give here the equations for calculating thermal transport coefficients in the microcanonical ensemble, for a fluid composed of N identical molecules.

The diffusion coefficient D is given (in three dimensions) by

$$D = \frac{1}{3} \int_0^\infty dt \left\langle \mathbf{v}_i(t) \cdot \mathbf{v}_i(0) \right\rangle \tag{2.116}$$

where $\mathbf{v}_i(t)$ is the centre-of-mass velocity of a single molecule. The corresponding Einstein relation, valid at long times, is

$$D = \lim_{t \to \infty} \frac{d}{dt} \frac{1}{6} \left\langle \left| \mathbf{r}_i(t) - \mathbf{r}_i(0) \right|^2 \right\rangle \tag{2.117}$$

where $\mathbf{r}_i(t)$ is the molecule position. There is also an equally valid 'intermediate' form:

$$D = \lim_{t \to \infty} \frac{1}{3} \left\langle \mathbf{v}_i(0) \cdot \left(\mathbf{r}_i(t) - \mathbf{r}_i(0) \right) \right\rangle. \tag{2.118}$$

In practice, these averages would be computed for each of the N particles in the simulation, the results added together, and divided by N, to improve statistical accuracy. Note that in the computation of eqns (2.117), (2.118), it is important not to switch attention from one periodic image to another, which is why it is sometimes useful to have available a set of particle coordinates which have not been subjected to periodic boundary conditions during the simulation (see Section 1.6 and Chapter 8).

The shear viscosity is given by

$$\eta = \frac{V}{k_B T} \int_0^\infty dt \left\langle \mathcal{P}_{\alpha\beta}(t) \mathcal{P}_{\alpha\beta}(0) \right\rangle \tag{2.119}$$

or

$$\eta = \lim_{t \to \infty} \frac{d}{dt} \frac{1}{2} \frac{V}{k_B T} \left\langle \left(Q_{\alpha\beta}(t) - Q_{\alpha\beta}(0) \right)^2 \right\rangle. \tag{2.120}$$

Here

$$\mathcal{P}_{\alpha\beta} = \frac{1}{V} \left(\sum_i p_{i\alpha} p_{i\beta} / m_i + \sum_i r_{i\alpha} f_{i\beta} \right) \tag{2.121}$$

or

$$\mathcal{P}_{\alpha\beta} = \frac{1}{V} \left(\sum_i p_{i\alpha} p_{i\beta} / m_i + \sum_i \sum_{j > i} r_{ij\alpha} f_{ij\beta} \right) \tag{2.122}$$

is an off-diagonal ($\alpha \neq \beta$) element of the pressure tensor (compare the virial expression for the pressure function eqns (2.61) and (2.65)) and

$$Q_{\alpha\beta} = \frac{1}{V} \sum_i r_{i\alpha} p_{i\beta}. \tag{2.123}$$

The negative of $\mathcal{P}_{\alpha\beta}$ is often called the stress tensor. These quantities are multi-particle properties, properties of the system as a whole, and no additional averaging over N particles is possible. Consequently η is subject to much greater statistical imprecision than

D. Some improvement is possible by averaging over different components $\alpha\beta = xy, yz, zx$, of $\mathcal{P}_{\alpha\beta}$. Just as for eqn (2.65), the origin-independent form, eqn (2.122), should be used rather than eqn (2.121), in periodic boundaries, and similar care needs to be taken in the calculation of $Q_{\alpha\beta}(t) - Q_{\alpha\beta}(0)$ in eqn (2.120).

The bulk viscosity is given by a similar expression:

$$\eta_V = \frac{V}{9k_{\mathrm{B}}T} \sum_{\alpha\beta} \int_0^\infty dt \left\langle \delta\mathcal{P}_{\alpha\alpha}(t)\delta\mathcal{P}_{\beta\beta}(0) \right\rangle = \frac{V}{k_{\mathrm{B}}T} \int_0^\infty dt \left\langle \delta\mathcal{P}(t)\delta\mathcal{P}(0) \right\rangle \qquad (2.124a)$$

where we sum over $\alpha, \beta = x, y, z$ and note that $\mathcal{P} = \frac{1}{3}\mathrm{Tr}\,\mathcal{P} = \frac{1}{3}\sum_\alpha \mathcal{P}_{\alpha\alpha}$. Rotational invariance leads to the equivalent expression

$$\eta_V + \tfrac{4}{3}\eta = \frac{V}{k_{\mathrm{B}}T} \int_0^\infty dt \left\langle \delta\mathcal{P}_{\alpha\alpha}(t)\delta\mathcal{P}_{\alpha\alpha}(0) \right\rangle. \qquad (2.124b)$$

Here the diagonal stresses must be evaluated with care, since a non-vanishing equilibrium average must be subtracted:

$$\delta\mathcal{P}_{\alpha\alpha}(t) = \mathcal{P}_{\alpha\alpha}(t) - \langle\mathcal{P}_{\alpha\alpha}\rangle = \mathcal{P}_{\alpha\alpha}(t) - P \qquad (2.125a)$$
$$\delta\mathcal{P}(t) = \mathcal{P}(t) - \langle\mathcal{P}\rangle = \mathcal{P}(t) - P \qquad (2.125b)$$

with $\mathcal{P}_{\alpha\beta}$ given by an expression like eqn (2.122). The corresponding Einstein relation is (Alder et al., 1970)

$$\eta_V + \tfrac{4}{3}\eta = \lim_{t\to\infty} \frac{d}{dt} \frac{1}{2} \frac{V}{k_{\mathrm{B}}T} \left\langle \left(Q_{\alpha\alpha}(t) - Q_{\alpha\alpha}(0) - Pt \right)^2 \right\rangle. \qquad (2.126)$$

The thermal conductivity λ_T can be written (Hansen and McDonald, 2013)

$$\lambda_T = \frac{V}{k_{\mathrm{B}}T^2} \int_0^\infty dt \left\langle j_\alpha^\epsilon(t)j_\alpha^\epsilon(0) \right\rangle \qquad (2.127)$$

or

$$\lambda_T = \lim_{t\to\infty} \frac{d}{dt} \frac{1}{2} \frac{V}{k_{\mathrm{B}}T^2} \left\langle \left(\delta\epsilon_\alpha(t) - \delta\epsilon_\alpha(0) \right)^2 \right\rangle. \qquad (2.128)$$

Here j_α^ϵ is a component of the energy current, that is, the time derivative of

$$\delta\epsilon_\alpha = \frac{1}{V} \sum_i r_{i\alpha}\left(\epsilon_i - \langle\epsilon_i\rangle\right). \qquad (2.129)$$

The term $\sum_i r_{i\alpha}\langle\epsilon_i\rangle$ makes no contribution if $\sum_i r_{i\alpha} = 0$, as is the case in a normal one-component MD simulation. In calculating the energy per molecule ϵ_i, the potential energy of two molecules (assuming pairwise additive potentials) is taken to be divided equally between them:

$$\epsilon_i = p_i^2/2m_i + \tfrac{1}{2}\sum_{j\neq i} v(r_{ij}). \qquad (2.130)$$

These expressions for η_V and λ_T are ensemble-dependent and the preceding equations hold for the microcanonical case only. A fuller discussion may be found in the standard texts (McQuarrie, 1976; Zwanzig, 1965).

Transport coefficients are related to the long-time behaviour of correlation functions. Short-time correlations, on the other hand, may be linked with static equilibrium ensemble averages by expanding in a Taylor series. For example, the velocity of particle i may be written

$$\mathbf{v}_i(t) = \mathbf{v}_i(0) + \dot{\mathbf{v}}_i(0)t + \tfrac{1}{2}\ddot{\mathbf{v}}_i(0)t^2 + \cdots .\tag{2.131}$$

Multiplying by $\mathbf{v}_i(0)$ and ensemble averaging yields

$$\left\langle \mathbf{v}_i(t) \cdot \mathbf{v}_i(0) \right\rangle = \langle v_i^2 \rangle + \tfrac{1}{2}\langle \ddot{\mathbf{v}}_i \cdot \mathbf{v}_i \rangle t^2 + \cdots = \langle v_i^2 \rangle - \tfrac{1}{2}\langle \dot{v}_i^2 \rangle t^2 + \cdots .\tag{2.132}$$

The vanishing of the term linear in t, and the last step, where we set $\langle \ddot{\mathbf{v}}_i \cdot \mathbf{v}_i \rangle = -\langle \dot{\mathbf{v}}_i \cdot \dot{\mathbf{v}}_i \rangle$, follow from time-reversal symmetry and stationarity (McQuarrie, 1976). Thus, the short-time velocity autocorrelation function is related to the mean-square acceleration, that is, to the mean-square force. This behaviour may be used to define the Einstein frequency ω_E

$$\langle \mathbf{v}_i(t) \cdot \mathbf{v}_i(0) \rangle = \langle v_i^2 \rangle \left(1 - \tfrac{1}{2}\omega_E^2 t^2 + \cdots \right).\tag{2.133}$$

The analogy with the Einstein model, of an atom vibrating in the mean force field of its neighbours, with frequency ω_E in the harmonic approximation, becomes clear when we replace the mean-square force by the average potential curvature using

$$\langle f_{i\alpha}^2 \rangle = -\left\langle f_{i\alpha}\partial\mathcal{V}/\partial r_{i\alpha} \right\rangle = -k_B T\left\langle \partial f_{i\alpha}/\partial r_{i\alpha} \right\rangle = k_B T\left\langle \partial^2\mathcal{V}/\partial r_{i\alpha}^2 \right\rangle\tag{2.134}$$

which is another application of $\langle \mathcal{A}\, \partial\mathcal{H}/\partial q_k \rangle = k_B T\langle \partial\mathcal{A}/\partial q_k \rangle$. The result is

$$\omega_E^2 = \frac{\langle f_i^2 \rangle}{m_i^2\langle v_i^2 \rangle} = \frac{1}{3m_i}\langle \nabla_{\mathbf{r}_i}^2 \mathcal{V} \rangle.\tag{2.135}$$

This may be easily evaluated for, say, a pairwise additive potential. Short-time expansions of other time correlation functions may be obtained using similar techniques. The temporal Fourier transform (see Appendix D) of the velocity autocorrelation function is proportional to the density of normal modes in a purely harmonic system, and is often loosely referred to as the 'density of states' in solids and liquids. The velocity autocorrelation function for the Lennard-Jones liquid near the triple point is illustrated in Fig. 2.3.

We can only mention briefly some other autocorrelation functions of interest in computer simulations. The generalization of eqn (2.109) to the time domain yields the intermediate scattering function $I(k, t)$.

$$I(k, t) = N^{-1}\left\langle \rho(\mathbf{k}, t)\rho(-\mathbf{k}, 0) \right\rangle\tag{2.136}$$

with $\rho(\mathbf{k}, t)$ defined by eqn (2.108). The temporal Fourier transform of this, the dynamic structure factor $S(k, \omega)$, in principle may be measured by inelastic neutron scattering. Spatially Fourier transforming $I(k, t)$ yields the van Hove function $G(r, t)$, a generalization of $g(r)$ which measures the probability of finding a particle at position r at time t, given that a particle was at the origin of coordinates at time 0. All of these functions may be divided into parts due to 'self' (i.e. single-particle) motion and due to 'distinct' (i.e. collective) effects. Other k-dependent variables may be defined, and their time correlation

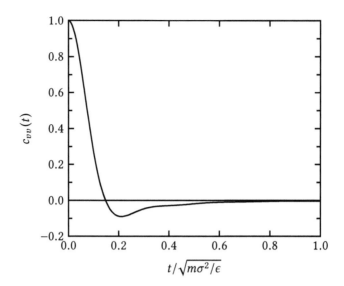

Fig. 2.3 The velocity autocorrelation function for the Lennard-Jones fluid (shifted potential, $r_c = 2.5\sigma$) close to its triple point ($T^* = 0.8$, $\rho^* = 0.8$).

functions are of great interest (Hansen and McDonald, 2013). For example, longitudinal and transverse momentum components may be defined

$$p^{\parallel}(\mathbf{k}, t) = \frac{1}{V} \sum_i p_{ix}(t) \exp\left(ikx_i(t)\right) \tag{2.137a}$$

$$p_1^{\perp}(\mathbf{k}, t) = \frac{1}{V} \sum_i p_{iy}(t) \exp\left(ikx_i(t)\right) \tag{2.137b}$$

$$p_2^{\perp}(\mathbf{k}, t) = \frac{1}{V} \sum_i p_{iz}(t) \exp\left(ikx_i(t)\right) \tag{2.137c}$$

where for convenience we take $\mathbf{k} = (k, 0, 0)$ in the x direction, and $x_i = r_{ix}$. These quantities are useful for discussing hydrodynamic modes in liquids. These functions may all be computed routinely in simulations, although, as always, the allowed k-vectors are restricted by small system sizes and periodic boundary conditions.

For systems of rigid molecules, the angular velocity ω_i plays a role in reorientational dynamics analogous to that of $\mathbf{v}_i$ in translation (see Chapter 3). The angular velocity correlation function $\langle \omega_i(t) \cdot \omega_i(0) \rangle$ may be used to describe rotation. Time-dependent orientational correlations may be defined (Gordon, 1968; Steele, 1969; 1980) as straightforward generalizations of the quantities seen earlier. For a linear molecule, the time correlation function of rank-ℓ spherical harmonics is

$$c_\ell(t) = 4\pi \left\langle Y_{\ell m}\big(\Omega_i(t)\big) Y_{\ell m}^*\big(\Omega_i(0)\big) \right\rangle = \left\langle P_\ell\big(\cos \delta\gamma(t)\big) \right\rangle \tag{2.138}$$

where $\delta\gamma(t)$ is the magnitude of the angle turned through in time t. Note that, assuming an isotropic system, there are $2\ell + 1$ rank-ℓ functions, all identical in form, corresponding to the different values of m.

Analogous formulae for non-linear molecules involve the Wigner rotation matrices. The starting point is the expansion of the probability density for rotation of the molecule through a set of Euler angles $\delta\Omega_i$ in time t:

$$\rho(\delta\Omega_i;t) = \sum_{\ell m m'} \frac{2\ell+1}{8\pi^2} c_{\ell m m'}(t) \mathcal{D}_{mm'}^{\ell*}(\delta\Omega_i)$$

which is equivalent to defining the key correlation functions

$$c_{\ell m m'}(t) = \left\langle \mathcal{D}_{mm'}^{\ell}\left(\delta\Omega_i(t)\right) \right\rangle.$$

The term on the right may be expressed in terms of the molecular orientations at time 0 and time t:

$$\mathcal{D}_{mm'}^{\ell}\left(\delta\Omega_i(t)\right) = \sum_{n=-\ell}^{\ell} \mathcal{D}_{nm}^{\ell*}\left(\Omega_i(0)\right) \mathcal{D}_{nm'}^{\ell}\left(\Omega_i(t)\right).$$

However, in an isotropic fluid, averaging over the initial distribution gives an identical result for each of these terms

$$\left\langle \mathcal{D}_{mm'}^{\ell}\left(\delta\Omega_i(t)\right) \right\rangle = (2\ell+1)\left\langle \mathcal{D}_{nm}^{\ell*}\left(\Omega_i(0)\right) \mathcal{D}_{nm'}^{\ell}\left(\Omega_i(t)\right) \right\rangle$$

independent of n, and so (Steele, 1980)

$$c_{\ell m m'}(t) = (2\ell+1)\left\langle \mathcal{D}_{nm}^{\ell*}\left(\Omega_i(0)\right) \mathcal{D}_{nm'}^{\ell}\left(\Omega_i(t)\right) \right\rangle. \tag{2.139}$$

When the symmetry of the phase is reduced, for example in liquid crystals, many more non-equivalent time correlation functions exist (Zannoni, 1994). These quantities are experimentally accessible and the relationships between them are of great theoretical interest. For example, first-rank autocorrelation functions may be related to infrared absorption, and second-rank functions to light scattering. Functions of all ranks contribute to inelastic neutron scattering spectra from molecular liquids.

2.8 Long-range corrections

As explained in Section 1.6, computer simulations frequently use a pair potential with a spherical cutoff at a distance r_c. It becomes useful to correct the results of simulations to compensate for the missing long-range part of the potential. Contributions to the energy, pressure, etc. for $r > r_c$ are frequently estimated by assuming that $g(r) \approx 1$ in this region, and using eqns (2.103)–(2.105)

$$E_{\text{full}} \approx E_c + E_{\text{LRC}} = E_c + 2\pi N\rho \int_{r_c}^{\infty} r^2 v(r)\, dr \tag{2.140}$$

$$(PV)_{\text{full}} = (PV)_c + (PV)_{\text{LRC}} = (PV)_c - \tfrac{2}{3}\pi N\rho \int_{r_c}^{\infty} r^2 w(r)\, dr \tag{2.141}$$

$$\mu_{\text{full}} = \mu_c + \mu_{\text{LRC}} = \mu_c + 4\pi\rho \int_{r_c}^{\infty} r^2 v(r)\, dr \tag{2.142}$$

where E_{full}, $(PV)_{\text{full}}$, μ_{full} are the desired values for a liquid with the full potential, and E_c, $(PV)_c$, μ_c are the values actually determined from a simulation using a potential with a cutoff. For the Lennard-Jones potential, eqn (1.6), these equations become

$$E^*_{\text{LRC}} = \tfrac{8}{9}\pi N \rho^* r_c^{*-9} - \tfrac{8}{3}\pi N \rho^* r_c^{*-3} \tag{2.143}$$

$$P^*_{\text{LRC}} = \tfrac{32}{9}\pi \rho^{*2} r_c^{*-9} - \tfrac{16}{3}\pi \rho^{*2} r_c^{*-3} \tag{2.144}$$

$$\mu^*_{\text{LRC}} = \tfrac{16}{9}\pi \rho^* r_c^{*-9} - \tfrac{16}{3}\pi \rho^* r_c^{*-3} \tag{2.145}$$

where we use Lennard-Jones reduced units (see Appendix B). In the case of the constant-NVE and constant-NVT ensembles, these corrections can be applied to the results after a simulation has run. However, if the volume or the number of particles is allowed to fluctuate (e.g. in a constant-NPT or constant-μVT simulation) it is important to apply the corrections to the calculated instantaneous energies, pressures, etc. during the course of a simulation, since they will change as the density fluctuates: it is far more tricky to attempt to do this when the simulation is over.

For three-body potentials such as the Axilrod–Teller potential, the potential energy from the three-body term in a homogeneous fluid is

$$\langle V_3 \rangle = \frac{\rho^3}{6} \iiint d\mathbf{r}_1 d\mathbf{r}_2 d\mathbf{r}_3 \, v^{(3)}(r_{12}, r_{13}, r_{23}) \, g^{(3)}(\mathbf{r}_1, \mathbf{r}_2, \mathbf{r}_3). \tag{2.146}$$

In a simulation the three-body energy is calculated explicitly for

$$r_{12} \leq r_c \text{ and } r_{13} \leq r_c \text{ and } r_{23} \leq r_c. \tag{2.147}$$

For the rest of the space, the long-range part, the three-body distribution function can be approximated using the superposition approximation

$$g^{(3)}(\mathbf{r}_1, \mathbf{r}_2, \mathbf{r}_3) \approx g^{(2)}(r_{12}) \, g^{(2)}(r_{13}) \, g^{(2)}(r_{23}). \tag{2.148}$$

There are six equivalent parts to this region, and the long-range correction is thus

$$\langle V_3 \rangle_{\text{LRC}} = \rho^3 \iiint_{\substack{r_{12} > r_c \\ r_{12} > r_{13} > r_{23}}} d\mathbf{r}_1 d\mathbf{r}_2 d\mathbf{r}_3 \, v^{(3)}(r_{12}, r_{13}, r_{23}) \, g^{(2)}(r_{13}) \, g^{(2)}(r_{23}) \tag{2.149}$$

where we have explicitly considered the region in which $r_{12} > r_c$ and thus $g(r_{12}) = 1$ (Barker et al., 1971). Then transforming to bipolar coordinates we have

$$E_{3,\text{LRC}} = \langle V_3 \rangle_{\text{LRC}} = 8\pi^2 \rho^2 N \int_{r_c}^{\infty} dr_{12}\, r_{12} \int_0^{r_{12}} dr_{13}\, r_{13}$$
$$\int_{r_{12}-r_{13}}^{r_{13}} dr_{23}\, r_{23}\, v^{(3)}(r_{12}, r_{13}, r_{23}) \, g^{(2)}(r_{13}) g^{(2)}(r_{23}) \tag{2.150}$$

which can be evaluated accurately using the kind of Monte Carlo integration method described in Section 4.2. Note that the approximate estimate of the three-body long-range correction requires the two-body radial distribution function for the fluid. This cannot simply be set to one since both r_{13} and r_{23} can be less than r_c. For the specific case of the Axilrod–Teller potential, which is a homogeneous function of order 9 (Graben and Fowler, 1969)

$$(PV)_{3,\text{LRC}} = 3E_{3,\text{LRC}}. \tag{2.151}$$

For more general potentials the Monte Carlo evaluation of eqn (2.149) can be adapted to calculate the long-range correction for the pressure and the chemical potential.

2.9 Quantum corrections

Most of this book will deal with the computer simulation of systems within the classical approximation, although we turn in Chapter 13 to the attempts which have been made to incorporate quantum effects in simulations. Quantum effects in thermodynamics may be measured experimentally via the isotope separation factor, while tunnelling, superfluidity, etc. are clear manifestations of quantum mechanics.

Even within the limitations of a classical simulation, it is still possible to estimate quantum corrections of thermodynamic functions. This is achieved by expanding the partition function in powers of Planck's constant, $\hbar = h/2\pi$ (Wigner, 1932; Kirkwood, 1933). For a system of N atoms we have

$$Q_{NVT} = \frac{1}{\Lambda^{3N} N!} \int \mathrm{d}\mathbf{r} \left(1 - \frac{\beta \hbar^2}{24m} \sum_{i=1}^{N} \left[\nabla_{\mathbf{r}_i} \beta \mathcal{V}(\mathbf{r})\right]^2\right) \exp[-\beta \mathcal{V}(\mathbf{r})] \tag{2.152}$$

where Λ is defined in eqn (2.24). The expansion accounts for the leading quantum-mechanical diffraction effects; other effects, such as exchange, are small for most cases of interest. Additional details may be found elsewhere (Landau and Lifshitz, 1980; McQuarrie, 1976). This leads to the following correction to the Helmholtz free energy, $\Delta A = A^{\text{qu}} - A^{\text{cl}}$

$$\Delta A_{\text{trans}} = \tfrac{1}{24} N \hbar^2 \beta^2 \langle f_i^2 \rangle / m. \tag{2.153}$$

Here, as in Section 2.7, $\langle f_i^2 \rangle$ is the mean-square force on one atom in the simulation. Obviously, better statistics are obtained by averaging over all N atoms. An equivalent expression is

$$\Delta A_{\text{trans}} = \frac{N \Lambda^2 \rho}{48\pi} \int \mathrm{d}\mathbf{r}\, g(r) \nabla^2 v(r)$$

$$= \frac{N \Lambda^2 \rho}{12} \int_0^\infty r^2 g(r) \left(\frac{\mathrm{d}^2 v(r)}{\mathrm{d}r^2} + \frac{2}{r}\frac{\mathrm{d}v(r)}{\mathrm{d}r}\right) \mathrm{d}r \tag{2.154}$$

assuming pairwise additive interactions. Additional corrections of order $\hbar^4$ can be estimated if the three-body distribution function g_3 can be calculated in a simulation (Hansen and Weis, 1969). Note that for hard systems, the leading quantum correction is of order $\hbar$: for hard spheres it amounts to replacing the hard sphere diameter σ by $\sigma + \Lambda/\sqrt{8}$ (Hemmer, 1968; Jancovici, 1969). By differentiating these equations, quantum corrections for the energy, pressure, etc. can easily be obtained.

Equation (2.153) is also the translational correction for a system of N molecules where it is understood that m now stands for the molecular mass and $\langle f_i^2 \rangle$ is the mean-square force acting on the molecular centre of mass. Additional corrections must be applied for a molecular system, to take account of rotational motion (St. Pierre and Steele, 1969; Powles and Rickayzen, 1979). For linear molecules, with moment of inertia I, the additional term is (Gray and Gubbins, 1984)

$$\Delta A_{\text{rot}} = \tfrac{1}{24} N \hbar^2 \beta^2 \langle \tau_i^2 \rangle / I - N \hbar^2 / 6I \tag{2.155}$$

where $\langle \tau_i^2 \rangle$ is the mean-square torque acting on a molecule. The correction for the general asymmetric top, with three different moments of inertia I_{xx}, I_{yy}, and I_{zz}, is rather more complicated

$$\Delta A_{\text{rot}} = \tfrac{1}{24} N \hbar^2 \beta^2 \left(\frac{\langle \tau_{ix}^2 \rangle}{I_{xx}} + \frac{\langle \tau_{iy}^2 \rangle}{I_{yy}} + \frac{\langle \tau_{iz}^2 \rangle}{I_{zz}} \right) - \left[\frac{N \hbar^2}{24} \sum_{\text{cyclic}} \frac{2}{I_{xx}} - \frac{I_{xx}}{I_{yy} I_{zz}} \right] \tag{2.156}$$

where the sum is over the three cyclic permutations of x, y, and z. These results are independent of ensemble, and from them the quantum corrections to any other thermo-dynamic property can be calculated. Moreover, it is easy to compute the mean-square force and mean-square torque in a simulation.

Another, possibly more accurate, way of estimating quantum corrections has been proposed (Berens et al., 1983). In this approach, the velocity autocorrelation function is calculated and is Fourier transformed to obtain a spectrum, or density of states,

$$\hat{c}_{vv}(\omega) = \int_{-\infty}^{\infty} dt \, \exp(i\omega t) \langle \mathbf{v}_i(t) \cdot \mathbf{v}_i(0) \rangle / \langle v_i^2 \rangle$$

$$= \frac{m}{3 k_{\text{B}} T} \int_{-\infty}^{\infty} dt \, \exp(i\omega t) \langle \mathbf{v}_i(t) \cdot \mathbf{v}_i(0) \rangle. \tag{2.157}$$

Then, quantum corrections are applied to any thermodynamic quantities of interest, using the approximation that the system behaves as a set of harmonic oscillators, whose frequency distribution is dictated by the measured velocity spectrum. For each thermo-dynamic function a correction function, which would apply to a harmonic oscillator of frequency ω, may be defined. The total correction is then obtained by integrating over all frequencies. For the Helmholtz free energy, the correction is given by

$$\Delta A = 3 N k_{\text{B}} T \int_{-\infty}^{\infty} \frac{d\omega}{2\pi} \, \hat{c}_{vv}(\omega) \ln \left(\frac{\exp\left(\tfrac{1}{2} \hbar\omega / k_{\text{B}} T\right) - \exp\left(-\tfrac{1}{2} \hbar\omega / k_{\text{B}} T\right)}{\hbar\omega / k_{\text{B}} T} \right) \tag{2.158}$$

which agrees with eqn (2.153) to $O(\hbar^2)$. The rationale here is that the harmonic approxima-tion is most accurate for the high-frequency motions that contribute the largest quantum corrections, whereas the anharmonic motions are mainly of low frequency, and thus their quantum corrections are less important. Simulations comparing the simulated and experimental heat capacity (Waheed and Edholm, 2011) seem to confirm this for liquid water, but anharmonicity for ice near the melting point is significant, and the method overestimates the heat of vaporization. This approach has also been applied to liquid methanol (Hawlicka et al., 1989) and ammonia under extreme conditions (Bethkenhagen et al., 2013). An alternative approach, however, is to incorporate quantum effects associated with light nuclei, such as hydrogen, into the simulation method directly using the path-integral approach (see Section 13.4).

Quantum corrections may also be applied to structural quantities such as $g(r)$. The formulae are rather complex and will not be given here, but they are based on the same formula eqn (2.152) for the partition function (Gibson, 1974). Again, the result is different for hard systems (Gibson, 1975a,b).

When it comes to time-dependent properties, there is one quantum correction which is essential to bring the results of classical simulation in line with experiment. Quantum mechanical autocorrelation functions obey the detailed balance condition

$$\hat{C}_{\mathcal{A}\mathcal{A}}(\omega) = \exp(\beta\hbar\omega)\hat{C}_{\mathcal{A}\mathcal{A}}(-\omega) \qquad (2.159)$$

whereas, of course, classical autocorrelation functions are even in frequency (Berne and Harp, 1970). The effects of detailed balance are clearly visible in experimental spectra, for example in inelastic neutron scattering, which probes $S(k, \omega)$; in fact experimental results are often converted to the symmetrized form $\exp(\frac{1}{2}\hbar\beta\omega)S(k, \omega)$ for comparison with classical theories. Simple empirical measures have been advocated to convert classical time correlation functions into approximate quantum-mechanical ones. Both the 'complex time' substitutions

$$C_{\mathcal{A}\mathcal{A}}(t) \rightarrow C_{\mathcal{A}\mathcal{A}}(t - \tfrac{1}{2}i\hbar\beta) \qquad \text{(Schofield, 1960)} \qquad (2.160)$$

$$\text{and} \quad C_{\mathcal{A}\mathcal{A}}(t) \rightarrow C_{\mathcal{A}\mathcal{A}}\big((t^2 - i\hbar\beta t)^{1/2}\big) \qquad \text{(Egelstaff, 1961)} \qquad (2.161)$$

result in functions which satisfy detailed balance. The former is somewhat easier to apply, since it equates the symmetrized experimental spectrum with the classical simulated one, while the latter satisfies some additional frequency integral relations.

2.10 Constraints

In modelling large molecules such as proteins it may be necessary to include constraints in the potential model, as discussed in Section 1.3.3. This introduces some subtleties into the statistical mechanical description. The system of constrained molecules moves on a well-defined hypersurface in phase space. The generalized coordinates corresponding to the constraints and their conjugate momenta are removed from the Hamiltonian. This affects the form of the distribution function, and expressions for ensemble averages, in Cartesian coordinates (see Ryckaert and Ciccotti, 1983, Appendix). This system is not equivalent to a fluid where the constrained degrees of freedom are replaced by harmonic springs, even in the limit of infinitely strong force constants (Fixman, 1974; Pear and Weiner, 1979; Chandler and Berne, 1979).

To explore this difference more formally, we consider a set of N atoms grouped into molecules in some arbitrary way by harmonic springs. The Cartesian coordinates of the atoms are the $3N$ values $\mathbf{r} = \{r_{i\alpha}\}, i = 1, 2, \cdots N, \alpha = x, y, z$. The system can be described by $3N$ generalized coordinates $\mathbf{q}$ (i.e. the positions of the centre of mass of each molecule, their orientations, and vibrational coordinates). The potential energy of the system can be separated into a part, $\mathcal{V}_s$, associated with the 'soft' coordinates $\mathbf{q}^s$ (the translations, rotations, and internal conversions) and a part $\mathcal{V}_h$ associated with the 'hard' coordinates $\mathbf{q}^h$ (bond stretching and possibly bond angle vibrations)

$$\mathcal{V}(\mathbf{q}) = \mathcal{V}_s(\mathbf{q}^s) + \mathcal{V}_h(\mathbf{q}^h). \tag{2.162}$$

If the force constants of the hard modes are independent of $\mathbf{q}^s$ then the canonical ensemble average of some configurational property $\mathcal{A}(\mathbf{q}^s)$ is (Berendsen and van Gunsteren, 1984)

$$\langle \mathcal{A} \rangle_{NVT} = \frac{\int \mathcal{A}(\mathbf{q}^s) \sqrt{\det(\mathbf{G})} \exp\left[-\beta \mathcal{V}_s(\mathbf{q}^s)\right] d\mathbf{q}^s}{\int \sqrt{\det(\mathbf{G})} \exp\left[-\beta \mathcal{V}_s(\mathbf{q}^s)\right] d\mathbf{q}^s} \tag{2.163}$$

where $\det(\mathbf{G})$ is the determinant of the mass-weighted metric tensor $\mathbf{G}$, which is associated with the transformation from Cartesian to generalized coordinates

$$G_{k\ell} = \sum_{i=1}^{N} \sum_{\alpha} m_i \frac{\partial r_{i\alpha}}{\partial q_k} \frac{\partial r_{i\alpha}}{\partial q_\ell}. \tag{2.164}$$

$\mathbf{G}$ involves all the generalized coordinates and is a $3N \times 3N$ matrix. If the hard degrees of freedom are actually constrained they are removed from the matrix $\mathbf{G}$:

$$\langle \mathcal{A} \rangle_{NVT}^s = \frac{\int \mathcal{A}(\mathbf{q}^s) \sqrt{\det(\mathbf{G}^s)} \exp\left[-\beta \mathcal{V}_s(\mathbf{q}^s)\right] d\mathbf{q}^s}{\int \sqrt{\det(\mathbf{G}^s)} \exp\left[-\beta \mathcal{V}_s(\mathbf{q}^s)\right] d\mathbf{q}^s} \tag{2.165}$$

where

$$G_{k\ell}^s = \sum_{i=1}^{N} \sum_{\alpha} m_i \frac{\partial r_{i\alpha}}{\partial q_k^s} \frac{\partial r_{i\alpha}}{\partial q_\ell^s}. \tag{2.166}$$

$\mathbf{G}^s$ is a sub-matrix of $\mathbf{G}$ and has the dimensions of the number of soft degrees of freedom. The simulation of a constrained system does not yield the same average as the simulation of an unconstrained system unless $\det(\mathbf{G}) / \det(\mathbf{G}^s)$ is independent of the soft modes. In the simulation of large flexible molecules, it may be necessary to constrain some of the internal degrees of freedom, and in this case we would probably require an estimate of $\langle \mathcal{A} \rangle_{NVT}$ rather than $\langle \mathcal{A} \rangle_{NVT}^s$. Fixman (1974) has suggested a solution to the problem of obtaining $\langle \mathcal{A} \rangle_{NVT}$ in a simulation employing constrained variables. A term,

$$\mathcal{V}_c = \tfrac{1}{2} k_B T \ln \det(\mathbf{H}) \tag{2.167}$$

is added to the potential $\mathcal{V}_s$. $\det(\mathbf{H})$ is given by

$$\det(\mathbf{H}) = \det(\mathbf{G}) / \det(\mathbf{G}^s). \tag{2.168}$$

Substituting $\mathcal{V}_s + \mathcal{V}_c$ as the potential in eqn (2.165) we recover the unconstrained average of eqn (2.163). The separate calculation of $\mathbf{G}$ and $\mathbf{G}^s$ to obtain their determinants is difficult. However, $\det(\mathbf{H})$ is the determinant of a simpler matrix

$$H_{k\ell} = \sum_{i=1}^{N} \sum_{\alpha} m_i \frac{\partial q_k^h}{\partial r_{i\alpha}} \frac{\partial q_\ell^h}{\partial r_{i\alpha}} \tag{2.169}$$

which has the dimensions of the number of constrained (hard) degrees of freedom.

As a simple example of the use of eqn (2.169) consider the case of a butane molecule (see Fig. 1.10). In our simplified butane, the four united atoms have the same mass m, the bond angles and torsional angles are free to change but the three bond lengths, C_1-C_2, C_2-C_3, and C_3-C_4 are fixed. The 3×3 matrix H is

$$
\begin{pmatrix}
2m & -m\cos\theta & 0 \\
-m\cos\theta & 2m & -m\cos\theta' \\
0 & -m\cos\theta' & 2m
\end{pmatrix}
$$

and

$$
\det(H) \propto \left(2 + \sin^2\theta + \sin^2\theta'\right). \tag{2.170}
$$

Since θ and θ' can change, H should be included through eqn (2.167). However, it is possible to use a harmonic bond-angle potential, which keeps the bond angles very close to their equilibrium values. In this case H is approximately constant and might be neglected without seriously affecting $\langle \mathcal{A} \rangle_{NVT}$. If we had also constrained the bond angles in our model of butane, then $\det(H)$ would have been a function of the torsional angle ϕ as well as the θ angles. Thus H can change significantly when the molecule converts from the *trans* to the *gauche* state and $\mathcal{V}_c$ must be included in the potential (van Gunsteren, 1980). In the case of a completely rigid molecule, $\det(H)$ is a constant and need not be included. We shall discuss the consequences of constraining degrees of freedom at the appropriate points in Chapters 3 and 4.

2.11 Landau free energy

The idea of a free energy which depends on certain constrained 'order parameters', often termed a 'Landau' free energy because of its prominence in the Landau theory of phase transitions, is very common in molecular simulation. Consider a single parameter $q(\mathbf{r})$, a generalized coordinate, which can be written as a function of all the atomic positions $\mathbf{r}$. (In the most general case, it might also depend on momenta, but the configurational case is by far the most common). Then, quite generally, we may write the probability density function for q in the canonical ensemble

$$
\rho(q) = \left\langle \delta[q - q(\mathbf{r})] \right\rangle = \frac{\int d\mathbf{r}\, \delta[q - q(\mathbf{r})] \exp[-\beta \mathcal{V}(\mathbf{r})]}{\int d\mathbf{r}\, \exp[-\beta \mathcal{V}(\mathbf{r})]} \equiv \frac{Q_{NVT}^{\mathrm{ex}}(q)}{Q_{NVT}^{\mathrm{ex}}}. \tag{2.171}
$$

Here we have defined the numerator $Q_{NVT}^{\mathrm{ex}}(q)$ as the excess partition function of a system which is restricted to configurations for which the generalized coordinate $q(\mathbf{r})$ takes the value q. The function $\rho(q)$ is, of course, normalized such that $\int dq\, \rho(q) = 1$. Taking logarithms, and identifying $A = -k_B T \ln Q_{NVT}^{\mathrm{ex}}$, we may use $Q_{NVT}^{\mathrm{ex}}(q)$ to define a q-dependent Helmholtz free energy, which we usually write as $\mathcal{F}(q)$:

$$
\mathcal{F}(q) = A - k_B T \ln \rho(q) = A - k_B T \ln\langle \delta[q - q(\mathbf{r})]\rangle. \tag{2.172}
$$

The thermodynamic Helmholtz free energy is frequently omitted from this equation, since usually one is interested in changes in $\mathcal{F}(q)$ as a function of q. For example, in discussing phase transitions in the Ising model of ferromagnetic systems, below the critical point, q might be the overall magnetization of the system, and $\mathcal{F}(q)$ would typically have a double-minimum structure corresponding to stable states with positive and negative values of q (Landau and Lifshitz, 1980; Chandler, 1987). As another example, q might be the coordinate of a molecule as it is moved from an aqueous phase into a non-aqueous phase, or a measure of 'crystallinity' in a nucleus of solid phase growing within a melt (Lynden-Bell, 1995). In many cases, the free-energy barrier between two states characterized by different values of q, is of primary interest. It is easy to extend these definitions to include several order parameters $\{q_i\}$ characterizing the system, in which case one would be interested in a Landau free-energy surface $\mathcal{F}(\{q_i\})$, and possibly in the typical trajectories followed by a system as it makes transitions between basins in this surface. Later chapters will give several examples of the use of simulations to measure these quantities, sometimes using special sampling methods to improve the efficiency. An interesting consequence of eqns (2.171) and (2.172) is that one can define a thermodynamic 'force'

$$-\frac{\mathrm{d}\mathcal{F}}{\mathrm{d}q} = \frac{k_B T}{\rho(q)}\frac{\mathrm{d}\rho(q)}{\mathrm{d}q} = \left\langle -\frac{\mathrm{d}\mathcal{V}}{\mathrm{d}q}\right\rangle_q$$

where the average is to be taken in the restricted $q(\mathbf{r}) = q$ ensemble, and the quantity being averaged is the 'mechanical' force (the negative of the derivative of the potential energy $\mathcal{V}$ with respect to the coordinate q). Essentially the same definitions appear in the development of coarse-grained effective potentials, discussed in Section 12.7. In that case, the aim is to use the Landau free energy, written as a function of a reduced set of coordinates **q**, in place of the full potential-energy function.

Often, phenomenological theories predict a particular form for $\mathcal{F}(q)$, which may be tested by simulation. Taking this further, theories may be based on a proposed free-energy *functional* $\mathcal{F}[q(\mathbf{r})]$ depending on an order parameter $q(\mathbf{r})$ which varies from place to place. Here, **r** is a position in the fluid, and to make contact with the microscopic description, it is necessary to have a suitable definition of q. For instance, a local property q_i of the molecules may be used to define a collective quantity

$$q(\mathbf{r}) \propto \sum_i q_i \delta(\mathbf{r} - \mathbf{r}_i).$$

Statistical mechanical theories of inhomogeneous systems often suppose that $\mathcal{F}$ depends on $q(\mathbf{r})$ through a local free-energy density $f(\mathbf{r}) = f(q(\mathbf{r}))$, together with terms involving gradients of q with respect to position

$$\mathcal{F}[q(\mathbf{r})] = \int \mathrm{d}\mathbf{r}\, f\big(q(\mathbf{r})\big) + \tfrac{1}{2}\kappa |\nabla q(\mathbf{r})|^2$$

where κ is a phenomenological constant, and we shall see examples of this kind in the following sections.

2.12   Inhomogeneous systems

In most of this book, we will deal with homogeneous fluids, but the methods of liquid-state simulation have been extended with great advantage to consider the gas–liquid and solid–liquid interfaces in a variety of geometries. In this section we will illustrate the

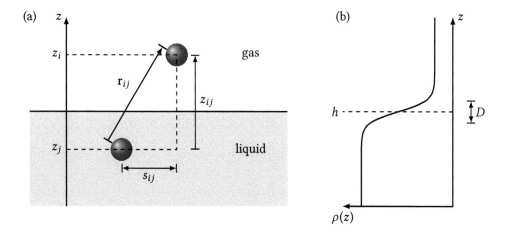

Fig. 2.4 (a) The geometry of the planar interface. (b) Density profile: h is the position (Gibbs dividing surface), and D the thickness, of the interface.

basic ideas with reference to the planar gas–liquid interface (see Fig. 2.4). For a more detailed review of the statistical mechanics of inhomogeneous systems see Rowlinson and Widom (1982), Nicholson and Parsonage (1982), Croxton (1980), and Safran (1994).

For an inhomogeneous fluid in the canonical ensemble, the singlet and pair density distribution functions for a fluid of N atoms at a temperature T are

$$\rho^{(1)}(\mathbf{r}_1) = N \frac{\int \cdots \int \exp(-\beta\mathcal{V})\,d\mathbf{r}_2 d\mathbf{r}_3 \cdots \mathbf{r}_N}{\int \cdots \int \exp(-\beta\mathcal{V})\,d\mathbf{r}_1 d\mathbf{r}_2 d\mathbf{r}_3 \cdots \mathbf{r}_N} \tag{2.173}$$

and

$$\rho^{(2)}(\mathbf{r}_1, \mathbf{r}_2) = N(N-1) \frac{\int \cdots \int \exp(-\beta\mathcal{V})\,d\mathbf{r}_3 \cdots \mathbf{r}_N}{\int \cdots \int \exp(-\beta\mathcal{V})\,d\mathbf{r}_1 d\mathbf{r}_2 d\mathbf{r}_3 \cdots \mathbf{r}_N}. \tag{2.174}$$

For the planar gas–liquid interface, the cylindrical symmetry allows us to express these distribution functions in terms of the height of an atom, z_i, and the distance s_{ij} parallel to the planar interface between the atoms (see Fig. 2.4). Unlike the homogeneous fluid, the singlet particle density depends now on the position in the fluid.

In this case, $\rho^{(1)}$ and $\rho^{(2)}$ can be expressed in terms of the Dirac delta function as

$$\rho^{(1)}(z) = \frac{1}{\mathcal{A}} \left\langle \sum_{i=1}^{N} \delta(z - z_i) \right\rangle \tag{2.175}$$

and

$$\rho^{(2)}(z, z', s) = \frac{1}{\mathcal{A}} \left\langle \sum_{i=1}^{N} \sum_{j \neq i} \delta(z - z_i)\, \delta(z' - z_j)\delta(s - s_{ij}) \right\rangle \tag{2.176}$$

where $\delta(s)$ is a two-dimensional delta function in the surface vector $s = (x, y)$, $s = |s|$, and $\mathcal{A}$ is the surface area. Note that from this point, we omit the superscript (1) for the single particle density. $\rho^{(2)}$ is simply related to the radial distribution function

$$\rho^{(2)}(z_1, z_2, s_{12}) = \rho(z_1)\,\rho(z_2)\,g^{(2)}(z_1, z_2, s_{12}). \tag{2.177}$$

In this geometry, $\rho^{(2)}(\mathbf{r}_1, \mathbf{r}_2)$ is often defined in terms of the variables $\{z_1, z_2, r_{12}\}$ where r_{12} is the interatomic separation. In this case (Nicholson and Parsonage, 1982),

$$\rho^{(2)}(z_1, z_2, s_{12})s_{12}\mathrm{d}s_{12}\mathrm{d}z_1\mathrm{d}z_2 = \rho^{(2)}(z_1, z_2, r_{12})r_{12}\mathrm{d}r_{12}\mathrm{d}z_1\mathrm{d}z_2. \tag{2.178}$$

The symmetry of the planar interface and the condition of hydrostatic equilibrium for the fluid

$$\nabla \cdot \mathbf{P} = 0 \tag{2.179}$$

means that the pressure tensor $\mathbf{P}$ is diagonal and is a function only of z. The independent components are

$$P_{zz}(z) = P_{\mathrm{N}}, \quad \text{normal component, independent of } z, \text{ and} \tag{2.180a}$$
$$P_{xx}(z) = P_{yy}(z) = P_{\mathrm{T}}(z), \quad \text{tangential component, dependent on } z. \tag{2.180b}$$

Usually we define $P_{\mathrm{T}}(z) = \frac{1}{2}\big(P_{xx}(z) + P_{yy}(z)\big)$ in this geometry.

Extending the well-known virial equation for the pressure of a homogeneous fluid is not straightforward. There is no unique way of deciding whether the force between two atoms 1 and 2 contributes to the stress across a microscopic element of the fluid at a particular z. An obvious choice (Irving and Kirkwood, 1950) is to say that the forces contribute if the vector $\mathbf{r}_{12}$ intersects the element. There are infinitely many other different possible choices for the contour joining two atoms (Schofield and Henderson, 1982). They all lead to the same value of the normal pressure, but they give rise to different values of the tangential pressure.

With the Irving–Kirkwood (IK) choice, for a fluid with a pair additive potential $v(r)$, an atom at $\mathbf{r}_1$ experiences a force $-(\mathbf{r}_{12}/r_{12})v'(r_{12})$ from an atom at $\mathbf{r}_2$. The probability of there being two atoms at these positions is $\rho^{(2)}(\mathbf{r}_1, \mathbf{r}_2)$, which may be rewritten as $\rho^{(2)}(\mathbf{r} + \alpha\mathbf{r}_{12}, \mathbf{r} + (\alpha - 1)\mathbf{r}_{12})$ where $\alpha = |\mathbf{r} - \mathbf{r}_2|/r_{12}$. Then

$$\mathbf{P}(\mathbf{r}) = \rho(\mathbf{r})k_{\mathrm{B}}T\mathbf{1} - \frac{1}{2}\int \mathrm{d}\mathbf{r}_{12}\frac{\mathbf{r}_{12}\,\mathbf{r}_{12}}{r_{12}}v'(r_{12})\int_0^1 \mathrm{d}\alpha\,\rho^{(2)}\big(\mathbf{r} + \alpha\mathbf{r}_{12}, \mathbf{r} + (\alpha - 1)\mathbf{r}_{12}\big)$$

$$= \rho(\mathbf{r})k_{\mathrm{B}}T\mathbf{1} - \frac{1}{2}\int \mathrm{d}\mathbf{r}_{12}\int \mathrm{d}\mathbf{r}_1\frac{\mathbf{r}_{12}\,\mathbf{r}_{12}}{r_{12}}v'(r_{12})\int_0^1 \mathrm{d}\alpha\,\rho^{(2)}(\mathbf{r}_1, \mathbf{r}_2)\delta(\mathbf{r} + \alpha\mathbf{r}_{12} - \mathbf{r}_1)$$

$$\tag{2.181}$$

where the first term is the ideal gas contribution and $\mathbf{r}_{12} = \mathbf{r}_1 - \mathbf{r}_2$. Eqn (2.181) is general, and for the planar interface, defining $z_{12} = z_1 - z_2$, it can be simplified to

$$\mathbf{P}(z) = \rho(z)k_{\mathrm{B}}T\mathbf{1} - \frac{1}{2\mathcal{A}}\int \mathrm{d}\mathbf{r}_1 \int \mathrm{d}\mathbf{r}_{12}\frac{\mathbf{r}_{12}\,\mathbf{r}_{12}}{r_{12}}v'(r_{12})\int_0^1 \mathrm{d}\alpha\,\delta(z + \alpha z_{12} - z_1)\rho^{(2)}(z_1, \mathbf{r}_{12}).$$

$$\tag{2.182}$$

The surface tension, γ, and the position of the surface of tension, z_s, can be calculated from the pressure tensor.

$$\gamma = \int_{-\infty}^{\infty} dz \left[P_N(z) - P_T(z) \right] \tag{2.183a}$$

$$\gamma z_s = \int_{-\infty}^{\infty} dz\, z \left[P_N(z) - P_T(z) \right] \tag{2.183b}$$

where the integral will be across one interface from far inside the gas to far inside the liquid. The surface tension is independent of the choice of contour discussed earlier in this section, whereas the surface of tension z_s depends on this choice and is therefore ill-defined from a microscopic perspective.

The surface tension can be obtained directly by calculating the change in the Helmholtz free energy, in the canonical ensemble (Buff, 1952), or the grand potential, in the grand canonical ensemble (Buff, 1955), with surface area at constant volume. In these derivations, all the particle coordinates are scaled as follows

$$x_i \to x_i' = (1+\epsilon)^{1/2} x_i \approx (1 + \tfrac{1}{2}\epsilon) x_i,$$
$$y_i \to y_i' = (1+\epsilon)^{1/2} y_i \approx (1 + \tfrac{1}{2}\epsilon) y_i,$$
$$z_i \to z_i' = (1+\epsilon)^{-1} z_i \approx (1 - \epsilon) z_i,$$

where $\epsilon \ll 1$. This means that, to first order, the volume is constant and the change in interfacial area is $\Delta\mathcal{A} = \epsilon\mathcal{A}$. For a pair potential the result for the surface tension is

$$\gamma = \frac{1}{4} \int_{-\infty}^{+\infty} dz_1 \int dr_{12} \left(r_{12} - \frac{3z_{12}^2}{r_{12}} \right) v'(r_{12})\, \rho^{(2)}(\mathbf{r}_1, \mathbf{r}_2)$$
$$= \frac{1}{2\mathcal{A}} \left\langle \sum_i \sum_{j>i} \left(r_{ij} - \frac{3z_{ij}^2}{r_{ij}} \right) v'(r_{ij}) \right\rangle. \tag{2.184}$$

Eqn (2.184) is formally equivalent to eqn (2.183a) for the surface tension.

Finally the chemical potential, μ, in the interface can be calculated by the particle insertion method of Widom (1963)

$$\mu = k_B T \ln \Lambda^3 \rho(z) - k_B T \ln\langle \exp[-\beta v_{test}(z)] \rangle \tag{2.185}$$

where Λ is the de Broglie wavelength of the atom and the first term is the ideal gas contribution to μ (compare with eqn (2.75)). $v_{test}(z)$ is the potential energy of a test particle that is inserted into the fluid at a particular value of z and a random position in the xy plane. This particle does not influence the behaviour of other atoms in the system. Since the system is at equilibrium, μ must be independent of z (the potential distribution theorem). The density varies sharply around the interface, so that changes in the first and second terms in eqn (2.185) with z must cancel.

The interface between two phases will, in general, not be planar, due to fluctuations. Its height, that is, its z-coordinate, will be a function of x and y: we denote this $h(x, y)$. This is illustrated schematically in Fig. 14.3 and, as we shall see in Chapter 14, defining

$h(x, y)$ can be subtle. Suppose that this can be done. Capillary wave theory (Buff et al., 1965; Rowlinson and Widom, 1982; Safran, 1994) is based on a coarse-grained free energy, reflecting the increase in surface area associated with gradients of h:

$$\mathcal{F} = \tfrac{1}{2}\gamma \iint dxdy \left[\left(\frac{\partial h}{\partial x}\right)^2 + \left(\frac{\partial h}{\partial y}\right)^2 \right]$$

(2.186)

where γ is the surface tension; this is expected to be valid for small gradients. Expanding in a Fourier series in the xy-plane, with $\mathbf{k} = (k_x, k_y)$, using Parseval's theorem (see Appendix D), and applying the equipartition principle to the resulting sum of independent Fourier modes gives:

$$\mathcal{F} = \frac{\gamma}{2\mathcal{A}} \sum_{\mathbf{k}} k^2 |\hat{h}(\mathbf{k})|^2 \quad \Rightarrow \quad \left\langle |\hat{h}(\mathbf{k})|^2 \right\rangle = \frac{\mathcal{A}k_B T}{\gamma k^2}.$$

(2.187)

where $\mathcal{A}$ is the cross-sectional area. The resulting distribution of heights is Gaussian and the mean-square deviation is

$$\left\langle \delta h(x, y)^2 \right\rangle = \left\langle h(x, y)^2 \right\rangle - \left\langle h(x, y) \right\rangle^2 = \frac{1}{\mathcal{A}} \iint_{\mathcal{A}} dxdy \left\langle h(x, y)^2 \right\rangle = \frac{1}{\mathcal{A}^2} \sum_{k_x, k_y} \left\langle |\hat{h}(\mathbf{k})|^2 \right\rangle$$

$$\approx \frac{1}{(2\pi)^2} \iint dk_x dk_y \frac{k_B T}{\gamma k^2} \approx \frac{k_B T}{2\pi\gamma} \int_{2\pi/L}^{2\pi/a} \frac{dk}{k} = \frac{k_B T}{2\pi\gamma} \ln\left(\frac{L}{a}\right).$$

(2.188)

In the second line, the sum over wavevectors is replaced by an integral. It is assumed that the coarse-grained description applies from a small length scale, a, usually taken to be of the order of molecular dimensions, up to L, where $\mathcal{A} = L^2$ (we have assumed a square cross-section). This capillary-wave variation is a major contribution to the observed density profile width D schematically indicated in Fig. 2.4(b); there is also an intrinsic contribution D_0 discussed further in Chapter 14. If one assumes that the two contributions are convoluted together (Semenov, 1993; 1994) the result is

$$\langle D \rangle^2 = D_0^2 + \frac{\pi}{2} \left\langle \delta h(x, y)^2 \right\rangle = D_0^2 + \frac{k_B T}{4\gamma} \ln\left(\frac{L}{a}\right).$$

(2.189)

Equation (2.189) shows how this width increases with increasing transverse box dimensions. Observing these fluctuations gives us a way of determining the surface tension, as an alternative to eqn (2.184).

The ideas described in this section are readily extended to the solid–liquid interface, where the solid can be represented as a static external field or in its full atomistic detail using pair potentials, and for interfaces of different symmetry such as the spherical droplet. We return to this in Chapter 14.

2.13   Fluid membranes

Amphiphilic molecules, consisting of both hydrophobic and hydrophilic sections, may spontaneously self-assemble in water. Often the hydrophobic part consists of one or two hydrocarbon tails, and the hydrophilic part is a charged or polar head group. A variety of phases may be observed, dependent upon composition and thermodynamic state: micelles, hexagonal arrangements of cylindrical assemblies, bicontinuous cubic phases, and the lamellar phase (Safran, 1994; Jones, 2002). The building block in this last case is the bilayer, consisting of two sheets of amphiphiles arranged tail-to-tail with the head groups on the outside, facing into layers of water. The bilayer itself is also found as the containing wall of spherical vesicles, which have water inside as well as outside. These are sometimes used as very simple models of biological cells, whose surrounding fluid membrane is typically a bilayer of phospholipid molecules, and simulation of these systems is a very active area (Sansom and Biggin, 2010).

The properties of planar bilayer membranes are, therefore, of interest, particularly their elasticity. Usually, a membrane is in a state of zero tension, and deformations away from planarity may be described by a coarse-grained Helfrich free energy (Helfrich, 1973), an integral over the membrane area $\mathcal{A}$

$$\mathcal{F} = \iint_{\mathcal{A}} \mathrm{d}\mathcal{A} \, \tfrac{1}{2}\kappa\big(C_1 + C_2 - C_0\big)^2 + \bar{\kappa} C_1 C_2 \qquad (2.190)$$

where C_0 is the spontaneous curvature (zero for a symmetrical bilayer), C_1 and C_2 are the two principal curvatures of the membrane surface at a given point, κ is the bending modulus, and $\bar{\kappa}$ is the saddle splay modulus. In the limit of small gradients, a description in terms of the membrane height $h(x, y)$ may be adopted, and

$$\mathcal{F} = \iint \mathrm{d}x\mathrm{d}y \, \tfrac{1}{2}\kappa\big(\nabla^2 h(x, y)\big)^2 = \frac{\kappa}{2\mathcal{A}} \sum_{\mathbf{k}} k^4 \big|\hat{h}(\mathbf{k})\big|^2$$

where the Laplacian, and the wavevector, are both taken in the xy-plane. Hence, by equipartition of energy

$$\left\langle \big|\hat{h}(\mathbf{k})\big|^2 \right\rangle = \frac{\mathcal{A} k_\mathrm{B} T}{\kappa k^4}. \qquad (2.191)$$

If the membrane is under tension, an appropriate term in k^2 must be included, as in eqn (2.187): however, this term would dominate at low k, so it is important that the tensionless state be maintained (see Section 14.6).

The moduli may also, in principle, be related to the pressure tensor profiles, in a manner analogous to equations (2.183) for the surface tension (Szleifer et al., 1990; Marsh, 2007; Ollila and Vattulainen, 2010). For the bilayer as a whole

$$\gamma = \int_{-\infty}^{\infty} \mathrm{d}z \left[P_\mathrm{N} - P_\mathrm{T}(z) \right] = 0, \qquad (2.192a)$$

$$-\kappa C_0 = \int_{-\infty}^{\infty} \mathrm{d}z \, z \left[P_\mathrm{N} - P_\mathrm{T}(z) \right] = 0, \qquad (2.192b)$$

$$\bar{\kappa} = \int_{-\infty}^{\infty} \mathrm{d}z \, z^2 \left[P_\mathrm{N} - P_\mathrm{T}(z) \right], \qquad (2.192c)$$

where the origin of coordinates $z = 0$ is chosen at the centre of the bilayer, and the pressure components are defined in eqns (2.180). (In the literature, the negative of the pressure, that is, the stress, is frequently used in these equations). The zeroth moment of the pressure difference vanishes in the tensionless state, while the first moment vanishes by symmetry. The following equations apply to each monolayer constituting the bilayer

$$-\kappa^m C_0^m = \int_0^\infty dz\, (z - z^m) \left[P_N - P_T(z) \right],$$ (2.193a)

$$\tilde{\kappa}^m = \int_0^\infty dz\, (z - z^m)^2 \left[P_N - P_T(z) \right].$$ (2.193b)

The superscript 'm' indicates that these are properties of the separate monolayers. Here, z^m is the position of the monolayer 'neutral plane' relative to the mid-plane. The integrals may, in practice, be truncated once $|z|$ exceeds the monolayer thickness, since $P_T(z) = P_N$ in the bulk. The use of these formulae is limited by ambiguities in defining z^m, approximations needed to relate monolayer and bilayer properties, and by the fact that κ and C_0 appear together, not separately. Also, as mentioned earlier, the convention adopted for the pressure tensor will affect the profiles. As discussed further in Section 14.6, alternative approaches to the membrane elastic moduli are more reliable.

2.14   Liquid crystals

For liquid crystalline phases, extra care may be needed in measuring some of the standard properties discussed in previous sections, and there are some additional properties that uniquely characterize these phases. For recent reviews see Zannoni (2000), Care and Cleaver (2005), and Wilson (2005).

Liquid crystals have long-range orientational order, and some liquid crystal phases have long-range positional order in certain directions. The simplest case is the *nematic* phase, typically formed by highly elongated (rod-like) or very flat (disk-like) molecules, or by similarly shaped particles in colloidal suspension. The orientational order is characterized by both a magnitude S, the order parameter, and a direction, $\mathbf{n}$, usually called the *director*. The molecules may be irregularly shaped, but we assume that it is possible to identify a principal axis for each one, perhaps related to the inertia tensor, the polarizability tensor, or some other molecular property, and that S and $\mathbf{n}$ will be insensitive to this choice. Let $\mathbf{e}_i$ be the unit vector pointing along this axis. Irrespective of whether the molecules have head–tail symmetry or not, nematic ordering is a second-rank property, and S is given by the average second Legendre polynomial of the cosine of the angle between a typical molecule and the director:

$$S = \frac{1}{N} \sum_{i=1}^N \langle P_2(\mathbf{e}_i \cdot \mathbf{n}) \rangle = \left\langle \frac{1}{N} \sum_{i=1}^N \tfrac{3}{2}(\mathbf{e}_i \cdot \mathbf{n})^2 - \tfrac{1}{2} \right\rangle = \mathbf{n} \cdot \langle \mathbf{Q} \rangle \cdot \mathbf{n},$$

where we have defined the second-rank orientational order tensor

$$Q_{\alpha\beta} = \frac{1}{N} \sum_{i=1}^N Q_{\alpha\beta}^i, \quad \text{where} \quad Q_{\alpha\beta}^i = \tfrac{3}{2} e_{i\alpha} e_{i\beta} - \tfrac{1}{2}\delta_{\alpha\beta}, \quad \text{and } \alpha, \beta = x, y, z. \quad (2.194)$$

Typically $\mathbf{n}$ is unknown at the start of the calculation, but can be defined as the direction that maximizes the value of S. The problem is then reduced to a variational one, which is solved by diagonalizing Q (Zannoni, 1979): S turns out to be the largest of the three eigenvalues, and n is the corresponding director. Because of the second-rank nature of the order, $-n$ is equivalent to n. Perfect orientational order corresponds to $S = 1$. For an isotropic liquid, $S = 0$, although in practice finite-size effects will result in fluctuations which make $S \sim N^{-1/2}$. Since Q is traceless, in a uniaxial nematic phase the other two eigenvalues will both be equal to $-\frac{1}{2}S$; once more, we expect finite-size effects to cause some small deviations (Eppenga and Frenkel, 1984).

Once the director has been determined, the single-particle orientational distribution function

$$\rho(\cos\theta) = \rho(e \cdot n) = \langle\delta(\cos\theta - \cos\theta_i)\rangle, \quad \text{where} \quad \int_{-1}^{1} d\cos\theta\, \rho(\cos\theta) = 1$$

may be calculated by constructing a histogram of $e_i \cdot n$ values, averaged over particle index i. Alternatively, the function may be expanded in Legendre polynomials P_ℓ

$$\rho(\cos\theta) = \sum_{\ell=0,2,\ldots}^{\infty} \frac{2\ell+1}{2} S_\ell P_\ell(\cos\theta) \quad \text{where} \quad S_\ell = \langle P_\ell(\cos\theta)\rangle;$$

the coefficients are simulation averages of the indicated quantities. Only components with even values of ℓ are non-vanishing.

The pressure tensor in a nematic liquid crystal is isotropic, that is, diagonal with $\langle\mathcal{P}_{\alpha\beta}\rangle = P\delta_{\alpha\beta}$ (Allen and Masters, 1993, appendix). Most other properties are affected by the reduced symmetry. The molecular centre–centre pair distribution function will be a function of components of the separation vector r_{ij} resolved along and perpendicular to the director. The orientation dependence of pair correlations may be expanded in appropriate angular functions (Zannoni, 2000); the formula is more complicated than eqn (2.106), for isotropic fluids, but similar in concept.

Director fluctuations can be described by a coarse-grained free energy (Oseen, 1933; Frank, 1958) which involves the gradients of n:

$$\mathcal{F} = \int_V dr\, \tfrac{1}{2}K_1(\nabla \cdot n)^2 + \tfrac{1}{2}K_2(n \cdot \nabla \times n)^2 + \tfrac{1}{2}K_3|n \times (\nabla \times n)|^2.$$

The quantities of interest are the Frank elastic constants K_1, K_2, and K_3. In a frame of reference where the equilibrium director is $n = (0, 0, 1)$, this may be rewritten in terms of the deviations from equilibrium, $n_x(r)$ and $n_y(r)$. Taking Fourier components (as for the capillary waves of Section 2.12) with $k = (k_x, 0, k_z)$ chosen in the xz-plane:

$$\mathcal{F} = \frac{1}{V}\sum_{k_x,k_y} \tfrac{1}{2}\left(K_1 k_x^2 + K_3 k_z^2\right)\left|\hat{n}_x(k_x, k_z)\right|^2 + \tfrac{1}{2}\left(K_2 k_x^2 + K_3 k_z^2\right)\left|\hat{n}_y(k_x, k_z)\right|^2.$$

Applying the equipartition principle

$$\langle|\hat{n}_x|^2\rangle = \frac{V k_B T}{K_1 k_x^2 + K_3 k_z^2}, \qquad \langle|\hat{n}_y|^2\rangle = \frac{V k_B T}{K_2 k_x^2 + K_3 k_z^2}. \qquad (2.195)$$

In practice, these fluctuations are determined by measuring fluctuations in $\mathbf{Q}$. We define

$$Q_{\alpha\beta}(\mathbf{r}) = \frac{V}{N} \sum_{i=1}^{N} Q_{\alpha\beta}^i \, \delta(\mathbf{r} - \mathbf{r}_i), \qquad \hat{Q}_{\alpha\beta}(\mathbf{k}) = \frac{V}{N} \sum_{i=1}^{N} Q_{\alpha\beta}^i \, \exp(-i\mathbf{k} \cdot \mathbf{r}_i).$$

For an axially symmetric phase it is easy to show $Q_{\alpha\beta}(\mathbf{r}) = \frac{3}{2}S[n_\alpha(\mathbf{r})n_\beta(\mathbf{r}) - \frac{1}{3}\delta_{\alpha\beta}]$, so for the off-diagonal elements of interest here $\hat{n}_\alpha(\mathbf{k}) = \hat{Q}_{z\alpha}(\mathbf{k})/(\frac{3}{2}S)$. The Frank elastic constants are obtained by fitting the director fluctuations to eqn (2.195) in the low-k limit (Allen et al., 1996; Humpert and Allen, 2015a).

Transport coefficients may be evaluated using time correlation functions or Einstein relations in the usual way, but taking into account the different symmetry cases (Sarman and Laaksonen, 2011); it is also possible to use nonequilibrium methods, of the general kind discussed in Chapter 11 (Sarman and Laaksonen, 2009a; 2015). The hydrodynamics of nematics is more complicated than that of simple liquids, and in particular the coupling between shear flow and director rotation is of great interest (Sarman and Laaksonen, 2009b; Humpert and Allen, 2015a,b).

3

Molecular dynamics

3.1 Equations of motion for atomic systems

In this chapter, we deal with the techniques used to solve the classical equations of motion for a system of N molecules interacting via a potential $\mathcal{V}$ as in eqn (1.4). These equations may be written down in various ways (Goldstein, 1980). Perhaps the most fundamental form is the Lagrangian equation of motion

$$\frac{\mathrm{d}}{\mathrm{d}t}(\partial\mathcal{L}/\partial\dot{q}_k) - (\partial\mathcal{L}/\partial q_k) = 0 \tag{3.1}$$

where the Lagrangian function $\mathcal{L}(\mathbf{q}, \dot{\mathbf{q}})$ is defined in terms of kinetic and potential energies

$$\mathcal{L} = \mathcal{K} - \mathcal{V} \tag{3.2}$$

and is considered to be a function of the generalized coordinates q_k and their time derivatives $\dot{q}_k$. If we consider a system of atoms, with Cartesian coordinates $\mathbf{r}_i$ and the usual definitions of $\mathcal{K}$ and $\mathcal{V}$ (eqns (1.3) and (1.4)) then eqn (3.1) becomes

$$m_i\ddot{\mathbf{r}}_i = \mathbf{f}_i \tag{3.3}$$

where m_i is the mass of atom i and

$$\mathbf{f}_i = \nabla_{\mathbf{r}_i}\mathcal{L} = -\nabla_{\mathbf{r}_i}\mathcal{V} \tag{3.4}$$

is the force on that atom. These equations also apply to the centre of mass motion of a molecule, with $\mathbf{f}_i$ representing the total force on molecule i; the equations for rotational motion may also be expressed in the form of eqn (3.1), and will be dealt with in Sections 3.3 and 3.4.

The generalized momentum p_k conjugate to q_k is defined as

$$p_k = \partial\mathcal{L}/\partial\dot{q}_k. \tag{3.5}$$

The momenta feature in the Hamiltonian form of the equations of motion

$$\dot{q}_k = \partial\mathcal{H}/\partial p_k \tag{3.6a}$$

$$\dot{p}_k = -\partial\mathcal{H}/\partial q_k. \tag{3.6b}$$

Computer Simulation of Liquids. Second Edition. M. P. Allen and D. J. Tildesley.
© M. P. Allen and D. J. Tildesley 2017. Published in 2017 by Oxford University Press.

The Hamiltonian is strictly defined by the equation

$$\mathcal{H}(\mathbf{p}, \mathbf{q}) = \sum_k \dot{q}_k p_k - \mathcal{L}(\mathbf{q}, \dot{\mathbf{q}}) \tag{3.7}$$

where it is assumed that we can write $\dot{q}_k$ on the right as some function of the momenta $\mathbf{p}$. For our immediate purposes (involving a potential $\mathcal{V}$ which is independent of velocities and time) this reduces to eqn (1.2), and $\mathcal{H}$ is automatically equal to the energy (Goldstein, 1980, Chapter 8). For Cartesian coordinates, Hamilton's equations become

$$\dot{\mathbf{r}}_i = \mathbf{p}_i / m_i \tag{3.8a}$$
$$\dot{\mathbf{p}}_i = -\nabla_{\mathbf{r}_i} \mathcal{V} = \mathbf{f}_i. \tag{3.8b}$$

Computing centre of mass trajectories, then, involves solving either a system of $3N$ second-order differential equations, eqn (3.3), or an equivalent set of $6N$ first-order differential equations, eqns (3.8a), (3.8b). Before considering how to do this, we can make some very general remarks regarding the equations themselves.

A consequence of eqn (3.6b), or equivalently eqns (3.5) and (3.1), is that in certain circumstances a particular generalized momentum p_k may be conserved, that is, $\dot{p}_k = 0$. The requirement is that $\mathcal{L}$, and hence $\mathcal{H}$ in this case, shall be independent of the corresponding generalized coordinate q_k. For any set of particles, it is possible to choose six generalized coordinates, changes in which correspond to translations of the centre of mass, and rotations about the centre of mass, for the system as a whole (changes in the remaining $3N - 6$ coordinates involving motion of the particles relative to one another). If the potential function $\mathcal{V}$ depends only on the magnitudes of particle separations (as is usual) and there is no external field applied (i.e. the term v_1 in eqn (1.4) is absent) then $\mathcal{V}, \mathcal{H}$ and $\mathcal{L}$ are manifestly independent of these six generalized coordinates. The corresponding conjugate momenta, in Cartesian coordinates, are the total linear momentum

$$\mathbf{P} = \sum_i \mathbf{p}_i \tag{3.9}$$

and the total angular momentum

$$\mathbf{L} = \sum_i \mathbf{r}_i \times \mathbf{p}_i = \sum_i m_i \mathbf{r}_i \times \dot{\mathbf{r}}_i \tag{3.10}$$

where we take the origin at the centre of mass of the system. Thus, these are conserved quantities for a completely isolated set of interacting molecules. In practice, we rarely consider completely isolated systems. A more general criterion for the existence of these conservation laws is provided by symmetry considerations (Goldstein, 1980, Chapter 8). If the system (i.e. $\mathcal{H}$) is invariant to translation in a particular direction, then the corresponding momentum component is conserved. If the system is invariant to rotation about an axis, then the corresponding angular momentum component is conserved. Thus, we occasionally encounter systems enclosed in a spherical box, and so a spherically symmetrical v_1 term appears in eqn (1.4); all three components of total angular momentum about the centre of symmetry will be conserved, but total translational momentum will not be. If the surrounding walls formed a cubical box, none of these quantities would be conserved.

In the case of the periodic boundary conditions described in Chapter 1, it is easy to see that translational invariance is preserved, and hence total linear momentum is conserved. Several different box geometries were considered in Chapter 1, but none of them were spherically symmetrical; in fact it is impossible (in Euclidean space) to construct a spherically symmetric periodic system. Hence, despite the fact that there may be no v_1-term in eqn (1.4), total angular momentum is not conserved in most molecular dynamics simulations. In the case of the spherical boundary conditions discussed in Section 1.6.5, a kind of angular momentum conservation law does apply. When we embed a two-dimensional system in the surface of a sphere, the three-dimensional spherical symmetry is preserved. Similarly, for a three-dimensional system, there should be a four-dimensional conserved 'hyper-angular momentum'.

We have left until last the most important conservation law. Assuming that $\mathcal{H}$ does not depend explicitly on time (so that $\partial \mathcal{H}/\partial t = 0$), the total derivative $\dot{\mathcal{H}}$ may be written

$$\frac{\mathrm{d}\mathcal{H}}{\mathrm{d}t} = \sum_k \left(\frac{\partial \mathcal{H}}{\partial q_k} \dot{q}_k + \frac{\partial \mathcal{H}}{\partial p_k} \dot{p}_k \right) = 0$$

by virtue of eqns (3.6). Hence the Hamiltonian is a constant of the motion. This energy conservation law applies whether or not an external potential exists: the essential condition is that no explicitly time-dependent (or velocity-dependent) forces shall act on the system.

The second point concerning the equations of motion is that they are reversible in time. By changing the signs of all the velocities or momenta, we will cause the molecules to retrace their trajectories. If the equations of motion are solved correctly, the computer-generated trajectories will also have this property.

Our final observation concerning eqns (3.3), (3.4), and (3.6) is that the spatial derivative of the potential appears. This leads to a qualitative difference in the form of the motion, and the way in which the equations are solved, depending upon whether or not $\mathcal{V}$ is a continuous function of particle positions. To use the finite-timestep method of solution to be described in the next section, it is essential that the particle positions vary smoothly with time: a Taylor expansion of $\mathbf{r}(t)$ about time t may be necessary, for example. Whenever the potential varies sharply (as in the hard-sphere and square-well cases) impulsive 'collisions' between particles occur at which the velocities (typically) change discontinuously. The particle dynamics at the moment of each collision must be treated explicitly, and separately from the smooth inter-collisional motion. The identification of successive collisions is the key feature of a molecular dynamics program for such systems, and we shall discuss this in Section 3.7.

3.2   Finite-difference methods

A standard method for solution of ordinary differential equations such as eqns (3.3) and (3.8) is the finite-difference approach. The general idea is as follows. Given the molecular positions, velocities, and other dynamic information at time t, we attempt to obtain the positions, velocities, etc. at a later time $t + \delta t$, to a sufficient degree of accuracy. The equations are solved on a step-by-step basis; the choice of the time interval δt will depend somewhat on the method of solution, but δt will be significantly smaller than the typical time taken for a molecule to travel its own length. Many different algorithms fall into the general finite-difference pattern. Historically, standard approaches such as predictor–corrector algorithms (Gear, 1966; 1971) and general-purpose approaches such as Runge–Kutta (Press et al., 2007) have been used in molecular dynamics simulations, and there have been several comparisons of different methods (Gear, 1971; van Gunsteren and Berendsen, 1977; Hockney and Eastwood, 1988; Berendsen and van Gunsteren, 1986; Gray et al., 1994; Leimkuhler and Reich, 2004). Which shall we choose?

A shortlist of desirable qualities for a successful simulation algorithm might be as follows.

(a) It should be fast, and require little memory.

(b) It should permit the use of a long timestep δt.

(c) It should duplicate the classical trajectory as closely as possible.

(d) It should satisfy the known conservation laws for energy and momentum, and be time-reversible.

(e) It should be simple in form and easy to program.

For molecular dynamics, the first point is generally less critical than the others, when it comes to choosing between algorithms. The memory required to store positions, velocities, accelerations, etc. is very small compared with that available on most computers, although this might become a consideration when taking advantage of special features of the architecture, such as graphics processing units (GPUs). Compared with the time-consuming force calculation, which is carried out at every timestep, the raw speed of the integration algorithm is not crucial. It is far more important to be able to employ a long timestep δt: in this way, a given period of 'simulation' time can be covered in a modest number of integration steps, that is, in an acceptable amount of computer time. Clearly, the larger δt, the less accurately will our solution follow the correct classical trajectory. How important are points (c) and (d) in the list?

It is unreasonable to expect that any approximate method of solution will dutifully follow the exact classical trajectory indefinitely. Any two classical trajectories which are initially very close will eventually diverge from one another exponentially with time (according to the 'Lyapunov exponents'), irrespective of the algorithm used to approximate the equations of motion. In the same way, any small perturbation, even the tiny error associated with finite precision arithmetic, will tend to cause a computer-generated trajectory to diverge from the true classical trajectory with which it is initially coincident. We illustrate the effect in Fig. 3.1: using one simulation as a reference, we show that a small perturbation applied at time $t = 0$ causes the trajectories in the perturbed simulation to diverge from the reference trajectories and become statistically uncorrelated, within a few hundred timesteps (see also Stoddard and Ford, 1973; Erpenbeck and Wood, 1977). In this example, we show the growing average 'distance in configuration space', defined as Δr where $\Delta r^2 = |\Delta \mathbf{r}|^2 = (1/N) \sum |\mathbf{r}_i(t) - \mathbf{r}_i^0(t)|^2$, $\mathbf{r}_i^0(t)$ being the position of molecule i at time t in a reference simulation, and $\mathbf{r}_i(t)$ being the position of the same molecule at the same time in the perturbed simulation. In the three cases illustrated here, all the molecules in the perturbed runs are initially displaced in random directions from their reference positions at $t = 0$, by $10^{-3}\sigma$, $10^{-6}\sigma$, and $10^{-9}\sigma$ respectively, where σ is the molecular diameter. In all other respects, the runs are identical; in particular, each corresponds

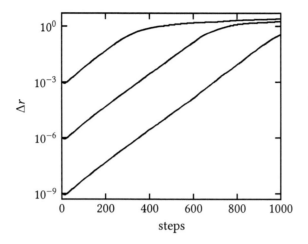

Fig. 3.1 The divergence of trajectories in molecular dynamics. Atoms interacting through the potential $v^{\mathrm{RLJ}}(r)$, eqn (1.10a), were used, and a dense fluid state was simulated ($\rho^* = 0.6$, $T^* = 1.05$, $\delta t^* = 0.005$). Δr is the phase space separation between perturbed and reference trajectories. These simulations used the velocity Verlet algorithm, eqn (3.11), but the results are essentially determined by the equations of motion rather than the integration algorithm.

to essentially the same total energy. As the runs proceed, however, other mechanical quantities eventually become statistically uncorrelated. Typically, properties such as the kinetic energy or pressure remain very close for a period whose length depends on the size of the initial perturbation; after this point the differences become noticeable very rapidly. Presumably, both the reference trajectory and the perturbed trajectory are diverging from the true solution of Newton's equations.

Clearly, no integration algorithm will provide an essentially exact solution for a very long time. Fortunately, we do not need to do this. Remember that molecular dynamics serves two roles. First, we need essentially exact solutions of the equations of motion for times comparable with the correlation times of interest so that we may accurately calculate time correlation functions. Second, we use the method to generate states sampled from the microcanonical ensemble. We do not need exact classical trajectories to do this but must lay great emphasis on energy conservation as being of primary importance for this reason. Momentum conservation is also important, but this can usually be easily arranged. The point is that the particle trajectories must stay on the appropriate constant-energy hypersurface in phase space, otherwise correct ensemble averages will not be generated. Energy conservation is degraded as the timestep is increased, and so all simulations involve a trade-off between economy and accuracy: a good algorithm permits a large timestep to be used while preserving acceptable energy conservation. Other factors dictating the energy-conserving properties are the shape of the potential-energy curves and the typical particle velocities. Thus, shorter timesteps are used at high temperatures, for light molecules and for rapidly varying potential functions.

The final quality an integration algorithm should possess is simplicity. A simple algorithm will involve the storage of only a few coordinates, velocities, etc., and will be

Code 3.1   Velocity Verlet algorithm

This snippet shows a direct translation from the so-called split-operator form of the algorithm (see Section 3.2.2). We have inserted a reminder that the arrays r, v, and f, are dimensioned to contain the entire set of $3N$ positions, velocities, and accelerations, and so the assignment statements apply to the entire array in each case. The (optional) syntax r(:,:) emphasizes this, but here, and henceforth, we omit it for brevity.

```
REAL , DIMENSION(3,n) :: r, v, a
v = v + 0.5 * dt * a
r = r + dt * v
! ... evaluate forces and hence a=f/m from r
v = v + 0.5 * dt * a
```

easy to program. Bearing in mind that solution of ordinary differential equations is a fairly routine task, there is little point in wasting valuable man-hours on programming a very complicated algorithm when the time might be better spent checking and optimizing the calculation of forces (see Chapter 5). Little computer time is to be gained by increases in algorithm speed, and the consequences of making a mistake in coding a complicated scheme might be significant.

For all these reasons, most molecular dynamics programs use a variant of the algorithm initially adopted by Verlet (1967) and attributed to Störmer (Gear, 1971). We describe this method in the following section.

3.2.1   The Verlet algorithm

Perhaps the most revealing way of writing the Verlet algorithm is in the so-called velocity Verlet form (Swope et al., 1982), which acts over a single timestep from t to $t + \delta t$ as follows:

$$\mathbf{v}(t + \tfrac{1}{2}\delta t) = \mathbf{v}(t) + \tfrac{1}{2}\delta t\, \mathbf{a}(t) \tag{3.11a}$$

$$\mathbf{r}(t + \delta t) = \mathbf{r}(t) + \delta t\, \mathbf{v}(t + \tfrac{1}{2}\delta t) \tag{3.11b}$$

$$\mathbf{v}(t + \delta t) = \mathbf{v}(t + \tfrac{1}{2}\delta t) + \tfrac{1}{2}\delta t\, \mathbf{a}(t + \delta t). \tag{3.11c}$$

The first step (3.11a) can be thought of as 'half-advancing' the velocities $\mathbf{v}$ to an intermediate time $t + \tfrac{1}{2}\delta t$, using the values of the accelerations $\mathbf{a}$ at time t; these mid-step velocities are then used to propel the coordinates from time t to $t + \delta t$ in step (3.11b). After this, a force evaluation is carried out to give $\mathbf{a}(t + \delta t)$ for the last step (3.11c) which completes the evolution of the velocities. The equations translate almost directly into computer code, as shown in Code 3.1. At the end of the step, we can calculate quantities such as the kinetic energy by summing the squares of the velocities, or the total momentum vector, by summing the different Cartesian components of the velocity. The potential energy at time $t + \delta t$ will have been computed in the force loop.

This method is numerically stable, convenient, and simple. It is exactly reversible in time and, given conservative forces, is guaranteed to conserve linear momentum. The

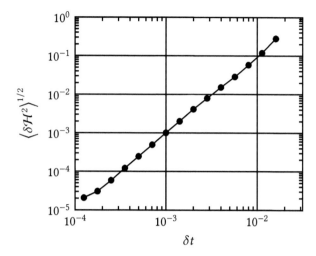

Fig. 3.2 Energy conservation of the Verlet algorithm. The system studied is as for Fig. 3.1. We calculate RMS energy fluctuations $\langle \delta \mathcal{H}^2 \rangle^{1/2}$ for various runs starting from the same initial conditions, and proceeding for the same total simulation time t_{run}, but using different timesteps δt and corresponding numbers of steps $\tau_{\text{run}} = t_{\text{run}}/\delta t$. The plot uses log–log scales.

method has been shown to have excellent energy-conserving properties even with long timesteps. As an example, for simulations of liquid argon near the triple point, RMS energy fluctuations $\langle \delta \mathcal{H}^2 \rangle^{1/2}$ of the order 0.01 % of the potential well depth are observed using $\delta t \approx 10^{-14}$ s, and these increase to 0.2 % for $\delta t \approx 4 \times 10^{-14}$ s (Verlet, 1967; Fincham and Heyes, 1982; Heyes and Singer, 1982). In fact, $\langle \delta \mathcal{H}^2 \rangle^{1/2}$ is closely proportional to δt^2 for Verlet-equivalent algorithms, as shown in Fig. 3.2. As we shall see in the next section, there is an interesting theoretical derivation of this version of the algorithm, which clarifies the reason for this dependence.

In the original 'velocity Verlet' paper (Swope et al., 1982), the previous equations were written in the slightly different form

$$r(t + \delta t) = r(t) + \delta t v(t) + \tfrac{1}{2} \delta t^2 a(t) \tag{3.12a}$$

$$v(t + \delta t) = v(t) + \tfrac{1}{2} \delta t \big[a(t) + a(t + \delta t) \big]. \tag{3.12b}$$

These are easily obtained from eqns (3.11) by eliminating the mid-step velocity. However, in practice, the velocity is still incremented in two steps, as the alternative is to (needlessly) store accelerations at both the start and end of the step. This is shown in Code 3.2.

Two other versions of the Verlet algorithm are worth mentioning at this point. The original implementation (Verlet, 1967) makes no direct use of the velocities at all but instead is directly related to the second-order equations (3.3). Consider addition of the equations obtained by Taylor expansion about $r(t)$

$$r(t + \delta t) = r(t) + \delta t v(t) + \tfrac{1}{2} \delta t^2 a(t) + \cdots$$

$$r(t - \delta t) = r(t) - \delta t v(t) + \tfrac{1}{2} \delta t^2 a(t) - \cdots \tag{3.13}$$

Code 3.2   Velocity Verlet algorithm (original)

Here dt_sq stores the value of δt^2. The algorithm is equivalent to that of Code 3.1, differing only in that the positions are updated before the mid-step velocities are calculated.

```
r = r + dt * v + 0.5 * dt_sq * a
v = v + 0.5 * dt * a
! ... evaluate forces and hence a=f/m from r
v = v + 0.5 * dt * a
```

to give

$$r(t + \delta t) = 2r(t) - r(t - \delta t) + \delta t^2 a(t). \tag{3.14}$$

The method is based on positions $r(t)$, accelerations $a(t)$, and the positions $r(t - \delta t)$ from the previous step. The Verlet algorithm is 'centered' (i.e. $r(t - \delta t)$ and $r(t + \delta t)$ play symmetrical roles in eqn (3.14)), making it time-reversible. It is straightforward to show that eqn (3.14) is equivalent to eqns (3.11), by considering two successive steps and eliminating the velocities.

The velocities are not needed to compute the trajectories, but they are useful for estimating the kinetic energy (and hence the total energy), as well as other interesting properties of the system. They may be obtained from the formula

$$v(t) = \frac{r(t + \delta t) - r(t - \delta t)}{2\delta t}. \tag{3.15}$$

Whereas eqn (3.14) is correct except for errors of order δt^4 (the local error) the velocities from eqn (3.15) are subject to errors of order δt^2. More accurate estimates of $v(t)$ can be made if more variables are stored, but this adds to the inconvenience already implicit in eqn (3.15), namely that $v(t)$ can only be computed once $r(t + \delta t)$ is known.

One implementation of the 'classic' Verlet algorithm is indicated in Code 3.3. It should be clear that the 'classic' Verlet algorithm has identical stability properties to the velocity form, and is very simple. Against it, we must say that the handling of velocities is rather awkward and that the form of the algorithm may needlessly introduce some numerical imprecision (Dahlquist and Björk, 1974). This arises because, in eqn (3.14), a small term ($O(\delta t^2)$) is added to a difference of large terms ($O(\delta t^0)$), in order to generate the trajectory.

Another alternative is the so-called half-step 'leapfrog' scheme (Hockney, 1970; Potter, 1972, Chapter 5). The origin of the name becomes apparent when we write the algorithm down:

$$v(t + \tfrac{1}{2}\delta t) = v(t - \tfrac{1}{2}\delta t) + \delta t a(t) \tag{3.16a}$$

$$r(t + \delta t) = r(t) + \delta t v(t + \tfrac{1}{2}\delta t). \tag{3.16b}$$

The stored quantities are the current positions $r(t)$ and accelerations $a(t)$ together with the mid-step velocities $v(t - \tfrac{1}{2}\delta t)$. The velocity equation eqn (3.16a) is implemented first,

Code 3.3 Classic Verlet algorithm

The 'classic' Verlet algorithm evaluates accelerations from the current positions, then uses these together with the old positions in the advancement step. The variable dt_sq stores δt^2 as usual. During this step, it is possible to calculate the current velocities. We handle this using a temporary array r_new to store the new positions. Then, a shuffling operation takes place in the last two statements. At the end of the step, the positions have been advanced, but the 'current' (now 'old') potential energy can be combined with the kinetic energy, calculated from the 'current' velocities. Following the particle move, we are ready to evaluate the forces at the start of the next step.

```
REAL, DIMENSION(3,n) :: r, r_old, r_new, a, v
! ... evaluate forces and hence a=f/m from r
r_new  = 2.0 * r - r_old + dt_sq * a
v      = ( r_new - r_old ) / ( 2.0 * dt )
r_old  = r
r      = r_new
```

and the velocities leap over the coordinates to give the next mid-step values $v(t + \frac{1}{2}\delta t)$. During this step, the current velocities may be calculated

$$v(t) = \tfrac{1}{2}\left(v(t + \tfrac{1}{2}\delta t) + v(t - \tfrac{1}{2}\delta t)\right). \tag{3.17}$$

This is necessary so that the energy $(\mathcal{H} = \mathcal{V} + \mathcal{K})$ at time t can be calculated, as well as any other quantities that require positions and velocities at the same instant. Following this, eqn (3.16b) is used to propel the positions once more ahead of the velocities. After this, the new accelerations may be evaluated ready for the next step.

Elimination of the velocities from these equations shows that the method is algebraically equivalent to Verlet's algorithm. In fact, eqn (3.16b) is identical to eqn (3.11b), while eqn (3.16a) is obtained by combining (3.11a) with (3.11c) *for the previous step.* Numerical benefits derive from the fact that at no stage do we take the difference of two large quantities to obtain a small one; this minimizes loss of precision on a computer. As is clear from eqn (3.17), the leapfrog method still does not handle the velocities in a completely satisfactory manner, and velocity Verlet is generally preferable. A complete molecular dynamics program for Lennard-Jones atoms, using the velocity Verlet method, is given in Code 3.4.

3.2.2 Formal basis of Verlet algorithm

To an extent, there is no need to understand 'where algorithms come from', as long as they work. Nonetheless, an understanding of the formal background to molecular dynamics algorithms, and particularly the Verlet algorithm, has been extremely useful in terms of knowing their limitations and how they may be extended to different situations. We shall only scratch the surface: for more details the reader is referred to the books of Leimkuhler and Reich (2004) and Tuckerman (2010). The following equations make use of the Liouville operator, introduced in eqn (2.4), and its exponential $\exp(iLt) \equiv U(t)$, which

Code 3.4 Molecular dynamics, NVE-ensemble, Lennard-Jones

These files are provided online. The program md_nve_lj.f90 controls the simulation, reads in the run parameters, implements the velocity Verlet algorithm, and writes out the results. It uses the routines in md_lj_module.f90 to evaluate the Lennard-Jones forces, and various utility modules (see Appendix A) for input/output and simulation averages. Code to set up an initial configuration is provided in initialize.f90.

```
! md_nve_lj.f90
! Molecular dynamics, NVE ensemble
PROGRAM md_nve_lj


! md_lj_module.f90
! Force routine for MD simulation, Lennard-Jones atoms
MODULE md_module
```

is often called the *propagator*: it has the effect of moving the system (i.e. the coordinates, momenta, and all the dynamical variables that depend on them) forward through time.

In the Verlet algorithm we use an approximate form of the propagator, which arises from splitting iL in two (Tuckerman et al., 1992):

$$iL = \dot{\mathbf{r}} \cdot \frac{\partial}{\partial \mathbf{r}} + \dot{\mathbf{p}} \cdot \frac{\partial}{\partial \mathbf{p}} = \mathbf{v} \cdot \frac{\partial}{\partial \mathbf{r}} + \mathbf{f} \cdot \frac{\partial}{\partial \mathbf{p}} \equiv iL_1 + iL_2, \tag{3.18}$$

where, as before, we abbreviate $\mathbf{r}$, $\mathbf{v}$ for the complete set of positions, velocities, etc. It is not hard to see that the 'propagators' corresponding to each of the separate parts will only affect the corresponding coordinate

$$\exp(iL_1 \delta t)\,\mathbf{r} = \mathbf{r} + \mathbf{v}\delta t \qquad \exp(iL_1 \delta t)\,\mathbf{p} = \mathbf{p} \qquad \text{drift,} \tag{3.19}$$

$$\exp(iL_2 \delta t)\,\mathbf{r} = \mathbf{r} \qquad \exp(iL_2 \delta t)\,\mathbf{p} = \mathbf{p} + \mathbf{f}\delta t \qquad \text{kick.} \tag{3.20}$$

The first of these is termed the 'drift' because it advances coordinates without changing momenta, rather like drifting in free flight, with the forces switched off. The second is called the 'kick' since it impulsively changes momenta without altering positions. It is important to realize that each of these separate propagation steps has been derived from a corresponding part of the Hamiltonian: the 'drift' arises from the kinetic-energy part, while the 'kick' comes from the potential-energy part.

Now, much like operators in quantum mechanics, iL_1 and iL_2 do not commute with each other, and this means that the following relation

$$\exp(iL\delta t) = \exp[(iL_1 + iL_2)\delta t] \approx \exp(iL_1 \delta t)\exp(iL_2 \delta t)$$

is only an approximation, not an exact relation. The error associated with the approximation is, however, 'small', that is, it becomes asymptotically exact in the limit $\delta t \to 0$. A slightly different approximation would result from combining the two partial propagators in the opposite order, and the following arrangement has the additional merit of being exactly time-reversible:

$$\exp(iL\delta t) \approx \exp(iL_2\delta t/2) \; \exp(iL_1\delta t) \; \exp(iL_2\delta t/2). \tag{3.21}$$

The operators act in turn, reading from right to left, upon the phase space variables $\mathbf{r}$ and $\mathbf{p}$, initially at time t, converting them into the new variables at $t + \delta t$. An attractive feature of this formalism is that the three successive steps embodied in eqn (3.21), with the operators defined by eqns (3.19) and (3.20), translate directly (Martyna et al., 1996) into the velocity Verlet algorithm of eqn (3.11):

(a) 'half-kick', $\mathbf{r}$ constant, $\mathbf{p}(t) \rightarrow \mathbf{p}(t + \frac{1}{2}\delta t) = \mathbf{p}(t) + \frac{1}{2}\delta t\,\mathbf{f}(t)$;

(b) 'drift', free flight with $\mathbf{p}$ constant, $\mathbf{r}(t) \rightarrow \mathbf{r}(t + \delta t) = \mathbf{r}(t) + \delta t\,\mathbf{p}(t + \frac{1}{2}\delta t)/m$;

(c) 'half-kick', $\mathbf{r}$ constant, $\mathbf{p}(t + \frac{1}{2}\delta t) \rightarrow \mathbf{p}(t + \delta t) = \mathbf{p}(t + \frac{1}{2}\delta t) + \frac{1}{2}\delta t\,\mathbf{f}(t + \delta t)$.

This particular splitting is quite simple; possible advantages of a higher-order decomposition have been discussed by Ishida et al. (1998).

A key consequence of the propagators when split in this way (the so-called symplectic property) is that, although the trajectories are approximate and will not conserve the true energy $\mathcal{H}$, they do exactly conserve a 'shadow Hamiltonian' $\mathcal{H}^{\ddagger}$ (Toxvaerd, 1994), where $\mathcal{H}$ and $\mathcal{H}^{\ddagger}$ differ from each other by a small amount, vanishing as $\delta t \rightarrow 0$. More precisely, it may be shown that the difference $\mathcal{H} - \mathcal{H}^{\ddagger}$ can be written as a Taylor expansion in δt, where the coefficients involve derivatives of $\mathcal{H}$ with respect to the coordinates. The consequence is that the system will remain on a hypersurface in phase space which is 'close' to the true constant-energy hypersurface. Such a stability property is extremely useful in molecular dynamics, since we wish to sample constant-energy states. It essentially eliminates any long-term 'drift' in the total energy.

We can illustrate this with the example of the simple one-dimensional harmonic oscillator for which the trajectory generated by the velocity Verlet algorithm and the corresponding shadow Hamiltonian may be written down explicitly (Venneri and Hoover, 1987; Toxvaerd, 1994). For natural frequency ω, the exact equations of motion and conserved Hamiltonian are

$$\dot{r} = p/m, \quad \dot{p} = -m\omega^2 r, \quad \mathcal{H}(r,p) = p^2/2m + \tfrac{1}{2}m\omega^2 r^2. \tag{3.22}$$

The shadow Hamiltonian depends on timestep through a quantity $\zeta = 1 - (\omega\delta t/2)^2$ and may be written

$$\mathcal{H}^{\ddagger}(r,p) = p^2/2m + \tfrac{1}{2}m\omega^2\zeta r^2. \tag{3.23}$$

It would be equally valid to divide the right-hand side by the factor ζ, in which case the timestep dependence would be associated with the kinetic-energy term, or by $\sqrt{\zeta}$, when it would appear in both terms; for all these choices the difference is $O((\omega\delta t)^2)$. The present choice allows us to compare trajectories which initially coincide at $r = 0$ and have the same momentum and energy, and we do this in the (r, p) 'phase portraits' of Fig. 3.3. The true dynamics follows an elliptical trajectory defined by $\mathcal{H} = $ constant (in the figure this is a circle). The equation $\mathcal{H}^{\ddagger} = $ constant also describes an ellipse, differing only slightly (for small $\omega\delta t$) from the true one. On this diagram, the 'kicks' are vertical line segments, and the 'drifts' are horizontal ones. At the end of each velocity Verlet step the discrete

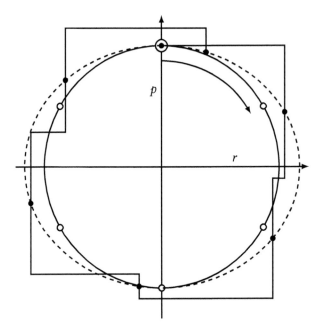

Fig. 3.3 Exact trajectory (circle) and velocity Verlet trajectory (straight line segments) for the harmonic oscillator. The shadow-Hamiltonian conservation law is shown by a dashed ellipse. The open and closed circles mark out phase points at the end of each timestep for the corresponding trajectories. The starting point is marked with a double circle. The timestep is chosen such that $\omega\delta t = \pi/3$, so the exact trajectory returns to its starting point after six timesteps.

trajectory lands exactly on the constant-$\mathcal{H}^{\ddagger}$ ellipse, although the intermediate stages lie off the ellipse. Therefore, the long-time trajectory is very stable in this sense: it will never leave the ellipse $\mathcal{H}^{\ddagger} \approx \mathcal{H} + O((\omega\delta t)^2) = $ constant. However, it can also be seen that the positions and coordinates at the end of each timestep are quickly losing their phase, relative to the corresponding points on exact trajectory. Of course, the example shown uses a very large timestep, to emphasize all these effects.

3.3   Molecular dynamics of rigid non-spherical bodies

Molecular systems are not rigid bodies in any sense: they consist of atoms interacting via intra- and inter-molecular forces. In principle, we should not distinguish between these forces, but as a practical definition we take the forces acting within molecules to be at least an order of magnitude greater than those acting between molecules. If treated classically, as in the earliest molecular simulations (Harp and Berne, 1968; 1970; Berne and Harp, 1970), molecular bond vibrations would occur so rapidly that an extremely short timestep would be required to solve the equations of motion. We return to this approach in Section 3.5; however, we must bear in mind that the classical approach is highly questionable for bond vibrations. A common solution to these problems is to take the intramolecular bonds to be of fixed length. This is not an inconsequential step, but seems reasonable if, as is commonly true at normal temperatures, the amplitude of vibration (classical or quantum) is small compared with molecular dimensions. For polyatomic molecules, we must also consider whether all bond angles should be assumed to be fixed. This is less reasonable in molecules with low-frequency torsional degrees of freedom, or indeed where conformer interconversion is of interest. In this section, we consider the molecular dynamics of molecules in which all bond lengths and internal angles are taken to be constant, that is, in which the molecule is a single rigid unit. In Section 3.4 we discuss the simulation of flexible polyatomic molecules.

In classical mechanics, it is natural to divide molecular motion into translation of the centre of mass and rotation about the centre of mass (Goldstein, 1980). The former motion is handled by the methods of the previous sections: we simply interpret the force $\mathbf{f}_i$ in the equation $m\ddot{\mathbf{r}}_i = \mathbf{f}_i$ as being the vector sum of all the forces acting on molecule i at the centre of mass $\mathbf{r}_i$. The rotational motion is governed by the torque τ_i about the centre of mass. When the interactions have the form of forces $\mathbf{f}_{ia}$ acting on sites at positions $\mathbf{r}_{ia}$ in the molecule, the torque is simply defined

$$\tau_i = \sum_a (\mathbf{r}_{ia} - \mathbf{r}_i) \times \mathbf{f}_{ia}. \tag{3.24}$$

When multipolar terms appear in the potential, the expression for the torque is more complicated (Price et al., 1984), but it may still be calculated from the molecular positions and orientations (see Appendix C). The torque enters the rotational equations of motion in the same way that the force enters the translational equations; the nature of orientation space, however, guarantees that the equations of reorientational motion will not be as simple as the translational equations. In this section, we consider the rotational motion of a molecule under the influence of external torques, taking our origin of coordinates to lie at the centre of mass. To simplify the notation, we drop the suffix i in this section, understanding that all the vectors and matrices refer to a single molecule.

3.3.1   Rotational equations of motion

The orientation of a rigid body, such as a molecule, specifies the relation between an axis system fixed in space and one fixed with respect to the body. We define these systems so that they coincide when the molecule is in an unrotated, reference, orientation. Any vector $\mathbf{e}$ may be expressed in terms of components in the body-fixed or space-fixed frames: we use the notation $\mathbf{e}^b$ and $\mathbf{e}^s$, respectively. These components are related by the 3×3 rotation matrix $\mathbf{A}$

$$\mathbf{e}^b = \mathbf{A} \cdot \mathbf{e}^s, \tag{3.25a}$$

$$\mathbf{e}^s = \mathbf{A}^T \cdot \mathbf{e}^b. \tag{3.25b}$$

Note that $\mathbf{A}$ is *orthogonal*, that is, its inverse $\mathbf{A}^{-1}$ is the same as its transpose, which we denote $\mathbf{A}^T$. These equations relate the components of the *same vector* in two *different coordinate systems*, which is usually called a *passive* rotation. We also need to consider an *active* rotation, which transforms a vector into a *different vector*, with the components expressed in the *same coordinate system*. We express this as

$$\mathbf{e}' = \mathbf{A}^T \cdot \mathbf{e}. \tag{3.26}$$

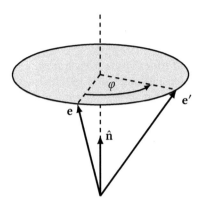

Fig. 3.4   Active rotation of a vector $e \rightarrow e'$ through an angle φ about an axis $\hat{n}$.

To relate these two conventions, consider a vector $\mathbf{e}$ fixed in the molecule. In the unrotated, reference orientation, and in the rotated new orientation, the body-fixed components of this vector will be unchanged because this coordinate system also rotates with the molecule. Therefore $\mathbf{e} = \mathbf{e}^b$. The space-fixed components, however, will change from $\mathbf{e}$ to $\mathbf{e}' = \mathbf{e}^s$. Therefore, eqns (3.26) and (3.25b) are equivalent. Taking this further, if we choose $\mathbf{e}^b$ to be, in turn, each of the body-fixed basis vectors, $(1, 0, 0)$, $(0, 1, 0)$, and $(0, 0, 1)$, we can see from eqns (3.25) and (3.26) that the columns of $\mathbf{A}^{\mathrm{T}}$ (and the rows of $\mathbf{A}$) are these same vectors, in the space-fixed frame, and they completely define the molecular orientation.

The same active transformation of a vector may be specified by an angle φ of rotation about an axis defined by a unit vector $\hat{n}$:

$$\mathbf{e}' = \mathbf{e} \cos \varphi + (1 - \cos \varphi)(\mathbf{e} \cdot \hat{n})\, \hat{n} + (\hat{n} \times \mathbf{e}) \sin \varphi \tag{3.27}$$

where the sense of rotation is shown in Fig. 3.4.

The time evolution of the rotation matrix is obtained by considering an infinitesimally small rotation $\varphi = \omega dt$, about a unit vector $\hat{n} = \hat{\omega}^s = \omega^s/\omega$, where ω is the magnitude of the angular velocity. Expanding the trigonometric functions in eqn (3.27) to first order in dt, in the space-fixed frame, this gives

$$\dot{\mathbf{e}}^s = \omega^s \times \mathbf{e}^s. \tag{3.28}$$

As discussed earlier, this gives an equation of motion for each of the rows of $\mathbf{A}$.

This space-fixed time derivative is only correct in an inertial (i.e. non-rotating) frame. Clearly, if $\mathbf{e}$ is a vector fixed in the molecular frame (e.g. a bond vector) then $\mathbf{e}^b$ will not change with time, that is, $\dot{\mathbf{e}}^b = 0$. In space-fixed coordinates, though, the components $\mathbf{e}^s$ will vary. In fact, eqn (3.28) is a specific case of the general equation linking time derivatives in the two systems

$$\left(\frac{d\mathbf{e}}{dt} \right)^s = \left(\frac{d\mathbf{e}}{dt} \right)^b + \omega \times \mathbf{e}, \tag{3.29}$$

and this is another way of looking at it.

Now we have the first ingredient of the equations of rotational motion: three equations, of the form of eqn (3.28), for the three basis vectors of the molecular frame, expressed in space-fixed coordinates. These three vectors correspond to the rows of the rotation matrix A, from which the components, in the space-fixed frame, of any other body-fixed vector may also be obtained.

To complete the picture, we need to relate the time evolution of the angular velocity vector ω to the angular momentum ℓ, through the inertia tensor I:

$$\ell = \mathbf{I} \cdot \omega. \tag{3.30}$$

The rotational part of the kinetic energy takes the form

$$\mathcal{K} = \tfrac{1}{2}\omega \cdot \mathbf{I} \cdot \omega = \tfrac{1}{2}\ell \cdot \mathbf{I}^{-1} \cdot \ell. \tag{3.31}$$

In eqns (3.30) and (3.31), the quantities may be expressed in whichever set of coordinates is convenient, provided the choice is made consistently throughout. The body-fixed frame is usually chosen to make the inertia tensor diagonal,

$$\mathbf{I}^b = \mathrm{diag}(I^b_{xx}, I^b_{yy}, I^b_{zz}) = \begin{pmatrix} I^b_{xx} & 0 & 0 \\ 0 & I^b_{yy} & 0 \\ 0 & 0 & I^b_{zz} \end{pmatrix}$$

where the three non-zero elements are the *principal* moments of inertia. In this frame, the connection between ω and ℓ is simply $\ell^b_x = I^b_{xx}\omega^b_x$ (and similarly for y and z), while the kinetic energy is a sum of three terms, one for each principal axis

$$\mathcal{K} = \sum_{\alpha=x,y,z} \tfrac{1}{2} I^b_{\alpha\alpha}\omega^{b\,2}_{\alpha} = \sum_{\alpha=x,y,z} \frac{\ell^{b\,2}_{\alpha}}{2I^b_{\alpha\alpha}}. \tag{3.32}$$

The torque τ is most easily evaluated in space-fixed axes, and is equal to the time derivative of the angular momentum in those same axes; in the body-fixed frame, however, the corresponding equation has an extra term, as per eqn (3.29):

$$\dot{\ell}^s = \tau^s, \tag{3.33a}$$

$$\dot{\ell}^b + \omega^b \times \ell^b = \tau^b. \tag{3.33b}$$

Note that we have taken the components of $\omega \times \ell$ in the body-fixed frame, since $\dot{\ell}^b$ and τ^b are expressed in that frame. The resulting equations for the components of ω in the body-fixed frame are

$$\dot{\omega}^b_x = \frac{\tau^b_x}{I^b_{xx}} + \left(\frac{I^b_{yy} - I^b_{zz}}{I^b_{xx}}\right)\omega^b_y\omega^b_z, \tag{3.34a}$$

$$\dot{\omega}^b_y = \frac{\tau^b_y}{I^b_{yy}} + \left(\frac{I^b_{zz} - I^b_{xx}}{I^b_{yy}}\right)\omega^b_z\omega^b_x, \tag{3.34b}$$

$$\dot{\omega}^b_z = \frac{\tau^b_z}{I^b_{zz}} + \left(\frac{I^b_{xx} - I^b_{yy}}{I^b_{zz}}\right)\omega^b_x\omega^b_y. \tag{3.34c}$$

This is the second ingredient of the rotational equations of motion. Eqns (3.34) are combined with eqns (3.28); conversion from space-fixed to body-fixed systems and back is handled by eqn (3.25), that is, $\boldsymbol{\tau}^b = \mathbf{A} \cdot \boldsymbol{\tau}^s$ and $\boldsymbol{\omega}^s = \mathbf{A}^T \cdot \boldsymbol{\omega}^b$.

In fact there is substantial redundancy in the rotation matrix $\mathbf{A}$ of eqn (3.25) and the equations of motion which apply to it: only three independent quantities (generalized coordinates) are needed to define $\mathbf{A}$. These are often taken to be the Euler angles ϕ, θ, ψ in a suitable convention (Goldstein, 1980). However, the equations of motion in Euler angles contain some singularities, and so in molecular dynamics it is far more common to use the rotation matrix directly, as before, or to employ a set of four *quaternion* parameters, as suggested by Evans (1977). Quaternions fulfil the requirements of having well-behaved equations of motion. The four quaternions are linked by one algebraic equation, so there is just one 'redundant' variable. The basic simulation algorithm has been described by Evans and Murad (1977). A quaternion $\mathbf{a}$ is a set of four scalar quantities

$$\mathbf{a} = (a_0, a_1, a_2, a_3) = (a_0, \mathbf{a}) \tag{3.35}$$

where, as indicated, it is often useful to think of the last three elements (a_1, a_2, a_3) as constituting a vector $\mathbf{a}$.[1] Sometimes these four quantities are called Euler parameters. The equations that follow are formally simplified by defining quaternion multiplication

$$\mathbf{a} \otimes \mathbf{b} = \left(a_0 b_0 - \mathbf{a} \cdot \mathbf{b}, \, a_0 \mathbf{b} + b_0 \mathbf{a} + \mathbf{a} \times \mathbf{b} \right) \tag{3.36a}$$

$$= \begin{pmatrix} a_0 & -a_1 & -a_2 & -a_3 \\ a_1 & a_0 & -a_3 & a_2 \\ a_2 & a_3 & a_0 & -a_1 \\ a_3 & -a_2 & a_1 & a_0 \end{pmatrix} \begin{pmatrix} b_0 \\ b_1 \\ b_2 \\ b_3 \end{pmatrix}. \tag{3.36b}$$

This operation is *not* commutative, $\mathbf{a} \otimes \mathbf{b} \neq \mathbf{b} \otimes \mathbf{a}$, but it *is* associative: $(\mathbf{a} \otimes \mathbf{b}) \otimes \mathbf{c} = \mathbf{a} \otimes (\mathbf{b} \otimes \mathbf{c})$.

Quaternions for rotations satisfy the normalization constraint

$$a_0^2 + a_1^2 + a_2^2 + a_3^2 = 1. \tag{3.37}$$

A rotation about an axis defined by a unit vector $\hat{\mathbf{n}}$, through an angle φ, is represented by a quaternion (Goldstein, 1980)

$$\mathbf{a} = \left(\cos \tfrac{1}{2}\varphi, \, \hat{\mathbf{n}} \sin \tfrac{1}{2}\varphi \right). \tag{3.38}$$

Note that $\mathbf{a}$ and $-\mathbf{a}$ represent the *same* rotation. The complement (inverse) of $\mathbf{a}$, corresponding to $\varphi \rightarrow -\varphi$, is $\mathbf{a}^{-1} = (a_0, -\mathbf{a})$; this satisfies $\mathbf{a}^{-1} \otimes \mathbf{a} = \mathbf{I} = (1, 0, 0, 0)$, the unit quaternion. Using eqns (3.36a) and (3.38), the active rotation formula (3.27) may be written

$$\mathbf{e}' = \mathbf{a} \otimes \mathbf{e} \otimes \mathbf{a}^{-1}, \tag{3.39}$$

where we have defined a quaternion $\mathbf{e} = (0, \mathbf{e})$. This may be rewritten, using eqn (3.36b), in the form $\mathbf{e}' = \mathbf{A}^T \cdot \mathbf{e}$ (3.26), with the rotation matrix given by

$$\mathbf{A} = \begin{pmatrix} a_0^2 + a_1^2 - a_2^2 - a_3^2 & 2(a_1 a_2 + a_0 a_3) & 2(a_1 a_3 - a_0 a_2) \\ 2(a_1 a_2 - a_0 a_3) & a_0^2 - a_1^2 + a_2^2 - a_3^2 & 2(a_2 a_3 + a_0 a_1) \\ 2(a_1 a_3 + a_0 a_2) & 2(a_2 a_3 - a_0 a_1) & a_0^2 - a_1^2 - a_2^2 + a_3^2 \end{pmatrix}. \tag{3.40}$$

---

[1] The use of vector notation for the components (a_1, a_2, a_3) is not technically correct, as explained elsewhere (Silva and Martins, 2002), but for our purposes no confusion should arise.

The expressions relating the components of a vector in two different frames, eqn (3.25), take a similar form:

$$\mathbf{e}^b = \mathbf{a}^{-1} \otimes \mathbf{e}^s \otimes \mathbf{a}, \quad \mathbf{e}^s = \mathbf{a} \otimes \mathbf{e}^b \otimes \mathbf{a}^{-1}.$$

A small rotation $(\cos\frac{1}{2}\varphi, \hat{\mathbf{n}}\sin\frac{1}{2}\varphi)$ with $\hat{\mathbf{n}} = \hat{\boldsymbol{\omega}}^s = \boldsymbol{\omega}^s/\omega$, and $\varphi = \omega\delta t$, becomes $(1, \frac{1}{2}\boldsymbol{\omega}^s\delta t)$. Applying this to the molecular orientation $\mathbf{a}$ gives the equation of motion for the quaternions representing the orientation of each molecule:

$$\dot{\mathbf{a}} = \tfrac{1}{2}\mathbf{w}^s \otimes \mathbf{a} = \tfrac{1}{2}\mathbf{a} \otimes \mathbf{w}^b,$$

where we have defined $\mathbf{w}^s = (0, \boldsymbol{\omega}^s)$ and $\mathbf{w}^b = (0, \boldsymbol{\omega}^b)$. In matrix form these equations become

$$\begin{pmatrix}\dot{a}_0\\\dot{a}_1\\\dot{a}_2\\\dot{a}_3\end{pmatrix} = \frac{1}{2}\begin{pmatrix}a_0 & -a_1 & -a_2 & -a_3\\a_1 & a_0 & -a_3 & a_2\\a_2 & a_3 & a_0 & -a_1\\a_3 & -a_2 & a_1 & a_0\end{pmatrix}\begin{pmatrix}0\\\omega_x^b\\\omega_y^b\\\omega_z^b\end{pmatrix}, \tag{3.41a}$$

$$= \frac{1}{2}\begin{pmatrix}a_0 & -a_1 & -a_2 & -a_3\\a_1 & a_0 & a_3 & -a_2\\a_2 & -a_3 & a_0 & a_1\\a_3 & a_2 & -a_1 & a_0\end{pmatrix}\begin{pmatrix}0\\\omega_x^s\\\omega_y^s\\\omega_z^s\end{pmatrix}. \tag{3.41b}$$

In the quaternion formulation, these equations replace eqn (3.28) for each of the basis vectors constituting the rotation matrix. Equations (3.41a) with (3.34), using the matrix of eqn (3.40) to transform between space-fixed and body-fixed coordinates, contain no unpleasant singularities.

Before considering the numerical solution of these equations, we mention the special case of linear molecules, where the orientation is completely specified by a single axis $\mathbf{e}^s$. Formally taking this to be the body-fixed z-axis, the definition of a linear rotor implies that the moment of inertia tensor has $I_{zz} = 0$, and $I_{xx} = I_{yy} = I$. The angular velocity, angular momentum, and torque vectors are all perpendicular to this axis. The time evolution of the axis vector is given by eqn (3.28), and the time evolution of the angular velocity by eqn (3.33a) together with the simple relation $\boldsymbol{\ell} = I\boldsymbol{\omega}$.

3.3.2 Rotational algorithms

How shall we solve these equations in a step-by-step manner? As for the translational equations of motion, a standard predictor–corrector approach may be applied. However, methods inspired by the Verlet-leapfrog approach, and especially the operator-splitting idea, are more stable. Several modifications of the leapfrog and Verlet algorithms, based on the equations of motion, seem to show good stability properties (Fincham, 1981; 1992; Omelyan, 1998; 1999; Hiyama et al., 2008), but we shall concentrate here on the algorithms based on more formal Hamiltonian splitting, for the same reasons outlined in Section 3.2.2: it can be guaranteed that the energy will be well conserved, because of the existence of a shadow Hamiltonian.

Once more, we concentrate on the rotational motion about the centre of mass, assuming that the translational motion is handled using the methods of Section 3.2. We focus on a single molecule to avoid too many indices, but of course the algorithm is applied to all the molecules simultaneously. Since the Hamiltonian splits naturally into a rotational kinetic-energy part, and a potential-energy part that depends on orientation, exactly the same kind of factorization of the Liouville operator may be used as in Section 3.2.2, and this leads to:

(a) 'half-kick' $\ell^s(t) \rightarrow \ell^s(t + \frac{1}{2}\delta t) = \ell^s(t) + \frac{1}{2}\delta t \, \tau^s(t)$ (fixed orientation);
(b) 'drift' $\mathbf{a}(t) \rightarrow \mathbf{a}(t + \delta t)$ or $\mathbf{A}(t) \rightarrow \mathbf{A}(t + \delta t)$ (free rotation with ℓ^s constant);
(c) 'half-kick' $\ell^s(t + \frac{1}{2}\delta t) \rightarrow \ell^s(t + \delta t) = \ell^s(t + \frac{1}{2}\delta t) + \frac{1}{2}\delta t \, \tau^s(t + \delta t)$ (fixed orientation).

The first and third steps are straightforward: our attention needs to focus on the middle one, which is the exact solution of the problem of a freely rotating body, with constant angular momentum (in the space-fixed frame), over one timestep.

A molecule whose principal moments of inertia are all equal, $I_{xx} = I_{yy} = I_{zz} = I$, is called a *spherical top*. Examples are CCl_4 and SF_6. In this case, $\ell = I\omega$, so the angular velocity vector is constant, in both space-fixed and body-fixed frames. This means that the solution is a simple rotation about the direction $\hat{\mathbf{n}} = \hat{\boldsymbol{\omega}}^s = \boldsymbol{\omega}^s/\omega$, through an angle $\varphi = \omega\delta t$, where $\omega = |\boldsymbol{\omega}^s|$. It is simply a matter of convenience whether to use the rotation matrix, or quaternion parameters, to represent the molecular orientation. In vector notation, the solution of eqn (3.28) for any of the basis vectors fixed in the molecule is given by eqn (3.27):

$$\mathbf{e}^s(t + \delta t) = \mathbf{e}^s(t) \cos \omega\delta t + (1 - \cos \omega\delta t)\left(\mathbf{e}^s(t) \cdot \hat{\boldsymbol{\omega}}^s\right)\hat{\boldsymbol{\omega}}^s + \left(\hat{\boldsymbol{\omega}}^s \times \mathbf{e}^s(t)\right) \sin \omega\delta t. \quad (3.42)$$

As explained before, this applies to each of the rows of $\mathbf{A}$. This equation is also the solution of the free-rotation problem for *linear* molecules. In quaternion form, using eqn (3.38) in a similar way

$$\mathbf{a}(t + \delta t) = \left(\cos \tfrac{1}{2}\omega\delta t, \hat{\boldsymbol{\omega}}^s \sin \tfrac{1}{2}\omega\delta t\right) \otimes \mathbf{a}(t). \quad (3.43)$$

More commonly, small rigid molecules have three unequal moments of inertia, in which case they are called *asymmetric tops* (H_2O is one example), or, more rarely, two equal values and one different, when they are termed *symmetric tops* (e.g. $CHCl_3$). Now, the exact solution of the free-rotation problem is not so straightforward, although it has been presented in a form suitable for molecular dynamics (van Zon and Schofield, 2007a,b). It consists of a sequence of rotations, and involves Jacobi elliptic functions and the numerical evaluation of an integral. Although this introduces some computational complexity into the problem, it should be remembered that the most CPU-intensive parts of the MD program most likely lie in the force loop, so this approach is quite feasible (Celledoni et al., 2008).

Alternatively, one can make approximations to the free-rotation propagator, giving a simpler implementation (Dullweber et al., 1997; Miller et al., 2002). One method, which has been extensively studied (Dullweber et al., 1997) is to take advantage of the splitting of the kinetic energy into three separate parts, eqn (3.32), and correspondingly split the propagator into successive rotations, each taken around one of the body-fixed principal axes. A symmetric splitting scheme of this kind may be written

$$U_x(\tfrac{1}{2}\delta t)\, U_y(\tfrac{1}{2}\delta t)\, U_z(\delta t)\, U_y(\tfrac{1}{2}\delta t)\, U_x(\tfrac{1}{2}\delta t),$$

where $U_\alpha(t) = \exp(iL_\alpha t)$ for $\alpha = x, y, z$. Each symbol represents a rotation about the corresponding principal axis, for the indicated time, which is implemented using an equation like (3.42) or (3.43). The first step, $U_x(\frac{1}{2}\delta t)$, for example, consists of selecting the vector $\mathbf{x}^s$ corresponding to the body-fixed x-axis, calculating the component of $\boldsymbol{\ell}$ along this axis, which is, of course, equal to ℓ_x^b, evaluating the corresponding angular velocity component $\omega_x^b = \ell_x^b/I_{xx}$, and replacing $\omega \to \omega_x^b$, $\hat{\boldsymbol{\omega}}^s \to \mathbf{x}^s$, and $\delta t \to \frac{1}{2}\delta t$ in eqn (3.42) (which applies to each of the basis vectors) or eqn (3.43). In this case, the principal x-axis will remain constant, while the y and z axes rotate around it. Then the same procedure is repeated for the other axes in sequence. Throughout the whole process, the space-fixed angular momentum $\boldsymbol{\ell}^s$ is constant, while the components $\boldsymbol{\ell}^b$ in body-fixed axes will change as they rotate. Of course, the particular sequence of axes given here is just one choice; there is scope to optimize the performance of the algorithm, depending on the relative values of the moments of inertia (Fassò, 2003).

Whichever method is used to integrate forward the rotational equations of motion, in principle the normalization constraint eqn (3.37), or the orthonormality of the rotation matrix, should be preserved. It is easy to show that the aforementioned algorithms do this, to the precision allowed by floating-point numbers. In practice, of course, small errors may build up over a period of time. To avoid this, it seems natural to 'renormalize' $\mathbf{a}$, or 're-orthonormalize' $\mathbf{A}$, at frequent intervals. However, it has been pointed out (Matubayasi and Nakahara, 1999) that imposing an additional step of this kind will violate the exact time reversibility of the algorithm. Therefore, provided an appropriate algorithm is being employed, it is best to avoid doing this.

3.4 Constraint dynamics

In polyatomic systems, it becomes necessary to consider not only the stretching of interatomic bonds but also bending motions, which change the angle between bonds, and twisting motions, which alter torsional angles (see Fig. 1.10). These torsional motions are, typically, of much lower frequency than bond vibrations, and are very important in long-chain organic molecules and biomolecules: they lead to conformational interconversion and have a direct influence on polymer dynamics. Clearly, these effects must be treated properly in molecular dynamics, within the classical approximation. It would be quite unrealistic to assume total rigidity of such a molecule, although bond lengths can be thought of as fixed, and a case might be made out for a similar approximation in the case of bond bending angles.

Of course, for any system with such holonomic constraints applied (i.e. a set of algebraic equations connecting the coordinates) it is possible to construct a set of generalized coordinates obeying constraint-free equations of motion (i.e. ones in which the constraints appear implicitly). For any molecule of moderate complexity, such an approach would be very complicated, although it was used in the first simulations of butane (Ryckaert and Bellemans, 1975). The equations of motion in such a case are derived from first principles, starting with the Lagrangian (eqns (3.1) and (3.2)). In practice, we will want to handle the dynamics in terms of Cartesian coordinates, not generalized ones. The definition of the microcanonical ensemble in this situation is discussed by Ryckaert and Ciccotti (1983, Appendix), and some of the consequences are outlined in Section 2.10. Here we examine the practical implementation of constraint dynamics using Cartesian coordinates.

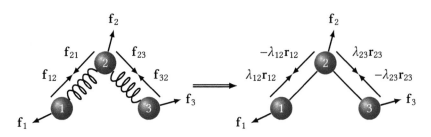

Fig. 3.5 Illustration of constraint scheme applied to triatomic molecule such as H_2O. Intramolecular forces $f_{12} = -f_{21}$ and $f_{23} = -f_{32}$, derived from bond-stretching potentials, are replaced by constraint forces $\pm\lambda_{12}r_{12}$ and $\pm\lambda_{23}r_{23}$, directed along the bonds. The multipliers λ_{12} and λ_{23} are determined by the constraint conditions at the end of the timestep, and for convenience a factor involving δt is often included in their definition. In both cases, the forces f_1, f_2, f_3, are due to *intermolecular* interactions, and to any other intramolecular terms such as the θ_{123} angle-bending potential.

A special technique has been developed to handle the dynamics of a molecular system in which certain arbitrarily selected degrees of freedom (such as bond lengths) are constrained while others remain free to evolve under the influence of intermolecular and intramolecular forces. This constraint dynamics approach (Ryckaert et al., 1977) in effect uses a set of undetermined multipliers to represent the magnitudes of forces directed along the bonds which are required to keep the bond lengths constant. The technique is to solve the equations of motion for one timestep in the absence of the constraint forces, and subsequently determine their magnitudes and correct the atomic positions. The method can be applied equally well to totally rigid and non-rigid molecules. Its great appeal is that it reduces even a complex polyatomic liquid simulation to the level of difficulty of an atomic calculation plus a constraint package based on molecular geometry. The original method, called SHAKE, designed to work with the Verlet algorithm, is described in detail by Ryckaert et al. (1977); a version built around the velocity Verlet algorithm (Section 3.2.1) was proposed by Andersen (1983), and called RATTLE. We describe this method here.

We shall illustrate the constraint method with a simple example. Consider a bent triatomic molecule such as H_2O, in which we wish to constrain two of the bonds to be of fixed length but allow the remaining bond, and hence the inter-bond angle, to vary under the influence of the intra-molecular potential. In this section, we drop the molecular indices, so r_a will represent the position of atom a in a specific molecule. Numbering the central (oxygen) atom 2, and the two outer (hydrogen) atoms 1 and 3, as shown in Fig. 3.5, we write the constraint equations

$$\chi_{12}^{(r)}(t) = \left|r_{12}(t)\right|^2 - d_{12}^2 = 0, \quad \text{and} \quad \chi_{23}^{(r)}(t) = \left|r_{23}(t)\right|^2 - d_{23}^2 = 0, \tag{3.44a}$$

and by time differentiation it follows that the following equations are also satisfied

$$\chi_{12}^{(v)}(t) = v_{12}(t) \cdot r_{12}(t) = 0, \quad \text{and} \quad \chi_{23}^{(v)}(t) = v_{23}(t) \cdot r_{23}(t) = 0, \tag{3.44b}$$

where d_{12} and d_{23} are the bond lengths, $r_{12} = r_1 - r_2$, $v_{12} = \dot{r}_{12}$, etc. The equations of motion take the form

$$m_1\ddot{r}_1 = f_1 + g_1, \quad m_2\ddot{r}_2 = f_2 + g_2, \quad m_3\ddot{r}_3 = f_3 + g_3. \tag{3.45}$$

Here $\mathbf{f}_1$, $\mathbf{f}_2$, and $\mathbf{f}_3$ are the forces due to intermolecular interactions as well as those intramolecular effects that are explicitly included in the potential. The remaining terms $\mathbf{g}_1$ etc. are the constraint forces: their role is solely to ensure that the constraint equations (3.44) are satisfied at all times.

The Lagrangian equations of motion are derived from the constraints (3.44a) (Bradbury, 1968, Chapter 11); they are eqns (3.45) with

$$\mathbf{g}_a = \Lambda_{12} \nabla_{\mathbf{r}_a} \chi_{12}^{(r)} + \Lambda_{23} \nabla_{\mathbf{r}_a} \chi_{23}^{(r)} \tag{3.46}$$

where Λ_{12} and Λ_{23} are undetermined (Lagrangian) multipliers. So far, we have made no approximations, and, in principle, could solve for the constraint forces (Orban and Ryckaert, 1974). However, because we are only able to solve the equations of motion approximately, using finite-difference methods, in practice this will lead to bond lengths that steadily diverge from the desired values. Instead, Ryckaert et al. (1977) suggested an approach in which the constraint forces are calculated so as to guarantee that the constraints are satisfied at each timestep; by implication, the constraint forces themselves are only correct to the same order of accuracy as the integration algorithm. In fact, we will be considering two different approximations to each $\mathbf{g}_a$: the first, which we call $\mathbf{g}_a^{(r)}$, ensures that eqns (3.44a) are satisfied, while the second, $\mathbf{g}_a^{(v)}$, guarantees that we satisfy eqns (3.44b). Both these approximations, of course, will be within $O(\delta t^2)$ of the true constraint forces.

Consider the first 'kick' and the 'drift' stage of the velocity Verlet algorithm, generating the half-advanced velocities, and the fully advanced positions. Let $\mathbf{r}_a'(t+\delta t)$ and $\mathbf{v}_a'(t+\tfrac{1}{2}\delta t)$ be the results of doing this, including all the physical forces $\mathbf{f}_a$, but *without* including the constraint forces $\mathbf{g}_a$. Referring to eqns (3.11a), (3.11b), and (3.12a):

$$\mathbf{v}_a'(t + \tfrac{1}{2}\delta t) = \mathbf{v}_a(t) + \tfrac{1}{2}(\delta t/m_a)\,\mathbf{f}_a(t)$$

$$\mathbf{r}_a'(t + \delta t) = \mathbf{r}_a(t) + \delta t\,\mathbf{v}_a'(t + \tfrac{1}{2}\delta t) = \mathbf{r}_a(t) + \delta t\,\mathbf{v}_a(t) + \tfrac{1}{2}\left(\delta t^2/m_a\right)\mathbf{f}_a(t).$$

Concentrating on the positions, the effects of subsequently including the constraint forces may be written

$$\mathbf{r}_a(t + \delta t) = \mathbf{r}_a'(t + \delta t) + \tfrac{1}{2}\left(\delta t^2/m_a\right)\mathbf{g}_a^{(r)}(t) \tag{3.47}$$

for each of the three atoms in the H_2O molecule. The constraint forces for this stage, $\mathbf{g}_a^{(r)}$, must be directed along the bond vectors $\mathbf{r}_{ab}(t)$, and must conform to Newton's third law, so we may write them in the form

$$\left(\tfrac{1}{2}\delta t^2\right)\mathbf{g}_1^{(r)}(t) = \lambda_{12}^{(r)}\mathbf{r}_{12}(t) \tag{3.48a}$$

$$\left(\tfrac{1}{2}\delta t^2\right)\mathbf{g}_2^{(r)}(t) = \lambda_{23}^{(r)}\mathbf{r}_{23}(t) - \lambda_{12}^{(r)}\mathbf{r}_{12}(t) \tag{3.48b}$$

$$\left(\tfrac{1}{2}\delta t^2\right)\mathbf{g}_3^{(r)}(t) = -\lambda_{23}^{(r)}\mathbf{r}_{23}(t) \tag{3.48c}$$

where $\lambda_{12}^{(r)}$ and $\lambda_{23}^{(r)}$ are the undetermined multipliers (one for each constraint), and we have included a factor of $\tfrac{1}{2}\delta t^2$ for convenience. Inserting these expressions into the corresponding three equations (3.47), we can calculate the bond vectors at $t + \delta t$:

$$\mathbf{r}_{12}(t + \delta t) = \mathbf{r}'_{12}(t + \delta t) + \left(m_1^{-1} + m_2^{-2}\right)\lambda_{12}^{(r)}\mathbf{r}_{12}(t) - m_2^{-1}\lambda_{23}^{(r)}\mathbf{r}_{23}(t) \tag{3.49a}$$

$$\mathbf{r}_{23}(t + \delta t) = \mathbf{r}'_{23}(t + \delta t) - m_2^{-1}\lambda_{12}^{(r)}\mathbf{r}_{12}(t) + \left(m_2^{-1} + m_3^{-2}\right)\lambda_{23}^{(r)}\mathbf{r}_{23}(t). \tag{3.49b}$$

Now we can take the square modulus of both sides and apply our desired constraints: $|\mathbf{r}_{12}(t + \delta t)|^2 = |\mathbf{r}_{12}(t)|^2 = d_{12}^2$ and similarly for $\mathbf{r}_{23}$. The result is a pair of quadratic equations in the two unknowns, $\lambda_{12}^{(r)}$ and $\lambda_{23}^{(r)}$, the coefficients in which are all known (given that we already have the 'unconstrained' bond vectors $\mathbf{r}'_{ab}$) and which can be solved for the undetermined multipliers. We turn to the method of solution shortly.

These values are used in eqns (3.48) and (3.47). At the same time, the half-step velocities $\mathbf{v}'_a(t + \frac{1}{2}\delta t)$ are adjusted according to

$$\mathbf{v}_a(t + \tfrac{1}{2}\delta t) = \mathbf{v}'_a(t + \tfrac{1}{2}\delta t) + \tfrac{1}{2}(\delta t/m_a)\,\mathbf{g}_a^{(r)}(t). \tag{3.50}$$

The second part of the algorithm follows evaluation of the non-constraint forces $\mathbf{f}_a(t + \delta t)$, which are used in eqn (3.11c) to give $\mathbf{v}'_a(t + \delta t)$:

$$\mathbf{v}'_a(t + \delta t) = \mathbf{v}_a(t + \tfrac{1}{2}\delta t) + \tfrac{1}{2}(\delta t/m_a)\,\mathbf{f}_a(t + \delta t).$$

The inclusion of constraints in the second stage is written

$$\mathbf{v}_a(t + \delta t) = \mathbf{v}'_a(t + \delta t) + \tfrac{1}{2}(\delta t/m_a)\mathbf{g}_a^{(v)}(t + \delta t). \tag{3.51}$$

These constraint forces $\mathbf{g}_a^{(v)}(t + \delta t)$, are directed along the bonds $\mathbf{r}_{ab}(t + \delta t)$, so a set of equations similar to (3.48) may be written

$$\left(\tfrac{1}{2}\delta t\right)\mathbf{g}_1^{(v)}(t + \delta t) = \lambda_{12}^{(v)}\mathbf{r}_{12}(t + \delta t) \tag{3.52a}$$

$$\left(\tfrac{1}{2}\delta t\right)\mathbf{g}_2^{(v)}(t + \delta t) = \lambda_{23}^{(v)}\mathbf{r}_{23}(t + \delta t) - \lambda_{12}^{(v)}\mathbf{r}_{12}(t + \delta t) \tag{3.52b}$$

$$\left(\tfrac{1}{2}\delta t\right)\mathbf{g}_3^{(v)}(t + \delta t) = -\lambda_{23}^{(v)}\mathbf{r}_{23}(t + \delta t) \tag{3.52c}$$

where once again it is convenient to introduce a factor involving the timestep. This reduces the problem to determining a new pair of undetermined multipliers, $\lambda_{12}^{(v)}$ and $\lambda_{23}^{(v)}$. These are chosen so that the velocities satisfy the constraint equations (3.44b) exactly at time $t + \delta t$. Those equations are *linear* in the unknowns, since they are obtained by taking the scalar products of the relative velocities:

$$\mathbf{v}_{12}(t + \delta t) = \mathbf{v}'_{12}(t + \delta t) + \left(m_1^{-1} + m_2^{-2}\right)\lambda_{12}^{(v)}\mathbf{r}_{12}(t) - m_2^{-1}\lambda_{23}^{(v)}\mathbf{r}_{23}(t) \tag{3.53a}$$

$$\mathbf{v}_{23}(t + \delta t) = \mathbf{v}'_{23}(t + \delta t) - m_2^{-1}\lambda_{12}^{(v)}\mathbf{r}_{12}(t) + \left(m_2^{-1} + m_3^{-2}\right)\lambda_{23}^{(v)}\mathbf{r}_{23}(t) \tag{3.53b}$$

with the corresponding, already determined, bond vectors $\mathbf{r}_{12}(t + \delta t)$, $\mathbf{r}_{23}(t + \delta t)$. Once determined, the values of $\lambda_{12}^{(v)}$ and $\lambda_{23}^{(v)}$ are used in eqns (3.52) to give the constraint forces, which are substituted into eqns (3.51). Note that, in the next integration step, a different approximation to these same constraint forces, namely $\mathbf{g}_a^{(r)}(t + \delta t)$, will be used. This step follows immediately.

The scheme for the original SHAKE method is somewhat simpler than the one just described, since the Verlet algorithm only involves positions, with velocities determined afterwards: therefore, only a single set of Lagrange multipliers, $\lambda_{ab}^{(r)}$, is needed. Nonetheless, the same ideas apply, and the equations are very similar.

We have examined this case in some detail so as to bring out the important features in a more general scheme. Bond angle (as opposed to bond length) constraints present no fundamental difficulty, and may be handled by introducing additional length constraints. For example, the H–O–H bond angle in water may be fixed by constraining the H–H distance, in addition to the O–H bond lengths. Instead of eqn (3.48) we would then have

$$\left(\tfrac{1}{2}\delta t^2\right) g_1^{(r)}(t) = \lambda_{12}^{(r)} r_{12}(t) - \lambda_{31}^{(r)} r_{31}(t)$$

$$\left(\tfrac{1}{2}\delta t^2\right) g_2^{(r)}(t) = \lambda_{23}^{(r)} r_{23}(t) - \lambda_{12}^{(r)} r_{12}(t)$$

$$\left(\tfrac{1}{2}\delta t^2\right) g_3^{(r)}(t) = \lambda_{31}^{(r)} r_{31}(t) - \lambda_{23}^{(r)} r_{23}(t)$$

and eqn (3.49) would be replaced by

$$r_{12}(t + \delta t) = r_{12}'(t + \delta t) + \left(m_1^{-1} + m_2^{-1}\right)\lambda_{12}^{(r)} r_{12}(t) - m_2^{-1}\lambda_{23}^{(r)} r_{23}(t) - m_1^{-1}\lambda_{31}^{(r)} r_{31}(t)$$

$$r_{23}(t + \delta t) = r_{23}'(t + dt) - m_3^{-1}\lambda_{31}^{(r)} r_{31}(t) + (m_2^{-1} + m_3^{-1})\lambda_{23}^{(r)} r_{23}(t) - m_2^{-1}\lambda_{12}^{(r)} r_{12}(t)$$

$$r_{31}(t + \delta t) = r_{31}'(t + \delta t) - m_1^{-1}\lambda_{12}^{(r)} r_{12}(t) - m_3^{-1}\lambda_{23}^{(r)} r_{23}(t) + (m_3^{-1} + m_1^{-1})\lambda_{31}^{(r)} r_{31}(t)$$

with a similar set of equations applying to the second-stage velocity constraints. This process of 'triangulating' the molecule by introducing fictitious bonds is straightforwardly applied to more complex systems. Figure 3.6 shows bond length constraints applied to the carbon units in a simple model of butane, which leaves just one internal parameter (the torsion angle ϕ) free to evolve under the influence of the potential. The extension to n-alkanes is discussed by Ryckaert et al. (1977) and an application to the case of n-decane has been described (Ryckaert and Bellemans, 1978).

Now we turn to the method of solving the constraint equations. For the velocity constraints, λ_{12}^v, λ_{23}^v, these equations are linear. For the position constraints, since λ_{12}^r, λ_{23}^r are proportional to δt^2, the quadratic terms are relatively small. These equations may be solved in an iterative fashion: the quadratic terms are dropped and the remaining linear equations solved; these approximate solutions are substituted into the quadratic terms to give new linear equations, which yield improved estimates of λ_{12}^r and λ_{23}^r, and so on. For very small molecules, as in the previous example, the (linearized) constraint equations may be solved by straightforward algebra. For a larger polyatomic molecule, with n_c constraints, the solution of these equations essentially requires inversion of an $n_c \times n_c$ matrix at each timestep. This could become time-consuming for very large molecules, such as proteins. Assuming, however, that only near-neighbour atoms and bonds are related by constraint equations, the constraint matrix will be sparse, and special inversion techniques might be applicable. An alternative procedure is to go through the constraints one by one, cyclically, adjusting the coordinates so as to satisfy each in turn. The procedure may be iterated until all the constraints are satisfied to within a given tolerance. To be precise, it is this algorithm that is termed SHAKE (Ryckaert et al., 1977) or RATTLE (Andersen, 1983). It is most useful when large molecules are involved, and is a standard part of most

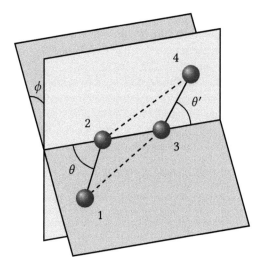

Fig. 3.6 Triangulation scheme for the simple model of butane illustrated in Fig. 1.10. Bonds 1–2, 2–3, and 3–4 are constrained to have specified lengths. In addition, constraining the 1–3 and 2–4 distances will fix the angles θ, θ', leaving the torsion angle ϕ as the only internal coordinate that is free to vary.

molecular dynamics packages. If the process is iterated to convergence, then it preserves the symplectic nature of the Verlet, or velocity Verlet, algorithm (Leimkuhler and Skeel, 1994). The process may be accelerated by a successive overrelaxation approach (Barth et al., 1995).

Problems may arise in the construction of a constraint scheme for certain molecules. Consider the linear molecule CS_2: it has three atoms and five degrees of freedom (two rotational and three translational) so we require $n_c = 3 \times 3 - 5 = 4$ constraints. This is impossible with only three bond lengths available to be specified. A more subtle example is that of benzene, modelled as six united CH atoms in a hexagon. For six degrees of freedom (three rotational and three translational) we require $n_c = 3 \times 6 - 6 = 12$ constraints, and this number may indeed be accommodated. However, the constraint matrix is then found to be singular, that is, its determinant vanishes. Physically, the problem is that all the constraints act in the plane of the molecule, and none of them act to preserve planarity. The solution to both these problems is to choose a subset of atoms sufficient to define the molecular geometry, apply constraints to those atoms, and express the coordinates of the remaining atoms as linear combinations of those of the primary 'core' (Ciccotti et al., 1982). In computing the dynamics of the core, there is a simple prescription for transferring the forces acting on the 'secondary' atoms to the core atoms so as to generate the correct linear and angular accelerations. The SHAKE method has been extended to handle more general geometrical constraints needed to specify (for example) the arrangement of side-chains or substituent atoms in flexible hydrocarbons (Ryckaert, 1985; Ciccotti and Ryckaert, 1986).

Having discussed SHAKE and RATTLE, we briefly summarize some of the alternatives. For small molecules, an algebraic solution of the constraint equations may still be preferable to iteration, and indeed this is the approach used in the SETTLE algorithm (Miyamoto

Code 3.5 Constraint algorithms for a chain molecule

These files are provided online. The code illustrates the RATTLE and MILC-SHAKE constraint algorithms for a chain of atoms. The program md_chain_nve_lj.f90 offers a run-time choice between these two methods, the routines for which are in the file md_chain_lj_module.f90. MILC-SHAKE requires a tridiagonal solver, and there are many options for this; the current implementation calls a routine dgtsv from the LAPACK library which is assumed to be installed. A more general example routine, using a similar approach, is referenced in Bailey et al. (2008), which also provides more details about the method. Various routines for input/output and simulation averages are provided in utility modules (see Appendix A). Code to set up an initial configuration is provided in initialize.f90.

```
! md_chain_nve_lj.f90
! Molecular dynamics, NVE ensemble, chain molecule
PROGRAM md_chain_nve_lj

! md_chain_lj_module.f90
! Force & constraint routines for MD, LJ chain
MODULE md_module
```

and Kollman, 1992). For linear or ring molecules, where each atom is connected to at most two others, the constraint matrix is tridiagonal, and again an efficient non-iterative solution of the constraint problem, MILC-SHAKE, may be implemented (Bailey et al., 2008; 2009). The linear constraint solver (LINCS) approach (Hess et al., 1997) is based on re-writing the constrained equations of motion using the principle of least action. The method works by applying a matrix correction to the unconstrained positions so as to set the bond lengths, projected along the direction of the 'old' bonds, to the correct values. Including a correction for bond rotation, the method is found to be quite accurate without iteration, and is claimed to be several times faster than SHAKE, as well as showing better convergence characteristics at large timesteps. WIGGLE (Lee et al., 2005) is derived by applying a projection operator method to the equations of motion, and considers atomic accelerations rather than forces. Some examples are given in Code 3.5.

For completely rigid molecules, symplectic integration algorithms based on quaternion parameters or rotation matrices seem to be the simplest approach. On the other hand, as soon as any non-rigidity is introduced into the molecular model, constraint dynamics as typified by RATTLE and SHAKE provide an attractive option. As we shall see in the next section, the only reasonable alternative is to allow the high-frequency motions to occur, and employ a multiple-timestep integrator to handle them properly. For flexible molecules, we have ample choice as to where to apply constraints, and it is generally believed that, while constraining bond lengths is worthwhile, it is best to leave bond angles (and certainly torsion angles) free to evolve under the influence of appropriate terms in the potential energy. This is partly on the grounds of program efficiency: the RATTLE/SHAKE algorithm iterations converge very slowly when rigid 'triangulated' molecular units are involved, often necessitating a reduced timestep, which might as well be used in a proper integration of the bond 'wagging' motions instead (van Gunsteren and Berendsen, 1977; van Gunsteren, 1980). The other reason is that the relatively low frequencies of these motions makes the constraint approximation less valid. As discussed in Section 2.10, a model with a strong harmonic potential is different from one in which the potential is replaced by a rigid constraint. This point has been recognized for some time in the field of polymer dynamics, and has been tested by computer simulation (Fixman, 1974; 1978a,b; Gō and Scheraga, 1976; Helfand, 1979; Pear and Weiner, 1979). In practical terms, for a model of a protein molecule, van Gunsteren and Karplus (1982) have shown that the introduction of bond length constraints into a model based otherwise on realistic intramolecular potential functions has little effect on the structure and dynamics, but the further introduction of constraints on bond angles seriously affects the torsion angle distributions and the all-important conformational interconversion rates. This effect can be countered by adding the additional constraint potential of eqn (2.161), which involves the calculation of the metric determinant $\det(\mathbf{H})$. This is time-consuming and algebraically complicated for all but the simplest flexible molecules, and the lesson seems to be that, for realistic molecular dynamics simulations, bond length constraints are permissible, but bond angle constraints should not be introduced without examining their effects.

Two final points should be made, in relation to the calculation of thermodynamic properties of model systems incorporating constraints. The calculation of the total kinetic energy of such a system is a simple matter of summing the individual atomic contributions in the usual way. When using this quantity to estimate the temperature, according to eqn (2.55), we must divide by the number of degrees of freedom. It should be clear from the specification of the molecular model how many independent constraints have been applied, and hence what the number of degrees of freedom is. Second, in molecular systems, quantities such as the pressure may be calculated in several ways, the two most important of which focus on the component atoms, and on the molecular centres of mass, respectively. Consider the evaluation of the virial function (eqn (2.65)) interpreting the sum as being taken over all atom–atom separations r_{ab} and forces $\mathbf{f}_{ab}$. In this case, all intramolecular contributions to $\mathcal{W}$ including the constraint forces should be taken into account. Now consider the alternative interpretation of eqn (2.67), in which we take the $\mathbf{f}_{ij}$ to represent the sum of all the forces acting on a molecule i, due to its interactions with molecule j, and take each such force to act at the centre of mass. In this case, all the intramolecular forces, including the constraint forces, cancel out and can be omitted from the sum. It is easy to show that, at equilibrium, the average pressure computed by either route is the same.

3.5   Multiple-timestep algorithms

An alternative to constraining some of the intramolecular degrees of freedom is to allow them to evolve according to the classical equations, but use a timestep small enough to cope with the high-frequency motion. The formal development of Section 3.2.2 allows this to be done in a relatively straightforward way (Tuckerman et al., 1992; Grubmüller et al., 1991; Martyna et al., 1996). Suppose, to illustrate this, that there are 'slow' $\mathbf{f}^{\text{slow}}$, and 'fast' $\mathbf{f}^{\text{fast}}$, forces. Therefore, the momentum equation is $\dot{\mathbf{p}} = \mathbf{f}^{\text{slow}} + \mathbf{f}^{\text{fast}}$. We may break up the

Liouville operator $iL = iL_1 + iL_2 + iL_3$ where the separate terms are defined:

$$iL_1 = \mathbf{v} \cdot \frac{\partial}{\partial \mathbf{r}}, \quad iL_2 = \mathbf{f}^{\text{fast}} \cdot \frac{\partial}{\partial \mathbf{p}}, \quad iL_3 = \mathbf{f}^{\text{slow}} \cdot \frac{\partial}{\partial \mathbf{p}}. \tag{3.54}$$

The propagator $U(\Delta t) = \exp\left(iL\Delta t\right)$ approximately factorizes

$$U(\Delta t) \approx U_3\left(\tfrac{1}{2}\Delta t\right) \exp\left((iL_2 + iL_1)\Delta t\right) U_3\left(\tfrac{1}{2}\Delta t\right), \tag{3.55}$$

where Δt represents a long timestep. The middle part is then split again, using the conventional separation as usual, iterating over small timesteps $\delta t = \Delta t / n_{\text{short}}$:

$$\exp\left((iL_2 + iL_1)\Delta t\right) \approx \left[U_2\left(\tfrac{1}{2}\delta t\right) U_1(\delta t) U_2\left(\tfrac{1}{2}\delta t\right)\right]^{n_{\text{short}}}. \tag{3.56}$$

The fast-varying forces are computed at short intervals, while the slow forces are computed once per long timestep. This algorithm translates naturally into computer code, as shown in Code 3.6. As we shall see in Section 5.4, this approach may be extended to speed up the calculation of non-bonded interactions, as well as bond vibrations and the like.

3.6 Checks on accuracy

Is it working properly? This is the first question that must be asked when a simulation is run for the first time, and the answer is frequently in the negative. Here, we discuss the tell-tale signs of a non-functioning MD program.

The first check must be that the conservation laws are properly obeyed, and in particular that the energy should be 'constant'. As we shall see later (Section 3.8), it is common practice to conduct MD simulations at constant temperature, in which energy fluctuations are expected. This can conceal some problems, so before carrying out such runs it is important to switch off the thermostat and conduct tests in the microcanonical ensemble.

In fact small changes in the energy will occur (see Fig. 3.2). For a simple Lennard-Jones system, fluctuations of order 1 part in 10^4 are generally considered to be acceptable, although some workers are less demanding, and some more so. Energy fluctuations may be reduced by decreasing the timestep. Assuming that a symplectic algorithm is being used, as explained earlier, the way these fluctuations scale with timestep is almost precisely known, because of the exact conservation of a shadow Hamiltonian. For the Verlet algorithm, $\mathcal{H}^{\ddagger} - \mathcal{H} = O(\delta t^2)$, and a suggestion due to Andersen (Berens et al., 1983) may be useful: plot RMS energy fluctuations $\sqrt{\langle \mathcal{H}^2 \rangle - \langle \mathcal{H} \rangle^2}$ against δt^2, and check that the resulting graph is linear (see also Fig. 3.2). To avoid systematic differences, these calculations should cover essentially the same MD trajectories. Several short runs should be undertaken, each starting from the same initial configuration and covering the same total time t_{run}: each run should employ a different timestep δt, and hence consist of a different number of steps $\tau_{\text{run}} = t_{\text{run}} / \delta t$.

A good initial estimate of δt is that it should be roughly an order of magnitude less than the Einstein period $t_{\text{E}} = 2\pi/\omega_{\text{E}}$, where the Einstein frequency ω_{E} is given by eqn (2.135). This gives a guide to the typical frequencies of atomic motion that the algorithm is attempting to reproduce faithfully. If a typical liquid starting configuration is available,

Code 3.6   Multiple-timestep algorithm

We show, schematically, the important parts of the algorithm for a simulation of n_{long} 'long' timesteps, each of length Δt (stored in dt_long), each step consisting of n_{short} 'short' timesteps δt (stored in dt_short). We take the atoms to have unit mass, so forces and accelerations are the same.

```
REAL, DIMENSION(3,n) :: r, f_fast, f_slow, v
! begin with these arrays all calculated

DO long_step = 1, n_long
   v = v + 0.5 * dt_long * f_slow
   DO short_step = 1, n_short
      v = v + 0.5 * dt_short * f_fast
      r = r + dt_short * v
      ! ... calculate f_fast from r
      v = v + 0.5 * dt_short * f_fast
   END DO
   ! ... calculate f_slow from r
   v = v + 0.5 * dt_long * f_slow
END DO
```

A program illustrating this for a chain molecule, with strong intramolecular springs giving rise to the 'fast' forces, is provided online in md_chain_mts_lj.f90 with the force routines in the same file md_chain_lj_module.f90 used in Code 3.5. As usual, routines for input/output and simulation averages are provided in utility modules (see Appendix A), while code to set up an initial configuration is provided in initialize.f90.

```
! md_chain_mts_lj.f90
! Molecular dynamics, multiple timesteps, chain molecule
PROGRAM md_chain_mts_lj
```

then ω_E may be obtained by averaging over all the molecules in the system. Otherwise, an approximate calculation may be made by considering a hypothetical solid phase of the same density as the system of interest. As can be deduced from the analysis of the harmonic oscillator in Section 3.2.2, the quantity ζ in eqn (3.23) switches from positive to negative when $\omega \delta t = 2$. This causes the trajectories for this model system to change from ellipses in the phase plane (Fig. 3.3) to hyperbolae, which diverge from the origin. Needless to say, the trajectories will become noticeably unphysical well before this limit is reached. In a simulation, too large a timestep will generate high velocities and atomic overlaps, with disastrous results: most simulation packages will test for this and stop before numerical overflows occur.

Although symplectic algorithms should generate no systematic drift in energy, there is a loophole in the preceding analysis: the connection between $\mathcal{H}^{\ddagger}$ and $\mathcal{H}$ relies on

Example 3.1 Protein folding on a special-purpose machine

Anton is a massively parallel computer that has been specifically designed to run MD calculations (Shaw et al., 2009). It consists of a set of 512 nodes connected in a toroidal topology. Each node has two sub-systems. The first consists of an array of 32 special pairwise point interaction pipelines (PPIPS) for calculating the short range interactions for a variety of potentials. The second flexible sub-system contains eight geometry cores for fast numerical calculations, four processors to control overall data flow in the system, and four data-transfer processors that allow the communication to be hidden behind the computation. The inner loops of a standard MD package such as GROMACS can be mapped onto Anton's PPIPS. A new Gaussian-split Ewald method (Shan et al., 2005) was specifically developed to handle the long-range part of the electrostatic interactions on this machine.

Once constructed and tested against results from more general-purpose computers, Anton was used to study the folding of two proteins in water (Shaw et al., 2010): the first of these, a villin fragment (m = 4.07 kDa) is from an actin binding protein; the second, FIP35 (m = 4.14 kDa) is a 35-residue mutant of the human pin1 protein. The starting configuration for both simulations was an extended protein surrounded by 4000 TIP3P water molecules and a small number of Na^+ and Cl^- ions at 30 mM ionic concentration. At a temperature of 300 K the villin folded to its natural state in 68 µs. At 337 K, FIP35 folded in 38 µs. Both simulations were run for a further 20 µs to confirm the final structures. These very long simulations were performed using constant-NVT MD with a Nosé–Hoover thermostat and a multiple-timestep approach: δt = 2.5 fs for the bonded and short-range interactions, and Δt = 5.0 fs for the long-range electrostatic forces. The simulated structures of the folded proteins show a backbone RMS deviation of ~ 0.1 nm with respect to the X-ray crystal structures. The MD allows us to follow the folding process in detail. Although the folding rates and native structures of these proteins are accurately described by the CHARMM and AMBER force fields used in the calculations, the simulated enthalpy of the folded state is often lower than experimental estimates (Piana et al., 2014). This is ascribed to a set of small, uncorrelated errors in the force-field parameters rather than to the overall structure of the force field itself. Subsequently, other fast-folding proteins were studied (Lindorff-Larsen et al., 2011), with simulations reaching milliseconds in length.

Protein folding has long been a stimulus for many aspects of MD simulation (for reviews see Lane et al., 2013; Towse and Daggett, 2015). It would be wrong to think that special-purpose hardware is the only solution: conventional parallel computers may still be the most convenient, accessible, or cost-effective route for many research groups. Advanced simulation techniques may greatly accelerate the sampling (see e.g. Du and Bolhuis, 2014; Miao et al., 2015; Pan et al., 2016) and we shall cover some of these in later chapters. Nonetheless, the design and construction of Anton, and its further development (Shaw et al., 2014), are noteworthy in the field.

the differentiability of the Hamiltonian. We usually employ a cutoff which introduces discontinuities in the potential or its higher derivatives (Engle et al., 2005), which can cause a drift. Depending on the nature of the discontinuity (i.e. in which derivative of the potential it occurs) and its magnitude, this may or may not be a small effect. It has been suggested that at least the potential, and its first four derivatives should be continuous at the cutoff for long-time energy conservation in practice (Toxvaerd et al., 2012).

A slow upward drift of energy may also indicate a program error. Effects with a 'physical' origin can be identified by the procedure outlined earlier, that is, duplicating a short run but using a larger number of smaller timesteps. If the drift as a function of simulation time is unchanged, then it is presumably connected with the system under study, whereas if it is substantially reduced, the method used to solve the equations of motion (possibly the size of the timestep) is responsible. In the category of program error, we should mention the possibility that the wrong quantity is being calculated. If the total energy varies significantly but the simulation is 'stable' in the sense that no inexorable climb in energy occurs, then the way in which the energy is calculated should be examined. Are potential and kinetic contributions added together correctly? Is the pairwise force (appearing in the double loop) in fact correctly derived from the potential? This last possibility may be tested by including statements that calculate the force on a given particle numerically, from the potential energy, by displacing it slightly in each of the three coordinate directions. The result may be compared with the analytical formula encoded in the program. We illustrate this for various examples in Appendix C. As emphasized earlier, although small fluctuations are permissible, it is essential to eliminate any traces of a drift in the total energy over periods of tens of thousands of timesteps, if the simulation is to probe the microcanonical ensemble correctly.

Rather than a slow drift, a very rapid, even catastrophic increase in energy may occur within the first few timesteps. There are two possibilities here: either a starting configuration with particle overlaps has been chosen (so that the intermolecular forces are unusually large) or there is a serious program error. The starting configuration may be tested simply by visualizing it, or writing out the coordinates and inspecting the numbers. Alternatively, particularly when the number of particles is large, statements may temporarily be incorporated into the force loop so as to test each of the pair separations and write out particle coordinates and identifiers whenever a very close pair is detected.

Tracking down a serious program error may be a difficult task. It is a favourite mistake, particularly when reading in the potential parameters in real (e.g. SI) units, to make a small, but disastrous, error in unit conversion. There is much to be said for testing out a program on a small number of particles before tackling the full-size system, but beware! Is the potential cutoff distance still smaller than half of the box length? Frequent program errors involve mismatching of number, length, or type of variables passed between routines. Simple typographical errors, while hard to spot, may have far-reaching effects. It is hard to overemphasize how useful modern software development tools can be in locating and eliminating mistakes of this kind. A good editor may be used to check the source code much more efficiently than simple visual inspection. Most Fortran compilers will detect mismatched variables, provided the program uses modules and/or interface blocks in a consistent way. On modern computers, excellent interactive Fortran debugging facilities exist, which allow the program to be run under user control, with constant monitoring of the program flow and the values of variables of interest. Needless to say, a program written in a simple, logical, and modular fashion will be easier to debug (and will contain fewer errors!) than one which has not been planned in this way. Some programming considerations appear in Appendix A.

For molecular simulations, errors may creep into the program more easily than in the simple atomic case. Energy should be conserved just as for atomic simulations although, for small molecules, a rather short timestep may be needed to achieve this, since rotational motion occurs so rapidly. If non-conservation is a problem, several points may need checking. Incorrectly differentiating the potential on the way to the torques may be a source of error: this is more complicated for potentials incorporating multipolar terms (see Appendix C). Again, this may be tested numerically by subjecting selected molecules to small rotations and observing the change in potential energy. If the angular part of the motion is suspect, the rest of the program may be tested by 'freezing out' the rotation. This is accomplished by disengaging the rotational algorithm; physically this corresponds to giving the molecules an infinite moment of inertia and zero angular velocity. Energy should still be conserved under these conditions. Conversely, the angular motion may be tested out by omitting, temporarily, the translational algorithm, thus fixing the molecular centres at their initial positions.

Two final points should be made. When the program appears to be running correctly, the user should check that the monitored quantities are in fact evolving in time. Even conserved variables will fluctuate a little if only due to round-off errors, and any quantity that appears to be constant to ten significant figures should be regarded with suspicion: it is probably not being updated at all. Excellent conservation, but no science, will result from a program that does not, in fact, move the particles for any reason. A timestep that is too small (or that has been accidentally set to zero) will be very wasteful of computer time, and the extent to which δt can be increased without prejudicing the stability of the simulation should be investigated. Finally, the problems just discussed are all 'mechanical' rather than 'thermodynamic'; that is, they are associated with the correct solution of the equations of motion. The quite separate question of attaining thermodynamic equilibrium will be discussed in Chapter 5. If a well-known system is being simulated (e.g. Lennard-Jones, soft-sphere potentials, etc.) then it is obviously sensible to compare the simulation output, when equilibrium has been attained, with the known thermodynamic properties reported in the literature.

3.7   Molecular dynamics of hard particles

The molecular dynamics of molecules interacting via hard potentials (i.e. discontinuous functions of distance) must be solved in a way which is qualitatively different from the molecular dynamics of soft bodies. Whenever the distance between two particles becomes equal to a point of discontinuity in the potential, then a 'collision' (in a broad sense) occurs: the particle velocities will change suddenly, in a specified manner, depending upon the particular model under study. Thus, the primary aim of a simulation program here is to locate the time, collision partners, and all impact parameters, for every collision occurring in the system, in chronological order. Instead of a regular, step-by-step, approach, as for soft potentials, hard potential programs evolve on a collision-by-collision basis, computing the collision dynamics and then searching for the next collision. The general scheme may be summarized as follows:

(a) locate next collision;
(b) move all particles forward until collision occurs;
(c) implement collision dynamics for the colliding pair;
(d) calculate collisional properties, ready for averaging, before returning to (a).

Because of the need to locate accurately future collision times, simulations have been restricted in the main to systems in which force-free motion occurs between collisions. In the simple case of spherical particles such as hard spheres and square wells (Alder and Wainwright, 1959; 1960), location of the time of collision between any two particles requires the solution of a quadratic equation. We examine this in detail in the next section. The computational problems become more daunting when the hard cores are supplemented with long-range soft potentials. An example is the primitive model of electrolytes, consisting of hard spheres plus Coulomb interactions. By contrast, such systems may be handled easily using Monte Carlo simulation (see Chapter 4). However, it is possible to treat these 'hybrid' hard-plus-soft systems, as well as non-spherical hard particles, by returning to an approximate 'step-by-step' approach; we consider this briefly in Section 3.7.2.

Although event-driven simulations of this kind are not as widespread as those using continuous potentials, the DYNAMO package provides a platform for carrying out a range of simulations using such models (Bannerman et al., 2011).

3.7.1 Hard spheres

A program to solve hard-sphere molecular dynamics has two functions to perform: the calculation of collision times and the implementation of collision dynamics. We illustrate the methods using a simple program in Code 3.7.

The collision time calculation is the expensive part of the program, since, in principle, all possible collisions between distinct pairs must be considered. Consider two spheres, i and j, of diameter σ, whose positions at time t are $\mathbf{r}_i$ and $\mathbf{r}_j$, and whose velocities are $\mathbf{v}_i$ and $\mathbf{v}_j$. If these particles are to collide at time $t + t_{ij}$ then the following equation will be satisfied:

$$\left| \mathbf{r}_{ij}(t + t_{ij}) \right| = \left| \mathbf{r}_{ij} + \mathbf{v}_{ij} t_{ij} \right| = \sigma \tag{3.57}$$

where $\mathbf{r}_{ij} = \mathbf{r}_i - \mathbf{r}_j$ and $\mathbf{v}_{ij} = \mathbf{v}_i - \mathbf{v}_j$. If we define $b_{ij} = \mathbf{r}_{ij} \cdot \mathbf{v}_{ij}$, then this equation becomes

$$v_{ij}^2 t_{ij}^2 + 2b_{ij} t_{ij} + r_{ij}^2 - \sigma^2 = 0. \tag{3.58}$$

This is a quadratic equation in t_{ij}. If $b_{ij} > 0$, then the molecules are going away from each other and they will not collide. If $b_{ij} < 0$, it may still be true that $b_{ij}^2 - v_{ij}^2(r_{ij}^2 - \sigma^2) < 0$, in which case eqn (3.58) has complex roots and again no collision occurs. Otherwise (assuming that the spheres are not already overlapping) two positive roots arise, the smaller of which corresponds to impact

$$t_{ij} = \frac{-b_{ij} - \left(b_{ij}^2 - v_{ij}^2 (r_{ij}^2 - \sigma^2) \right)^{1/2}}{v_{ij}^2}. \tag{3.59}$$

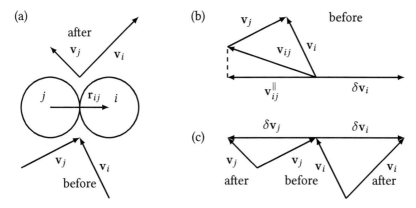

Fig. 3.7 A smooth hard-sphere collision. For illustrative purposes, we have taken all the vectors to be coplanar. We assume equal masses. (a) Collision geometry, and velocities before and after collision. (b) The vector construction gives the change in velocities $\delta\mathbf{v}_i = -\delta\mathbf{v}_j$ for each particle. (c) Relation between velocities before and after collision.

The program should store, in a convenient way, the earliest upcoming collision involving each atom. Methods for handling collision lists can be quite sophisticated (Rapaport, 1980; Bannerman et al., 2011) but a simple approach is to identify the collision by the index of the first particle, i, and store the collision time in an array element `coltime(i)`, and the collision partner, j, in another array `partner(i)`. In principle, the search need only consider distinct pairs with $i < j$ (except as mentioned later in this section), as in a conventional force loop. Also, for a reasonable liquid density, we may assume that we only need to examine the nearest images of any two particles in order to pick out the collision between them. Relaxing this assumption makes the simulation program a little more complicated; it should only break down at very low densities.

The next stage of the program is to locate the earliest collision time t_{ij}, and the colliding pair i and j; this may be accomplished with the Fortran `MINLOC` function. All molecules are moved forward by the time t_{ij}, the periodic boundary conditions are applied, and the table of future collision times is adjusted accordingly.

Now we are ready to carry through the second part of the calculation, namely the collision dynamics themselves. The changes in velocities of the colliding pair are completely dictated by the requirements that energy and linear momentum be conserved and (for smooth hard spheres) that the impulse act along the line of centres, as shown in Fig. 3.7.

Using conservation of total linear momentum and (kinetic) energy, and assuming equal masses, the velocity change $\delta\mathbf{v}_i$, such that

$$\mathbf{v}_i(\text{after}) = \mathbf{v}_i(\text{before}) + \delta\mathbf{v}_i, \quad \mathbf{v}_j(\text{after}) = \mathbf{v}_j(\text{before}) - \delta\mathbf{v}_i, \tag{3.60}$$

is given by

$$\delta\mathbf{v}_i = -(b_{ij}/\sigma^2)\mathbf{r}_{ij} = -\mathbf{v}_{ij}^{\parallel} \tag{3.61}$$

with $b_{ij} = \mathbf{r}_{ij} \cdot \mathbf{v}_{ij}$ evaluated now at the moment of impact (it is still a negative number). Thus, $\delta\mathbf{v}_i$ is simply the negative of the projection of $\mathbf{v}_{ij}$ along the $\mathbf{r}_{ij}$ direction, which we

Code 3.7 Molecular dynamics of hard spheres

These files are provided online. The program md_nve_hs.f90 controls the simulation, reads in the run parameters, implements the free-flight dynamics, and writes out the results. It uses the routines in md_nve_hs_module.f90 to update the collision lists and implement the collision dynamics, and utility module routines (see Appendix A) for input/output and simulation averages.

```
! md_nve_hs.f90
! Molecular dynamics, NVE ensemble, hard spheres
PROGRAM md_nve_hs
```

```
! md_nve_hs_module.f90
! Collisions and overlap for MD of hard spheres
MODULE md_module
```

denote $\mathbf{v}_{ij}^{\|}$ (see Fig. 3.7). The code for the collision dynamics is simply a transcription of eqn (3.61) followed by eqn (3.60).

Now we could return to the initial loop and recalculate all collision times afresh. In fact, there is no need to carry out this calculation in entirety, since many of the details in coltime and partner will have been unaffected by the collision between i and j. Obviously, we must look for the next collision partners of i and j; to be certain of finding them, we must examine all other atoms (with indices greater than *or* less than i and j respectively). Also we have to discover the fate of any other atoms which *were* due to collide with i and j, had these two not met each other first; for this it is sufficient to search only for partners with (say) higher indices, as before. Apart from these, the information in our collision lists is still quite valid. The 'update' procedure is given in Code 3.7. Following this, the smallest time in coltime is located, the particles are moved on, and the whole procedure is repeated.

The generalization of this program to the case of the square-well potential is straight-forward. Now, for each pair, there are two distances at which 'collisions' occur, so the algorithm for determining collision times is slightly more involved. Collisions at the inner sphere obey normal hard-sphere dynamics. At the outer boundary, where a finite change in potential energy occurs, the change in momentum is determined by the usual conservation laws. For molecules approaching each other, the potential energy drops on crossing the boundary, and so the kinetic energy shows a corresponding increase. If the molecules are within the well, and separating from each other as they approach the boundary, two possibilities arise. If the total kinetic energy is sufficient, the molecules cross the boundary with a loss in $\mathcal{K}$ to compensate the rise in $\mathcal{V}$, and continue separating more slowly. Alternatively, if $\mathcal{K}$ is insufficient, reflection at the outer boundary occurs: the particles remain 'bound' within the attractive well, and start approaching each other again.

We should mention one elegant approach to the model of a flexible chain of hard spheres (Rapaport, 1978; 1979; Bellemans et al., 1980) which once more reduces the complexity of a polyatomic simulation to the level of a simple atomic simulation. In the Rapaport model, the length of the bond between two adjacent atoms in the chain is not fixed, but is constrained to lie between two values σ and $\sigma + \delta\sigma$. Interactions between non-bonded atoms, and between atoms on different polymer molecules, are of the usual hard-sphere form. The spherical atoms undergo free flight between collisions that are of the usual kind: in fact the 'bonds' are no more than extreme examples of the square-well potential (with infinite walls on both sides of the well). By choosing $\delta\sigma$ to be small, the bond lengths may be constrained as closely as desired, at the expense (of course) of there being more 'bond collisions' per unit time. The model can be extended so that we can construct nearly rigid, as well as more complicated flexible molecules from the basic building blocks (Chapela et al., 1984).

More complicated potentials involving several 'steps' can be treated in the same way; a quite realistic potential can be constructed from a large number of vertical and horizontal segments, but of course the simulation becomes more expensive as more 'collisions' have to be dealt with per unit time (Chapela et al., 1984).

A further extension of the model, which preserves spherical symmetry, is the introduction of roughness. Rough spheres (Subramanian and Davis, 1975; O'Dell and Berne, 1975), differ from simple hard spheres only in their collision dynamics: the free-flight dynamics between collisions, and hence the techniques used to locate future collisions, are identical. They are characterized by a moment of inertia, and carry an angular or 'spin' velocity ω: the collision rules guarantee conservation of total energy, total linear momentum, and total angular momentum of the colliding pair about an arbitrarily chosen point. (We remind the reader, in passing, that periodic boundary conditions will destroy the conservation law for the total angular momentum of the whole system; see Chapter 1). Attempts have been made to introduce 'partial roughness' into the basic hard-sphere model (Lyklema, 1979a,b) but we shall not discuss them here.

3.7.2   Hard non-spherical bodies

For any non-spherical rigid-body model, calculating the collision point for two molecules, even in the case of free flight between collisions, is numerically more taxing than simply solving a quadratic equation. First, it is necessary to express the contact condition as a function of the orientations and relative positions of each candidate pair of molecules. Then, assuming free flight, expressions for these, in terms of the individual momenta and angular momenta, as well as the positions and orientations at the current time t, must be obtained. In favourable cases, the equation will take the form $\Phi(t_{ij}; \mathbf{r}_{ij}(t), \mathbf{e}_i(t), \mathbf{e}_j(t)) = 0$ at contact, with $\Phi > 0$ applying when the particles do not overlap, and $\Phi < 0$ indicating that the particles do overlap. This opens up the possibility of checking for overlap at regular intervals δt; on detecting overlap between a pair, it is possible to solve for the collision time numerically, knowing that the root of the equation is bracketed within the time interval: efficient methods exist for doing this, particularly if the time derivative $\dot{\Phi}$ can be written down analytically (Press et al., 2007). If several collisions are predicted to occur, then the earliest one is selected and dealt with; there may then be some recalculation of future collisions, as in the case of hard spheres. The method will miss a certain fraction of collisions, if two particles enter and then leave the overlap region within the course of one timestep. This approach has been applied to a range of hard particles (Rebertus and

Sando, 1977; Bellemans et al., 1980; Stratt et al., 1981; Allen and Imbierski, 1987; Allen et al., 1989; 1993). It is also possible to extend the approach to handle models in which there is never any overlap (Frenkel and Maguire, 1983; Allen and Cunningham, 1986), and hybrid systems of hard cores plus soft attractive (or repulsive) potentials (McNeil and Madden, 1982).

Checks on the working of a program which simulates hard molecular systems must include tests of the basic conservation laws on collision, and periodic examination of the configuration for unphysical overlaps. It is also sensible to conduct preliminary runs for any special cases of the molecular model (e.g. hard spheres) whose properties are well known.

3.8   Constant-temperature molecular dynamics

Quite often, we wish to conduct MD in the canonical ensemble, which means implementing a thermostat of some kind. In this section, we describe some of the approaches in common use, and discuss practical points in their implementation. Some of these approaches are discussed in the review of Hünenberger (2005). For some of these algorithms, a formal proof that they sample the desired ensemble requires ideas discussed in the next chapter, such as Markov chains.

All the methods involve modifying the equations of motion, whether by introducing stochastic terms or adding deterministic effects through constraints or additional dynamical variables. This naturally leads to the question: how will this affect the time correlation functions and related dynamical properties? The answer usually depends on the system studied, and the parameters of the method; a study of most of the methods discussed in this section (Basconi and Shirts, 2013) gives some guidance on this issue.

More seriously, the application of a thermostat does not guarantee that the system is properly at equilibrium. There are some celebrated examples of different degrees of freedom apparently having different temperatures (Harvey et al., 1998; Mor et al., 2008); this is sometimes due to poor choice of thermostat parameters, and in some cases the origins of the error are rather subtle (Eastwood et al., 2010).

3.8.1   Stochastic methods

A physical picture of a system corresponding to the canonical ensemble involves 'stray interactions' between the molecules of the system and the particles of a heat bath at a specified temperature (Tolman, 1938). This leads to a straightforward adaptation of the MD method due to Andersen (1980). At intervals, the velocity of a randomly selected molecule is chosen afresh from the Maxwell–Boltzmann distribution (see Appendix E). This corresponds to a collision with an imaginary heat-bath particle. The system follows standard energy-conserving dynamics in between collisions, jumping between constant-energy surfaces whenever a collision occurs. In this way, the system samples the canonical ensemble.

In the original description of the method (Andersen, 1980) times between collisions are chosen from a Poisson distribution with a specified mean collision time, but this detail does not affect the final phase-space distribution. A practical alternative, having specified a collision rate per particle, ν, is to refresh the velocity of every atom with probability $P = \nu \delta t$ at each timestep δt. If the collisions take place infrequently, energy fluctuations will occur slowly, but kinetic-energy (temperature) fluctuations will occur much as they do in conventional MD. If the collisions occur very frequently, then velocities and kinetic-energy fluctuations are dominated by them, rather than by the deterministic dynamics. Too high a collision rate will slow down the speed at which the molecules in the system explore configuration space, whereas too low a rate means that the canonical distribution of energies will only be sampled slowly. If it is intended that the system mimic a volume element in a liquid, in thermal contact with its surroundings, then Andersen suggests

$$\nu \propto \lambda_T \rho^{-1/3} N^{-2/3}$$

where λ_T is the thermal conductivity. Note that this decreases as the system size goes up; in other words the overall rate of collisions in the whole system increases slowly, $N\nu \propto N^{1/3}$. In suitable circumstances, the collisions have only a small effect on single-particle time correlation functions (Haile and Gupta, 1983) but too high a collision rate will lead to velocity correlation functions that decay exponentially in time (Evans and Morriss, 1984a). Whether or not the dynamics are of interest, it is sensible to compare the single-particle velocity autocorrelation functions with and without the thermostat, to assess its impact on particle motion and hence (one aspect of) the efficiency of phase space exploration.

The Andersen thermostat is simple to implement and requires no changes to the dynamical algorithm. It is relatively straightforward to combine it with constraint algorithms (Ryckaert and Ciccotti, 1986). For systems with strong intramolecular forces, such as bond stretches, it does not introduce any extra concerns regarding ergodicity, and indeed it may be quite effective at equilibrating such degrees of freedom. Variants exist, in which all the particle velocities are resampled, or scaled up and down, simultaneously (Andrea et al., 1983; Heyes, 1983b; Bussi et al., 2007), allowing a gentler perturbation of the trajectories.

An alternative approach is the Langevin thermostat, which also mimics the coupling of the system of interest to a thermal bath. Here, the equations of motion are modified in two ways: first, a 'random force' term is introduced: this is the stochastic element of the algorithm. Second, a deterministic 'frictional force' is added, proportional to particle velocities. The strength of these terms, and the prescribed temperature, are connected by the fluctuation–dissipation theorem. We discuss the Langevin equation in detail in Section 12.2. It can also be very effective in equilibrating stiff degrees of freedom, although the same care should be taken in choosing the friction constant as for the collision rate in the Andersen method. It is obviously sensible to conduct some preliminary tests, monitoring dynamical properties, to assess the effect of the thermostat.

A feature of the thermostats described so far is that they do not conserve the total momentum of the system, and so are not Galilean-invariant: the equations of motion would change if the underlying coordinate system were made to move at some constant velocity. One minor consequence of this is that the number of degrees of freedom associated with the velocities is $g = 3N$ instead of $g = 3(N-1)$ as in the usual MD ensemble. However, this non-conservation may become problematic in the study of fluid flow. To avoid this, a momentum-conserving 'pairwise' stochastic thermostat was proposed by Lowe (1999). It extends the idea of Andersen (1980), and is usually referred to as the Lowe–Andersen thermostat. At each timestep δt, after advancing the positions and momenta in the usual way, with a conventional velocity Verlet algorithm, pairs ij are examined in random order. For each pair, with probability $P = v\delta t$, the momenta are updated in a conservative way

$$\mathbf{p}'_i = \mathbf{p}_i + \delta\mathbf{p}_{ij}, \qquad \mathbf{p}'_j = \mathbf{p}_j - \delta\mathbf{p}_{ij},$$
$$\delta\mathbf{p}_{ij} = \tfrac{1}{2}m \left[\zeta_{ij}\sqrt{2k_\mathrm{B}T/m} - (\mathbf{v}_{ij} \cdot \hat{\mathbf{r}}_{ij}) \right] \hat{\mathbf{r}}_{ij}$$

where ζ_{ij} is a Gaussian random variable, with zero mean and unit variance. Just as in the Andersen case, the parameter v affects the properties of the fluid: high (low) v gives strong (weak) T control, and high (low) viscosity. Usually, the list of pairs would be restricted to those lying within a predefined separation of each other. The probability of selecting a given pair may include a separation-dependent weight function. The algorithm for selecting pairs in a random order is discussed in Appendix E.6. This thermostat was originally proposed as an alternative to the dissipative particle dynamics (DPD) method discussed in Section 12.4. The DPD algorithm itself may be used as a thermostat, being essentially a pairwise version of the Langevin equation.

3.8.2   Deterministic methods

Starting with a few key papers in the early 1980s (Andersen, 1980; Hoover et al., 1982; Evans, 1983; Nosé, 1984), there has been a tremendous amount of work on developing deterministic MD algorithms for sampling at constant temperature and pressure. In most cases, the process involves establishing modified equations of motion, for which the desired distribution is a stationary solution, usually assumed to be the equilibrium solution. The next stage is to develop a stable numerical algorithm to integrate these equations. As we have seen, a Liouville-operator-splitting approach is very attractive, since it provides such a direct link between the formal theory of the dynamics, and the numerical method. However, the situation is complicated because often the equations are not Hamiltonian, and do not have a symplectic structure. This has stimulated interest in the statistical mechanics of non-Hamiltonian systems (Tuckerman et al., 1999; 2001), as well as schemes for transforming the dynamical equations into a Hamiltonian form (Bond et al., 1999) which may be treated in the conventional way.

Controlling the temperature deterministically is most crudely done, within the MD algorithm, by rescaling the velocities at each step by a factor $\sqrt{T/\mathcal{T}}$ where $\mathcal{T}$ is the current kinetic temperature and T is the desired thermodynamic temperature. This is better regarded as the solution of the isokinetic equations of motion derived using Gauss' principle of least constraint (Hoover et al., 1982; Evans, 1983; Hoover, 1983b; Evans and Morriss, 1984a)

$$\dot{\mathbf{r}} = \mathbf{v} = \mathbf{p}/m \tag{3.62a}$$
$$\dot{\mathbf{p}} = \mathbf{f} - \xi(\mathbf{r}, \mathbf{p})\mathbf{p} \tag{3.62b}$$
$$\xi = \frac{\mathbf{p} \cdot \mathbf{f}/m}{\mathbf{p} \cdot \mathbf{p}/m}. \tag{3.62c}$$

We adopt the usual shorthand, $\mathbf{r} \equiv \{\mathbf{r}_i\}$, $\mathbf{p} \cdot \mathbf{f}/m \equiv \sum_i \mathbf{p}_i \cdot \mathbf{f}_i/m_i = \sum_i \mathbf{v}_i \cdot \mathbf{f}_i$, etc. The formula for the 'friction coefficient' ξ, which acts as a Lagrange multiplier enforcing the constraint, guarantees that $\dot{\mathcal{T}} = 0$. (Note, however, that this term is not derived in the same way as those of Section 3.4). Evans and Morriss (1983b) have shown that the ensemble distribution function is proportional to

$$\delta(\mathcal{K}(\mathbf{p}) - \mathcal{K}_0)\delta(\mathbf{P} - \mathbf{P}_0) \, \exp(-\mathcal{V}(\mathbf{r})/k_B \mathcal{T}) \tag{3.63}$$

where $\mathcal{K}_0$ and $\mathbf{P}_0$ are the initial (and conserved) kinetic energy and total momentum, respectively. Because of these four constraints, the connection between temperature and kinetic energy is

$$k_B \mathcal{T} = \frac{2\mathcal{K}_0}{g}, \quad \text{where} \quad g = 3N - 4$$

for an atomic system in 3D. Because of the factorization of eqn (3.63) into $\mathbf{p}$- and $\mathbf{r}$-dependent parts, we see that the configurational distribution is canonical. Incidentally, although it is not obvious from eqn (3.63), the momentum distribution is still very close to Maxwellian for a many-particle system, just as it is in the microcanonical ensemble.

What algorithm shall we use to solve eqn (3.62)? Based on the statistical mechanics of non-Hamiltonian systems (Tuckerman et al., 1999; 2001), Minary et al. (2003) have proposed a variety of algorithms and revisit the operator-splitting approach due to Zhang (1997), which we describe here. This uses the same kind of decomposition as eqn (3.18): $iL = iL_1 + iL_2$. The first component corresponds to the equations

$$\begin{pmatrix} \dot{\mathbf{r}} \\ \dot{\mathbf{p}} \end{pmatrix} = iL_1 \begin{pmatrix} \mathbf{r} \\ \mathbf{p} \end{pmatrix} = \mathbf{v} \cdot \frac{\partial}{\partial \mathbf{r}} \begin{pmatrix} \mathbf{r} \\ \mathbf{p} \end{pmatrix} = \begin{pmatrix} \mathbf{v} \\ \mathbf{0} \end{pmatrix}$$

and so advances the positions at constant momenta

$$U_1(t)\mathbf{r} = \exp\big(iL_1 t\big)\mathbf{r} = \mathbf{r} + \mathbf{v}\,t,$$

leaving $\mathbf{p}$ unchanged. The second component corresponds to the equations

$$\begin{pmatrix} \dot{\mathbf{r}} \\ \dot{\mathbf{p}} \end{pmatrix} = iL_2 \begin{pmatrix} \mathbf{r} \\ \mathbf{p} \end{pmatrix} = \big(\mathbf{f} - \xi(\mathbf{r}, \mathbf{p})\mathbf{p}\big) \cdot \frac{\partial}{\partial \mathbf{p}} \begin{pmatrix} \mathbf{r} \\ \mathbf{p} \end{pmatrix} = \begin{pmatrix} \mathbf{0} \\ \mathbf{f} - \xi(\mathbf{r}, \mathbf{p})\mathbf{p} \end{pmatrix}.$$

The equation for $\dot{\mathbf{p}}$ is exactly soluble: the positions $\mathbf{r}$ are fixed, so the forces $\mathbf{f}$ are constant, as is the denominator in the definition of ξ. The solution may be written (Zhang, 1997; Minary et al., 2003) in terms of two (related) functions of time

$$\alpha(t) = \cosh \omega_0 t + \xi_0 t \, \frac{\sinh \omega_0 t}{\omega_0 t} \qquad \approx 1 + \xi_0 t + \tfrac{1}{2}\big(1 + \tfrac{1}{3}\xi_0 t\big)\omega_0^2 t^2,$$

$$\beta(t) = \frac{\sinh \omega_0 t}{\omega_0 t} + \xi_0 t \, \frac{\cosh \omega_0 t - 1}{\omega_0^2 t^2} \qquad \approx 1 + \tfrac{1}{2}\xi_0 t + \tfrac{1}{6}\big(1 + \tfrac{1}{4}\xi_0 t\big)\omega_0^2 t^2,$$

where the quantities

$$\omega_0^2 = \frac{\mathbf{f} \cdot \mathbf{f}/m}{\mathbf{p} \cdot \mathbf{p}/m} \quad \text{and} \quad \xi_0 = \frac{\mathbf{p} \cdot \mathbf{f}/m}{\mathbf{p} \cdot \mathbf{p}/m}$$

are evaluated at the start of the U_2 propagation step. To avoid numerical inaccuracy at small $\omega_0 t$, the indicated expansions can be used. The resulting propagator is

$$U_2(t)\mathbf{p} = \exp\big(iL_2 t\big)\mathbf{p} = \frac{\mathbf{p} + \beta(t)\mathbf{f}\,t}{\alpha(t)},$$

leaving **r** unchanged. The propagators are combined in the familiar pattern

$$U(\delta t) = \exp(iL\delta t) \approx U_2\left(\tfrac{1}{2}\delta t\right) U_1\left(\delta t\right) U_2\left(\tfrac{1}{2}\delta t\right)$$

and this translates straightforwardly into the same kind of kick–drift–kick code as the velocity Verlet algorithm. The result is (by design) time-reversible; since the U_2 steps exactly conserve the kinetic energy, so does the overall algorithm. Minary et al. (2003) discuss multiple-timestep and constraint versions of the method. An extension to include rigid body rotation has been presented by Terada and Kidera (2002). There is also a different approach, using a transformation into Hamiltonian form (Dettmann and Morriss, 1996) in which case the standard velocity Verlet method can be used.

A second way to treat the dynamics of a system in contact with a thermal bath is to include one or more degrees of freedom which represent the reservoir, and carry out a simulation of this 'extended system'. Energy is allowed to flow dynamically from the reservoir to the system and back; the reservoir has a certain 'thermal inertia' Q associated with it, and the whole technique is rather like controlling the volume of a sample by using a piston (see Section 3.9). Nosé (1984) proposed a method for doing this, based on a Hamiltonian for an extended system including an additional coordinate s, and associated momentum p_s, together with the prescribed temperature T as a parameter. He demonstrated that the microcanonical equations of motion for the extended system would generate a canonical distribution of the variables **r** and **p**/s. This means that a slightly awkward time scaling, which varies during the simulation, must be applied to convert the results back to the physical set of variables **r** and **p**. Most commonly nowadays, the modified equations due to Hoover (1985) are used instead, and are generally referred to as the Nosé–Hoover equations:

$$\dot{\mathbf{r}} = \mathbf{p}/m, \qquad\qquad\qquad \dot{\mathbf{p}} = \mathbf{f} - (p_\eta/Q)\mathbf{p}, \qquad\qquad (3.64a)$$

$$\dot{\eta} = p_\eta/Q, \qquad\qquad\qquad \dot{p}_\eta = \mathbf{p}\cdot\mathbf{p}/m - gk_BT. \qquad\qquad (3.64b)$$

Here, T is the prescribed temperature and g the number of degrees of freedom; typically $g = 3N - 3$. The combination $p_\eta/Q = \dot{\eta} \equiv \xi$ may be recognized as a dynamical friction coefficient: its equation of motion drives it towards higher values if the system is too hot, and lower (possibly negative) values if too cold.

It is important to realize that these equations are non-Hamiltonian. Nonetheless, they do conserve an energy-like variable

$$\mathcal{H} = \mathcal{K}(\mathbf{p}) + \mathcal{V}(\mathbf{r}) + \frac{p_\eta^2}{2Q} + gk_BT\,\eta. \qquad\qquad (3.65)$$

The variable η is only of interest inasmuch as it appears in this quantity, and hence can be used to check the correct solution of the dynamics. The stationary distribution is (Hoover, 1985; Nosé, 1991)

$$\exp\left(-\mathcal{V}(\mathbf{r})/k_BT\right) \exp\left(-\mathcal{K}(\mathbf{p})/k_BT\right) \exp\left(-p_\eta^2/2Qk_BT\right)$$

that is, it is canonical, at temperature T, in the variables **r**, **p**.

The operator-splitting approach, similar to that seen before, may be applied to the Nosé–Hoover equations (Tuckerman et al., 1992; Martyna et al., 1996). Although they are non-Hamiltonian, it turns out to be possible to formulate such an algorithm so as to conserve the appropriate invariant measure or phase space volume (Ishida and Kidera, 1998; Sergi and Ferrario, 2001; Legoll and Monneau, 2002; Ezra, 2006). Here we describe one of these schemes. The splitting is

$$iL = iL_1 + iL_2 + iL_3 + iL_4$$

where

$$iL_1 = \mathbf{v} \cdot \frac{\partial}{\partial \mathbf{r}}, \qquad\qquad iL_2 = \mathbf{f} \cdot \frac{\partial}{\partial \mathbf{p}},$$

$$iL_3 = \xi \frac{\partial}{\partial \eta} - \xi \mathbf{p} \cdot \frac{\partial}{\partial \mathbf{p}}, \qquad iL_4 = \left(\mathbf{p} \cdot \mathbf{p}/m - g k_B T\right)\frac{\partial}{\partial p_\eta},$$

where again $\xi = p_\eta/Q$. The effect of each of the propagators is as follows (for simplicity we only show the action on those coordinates and momenta that are changed):

$$U_1(t)\mathbf{r} = \exp(iL_1 t)\mathbf{r} \quad = \mathbf{r} + \mathbf{v}\,t$$

$$U_2(t)\mathbf{p} = \exp(iL_2 t)\mathbf{p} \quad = \mathbf{p} + \mathbf{f}\,t$$

$$U_3(t)\begin{pmatrix}\eta\\\mathbf{p}\end{pmatrix} = \exp(iL_3 t)\begin{pmatrix}\eta\\\mathbf{p}\end{pmatrix} = \begin{pmatrix}\eta + \xi\,t\\\mathbf{p}\exp\!\left(-\xi t\right)\end{pmatrix}$$

$$U_4(t)p_\eta = \exp(iL_4 t)p_\eta \quad = p_\eta + \left(\mathbf{p} \cdot \mathbf{p}/m - g k_B T\right)t.$$

Notice that the two parts of the iL_3 operator commute with each other, so the order in which the corresponding propagators are implemented is unimportant.

One simple approach (Martyna et al., 1996) is to split the propagator into non-thermostat and thermostat parts, and further split each of these into its two components:

$$\exp(iL\delta t) = U(\delta t) \approx U_{3\&4}\!\left(\tfrac{1}{2}\delta t\right) U_{1\&2}\!\left(\delta t\right) U_{3\&4}\!\left(\tfrac{1}{2}\delta t\right) \tag{3.66a}$$

$$U_{1\&2}\!\left(\delta t\right) = \exp\!\left((iL_1 + iL_2)\delta t\right) \approx U_2\!\left(\tfrac{1}{2}\delta t\right) U_1\!\left(\delta t\right) U_2\!\left(\tfrac{1}{2}\delta t\right) \tag{3.66b}$$

$$U_{3\&4}\!\left(\tfrac{1}{2}\delta t\right) = \exp\!\left((iL_3 + iL_4)\tfrac{1}{2}\delta t\right) \approx U_4\!\left(\tfrac{1}{4}\delta t\right) U_3\!\left(\tfrac{1}{2}\delta t\right) U_4\!\left(\tfrac{1}{4}\delta t\right). \tag{3.66c}$$

The algorithm consists of a sequence of nine steps: the last three are just a repeat of the first three, updating the thermostat variables, while the middle three are the usual (constant-energy) velocity Verlet algorithm. However, this is not the only choice. A quite reasonable alternative ordering (Itoh et al., 2013) is

$$U(\delta t) \approx U_4\!\left(\tfrac{1}{2}\delta t\right) U_3\!\left(\tfrac{1}{2}\delta t\right) U_2\!\left(\tfrac{1}{2}\delta t\right) U_1\!\left(\delta t\right) U_2\!\left(\tfrac{1}{2}\delta t\right) U_3\!\left(\tfrac{1}{2}\delta t\right) U_4\!\left(\tfrac{1}{2}\delta t\right) \tag{3.67}$$

and it is also possible to group the U_4 step together with the central U_1 step (since the corresponding Liouville operators commute, they can both be implemented over the full δt). Itoh et al. (2013) have discussed the consequences of applying the component parts in different orders. As for the velocity Verlet algorithm, each individual step may be expressed in exact form, which is easily translated into computer code, as illustrated in Code 3.8.

Code 3.8 Measure-preserving constant-*NVT* MD algorithm

These files are provided online. The program `md_nvt_lj.f90` controls the simulation, reads in the run parameters, implements an algorithm for the Nosé–Hoover equations, and writes out the results. Routines in the file `md_lj_module.f90` (see Code 3.4) are used for the forces, and various utility module routines (see Appendix A) handle input/output and simulation averages.

```
! md_nvt_lj.f90
! Molecular dynamics , NVT ensemble
PROGRAM md_nvt_lj
```

It is also possible to use a different splitting, for example (Ezra, 2006)

$$\mathrm{i}L_1 = \mathbf{v}\cdot\left(\frac{\partial}{\partial\mathbf{r}} + \mathbf{p}\frac{\partial}{\partial p_\eta}\right), \qquad \mathrm{i}L_2 = \mathbf{f}\cdot\frac{\partial}{\partial\mathbf{p}},$$

$$\mathrm{i}L_3 = \xi\frac{\partial}{\partial\eta} - \xi\mathbf{p}\cdot\frac{\partial}{\partial\mathbf{p}}, \qquad \mathrm{i}L_4 = -gk_\mathrm{B}T\frac{\partial}{\partial p_\eta}.$$

Although this is not just a simple identification of terms in the equations of motion, each of the operators also preserves the key volume elements (Ezra, 2006). Finally, the list of possible algorithms grows enormously once one considers the possibility of higher-order decompositions (Ishida and Kidera, 1998).

The thermal inertia Q governs the rate of flow of energy between the physical system and the reservoir: in the limit $Q \to \infty$ we regain conventional MD, while too low a value will result in long-lived, weakly damped, oscillations of the energy. Q has units of energy $\times$ time^2. Rewriting eqn (3.64b) in the form

$$\dot{\xi} = \frac{\dot{p}_\eta}{Q} = \frac{gk_\mathrm{B}T}{Q}\left(\frac{\mathbf{p}\cdot\mathbf{p}/m}{gk_\mathrm{B}T} - 1\right)$$

shows that the combination $Q/gk_\mathrm{B}T \equiv \tau^2$ gives the square of a characteristic timescale, and Martyna et al. (1992) suggest matching τ, roughly, to the physical timescales of motions in the system to achieve the most effective thermostatting. However, clearly an empirical approach, varying Q and observing the effects on the system of interest, is advisable.

A possible deficiency of the Nosé–Hoover equations is their lack of ergodicity, or at least very slow sampling of the full equilibrium distribution, under certain circumstances. This is dramatically clear for a single harmonic oscillator (Hoover, 1985), but can occur for other small and/or stiff systems. A comparative example is the simulation of a single butane molecule (D'Alessandro et al., 2002), where the simple Gaussian isokinetic method is demonstrably better. To improve this situation, a 'chain' of thermostatting variables may be introduced (Martyna et al., 1992). The equations of motion for M such variables are written

$$\dot{\mathbf{r}} = \mathbf{p}/m, \tag{3.68a}$$

$$\dot{\mathbf{p}} = \mathbf{f} - \left(\frac{p_{\eta_1}}{Q_1}\right)\mathbf{p}, \tag{3.68b}$$

$$\dot{\eta}_j = \left(\frac{p_{\eta_j}}{Q_j}\right), \qquad j = 1, \ldots, M, \tag{3.68c}$$

$$\dot{p}_{\eta_1} = G_1 - \left(\frac{p_{\eta_2}}{Q_2}\right)p_{\eta_1}, \tag{3.68d}$$

$$\dot{p}_{\eta_j} = G_j - \left(\frac{p_{\eta_{j+1}}}{Q_{j+1}}\right)p_{\eta_j}, \qquad j = 2, \ldots, M-1, \tag{3.68e}$$

$$\dot{p}_{\eta_M} = G_M, \tag{3.68f}$$

where for brevity we have introduced the driving forces

$$G_1 = G_1(\mathbf{p}) = \mathbf{p} \cdot \mathbf{p}/m - gk_{\mathrm{B}}T, \tag{3.69a}$$

$$G_j = G_j(p_{\eta_{j-1}}) = \frac{p_{\eta_{j-1}}^2}{Q_{j-1}} - k_{\mathrm{B}}T, \qquad j > 1. \tag{3.69b}$$

The conserved quantity is now

$$\mathcal{H} = \mathcal{K}(\mathbf{p}) + \mathcal{V}(\mathbf{r}) + \sum_{j=1}^{M} \frac{p_{\eta_j}^2}{2Q_j} + gk_{\mathrm{B}}T\,\eta_1 + \sum_{j=2}^{M} k_{\mathrm{B}}T\,\eta_j. \tag{3.70}$$

and the equilibrium distribution is

$$\exp\left(-\mathcal{V}(\mathbf{r})/k_{\mathrm{B}}T\right) \exp\left(-\mathcal{K}(\mathbf{p})/k_{\mathrm{B}}T\right) \prod_{j=1}^{M} \exp\left(-p_{\eta_j}^2/2Q_j k_{\mathrm{B}}T\right).$$

The recommended thermostat masses reflect the expected timescale τ of fluctuations in the physical system: if we choose $Q_1 \approx gk_{\mathrm{B}}T\tau^2$, then it is sensible to make $Q_j \approx k_{\mathrm{B}}T\tau^2$, for $j = 2, \ldots, M$. As before, some experimentation with these parameters is advisable.

The additional variables add very little to the cost of the simulation, and can readily be incorporated into the integration algorithm (Martyna et al., 1996). The Liouville operator contains terms iL_1 and iL_2 defined as before, plus modified thermostat terms

$$iL_3 = \sum_{j=1}^{M} \xi_j \frac{\partial}{\partial \eta_j} - \xi_1 \mathbf{p} \cdot \frac{\partial}{\partial \mathbf{p}}, \tag{3.71a}$$

$$iL_4 = \sum_{j=1}^{M} iL_{4,j} \tag{3.71b}$$

$$iL_{4,j} = \begin{cases} \left(G_j - \xi_{j+1}p_{\eta_j}\right)\dfrac{\partial}{\partial p_{\eta_j}} & j < M \\[2ex] G_M \dfrac{\partial}{\partial p_{\eta_M}} & j = M \end{cases} \tag{3.71c}$$

where, again for brevity, we introduce $\xi_j = p_{\eta_j}/Q_j$. The propagator corresponding to $\mathrm{i}L_3$ acts in a similar way to before

$$U_3(t)\eta_j = \exp(\mathrm{i}L_3 t)\eta_j = \eta_j + \xi_j\, t \qquad j = 1, \ldots, M \tag{3.72a}$$

$$U_3(t)\mathbf{p} = \exp(\mathrm{i}L_3 t)\mathbf{p} = \mathbf{p}\exp\left(-\xi_1 t\right). \tag{3.72b}$$

All the different operators in $\mathrm{i}L_3$ commute with each other (the variables η_j have no effect on the other variables) so these updates can be conducted in any order. The propagator corresponding to each $\mathrm{i}L_{4,j}$ is exactly soluble in each case. For the case $j = M$ we have simply

$$U_{4,M}(t)p_{\eta_M} = \exp(\mathrm{i}L_4 t)p_{\eta_M} = p_{\eta_M} + G_M t \tag{3.73}$$

while for $j < M$

$$U_{4,j}(t)p_{\eta_j} = \exp(\mathrm{i}L_4 t)p_{\eta_j} = p_{\eta_j}\exp(-\xi_{j+1}t) + G_j t \left(\frac{1 - \exp(-\xi_{j+1}t)}{\xi_{j+1}t}\right) \tag{3.74a}$$

$$\approx p_{\eta_j}\exp(-\xi_{j+1}t) + G_j t\,\exp(-\xi_{j+1}t/2). \tag{3.74b}$$

Equation (3.74b) is an approximate form resulting from a further factorization of the propagator, or by considering a Taylor expansion, which is recommended at small $\xi_{j+1}t$ (Martyna et al., 1996). Finally, all these propagators may be put together as in eqns (3.66), but with (3.66c) taking the form

$$U_{3\&4}\left(\tfrac{1}{2}\delta t\right) \approx U_{4,M}\left(\tfrac{1}{4}\delta t\right) \cdots U_{4,1}\left(\tfrac{1}{4}\delta t\right) U_3\left(\tfrac{1}{2}\delta t\right) U_{4,1}\left(\tfrac{1}{4}\delta t\right) \cdots U_{4,M}\left(\tfrac{1}{4}\delta t\right). \tag{3.75}$$

Alternative patterns, based for instance on eqn (3.67), might also be possible. However, if the chain momenta vary quickly, a higher-order method may be needed: Tuckerman et al. (2006) recommend using a Suzuki–Yoshida decomposition of the thermostat propagator (essentially a multiple-timestep scheme) and then factorize each part (for details, see Tuckerman et al., 2006). There is no particular concern about the computational cost of this part, since the expensive force evaluation happens only once. In any case, the equations translate straightforwardly into computer code. Once more, alternative measure-preserving splittings of the Liouville operator for Nosé–Hoover chains are possible (Ezra, 2006).

Nosé–Hoover chains provide a flexible approach to thermostatting, in which the precise values of the inertia parameters Q_j are not thought to be critical to ensuring good sampling. However, this is clearly system-dependent, and a mixture of different types of molecule may require some thought to be given to these parameters. A further option is to apply different thermostat chains to different species (to reflect, for example, their different masses). Once more, the aim is to equilibrate all the degrees of freedom at approximately the same rate; however, a poor or unlucky choice of parameters can give rise to the artefacts mentioned at the start of Section 3.8: an observation of solvent and solute having different temperatures, for instance, might indicate a problem of this kind.

As mentioned, the Nosé–Hoover equations are not Hamiltonian, and hence do not have a symplectic structure. An alternative to the approach of this section is to transform the Nosé–Hoover equations, including reparameterizing the time variable, so as to give a truly Hamiltonian form (Bond et al., 1999). In this case, a symplectic algorithm may be applied. This is usually called the Nosé–Poincaré approach. It is possible to extend this algorithm, as well, by using a chain of thermostats. This is explained, along with a more general discussion of the role of the thermal inertia, by Leimkuhler and Sweet (2004; 2005). Okumura et al. (2007) have combined Nosé–Poincaré with the symplectic integrator for rotational motion due to Miller et al. (2002). A method to improve conservation in both Nosé–Hoover and Nosé–Poincaré dynamics has been proposed by Okumura et al. (2014).

Because of the way the momenta are scaled by the thermostatting, the Nosé–Hoover equations are not Galilean-invariant, and they only conserve the total momentum if it is set to zero initially. Actually, the situation is rather more complicated: if the total momentum is not zero, the equations do not generate the canonical ensemble (Cho et al., 1993). The reasons for this are discussed in detail by Tuckerman et al. (2001); the problem is not present for Nosé–Hoover chains (Martyna, 1994). In passing, we note that a Galilean-invariant 'pairwise' analogue of the Nosé–Hoover thermostat has also been derived (Allen and Schmid, 2007).

There are thermostats based on the configurational temperature (Braga and Travis, 2005; Travis and Braga, 2006; 2008; Pieprzyk et al., 2015), which are Galilean-invariant in the absence of external forces. An example (Braga and Travis, 2005) is

$$\dot{\mathbf{r}} = \mathbf{p}/m + \frac{p_\eta}{Q}\mathbf{f}, \qquad \dot{\mathbf{p}} = \mathbf{f}, \tag{3.76a}$$

$$\dot{\eta} = \frac{p_\eta}{Q}, \qquad \dot{p}_\eta = \boldsymbol{\nabla}\mathcal{V} \cdot \boldsymbol{\nabla}\mathcal{V} - k_{\mathrm{B}}T\nabla^2\mathcal{V}. \tag{3.76b}$$

An extra momentum p_η appears, which is driven by the difference between two terms. The squared force $\boldsymbol{\nabla}\mathcal{V} \cdot \boldsymbol{\nabla}\mathcal{V} = \mathbf{f} \cdot \mathbf{f} = \sum_{i\alpha} f_{i\alpha}^2$, and the Laplacian $\nabla^2\mathcal{V} = \sum_{i\alpha} \partial^2\mathcal{V}/\partial r_{i\alpha}^2$, are sums over atoms and Cartesian components of the terms which appear in the definition of the configurational temperature, eqn (2.56). Q as usual represents a thermostat mass, and the desired temperature is T; Braga and Travis (2005) show that the canonical ensemble is generated by these equations and present an integration scheme based on Liouville operator splitting.

Our final deterministic method for temperature control is the 'weak coupling' algorithm of Berendsen et al. (1984), more usually termed the 'Berendsen thermostat'. At each timestep, momenta are scaled as follows

$$\mathbf{p}' = \mathbf{p}\sqrt{1 + \frac{\delta t}{\tau}\left(\frac{T}{\mathcal{T}} - 1\right)} \tag{3.77}$$

where τ is a preset time constant. The aim is to make the temperature relax towards the desired value, with the prescribed time constant. As pointed out by Hoover (1985), this can be regarded as the solution of eqns (3.62a), (3.62b), with friction coefficient given by

$$\xi = \frac{1}{2\tau}\left(1 - \frac{T}{\mathcal{T}}\right).$$

This means that the equations of motion are no longer time-reversible. Also, the distribution is not canonical, and in fact depends on the parameter τ: an analysis due to Morishita (2000) suggests that it tends to isokinetic as $\tau \to 0$ and microcanonical as $\tau \to \infty$. Hence, fluctuations in temperature can never be as large as those in the canonical ensemble. The Berendsen thermostat is quite widely used, but its non-canonical nature must be borne in mind.

3.9   Constant-pressure molecular dynamics

To simulate a system at a prescribed pressure, it is inevitable that the system box must change its volume. Andersen (1980) originally proposed a method which included V as a dynamical variable, associated with an inertia, or 'piston mass', W, and an additional potential-energy term PV. Equations of motion, employing reduced positional variables, were obtained from a Lagrangian: there is a conserved, extended Hamiltonian, and the trajectories were shown to sample the constant-NPH (isobaric–isoenthalpic) ensemble. Combination with one of the thermostats of the previous section would generate the isothermal–isobaric NPT ensemble.

These equations were reformulated in real-space variables by Hoover (1985; 1986), Melchionna et al. (1993), and Martyna et al. (1994), amongst others. As discussed in detail by Tuckerman et al. (2001), only the last formulation generates the correct ensemble in all circumstances (although the others generate a closely related distribution, if the total momentum is set to zero). The NPT equations of Martyna et al. (1994), in d dimensions, are

$$\dot{\mathbf{r}} = \mathbf{p}/m + \left(\frac{p_\varepsilon}{W}\right)\mathbf{r}, \tag{3.78a}$$

$$\dot{\mathbf{p}} = \mathbf{f} - \alpha\left(\frac{p_\varepsilon}{W}\right)\mathbf{p} - \left(\frac{p_{\eta_1}}{Q_1}\right)\mathbf{p}, \tag{3.78b}$$

$$\dot{V} = d\left(\frac{p_\varepsilon}{W}\right)V \quad \text{or} \quad \dot{\varepsilon} = \frac{p_\varepsilon}{W}, \tag{3.78c}$$

$$\dot{p}_\varepsilon = dV(\mathcal{P}' - P) - \left(\frac{p_{\eta_1'}}{Q_1'}\right)p_\varepsilon, \tag{3.78d}$$

together with a thermostatting scheme which we will come to shortly. Here we have introduced

$$\alpha = 1 + \frac{d}{g} = 1 + \frac{1}{N}$$

where the last equation applies for a system with no constraints, when the number of degrees of freedom is $g = dN$. In these equations, the positions and (in a consistent way) the volume, are dynamically scaled by a factor depending on the new 'momentum' variable p_ε. This corresponds to a 'rate of strain'

$$\xi_\varepsilon = \frac{p_\varepsilon}{W} = \dot{\varepsilon} = \frac{d}{dt}\ln\left(V^{1/d}\right) = \frac{d}{dt}\ln L,$$

where the strain is $\quad \varepsilon(t) = \frac{1}{d}\ln\left(\frac{V(t)}{V(0)}\right) = \ln\left(\frac{L(t)}{L(0)}\right),$

if we specialize to the case of a cubic box with $V = L^d$. There is an associated 'piston mass' W, which can control the damping of the volume fluctuations; W has units of energy $\times$ time^2. The driving force for p_ε is the difference between the desired pressure P and the instantaneous pressure. We define this, purely for convenience in writing down the algorithm,

$$\mathcal{P}' = \mathcal{P} + (d/g)\mathbf{p} \cdot \mathbf{p}/m = \frac{1}{dV}\left(\alpha\mathbf{p} \cdot \mathbf{p}/m - \mathbf{r} \cdot \mathbf{f}\right) - \frac{\partial\mathcal{V}}{\partial V}.$$

Note the inclusion of a (small) extra kinetic term; also, any explicit volume dependence of the potential-energy function has been included. In the common case of periodic boundaries with no external forces, the $\mathbf{r} \cdot \mathbf{f}$ term should be re-expressed as usual in a translationally invariant form. Martyna et al. (1994) and Tuckerman (2010) discuss some subtle dependencies of the conserved variables and distributions on the conservation, or non-conservation, of total momentum.

Two Nosé–Hoover thermostat chains act on, respectively, the particle momenta and the barostat:

$$\dot{\eta}_j = \frac{p_{\eta_j}}{Q_j}, \qquad \dot{\eta}'_j = \frac{p_{\eta'_j}}{Q'_j}, \qquad j = 1, \ldots, M$$

$$\dot{p}_{\eta_j} = G_j - \left(\frac{p_{\eta_{j+1}}}{Q_{j+1}}\right)p_{\eta_j}, \qquad \dot{p}_{\eta'_j} = G'_j - \left(\frac{p_{\eta'_{j+1}}}{Q'_{j+1}}\right)p_{\eta'_j}, \qquad j = 1, \ldots, M-1$$

$$\dot{p}_{\eta_M} = G_M, \qquad \dot{p}_{\eta'_M} = G'_M.$$

The thermostat driving forces G_j are given by eqns (3.69), and the G'_j terms are similarly given by

$$G'_1 = G'_1(p_\varepsilon) = \frac{p_\varepsilon^2}{W} - k_\mathrm{B}T, \qquad\qquad (3.79a)$$

$$G'_j = G'_j(p_{\eta'_{j-1}}) = \frac{p_{\eta'_{j-1}}^2}{Q'_{j-1}} - k_\mathrm{B}T, \qquad j > 1. \qquad\qquad (3.79b)$$

The thermostat masses Q_j and Q'_j are chosen along the same lines as discussed previously; it is recommended to use two different thermostats, because the natural timescales associated with particle motions and volume fluctuations may be significantly different.

The equations of this section conserve the energy-like function

$$\mathcal{H} = \mathcal{K}(\mathbf{p}) + \mathcal{V}(\mathbf{r}) + PV + \frac{p_\varepsilon^2}{2W} + \sum_{j=1}^{M}\left(\frac{p_{\eta_j}^2}{2Q_j} + \frac{p_{\eta'_j}^2}{2Q'_j} + k_\mathrm{B}T\eta'_j\right) + gk_\mathrm{B}T\eta_1 + \sum_{j=2}^{M}k_\mathrm{B}T\eta_j.$$

They correctly generate the isothermal–isobaric distribution for $\mathbf{r}$, $\mathbf{p}$, and V, multiplied by Gaussian distributions in the physically uninteresting variables p_ε, p_{η_j} and $p_{\eta'_j}$. If the thermostatting is completely removed, the scheme generates the constant-NPH ensemble, apart from the small fluctuating term $p_\varepsilon^2/2W$.

An integration scheme for these equations of motion was originally proposed by Martyna et al. (1996), but did not satisfactorily conserve the appropriate measure. A corrected scheme has been presented by Tuckerman et al. (2006). Consider the following operator splitting

$$iL = iL_1 + iL'_1 + iL_2 + iL'_2 + iL_3 + iL'_3 + iL_4 + iL'_4.$$

The first four operators act on the particle coordinates and volume:

$$iL_1 = (\mathbf{v} + \xi_\varepsilon \mathbf{r}) \cdot \frac{\partial}{\partial \mathbf{r}}, \qquad\qquad iL'_1 = \xi_\varepsilon \frac{\partial}{\partial \varepsilon},$$

$$iL_2 = (\mathbf{f} - \alpha \xi_\varepsilon \mathbf{p}) \cdot \frac{\partial}{\partial \mathbf{p}}, \qquad\qquad iL'_2 = (\mathcal{P}' - P) V \frac{\partial}{\partial p_\varepsilon},$$

where, for short, $\mathbf{v} = \mathbf{p}/m$ (not equal to $\dot{\mathbf{r}}$, note) and $\xi_\varepsilon = p_\varepsilon/W$. The remaining contributions are the thermostats. iL_3 and iL_4 are defined as in eqn (3.71), while iL'_3 and iL'_4 are defined in exactly the same way, in terms of the primed variables η'_j, $p_{\eta'_j}$, G'_j, etc. The propagators are nested together in exactly the same manner as eqn (3.66), with the thermostat parts $U_{3\&4}\left(\frac{1}{2}\delta t\right)$ on the 'outside' and the particle parts $U_{1\&2}(\delta t)$ on the 'inside'. Consider $U_{3\&4}$ first. The two thermostat chains are independent of each other (the corresponding Liouville operators commute), so each can be updated separately, and the order does not matter. As in the case of *NVT* Nosé–Hoover chains, eqn (3.75), the $U_{4,M}$ propagator is outermost, and we count inwards to $U_{4,1}$, updating each thermostat momentum p_{η_j} according to eqns (3.73), (3.74). The U'_4 propagator acts similarly on the $p_{\eta'_j}$. For the innermost thermostat stage, U_3 acts on the variables η_j and the momenta $\mathbf{p}$ according to eqn (3.72), and U'_3 acts on η'_j and p_ε according to

$$U'_3(t)\eta'_j = \exp(iL'_3 t)\eta'_j = \eta'_j + \xi'_j\, t \qquad j = 1, \dots, M, \tag{3.80a}$$

$$U'_3(t)p_\varepsilon = \exp(iL'_3 t)p_\varepsilon = p_\varepsilon \exp\left(-\xi'_1 t\right). \tag{3.80b}$$

Once more, depending on the speed of the thermostat variables, it is possible to use a Suzuki–Yoshida decomposition (Tuckerman et al., 2006) to improve the accuracy. This is more critical for the particle thermostats, since the box volume typically evolves much more slowly, and the corresponding thermostat will ideally match this timescale.

Now we turn to the inner, particle and volume, propagator, $U_{1\&2}$. The two operators iL_1 and iL'_1 commute, and so we may advance both the coordinates $\mathbf{r}$ and ε (and hence the volume V), without worrying about the order; both parts are exactly soluble:

$$U_1(t)\mathbf{r} = \mathbf{r}\exp(\xi_\varepsilon t) + \mathbf{v}\, t \frac{\exp(\xi_\varepsilon t) - 1}{\xi_\varepsilon t}$$

$$U'_1(t)\varepsilon = \varepsilon + \xi_\varepsilon\, t.$$

As seen before, a Taylor expansion is advisable for small values of $\xi_\varepsilon t$. The two operators iL_2 and iL'_2 do not commute (remember, $\xi_\varepsilon = p_\varepsilon/W$, and the momenta $\mathbf{p}$ appear inside $\mathcal{P}'$), but each part is exactly soluble; Tuckerman et al. (2006) write

$$U_{1\&2}(\delta t) \approx U'_2\left(\tfrac{1}{2}\delta t\right)U_2\left(\tfrac{1}{2}\delta t\right)U_1(\delta t)U'_1(\delta t)U_2\left(\tfrac{1}{2}\delta t\right)U'_2\left(\tfrac{1}{2}\delta t\right),$$

Code 3.9 Measure-preserving constant-NPT MD algorithm

These files are provided online. The program `md_npt_lj.f90` controls the simulation, reads in the run parameters, implements the algorithm of Tuckerman et al. (2006), and writes out the results. Routines in the file `md_lj_module.f90` (see Code 3.4) are used for the forces, and various utility module routines (see Appendix A) handle input/output and simulation averages.

```
! md_npt_lj.f90
! Molecular dynamics, NPT ensemble
PROGRAM md_npt_lj
```

where

$$U_2(t)\mathbf{p} = \mathbf{p}\exp(-\alpha\xi_\varepsilon t) + \mathbf{f}\,t\frac{1 - \exp(-\alpha\xi_\varepsilon t)}{\alpha\xi_\varepsilon t} \tag{3.81a}$$

$$U_2'(t)p_\varepsilon = p_\varepsilon + (\mathcal{P}' - P)V\,t, \tag{3.81b}$$

with the usual comment about Taylor expanding for small $\xi_\varepsilon t$. Although the prescription may seem lengthy, it is very easy to translate into computer code, as shown in Code 3.9.

All of these equations consider isotropic changes in the simulation box. An important extension of this approach is to allow the box shape, as well as size, to vary. This is vital to allow the relaxation of stress, as well as crystal structure transformations in solid-state simulations, and the relevant equations were originally formulated by Parrinello and Rahman (1980; 1981; 1982) and further discussed by Nosé and Klein (1983). Measure-preserving integrators for this case are discussed by Yu et al. (2010), who also describe how to combine the algorithm with holonomic constraints. Fully flexible boxes are of limited use in liquid state simulations, since they may become extremely thin in one or more dimensions, in the absence of elastic restoring forces. However, when studying liquid–solid coexistence in slab geometry (see Chapter 14) or bilayer membranes which span the cross-section of the box, it may be important to allow anisotropic variations. Sometimes, it is desired to hold the area in the xy-plane fixed, while prescribing a pressure P_{zz} in the z-direction; a suitable algorithm for this is described by Romero-Bastida and López-Rendón (2007). Ikeguchi (2004) has also discussed integrators for this case, and other constant-P ensembles, as well as combining with algorithms for rotational motion.

As in the NVT case, there is an alternative to the non-Hamiltonian equations just described. The Nosé–Poincaré approach may also be applied to the isothermal–isobaric ensemble (Sturgeon and Laird, 2000), and the resulting equations of motion are Hamiltonian in form. A configurational constant-pressure algorithm has also been proposed (Braga and Travis, 2006).

A deterministic thermostat is not the only choice for performing simulations in the constant-NPT ensemble. As an extreme alternative, one might combine Monte Carlo volume moves (see Chapter 4), and the stochastic Andersen thermostat, with conventional constant-NVE dynamics; this is actually a sensible idea for hard-sphere MD, where the simplicity of free-flight motion is compromised by modifying the equations of motion.

A less dramatic approach is to replace the Nosé–Hoover thermostat by the Andersen thermostat, as originally proposed by Andersen (1980). Quigley and Probert (2004) have advocated using Langevin dynamics to thermostat constant-pressure simulations.

We note that equations of motion have been devised (Evans and Morriss, 1983a,b; 1984a) which constrain both the kinetic energy and the instantaneous (virial) pressure to fixed values. The ensemble is well defined, but it is not generally as useful as the constant-NPT ensemble discussed in this section. Also, there is a 'weak coupling' barostat, which relaxes the pressure towards a desired value, in the same way as the Berendsen thermostat (Berendsen et al., 1984). This is a widely available option in simulation packages, but the user should be aware that the ensemble (just as for the thermostat) is not the isothermal–isobaric one, and is not known.

3.10   Grand canonical molecular dynamics

In the grand canonical ensemble, the number of particles may vary, so it is natural to use a Monte Carlo approach in which creation or destruction of a particle is one of the options. We turn to this in Chapter 4. Is it possible to use MD to sample states in this ensemble? One can imagine a hybrid method, combining dynamical trajectories with particle creation and destruction, although such moves would have a dramatic effect when they occurred. They may be confined to a designated region of the simulation box (Papadopoulou et al., 1993), and two such control regions may be used to generate a steady diffusive flux (Heffelfinger and van Swol, 1994).

Pure MD simulation of systems in this ensemble rely on extending it to allow continuous variation of N in some way, and writing down a corresponding Lagrangian and Hamiltonian (Çağin and Pettitt, 1991a,b; Weerasinghe and Pettitt, 1994; Palmer and Lo, 1994; Lo and Palmer, 1995; Lynch and Pettitt, 1997; Boinepalli and Attard, 2003). The most recent version of this kind (Eslami and Müller-Plathe, 2007) uses a scaling variable $0 \le \nu \le 1$ for one of the particles in the system: this variable affects both the mass and the interaction potential of the particle with the rest of the system. An equation of motion is developed for ν. Whenever ν approaches 0, the scaled particle is deleted, a new particle is chosen to be the scaled particle with $\nu = 1$, and N is replaced by $N-1$. When ν approaches 1, the scaled particle is converted into a regular particle, N is replaced by $N+1$, and a new scaled particle with $\nu = 0$ is created at a random location in space. Although the method has been tested on model systems such as Lennard-Jones atoms and water (Eslami and Müller-Plathe, 2007) it is not widely used.

We mention in passing the simpler approach of actually including a particle reservoir in the simulation, allowing physical transfer into and out of the sub-system of interest. Otherwise, the simulation algorithm is conventional MD; the method has been applied to the study of confined systems (Gao et al., 1997a,b). Potential drawbacks are the expense of simulating a relatively large reservoir, which is not itself of interest, or the hysteresis that may affect transfer of particles between reservoir and sub-system, depending on the geometry. The chemical potential is controlled indirectly by adjusting the thermodynamic state of the reservoir in some way. Recently, it has been suggested to combine this with a coarse-grained model of the particles in the reservoir (Wang et al., 2013; Perego et al., 2015); this general type of 'adaptive resolution' simulation is discussed in Chapter 12.

3.11 Molecular dynamics of polarizable systems

We have already discussed the inclusion of polarization in the molecular model, via induced atomic dipoles, in Section 1.3.3. How is this handled, in practice, in an MD simulation? A useful summary of models and methods is given by Rick and Stuart (2003).

Souaille et al. (2009) have described an implementation of the induced atomic dipole model, with Thole screening. The traditional approach is to iterate the self-consistent equations (1.20) until a convergence criterion is met. A typical scheme sets

$$\left[\Delta\mu_\gamma^a\right]^{n+1} = (1-\omega)\left[\Delta\mu_\gamma^a\right]^n + \omega\alpha_{\alpha\gamma}^a\left(\mathcal{E}_\alpha^a + \sum_{b\neq a}T_{\alpha\beta}\left[\Delta\mu_\beta^b\right]^n\right),$$

where $[\cdots]^n$ indicates iteration number n. This procedure is carried out every step; the iterations begin with the values of μ^a obtained in the previous step. Souaille et al. (2009) give $\omega = 0.7$ as a typical mixing factor.

An alternative approach was pioneered by Sprik and Klein (1988) in connection with a polarizable water model. It is based on the idea of relaxing degrees of freedom through an extended Lagrangian formalism, as originally applied by Car and Parrinello to electronic structure (this will be discussed in Section 13.2). The dipoles μ^a are added as dynamical variables alongside the atomic coordinates, and are given momenta $\mathbf{p}_{\mu^a} = m_{\mu^a}\dot{\mu}^a$. An additional, fictitious, kinetic-energy term appears in the Hamiltonian

$$\sum_a \left|\mathbf{p}_{\mu^a}\right|^2/m_{\mu^a},$$

along with the electrostatic potential terms, which also depend on the μ^a. The extended equations of motion are solved by standard algorithms, in which the dipole momenta have a Nosé–Hoover thermostat applied to keep their temperature extremely low (of order 1 K). The idea is that the dynamics keeps all the dipole values very close to the values that they would have by minimizing the corresponding potential energy, which are in turn the values given by the self-consistent equations (1.20). Of course, in principle, maintaining their temperature at a low value is not sufficient: energy will tend to flow into the dipoles from the other degrees of freedom, as the system is not at equilibrium. To prevent this, there needs to be only weak coupling between the dipolar variables and the nuclear motion. The fictitious masses are chosen to make this the case, in other words, to guarantee a time scale separation. Souaille et al. (2009) report the extended dynamics method to be an order of magnitude faster than solving the self-consistent equations, while being just as accurate, for a test using a polarizable water potential.

The induced atomic dipole approach is not the only way of introducing polarization. The charges themselves may be allowed to vary, without introducing any more variables. The underlying idea of fluctuating charges is underpinned by the idea of electronegativity equalization, which arises from minimization of the electrostatic energy with respect to variation of the charges, subject to constraints on the total charge on various molecules or ions. Instead of treating this as a minimization problem, however, the charges may be taken as dynamical variables, in an extended Lagrangian formalism (Rick et al., 1994), similar to the scheme described earlier. They are given a fictitious kinetic energy, and a self-energy potential term controls their fluctuations; electroneutrality is imposed through Lagrangian constraints.

Wilson and Madden (1993) describe a combination of approaches, in the spirit of Sprik and Klein (1988), involving additional charges which are allowed to vary so as to represent multipoles. Each multipole is represented as a set of variable charges located on a fixed framework rather than as a point entity: for example, a dipole would be represented by a pair of charges separated by a vector of fixed length. Once more, an extended Lagrangian is introduced; the rotation of the charge framework, as well as the fluctuation of the charges, is handled by the dynamical equations.

Models of the shell or Drude type involve allowing the separation of charges to vary, under the influence of a 'spring' potential, while the charges themselves are fixed. In the limit of strong springs (i.e. small separations) and correspondingly high charges, this becomes identical to the point induced dipole approach. Once more, the problem may be treated on the computer by iterative minimization or dynamically; in the latter case, the degrees of freedom are conventional 'mechanical' quantities (bond lengths and masses), but the same ideas apply as in the extended Lagrangian methods. For an early example see Mitchell and Fincham (1993), and for a recent review see Lamoureux and Roux (2003). A simple version of this, the charge-on-spring model (Straatsma and McCammon, 1990), has been implemented in the Groningen molecular simulation (GROMOS) package (Yu and van Gunsteren, 2005). There is a considerable literature on developing polarizable force fields (see e.g. Halgren and Damm, 2001; Rick and Stuart, 2003; Warshel et al., 2007; Baker, 2015). They are widely implemented in a range of simulation packages.

4

Monte Carlo methods

4.1 Introduction

The Monte Carlo method was first developed by von Neumann, Ulam, and Metropolis at the end of the Second World War to study the diffusion of neutrons in fissionable material. The name 'Monte Carlo', coined by Metropolis in 1947 and used in the title of a paper describing the early work at Los Alamos (Metropolis and Ulam, 1949), derives from the extensive use of random numbers in the approach.

The method is based on the idea that a determinate mathematical problem can be treated by finding a probabilistic analogue which is then solved by a stochastic sampling experiment (von Neumann and Ulam, 1945). For example, the configurational energy of a liquid can be calculated by solving the coupled equations of motion of the atoms and averaging over time. Alternatively, one can set up an ensemble of states of the liquid, choosing individual states with the appropriate probability, and calculating the configurational energy by averaging uniformly over the ensemble. These sampling experiments involve the generation of random numbers followed by a limited number of arithmetic and logical operations, which are often the same at each step. These are tasks that are well suited to a computer and the growth in the importance of the method can be linked to the rapid development of these machines. The arrival of the MANIAC computer at Los Alamos in 1952 prompted the study of the many-body problem by Metropolis et al. (1953) and the development of the Metropolis Monte Carlo method (Wood, 1986), which is the subject of this chapter. Today, the Monte Carlo method is widely applied in all branches of the natural and social sciences and is, arguably, 'the most powerful and commonly used technique for analysing complex problems' (Rubinstein, 1981).

4.2 Monte Carlo integration

As outlined in Chapter 2, the Metropolis Monte Carlo method aims to generate a trajectory in phase space which samples from a chosen statistical ensemble. There are several difficulties involved in devising such a prescription and making it work for a system of molecules in a liquid. So we take care to introduce the Monte Carlo method with a simple example.

Consider the problem of finding the volume, V, of the solid bounded by the coordinate axes, and the planes $z = 1 + y$ and $2x + y = 2$. This is the volume below the dark-grey

Computer Simulation of Liquids. Second Edition. M. P. Allen and D. J. Tildesley.
© M. P. Allen and D. J. Tildesley 2017. Published in 2017 by Oxford University Press.

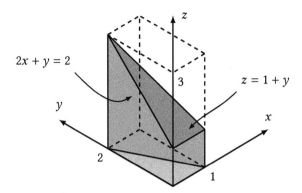

Fig. 4.1 The solid volume V, below the dark-grey triangle and above the light-grey triangle can be evaluated from the integral in eqn (4.1).

triangle in Fig. 4.1. The volume is exactly given by

$$V = \int_0^1 dx \int_0^{2-2x} dy\,(1+y) = \frac{5}{3}. \tag{4.1}$$

Here we consider two simple Monte Carlo methods to evaluate it numerically.

4.2.1   Hit and miss

In a Monte Carlo evaluation of the integral, the volume of interest, V, would be surrounded by a sampling region of a simple geometry whose volume is known. In this case we choose the rectangular box of volume $V_0 = 6$ indicated by the dashed lines in Fig. 4.1. A random position is chosen within the rectangular box $\mathbf{r}_\tau = (x_\tau, y_\tau, z_\tau)$; this is a shot, τ. If this shot is within the required volume, V, it is a hit. If a total of τ_{shot} shots are fired and τ_{hit} hits scored then

$$V = \frac{V_0\,\tau_{\text{hit}}}{\tau_{\text{shot}}}. \tag{4.2}$$

The key to this method is the generation of $3\tau_{\text{shot}}$ random numbers from a uniform distribution. A sample program to perform this integration is given in Code 4.1. RANDOM_SEED() and RANDOM_NUMBER() are built-in Fortran functions for generating uniform random numbers on $(0, 1)$. Random number generators are discussed briefly in Appendix E.

4.2.2   Sample mean integration

Hit-and-miss integration is conceptually easy to understand but the sample mean method is more generally applicable and offers a more accurate estimate for most integrals (Hammersley and Handscomb, 1964; Rubinstein, 1981). In this case the integral of interest

$$F = \int_{x_1}^{x_2} dx\, f(x) \tag{4.3}$$

is rewritten as

$$F = \int_{x_1}^{x_2} dx \left(\frac{f(x)}{\rho(x)} \right) \rho(x) \tag{4.4}$$

Code 4.1 Hit-and-miss integration

This program is also available online in the file hit_and_miss.f90.

```fortran
PROGRAM hit_and_miss
  USE, INTRINSIC :: iso_fortran_env, ONLY : output_unit
  IMPLICIT NONE
  REAL                                :: v
  REAL, DIMENSION(3)                  :: r, zeta
  REAL, DIMENSION(3), PARAMETER :: r_0 = [1.0, 2.0, 3.0]
  REAL,               PARAMETER :: v_0 = PRODUCT(r_0)
  INTEGER                             :: tau, tau_shot, tau_hit

  CALL RANDOM_SEED()
  tau_hit  = 0
  tau_shot = 1000000

  DO tau = 1, tau_shot
     CALL RANDOM_NUMBER ( zeta(:) ) ! uniform in range (0,1)
     r = zeta * r_0                 ! uniform in v_0
     IF (    r(2) < ( 2.0 - 2.0*r(1) ) .AND. &
          &  r(3) < ( 1.0 + r(2) )       ) THEN ! in polyhedron
        tau_hit = tau_hit + 1
     END IF
  END DO
  v = v_0 * REAL ( tau_hit ) / REAL ( tau_shot )
  WRITE (UNIT=output_unit,FMT='(a,f10.5)') 'Estimate_=_', v
END PROGRAM hit_and_miss
```

where $\rho(x)$ is an arbitrary probability density function. Consider performing a number of trials τ, each consisting of choosing a random number ζ_τ, from the distribution $\rho(x)$ in the range (x_1, x_2). Then

$$F = \left\langle \frac{f(\zeta_\tau)}{\rho(\zeta_\tau)} \right\rangle_{\text{trials}} \tag{4.5}$$

where the brackets represent an average over all trials. A simple application would be to choose $\rho(x)$ to be uniform, that is,

$$\rho(x) = \frac{1}{(x_2 - x_1)} \qquad x_1 \le x \le x_2 \tag{4.6}$$

and then the integral F can be estimated as

$$F \approx \frac{(x_2 - x_1)}{\tau_{\max}} \sum_{\tau=1}^{\tau_{\max}} f(\zeta_\tau). \tag{4.7}$$

Code 4.2   Sample mean integration

This program is also available online in the file sample_mean.f90.

```fortran
PROGRAM sample_mean
  IMPLICIT NONE
  REAL                              :: v, f
  REAL , DIMENSION(2)               :: r, zeta
  REAL , DIMENSION(2), PARAMETER :: r_0 = [1.0, 2.0]
  REAL ,               PARAMETER :: a_0 = PRODUCT(r_0)
  INTEGER                           :: tau, tau_max

  CALL RANDOM_SEED ()
  tau_max = 1000000

  f = 0.0
  DO tau = 1, tau_max
     CALL RANDOM_NUMBER ( zeta ) ! uniform in (0,1)
     r = zeta * r_0 ! uniform in xy rectangle
     IF ( r(2) < 2.0-2.0*r(1) ) THEN ! in xy triangle
        f = f + ( 1.0 + r(2) )         ! value of z
     END IF
  END DO
  v = a_0 * f / REAL ( tau_max )
  WRITE (UNIT=output_unit,FMT='(a,f10.5)') 'Estimate =_', v
END PROGRAM sample_mean
```

The method can be readily generalized to multiple integrals. A Monte Carlo sample mean evaluation of the volume in Fig. 4.1 can be performed with the program in Code 4.2. In this case, the integration is carried out by selecting points in a rectangle in the xy plane. The function f to be summed is zero if the points lie outside the light-grey triangle, and equal to $z = 1 + y$ inside. The sample mean method can be used to calculate many of the multiple integrals of liquid state theory, for example the long-range correction to the three-body potential energy in eqn (2.149).

4.2.3   A direct evaluation of the partition function?

For the multidimensional integrals of statistical mechanics, the sample mean method, with a suitable choice of $\rho(x)$, is the only sensible solution. To understand this, we consider the evaluation of the configurational integral $Z_{NVT} = \int d\mathbf{r} \exp(-\beta \mathcal{V})$, eqn (2.26), for a system of, say, $N = 100$ molecules in a cube of side L. The sample mean approach to this integral, using a uniform distribution could involve the following trials:

(a) pick a point at random in the 300-dimensional configuration space, by generating 300 random numbers, on $(-\frac{1}{2}L, \frac{1}{2}L)$, which, taken in triplets, specify the coordinates of each molecule;

(b) calculate the potential energy, $\mathcal{V}(\tau)$, and hence the Boltzmann factor for this configuration.

This procedure is repeated for many trials and the configurational integral is estimated
using

$$Z_{NVT} \approx \frac{V^N}{\tau_{\max}} \sum_{\tau=1}^{\tau_{\max}} \exp\left(-\beta \mathcal{V}(\tau)\right). \tag{4.8}$$

In principle, the number of trials $\tau_{\max}$ may be increased until Z_{NVT} is estimated to the
desired accuracy. Unfortunately, a large number of the trials would give a very small
contribution to the average. In such a random configuration, molecules would overlap,
$\mathcal{V}(\tau)$ would be large and positive, and the Boltzmann factor vanishingly small. An accurate
estimation of Z_{NVT} for a dense liquid using a uniform sample mean method is not possible,
although methods of this type have been used to examine the structural properties of the
hard-sphere fluid at low densities (Alder et al., 1955). The difficulties in the calculation of
Z_{NVT} apply equally to the calculation of ensemble averages such as

$$\langle \mathcal{A} \rangle_{NVT} = \frac{\int d\mathbf{r} \mathcal{A} \exp\left(-\beta \mathcal{V}\right)}{\int d\mathbf{r} \exp\left(-\beta \mathcal{V}\right)} \approx \frac{\sum_{\tau=1}^{\tau_{\max}} \mathcal{A}(\tau) \exp\left(-\beta \mathcal{V}(\tau)\right)}{\sum_{\tau=1}^{\tau_{\max}} \exp\left(-\beta \mathcal{V}(\tau)\right)}, \tag{4.9}$$

if we attempt to estimate the numerator and denominator separately by using the uniform
sample mean method. However, at realistic liquid densities the problem might be solved
using a sample mean integration where the random coordinates are chosen from a nonuniform distribution. This method of 'importance sampling' is discussed in the next
section.

4.3   Importance sampling

Importance sampling techniques choose random numbers from a distribution $\rho(x)$, which
allows the function evaluation to be concentrated in the regions of space that make
important contributions to the integral. Consider the canonical ensemble. In this case the
desired integral is

$$\langle \mathcal{A} \rangle = \int d\Gamma \rho_{NVT}(\Gamma) \mathcal{A}(\Gamma)$$

that is, the integrand is $f = \rho_{NVT} \mathcal{A}$. By sampling configurations at random, from a
chosen distribution $\rho(\Gamma)$ we can estimate the integral as

$$\langle \mathcal{A} \rangle_{NVT} = \langle \mathcal{A} \rho_{NVT}/\rho \rangle_{\text{trials}}. \tag{4.10}$$

For most functions $\mathcal{A}(\Gamma)$, the integrand will be significant where $\rho_{NVT}(\Gamma)$ is significant.
In these cases choosing $\rho(\Gamma) = \rho_{NVT}(\Gamma)$ should give a good estimate of the integral. In
this case

$$\langle \mathcal{A} \rangle_{NVT} = \langle \mathcal{A} \rangle_{\text{trials}}. \tag{4.11}$$

(This is not always the best choice, and sometimes we choose alternative distributions
$\rho(\Gamma)$; see Section 9.2.3.)

Such a method, with $\rho(\Gamma) = \rho_{NVT}(\Gamma)$ was originally developed by Metropolis et al.
(1953). The problem is not solved, simply rephrased. The difficult job is finding a method

of generating a sequence of random states so that by the end of the simulation each state has occurred with the appropriate probability. It turns out that it is possible to do this without ever calculating the normalizing factor for ρ_{NVT}, that is, the partition function (see eqns (2.11)–(2.13)).

The solution is to set up a Markov chain of states of the liquid, which is constructed so that it has a limiting distribution of $\rho_{NVT}(\Gamma)$. A Markov chain is a sequence of trials that satisfies two conditions:

(a) The outcome of each trial belongs to a finite set of outcomes, $\{\Gamma_1, \Gamma_2, \ldots\}$, called the state space.
(b) The outcome of each trial depends only on the outcome of the trial that immediately precedes it.

Two such states Γ_m and Γ_n are linked by a transition probability π_{mn} which is the probability of going from state m to state n. The properties of a Markov chain are best illustrated with a simple example. Suppose the reliability of your computer follows a certain pattern. If it is up and running on one day it has a 60 % chance of running correctly on the next. If however, it is down, it has a 70 % chance of also being down the next day. The state space has two components, up ($\uparrow$) and down ($\downarrow$), and the transition matrix has the form

$$\pi = \begin{matrix} \uparrow \\ \downarrow \end{matrix} \begin{pmatrix} \overset{\uparrow}{0.6} & \overset{\downarrow}{0.4} \\ 0.3 & 0.7 \end{pmatrix}. \tag{4.12}$$

If the computer is equally likely to be up or down to begin with, then the initial probability can be represented as a vector, which has the dimensions of the state space

$$\rho^{(1)} = \begin{pmatrix} \overset{\uparrow}{0.5} & \overset{\downarrow}{0.5} \end{pmatrix}. \tag{4.13}$$

The probability that the computer is up on the second day is given by the matrix equation

$$\rho^{(2)} = \rho^{(1)}\pi = (0.45, 0.55) \tag{4.14}$$

that is, there is a 45 % chance of running a program. The next day would give

$$\rho^{(3)} = \rho^{(2)}\pi = \rho^{(1)}\pi\pi = \rho^{(1)}\pi^2 = (0.435, 0.565), \tag{4.15}$$

and a 43.5 % chance of success. If you are anxious to calculate your chances of running a program in the long run, then the limiting distribution is given by

$$\rho = \lim_{\tau \to \infty} \rho^{(1)}\pi^\tau. \tag{4.16}$$

A few applications of eqn (4.16) show that the result converges to $\rho = (0.4286, 0.5714)$. It is clear from eqn (4.16) that the limiting distribution, ρ, must satisfy the eigenvalue equation

$$\rho\pi = \rho, \quad \text{or} \quad \sum_m \rho_m \pi_{mn} = \rho_n \quad \forall n, \tag{4.17}$$

with eigenvalue unity. π is termed a stochastic matrix since its rows add to 1

$$\sum_n \pi_{mn} = 1 \quad \forall m. \tag{4.18}$$

It is the transition matrix for an irreducible Markov chain. (An irreducible or ergodic chain is one where every state can eventually be reached from another state.) More formally, we note that the Perron–Frobenius theorem (Chung, 1960; Feller, 1957) states that an irreducible stochastic matrix has one left eigenvalue which equals unity, and the corresponding eigenvector is the limiting distribution of the chain. The other eigenvalues are less than unity and they govern the rate of convergence of the Markov chain. The limiting distribution, ρ, implied by the chain is quite independent of the initial condition $\rho^{(1)}$ (in the long run, it matters nothing if your computer is down today). In the case of a liquid, we must construct a much larger transition matrix, which is stochastic and ergodic (see Chapter 2). In contrast to the previous problem, the elements of the transition matrix are unknown, but the limiting distribution of the chain is the vector with elements $\rho_m = \rho_{NVT}(\Gamma_m)$ for each point Γ_m in phase space. It is possible to determine elements of π which satisfy eqns (4.17) and (4.18) and thereby generate a phase space trajectory in the canonical ensemble. We have considerable freedom in finding an appropriate transition matrix, with the crucial constraint that its elements can be specified without knowing Q_{NVT}. A useful trick in searching for a solution of eqn (4.17) is to replace it by the unnecessarily strong condition of 'microscopic reversibility':

$$\rho_m \pi_{mn} = \rho_n \pi_{nm}. \tag{4.19}$$

Summing over all states m and making use of eqn (4.18) we regain eqn (4.17)

$$\sum_m \rho_m \pi_{mn} = \sum_m \rho_n \pi_{nm} = \rho_n \sum_m \pi_{nm} = \rho_n. \tag{4.20}$$

A suitable scheme for constructing a phase space trajectory in the canonical ensemble involves choosing a transition matrix which satisfies eqns (4.18) and (4.19). The first such scheme was suggested by Metropolis et al. (1953) and is often known as the asymmetrical solution. If the states m and n are distinct, this solution considers two cases

$$\pi_{mn} = \alpha_{mn} \qquad\qquad \rho_n \geq \rho_m \qquad m \neq n \tag{4.21a}$$
$$\pi_{mn} = \alpha_{mn}(\rho_n/\rho_m) \qquad \rho_n < \rho_m \qquad m \neq n. \tag{4.21b}$$

It is also important to allow for the possibility that the liquid remains in the same state,

$$\pi_{mm} = 1 - \sum_{n \neq m} \pi_{mn}. \tag{4.21c}$$

In this solution α is a symmetrical stochastic matrix, $\alpha_{mn} = \alpha_{nm}$, often called the underlying matrix of the Markov chain. The symmetric properties of α can be used to show that for the three cases ($\rho_m = \rho_n$, $\rho_m < \rho_n$, and $\rho_m > \rho_n$) the transition matrix defined in eqn (4.21) satisfies eqns (4.18) and (4.19). It is worth stressing that it is the symmetric property of α that is essential in satisfying microscopic reversibility in this case. Non-symmetrical α matrices which satisfy microscopic reversibility or just the weaker condition, eqn (4.17), can be constructed but these are not part of the basic Metropolis recipe (Owicki and Scheraga, 1977b). These cases are considered in more detail in Chapter 9. This Metropolis solution only involves the ratio ρ_n/ρ_m and is therefore independent of Q_{NVT} which is not required to perform the simulations.

There are other solutions to eqns (4.18) and (4.19). The symmetrical solution (Wood and Jacobson, 1959; Flinn and McManus, 1961; Barker, 1965) is often referred to as Barker sampling:

$$\pi_{mn} = \alpha_{mn}\,\rho_n/(\rho_n + \rho_m) \qquad m \neq n \tag{4.22a}$$

$$\pi_{mm} = 1 - \sum_{n\neq m} \pi_{mn}. \tag{4.22b}$$

Equation (4.22) also satisfies the condition of microscopic reversibility.

If states of the fluid are generated using transition matrices such as eqns (4.21) and (4.22), then a particular property, $\langle \mathcal{A}\rangle_{\mathrm{run}}$, obtained by averaging over the τ_{run} trials in the Markov chain, is related to the average in the canonical ensemble (Chung, 1960; Wood, 1968b)

$$\langle \mathcal{A}\rangle_{NVT} = \langle \mathcal{A}\rangle_{\mathrm{run}} + O\!\left(\tau_{\mathrm{run}}^{-1/2}\right). \tag{4.23}$$

As mentioned in Chapter 2, we usually restrict simulations to the configurational part of phase space, calculate average configurational properties of the fluid, and add the ideal gas parts after the simulation.

Since there are a number of suitable transition matrices, it is useful to choose a particular solution which minimizes the variance in the estimate of $\langle \mathcal{A}\rangle_{\mathrm{run}}$. Suitable prescriptions for defining the variance in the mean, $\sigma^2(\langle \mathcal{A}\rangle_{\mathrm{run}})$ are discussed in Chapter 8. In particular, the statistical inefficiency (Section 8.4.1)

$$s = \lim_{\tau_{\mathrm{run}}\to\infty} \tau_{\mathrm{run}}\sigma^2(\langle \mathcal{A}\rangle_{\mathrm{run}})/\sigma^2(\mathcal{A}) \tag{4.24}$$

measures how slowly a run converges to its limiting value. Peskun (1973) has shown that it is reasonable to order two transition matrices,

$$\pi_1 \leq \pi_2 \tag{4.25}$$

if each off-diagonal element of π_1 is less than the corresponding element in π_2. If this is the case, then

$$s(\langle \mathcal{A}\rangle, \pi_1) \geq s(\langle \mathcal{A}\rangle, \pi_2) \tag{4.26}$$

for any property $\mathcal{A}$. If the off-diagonal elements of π are large then the probability of remaining in the same state is small and the sampling of phase space will be improved. With the restriction that ρ_m and ρ_n are positive, eqns (4.21) and (4.22) show that the Metropolis solution leads to a lower statistical inefficiency of the mean than the Barker solution.

Valleau and Whittington (1977b) stress that a low statistical inefficiency is not the only criterion for choosing a particular π. Since the simulations are of finite length, it is essential that the Markov chain samples a representative portion of phase space in a reasonable

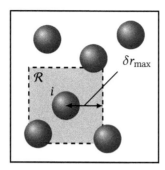

Fig. 4.2 State n is obtained from state m by moving atom i with a uniform probability to any point in the shaded region $\mathcal{R}$.

number of moves. All the results derived in this section depend on the ergodicity of the chain (i.e. that there is some non-zero multi-step transition probability of moving between any two allowed states of the fluid). If these allowed states are not connected, the MC run may produce a low s but in addition a poor estimate of the canonical average. When the path between two allowed regions of phase space is difficult to find, the situation is described as a bottleneck. These bottlenecks are always a worry in MC simulations but are particularly troublesome in the simulation of two-phase coexistence (Lee et al., 1974), in the simulation of phase transitions (Evans et al., 1984), and in simulations of ordinary liquids at unusually high density.

Where a comparison has been made between the two common solutions to the transition matrix, eqns (4.21) and (4.22), the Metropolis solution appears to lead to a faster convergence of the chain (Valleau and Whittington, 1977b). The Metropolis method becomes more favourable as the number of available states at a given step increases and as the energy difference between the states increases. (For two-state problems such as the Ising model the symmetric algorithm may be favourable (Cunningham and Meijer, 1976).) In the next section we describe the implementation of the asymmetric solution.

4.4   The Metropolis method

To implement the Metropolis solution to the transition matrix, it is necessary to specify the underlying stochastic matrix $\boldsymbol{\alpha}$. This matrix is designed to take the system from state m into any one of its neighbouring states n. In this chapter, we normally consider the use of a symmetric underlying matrix, that is, $\alpha_{mn} = \alpha_{nm}$. A useful but arbitrary definition of a neighbouring state is illustrated in Fig. 4.2. This diagram shows six atoms in a state m; to construct a neighbouring state n, one atom (i) is chosen at random and displaced from its position $\mathbf{r}_i^m$ with equal probability to any point $\mathbf{r}_i^n$ inside the square $\mathcal{R}$. This square is of side $2\delta r_{\max}$ and is centred at $\mathbf{r}_i^m$. In a three-dimensional example, $\mathcal{R}$ would be a small cube. On the computer there is a large but finite number of new positions, $N_{\mathcal{R}}$, for the atom i and in this case α_{mn} can be simply defined as

$$\alpha_{mn} = \begin{cases} 1/N_{\mathcal{R}} & \mathbf{r}_i^n \in \mathcal{R} \\ 0 & \mathbf{r}_i^n \notin \mathcal{R}. \end{cases} \tag{4.27}$$

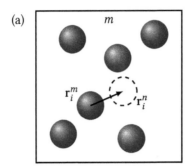 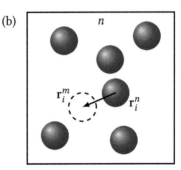

Fig. 4.3 (a) State n is generated from state m by displacing atom i from $\mathbf{r}_i^m$ to $\mathbf{r}_i^n$ (dashed circle). (b) The reverse move. To ensure microscopic reversibility, in the simple Metropolis method the probabilities of attempting the forward and reverse moves should be equal, $\alpha_{mn} = \alpha_{nm}$.

With this choice of α, eqn (4.21) is readily implemented. At the beginning of an MC move an atom is picked at random and given a uniform random displacement along each of the coordinate directions. The maximum displacement, δr_{max} is an adjustable parameter that governs the size of the region and controls the convergence of the Markov chain. The new position is obtained with the following code; dr_max is the maximum displacement δr_{max}, and the simulation box has unit length.

```
REAL , DIMENSION(3,n) :: r
REAL , DIMENSION(3)    :: ri, zeta

CALL RANDOM_NUMBER ( zeta )        ! uniform in range (0,1)
zeta = 2.0*zeta - 1.0              ! now in range (-1,+1)
ri(:) = r(:,i) + zeta * dr_max     ! trial move to new position
ri(:) = ri(:) - ANINT ( ri(:) )   ! periodic boundaries
```

The appropriate element of the transition matrix depends on the relative probabilities of the initial state m and the final state n. There are two cases to consider. If $\delta \mathcal{V}_{nm} = \mathcal{V}_n - \mathcal{V}_m \leq 0$ then $\rho_n \geq \rho_m$ and eqn (4.21a) applies. If $\delta \mathcal{V}_{nm} > 0$ then $\rho_n < \rho_m$ and eqn (4.21b) applies. (The symbol $\mathcal{V}_m$ is used as a shorthand for $\mathcal{V}(\Gamma_m)$.) The next step in an MC move is to determine $\delta \mathcal{V}_{nm}$. The determination of $\delta \mathcal{V}_{nm}$ does not require a complete recalculation of the configurational energy of state m, just the changes associated with the moving atom. For example (see Fig. 4.3), the change in potential energy is calculated by computing the energy of atom i with all the other atoms before and after the move

$$\delta \mathcal{V}_{nm} = \left(\sum_{j=1}^{N} v(r_{ij}^n) - \sum_{j=1}^{N} v(r_{ij}^m) \right) \tag{4.28}$$

where the sum over the atoms excludes atom i. In calculating the change of energy, the explicit interaction of atom i with all its neighbours out to a cutoff distance r_c is considered. The contribution from atoms beyond the cutoff could be estimated using a mean field correction (see Section 2.8), but in fact the correction for atom i in the old and new positions is exactly the same in a homogeneous fluid, and does not need to be included explicitly in the calculation of $\delta \mathcal{V}_{nm}$.

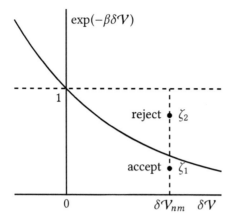

Fig. 4.4 Accepting uphill moves in the MC simulation

If the move is downhill in energy $\delta V_{nm} \leq 0$, then the probability of state n is greater than state m and the new configuration is accepted. The method of choosing trial moves ensures that the transition probability $\pi_{mn} = \alpha_{nm}$, the value required by eqn (4.21a).

If the move is uphill in energy $\delta V_{nm} > 0$, then the move is accepted with a probability ρ_n/ρ_m according to eqn (4.21b). Again the factor α_{mn} is automatically included in making the move. This ratio can be readily expressed as the Boltzmann factor of the energy difference:

$$\frac{\rho_n}{\rho_m} = \frac{Z_{NVT}^{-1}\exp\left(-\beta V_n\right)}{Z_{NVT}^{-1}\exp\left(-\beta V_m\right)} = \frac{\exp\left(-\beta V_n\right)\exp\left(-\beta\delta V_{nm}\right)}{\exp\left(-\beta V_n\right)} = \exp\left(-\beta\delta V_{nm}\right). \quad (4.29)$$

To accept a move with a probability of $\exp(-\beta\delta V_{nm})$, a random number ζ is generated uniformly on $(0, 1)$. The random number is compared with $\exp(-\beta\delta V_{nm})$. If it is less than $\exp(-\beta\delta V_{nm})$ the move is accepted. This procedure is illustrated in Fig. 4.4. During the run, suppose that a particular uphill move, δV_{nm} is attempted. If at that point a random number ζ_1 is chosen (see Fig. 4.4), the move is accepted. If ζ_2 is chosen the move is rejected. Over the course of the run the net result is that energy changes such as δV_{nm} are accepted with a probability $\exp(-\beta\delta V_{nm})$. If the uphill move is rejected, the system remains in state m in accord with the finite probability π_{mm} of eqn (4.21c). In this case, the atom is retained at its old position and the old configuration is recounted as a new state in the chain. This procedure can be summarized by noting that we accept any move (uphill or downhill) with probability $\min[1, \exp(-\beta\delta V_{nm})]$, where the min function returns a value equal to the minimum of its arguments (as does the Fortran function with the same name).

A complete MC program for a fluid of Lennard-Jones atoms is given in Code 4.3. Here, we show the typical code for the heart of the program, the acceptance and rejection of moves. In this code, pot_old and pot_new are the potential energies of atom i summed over all its neighbours j, as in eqn (4.28). We also expect that, in calculating pot_new, a logical flag overlap is set if a significant molecular overlap is detected, that is, an interaction with a very high potential energy, which may be regarded as infinite for

Code 4.3   Monte Carlo *NVT*-ensemble for Lennard-Jones atoms

These files are provided online. The program mc_nvt_lj.f90 controls the simulation, reads in the run parameters, selects moves, and writes out the results. It uses the routines in mc_lj_module.f90 to evaluate the Lennard-Jones potential, and actually implement the moves, and utility modules (see Appendix A) for the Metropolis function, input/output and simulation averages.

```
! mc_nvt_lj.f90
! Monte Carlo, NVT ensemble
PROGRAM mc_nvt_lj

! mc_lj_module.f90
! Energy and move routines for MC simulation, LJ potential
MODULE mc_module
```

practical purposes. We use this to guard against a trial move with a very large value of δV_{nm} which might cause underflow problems in the computation of $\exp(-\beta \delta V_{nm})$. The threshold should be high enough to guarantee that $\exp(-\beta \delta V_{nm})$ is negligibly small at the chosen temperature; an advantage of including this in the energy calculation is that the program can immediately save time by setting the flag and leaping out of the energy loop.

```
IF ( .NOT. overlap ) THEN ! consider non-overlap only
   delta = ( pot_new - pot_old ) / temperature
   IF ( metropolis ( delta ) ) THEN    ! accept metropolis
      pot     = pot + pot_new - pot_old ! update potential
      r(:,i) = ri(:)                    ! update position
      moves   = moves + 1               ! update move counter
   END IF ! reject metropolis test
END IF ! reject overlap without calculation
```

Here pot holds the *total* potential energy of the system, V, which changes by δV_{nm} if the move is accepted. The metropolis function simply returns a .TRUE. or .FALSE. result, using code like

```
REAL :: zeta
IF ( delta < 0.0 ) THEN ! downhill, accept
   metropolis = .TRUE.
ELSE
   CALL RANDOM_NUMBER ( zeta )      ! uniform in range (0,1)
   metropolis = EXP(-delta) > zeta ! metropolis test
END IF
```

In the function referred to by the program of Code 4.3 and other online programs, we include a further guard against underflow.

So far we have said little about the maximum allowed displacement of the atom, δr_{max}, which governs the size of the trial move. If this parameter is too small then a large fraction of moves are accepted but the phase space of the liquid is explored slowly, that is, consecutive states are highly correlated. If δr_{max} is too large then nearly all the trial moves are rejected and again there is little movement through phase space. In fact δr_{max} is often adjusted during the simulation so that about half the trial moves are rejected. This adjustment can be handled automatically using code similar to the following, at predefined intervals, for example, at the end of every sweep, assuming that move_ratio is the ratio of accepted to attempted moves during the sweep.

```
IF ( move_ratio > 0.55 ) THEN
   dr_max = dr_max * 1.05
ELSE IF ( move_ratio < 0.45 ) THEN
   dr_max = dr_max * 0.95
END IF
```

It is not clear that an acceptance ratio of 0.5 is optimum. A reported study of the parameter δr_{max} (Wood and Jacobson, 1959) suggests that an acceptance ratio of only 0.1 maximizes the root-mean-square displacement of atoms as a function of computer time. The root-mean-square displacement is one possible measure of the movement through phase space and the work suggests that a small number of large moves is most cost-effective. Few simulators would have the courage to reject nine out of ten moves on this limited evidence and an acceptance ratio of 0.5 is still common. This issue highlights a difficulty in assessing particular simulation methods. The work of Wood and Jacobson was performed on 32 hard spheres, at a particular packing fraction, on a first-generation computer. There is no reason to believe that their results would be the same for a different potential, at a different state point, on a different machine. The MC technique is time-consuming and since most researchers are more interested in new results rather than methodology there has been little work on the optimization of parameters such as δr_{max} and the choice of transition matrix.

In the original Metropolis method one randomly chosen atom is moved to generate a new state. The underlying stochastic matrix can be changed so that several or all of the atoms are moved simultaneously (Ree, 1970; Ceperley et al., 1977). $\delta \mathcal{V}_{nm}$ is calculated using a straightforward extension of eqn (4.28) and the move is accepted or rejected using the normal criteria. Chapman and Quirke (1985) have performed a simulation of 32 Lennard-Jones atoms at a typical liquid density and temperature. In this study, all 32 atoms were moved simultaneously, and an acceptance ratio of $\approx 30\,\%$ was obtained using $\delta r_{max} \approx 0.3\sigma$. Chapman and Quirke found that equilibration (see Chapter 5) was achieved more rapidly by employing multi-particle moves rather than single-particle moves. The relative efficiency of multi-particle and single-particle moves, as measured by their ability to sample phase space in a given amount of computer time, has not been subjected to systematic study.

A common practice in MC simulation is to select the atoms to move sequentially (i.e. in order of atom index) rather than randomly. This cuts down on the amount of random number generation and is an equally valid method of generating the correctly weighted states (Hastings, 1970). The length of an MC simulation is conveniently measured in 'cycles'; that is, N trial moves whether selected sequentially or randomly. The computer

Code 4.4   Monte Carlo of hard spheres

These files are provided online. The program `mc_nvt_hs.f90` controls the simulation, reads in the run parameters, selects moves, and writes out the results. It uses the overlap routines in `mc_hs_module.f90`, and utility module routines (see Appendix A) for input/output and simulation averages.

```
! mc_nvt_hs.f90
! Monte Carlo, NVT ensemble, hard spheres
PROGRAM mc_nvt_hs
```

```
! mc_hs_module.f90
! Overlap routines for MC simulation, hard spheres
MODULE mc_module
```

time involved in an MC cycle is comparable (although obviously not equivalent) to that in an MD timestep.

The simulation of hard spheres is particularly easy using the MC method. The same Metropolis procedure is used, except that, in this case, the overlap of two spheres results in an infinite positive energy change and $\exp(-\beta\delta\mathcal{V}_{nm}) = 0$. All trial moves involving an overlap are immediately rejected since $\exp(-\beta\delta\mathcal{V}_{nm})$ would be smaller than any random number generated on $(0, 1)$. Equally, all moves that do not involve overlap are immediately accepted. As before in the case of a rejection, the old configuration is recounted in the average. As discussed in Section 2.4, one minor complication is that the pressure must be calculated by a box-scaling (or related) method. An example program is given in Code 4.4.

The importance sampling technique, as described, only generates states that make a substantial contribution to ensemble averages such as the energy. In practice we cannot sum over all the possible states of the fluid and so cannot calculate Z_{NVT}. Consequently, this is not a direct route to the 'statistical' properties of the fluid such as A, S, and μ. In the canonical ensemble there are a number of ways around this problem, such as thermodynamic integration and the particle insertion methods (see Section 2.4). It is also possible to estimate the free energy difference between the simulated state and a neighbouring state point, and a modification of the sampled distribution, so-called umbrella sampling or non-Boltzmann sampling can make this more efficient. A process of iterative refinement may allow the simulation to sample a greatly extended range of energies, and hence estimate the entropy. We return to these approaches in Chapter 9. Alternatively the problem can be tackled at root by conducting simulations in the grand canonical ensemble (Section 4.6), but as we shall see, this approach may have limited application to dense liquids.

4.5   Isothermal–isobaric Monte Carlo

An advantage of the MC method is that it can be readily adapted to the calculation of averages in any ensemble. Wood (1968a,b; 1970) first showed that the method could be extended to the isothermal–isobaric ensemble. This ensemble was introduced in Section 2.2, and in designing a simulation method we should recall that the number of molecules, the temperature, and the pressure are fixed while the volume of the simulation box is allowed to fluctuate. The original constant-NPT simulations were performed on hard spheres and disks, but McDonald (1969; 1972) extended the technique to cover continuous potentials in his study of Lennard-Jones mixtures. This ensemble was thought to be particularly appropriate for simulating mixtures since experimental measurements of excess properties are recorded at constant pressure and theories of mixing are often formulated with this assumption. The method has also been used in the simulation of single-component fluids (Voronstov-Vel'Yaminov et al., 1970), and in the study of phase transitions (Abraham, 1982). It is worth recalling that at constant N, P, T we should not see two phases coexisting in the same simulation cell, a problem which bedevils the simulation of phase transitions in the canonical ensemble.

In the constant-NPT ensemble the configurational average of a property $\mathcal{A}$, is given by

$$\langle \mathcal{A} \rangle_{NPT} = \frac{\int_0^\infty dV \exp(-\beta PV)V^N \int ds\, \mathcal{A}(s) \exp\left(-\beta \mathcal{V}(s)\right)}{Z_{NPT}}. \tag{4.30}$$

In eqn (4.30), Z_{NPT} is the appropriate configurational integral eqn (2.30) and V is the volume of the fluid. Note that in this equation we use a set of scaled coordinates, $s = (s_1, s_2, \ldots s_N)$, where

$$s = L^{-1}r. \tag{4.31}$$

In this case the configurational integral in eqn (4.30) is over the $3N$-dimensional unit cube and the additional factor of V^N comes from the volume element dr. (In this section the simulation box is assumed to be a cube of side $L = V^{-1/3}$; the arguments can be easily extended to non-cubic boxes.)

The Metropolis scheme is implemented by generating a Markov chain of states which has a limiting distribution proportional to

$$\exp\left[-\beta\left(PV + \mathcal{V}(s)\right) + N \ln V\right]$$

and the method used is a direct extension of the ideas discussed in Section 4.4.

A new state is generated by displacing a molecule randomly and/or making a random volume change from V_m to V_n

$$s_i^n = s_i^m + \delta s_{\max}(2\zeta - 1) \tag{4.32}$$
$$V_n = V_m + \delta V_{\max}(2\zeta - 1).$$

Here, as usual, ζ is a random number generated uniformly on $(0, 1)$, while $\boldsymbol{\zeta}$ is a vector whose components are also uniform random numbers on $(0, 1)$ and $\mathbf{1}$ is the vector $(1, 1, 1)$. $\delta s_{\max}$ and $\delta V_{\max}$ govern the maximum changes in the scaled coordinates of the particles, and in the volume of the simulation box, respectively. Their precise values will depend on the state point studied and they are chosen to produce an acceptance ratio of 35 %–50 % (McDonald, 1972). These values are initial guesses and can be automatically adjusted by the program, although in this case there are two independent maximum displacements and many different combinations will produce a given acceptance ratio.

Once the new state n has been produced the quantity δH is calculated,

$$\delta H_{nm} = \delta \mathcal{V}_{nm} + P(V_n - V_m) - N\beta^{-1}\ln(V_n/V_m). \tag{4.33}$$

δH_{nm} is closely related to the enthalpy change in moving from state m to state n. Moves are accepted with a probability equal to $\min[1, \exp(-\beta\delta H_{nm})]$ using the techniques discussed in Section 4.4. A move may proceed with a change in particle position or a change in volume or a combination of both.

Eppenga and Frenkel (1984) have pointed out that it may be more convenient to make random changes in $\ln V$ rather than in V itself. A random number $\delta(\ln V)$ is chosen uniformly in some range $[-\delta(\ln V)_{max}, \delta(\ln V)_{max}]$, the volume multiplied by $\exp[\delta(\ln V)]$ and the molecular positions scaled accordingly. The only change to the acceptance/rejection procedure is that the factor N in eqn (4.33) is replaced by $N + 1$.

One important difference between this ensemble and the canonical ensemble is that when a move involves a change in volume, the density of the liquid changes. In this case the long-range corrections to the energy in states m and n are different and must be included directly in the calculation of $\delta \mathcal{V}_{nm}$ (see Section 2.8).

In the general case, changing the volume is computationally more expensive than displacing a molecule. For a molecule displacement there are at most $2(N-1)$ calculations of the pair potential in calculating $\delta \mathcal{V}_{nm}$. In general, a volume change in a pair-additive fluid requires the recalculation of all the $\frac{1}{2}N(N-1)$ interactions. Fortunately, for the simplest potentials, the change in $\mathcal{V}$ with volume can be calculated by scaling. As an example, consider the configurational energy of a Lennard-Jones fluid in state m:

$$\mathcal{V}_m = 4\epsilon \sum_i \sum_{j>i} \left(\frac{\sigma}{L_m s_{ij}^m}\right)^{12} - 4\epsilon \sum_i \sum_{j>i} \left(\frac{\sigma}{L_m s_{ij}^m}\right)^6$$

$$= \mathcal{V}_m^{(12)} - \mathcal{V}_m^{(6)}. \tag{4.34}$$

Here we have divided up the potential into its separate twelfth-power and sixth-power components. If the only change between the states m and n is the length of the box then the energy of the new state is

$$\mathcal{V}_n = \mathcal{V}_m^{(12)} \left(\frac{L_m}{L_n}\right)^{12} + \mathcal{V}_m^{(6)} \left(\frac{L_m}{L_n}\right)^6$$

and

$$\delta \mathcal{V}_{nm} = \delta \mathcal{V}_{nm}^{vol} = \mathcal{V}_m^{(12)} \left[\left(\frac{L_m}{L_n}\right)^{12} - 1\right] + \mathcal{V}_m^{(6)} \left[\left(\frac{L_m}{L_n}\right)^6 - 1\right]. \tag{4.35}$$

This calculation is extremely rapid and only requires that the two components of the potential energy, $\mathcal{V}^{(12)}$ and $\mathcal{V}^{(6)}$, be stored separately. If the potential cutoff is taken to scale with the box length (i.e. $r_c = s_c L$ with s_c constant) then the separate terms $\mathcal{V}_{LRC}^{(12)}$ and $\mathcal{V}_{LRC}^{(6)}$ scale just like $\mathcal{V}^{(12)}$ and $\mathcal{V}^{(6)}$ respectively. If in addition to a box-length change a molecule is simultaneously displaced, then there are two contributions

$$\delta \mathcal{V}_{nm} = \delta \mathcal{V}_{nm}^{dis} + \delta \mathcal{V}_{nm}^{vol} \tag{4.36}$$

Code 4.5 Monte Carlo *NPT*-ensemble for Lennard-Jones atoms

These files are provided online. The program `mc_npt_lj.f90` controls the simulation, reads in the run parameters, selects moves, and writes out the results. It uses the routines in `mc_lj_module.f90` (see Code 4.3) to evaluate the Lennard-Jones potential and actually implement the moves. It also uses utility modules (see Appendix A) for the Metropolis function, input/output, and simulation averages.

```
! mc_npt_lj.f90
! Monte Carlo, NPT ensemble
PROGRAM mc_npt_lj
```

where $\delta V_{nm}^{\text{vol}}$ is given by eqn (4.35) and

$$\delta V_{nm}^{\text{dis}} = \mathcal{V}_n(L_n) - \mathcal{V}_m(L_n). \tag{4.37}$$

Thus the energy change on displacement is obtained using the new box length L_n (think of scaling the box, followed by moving the molecule).

This simple prescription for the calculation of δV_{nm} relies on there being just one characteristic length in the potential function. This may not be the case for some complicated pair potentials, and it is also not true for most molecular models, where intramolecular bond lengths as well as site–site potentials appear. For an interaction site model, simple scaling would imply a non-physical change in the molecular shape. Note that for non-conformal potentials, such as the site–site model, the energy change on scaling can be estimated quickly and accurately (but not exactly) using a Taylor series expansion of $\mathcal{V}_m$ in the scaling ratio, L_n/L_m (Brennan and Madden, 1998). The code for a constant-*NPT* simulation is given in Code 4.5.

By averaging over the states in the Markov chain it is possible to calculate mechanical properties such as the volume and the enthalpy as well as various properties related to their fluctuations. In common with the constant-*NVT* simulation, this method only samples important regions of phase space and it is not possible to calculate the 'statistical' properties such as the Gibbs free energy. During the course of a particular run the virial can be calculated in the usual manner to produce an estimate of the pressure. This calculated pressure (including the long-range correction) should be equal to the input pressure, *P*, used in eqn (4.33) to generate the Markov chain. This test is a useful check of a properly coded constant-*NPT* program.

From the limited evidence available, it appears that the fluctuations of averages calculated in a constant-*NPT* MC simulation are greater than those associated with the averages in a constant-*NVT* simulation. However, the error involved in calculating excess properties of mixtures in the two ensembles is comparable, since they can be arrived at more directly in a constant-*NPT* calculation (McDonald, 1972).

Finally, constant-pressure simulations of hard disks and spheres (Wood, 1968a; 1970) can be readily performed using the methods described in this section. Wood (1968a) has also developed an elegant method for hard-core systems in which the integral over $\exp(-\beta PV)$ in eqn (4.30) is used to define a Laplace transform. The simulation is performed by generating a Markov chain in the transform space using a suitably defined pseudo-potential. This method avoids direct scaling of the box; details can be found in the original paper.

Constant-pressure MC simulation can be particularly difficult to apply to atomic and molecular solids. In this case, even a small, random scaling of the box can change the atom separations far from those associated with a perfect lattice, causing a large change in δV_{nm} and a likely rejection of the trial move. Schultz and Kofke (2011) have suggested a new approach which involves a scaling of the perfect lattice positions, $r_{i,\text{latt}}$, during a volume change δV_{nm} and a separate scaling of the relative positions of atoms with respect to their associated lattice sites. This second scaling depends on the imposed value of the pressure. It is necessary to calculate both δV_{nm} for the atoms and $\delta V_{nm}^{\text{latt}}$ for the atoms on a perfect lattice, during the trial volume change. This method improves the convergence of the Markov chain for solids of Lennard-Jones atoms, hard spheres, and hard dumbbells (Schultz and Kofke, 2011).

4.6   Grand canonical Monte Carlo

In grand canonical Monte Carlo (GCMC) the chemical potential is fixed while the number of molecules fluctuates. The simulations are carried out at constant μ, V, and T, and the average of some property $\mathcal{A}$, is given by

$$\langle \mathcal{A} \rangle_{\mu VT} = \frac{\sum_{N=0}^{\infty} (N!)^{-1} V^N z^N \int \mathrm{d}\mathbf{s}\, \mathcal{A}(\mathbf{s}) \exp\left(-\beta \mathcal{V}(\mathbf{s})\right)}{Q_{\mu VT}} \tag{4.38}$$

where $z = \exp(\beta\mu)/\Lambda^3$ is the activity, Λ is defined in eqn (2.24), and $Q_{\mu VT}$ in eqn (2.32). Again it is convenient to use a set of scaled coordinates $\mathbf{s} = (\mathbf{s}_1, \mathbf{s}_2, \ldots \mathbf{s}_N)$ defined as in eqn (4.31) for each particular value of N. In common with the other ensembles discussed in this chapter only the configurational properties are calculated during the simulation and the ideal gas contributions are added at the end. A minor complication is that these contributions will depend on $\langle N \rangle_{\mu VT}$, which must be calculated during the run. N is not a continuous variable (the minimum change in N is one), and the sum in eqn (4.38) will not be replaced by an integral.

In GCMC the Markov chain is constructed so that the limiting distribution is proportional to

$$\exp\left[-\beta\left(\mathcal{V}(\mathbf{s}) - N\mu\right) - \ln N! - 3N \ln \Lambda + N \ln V\right]. \tag{4.39}$$

A number of methods of generating this chain have been proposed. A method applied in early studies of lattice systems (Salsburg et al., 1959; Chesnut, 1963), uses a set of variables $(c_1, c_2, \ldots)$, each taking the value 0 (unoccupied) or 1 (occupied), to define a configuration. In the simplest approach a trial move attempts to turn either a 'ghost' site ($c_i = 0$) into a real site ($c_i = 1$) or vice versa.

This method has been extended to continuous fluids by Rowley et al. (1975) and Yao et al. (1982). In this application real and ghost molecules are moved throughout the system using the normal Metropolis method for displacement. This means that 'ghost' moves are always accepted because no interactions are involved. In addition there are frequent conversion attempts between 'ghost' and real molecules. Unfortunately a 'ghost' molecule tends to remain close to the position at which its real precursor was destroyed, and is likely to rematerialize, at some later step in the simulation, in this same 'hole' in the liquid. This memory effect does not lead to incorrect results (Barker and Henderson, 1976), but may result in a slow convergence of the chain. The total number of real and ghost molecules, M, must be chosen so that if all the molecules became real, $\mathcal{V}$ would be very high for all possible configurations. In this case the sum in eqn (4.38) can be truncated at M. This analysis makes it clear that in GCMC simulations, we are essentially transferring molecules between our system of interest and an ideal gas system, each of which is limited to a maximum of M molecules. Thus the system properties are measured relative to those of this restricted ideal gas; if M is sufficiently large this should not matter.

Most workers now adopt the original method of Norman and Filinov (1969). In this technique there are three different types of move:

(a) a molecule is displaced;
(b) a molecule is destroyed (no record of its position is kept);
(c) a molecule is created at a random position in the fluid.

Displacement is handled using the normal Metropolis method. If a molecule is destroyed the ratio of the probabilities of the two states is

$$\frac{\rho_n}{\rho_m} = \exp(-\beta\delta\mathcal{V}_{nm})\exp(-\beta\mu)\frac{N\Lambda^3}{V} \tag{4.40}$$

where N is the number of molecules initially in state m. In terms of the activity this is

$$\frac{\rho_n}{\rho_m} = \exp\left[-\beta\delta\mathcal{V}_{nm} + \ln\left(\frac{N}{zV}\right)\right] \equiv \exp(-\beta\delta D_{nm}). \tag{4.41}$$

Here we have defined the 'destruction function' δD_{nm}. A destruction move is accepted with probability $\min[1, \exp(-\beta\delta D_{nm})]$ using the methods of Section 4.4. Finally, in a creation step, similar arguments give

$$\frac{\rho_n}{\rho_m} = \exp\left[-\beta\delta\mathcal{V}_{nm} + \ln\left(\frac{zV}{N+1}\right)\right] \equiv \exp(-\beta\delta C_{nm}) \tag{4.42}$$

(defining the 'creation function' δC_{nm}) and the move is accepted or rejected using the corresponding criterion.

The underlying matrix for the creation of an additional molecule in a fluid of N existing molecules is $\alpha_{nm} = \alpha^c/(N+1)$, where α^c is the probability of making an attempted creation. The underlying matrix for the reverse move is $\alpha_{mn} = \alpha^d/(N+1)$. The condition of microscopic reversibility for a creation/destruction attempt is satisfied if $\alpha_{mn} = \alpha_{nm}$, that is $\alpha_c = \alpha_d$ (Nicholson and Parsonage, 1982). The method outlined allows for the destruction or creation of only one molecule at a time. Except at low densities, moves which involve the addition or removal of more than one molecule would be highly improbable and such changes are not cost-effective (Norman and Filinov, 1969).

Although α^d must equal α^c there is some freedom in choosing between creation/destruction and a simple displacement, α^m. Again Norman and Filinov (1969) varied α^m and found that $\alpha^m = \alpha^d = \alpha^c = 1/3$ gave the fastest convergence of the chain, and these are the values commonly employed. Thus moves, destructions, and creations are selected at random, with equal probability.

Typically, the configurational energy, pressure, and density are calculated as ensemble averages during the course of the GCMC simulations. The beauty of this type of simulation is that the free energy can be calculated directly,

$$A/N = \mu - \langle \mathcal{P} \rangle_{\mu VT} V / \langle N \rangle_{\mu VT} \tag{4.43}$$

and using eqn (4.43) it is possible to determine all the 'statistical' properties of the liquid.

Variations on the method described in this section have been employed by a number of workers. The Metropolis method for creation and destruction can be replaced by a symmetrical algorithm. In this case the decisions for creation and destruction are respectively

$$\text{create if} \quad \left(1 + \frac{N+1}{zV} \exp(\beta \delta \mathcal{V}_{nm}) \right)^{-1} \geq \zeta$$

$$\text{destroy if} \quad \left(1 + \frac{zV}{N} \exp(\beta \delta \mathcal{V}_{nm}) \right)^{-1} \geq \zeta$$

with ζ generated uniformly on $(0, 1)$.

Adams (1974; 1975) has also suggested an alternative formulation which splits the chemical potential into the ideal gas and excess parts:

$$\mu = \mu^{\text{ex}} + \mu^{\text{id}}$$
$$= \left(\mu^{\text{ex}} + k_B T \ln \langle N \rangle_{\mu VT} \right) + k_B T \ln \left(\Lambda^3 / V \right)$$
$$= k_B T B + k_B T \ln(\Lambda^3 / V). \tag{4.44}$$

Adams performed the MC simulation at constant B, V, and T, where B is defined by eqn (4.44). μ can be obtained by calculating $\langle N \rangle_{\mu VT}$ during the run and using it in eqn (4.44). The technique is completely equivalent to the normal method at constant z, V, and T.

There are a number of technical points to be considered in performing GCMC. In common with the constant-NPT ensemble, the density is not constant during the run. In these cases the long-range corrections must be included directly in the calculation of $\delta \mathcal{V}_{nm}$. The corrections should also be applied during the run to other configurational properties such as the virial. If this is not done, difficulties may arise in correcting the pressure at the end of the simulation: this can affect the calculation of the free energy through eqn (4.43) (Barker and Henderson, 1976; Rowley et al., 1978).

A problem which is peculiar to GCMC is that when molecules are created or destroyed, the storage of coordinates needs to be properly updated. One approach is to label the molecules as being 'active' or 'inactive', and use an array to look up the indices of the active ones (Nicholson, 1984). In this way, the coordinates of a deleted molecule are left in place, but never referred to; subsequently they may be overwritten when a new molecule is created. A simpler approach is to replace the coordinates of the deleted molecule immediately with those of the molecule currently stored in the last position of the coordinate array, prior to reducing the number of molecules by one. In this case, newly created molecules are added to the end of the array.

Code 4.6   Monte Carlo μVT-ensemble for Lennard-Jones atoms

These files are provided online. The program mc_zvt_lj.f90 controls the simulation, reads in the run parameters, selects moves, and writes out the results. It uses the routines in mc_lj_module.f90 (see Code 4.3) to evaluate the Lennard-Jones potential, and actually implement the moves (including creation and destruction of atoms) and utility modules (see Appendix A) for the Metropolis function, input/output, and simulation averages.

```
!  mc_zvt_lj.f90
!  Monte Carlo, zVT (grand) ensemble
PROGRAM mc_zvt_lj
```

An example grand canonical simulation program is given in Code 4.6. Grand canonical simulations are more complicated to program than those in the canonical ensemble. The advantage of the method is that it provides a direct route to the 'statistical' properties of the fluid. For example, by determining the free energy of two different phases in two independent GCMC simulations we can say which of the two is thermodynamically stable at a particular μ and T. GCMC is particularly useful for studying inhomogeneous systems such as monolayer and multilayer adsorption near a surface (Whitehouse et al., 1983) or the electrical double-layer (Carnie and Torrie, 1984; Guldbrand et al., 1984). In these systems the surface often attracts the molecules strongly so that when a molecule diffuses into the vicinity of the surface it may tend to remain there throughout the simulation. GCMC additionally destroys particles in the dense region near the surface and creates them in the dilute region away from the surface. In this way it should encourage efficient sampling of some less likely but allowed regions of phase space as well as helping to break up metastable structures near the surface.

GCMC simulations of fluids have not been used widely. The problem is that as the density of the fluid is increased the probability of successful creation or destruction steps becomes small. Creation attempts fail because of the high risk of overlap. Destruction attempts fail because the removal of a particle without the subsequent relaxation of the liquid structure results in the loss of attractive interactions. Clearly this means that destructions in the vicinity of a surface may be infrequent and this somewhat offsets the advantage of GCMC in the simulation of adsorption (Nicholson, 1984). To address these problems, Mezei (1980) has extended the basic method to search for cavities in the fluid which are of an appropriate size to support a creation. Once they are located, creation attempts are made more frequently in the region of a cavity. In the Lennard-Jones fluid at $T^* = 2.0$, the highest density at which the system could be successfully studied was increased from $\rho^* = 0.65$ (conventional GCMC) to $\rho^* = 0.85$ (extended GCMC). Techniques for preferential sampling close to a molecule or a cavity are discussed in Section 9.3.

GCMC can be readily extended to simulate fluid mixtures. In a two-component mixture of atoms A and B, the two chemical potentials, μ_A and μ_B, are fixed and the mole fractions $\langle x_A \rangle$ and $\langle x_B \rangle$ are calculated during the course of simulation. A decision to create or destroy atoms A or B can be made with a fixed but arbitrary probability; once the species has been chosen, a creation or destruction attempt should be made with equal probability. In a two-component mixture, it is possible to swap the identities of atoms A and B without changing their positions. If we swap an atom from A to B, the trial move is accepted with a probability given by

$$\min\left[1, \frac{z_B N_A}{(N_B + 1)z_A} \exp(-\beta \delta V_{nm})\right]$$

where z_A and z_B are the respective activities (Cracknell et al., 1993; Lachet et al., 1997).

4.7   Semi-grand Monte Carlo

In the semi-grand ensemble the total number of atoms, N, is fixed but the number of atoms of a particular species i is allowed to vary at a fixed chemical potential difference, $\mu_i - \mu_1$. The partition function, eqn (2.37), is defined with the volume, V, and the temperature, T, fixed. For simulating mixtures it is often more convenient to work at a fixed pressure, P, and we can define a new semi-grand partition function using the Legendre transform of eqn (2.37) (Kofke and Glandt, 1988)

$$Q_{\{\mu_i | i \neq 1\}NPT} = \frac{1}{V_0} \int dV \, \exp(-\beta PV) \frac{V^N}{\Lambda^{3N} N!}$$

$$\times \sum_{i_1=1}^{c} \cdots \sum_{i_N=1}^{c} \left[\prod_{i=1}^{c} \exp\left((\mu_i - \mu_1)N_i/k_B T\right)\right] \int ds \, \exp\left(-\mathcal{V}(s)/k_B T\right) \tag{4.45}$$

where the sums are over the c possible identities for each atom and we assume each atom has the same mass. Eqn (4.45) can be simplified by introducing the fugacity, f_i

$$\mu_i(P, T, \{x_i\}) = \mu_i^{\ominus}(T) + k_B T \ln f_i \tag{4.46}$$

where $\mu_i^{\ominus}(T)$ is the chemical potential in the ideal gas reference state, $P = 1$ bar. It is more convenient to introduce the fugacity fraction ξ_i

$$\xi_i = \frac{f_i}{\sum_{i=1}^{c} f_i}. \tag{4.47}$$

This is a convenient quantity that varies from zero to unity as the mixture composition changes from pure species 1 to pure species i (Mehta and Kofke, 1994). Then the semi-grand partition function becomes

$$Q_{\{\mu_i | i \neq 1\}NPT} = \frac{1}{V_0} \int dV \, \exp(-\beta PV) \frac{V^N}{\Lambda^{3N} N!}$$

$$\times \sum_{i_1=1}^{c} \cdots \sum_{i_N=1}^{c} \left[\prod_{i=1}^{c} \left(\frac{\xi_i}{\xi_1}\right)^{N_i}\right] \int ds \, \exp\left(-\beta \mathcal{V}(s)\right). \tag{4.48}$$

The average of a property, $\mathcal{A}$, in the semi-grand ensemble is

$$\langle \mathcal{A} \rangle = \frac{\int dV \, \exp(-\beta PV) V^N \sum_{i_1=1}^{c} \cdots \sum_{i_N=1}^{c} \prod_{i=1}^{c} (\xi_i/\xi_1)^{N_i} \int ds \, \mathcal{A}(s) \exp\left(-\beta \mathcal{V}(s)\right)}{(V_0 \Lambda^{3N} N!) \, Q_{\{\mu_i | i \neq 1\}NPT}}. \tag{4.49}$$

In the semi-grand Monte Carlo method, the Markov chain is constructed so that the limiting distribution is proportional to

$$\exp\left[-\beta\left(\mathcal{V}(\mathbf{s}) + PV\right) + N\ln V + \sum_{i=1}^{c} N_i \ln \frac{\xi_i}{\xi_1}\right]. \tag{4.50}$$

In the course of the simulation, P, T, N, and $\{\xi_i | i \neq 1\}$ remain fixed, while V, $\{N_i\}$, and the atom positions, $\mathbf{s}$, are allowed to vary. This Markov chain is realized with three types of move: atom displacements, random volume changes, and a switch in the identity of two atoms. The first two types of move have already been discussed in detail. To switch the identity of a pair, an atom is chosen at random (this atom is of type i) and it is assigned a new identity, i', chosen randomly from all the atom types. In this way, only the terms involving i and i' in the sum in eqn (4.50) are involved in the identity switch. Then for an attempted switch, the ratio of the probabilities in the new trial state and the old state is

$$\frac{\rho_n}{\rho_m} = \frac{(\xi_i/\xi_1)^{N_i - 1}(\xi_{i'}/\xi_1)^{N_{i'} + 1}}{(\xi_i/\xi_1)^{N_i}(\xi_{i'}/\xi_1)^{N_{i'}}} \exp(-\beta\delta\mathcal{V}_{nm}) = \frac{\xi_{i'}}{\xi_i}\exp(-\beta\delta\mathcal{V}_{nm}) \tag{4.51}$$

where $\delta\mathcal{V}_{nm}$ is the change in configurational energy resulting from the identity switch. The trial switch is accepted with a probability given by $\min(1, \rho_n/\rho_m)$. The underlying stochastic transition matrix is symmetric, with $\alpha_{nm} = \alpha_{mn} = 1/(Nc)$.

The semi-grand ensemble has been used to simulate simple mixtures (Caccamo et al., 1998), liquid crystalline mixtures (Bates and Frenkel, 1998), and complex ionic mixtures (Madurga et al., 2011). The method is particular suited to the study of fluids with variable polydispersity. These are simulations in which the chemical potential distribution, $\mu(\sigma)$, is fixed, and in which the number or density distribution, $\rho(\sigma)$, is calculated (Bolhuis and Kofke, 1996; Kofke and Bolhuis, 1999). The variable σ, describing the polydispersity of the fluid, could, for example, be the diameter of a polydisperse mixture of hard spheres. In a semi-grand exchange move, σ is changed by selecting a particle j at random, changing its diameter by a small amount from σ_j to σ'_j, and calculating $\delta\mu_{nm} = \mu(\sigma'_j) - \mu(\sigma_j)$. The trial move is accepted with a probability given by

$$\min\left(1, \exp\left[+\beta(\delta\mu_{nm} - \delta\mathcal{V}_{nm})\right]\right).$$

The more difficult case of simulating at fixed polydispersity, $\rho(\sigma)$, has also been tackled in the grand canonical (Wilding and Sollich, 2004) and semi-grand (Wilding, 2009) ensembles.

4.8 Molecular liquids

4.8.1 Rigid molecules

In the MC simulation of a molecular liquid the underlying matrix of the Markov chain is altered to allow moves which usually consist of a combined translation and rotation of one molecule. Sequences involving a number of purely translational and purely rotational steps are perfectly proper but are not usually exploited in the simulation of molecular liquids. (There have been a number of simulations of idealized models of liquid crystals and plastic crystals where the centres of the molecules are fixed to a three-dimensional lattice. These simulations consist of purely rotational moves (see e.g. Luckhurst and Simpson, 1982; O'Shea, 1978)).

The translational part of the move is carried out by randomly displacing the centre of mass of a molecule along each of the space-fixed axes. As before, the maximum displacement is governed by the adjustable parameter δr_{max}. The orientation of a molecule is often described in terms of the Euler angles ϕ, θ, ψ mentioned in Section 3.3.1 (see Goldstein, 1980). A change in orientation can be achieved by taking small random displacements in each of the Euler angles of molecule i.

$$\phi_i^n = \phi_i^m + (2\zeta_1 - 1)\delta\phi_{max} \tag{4.52a}$$
$$\theta_i^n = \theta_i^m + (2\zeta_2 - 1)\delta\theta_{max} \tag{4.52b}$$
$$\psi_i^n = \psi_i^m + (2\zeta_3 - 1)\delta\psi_{max} \tag{4.52c}$$

where $\delta\phi_{max}$, $\delta\theta_{max}$ and $\delta\psi_{max}$ are the maximum displacements in the Euler angles.

In an MC step the ratio of the probabilities of the two states is given by

$$\frac{\rho_n}{\rho_m} = \frac{\exp\left(-\beta(\mathcal{V}_m + \delta\mathcal{V}_{nm})\right) dr^n d\Omega^n}{\exp(-\beta\mathcal{V}_m) dr^m d\Omega^m}. \tag{4.53}$$

The appropriate volume elements have been included to convert the probability densities into probabilities. $d\Omega^m = \prod_{i=1}^N d\Omega_i^m$, and $d\Omega_i^m = \sin\theta_i^m d\psi_i^m d\theta_i^m d\phi_i^m / \Omega$ for molecule i in state m. Ω is a constant which is $8\pi^2$ for non-linear molecules. In the case of linear molecules, the angle ψ is not required to define the orientation, and $\Omega = 4\pi$. The volume elements for states m and n have not previously been included in the ratio ρ_m/ρ_n (see eqn (4.29)), for the simple reason that they are the same in both states for a translational move, and cancel. For a move which only involves one molecule, i,

$$\frac{\rho_n}{\rho_m} = \exp\left(-\beta\delta\mathcal{V}_{nm}\right)\frac{\sin\theta_i^n}{\sin\theta_i^m}. \tag{4.54}$$

The ratio of the sines must appear in the transition matrix π_{mn} either in the acceptance/rejection criterion or in the underlying matrix element α_{mn}. This last approach is most convenient. It amounts to choosing random displacements in $\cos\theta_i$ rather than in θ_i, so eqn (4.52b) is replaced by

$$\cos\theta_i^n = \cos\theta_i^m + (2\zeta_2 - 1)\delta(\cos\theta)_{max}. \tag{4.55}$$

Then the usual Metropolis recipe of accepting with a probability of $\min[1, \exp(-\beta\delta\mathcal{V}_{nm})]$ is used. Including the $\sin\theta$ factor in the underlying chain avoids difficulties with $\theta_i^m = 0$ analogous to the problems mentioned in Section 3.3.1. Equations (4.52a), (4.52c), and (4.55) move a molecule from one orientational state into any one of its neighbouring orientational states with equal probability and fulfil the condition of microscopic reversibility.

It is useful to keep the angles which describe the orientation of a particular molecule in the appropriate range $(-\pi, \pi)$ for ψ and ϕ, and $(0, \pi)$ for θ. This is not essential, but avoids unnecessary work and possible overflow in the subsequent evaluation of any trigonometric functions. This can be done by a piece of code which is rather like that used to implement periodic boundary conditions. If dphi_max is the maximum change in ϕ and twopi stores the value 2π, one can use statements similar to

```
CALL RANDOM_NUMBER ( zeta )
phi = phi + 2.0 * ( zeta - 1.0 ) * dphi_max
phi = phi - ANINT ( phi / twopi ) * twopi
```

with similar code for ψ. In the case of eqn (4.55), it is necessary to keep $\cos\theta$ in the range $(-1, 1)$. One can either simply reject moves which would go outside this range, or implement the following correction (with dcos_max holding the value of $\delta(\cos\theta)_{\max}$)

```
cos_theta = cos_theta + 2.0 * ( zeta - 1.0 ) * dcos_max
cos_theta = cos_theta - ANINT ( cos_theta / 2.0 ) * 2.0
```

Note that when the ANINT function is not zero the molecule is rotated by π. For unsymmetrical molecules, this may result in a large trial energy change, which is likely to be rejected; nonetheless the method is correct.

More commonly, the orientation of a linear molecule is stored as a unit vector **e**, rather than in terms of angles. There are several ways of generating unbiased, small, changes in the orientation, suitable to use as trial moves in combination with an unmodified Metropolis formula based on the change in potential. One suggestion (Jansoone, 1974) is to randomly select new orientations $\mathbf{e}_i^n$, uniformly on the surface of a sphere, until one is generated that satisfies an inequality

$$\mathbf{e}_i^n \cdot \mathbf{e}_i^m > 1 - d_{\max}, \tag{4.56}$$

where $d_{\max} \ll 1$ controls the size of the maximum displacement. When such a trial orientation is obtained, the algorithm proceeds to calculate the potential-energy change and apply the Metropolis criterion. Another possibility is to set

$$\mathbf{e}_i^n = \mathbf{e}_i^m + \delta\mathbf{e}$$

where $\delta\mathbf{e}$ is a small vector whose orientation is chosen uniformly in space, and then renormalize $\mathbf{e}_i^n$ afterwards. A third approach is to rotate the orientation vector by a small, randomly chosen, angle, using eqn (3.27). The axis may be chosen randomly in space, or one of the three Cartesian axes may be chosen at random (Barker and Watts, 1969). Examples of code for these methods of generating a random direction in space, and a new orientation, are given in the utility modules of Appendix A.

Perhaps the most general approach to describing the orientation of polyatomic molecules is to use the quaternion parameters introduced in Section 3.3.1. A random orientation in space can be generated by choosing a random unit quaternion on $\mathbb{S}^3$ (Marsaglia, 1972; Vesely, 1982). This can be achieved by repeatedly selecting two random numbers, ζ_1, ζ_2 on $(-1, 1)$ until $s_1 = \zeta_1^2 + \zeta_2^2 < 1$. Then select ζ_3, ζ_4 on $(-1, 1)$ until $s_2 = \zeta_3^2 + \zeta_4^2 < 1$. The random quaternion is

$$\mathbf{a} = (\zeta_1, \zeta_2, \zeta_3\sqrt{(1-s_1)/s_2}, \zeta_4\sqrt{(1-s_1)/s_2}). \tag{4.57}$$

This is an extension of the technique described in Appendix E.4 for generating a random vector on the surface of a sphere. The quaternion, **a**, can be used to construct the rotation

Code 4.7   Monte Carlo program using quaternions

These files are provided online. The program `mc_nvt_poly_lj.f90` controls the simulation, reads in the run parameters, selects moves, and writes out the results. It uses various routines in `mc_poly_lj_module.f90`, and utility module routines (see Appendix A) for input/output and simulation averages. The model is a set of rigid polyatomic molecules, whose atoms interact through a shifted Lennard-Jones potential.

```
! mc_nvt_poly_lj.f90
! Monte Carlo, NVT ensemble, polyatomic molecule
PROGRAM mc_nvt_poly_lj

! mc_poly_lj_module.f90
! Routines for MC simulation, polyatomic molecule, LJ atoms
MODULE mc_module
```

matrix, eqn (3.40), and hence the positions, in a space-fixed frame, of the atoms within a molecule.

A random orientational displacement can be accomplished, just as in the linear-molecule case, by rotating $\mathbf{a}$ through an angle θ about some arbitrary axis $\mathbf{s}$, controlling the maximum size of the displacement with a parameter d. This can be readily accomplished by selecting $\mathbf{s}$ from an arbitrary but even distribution. Karney (2007) suggests that a useful choice is a three-dimensional Gaussian distribution (see Appendix E.3) of the form

$$\rho(\mathbf{s}) = \frac{\exp(-|\mathbf{s}|^2/2d^2)}{(2\pi)^{3/2}d^3},\tag{4.58}$$

where larger values of d correspond to larger orientational displacements. (Note that the use of a uniform distribution results in a singularity at $|\mathbf{s}| = 0$.) The new quaternion, $\mathbf{a}'$ is given by

$$\mathbf{a}' = (\cos\tfrac{1}{2}\theta, \hat{\mathbf{s}}\sin\tfrac{1}{2}\theta) \otimes \mathbf{a}\tag{4.59}$$

where $\theta = |\mathbf{s}|$ and $\hat{\mathbf{s}} = \mathbf{s}/|\mathbf{s}|$ and the quaternion multiplication is defined in eqn (3.36a). If $\theta = |\mathbf{s}| > \pi$, then the orientation will be identical to the wrapped orientation $\mathbf{s} - 2\pi\hat{\mathbf{s}}$. Karney (2007) recommends rejecting moves with $|\mathbf{s}| > \pi$ to avoid a subtlety associated with detailed balance.

One difficulty with MC methods for molecular fluids is that there are usually a number of parameters governing the maximum translational and orientational displacement of a molecule during a move. As usual, these parameters can be adjusted automatically to give an acceptance rate of ≈ 0.5, but there is not a unique set of maximum displacement parameters which will achieve this. A sensible set of values is best obtained by trial and error for the particular simulation in hand. An example of an MC program using quaternions is given in Code 4.7.

The MC method is particularly useful for simulating hard-core molecules. The complicated MD schemes mentioned in Section 3.7.2 can be avoided and the program consists

Code 4.8   Monte Carlo of hard spherocylinders

These files are provided online. The program `mc_nvt_sc.f90` controls the simulation, reads in the run parameters, selects moves, and writes out the results. It uses the spherocylinder overlap routines in `mc_sc_module.f90`, and utility module routines (see Appendix A) for input/output and simulation averages.

```
! mc_nvt_sc.f90
! Monte Carlo, NVT ensemble, linear hard molecules
PROGRAM mc_nvt_sc

! mc_sc_module.f90
! Overlap routines for MC simulation, hard spherocylinders
MODULE mc_module
```

simply of choosing one of the aforementioned schemes for moving a molecule and an algorithm for checking for overlap. A simple MC program for hard spherocylinders is given in Code 4.8. Again, as discussed in Section 2.4, the pressure must be calculated by a box-scaling or related method.

The MC method has been used successfully in the canonical ensemble for simulating hard-core molecules (Streett and Tildesley, 1978; Wojcik and Gubbins, 1983) and more realistic linear and non-linear molecules (Barker and Watts, 1969; Romano and Singer, 1979). Simulations of molecular fluids have also been attempted in the isothermal–isobaric ensemble (Owicki and Scheraga, 1977a; Eppenga and Frenkel, 1984). Simulations of molecular fluids in the grand canonical ensemble use the destruction and creation steps given by eqns (4.41) and (4.42). In this case the activity $z = \exp(\beta\mu)q_{id}$. For a molecular fluid q_{id} is a product of a translational term Λ^{-3} and terms corresponding to the rotation, vibration and electronic structure of the molecule depending on the precise model (Hill, 1956). In making an attempted creation it is necessary to insert the molecule at a random position and to choose a random orientation, possibly using the quaternion method of eqn (4.59). In dense systems of highly anisotropic molecules, these insertions are likely to fail without the use of some biasing technique and the method has not been widely used. For small molecules the method can been used to study the adsorption of gases and their mixtures into pores (Cracknell et al., 1993; Smit, 1995; Lachet et al., 1997) but the preferred approach for longer molecules would be the configurational-bias method discussed in Chapter 9.

4.8.2   Non-rigid molecules

Non-rigidity introduces new difficulties into the MC technique. The problem in this case is to find a suitable set of generalized coordinates to describe the positions and momenta of the molecules. Once these have been established, the integrations over the momenta can be performed analytically, which will leave just the configurational part of the ensemble average. However, the integration over momenta will produce complicated Jacobians in the configurational integral, one for each molecule (see Section 2.10). The Jacobian will

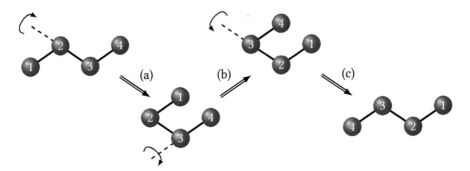

Fig. 4.5 A possible method for moving a chain molecule (butane), subject to bond length and angle constraints, in an MC simulation. For illustrative purposes, all the atoms are taken to be coplanar and the rotation angles are equal to π radians.

be some function, for example, of the angles θ, ϕ, which describe the overall orientation of the molecule, and the bond bending and torsion angles which describe the internal configuration. A simple example of this type of term is the $\sin \theta_i$ in the configurational integral for rigid molecules, which comes from the integration over the momenta $(p_\phi)_i$. As we have already seen in Section 4.8.1, these Jacobians are important in calculating the ratio ρ_n/ρ_m used in generating the Markov chain in the Metropolis method or, correspondingly, in designing the correct underlying stochastic matrix. For non-rigid molecules, correctly handling the Jacobian terms is more difficult.

This problem can be solved satisfactorily for the class of non-rigid molecules where the overall moment of inertia is independent of the coordinates of internal rotation (e.g. isobutane, acetone) (Pitzer and Gwinn, 1942). Generalized coordinates have also been developed for a non-rigid model of butane, which does not fall into this simple class (Ryckaert and Bellemans, 1975; Pear and Weiner, 1979), but the expressions are complicated and become increasingly so for longer molecules.

One way of working with generalized coordinates is as follows. In butane (see Section 1.3), it is possible to constrain bond lengths and bond bending angles, while allowing the torsional angle to change according to its potential function. The movement of the molecule in the simulation is achieved by random movements of randomly chosen atoms subject to the required constraints (Curro, 1974). An example of this is shown for butane in Fig. 4.5. A typical MC sequence might be (assuming that each move is accepted):

(a) atom 1 is moved by rotating around the 2–3 bond;
(b) atoms 1 and 2 are moved simultaneously by rotating around the 3–4 bond;
(c) atom 4 is moved by rotating around the 2–3 bond.

Moves (a) and (c) involve a random displacement of the torsional angle ϕ, in the range $(-\pi, \pi)$. The entire molecule is translated and rotated through space by making random rotations of atoms around randomly chosen bonds. We can also include an explicit translation of the whole molecule, and an overall rotation about one of the space-fixed axes. The disadvantage of this simple approach at high density is that a small rotation around the 1–2 bond can cause a substantial movement of atom 4, which is likely to result in overlap and a high rejection rate for new configurations.

If we consider the case of the simplified butane molecule introduced in Sections 1.3.3 and 2.10, then a trial MC move might consist of a translation and rotation of the whole molecule and a change in the internal configuration made by choosing random increments in $d(\cos \theta)$, $d(\cos \theta')$, and $d\phi$ (see Fig. 1.10). To avoid the artefacts associated with the constraint approximation, the Markov chain should be generated with a limiting distribution (see eqns (2.167), (2.170)) proportional to

$$\exp\left(-\beta(\mathcal{V} + \mathcal{V}_c)\right) = \exp\left(-\beta\left[\mathcal{V} + \tfrac{1}{2}k_B T \ln(2 + \sin^2 \theta + \sin^2 \theta')\right]\right). \qquad (4.60)$$

If θ and θ' stay close to their equilibrium values throughout, it might be possible to introduce only a small error by neglecting the constraint potential $\mathcal{V}_c$ in eqn (4.60). The constraint term becomes more complicated and important in the case of bond-angle constraints. Small non-rigid molecules can probably be best treated using the configurational-bias MC method (see Chapter 9).

There have been many volumes devoted to the study of polymer systems using the MC method (see e.g. Binder, 1984; Landau and Binder, 2009). Single chains can be simulated using crude MC methods. In this technique a polymer chain of specified length is built up randomly in space (Lal and Spencer, 1971) or on a lattice (Suzuki and Nakata, 1970). A chain is abandoned if a substantial overlap is introduced during its construction. When a large number N of chains of the required length have been produced, the average of a property (such as the end-to-end distance) is calculated from

$$\langle \mathcal{A} \rangle = \frac{\sum_{i=1}^{N} \mathcal{A}_i \exp(-\beta \mathcal{V}_i)}{\sum_{i=1}^{N} \exp(-\beta \mathcal{V}_i)} \qquad (4.61)$$

where the sums range over all the N polymer chains. The approach is inapplicable for a dense fluid of chains.

A more conventional MC method, which avoids this problem, was suggested by Wall and Mandel (1975). In a real fluid a chain is likely to move in a slithering fashion: the head of the chain moves to a new position and the rest of the chain follows like a snake or lizard. This type of motion is termed 'reptation' (de Gennes, 1971). Such 'reptation MC' algorithms have been applied to chains on a three-dimensional lattice (Wall et al., 1977) and to continuum fluids (Brender and Lax, 1983). Bishop et al. (1980) have developed a reptation algorithm which is suitable for a chain with arbitrary intermolecular and intramolecular potentials in a continuum fluid. The method exploits the Metropolis solution to the transition matrix to asymptotically sample the Boltzmann distribution. In the case studied by Bishop et al., the model consists of N chains each containing ℓ atoms. All the atoms in the fluid interact through the repulsive part of the Lennard-Jones potential, $v^{RLJ}(r)$, eqn (1.10a); this interaction controls the excluded volume of the chains. Adjacent atoms in the same chain interact additionally through the FENE potential of eqn (1.33) and Fig. 1.12, with typical parameters $R_0 = 1.95$ and $k = 20$. This gives a modified harmonic potential

$$v^H(r) = v^{FENE}(r) + v^{RLJ}(r). \qquad (4.62)$$

Any reasonable forms for the intermolecular and intramolecular potentials can be used in this approach.

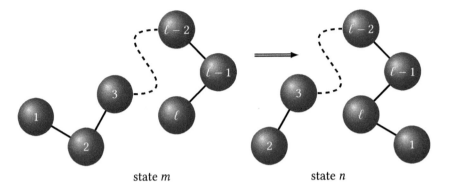

state *m*   state *n*

Fig. 4.6 The reptation of a polymer chain, where a single atom is transferred from the tail of the polymer to its head, while atoms 2 to ℓ are unchanged.

Each chain, see Fig. 4.6, is considered in turn and one end is chosen randomly as the head. The initial coordinates of the atoms in a particular chain, i, are

$$\{\mathbf{r}_{i,1}^m, \mathbf{r}_{i,2}^m, \ldots, \mathbf{r}_{i,\ell}^m\}.$$

A new position is selected for the head of chain i

$$\mathbf{r}_{i,1}^n = \mathbf{r}_{i,\ell}^m + \delta\mathbf{r}. \tag{4.63}$$

The direction of $\delta\mathbf{r}$ is chosen at random on the surface of a sphere, and the magnitude δr is chosen according to the probability distribution corresponding to $v^H(\delta r)$ using a rejection technique (see Appendix E). Thus, the intramolecular bonding potential is used in selecting the trial move (other examples of introducing bias in this way will be seen in Chapter 9). The chain has a new trial configuration

$$\{\mathbf{r}_{i,2}^n, \mathbf{r}_{i,3}^n, \ldots, \mathbf{r}_{i,\ell}^n, \mathbf{r}_{i,1}^n\},$$

where the positions of atoms $\{2 \ldots \ell\}$ are unchanged in moving from the old state m to the new state n. The atom at $\mathbf{r}_{i,1}^m$ has been transferred to the other end of the chain in the new state n. The change in non-bonded interactions in creating a new configuration is calculated by summing over all the atoms

$$\delta\mathcal{V}_{nm} = \sum_{a=2}^{\ell} v^{RLJ}\left(|\mathbf{r}_{i,a} - \mathbf{r}_{i,1}^n|\right) - v^{RLJ}\left(|\mathbf{r}_{i,a} - \mathbf{r}_{i,1}^m|\right)$$

$$+ \sum_{j\neq i}\sum_{a=1}^{\ell} v^{RLJ}\left(|\mathbf{r}_{j,a} - \mathbf{r}_{i,1}^n|\right) - v^{RLJ}\left(|\mathbf{r}_{j,a} - \mathbf{r}_{i,1}^m|\right). \tag{4.64}$$

Non-bonded interactions from chain i and from all other chains j are included here; we highlight the fact that all the $\mathbf{r}_{j,a}$ are unchanged, as well as all the $\mathbf{r}_{i,a}$ except $\mathbf{r}_{i,1}$, by omitting their superscripts. $\delta\mathcal{V}_{nm}$ is used to decide whether the move should be accepted or rejected according to the Metropolis criterion. As usual, the state following a rejected move is recounted in the calculation of simulation averages. The approach is simple to execute when there are no geometrical constraints to take into account. For example, in the work of Bishop et al. (1980), all the atoms are free to move under the influence of the potentials.

In general, the reptation method is not a particularly efficient algorithm for dense polymers because it does not induce significant changes in the internal structure of the chain. At high density it can produce a series of states in which an atom simply moves backwards and forwards between the head and the tail. More recently, polymer simulations have been performed using the configurational-bias method which attempts to rebuild the entire polymer by searching for regions of low energy (high probability) for each consecutive atom, and also by the introduction of a number of concerted moves which involve rotating or replacing complete segments of the chain. These lie beyond the scope of the normal Metropolis approach and we shall consider them in detail in Chapter 9.

4.9   Parallel tempering

The parallel tempering technique is a method of simulating a set of systems, each sampling an equilibrium ensemble, which frequently exchange configurations with each other. This is particularly useful for exploring a highly corrugated energy surface with deep local minima. In these circumstances, the application of a normal Metropolis Monte Carlo method with a value of $k_B T$ well below the barrier height, is likely to result in a system trapped in one of its local minima and the simulated phase space trajectory will be non-ergodic (see Section 2.1). One way to overcome this problem is to simulate many replicas of the system simultaneously, each at a different temperature. At the highest temperatures the system is likely to cross the barriers frequently and explore all of the regions of the energy surface. By exchanging the coordinates between the replicas, information relating to many different energy minima is transferred to the systems at lower temperature. For this reason the method is often referred to as replica exchange Monte Carlo (REMC).

This method employs the Metropolis algorithm for atom moves within an individual replica and introduces additional attempts to exchange the coordinates of two replicas (Swendsen and Wang, 1986; Geyer, 1991). Consider a set of M replicas, each containing the same number of atoms interacting through the same intermolecular potential. The temperatures are $T_{M-1} > T_{M-2} > \ldots T_I > T_J \ldots T_2 > T_1 > T_0$, and $T = T_0$ is often of most interest. In an exchange move, two adjacent replicas, I and J, with a small temperature difference, are chosen at random. Since the replicas are independent, the probability of the initial state is

$$\rho_m \propto \exp\left(-\beta_I \mathcal{V}(\mathbf{r}^I)\right) \exp\left(-\beta_J \mathcal{V}(\mathbf{r}^J)\right) \tag{4.65}$$

where $\mathbf{r}^I = \{\mathbf{r}_1^I, \mathbf{r}_2^I, \ldots, \mathbf{r}_N^I\}$ are the coordinates of the atoms in replica I, and similarly for J. After the attempted exchange of coordinates, the probability of the final state is

$$\rho_n \propto \exp\left(-\beta_I \mathcal{V}(\mathbf{r}^J)\right) \exp\left(-\beta_J \mathcal{V}(\mathbf{r}^I)\right). \tag{4.66}$$

The random choice of replicas ensures that $\alpha_{nm} = \alpha_{mn}$ and the exchange is accepted with a probability given by

$$\min\left(1, \exp\left[(\beta_I - \beta_J)\left(\mathcal{V}(\mathbf{r}^I) - \mathcal{V}(\mathbf{r}^J)\right)\right]\right). \tag{4.67}$$

This exchange of atomic coordinates does not upset the equilibrium distribution of a particular replica and, unlike the simulated annealing methods (Kirkpatrick et al., 1983), proper equilibrium Monte Carlo simulations are performed at each temperature. Since the total potential energy of a configuration is already calculated in the Metropolis method, exchange moves are relatively inexpensive. Typically in a simulation, 90 % of the trial moves would be attempted atom displacements and 10 % attempted replica exchanges (Falcioni and Deem, 1999); approximately 20 % of attempted exchanges should be accepted (Rathore et al., 2005). The basic method is well-suited to the simulation of polymer melts and conformational properties of proteins (Hansmann, 1997) where it opens up the possibility for efficient studies of the folding thermodynamics of more detailed protein models. After the simulations have been completed, it is possible to obtain an optimized estimate of properties at a range of temperatures using reweighted histogram techniques of the type described in Chapter 9 (Chodera et al., 2007).

The method can be readily extended to the grand canonical ensemble. In this case, the two replicas chosen at random are simulations performed at fixed chemical potentials, μ_I and μ_J, and (inverse) temperatures β_I and β_J, using the technique described in Section 4.6. An exchange move consists of swapping the atom coordinates in replica I with those in replica J. This move is accepted with a probability given by

$$\min\left(1, \left(\frac{\beta_I}{\beta_J}\right)^{3(N_I-N_J)/2} \exp\left[(\beta_I-\beta_J)\left(\mathcal{V}(\mathbf{r}^I)-\mathcal{V}(\mathbf{r}^J)\right)-(\beta_I\mu_I-\beta_J\mu_J)(N_I-N_J)\right]\right). \quad (4.68)$$

The factor in front of the exponential arises from the temperature dependence of the de Broglie wavelength. This technique has been used by Yan and de Pablo (1999) to construct a phase diagram for the Lennard-Jones fluid and the restricted primitive model of an ionic fluid. In these simulations the probability density $\rho_{\mu\beta}(N)$ is calculated, and the densities, corresponding to the coexisting liquid and gas, are obtained from the two peaks in this function. We note that at temperatures well below the critical point of the fluid, these two peaks would not be observed in a normal grand canonical MC simulation because the energy barrier for the system to move from the liquid to the gas phase would be too high. However, using the parallel tempering method, it is possible to move the system up towards the critical point, where such exchanges occur readily, and then back down to either the liquid or gas region, thus sampling the whole curve. In this case, the reweighted histogram techniques described in Chapter 9 are particularly useful.

The method can also be extended to the case where the replicas have the same values of N and T, but different values of the potential function between the atoms (Bunker and Dünweg, 2001; Fukunishi et al., 2002). The method has been used by Bunker and Dünweg (2001) to simulate polymer melts, each of which has a slightly different repulsive core potential for the interaction between monomers. The exchange moves connect the full excluded-volume potential to an ideal gas of chains. The parallel tempering, with swaps of the polymer configurations between adjacent potential functions, works with additional large-scale attempted pivot and translation moves for the polymer. These moves only have a realistic acceptance probability as the limit of the ideal gas chains is reached. For chains of length 200 monomers, the parallel tempering technique delivers a speedup of greater than a factor of eight, as measured by the integrated autocorrelation time for the first Rouse mode of the polymer.

Returning to the properties of parallel tempering for variable T, it is important to consider the optimum choice of temperatures for the set of replicas. If the probability distributions, $\rho_{NT}(\mathcal{V})$, of two replicas at two different temperatures do not exhibit a significant overlap then the acceptance ratio for the exchange moves between these replicas will be low and the advantages of the method lost. More quantitatively, Kofke (2002) has shown that the acceptance ratio of exchange moves is related to the entropy difference between the replicas

$$\langle p_{\text{acc}} \rangle \sim \exp(-\Delta S/k_B) \sim \left(\frac{T_J}{T_I} \right)^{C_V/k_B} \tag{4.69}$$

where $\Delta S > 0$, that is, the entropy of the high-temperature replica T_I minus that of the low-temperature replica, T_J, and C_V is the specific heat of the system at constant volume. Equation (4.69) confirms that the average transition probability decreases with increasing system size N, since ΔS and C_V are extensive properties. The larger the system, the greater the number of replicas required to connect the high-temperature and low-temperature states. The average transition probability also decreases with increasing temperature difference between replicas. Kofke's analysis demonstrates that the acceptance ratio for exchange can be made uniform across a set of replicas if the ratio of the temperatures of adjacent replicas is fixed. Kone and Kofke (2005) suggest that temperatures should be chosen such that approximately 20 % of swap attempts are accepted. Iterative methods of adjusting the temperature differences during the simulation are discussed in the review of Earl and Deem (2005).

Replica exchange is very commonly combined with MD packages (REMD) because the actual process of exchanging configurations can be handled by an external script rather than requiring a modification to the code. Some care is needed, however. Velocities or momenta need to be exchanged as well as positions. It is usual to run the individual simulations in the canonical ensemble, using one of the thermostatting algorithms discussed in Section 3.8, and to base the exchange acceptance–rejection criterion on the potential energies alone through eqn (4.67). On exchanging, the momenta are rescaled, to be consistent with the change in temperature (in other words, to satisfy detailed balance). The necessary prescriptions are (Sugita and Okamoto, 1999; Mori and Okamoto, 2010)

$$\mathbf{p}^I \rightarrow \mathbf{p}^I \sqrt{\frac{T_J}{T_I}}, \quad \mathbf{p}^J \rightarrow \mathbf{p}^J \sqrt{\frac{T_I}{T_J}}, \quad \text{or in general} \quad \mathbf{p} \rightarrow \mathbf{p} \sqrt{\frac{T_{\text{new}}}{T_{\text{old}}}}$$

where the momenta $\mathbf{p}^I$ are for the system going $T_I \rightarrow T_J$ and the $\mathbf{p}^J$ are going $T_J \rightarrow T_I$. This applies for Langevin, Andersen, and Nosé–Hoover thermostats. Of course, for the latter case, there is a thermostat velocity variable, which should also be rescaled

$$\zeta \rightarrow \zeta \sqrt{\frac{T_{\text{new}}/Q_{\text{new}}}{T_{\text{old}}/Q_{\text{old}}}}$$

to cover the case where the thermostat masses Q are different at the different temperatures. A similar prescription applies to Nosé–Hoover chains. This is one of those cases where it is essential that the MD thermostat correctly samples the canonical ensemble. Cooke and

Example 4.1 Sweetening the glass

The sugars sucrose and trehalose are used in the preservation of biological samples containing cells, proteins, and DNA. It is thought that the presence of a glassy matrix around the biological material strengthens the tertiary structure of the proteins and inhibits the diffusion of small molecules to and from the cells (Smith and Morin, 2005). Although sucrose and trehalose have the same molecular weight, the former contains one five-membered fructose ring and one six-membered glucose ring; the latter contains two glucose rings. There is considerable interest in the molecular mechanism of glass formation in concentrated aqueous solutions containing these two sugars.

Ekdawi-Sever et al. (2001) performed MC simulations of sucrose and trehalose using a parallel tempering constant-NPT algorithm. In this case all the systems are at a pressure $P = 1$ bar and the exchange of the atom coordinates and box volumes are accepted with a probability given by $\min[1, \exp(P\Delta\beta\Delta V + \Delta\beta\Delta\mathcal{V})]$. Simulations of the sugars were performed in aqueous solution with concentrations from 6 % to 100 % by mass, and temperatures between 300 K and 600 K. The simulated sucrose densities were calculated as a function of concentration, and for solutions above 80 % by mass, the replica exchange simulations predict densities that are lower than those found in conventional constant-NPT simulations, and that are closer to the experimental densities. A failure to employ parallel tempering in the simulations can result in the incorrect prediction of the glass transition temperatures. In comparing the two sugars at 1 bar, the calculated trehalose densities are consistently higher than those of sucrose over the temperature range. For example, at 500 K the simulated densities of sucrose and trehalose are 1.366 g cm^{-3} and 1.393 g cm^{-3} respectively. The higher trehalose density is associated with the hydrogen bonding in the solution. There are more sites available for intermolecular hydrogen bonding in trehalose than in sucrose, and for this reason trehalose has a higher hydration number than sucrose. This may account for the superior cryo- and lyo-protection afforded by trehalose.

Li et al. (2014) have used REMD to study the protection of insulin by trehalose. Starting from the crystal structure of insulin, the protein was solvated with TIP3P water and simulated using the GROMACS package. The drying process was modelled by the removal of water molecules. In the absence of trehalose, the drying resulted in a loss of the secondary structure of the insulin. In contrast, with the addition of the protectant trehalose, the number of amino acids in the helical conformation is maintained during drying, and both the secondary and tertiary structure of the protein are maintained.

Schmidler (2008) and Rosta et al. (2009) have demonstrate the artefacts that may result from a poor choice of thermostat (see also Lin and van Gunsteren, 2015); other practical points are emphasized by Sindhikara et al. (2010).

Normally neighbouring replicas are chosen at random for an attempted exchange, and with close enough temperatures, an acceptance ratio of ~ 20 % can be achieved. This method leads to a diffusive walk of replicas in the temperature space. The round trip rate for a replica is proportional to M^{-2} and since we typically choose $M \propto N^{1/2}$ (Hansmann, 1997), the round trip rate is inversely proportional to the system size.

For large systems, such as solvated proteins, the energy landscapes are sufficiently rugged that many round trips are required to explore all important configurations and the method struggles to sample the system properly. The ruggedness of the landscape is normally caused by the high barriers to internal rotation within the protein and not by the large bath of solvent molecules away from the protein. Yet both the protein and the bath contribute in an equal manner when attempting an exchange. The large number of degrees of freedom demands a close spacing of the replica temperatures. This problem has been addressed using the replica exchange with solute tempering (REST) technique (Liu et al., 2005).

The method considers a set of replicas with different temperatures, β_I, atom coordinates, $\mathbf{r}^I \equiv \{\mathbf{r}_1^I, \mathbf{r}_2^I, \dots, \mathbf{r}_N^I\}$, and potential functions, $\mathcal{V}_I$. The acceptance probability of exchanging the coordinates of two replicas I and J is given by $\min(1, \rho_n/\rho_m)$ where

$$\frac{\rho_n}{\rho_m} = \exp\left[-\beta_I\left(\mathcal{V}_I(\mathbf{r}^J) - \mathcal{V}_I(\mathbf{r}^I)\right) - \beta_J\left(\mathcal{V}_J(\mathbf{r}^I) - \mathcal{V}_J(\mathbf{r}^J)\right)\right]. \tag{4.70}$$

In the REST approach, the total potential energy for the protein in water is written as the sum of three terms, corresponding to the isolated protein (P), the solvent (W), and the coupling between them (WP)

$$\mathcal{V} = \mathcal{V}^P + \mathcal{V}^W + \mathcal{V}^{WP}. \tag{4.71}$$

However, this only applies at the lowest temperature T_0. Liu et al. (2005) define a temperature-dependent potential for replica I as

$$\mathcal{V}_I = \mathcal{V}^P + \left(\frac{\beta_0}{\beta_I}\right)\mathcal{V}^W + \left(\frac{\beta_0 + \beta_I}{2\beta_I}\right)\mathcal{V}^{WP}. \tag{4.72}$$

Substituting eqn (4.72) into eqn (4.70) gives

$$\frac{\rho_n}{\rho_m} = \exp\left[(\beta_I - \beta_J)\left(\mathcal{V}^P(\mathbf{r}^I) - \mathcal{V}^P(\mathbf{r}^J) + \tfrac{1}{2}\mathcal{V}^{WP}(\mathbf{r}^I) - \tfrac{1}{2}\mathcal{V}^{WP}(\mathbf{r}^J)\right)\right] \tag{4.73}$$

where the potential energy of the protein, $\mathcal{V}^P$, depends only on the subset of $\mathbf{r}$ corresponding to the protein. This judicious choice of scaling potential means that the $\mathcal{V}^W$ term has disappeared from the acceptance criterion. The number of replicas required in the calculation is significantly reduced as the solvent–solvent interactions have been removed from the attempted replica exchange. The Boltzmann factor for a particular replica I using the scaled potential is

$$\exp\left[-\beta_I\mathcal{V}^P - \beta_0\mathcal{V}^W - \tfrac{1}{2}(\beta_0 + \beta_I)\mathcal{V}^{WP}\right].$$

This is suggestive of a set of replicas in which the temperature of the protein of interest is being increased while the temperature of the bath is fixed at the lowest value, β_0; only the solute is tempered. The choice of scaled potential is not unique and Wang et al. (2011)

have also suggested simulating all of the replicas at a fixed temperature, β_0, with the scaled potential

$$\mathcal{V}_I = \frac{\beta_I}{\beta_0}\mathcal{V}^{\mathrm{P}} + \mathcal{V}^{\mathrm{W}} + \sqrt{\frac{\beta_I}{\beta_0}}\mathcal{V}^{\mathrm{WP}}. \tag{4.74}$$

This scaling is an improvement over eqn (4.72) when applied to the β-hairpin motif of a protein. This improvement in efficiency is thought to be due to the greater cancellation of scaled terms, $\mathcal{V}^{\mathrm{P}}$ and $\mathcal{V}^{\mathrm{WP}}$, in the acceptance criterion derived from eqn (4.74), and the better sampling between replica exchanges at high temperature.

The protein-in-solvent problem can also be tackled using a development of REMD called temperature intervals with global energy reassignment (TIGER) (Li et al., 2007b). This method uses a much smaller number of replicas, widely spaced in temperature, and constructs an overlap between the potential-energy distributions of each replica by using a series of heating–sampling–quenching cycles. Consider three replicas at increasing temperatures, $T_0 < T_1 < T_2$. T_0 is the baseline temperature where states will be generated according to a Boltzmann distribution. The three replicas are evolved with time at T_0 using constant-NVT MD. Replicas 1 and 2 are heated to their required temperatures by scaling the momenta over a period of 1 ps. The dynamics of the three replicas at the required temperatures proceeds for a further 1 ps in the sampling phase, and finally the replicas are quenched using 30 steps of an adopted basis Newton–Raphson energy minimization. This is followed by a heating and equilibration phase of 1.2 ps, for all of the replicas at the baseline temperature, T_0. The temperatures are fixed using a Nosé–Hoover thermostat (see Section 3.8.2). The total cycle time is 3.2 ps and a TIGER simulation might typically comprise ca. 5000 cycles. At the end of a particular cycle the three replicas are at the same time t in their evolution, at the same temperature T_0, and they have potential energies $\mathcal{V}_{I,T_0}(\mathbf{r}^I)$ for $I = 0 \ldots 2$. Thus, $\mathcal{V}_{1,T_0}(\mathbf{r}^1)$ is the potential energy of replica 1, quenched from T_1 to T_0, where the coordinates of the N atoms after the end of the quench and re-equilibration are $\mathbf{r}^1 \equiv \{\mathbf{r}_1^1, \ldots \mathbf{r}_N^1\}$. Li et al. (2007b) show that it is possible to swap the replicas I and J at this point by accepting the exchange with a probability given by

$$\min\left(1, \exp\left[(\beta_I - \beta_J)\left(\mathcal{V}_{I,T_0}(\mathbf{r}^I) - \mathcal{V}_{J,T_0}(\mathbf{r}^J)\right)\right]\right) \tag{4.75}$$

where we consider three possible swaps 0–1, 0–2, and 1–2 in that order. The replicas are then reheated to their required temperatures in the next cycle.

The method has been applied to an alanine dipeptide (Ace–Ala–Nme) in 566 TIP3P water molecules. All the replicas were started with the dipeptide in one particular conformation; the two dihedral angles ϕ and ψ were in the α_{R} region of the Ramachandran plot (the conformational distribution $\rho(\phi, \psi)$). A careful analysis of the evolution of the Ramachandran plot over the 16 ns of the run demonstrates that the three-level TIGER simulation is more effective than a REMD simulation with 24 replicas in evolving the equilibrium conformational distribution of the solvated dipeptide. Note that the number of replicas used in the TIGER method can be readily increased. However, there are two assumptions in the acceptance criterion, eqn (4.75): that the ratio of the conformational states of the peptide does not change significantly during quenching, and that the specific heat, C_V is not a strong function of the conformational state of the system (in the absence of a major conformational change). Li et al. (2007b) demonstrate that these approximations are accurate for the solvated dipeptide, and a more recent development, TIGER2 (Li et al., 2009), avoids these assumptions.

An extension of REMC known as convective replica exchange (Spill et al., 2013) has been developed to reduce the round trip time by deliberately moving one of the replicas up and down the temperature ladder. A replica is chosen at random and labelled as the walking replica; suppose this is initially in state I. The other $M - 1$ replicas are passive. When an exchange move is to be attempted, a trial exchange is made between the walker, I, and the higher-temperature state, $I + 1$. If the move is accepted, the new walking state is $I + 1$. If the attempt fails, a number of atom moves will be attempted in all replicas; at the next trial exchange, we return to the walker in state I and try to move it to $I + 1$ again. This strategy continues until the exchange is successful. When this happens, the underlying transition matrix is changed to allow for an exchange between $I + 1$ and $I + 2$, and in this way the walker moves all the way up to the highest-temperature state. At this point the direction of the walker is reversed, and exchange moves continue down the ladder of states until the walker reaches the lowest temperature. Then, the direction of moves is reversed again, and the process continues until the walker climbs back up the ladder to state I. The initial walker has accomplished its round trip, visiting every other state, and another random replica J is chosen to walk. The essential requirement for microscopic reversibility in this algorithm is that movements of a walker in one direction are balanced by movements in the opposite direction.

Spill et al. (2013) demonstrate that the convective replica exchange algorithm has two advantages over conventional replica exchange. First, the method prevents the formation of replica-exchange bottlenecks. By trying exchanges at the bottleneck until they succeed, the method makes sampling of the energy landscape more efficient. Second, in the simulation of complicated peptides, convective replica exchange is twice as fast as the conventional method in exploring large numbers of free-energy basins. In these systems, the use of convective replica exchange increased the number of round trips by a factor of between 8 and 48 over the conventional method, depending on the complexity of the peptide.

Parallel tempering has now been applied to polymers, proteins, ligand docking, spin glasses, solid state structure determinations, and in more general optimization problems such as image analysis, NMR interpretation, and risk analysis (Earl and Deem, 2005). Powerful extensions of the methodology are also possible (Coluzza and Frenkel, 2005; Kouza and Hansmann, 2011; Spill et al., 2013). One of the most useful features of the technique is that it can be implemented efficiently, and fairly easily, on a parallel computer. We return to this in Section 7.3.

4.10 Other ensembles

We have only discussed a few of the enormous number of MC algorithms that can be devised to tackle different simulation problems. Some advanced techniques will be described in Chapter 9. However, it is worth mentioning that MC methods are available to sample the constant-NVE and NPH ensembles (Ray, 1991; Lustig, 1998) to complement the MD algorithms described in Chapter 3. The basic idea is to include the ideal gas contribution to the energy through the specified ensemble constraint, e.g. $\mathcal{K} = E - \mathcal{V}(\mathbf{r})$ or $\mathcal{K} = H - PV - \mathcal{V}(\mathbf{r})$. Then Monte Carlo sampling of the configurational variables $\mathbf{r}$ uses a weight function which includes the ideal gas term (as is done in the NPT and μVT ensembles, although the formulae are rather different). Escobedo (2005) has described this approach for a range of ensembles, and given examples of their usefulness. For example, the NPH ensemble provides a very convenient way of traversing the two-phase region of a single-component system. Combining with multi-ensemble or replica-exchange methods is straightforward.

5

Some tricks of the trade

5.1 Introduction

The purpose of this chapter is to put flesh on the bones of the techniques that have been outlined in Chapters 3 and 4. There is a considerable gulf between understanding the ideas behind the MC and MD methods, and writing and running efficient programs. In this chapter, we describe some of the programming techniques commonly used in the simulation of fluids. There are a number of similarities in the structure of MC and MD programs. They involve a start-up from an initial configuration of molecules, the generation of new configurations in a particular ensemble, and the calculation of observable properties by averaging over a finite number of configurations. Because of the similarities, most of the ideas developed in this chapter are applicable to both techniques, and we shall proceed with this in mind, pointing out any specific exceptions. The first part of this chapter describes the methods used to speed up the evaluation of the interactions between molecules, which are at the heart of a simulation program. The second part describes the overall structure of a typical program and gives details of running a simulation.

5.2 The heart of the matter

In Chapter 1, we gave an example of the calculation of the potential energy for a system of particles interacting via the pairwise Lennard-Jones potential. At that point, we paid little attention to the efficiency of that calculation, although we have mentioned points such as the need to avoid the square root function, and the relative speeds of arithmetic operations (see Chapter 1 and Appendix A). The calculation of the potential energy of a particular configuration (and, in the case of MD, the forces acting on all molecules) is the heart of a simulation program, and is executed many millions of times. Great care must be taken to make this particular section of code as efficient as possible. In this section we return to the force/energy routine with the following questions in mind. Is it possible to avoid expensive function evaluations when we calculate the forces on a molecule? What can we do with much more complicated forms of pair potential?

5.2.1 Efficient calculation of forces, energies, and pressures

Consider, initially, an atomic system with a pairwise potential $v(r)$. Assume that we have identified a pair of atoms i and j. Using the minimum image separations, the squared interatomic distance is readily calculated. The force on atom i due to j is

$$\mathbf{f}_{ij} = -\nabla_{\mathbf{r}_i} v(r_{ij}) = -\nabla_{\mathbf{r}_{ij}} v(r_{ij}). \tag{5.1}$$

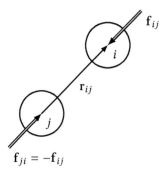

Fig. 5.1 The separation vector and force between two molecules. $\mathbf{r}_{ij}$ is the vector to i from j; $\mathbf{f}_{ij}$ is the force on i due to j; $\mathbf{f}_{ji}$ is the force on j due to i. Here the forces are drawn corresponding to an attractive interaction.

This force is directed along the interatomic vector $\mathbf{r}_{ij} = \mathbf{r}_i - \mathbf{r}_j$ (see Fig. 5.1) and it is easy to show that

$$\mathbf{f}_{ij} = -\frac{1}{r_{ij}}\left(\frac{dv(r_{ij})}{dr_{ij}}\right)\mathbf{r}_{ij} = -\frac{w(r_{ij})}{r_{ij}^2}\mathbf{r}_{ij}. \tag{5.2}$$

This equation makes it clear that if $v(r_{ij})$ is an even function of r_{ij}, then the force vector can be calculated without ever working out the absolute magnitude of $\mathbf{r}_{ij}$: r_{ij}^2 will do. The function $w(r_{ij})$ is the pair virial function introduced in eqns (2.65)–(2.69). If $v(r_{ij})$ is even in r_{ij}, then so is $w(r_{ij})$. Taking the Lennard-Jones potential, eqn (1.6), as our example, we have

$$\mathbf{f}_{ij} = \frac{24\epsilon}{r_{ij}^2}\left[2\left(\frac{\sigma}{r_{ij}}\right)^{12} - \left(\frac{\sigma}{r_{ij}}\right)^6\right]\mathbf{r}_{ij}. \tag{5.3}$$

In an MD simulation, $v(r_{ij})$, $w(r_{ij})$, and $\mathbf{f}_{ij}$ are calculated within a double loop over all pairs i and j as outlined in Chapter 1. The force on particle j is calculated from the force on i by exploiting Newton's third law. There are one or two elementary steps that can be taken to make this calculation efficient, and these appear in Code 3.4. In an MC calculation, $v(r_{ij})$ and $w(r_{ij})$ will typically be calculated in a loop over j, with i (the particle being given a trial move) specified. This is illustrated in Code 4.3.

The calculation of the configurational energy and the force can be readily extended to molecular fluids in the interaction site formalism. In this case the potential energy is given by eqn (1.12) and the virial by (for example) eqn (2.69). If required, the forces are calculated in a straightforward way. In this case, it may be simplest to calculate the virial by using the definitions (compare eqns (2.65), (2.67))

$$w(r_{ab}) = -\mathbf{r}_{ab} \cdot \mathbf{f}_{ab}, \tag{5.4}$$

summed over all distinct site–site separations $\mathbf{r}_{ab}$ and forces $\mathbf{f}_{ab}$ (including intramolecular ones) or

$$w(r_{ij}) = -\mathbf{r}_{ij} \cdot \mathbf{f}_{ij}, \tag{5.5}$$

summed over distinct pairs of molecules, where $\mathbf{f}_{ij}$ is the sum of site–site interactions $\mathbf{f}_{ab}$ acting between each pair. These equations translate easily into code. For more complicated intermolecular potentials, for example involving multipoles, the expressions given in Appendix C may be used.

There are some special considerations which apply to MC simulations, and which may improve the efficiency of the program. When a molecule i is subjected to a trial move, the new interactions with its neighbours j are calculated. It is possible to keep a watch for substantial overlap energies during this calculation: if one is detected, the remainder of the loop over j is immediately skipped and the move rejected. The method is particularly effective in the simulation of hard-core molecules, when a single overlap is sufficient to guarantee rejection. Note that it only makes sense to test the trial configuration in this way, since the current configuration is presumably free of substantial overlaps. For soft-core potentials, care should be taken not to set the overlap criterion at too low an energy: occasional significant overlaps may make an important contribution to some ensemble averages.

If no big overlaps are found, the result of the loops over j is a change in potential energy which is used in the MC acceptance/rejection test. If the move is accepted, this number can be used to update the current potential energy (as seen in Section 4.4): there is no need to recalculate the energy from scratch. It may be worth considering a similar approach when calculating the virial; that is, compute the change in this function which accompanies each trial move, and update $\mathcal{W}$ if it is accepted. Whether this is cost-effective compared with a less frequent complete recalculation of $\mathcal{W}$ depends on the acceptance ratio: it would not be worthwhile if a large fraction of moves were rejected. In any case, a complete recalculation of the energy and the virial should be carried out at the end of the simulation as a check that all is well.

The calculation of the pressure in systems of hard molecules (whether in MC or in MD) is carried out in a slightly different fashion, not generally within the innermost loop of the program, and we return to this in Section 5.5.

5.2.2   Table look-up and spline fit potentials

As the potentials used in simulations become more complicated, the repeated evaluation of algebraic expressions for the potential and forces can be avoided by using a prepared table. This technique has been used in the simulation of the Barker–Fisher–Watts potential for argon (Barker et al., 1971), which contains 11 adjustable parameters and an exponential.

The table is constructed once, at the beginning of the simulation program, and the potential and force are calculated as functions of $s = r_{ij}^2$. For example, we might set up a table to calculate the exponential-6 potential,

$$v^{\text{E6}}(r) = -A/r^6 + B\exp(-Cr) \quad \Rightarrow \quad v^{\text{E6}}(s) = -A/s^3 + B\exp(-Cs^{1/2}), \quad (5.6)$$

where A, B, and C are parameters. During the course of the run, values of $s = r_{ij}^2$ are calculated for a particular pair of molecules, and the potential is interpolated from the table. Polynomial or rational function interpolation from a set of tabulated values and methods for efficiently searching an ordered table are discussed by Press et al. (2007, Chapter 3). In a molecular dynamics simulation, of course, we also need to evaluate the forces. We may compute these by constructing a separate table of values of the function $w(r_{ij})/r_{ij}^2$, which enters into the force calculation through eqn (5.2). The success of the interpolation method depends on the careful choice of the table-spacing. Typically, we find that $\delta s = \delta r_{ij}^2 = 0.01 r_m^2$, where r_m is the position of the potential minimum, produces a sufficiently fine grid for use in MD and MC simulations.

Andrea et al. (1983) suggested an improvement to this method, using a spline fit to the potential. The function $v(s)$ (where once more $s = r_{ij}^2$) is divided into a number of regions by grid points or knots s_k. In each interval (s_k, s_{k+1}) the function is approximated by a fifth-order polynomial

$$v(s) \approx \sum_{n=0}^{5} c_n^{(k)} (s - s_k)^n.\tag{5.7}$$

The coefficients $c_0^{(k)} \cdots c_5^{(k)}$ are uniquely determined by the exact values of $v(s)$, $dv(s)/ds$, and $d^2 v(s)/ds^2$ at the two ends of the interval (Andrea et al., 1983, Appendix). Thus, we need to store the grid points s_k (which need not be evenly spaced) and six coefficients for each interval. In their simulation of water, Andrea et al. represented the O–O, O–H, and H–H potentials using 14, 16, and 26 intervals respectively. For MD, the forces are easily obtained by differentiating eqn (5.7) and using

$$\frac{w(r_{ij}^2)}{r_{ij}^2} = \frac{w(s)}{s} = 2\frac{dv}{ds}.\tag{5.8}$$

Another interesting example of a force represented by a polynomial fit is the construction of the force matched (FM) representation of a spherically truncated Coulomb interaction as discussed in Section 6.4 (Izvekov et al., 2008). In this case the truncated Coulomb force, $q_i q_j/s$ for $s^{1/2} < r_c$, is represented by the polynomial

$$f_{ij}^{FM}(s) = q_i q_j \left[\frac{1}{s} + \sum_{k=0}^{7} a_k s^{k/2}\right], \quad r_{core} < s^{1/2} < r_c,\tag{5.9}$$

where the coefficients a_k are determined by matching the trajectories from the FM force and the true long-range Coulomb interaction.

5.2.3 Shifted and shifted-force potentials

The truncation of the intermolecular potential at a cutoff introduces some difficulties in defining a consistent potential and force for use in the MD method. The function $v(r_{ij})$ used in a simulation contains a discontinuity at $r_{ij} = r_c$: whenever a pair of molecules crosses this boundary, the total energy will not be conserved. We can avoid this by shifting the potential function by an amount $v_c = v(r_c)$, that is, using instead the function

$$v^S(r_{ij}) = \begin{cases} v(r_{ij}) - v_c & r_{ij} \leq r_c \\ 0 & r_{ij} > r_c. \end{cases}\tag{5.10}$$

The small additional term is constant for any pair interaction, and does not affect the forces, and hence the equations of motion of the system. However, its contribution to the total energy varies from timestep to timestep, since the total number of pairs within cutoff range varies. This term should certainly be included in the calculation of the total energy, so as to check the conservation law. However, there is a further problem. The

Example 5.1 Learning the potentials with a neural network

For complicated intermolecular potentials, such as those developed from *ab initio* calculations, it can be cost-effective to teach the computer to estimate the potential energy of a particular atom from a knowledge of its environment in the liquid. This can be done by constructing an artificial neural network (Behler, 2011).

In this approach, the Cartesian coordinates of a particular atom, i, are used to calculate a set of μ symmetry functions, G_i^μ. The first of these functions, $\mu = 1$, might depend on the radial distribution of all the other atoms around i. The second might depend on the distribution of triplets around atom i through the variable $\cos\theta_{ijk}$, and so on (see e.g. Behler and Parrinello, 2007, eqns (4) and (5)). Symmetry functions of different order can be combined with weights $w_{i\mu}^s$ to produce an overall $G_i = \sum_\mu w_{i\mu}^s G_i^\mu$. The G_i are the input to a simple neural network of the kind shown.

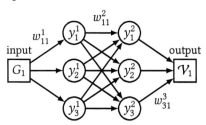

There is a separate network for each atom, containing a number of hidden layers (two in this example) each with a number of nodes (three in this example). The value y_m^ℓ held by the network at a particular node, m, in layer, ℓ, is calculated as a function of the weighted sum of the connected nodes in the previous layer, that is, $y_m^\ell = f(\sum_{n=1}^3 w_{nm}^\ell y_n^{\ell-1})$ for $\ell = 2$ in our example. The activation function is often taken to be $f(x) = \tanh(x)$, although different functions can be used for different layers. The output from the network is the potential energy, $\mathcal{V}_1$. The outputs from all the separate networks can be added to produce the total potential energy, $\mathcal{V}_{net}$. For a number of training configurations the full $\mathcal{V}_{QM}$ is also calculated. The weights, w, in the network, and symmetry functions, are adjusted to minimize $|\mathcal{V}_{net} - \mathcal{V}_{QM}|$ for the training set. The same connectivity and inter-layer weights are used in the different networks established for each atom. Once the optimum weights have been established, the network can be used to accurately estimate the potential and forces for an unknown configuration.

Morawietz et al. (2016) have employed neural network potentials in *ab initio* molecular dynamics simulations of water. The efficiency of the approach enabled them to show that the relatively weak, isotropic van der Waals forces are crucial in producing the density maximum and the negative volume of melting of the fluid.

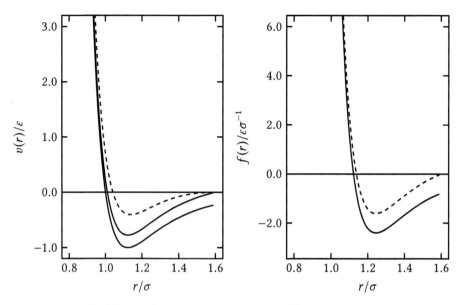

Fig. 5.2   Magnitude of the pair potential $v(r)$ and the force $f(r)$, for the Lennard-Jones and shifted Lennard-Jones potentials (solid lines), and the shifted-force modification (dashed lines). Note that, for illustration, we have chosen a very short cutoff distance, $r_c = 1.6\sigma$.

force between a pair of molecules is still discontinuous at $r_{ij} = r_c$. For example, in the Lennard-Jones case, the force is given by eqn (5.3) for $r_{ij} \leq r_c$, but is zero for $r_{ij} > r_c$. The magnitude of the discontinuity is $\approx 0.039\,\epsilon\sigma^{-1}$ for $r_c = 2.5\sigma$. It can cause instability in the numerical solution of the differential equations. To avoid this difficulty, a number of workers have used a 'shifted-force potential' (Stoddard and Ford, 1973; Streett et al., 1978; Nicolas et al., 1979; Powles et al., 1982). A small linear term is added to the potential, so that its derivative is zero at the cutoff distance

$$v^{\mathrm{SF}}(r_{ij}) = \begin{cases} v(r_{ij}) - v_c - (r_{ij} - r_c)\left(\dfrac{\mathrm{d}v(r_{ij})}{\mathrm{d}r_{ij}}\right)_{r_{ij}=r_c} & r_{ij} \leq r_c \\[2mm] 0 & r_{ij} > r_c. \end{cases} \tag{5.11}$$

The discontinuity now appears in the gradient of the force, not in the force itself. The shifted-force potential for the Lennard-Jones case is shown in Fig. 5.2. The force goes smoothly to zero at the cutoff r_c, removing problems in energy conservation and any numerical instability in the equations of motion. Making the additional term quadratic (Stoddard and Ford, 1973) avoids taking a square root. Of course, the difference between the shifted-force potential and the original potential means that the simulation no longer corresponds to the desired model liquid. However, the thermodynamic properties of a fluid of particles interacting with the unshifted potential can be recovered from the shifted-force potential simulation results, using a simple perturbation scheme (Nicolas et al., 1979; Powles, 1984).

5.2.4 Thermodynamic properties with a truncated potential

Comparing simulation results obtained with different versions of a potential can lead to considerable confusion. Here we discuss the most common situations. First, suppose that an MC simulation of a model fluid has been performed with a truncated (but not shifted) pair potential, which we write formally as

$$v_c(r) = v(r)\Theta(r_c - r) \tag{5.12}$$

where r_c is the cutoff distance and Θ is the unit step function. The corresponding virial function for the calculation of the pressure is

$$w_c(r) = w(r)\Theta(r_c - r) - rv(r)\delta(r - r_c). \tag{5.13}$$

The second term arises from the discontinuity of the potential at the cutoff. Using eqn (2.102)

$$\left\langle \sum_i \sum_{j>i} a(r_{ij}) \right\rangle = \tfrac{1}{2}N\rho \int_0^\infty a(r)g(r)4\pi r^2 \, dr. \tag{5.14}$$

and recalling $\mathcal{W} = -\tfrac{1}{3}\sum_i \sum_{j>i} w(r_{ij})$, we can evaluate the average of the delta function in calculating the pressure

$$P_c V = Nk_BT + \langle \mathcal{W} \rangle + \frac{2\pi N\rho}{3}r_c^3 g(r_c)v(r_c) \approx Nk_BT + \langle \mathcal{W} \rangle + \underbrace{\frac{2\pi N\rho}{3}r_c^3 v(r_c)}_{\Delta P_c V}, \tag{5.15}$$

thereby defining the correction ΔP_c for the delta function contribution. The approximation results from assuming that the cutoff is large enough that $g(r_c) \approx 1$. It is understood that $\mathcal{W}$ involves the sum over distinct pairs of the first term on the RHS of eqn (5.13), that is, those pairs within the cutoff. Each of these three terms contributes to the pressure associated with the truncated potential, hence the notation P_c. The value of ΔP_c can be calculated in advance of the simulation. For the Lennard-Jones potential, in reduced units, it is

$$\Delta P_c^* = \tfrac{8}{3}\pi\rho^{*2}\left(r_c^{*-9} - r_c^{*-3}\right). \tag{5.16}$$

Although the truncated potential is used to sample the canonical ensemble, it is the pressure corresponding to the full, untruncated, potential that is normally of interest. Now, the quantity to be averaged is the pairwise sum of $w(r)$ functions, that is, it involves no delta functions. Moreover, in a perturbative, mean-field, approach, the long-range term in the potential is assumed to have no effect on the structure. Therefore the pressure of the full system can be written, assuming $g(r) = 1$ for $r > r_c$

$$PV \approx Nk_BT + \langle \mathcal{W} \rangle - \frac{2\pi}{3}N\rho \int_{r_c}^\infty w(r)r^2 \, dr \equiv Nk_BT + \langle \mathcal{W} \rangle + P_{\mathrm{LRC}}V. \tag{5.17}$$

$\mathcal{W}$ here has exactly the same interpretation as in eqn (5.15); that is, the sum of the virial functions of distinct pairs within cutoff range, measured in the simulation using the truncated potential. The delta function correction does not appear at all, and eqn (5.17) can be used straightforwardly to compute the pressure during the simulation. The long-range correction, P_{LRC}, can be computed ahead of the simulation. For the example of the Lennard-Jones potential, it is given in eqn (2.144)

$$P^*_{\mathrm{LRC}} = \tfrac{16}{9}\pi\rho^{*2}\left(2r_c^{*-9} - 3r_c^{*-3}\right). \tag{5.18}$$

It may be desirable to measure the pressure P_c corresponding to the actual truncated potential used in the simulation, and express the pressure for the full potential in terms of this. In this case, the relevant formula is obtained by combining eqns (5.15) and (5.17):

$$P = P_c - \Delta P_c + P_{\mathrm{LRC}}. \tag{5.19}$$

In other words, the delta function correction used in the calculation of P_c must be subtracted off again, and the long-range correction added on. This is a very common approach to use in constant-pressure Monte Carlo simulations of the truncated potential, because then the specified input pressure corresponds to P_c, not P. The average of P_c may be calculated (as a consistency check) in the simulation, using eqn (5.15), but in any case the measured average density ρ will correspond to an equation of state $\rho(P_c)$, which we would normally invert to get $P_c(\rho)$. We can recover $P(\rho)$ afterwards using eqn (5.19). For the Lennard-Jones potential, the *combined* correction term is

$$P^* - P_c^* = P^*_{\mathrm{LRC}} - \Delta P_c^* = \pi\rho^{*2}\left(\tfrac{8}{9}r_c^{*-9} - \tfrac{8}{3}r_c^{*-3}\right). \tag{5.20}$$

This formula was used by Finn and Monson (1989) and Smit (1992), omitting the r_c^{*-9} term which is relatively small, to compare results from the truncated Lennard-Jones potential with the equation of state of Nicolas et al. (1979).

A little thought reveals that the same result will be obtained (on the grounds of ensemble equivalence) if one uses eqn (5.17) directly, even in a constant-pressure simulation of the truncated potential. However, one must be aware that the value of P obtained from it will not agree with the specified pressure of the simulation. In this case, the long-range correction term will vary with density.

Corresponding molecular dynamics simulations are performed with a truncated pair force

$$f_{cs}(r) = f(r)\Theta(r_c - r), \tag{5.21}$$

where the notation stands for 'cut and shifted'. In this case the virial function inside the cutoff is the same as that of the full potential, but there is no discontinuity at $r = r_c$. Therefore there is no extra term in the calculation of the pressure for this model

$$P_{cs}V = Nk_BT + \langle \mathcal{W} \rangle. \tag{5.22}$$

Making the same assumptions as before, the pressure for the full potential is straightforwardly given by the long-range correction

$$P = P_{cs} + P_{\mathrm{LRC}}. \tag{5.23}$$

The corresponding potential energy function, calculated by integrating the negative force from infinity to r is

$$v_{cs}(r) = \left[v(r) - v(r_c)\right]\Theta(r_c - r), \tag{5.24}$$

and the total potential energy is

$$\mathcal{V}_{cs} = \sum_i \sum_{j>i} v_{cs}(r_{ij}) = \mathcal{V} - N_c v(r_c), \tag{5.25}$$

where $\mathcal{V}$ is the total potential energy for all pairs within cutoff range (without shifting), and N_c is the total number of such pairs

$$N_c = \sum_i \sum_{j>i} \Theta(r_c - r_{ij}). \tag{5.26}$$

$\mathcal{V}_{cs}$ is the true potential energy for the truncated force and it is this potential that must be added to the kinetic energy to check for total energy conservation in the simulation. In estimating the potential energy for the full Lennard-Jones fluid, it is consistent to calculate the potential inside the cutoff using just the Lennard-Jones term without the shift and to add on the mean-field, long-range correction given by eqn (2.143).

Monte Carlo calculations may also be performed using the 'cut-and-shifted' potential, and the pressure for the full potential estimated using the long-range correction P_{LRC}. The discontinuity term ΔP_c does not arise in this case.

5.3 Neighbour lists

In the inner loops of the MD and MC programs, we consider a molecule i and loop over all molecules j to calculate the minimum image separations. If molecules are separated by distances greater than the potential cutoff, the program skips to the end of the inner loop, avoiding expensive calculations, and considers the next neighbour. In this method, the time to examine all pair separations is proportional to N^2. Verlet (1967) suggested a technique for improving the speed of a program by maintaining a list of the neighbours of a particular molecule, which is updated at intervals. Between updates of the neighbour list, the program does not check through all the j molecules, but just those appearing on the list. The number of pair separations explicitly considered is reduced. This saves time in looping through j, minimum imaging, calculating r_{ij}^2, and checking against the cutoff, for all those particles not on the list. Obviously, there is no change in the time actually spent calculating the energy and forces arising from neighbours within the potential cutoff. In this section, we describe some useful time-saving neighbour list methods. These methods are equally applicable to MC and MD simulations, and for convenience we use the MD method to illustrate them. There are differences concerning the relative sizes of the neighbour lists required in MC and MD and we return to this point at the end of the next section. Related techniques may be used to speed up MD of hard systems (Erpenbeck and Wood, 1977). In this case, the aim is to construct and maintain, as efficiently as possible, a table of future collisions between pairs of molecules. The scheduling of molecular collisions has been discussed in detail by Rapaport (1980) and Bannerman et al. (2011).

5.3.1 The Verlet neighbour list

In the original Verlet method, the potential cutoff sphere, of radius r_c, around a particular molecule is surrounded by a 'skin' to give a larger sphere of radius r_ℓ, as shown in Fig. 5.3. At the first step in a simulation, a list is constructed of all the neighbours of each molecule,

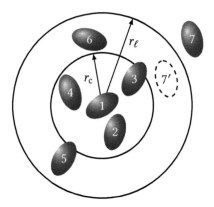

Fig. 5.3 The cutoff sphere, radius r_c, and its skin, radius r_ℓ, around a molecule 1. Molecules 2, 3, 4, 5, and 6 are on the list of molecule 1; molecule 7 is not. Only molecules 2, 3, and 4 are within the range of the potential at the time the list is constructed.

Code 5.1 Force routine using Verlet neighbour list

These files are provided online. md_lj_vl_module.f90 contains a Lennard-Jones force routine using a Verlet neighbour list, while verlet_list_module.f90 contains the routines for constructing and updating the list. These two files together act as a drop-in replacement for md_lj_module.f90 (see Code 3.4). They can be combined with, for example, md_nve_lj.f90 of Code 3.4 and the utility modules described in Appendix A to make a molecular dynamics program, as illustrated in the supplied SConstruct file.

```
! md_lj_vl_module.f90
! Force routine for MD, LJ atoms, Verlet neighbour list
MODULE md_module
```

```
! verlet_list_module.f90
! Verlet list handling routines for MD simulation
MODULE verlet_list_module
```

for which the pair separation is within r_ℓ. These neighbours are stored in an array, called, shall we say, list. list is quite large, of dimension roughly $4\pi r_\ell^3 \rho N/6$. At the same time, a second indexing array, point, of size N, is constructed. point(i) points to the position in the array list where the first neighbour of molecule i can be found. Since point(i+1) points to the first neighbour of molecule i+1, then point(i+1)-1 points to the last neighbour of molecule i. Thus, using point, we can readily identify the part of the large list array which contains neighbours of i. Routines for setting up the arrays list and point are given in Code 5.1.

Over the next few timesteps, the list is used in the force/energy evaluation routine. For each molecule i, the program identifies the neighbours j, by running over list from point(i) to point(i+1)-1. This loop should automatically handle the case when molecule i has no neighbours, and can be skipped, in which case point(i+1)-1 will be less than point(i). This is certainly possible in dilute systems. A sample force routine using the Verlet list is given in Code 5.1. From time to time, the neighbour list is reconstructed, and the cycle is repeated. The algorithm is successful because the skin around r_c is chosen to be thick enough so that between reconstructions a molecule, such as 7 in Fig. 5.3, which is not on the list of molecule 1, cannot penetrate through the skin into the important r_c sphere. Molecules such as 3 and 4 can move in and out of this sphere, but since they are on the list of molecule 1, they are always considered regardless, until the list is next updated.

The interval between updates of the table is often fixed at the beginning of the program, and intervals of 10–20 steps are quite common. An important refinement allows the program to update the neighbour list automatically. When the list is constructed, a vector for each molecule is set to zero. At subsequent steps, the vector is incremented with the displacement of each molecule. Thus it stores the total displacement for each molecule since the last update. When the sum of the magnitudes of the two largest displacements exceeds $r_\ell - r_c$, the neighbour list should be updated again (Fincham, 1981; Thompson, 1983). The code for automatic updating of the neighbour list is given in the routine of Code 5.1.

The list sphere radius, r_ℓ, is a parameter that we are free to choose. As r_ℓ is increased, the frequency of updates of the neighbour list will decrease. However, with a large list, the efficiency of the non-update steps will decrease. This trade-off is illustrated in Fig. 5.4, for different system sizes. For systems of a few hundred particles, the initial improvement reflects the interval between updates, which is when an all-pairs calculation is needed. The actual cost of using the list is relatively small, and a significant speed increase is seen up to $r_\ell^* \approx 2.8$. For larger values of r_ℓ^*, the cost of looping over all particles within the list becomes more significant. For systems $N > 500$, the improvement is dramatic and the turnover point shifts to larger r_ℓ^*, but as we shall see, the method of Section 5.3.2 becomes preferable. As the size of the system becomes larger, the size of the list array grows, approximately $\propto N$. If storage is a priority, then a binary representation of the list can be employed (O'Shea, 1983).

In the MC method, the array point has a size $N + 1$ rather than N, since the index i runs over all N atoms rather than $N - 1$ as in MD. In addition, the array list is roughly twice as large in MC as in a corresponding MD program. In the MD technique, the list for a particular molecule i contains only the molecules j with an index greater than i, since in this method we use Newton's third law to calculate the force on j from i at the same time as the force on i from j. In the MC method, particles i and j are moved independently and the list must contain separately the information that i is a neighbour of j and j a neighbour of i.

5.3.2 Cell structures and linked lists

As the size of the system increases towards 1000 molecules, the conventional neighbour list becomes too large to store easily, and the logical testing of every pair in the system is inefficient. An alternative method of keeping track of neighbours for large systems is the cell index method (Quentrec and Brot, 1973; Hockney and Eastwood, 1988). The cubic simulation box (extension to non-cubic cases is possible) is divided into a regular lattice of

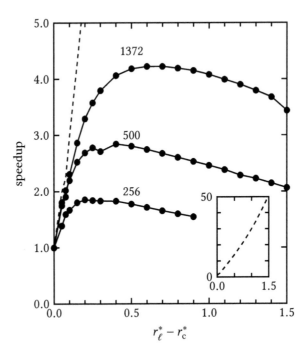

Fig. 5.4 Speedup with Verlet list. Example results are shown for the Lennard-Jones potential, with cutoff $r_c^* = 2.5$, and various values of skin thickness $r_\ell^* - r_c^*$, for the indicated system sizes N. The state point is $\rho^* = 0.78$, $T^* = 0.85$, and the timestep is $\delta t^* = 0.005$. The curves show timesteps per CPU-second, normalized by the speed for zero skin thickness (when the list is updated every step). The dashed line (also shown in the inset) gives the average number of steps between updates, which is almost independent of system size.

$s_c \times s_c \times s_c$ cells. A two-dimensional representation of this is shown in Fig. 5.5. These cells are chosen so that the side of the cell $\ell = L/s_c$ is greater than the cutoff distance for the forces, r_c. For the two-dimensional example of Fig. 5.5, the neighbours of any molecule in cell 8 are to be found in the cells 2, 3, 4, 7, 8, 9, 12, 13, and 14. If there is a separate list of molecules in each of those cells then searching through the neighbours is a rapid process. For the two-dimensional system illustrated, there are approximately $\rho\ell^2$ molecules in each cell (where ρ is the number density per unit area); the analogous result in three dimensions would be $\rho\ell^3$, where $\rho = N/V$. Using the cell structure in two dimensions, we need only examine $9N\rho\ell^2$ pairs (or just $4.5N\rho\ell^2$ if we take advantage of the third law in the MD method). This contrasts with N^2 (or $\frac{1}{2}N(N-1)$) for the brute-force approach. When the cell structure is used in three dimensions, then we compute $27N\rho\ell^3$ interactions ($13.5N\rho\ell^3$ for MD) as compared with N^2 (or $\frac{1}{2}N(N-1)$). The cost of the method scales with N rather than N^2, and the speedup over the brute-force approach is $\approx L^2/9\ell^2$ in 2D, or $\approx L^3/27\ell^3$ in 3D.

The cell structure may be set up and used by the method of linked lists (Knuth, 1973, Chapter 2; Hockney and Eastwood, 1988, Chapter 8). The first part of the method involves sorting all the molecules into their appropriate cells. This sorting is rapid, and may be

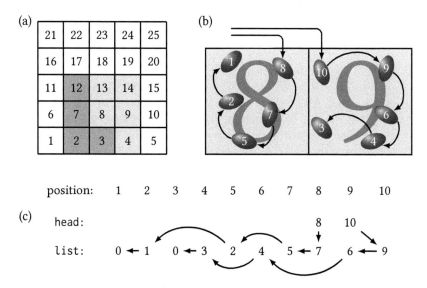

(a)

21	22	23	24	25
16	17	18	19	20
11	12	13	14	15
6	7	8	9	10
1	2	3	4	5

(b)

(c)

position: 1   2   3   4   5   6   7   8   9   10

head: 8   10

list: 0 ← 1   0 ← 3   2   4   5 ← 7   6 ← 9

Fig. 5.5 The cell method in two dimensions. (a) The central box is divided into $s_c \times s_c$ cells ($s_c = 5$). A molecule in cell 8 may interact with molecules in any of the shaded cells. In an MD program, only the light-shaded neighbours of cell 8 need be examined in the inner loop over cells. (b) A close-up of cells 8 and 9, showing the molecules and the link-list structure. (c) The head and list array elements corresponding to these two cells. Each entry gives the position of the next in the list array.

performed every step. Two arrays are created during the sorting process. The 'head-of-chain' array (head) has one element for each cell. This element contains the identification number of one of the molecules sorted into that cell. This number is used to address the element of a linked-list array (list), which contains the number of the next molecule in that cell. In turn, the list array element for that molecule is the index of the next molecule in the cell, and so on. If we follow the trail of linked-list references, we will eventually reach an element of list which is zero. This indicates that there are no more molecules in that cell, and we move on to the head-of-chain molecule for the next cell. To illustrate this, imagine a simulation of particles in two cells: 1, 2, 5, 7, and 8 in cell 8 and 3, 4, 6, 9, and 10 in cell 9 (see Fig. 5.5). The head and list arrays are illustrated in the figure. For cell 8, head(8) = 8 (the last particle found to lie in that cell), while for cell 9, head(9) = 10; in both cases the route through the linked list is arrowed. The construction of head and list is straightforward, and is illustrated in Code 5.2.

In MD, the calculation of the forces is performed by looping over all cells. For a given cell, the program traverses the linked list. A particular molecule on the list may interact with all molecules in the same cell that are further down the list. This avoids counting the *ij* interaction twice. A particular molecule also may interact with the molecules in the neighbouring cells. To avoid double counting of these forces, only a limited number of the neighbouring cells are considered. This idea is most easily explained with reference to our two-dimensional example shown in Fig. 5.5. A molecule in cell 8 interacts with other molecules in the same cell, and with eight neighbouring cells, but the program only

Code 5.2   Building a cell structure with linked lists

These are the essential statements for assigning each atom to a cell, and building the links between atoms lying in each cell. In the usual unit cube simulation box, the cell length is $1/s_c$. Here, we identify each cell by a three-dimensional index denoting its position in space, rather than a single index as in Fig. 5.5. This means that the cell c(1:3,i), in which an atom i lies, is essentially a discretized form of the coordinate vector r(1:3,i); it is convenient to make the indices lie in the range (0,sc-1) rather than (1,sc). We assume that periodic boundary corrections have already been applied to r. Correspondingly, the head array is of rank 3, and each element contains the particle index i of the last particle added to each cell. An expanded version of the code, including guards against roundoff error, appears in the online file link_list_module.f90 of Code 5.3.

```
REAL,     DIMENSION(3,n)                    :: r
INTEGER,  DIMENSION(3,n)                    :: c
INTEGER,  DIMENSION(n)                      :: list
INTEGER,  DIMENSION(0:sc-1,0:sc-1,0:sc-1)  :: head

head(:,:,:) = 0
DO i = 1, n
   c(:,i)   = FLOOR ( ( r(:,i) + 0.5 ) * REAL(sc) )
   list(i) = head(c(1,i),c(2,i),c(3,i))
   head(c(1,i),c(2,i),c(3,i)) = i
END DO
```

checks cells 8, 4, 9, 13, and 14. Interactions between cells 7 and 8 are checked when cell 7 is the focus of attention, and so on. In this way, we can make full use of Newton's third law in calculating the forces. We note that for a cell at the edge of the basic simulation box (15 for example in Fig. 5.5) it is necessary to consider the periodic cells (6, 11, and 16). An example of the code for constructing and searching through cells is given in Code 5.3. The linked-list method can also be used with the MC technique. In this case, a molecule move involves checking molecules in the same cell, and in all the neighbouring cells. This is illustrated in Code 5.4.

The cell structure may be used somewhat more efficiently, by avoiding unnecessary distance calculations, if the molecules in each cell are sorted into order of increasing (say) x-coordinate (Hockney and Eastwood, 1988). Extending this idea, Gonnet (2007) suggests sorting particles according to their projection onto the vector that connects the cell centres. For each pair of cells, the loops over particle pairs may then be restricted to those whose projected separation (a lower bound on the true separation) is less than r_c. Similar improvements are reviewed by Heinz and Hünenberger (2004) and Welling and Germano (2011).

The linked-list method has been used with considerable success in simulation of systems such as plasmas, galaxies, and ionic crystals, which require a large number of

Code 5.3   Force routine using linked lists

These files are provided online. md_lj_ll_module.f90 contains a Lennard-Jones force routine using a linked list, while link_list_module.f90 contains the routines for constructing and updating the list. These two files together act as a drop-in replacement for md_lj_module.f90 (see Code 3.4). They can be combined with, for example, md_nve_lj.f90 of Code 3.4 and the utility modules described in Appendix A, to make a molecular dynamics program, as illustrated in the supplied SConstruct file.

```
! md_lj_ll_module.f90
! Force routine for MD simulation, LJ atoms, linked lists
MODULE md_module
```

```
! link_list_module.f90
! Link list handling routines for MC or MD simulation
MODULE link_list_module
```

Code 5.4   Monte Carlo routines using linked lists

This file is provided online. mc_lj_ll_module.f90 contains Lennard-Jones energy routines using a linked list. Together with link_list_module.f90 (Code 5.3) it acts as a drop-in replacement for mc_lj_module.f90 (see Code 4.3). They can be combined with several of the supplied Lennard-Jones Monte Carlo programs in the common ensembles, for example mc_nvt_lj.f90 (Code 4.3), mc_zvt_lj.f90 (Code 4.6), plus the utility modules described in Appendix A, to make complete MC programs, as illustrated in the SConstruct file.

```
! mc_lj_ll_module.f90
! Energy and move routines for MC, LJ potential, linked lists
MODULE mc_module
```

particles (Hockney and Eastwood, 1988). In both the linked-list methods and the Verlet neighbour list, the computing time required to run a simulation tends to increase linearly with the number of particles N rather than quadratically. This rule of thumb does not take into account the extra time required to set up and manipulate the lists. For small systems ($N \approx 100$) the overheads involved make the use of lists unprofitable. Mazzeo et al. (2010) have developed a linked-list method for Monte Carlo simulations, and discuss several optimization strategies.

Parallel computers offer a more cost-effective route to high-performance computing than traditional single processor machines. Both the force calculation and integration steps of molecular dynamics are parallel in nature, and for that reason parallel algorithms based on the cell-structure, linked-list technique have been developed. This method is particularly efficient when the range of intermolecular potential is much smaller than the dimensions of the simulation box. This domain-decomposition approach was originally tested for fluids of up to 2×10^6 atoms in two and three dimensions (Pinches et al., 1991). A detailed implementation using the message passing interface (MPI) is provided by Griebel et al. (2007, Chapter 4).

One is not restricted to using cells of length r_c. A modification of this general approach employs cells that are sufficiently small that at most one particle can occupy each cell. In this case, a linked-list structure as described earlier is not required: the program simply loops over all cells, and conducts an inner loop over the cells that are within a distance r_c of the cell of interest. Advantages of this method are that the list of cells 'within range' of each given cell may be computed at the start of the program, remaining unaltered throughout, and that a simple decision (is the cell occupied or not?) is required at each stage. However, the number of cells is large, and, of course, many of them are empty. This version of the method has been analysed in detail (Mattson and Rice, 1999; Yao et al., 2004; Heinz and Hünenberger, 2004).

5.4   Non-bonded interactions and multiple timesteps

In Section 3.5 we introduced the idea of separating the Liouville operator into parts that vary at different rates in time, due to a division into 'fast' and 'slow' forces. One obvious such division is between intramolecular forces, especially bond stretching potentials which produce rapid vibration, and intermolecular non-bonded forces. This idea may be extended, and molecular dynamics packages quite commonly divide non-bonded potentials into a hierarchy of terms, from short-ranged contributions which have high gradients, to longer-ranged, more gently varying, terms. These are then handled by a corresponding hierarchy of timesteps. The rationale is that the number of long-range pair interactions is much larger than the number of short-range ones (there is a geometrical factor of r_{ij}^2) but they may be computed much less frequently.

A simplified example, following Procacci and Marchi (1996), illustrates the idea. Consider sub-dividing the Lennard-Jones potential according to a set of cutoff distances $R_1 < R_2 < \cdots < R_k < \cdots < R_K$ as follows

$$v_k^{\text{LJ}}(r) \equiv \left[S_k \left(\frac{r - R_k}{\lambda} \right) - S_{k-1} \left(\frac{r - R_{k-1}}{\lambda} \right) \right] v^{\text{LJ}}(r) \tag{5.27}$$

where the switching functions are defined $S_0(x) = 0$, $S_K(x) = 1$, and

$$S_k(x) = \begin{cases} 1 & x < -1 \\ (2x + 3)x^2 & -1 < x < 0 \\ 0 & x > 0 \end{cases} \qquad k = 1 \ldots K - 1.$$

It is understood that $R_0 = 0$, and R_K is the potential cutoff distance. The parameter λ defines a range of r over which S_k changes smoothly from 1 to 0, and the successive shells defined by the square brackets in eqn (5.27) overlap by this distance. It is easy to show that

$$v^{\text{LJ}}(r) = \sum_{k=1}^{K} v_k^{\text{LJ}}(r).$$

Code 5.5 Multiple-timestep algorithm, Lennard-Jones atoms

These files are provided online. md_lj_mts.f90 carries out MD using a three-timestep scheme, for three shells of pair interaction, which are computed using the force routine supplied in md_lj_mts_module.f90. Utility module routines described in Appendix A handle input/output and averages. The example is slightly contrived: for clarity, no neighbour lists are used.

```
! md_lj_mts.f90
! Molecular dynamics, NVE, multiple timesteps
PROGRAM md_lj_mts
```

```
! md_lj_mts_module.f90
! Force routine for MD, LJ atoms, multiple time steps
MODULE md_module
```

Then, for a suitable choice of the R_k, each contribution k may be treated using a different timestep, using a scheme similar to that described in Section 3.5, taking advantage of the slower variation in time of the longer-range contributions. Of course, as the atoms move around, the pairs belonging to each shell will change. Also, in calculating the forces from each separate contribution $v_k^{LJ}(r)$, the r-dependence of the switching functions must be taken into account. Code 5.5 illustrates this approach.

It should be emphasized that this example is oversimplified, and very little improvement in efficiency should be expected for a relatively short-ranged potential such as Lennard-Jones. The method comes into its own when electrostatic forces are present, for which the long-range parts may be handled with a long timestep, and when high-frequency intramolecular terms may be tackled with a short timestep, as discussed in Section 3.5. Procacci and Marchi (1996) propose a separation based not on atom pairs, but on pairs of neutral groups. It is worthwhile putting some effort into optimizing the efficiency of the arrangement, and in principle it is possible to handle a very large number of classes based on distance (Kräutler and Hünenberger, 2006).

5.5 When the dust has settled

At the end of the central loop of the program, a new configuration of molecules is created, and there are a number of important configurational properties that can be calculated. At this point in the program, the potential energy and the forces on particular molecules are available. These quantities, and functions derived from them, may be added to the accumulators used to eventually calculate simulation run averages. For example, the square of the configurational energy, $\mathcal{V}^2$, is calculated so that, at the end of the simulation, $\langle \mathcal{V}^2 \rangle - \langle \mathcal{V} \rangle^2$ can be used in eqns (2.80) or (2.89) to calculate the specific heat in the canonical or microcanonical ensemble. Although the average force and torque on a molecule in a fluid are zero, the mean-square values of these properties can be used to calculate the quantum corrections to the free energy given by eqns (2.153), (2.155). In an MD simulation, the force $\mathbf{f}_i$ on molecule i from its neighbours is calculated anyway, to move the molecules. The forces are not required in the implementation of the Metropolis MC method, so that they must be evaluated in addition to the potential energy if the quantum corrections are to be evaluated. The alternative of calculating the corrections via $g(r)$ (eqn (2.154)) is less accurate. The mean-square forces, as well as the average of the Laplacian of the potential, are needed to calculate the configurational temperature of eqn (2.56) and Appendix F.

This is the point in the simulation at which a direct calculation of the chemical potential can be carried out. A test particle, which is identical to the other molecules in the simulation, is inserted into the fluid at random (Widom, 1963). The particle does not disturb the phase trajectory of the fluid, but the energy of interaction with the other molecules, $\mathcal{V}_{test}$, is calculated. This operation is repeated many times, and the quantity $\exp(-\beta \mathcal{V}_{test})$ is used in eqn (2.75) to compute the chemical potential. In the MD method, the total kinetic temperature, $\mathcal{T}$, fluctuates, and it is essential to use eqn (2.76a). The insertion subroutine increases the running time somewhat; a figure of 20 % is typical.

The difficulty with this method is that a large number of substantial overlaps occur when particles are inserted. The exponential is then negligible, and we do not improve the statistics in the estimation of μ. Special techniques may be needed in such cases and we address these in Chapter 9. This is not a severe problem for the Lennard-Jones fluid, where Powles et al. (1982) have calculated μ close to the triple point with runs of less than 8000 timesteps. For molecular fluids, Romano and Singer (1979) calculated μ for a model of liquid chlorine up to a reduced density of $\rho\sigma^3 = 0.4$ (the triple point density is ≈ 0.52); Fincham et al. (1986) obtained an accurate estimate of μ by direct insertion at $\rho\sigma^3 = 0.45$ for the same model. In the case of mixtures, the chemical potential of each species can be determined by separately inserting molecules of that species.

The calculation of μ for a chain molecule, such as a small polymer in a polymer melt or solution, is more difficult. Once the first monomer of the ghost chain has been randomly inserted, without a significant overlap, the addition of subsequent monomers is likely to lead to such overlaps and, even for chains of modest length, the overall trial insertion will result in a zero contribution to μ. The configurational-bias method (see Section 9.3.4) can be used to thread the inserted chain through the fluid so that it makes a reasonable contribution to μ (Willemsen et al., 1998).

In calculating μ for an ionic fluid, charge neutrality can be preserved by inserting a pair of oppositely charged test particles and calculating the Boltzmann factor of the energy of the test ion pair with the ionic fluid. The two charged particles can be inserted at random positions in the fluid, but the accuracy of the calculation is improved if we use a Rosenbluth approach to increase the likelihood that the inserted pairs are separated by a short distance, that is, close to contact in the primitive model of the electrolyte (Orkoulas and Panagiotopoulos, 1993). More accurate calculations of the chemical potential of ions in aqueous environments can be made using methods such as MBAR (Mester and Panagiotopoulos, 2015); for further discussion see Section 9.2.4.

For systems of hard molecules, the pressure is generally not calculated in the conventional way: the potential energy is not differentiable, and so it is impossible to define forces. In MD of such systems, as described in Section 3.7, the program proceeds from collision to collision. At each event, the collisional virial may be calculated, from the impulse (rather than force) acting between the particles. This is accumulated in a time average and, at intervals, these averages are used to calculate the excess part of the pressure, as described in Section 2.4. However, for nonspherical hard particle models (Section 3.7.2), event-driven MD is a more complicated proposition than MC.

In MC simulations of hard-particle systems, the pressure may be calculated numerically by a box-scaling procedure, first introduced by Eppenga and Frenkel (1984). The problem has been revisited several times (Allen, 2006c; de Miguel and Jackson, 2006; Brumby et al., 2011; Jiménez-Serratos et al., 2012). Start with the canonical ensemble expression

$$\frac{P^{\text{ex}}}{k_{\text{B}}T} = -\frac{1}{k_{\text{B}}T}\frac{\partial A^{\text{ex}}}{\partial V} = \frac{\partial \ln Q^{\text{ex}}}{\partial V} = \frac{1}{Q^{\text{ex}}}\frac{\partial Q^{\text{ex}}}{\partial V}$$

$$= \lim_{\Delta V \to 0_+} \frac{1}{\Delta V}\frac{Q^{\text{ex}}(V) - Q^{\text{ex}}(V - \Delta V)}{Q^{\text{ex}}(V)}$$

$$= \lim_{\Delta V \to 0_+} \frac{1}{\Delta V}\left(1 - \frac{Q^{\text{ex}}(V - \Delta V)}{Q^{\text{ex}}(V)}\right).$$

Here we are considering a system that has been reduced in volume by a small amount ΔV relative to the system of interest. We can relate the ratio of partition functions to a single quantity, averaged over particle coordinates, if we take the latter to be scaled homogeneously, along with the box lengths

$$L' = fL, \qquad \mathbf{r}' = f\mathbf{r}, \qquad \text{where} \quad f = \left(\frac{V - \Delta V}{V}\right)^{1/3}.$$

The ratio $Q^{\text{ex}}(V - \Delta V)/Q^{\text{ex}}(V)$ may be interpreted as the relative statistical weights of these systems. For hard particles, the statistical weights of configurations are simply zero (if there is an overlap) or one (if there is no overlap): therefore this ratio is the average probability of *no overlap* being generated by the volume scaling. Eppenga and Frenkel (1984) argue that, for sufficiently small ΔV, this may be written as a product of independent probabilities of no overlap for all the pairs, each of which can be written as $1 - p$ where p is the probability that the volume scaling results in an overlap of a given pair; this value is the same for all pairs. Hence

$$\frac{P^{\text{ex}}}{k_{\text{B}}T} = \lim_{\Delta V \to 0_+} \frac{1 - \prod_{i<j}(1 - p)}{\Delta V} = \lim_{\Delta V \to 0_+} \frac{\sum_{i<j} p}{\Delta V} = \lim_{\Delta V \to 0_+} \frac{\langle N_{\text{overlap}} \rangle}{\Delta V},$$

where we use the fact that $p \ll 1$ for small ΔV. Therefore we may calculate P^{ex} by MC simulations in which, at intervals, we count the overlaps N_{overlap} (summed over all distinct pairs) that would result from an isotropic scaling of the box and particle coordinates by the factor f just defined, average them and divide by ΔV. Alternatively, N_{overlap} can be calculated by counting the overlaps that would result from scaling the particle dimensions up, isotropically, by a factor $1/f$, keeping the coordinates fixed. In practice, an extrapolation to low ΔV may be needed.

As just described, the method is restricted to *convex* hard molecules: an expansion (rather than a contraction) of the box cannot produce any overlaps in this case. For non-convex particles (for example, hard dumb-bells) an expansion can produce overlaps and the prescription must be modified: Brumby et al. (2011) discuss how to take this into account, as well as some of the subtleties of the statistical mechanics. Jiménez-Serratos et al. (2012) explain how to handle the case of finite discontinuities in the potential, as seen for square wells, for instance. It is possible to extend the procedure to calculate all the components of the pressure tensor, either by using the geometry of convex particles (Allen, 2006c) or by scaling separately in the x, y, and z directions (de Miguel and Jackson, 2006; Brumby et al., 2011). There are several ways of deriving the expression for P^{ex} (de Miguel and Jackson, 2006), and it is also possible to relate it to the pair distribution function at contact (Boublik, 1974), which is a well-known statistical mechanical route to the pressure. The use of this approach to estimate the surface tension of interfaces (Gloor et al., 2005) will be discussed in Chapter 14.

In this book, we have assumed that calculation of most other properties of interest will be carried out after the simulation, by analysis of output configurations stored on disk or other media. This analysis will be the subject of Chapter 8. In some cases, however, it may be preferable to calculate properties such as time correlation functions and structural distributions during the simulation itself. Against this, it must be said that the simulation program may be slowed down unnecessarily if too much calculation is included in it. For example, the pair distribution function $g(r)$ involves a sum over pairs of molecules, and this is sometimes included in the inner loop of an MD program. However, $g(r)$ is generally of interest for separations r much greater than the potential cutoff r_c, and so we need to examine many more pairs than would be required to calculate the energy and forces. Also, successive configurations are likely to be highly correlated. It is sensible to carry out this expensive summation less frequently than once per timestep. Similar observations apply to many other properties. The cleanest approach is to write a completely separate routine for calculating $g(r)$, and call it (say) every 10–20 steps or MC cycles; or alternatively to perform the analysis afterwards from stored configurations. In a similar way, time correlation functions may be calculated during an MD simulation, by methods very similar to some of those described in Chapter 8. However, this will require extra storage in the simulation program, and will make the program itself more complicated.

5.6   Starting up

In the remainder of this chapter, we consider the overall structure of the simulation programs. It is quite common to carry out sequences of runs at different state points, each run following on from the previous one. By this means, the starting configuration for most runs is obtained from a nearby state point, and will not require as long to equilibrate as one prepared from scratch. On the other hand, with the availability of parallel computing resources that can be used as 'task farms', the overall workflow may be more efficient if many runs can be conducted simultaneously. In this case, each one will require an independent starting point.

In both MD and MC techniques, therefore, it is necessary to design a starting configuration for the first simulation of a sequence. For MC, the molecular positions and orientations are specified, and for MD, in addition, the initial velocities and angular velocities must be chosen. Assuming that the liquid state is the target, it is important to choose a configuration that can relax quickly to the structure and velocity distribution appropriate to a fluid. This period of equilibration must be monitored carefully, since the disappearance of the initial structure may be quite slow. As a series of runs progresses, the coordinates and velocities from the last configuration of the previous run can be scaled (giving a new density, energy, etc.) to form the initial configuration for the next run. Again, with each change in state point, a period of equilibration must be set aside before attempting to compute proper simulation averages.

5.6.1 The initial configuration

The simplest method of constructing a liquid structure is to place molecules at random inside the simulation box (see Appendix E). The difficulty with this technique is that the configuration so constructed may contain substantial overlaps. This would be totally unphysical for a hard-core system, for which the potential energy would be infinite. For soft potentials, the energy for most random configurations, although high, can be calculated (provided no two molecules are centred at exactly the same point), so this type of configuration can be used to start Monte Carlo simulations, provided that the system is allowed to relax. In molecular dynamics, on the other hand, the large intermolecular potentials and the correspondingly large forces can cause difficulties in the solution of the differential equations of motion. A standard approach is to begin with a period of energy minimization: effectively, the velocities are set to zero at the start of each step, so the system evolves in the direction of the forces ('downhill'). During this period, a maximum limit can be set on the individual atom displacements, to avoid the creation of further overlaps. Another strategy is to replace the real potential by a version in which the repulsive core is reduced in size and/or strength, and slowly restore the full potential during equilibration.

It is more usual to start from a lattice. Almost any lattice is suitable, but historically the face-centred cubic structure, with its $4M^3$ (M = integer) lattice points has been the starting configuration for many simulations. This lattice is shown in Fig. 5.6. The lattice spacing is chosen so that the appropriate liquid state density is obtained. During the course of the simulation the lattice structure will disappear, to be replaced by a typical liquid structure. This process of 'melting' can be enhanced by giving each molecule a small random displacement from its initial lattice point along each of the space-fixed axes (Schofield, 1973).

In the case of a molecular fluid, it is also necessary to assign the initial orientations of the molecules. A model commonly used for linear molecules is the α-FCC lattice, which is the solid structure of CO_2 and one of the phases of N_2 (see Fig. 5.6). In this structure, there are four sublattices of molecules oriented along the four diagonals of the unit cell. A code for generating the α-FCC lattice is given in Code 5.6. For non-linear molecules, any suitable known crystal structure could be used. Small random displacements can also be applied to the lattice orientations so as to speed up melting. Some workers prefer to choose the orientations completely randomly given a centre of mass structure, although at high densities, with elongated molecules, random assignment of the directions can result in non-physical overlaps.

5.6.2 The initial velocities

For a molecular dynamics simulation, the initial velocities of all the molecules must be specified. It is usual to choose random velocities, with magnitudes conforming to the

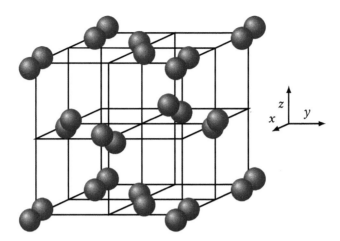

Fig. 5.6 Unit cell of the α-FCC structure for linear molecules. The centre-of-mass positions are as in the argon lattice. The orientations are in four sublattices: $(1, 1, 1)$, $(1, -1, -1)$, $(-1, 1, -1)$, $(-1, -1, 1)$.

Code 5.6 Initialization of a crystal lattice

These files are provided online. initialize.f90 contains a program to initialize the α-FCC lattice of linear molecules, or the simple FCC lattice of atoms, using routines in initialize_module.f90, and utility module routines (see Appendix A) to handle random numbers and file output. Optionally the velocities and angular velocities can also be initialized, for a chosen temperature. The program can also handle nonlinear molecules and chains of atoms.

```
! initialize.f90
! Sets up initial configuration for MD or MC
PROGRAM initialize
```

```
! initialize_module.f90
! Routines to initialize configurations and velocities
MODULE initialize_module
```

required temperature, corrected so that there is no overall momentum

$$\mathbf{P} = \sum_{i=1}^{N} m_i \mathbf{v}_i = 0. \tag{5.28}$$

The velocities may be chosen randomly from a Gaussian distribution (see Appendix E and Code 5.6). For example, in an atomic system

$$\rho(v_{ix}) = (m_i/2\pi k_{\mathrm{B}}T)^{1/2} \exp\left(-\tfrac{1}{2}m_i v_{ix}^2/k_{\mathrm{B}}T\right) \tag{5.29}$$

where $\rho(v_{ix})$ is the probability density for velocity component v_{ix}, and similar equations apply for the y and z components. The same equations apply to the centre-of-mass velocities in a molecular system. As a simple alternative, each velocity component may be chosen to be uniformly distributed in a range $(-v_{max}, +v_{max})$; the Maxwell–Boltzmann distribution is rapidly established by molecular collisions within (typically) 100 timesteps. The (re)selection of velocities, and the scaling needed to make them consistent with the chosen temperature, are both standard features of most MD packages.

For a molecular fluid, the angular velocity in the body-fixed frame is also chosen to be consistent with the required temperature

$$\frac{f}{2} N k_B \mathcal{T} = \frac{1}{2} \sum_{i=1}^{N} \omega_i^b \cdot \mathbf{I} \cdot \omega_i^b. \tag{5.30}$$

Here, $\mathbf{I}$ is the moment of inertia tensor and f the number of degrees of rotational freedom (two for a linear molecule, three for a nonlinear one). Because the total angular momentum is not conserved, it is not essential to set the initial value of this quantity to zero, but it is sensible to ensure that the molecular angular momenta roughly cancel each other. For linear molecules, each angular velocity ω_i must be chosen perpendicular to the molecular axis (see Appendix E). An example of this technique is given in Code 5.6. One method for initializing the angular velocity for a lattice configuration involves choosing pairs of molecules with identical orientations and assigning them equal and opposite angular velocities chosen at random. An alternative method is to set the angular velocity of every molecule to zero at the start of the run, and to choose the translational kinetic temperature to be greater than required. The normal process of equilibration will then redistribute the energy amongst the different degrees of freedom. Precise adjustments to the kinetic temperature are made by scaling velocities during equilibration.

5.6.3 Equilibration

If a simulation is started from a lattice, or from a disordered configuration at a different density and temperature, it is necessary to run for a period so that the system can come to equilibrium at the new state point. At the end of this equilibration period, all memory of the initial configuration should have been lost. A simple way to monitor system equilibration is to record the instantaneous values of the potential energy and pressure during this period. In the case of a lattice start, the potential energy rises from a large negative value to a value typical of a dense liquid, as shown in Fig. 5.7(a). The behaviour of the instantaneous pressure is also shown in Fig. 5.7(b). The equilibration period should be extended at least until these quantities have ceased to show a systematic drift and have started to oscillate about steady mean values.

Equilibration is especially important when the initial configuration is a lattice, and the state point of interest is in the liquid region of the phase diagram. There are a number of parameters that can be monitored to track the 'melting' of the lattice, and subsequent progress to equilibrium. The degree of translational order in the centres of mass is tested by evaluating the translational order parameter

$$s(\mathbf{k}) = |\rho(\mathbf{k})|^2 \quad \text{with} \quad \rho(\mathbf{k}) = \frac{1}{N} \sum_{i=1}^{N} \exp(i\mathbf{k} \cdot \mathbf{r}_i) \tag{5.31}$$

where $\mathbf{r}_i$ is the position vector of the centre of mass of the ith molecule and $\mathbf{k}$ is a reciprocal lattice vector of the initial lattice. This is equal to the structure factor $S(\mathbf{k})$ divided by the

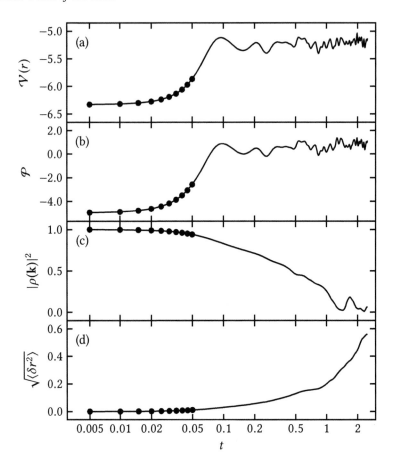

Fig. 5.7 The equilibration phase of an MD simulation. The horizontal scale is logarithmic; the first ten steps are shown explicitly. We show the evolution of: (a) the potential energy; (b) the instantaneous pressure; (d) the square modulus of the translational order parameter; and (c) the root-mean-square displacement from the initial positions. The system consists of 108 atoms interacting via the shifted Lennard-Jones pair potential, $r_c^* = 2.5$, no long-range corrections applied. The simulation starts from an FCC lattice with a Maxwell–Boltzman velocity distribution. The system is near the triple point ($\rho^* = 0.8442$, $T^* = 0.722$, $P^* = 0.610$, $E^*/N = -4.129$). The simulation is in the constant-energy ensemble, using the velocity Verlet algorithm, with $\delta t^* = 0.005$.

number of particles. For example,

$$\mathbf{k} = (2\pi/\ell)(-1,\ 1,\ -1) = \left[\left(\tfrac{1}{4}N\right)^{1/3} 2\pi/L\right](-1,\ 1,\ -1) \quad \text{for FCC,}$$

where ℓ is the unit cell size, which may be set equal to $L/(\tfrac{1}{4}N)^{1/3}$ in a cubic simulation box. It is, of course, possible to monitor several such components. For a solid, $\rho(\mathbf{k})$ is of order unity, apart from an origin-dependent phase factor which disappears on taking the square modulus to get $s(\mathbf{k})$. For a liquid, $s(\mathbf{k})$ will be positive, with amplitude $O(N^{-1/2})$. The translational order parameter for a simulation starting in the FCC lattice is shown in

Fig. 5.7(c). It is clear that, in this instance, $\rho(\mathbf{k})$ is a much more sensitive indicator of the persistence of a lattice structure, and of the need to extend the equilibration period, than the 'thermodynamic' quantities shown in Fig. 5.7(a) and (b).

The orientational order parameter, as introduced by Vieillard-Baron (1972) is given for linear molecules by the first Legendre polynomial

$$P_1 = \frac{1}{N} \sum_{i=1}^{N} P_1(\cos \gamma_i) = \frac{1}{N} \sum_{i=1}^{N} \cos \gamma_i \qquad (5.32)$$

where γ_i is the angle between the molecular axis of molecule i and the original axis direction in the perfect crystal. Several other parameters of this type (for example, higher-order Legendre polynomials) can be monitored. $P_1 = 1$ for the initial configuration and fluctuates around zero with amplitude $O(N^{-1/2})$ when the fluid is rotationally disordered. For non-linear molecules, several similar order parameters, based on different molecular axes, may be examined; they should all vanish simultaneously on 'melting'. An example of an order parameter routine is given in the utility modules of Appendix A.

An additional strategy involves monitoring the mean-squared displacements of the molecules from their initial lattice positions. This function increases during the course of a liquid simulation, as illustrated in Fig. 5.7(d) (see eqn (2.117)), but oscillates around a mean value in a solid. A useful rule of thumb is that when the root-mean-squared displacement per particle exceeds $0.5\,\sigma$ and is clearly increasing, then the system has 'melted' and the equilibration is complete. Care should be taken to exclude periodic boundary corrections in the computation of this quantity. This technique is useful for monitoring equilibration not only from a lattice but also from a disordered starting configuration, particularly when there is a danger that the system may become trapped in a glassy state rather than forming a liquid: eqn (5.31) would not be appropriate for these cases.

An additional danger during the equilibration period is that the system may enter a region of gas–liquid coexistence. If a study of a homogeneous fluid is attempted in the two-phase region, large, slow density fluctuations occur in the central simulation box. This is most clearly manifest in the radial distribution function (Jacucci and Quirke, 1980b). In the two-phase region, $g(r)$ has an unusually large first peak ($g(r) \approx 5$), it exhibits long-ranged, slow, oscillations, and does not decay to its correct long-range value of 1. The structure factor $S(k)$ diverges as $k = 0$, indicating long-wavelength fluctuations. Monitoring these structural quantities may give a warning that the system has entered a two-phase region, in which case extremely long equilibration times will be required.

One useful trick that may be used to increase the rate of equilibration from a lattice, is to raise the kinetic temperature to a high value (e.g. $T^* = 5$ for Lennard-Jones atoms) for the initial 500 steps (e.g. by scaling all the velocities). The temperature is reset to the desired value during the second part of the equilibration period. It is sometimes convenient to continually adjust the temperature and/or pressure of an MD simulation throughout the equilibration phase, using one of the methods described in Chapter 3.

It is difficult to say how long a run is needed for equilibration, but periods of 1000–10 000 timesteps or MC cycles are typical for small systems, $N \sim 1000$ atoms; remember (Section 4.4) that for an N-atom system, one MC cycle is N attempted moves. More time should be set aside for equilibration from an initial lattice, when large structural changes are anticipated, or whenever it is suspected that a phase transition is close; somewhat less time is required for high-temperature fluids. The golden rule is to examine carefully the aforementioned parameters, as the simulation proceeds. At the end of equilibration, they should have clearly reached the expected limiting behaviour. In an MD simulation, it is also worthwhile to check the proper partitioning of kinetic temperature for a molecular system (i.e. $\mathcal{T}_{\text{rot}} = \mathcal{T}_{\text{trans}}$) and that the kinetic temperature is equal to the configurational temperature, eqn (2.56), although it should be remembered that these instantaneous values are subject to significant fluctuations. At the end of the equilibration period, the accumulators for the ensemble averages are reset to zero, and the production phase of the simulation begins.

5.7   Organization of the simulation

Computer simulations are programs that may require a substantial amount of central processing unit (CPU) time, and produce significant amounts of data. It is not always clear, at the outset, how long a simulation should be, in order to produce the desired results with adequate statistical precision. For this reason, simulations should be designed so that they can be restarted or continued with the minimum difficulty. The restart facility enables the total simulation to be broken up into manageable chunks of computing time. In the event of an unexpected computer failure, including filling up the data storage, the program can be started again with a minimum loss of resources. It may even be possible to make the simulation self-starting, so that it can be run as a series of small jobs without human intervention. The details of job organization clearly depend on the particular computer being used.

5.7.1   Input/output and file handling

Ideally, manipulation of files by the user should be kept to a minimum. Often it is useful to write scripts to help set up many simulations at once, and to run the analysis programs afterwards. Python is very commonly used; alternatives include Perl, Tcl, and shells such as bash. The whole operation will be made easier by adopting a sensible, usually hierarchical, structure of directories and files. It also makes sense to keep all the files associated with a particular run, both input and output, in (or under) the same directory. This helps avoid the embarrassment of forgetting the parameters that were used to generate a large dataset.

Only a handful of parameters define the important features of a simulation: the run length, step size, desired temperature, etc. These numbers can be stored in a small input file, which can easily be accessed and altered by the user. Legibility and clarity are essential: a key–value system makes it easy to recognize which parameter is which, and this is the scheme used by many packages. The file can be prepared in a simple editor, or via a script or other user interface. Usually the simulation program or package will write these parameters out, somewhere near the start of the main output file, which will also typically contain some bare-bones information about the progress of the simulation, and a summary of results on successful completion.

A second, essential, body of information is the specification of the force field or interaction potential. For a complex system, this can be a significant amount of data, although it may be reduced if default parameters are built into the program itself, and simply selected as options in the data file. This information may be combined with the run parameters described in the previous paragraph, or may be stored separately, depending on the package and/or individual preference. It should also be user-readable, for convenience.

The starting configuration (molecular positions, velocities, accelerations, etc), and final configuration produced at the end of the run, may be either in machine-readable (unformatted, binary) or user-readable (formatted, e.g. ASCII) form. The formatted versions occupy more space, but can be compressed (zipped), and some programs (e.g. for visualization) can read the compressed forms directly. Most important, perhaps, is to adopt a *standard* file format where possible, for portability. There is no universally agreed standard, however, as the field is so broad. *De facto* standards are associated with the major packages such as Groningen machine for chemical simulations (GROMACS), chemistry at Harvard molecular mechanics (CHARMM), large-scale atomic/molecular massively parallel simulator (LAMMPS) (see Lundborg et al., 2013, and references therein) and attempts have been made to suggest more generally applicable models and formats (see e.g. Hinsen, 2014). Some of these incorporate structure, or connectivity, information as well as simply atomic coordinates. The protein data base (PDB) format is designed for proteins; the xyz format is a common and simple format in which, following a short header section, each atom is represented by a line of information containing its chemical symbol and coordinates. Toolkits for interconverting different file formats exist, and some visualization programs (for instance Jmol (2016) and VMD (Humphrey et al., 1996)) are capable of reading and writing a variety of formats. As the simulation proceeds, it may be convenient to write out at intervals the current configuration, in the same format as that used for the start and end configurations. By overwriting this file frequently (perhaps every 100–500 steps or cycles) we make it easy to restart the simulation in the event of program (or computer) failure.

There may be other specific quantities that should be written to files as the simulation proceeds, especially if they would be expensive to recompute from stored positions and velocities. A good example would be components of the stress tensor, required to compute shear viscosities and surface tensions.

Finally, we come to the very large file that stores molecular positions, velocities, and accelerations, taken at frequent intervals (e.g. every ten steps) during the run, for future analysis. This is usually called the *trajectory file*; alternatively it may consist of a succession of individual files, named according to the timestep, stored in a single directory. Eventually, this substantial amount of information must be archived. Often, to save space, these configurations are stored in machine-readable form (unformatted) or are compressed after the run is complete. The high degree of correlation between successive snapshots, as well as between the positions of neighbouring particles, make special-purpose compression algorithms quite attractive (Spångberg et al., 2011; Marais et al., 2012; Huwald et al., 2016). It is important to decide, before the run, how much information is needed in this file. Are molecular positions sufficient? Or are velocities required as well, or possibly even the forces on each atom?

It goes without saying that the user should become familiar with the format of the trajectory file, and the starting and finishing configurations, of any simulation package. In the simplest cases, the atomic positions and velocities may be written out in a consistent fashion by particle index, with particle 1 followed by particles 2, 3, etc. in order. This is natural if they are stored in contiguous arrays within the program, for instance. However, this is not always the case. In parallel simulation programs, as we shall see in Chapter 7, particle information is passed around between processors very frequently, and the particle identifier is just one more piece of information. It may be inefficient to gather all the particle coordinates and sort them by identifier before writing them out, so quite often they are output (with their identifiers) in an order which varies from step to step. Similarly, if the program is written in an object-oriented style, there is likely to be less emphasis on the use of the identifier as an array index. In either case, provided the particle identifier is written out together with coordinates and velocities, there should be no problem: but the user must be aware of this, for example, when computing single-particle time correlation functions and other properties!

5.7.2   Program structure

The overall structure of most simulation programs is fairly standard:

- Reading in parameters, starting configuration, and initialization.
- The main loop over timesteps or MC cycles.
- Writing out results, final configuration.

We will assume that all preparation of the configuration file is performed in separate utility programs. This includes the initial generation of a configuration at the start of a series of simulations (Section 5.6), and the scaling of coordinates or velocities to generate a different state point from a previous final configuration. In a sequence of runs, this can be handled by a script which calls the appropriate programs, in between calls to the main simulation program. In short, the utility programs produce a configuration file which is 'ready to go'. By separating these activities from the main simulation program, the latter is kept simple in structure: it accepts an initial configuration and, during the course of the run, maintains it in a state suitable for continuing the run.

Date and time routines are usually called, so a time-stamp can be written out at the start of the run, for easy reference. In the same way, it is most advisable to write out the basic run parameters, exactly as they are read in, at the start of the output file, thus avoiding any ambiguity about the nature of the simulation, when the results are studied later. It is important to record which algorithm is being used to move the molecules, and which ensemble is being sampled. Unless we are dealing with a very simple pair potential, we would want to record the potential or force-field parameters, the length of potential cutoffs, and perhaps the relative molecular masses in the output file. The units that apply in the simulation (Appendix B) should also be made clear. The essential point is that, by looking at the information at the start of the output file, it should be possible to recreate the run exactly, even if the details have been forgotten in the time since it was originally carried out.

If we are using potential tables and neighbour lists, we will also have to initialize them. Generally, it is advisable to delegate each separate task of this kind to a separate function or subroutine, so as to maintain a simple and clear, modular, program structure.

The initialization stage includes the setting to zero of run-average accumulators, and opening the trajectory file. It may be desirable, if a continuation run is being carried out, to open this file in 'append' mode, so as to give an uninterrupted succession of configurations as time proceeds. Obviously, one should guard against the danger of overwriting the old trajectory file, which should only be done if the run is being restarted from the beginning for some reason. On the other hand, if the restart is initiated from a configuration which was written before the last few records in the trajectory file, then those records will be generated afresh, and care must be taken not to duplicate them.

We are almost ready to deal with the body of the simulation: the main loop. Before doing so, for a molecular dynamics simulation, we may have to call some routine to compute the initial forces on the molecules. For Monte Carlo, there will be an initial calculation of the total energy, and perhaps other properties, of the system. Schematically the main loop takes the form

```
step = 0
DO
   IF ( step >= nstep ) EXIT
   IF ( CPU_TIME() > cpu_time_max ) EXIT

   CALL move ( r, v, ....)
   step = step + 1
   IF ( MOD(step, calc_step)   == 0 ) CALL calc
   IF ( MOD(step, output_step) == 0 ) CALL write_output
   IF ( MOD(step, traj_step)   == 0 ) CALL write_traj
   IF ( MOD(step, config_step) == 0 ) CALL write_config
END DO
```

At the start of each step we indicate possible conditions for exiting the loop: completing the desired number of steps, or detecting that the allocated CPU time has nearly been used up, as examples. The move subroutine is really shorthand for any of the MD or MC algorithms mentioned in the earlier chapters, complete with any force or energy evaluation routines that may be necessary. In the course of this, the current values of the potential and kinetic energies, virial function, etc. will normally be calculated. Any other instantaneous values of interest, such as order parameters and distribution function histograms, should be computed in subroutines immediately following the move routine (indicated schematically in the code by CALL calc). There may well be several of these routines, each handling different properties. They may require calculating at different intervals, which again we indicate schematically by calculating the step number, modulo the desired interval. It is assumed that each routine such as calc also increments the appropriate run-average accumulators. Following these calculations, at the appropriate intervals, information on the progress of the run is sent to the output channel, data are output to the trajectory file, and, lastly (for reasons of safety), the current coordinates, momenta, etc are output to the configuration file, which will be used for a restart if need be. At the same time, it may be convenient to write out all the accumulators, so that the calculation of run averages can be picked up again smoothly, if a restart is necessary.

This general scheme will only allow restarts, of course, if files (specifically the trajectory file) that were open at the time of a program crash remain uncorrupted and accessible afterwards. If this is not the case, then more complicated steps must be taken to maximize the safety of the information generated by the program. This may involve opening, appending to, and then closing the trajectory file at frequent intervals (perhaps every record).

Once the main loop is complete, the last stage of the program takes place: computing run averages, fluctuations, and possibly statistical error estimates, from the accumulators, and writing out the final configuration.

5.7.3   The scheme in action

The scheme outlined in the previous section would typically be run as a series of simulations alternating with configuration-modifying programs. Typically, the configuration handler would be used to generate the initial configuration, either from scratch (see Section 5.6) or by modifying the final configuration from an old run. This would be followed by an equilibration run. The output configuration file from the equilibration run would then be used by the configuration utility routine to generate an initial configuration for the production phase. All this assumes that, after each run, the user will examine the output before setting the subsequent run in motion. This is usually to be recommended: it is always desirable to check that, for example, equilibration has proceeded satisfactorily, and to investigate the reasons for any unusual behaviour (failure of the underlying computer hardware, filling up of the storage, or atom overlaps in the simulation). However, it is also quite possible, and may be desirable, to set up a sequence of jobs to run consecutively, without intervention; also, a large set of jobs, identical except for the input files, may be initiated in parallel. It is still worth emphasizing, however, the importance of looking at the raw results as soon as possible after the jobs have run.

It is notable that many of the key activities, such as file handling, job submission, and data analysis, are more conveniently handled through a script written in an appropriate language. Other aspects, especially the time-intensive advancement of the configuration, whether by MD or MC, really benefit from a compiled language. A powerful approach is to combine the two methods. Some packages are sufficiently modular that they can be treated as a library of routines, which are callable from, for example, a Python script. If this approach is adopted, there are two obvious problems to tackle. The first is to match the arguments of, say, a C or Fortran function, to those of the calling script. This is a very general issue, which has been addressed through 'wrapper' codes. The second is the efficiency of passing information back and forth, especially the atomic configurations, which may be quite large. The simplest method, to write the configuration out to a file, and read it back in again, although fairly foolproof, is likely to be much too slow.

Once a simulation is successfully completed, the output should be carefully filed and a backup copy made. Simulations tend to produce a wealth of information. The aim is to be able to recover, and understand, the information at a later date. Data archiving and curation are serious issues, especially in a collaborative research environment. Quite frequently, interesting scientific questions arise, long after the data were originally produced. Moreover, the people who produced the data may no longer be around. When considering backups, it is best to assume that disk or other media faults are a matter of 'when' rather than 'if'. For all these reasons, it is vital to have a simple, well-documented, data-recovery process.

5.8   Checks on self-consistency

We already mentioned, in Section 3.6, a few checks which should be performed to build some confidence in the proper working of an MD program. Once results start to appear, from either MD or MC programs, it makes sense to look critically at them, and try to spot any signs of inconsistency. There are two very common sources of error:

(a) the system may not be at equilibrium, or may have been simulated for insufficient time to generate well-sampled averages;

(b) the model may not be the one you think it is.

For the first of these, there is never a guarantee that a simulation is long enough. A serious attempt to calculate statistical errors, including estimating the correlation times or statistical inefficiencies, will be a good starting point. Much better, however, is to conduct several separate runs, from starting configurations that have been prepared completely independently, and compare the results. There are well-defined statistical tests to compute the likelihood that different estimates of any particular property, with their estimated errors, have been sampled from the same underlying distribution. Wherever possible, a property should be estimated in two different ways, which can be checked for consistency. For instance, heat capacities and compressibilities may be obtained from the fluctuation formulae given in Chapter 2, and by fitting an equation of state over a range of state points. Temperature can be estimated from kinetic energies, and from configurational variables. Pair distribution functions should tend to the correct limit at large separation: for instance, $g(r) \to 1$ (actually, in constant-N ensembles, $g(r) \to 1 - 1/N$) as $r \to \infty$. Finally, it bears repeating that measured properties (especially equations of state) should be compared wherever possible with published literature values.

With an ever-growing reliance on packaged force fields, it becomes more and more important to check the supplied 'default' values of interaction parameters, and to take care in specifying the particular version of a potential that is to be used in the simulation. As discussed in Chapter 6, there is a wealth of competing options for handling long-range forces, all of them depending on the selection of a few key parameters. Potential cutoff distances may vary significantly from one study to another, and it is not always obvious whether a potential is shifted to zero at the cutoff, or whether a shifted-force version, or some other tempering or smoothing near the cutoff, is being employed. If the package code is not sufficiently clear, it may be possible to compute the potential numerically as a function of a single variable (distance, orientation or internal coordinate), by calling the appropriate routine directly from an external script, and then plotting or comparing with the expected function in some other way.

In some situations, the study of system-size dependence is an integral part of the problem; in other cases, it is just a useful check that the system studied is large enough to avoid boundary effects. One should always take care to prepare the larger systems in a way that is as independent as possible from the smaller ones. It is tempting to simply replicate the smaller systems in the different Cartesian coordinate directions, to produce the larger ones, but this will build in the original system periodicity, which must be given ample time to disappear. It must also be borne in mind that the natural physical timescales of larger systems tend to be longer (sometimes much longer) than those of smaller ones, so they may need to be simulated for a longer time, not a shorter time!

6

Long-range forces

6.1 Introduction

So far in this book, we have discussed the core of the program when the forces are short-ranged. In this chapter, we turn our attention to the handling of long-range forces in simulations. A long-range force is often defined as one in which the spatial interaction falls off no faster than r^{-d} where d is the dimensionality of the system. In this category are the charge–charge interaction between ions $(v^{qq}(r) \sim r^{-1})$ and the dipole–dipole interaction between molecules $(v^{\mu\mu}(r) \sim r^{-3})$. These forces are a serious problem for the computer simulator since their range is greater than half the box length for simulations with many thousands of particles. The brute-force solution to this problem would be to increase the box size L to hundreds of nanometres so that the screening by neighbours would diminish the effective range of the potentials. Clearly, this is not a practical solution since the time required to run such a simulation is proportional to N^2, that is, L^6 in 3D.

How can this problem, which is particularly acute for $v^{qq}(r)$, be handled? One approach is to include more than the nearest or minimum image of a charge in calculating its energy. Lattice methods such as the Ewald sum, described in Section 6.2, include the interaction of an ion or molecule with all its periodic images. The Ewald method can be further optimized by assigning the charges in the simulation cells to a fine, regular mesh. This enables the long-range part of the force to be calculated efficiently using a fast Fourier transform (FFT). Approaches of this kind, such as the particle–particle particle–mesh (PPPM) method, are described in Section 6.3.

In the calculation of long-range forces, straightforward spherical truncation of the potential can be ruled out. The resulting sphere around a given ion could be charged, since the number of anions and cations need not balance at any instant. The tendency of ions to migrate back and forth across the spherical surface would create artificial effects at $r = r_c$. Methods for adding image charges and dipoles to the surface of the truncation sphere which ensure its charge neutrality are discussed in Section 6.4, along with methods to include periodic images of the truncation sphere and a technique for estimating the 'best' truncated potential. A charge distribution within a spherical cavity polarizes the surrounding medium. This polarization, which depends upon the relative permittivity of the medium, has an effect on the charge distribution in the cavity (see e.g. Fröhlich, 1949). The effect can be included in the simulation using the reaction field method for charged and dipolar fluids which is discussed in Section 6.5.

In Sections 6.6 and 6.8, we present two methods for calculating long-range forces for large systems with many thousands of charges. The first of these, the fast multipole method (FMM), uses the multipole expansion to calculate the field from charges in different regions of space. Using a hierarchical tree structure, individual fields are combined to produce the total field due to the sample and its surrounding images. The total field is then moved back down the tree using a local expansion to calculate the forces at individual charges. The second method, Maxwell equation molecular dynamics (MEMD), combines the classical equations of motion of the charges with Maxwell's equations for the electric field. By using a nominal value of the speed of light, it is possible to couple these equations despite their disparate timescales. Both these methods are local and have the advantage that the computer time scales linearly with the number of charges.

The terms 'electrostatic potential' and 'field' are used frequently throughout this chapter. This is a reminder that the electrostatic potential, ϕ, created at a distance r from a charge q is $\phi(r) = q/r$ and the corresponding electric field is $\mathcal{E} = -\nabla\phi(r)$. The field is simply the force per unit charge. For simplicity of notation, we are omitting all factors of $4\pi\epsilon_0$: this corresponds to adopting a non-SI unit of charge (see Appendix B), that is, the charge q is reduced by $(4\pi\epsilon_0)^{1/2}$ for convenience.

6.2 The Ewald sum

When applied to an ionic fluid, the basic minimum image method corresponds to cutting off the potential at the surface of a cube surrounding the ion in question (see Fig. 1.16). This cube will be electrically neutral. However, the drawback is that similarly charged ions will tend to occupy positions in opposite corners of the cube: the periodic image structure will be imposed directly on what should be an isotropic liquid, and this results in a distortion of the liquid structure. An alternative is to consider the interaction of an ion with all the other ions in the central box and with all the ions in all of its periodic images.

The Ewald sum is a technique for efficiently performing this sum. It was originally developed in the study of ionic crystals (Ewald, 1921; Madelung, 1918). In Fig. 1.13, ion 1 interacts with ions 2, 2_A, 2_B and all the other images of 2. The potential energy can be written as

$$\mathcal{V}^{qq} = \frac{1}{2}\sum_{\mathbf{m}\in\mathbb{Z}^3}{}' \left(\sum_{i=1}^{N} \sum_{j=1}^{N} q_i q_j |\mathbf{r}_{ij} + \mathbf{m}L|^{-1} \right) \tag{6.1}$$

where q_i, q_j, are the charges. As mentioned earlier, all factors of $4\pi\epsilon_0$ are omitted. The sum over $\mathbf{m} = (m_x, m_y, m_z)$ is over all triplets of integers, $\mathbb{Z}^3$. For a cubic box, $\mathbf{m}L$ represents the centre of each box in the periodic array. The prime indicates that we omit $i = j$ for $\mathbf{m} = 0$. For long-range potentials, this sum is conditionally convergent, that is, the result depends on the order in which we add up the terms. A natural choice is to take boxes in order of their proximity to the central box. The unit cells are added in sequence: the first term has $|\mathbf{m}| = 0$, that is, $\mathbf{m} = (0, 0, 0)$; the second term, $|\mathbf{m}| = 1$, comprises the six boxes centred at $(\pm L, 0, 0)$, $(0, \pm L, 0)$, $(0, 0, \pm L)$, etc. As we add further terms to the sum, we are building up our infinite system in roughly spherical layers (see Fig. 6.1). When we adopt this approach, we must specify the nature of the medium surrounding the sphere, in particular its relative permittivity (dielectric constant) ϵ_s. The results for

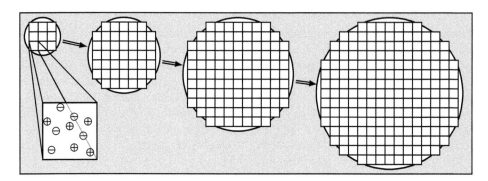

Fig. 6.1 Building up the sphere of simulation boxes. Each copy of the periodic box is represented by a small square. The shaded region represents the external dielectric continuum of relative permittivity ϵ_s.

a sphere surrounded by a good conductor such as a metal ($\epsilon_s = \infty$) and for a sphere surrounded by vacuum ($\epsilon_s = 1$) are different (de Leeuw et al., 1980):

$$\mathcal{V}^{qq}(\epsilon_s = \infty) = \mathcal{V}^{qq}(\epsilon_s = 1) - \frac{2\pi}{3L^3}\left|\sum_i q_i \mathbf{r}_i\right|^2 . \tag{6.2}$$

This equation applies in the limit of a very large sphere of boxes. In the vacuum, the sphere has a dipolar layer on its surface: the last term in eqn (6.2) cancels this. For the sphere in a conductor there is no such layer. The Ewald method is a way of efficiently calculating $\mathcal{V}^{qq}(\epsilon_s = \infty)$. Equation (6.2) enables us to use the Ewald sum in a simulation where the large sphere is in a vacuum, if this is more convenient. The mathematical details of the method are given by de Leeuw et al. (1980) and Heyes and Clarke (1981). Here we concentrate on the physical ideas. At any point during the simulation, the distribution of charges in the central cell constitutes the unit cell for a neutral lattice which extends throughout space. In the Ewald method, each point charge is surrounded by a charge distribution of equal magnitude and opposite sign, which spreads out radially from the charge. This distribution is conveniently taken to be Gaussian

$$\rho_i^q(\mathbf{r}) = q_i \kappa^3 \exp(-\kappa^2 r^2)/\pi^{3/2} \tag{6.3}$$

where the arbitrary parameter κ determines the width of the distribution, and $\mathbf{r}$ is the position relative to the centre of the distribution. This extra distribution acts like an ionic atmosphere, to screen the interaction between neighbouring charges. The screened interactions are now short-ranged, and the total screened potential is calculated by summing over all the molecules in the central cube and all their images in the real-space lattice of image boxes. This is illustrated in Fig. 6.2(a).

A charge distribution of the same sign as the original charge, and the same shape as the distribution $\rho_i^q(\mathbf{r})$ is also added (see Fig. 5.6(b)). This cancelling distribution reduces the overall potential to that due to the original set of charges. The cancelling distribution is summed in reciprocal space. In other words, the Fourier transforms of the cancelling distributions (one for each original charge) are added, and the total transformed back

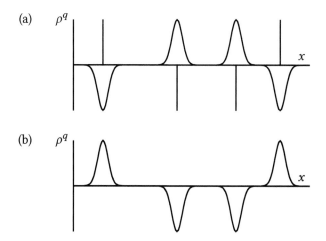

Fig. 6.2 Charge distribution in the Ewald sum, illustrated in 1D. (a) Original point charges plus screening distribution. (b) Cancelling distribution.

into real space. (Fourier transforms are discussed in Appendix D.) There is an important correction: the recipe includes the interaction of the cancelling distribution centred at r_i with itself, and this self term must be subtracted from the total. Thus, the final potential energy will contain a real-space sum plus a reciprocal-space sum minus a self-term plus the surface term already discussed. The final result is

$$
\mathcal{V}^{qq}(\epsilon_s = 1) = \frac{1}{2} \sum_{i=1}^{N} \sum_{j=1}^{N} q_i q_j \left(\sum_{|m|=0}^{\infty}{}' \frac{\mathrm{erfc}\left(\kappa|r_{ij} + Lm|\right)}{|r_{ij} + Lm|} \right.
$$

$$
+ (1/\pi L^3) \sum_{k \neq 0} (4\pi^2/k^2) \exp(-k^2/4\kappa^2) \exp(-ik \cdot r_{ij}) \Bigg)
$$

$$
- (\kappa/\pi^{1/2}) \sum_{i=1}^{N} q_i^2 + (2\pi/3L^3) \left| \sum_{i=1}^{N} q_i r_i \right|^2 .
$$

(6.4)

Here erfc(x) is the complementary error function (erfc$(x) = (2/\pi^{1/2}) \int_x^\infty \exp(-t^2)dt$) which falls to zero with increasing x. Thus, if κ is chosen to be large enough, the only term which contributes to the sum in real space is that with $m = 0$, and so the first term reduces to the normal minimum image convention.

The second term in eqn (6.4) is a sum over the reciprocal vectors, $k = 2\pi n/L$ where $n = (n_x, n_y, n_z)$ is a triplet of integers. A large value of κ corresponds to a sharp distribution of charge, so that we need to include many terms in the k-space summation to model it. In a simulation, the aim is to choose a value of κ and a sufficient number of k-vectors, so that eqn (6.4) (with the real-space sum truncated at $m = 0$) and eqn (6.2) give the same energy for typical liquid configurations.

Before considering the implementation of the Ewald method, we note that the Gaussian form of the charge distribution is not unique. Any function that smears the charge distribution can be used and Heyes (1981) lists the appropriate real- and reciprocal-space contributions for a number of different choices. The rigorous derivation of the Ewald sum requires that the system is neutral. In the presence of a net charge, a uniform background charge density, $-\sum_{i=1}^{N} q_i/L^3$, should be included to neutralize the fluid. This gives rise to an additional term in eqn (6.4) (Hub et al., 2014)

$$\mathcal{V}_{\text{charged}} = -\frac{\pi}{2L^3\kappa^2} \left| \sum_{i=1}^{N} q_i \right|^2. \tag{6.5}$$

For a homogeneous fluid, the calculation of the pressure resulting from the Ewald sum is straightforward since $\mathcal{W}^{qq} = \mathcal{V}^{qq}/3$.

In implementing the Ewald sum, the first step is to decide on a value of κ for the maximum number of wavevectors in the reciprocal-space sum. Kolafa and Perram (1992) have shown that it is possible to determine the best value of κ by balancing the errors in the real- and reciprocal-space parts of the sum. In practice, for a cubic box, we select a value of r_c, the cutoff in the real-space potential. This is normally set to $r_c = L/2$. Then starting with $\mathbf{n}_{\text{max}} = (10, 10, 10)$ calculate the average Coulomb energy for a few molecular dynamics steps as a function of κ around $\kappa = 3/r_c$. A suitable value of κ corresponds to the onset of a plateau in $\mathcal{V}^{qq}(\kappa)$. The parameter space can be explored by reducing $\mathbf{n}_{\text{max}}$ towards $(6, 6, 6)$ and repeating the test. In practice, κ is typically set to $6/L$ and 200–300 wavevectors are used in the k-space sum.

For the real-space part of the sum, the modified charge–charge interaction is calculated in the normal way in the main loop of the simulation program; $\text{erfc}(x)$ is an intrinsic function in Fortran 2008, and provided in most mathematical function libraries. The long-range (k-space) part of the potential energy, the second term in eqn (6.4), can be written as

$$\mathcal{V}_{\text{long}}^{qq} = \frac{1}{2L^3} \sum_{k\neq0} \frac{4\pi}{k^2} \exp(-k^2/4\kappa^2) \left| \hat{\rho}^q(\mathbf{k}) \right|^2, \tag{6.6}$$

where the Fourier transform of the charge density is

$$\hat{\rho}^q(\mathbf{k}) = \sum_{i=1}^{N} q_i \exp(-i\mathbf{k} \cdot \mathbf{r}_i). \tag{6.7}$$

The triple sum over $|\mathbf{k}|$, i, and j, in eqn (6.4) has been replaced by a double sum over $|\mathbf{k}|$ and i in eqn (6.6). The sum over $\mathbf{k}$-vectors is normally carried using complex arithmetic and a version of this subroutine is given in Code 6.1. The calculation of the force on a particular charge i is described in Appendix C.5. This part of the program can be quite expensive on conventional computers, but it may be efficiently vectorized, and so is well-suited for pipeline processing. It can also be efficiently evaluated on a P-processor parallel machine by dividing the charges across the processors:

$$\hat{\rho}^q(\mathbf{k}) = \sum_{p=1}^{P} \sum_{i\in p} q_i \exp(-i\mathbf{k} \cdot \mathbf{r}_i) = \sum_{p=1}^{P} \hat{\rho}_p^q(\mathbf{k}), \tag{6.8}$$

where the partial charge density, $\hat{\rho}_p^q(\mathbf{k})$, is calculated on each processor $p = 1, \ldots, P$ for the charges stored on that processor, and for all k vectors, and finally summed across the processors (Kalia et al., 1993).

Code 6.1 Force routine using the Ewald sum

This file is provided online. It contains two subroutines, to calculate the real-space and reciprocal-space parts of the charge–charge energy, respectively.

```
! ewald_module.f90
! r-space and k-space parts of Ewald sum for ions
MODULE ewald_module
```

How does the Ewald sum scale with the number of charges N in the simulation cell? At first glance, the sum over pairs in the first term of eqn (6.4) would suggest that the algorithm scales as $O(N^2)$. However, we have already seen that for sufficiently large systems, the linked-list method can be used to improve the efficiency of the real-space calculation. Following an analysis of Perram et al. (1988), the central box is divided into $S = s_c^3$ small cubic cells whose side L/s_c is at least the real-space cutoff. The number of operations in the real-space calculation is

$$N_{\text{real}} \propto S(N/S)^2 = N^2/S. \tag{6.9}$$

In this situation the number of operations in reciprocal space is

$$N_{\text{reciprocal}} \propto NS, \tag{6.10}$$

and the total time for computing the forces is

$$\tau = aN^2/S + bNS \tag{6.11}$$

where the constants a and b contain the constants of proportionality in eqns (6.9) and (6.10) and the overheads associated with the two contributions. τ achieves its minimum value for $S_{\text{min}} \propto N^{1/2}$ and for this value $\tau \propto N^{3/2}$. Thus for a careful choice of the number of cells we see that the algorithm scales as $O(N^{3/2})$.

The original method of Ewald can be readily extended to dipolar systems. In the derivation of eqn (6.4), q_i is simply replaced by $\mu_i \cdot \nabla_{r_i}$, where μ_i is the molecular dipole. The resulting expression is (Kornfeld, 1924; Adams and McDonald, 1976; de Leeuw et al., 1980)

$$
\mathcal{V}^{\mu\mu}(\epsilon_s = 1) = \tfrac{1}{2} \sum_{i=1}^{N} \sum_{j=1}^{N} \left(\sideset{}{'}\sum_{|\mathbf{m}|=0}^{\infty} (\boldsymbol{\mu}_i \cdot \boldsymbol{\mu}_j) B(\mathbf{r}_{ij} + \mathbf{Lm}) - (\boldsymbol{\mu}_i \cdot \mathbf{r}_{ij})(\boldsymbol{\mu}_j \cdot \mathbf{r}_{ij}) C(\mathbf{r}_{ij} + \mathbf{Lm}) \right.
$$
$$
\left. + \sum_{\mathbf{k} \neq 0} (1/\pi L^3)(\boldsymbol{\mu}_i \cdot \mathbf{k})(\boldsymbol{\mu}_j \cdot \mathbf{k})(4\pi^2/k^2) \exp(-k^2/4\kappa^2) \exp(-i\mathbf{k} \cdot \mathbf{r}_{ij}) \right)
$$
$$
- \sum_{i=1}^{N} 2\kappa^3 \mu_i^2/3\pi^{1/2} + \tfrac{1}{2} \sum_{i=1}^{N} \sum_{j=1}^{N} (4\pi/3L^3)\boldsymbol{\mu}_i \cdot \boldsymbol{\mu}_j \tag{6.12}
$$

where again factors of $4\pi\epsilon_0$ are omitted. In this equation, the sums over i and j are for dipoles in the central box and

$$B(r) = \mathrm{erfc}(\kappa r)/r^3 + \left(2\kappa/\pi^{1/2}\right)\exp(-\kappa^2 r^2)/r^2 \tag{6.13}$$

$$C(r) = 3\,\mathrm{erfc}(\kappa r)/r^5 + \left(2\kappa/\pi^{1/2}\right)\left(2\kappa^2 + 3/r^2\right)\exp(-\kappa^2 r^2)/r^2. \tag{6.14}$$

This expression can be used in the same way as the Ewald sum, with the real-space sum truncated at $|\mathbf{m}| = 0$ and a separate subroutine to calculate the k-vector sum.

The Ewald sum can be readily extended to higher-order multipoles. Aguado and Madden (2003) provide explicit expressions for the energy, forces, and stress tensor, when using the Ewald sum, for all the interactions up to quadrupole–quadrupole. The multipolar energy can be calculated using the interaction tensor $\mathbf{T}$ (see eqns (1.16) and eqn (1.17)). The real-space part of the Ewald sum can be calculated by replacing powers of r_{ij} by their screened counterparts

$$\frac{1}{r_{ij}} \;\rightarrow\; \frac{\widehat{1}}{r_{ij}} = \frac{\mathrm{erfc}(\kappa r_{ij})}{r_{ij}} \tag{6.15a}$$

$$\frac{1}{r_{ij}^{2n+1}} \;\rightarrow\; \frac{\widehat{1}}{r_{ij}^{2n+1}} = \frac{1}{r_{ij}^2}\left(\frac{\widehat{1}}{r_{ij}^{2n-1}} - \frac{(2\kappa^2)^n}{\sqrt{\pi}\kappa(2n-1)}\exp(-\kappa^2 r_{ij}^2)\right) \tag{6.15b}$$

for $n = 1, 2\ldots$, in eqns (1.17) to define a modified tensor, $\widehat{T}_{\alpha\beta}$, which can then be used in the calculation of the real-space energy and force. For example, the real-space contribution to the charge–dipole force is simply given by

$$(\mathbf{f}_{ij})_\alpha = -q_i \widehat{T}_{\alpha\beta}\mu_{j\beta} + q_j \widehat{T}_{\alpha\beta}\mu_{i\beta}. \tag{6.16}$$

Corresponding expressions have been developed for the reciprocal-space and self terms in the Ewald sum. The technique has been used to study the phonon dispersion curves of solid MgO and Al_2O_3 (Aguado and Madden, 2003) where there are small but significant differences in the simulations that include the quadrupolar terms using the Ewald method compared with a simple cutoff at $r_{ij} = r_c = L/2$. The formalism can be readily extended to the calculation of the energies for systems of polarizable ions and molecules.

A conceptually simple method for modelling dipoles and higher moments is to represent them as partial charges within the core of a molecule. In this case, the Ewald method may be applied directly to each partial charge at a particular site. The only complication in this case is in the self term. In a normal ionic simulation, we subtract the spurious term in the k-space summation that arises from the interaction of a charge at $\mathbf{r}_i$ with the distributed charge also centred at $\mathbf{r}_i$. In the simulation of partial charges within a molecule, it is necessary to subtract the terms that arise from the interaction of the charge at $\mathbf{r}_{ia}$ with the distributed charges centred at all the other sites within the same molecule ($\mathbf{r}_{ia}, \mathbf{r}_{ib}$, etc.) (Heyes, 1983a). This gives rise to a self-energy of

$$\mathcal{V}^{\mathrm{self}} = \tfrac{1}{2}\sum_i \sum_{a=1}^{n_s} q_{ia}\left(2\kappa q_{ia}/\pi^{1/2} + \sum_{b\neq a}^{n_s} q_{ib}\;\mathrm{erf}(\kappa d_{ab})/d_{ab}\right)$$

$$= \sum_i \left(\sum_{a=1}^{n_s} \kappa q_{ia}^2/\pi^{1/2} + \tfrac{1}{2}\sum_{a=1}^{n_s}\sum_{b\neq a}^{n_s} q_{ia}q_{ib}\;\mathrm{erf}(\kappa d_{ab})/d_{ab}\right) \tag{6.17}$$

where there are n_s sites on molecule i and the intramolecular separation of sites a and b is d_{ab}. This term must be subtracted from the potential energy.

The Ewald sum can also be used to calculate the energy resulting from the dispersion interaction $\propto r_{ij}^{-6}$ (Williams, 1971). Explicit expression for the real-space, reciprocal-space, and self terms are available for the energy, forces, and the stress (Karasawa and Goddard, 1989) and this method has been applied to crystal structure calculations. It has also been successfully applied to systems containing both dispersion and Coulombic interactions (in 't Veld et al., 2007). Since the dispersion interactions fall off rapidly with distance, one would guess that it is significantly faster to apply a spherical cutoff and add the standard long-range correction at the end of the simulation. However, Isele-Holder et al. (2013) have demonstrated that reducing the cutoff radius for local interactions and employing the particle–particle particle–mesh (PPPM) Ewald method for dispersion can be faster than truncating dispersion interactions. The Ewald approach to the dispersion interaction has also been used to calculate the full potential in slab simulations of the vapour–liquid interface (López-Lemus and Alejandre, 2002) (see Chapter 14).

Detailed theoretical studies (de Leeuw et al., 1980; Felderhof, 1980; Neumann and Steinhauser, 1983a,b; Neumann et al., 1984) have revealed the precise nature of the simulation when we employ a lattice sum. In this short section we present a simplified picture. For a review of the underlying ideas in dielectric theory, the reader is referred to Madden and Kivelson (1984) and for a more detailed account of their implementation in computer simulations to McDonald (1986).

The simulations which use the potential energy of eqn (6.4) are for a very large sphere of periodic replications of the central box *in vacuo*. As ϵ_s, the relative permittivity of the surroundings, is changed, the potential-energy function is altered. For a dipolar system, de Leeuw et al. (1980) have shown

$$\mathcal{V}^{\mu\mu}(\epsilon_s) = \mathcal{V}^{\mu\mu}(\epsilon_s = 1) - \frac{3k_BT}{N\mu^2} y \frac{(\epsilon_s - 1)}{(2\epsilon_s + 1)} \sum_{i=1}^{N} \sum_{j=1}^{N} \boldsymbol{\mu}_i \cdot \boldsymbol{\mu}_j \qquad (6.18)$$

where $y = 4\pi\rho\mu^2/9k_BT$, and we take $\mu = |\boldsymbol{\mu}_i|$ for all i. If, instead of vacuum, the sphere is considered to be surrounded by metal ($\epsilon_s \to \infty$), the last term in this equation exactly cancels the surface term in eqn (6.12). The potential functions of eqns (6.12) and (6.4) are often used without the final surface terms (Woodcock, 1971; Adams and McDonald, 1976) corresponding to a sphere surrounded by a metal. That the sum of the first three terms in these equations corresponds to the case $\epsilon_s = \infty$ can be traced to the neglect of the term $k = 0$ in the reciprocal-space summations.

The relative permittivity ϵ of the system of interest is, in general, not the same as that of the surrounding medium. The appropriate formula for calculating ϵ in a particular simulation does, however, depend on ϵ_s in the following way

$$\frac{1}{\epsilon - 1} = \frac{1}{3y\,g(\epsilon_s)} - \frac{1}{2\epsilon_s + 1} \qquad (6.19)$$

where g is related to the fluctuation in the total dipole moment of the central simulation box

$$g = g(\epsilon_s) = \frac{\left\langle \left| \sum_{i=1}^{N} \boldsymbol{\mu}_i \right|^2 \right\rangle - \left\langle \left| \sum_{i=1}^{N} \boldsymbol{\mu}_i \right| \right\rangle^2}{N\mu^2}. \qquad (6.20)$$

Note that the calculated value of g depends upon ϵ_s through the simulation Hamiltonian. For $\epsilon_s = 1$, eqn (6.19) reduces to the Clausius–Mosotti result

$$\frac{\epsilon - 1}{\epsilon + 2} = y\, g(1) \tag{6.21}$$

and for $\epsilon_s \to \infty$

$$\epsilon = 1 + 3y\, g(\infty). \tag{6.22}$$

A sensible way of calculating ϵ is to run the simulation using the potential energy of eqn (6.12) without the surface term, and to substitute the calculated value of g into eqn (6.22). The error magnification in using eqn (6.21) is substantial, and this route to ϵ should be avoided. In summary, the thermodynamic properties (E and P) and the permittivity are independent of ϵ_s, whereas the appropriate Hamiltonian and g-factor are not. There may also be a small effect on the structure. The best choice for ϵ_s would be ϵ, in which case eqn (6.19) reduces to the Kirkwood formula (Fröhlich, 1949)

$$\frac{(2\epsilon + 1)(\epsilon - 1)}{9\epsilon} = y\, g(\epsilon). \tag{6.23}$$

However, of course, we do not know ϵ in advance.

6.3   The particle–particle particle–mesh method

The PPPM algorithm (Eastwood et al., 1980) is a straightforward development of the Ewald method based on the use of the fast Fourier transform (FFT). The total potential is divided into a long-range and short-range part as in eqn (6.4). The evaluation of the short-range (r-space) part of the potential energy is the 'particle–particle' part of the method and for a large system this can be efficiently evaluated using the linked-list method described in Section 5.3.2.

The long-range (k-space) part of the potential energy, eqn (6.6), depends on the Fourier transform of the charge density

$$\hat{\rho}^q(\mathbf{k}) = \int d\mathbf{r}\, \rho^q(\mathbf{r}) \exp(-i\mathbf{k}\cdot\mathbf{r}) = \sum_{i=1}^{N} q_i \exp(-i\mathbf{k}\cdot\mathbf{r}_i). \tag{6.24}$$

In the PPPM method the charge density is interpolated onto a mesh in the simulation box. Once the charge density on the mesh is known, it is possible to calculate the electric field, $\mathcal{E}(\mathbf{r})$, and the potential, $\phi(\mathbf{r})$, using Poisson's equation

$$\nabla^2 \phi(\mathbf{r}) = -\nabla \cdot \mathcal{E}(\mathbf{r}) = -4\pi \rho^q(\mathbf{r}) \tag{6.25}$$

(the factor of 4π arises from our definition of the charges, see Appendix B). The field at the mesh can be interpolated to produce the force at a charge. This is the 'particle–mesh' part of the algorithm. Poisson's equation is most easily solved in reciprocal space

$$k^2 \hat{\phi}(\mathbf{k}) = 4\pi \hat{\rho}^q(\mathbf{k}) \tag{6.26}$$

where $\hat{\phi}(\mathbf{k})$ and $\hat{\rho}^q(\mathbf{k})$ are the Fourier transforms of the electric potential and the charge density respectively. An efficient fast Fourier transform (FFT) routine (see Appendix D.3) is used to calculate $\hat{\rho}^q(\mathbf{k})$. We consider each of these steps in more detail. First, the atomic charge density in the fluid is approximated by assigning the charges to a finely spaced mesh in the simulation box. Normally the mesh is taken to be cubic with a spacing ℓ. The number of cells $S = s_c^3$, where $s_c = L/\ell$ is normally chosen to be a power of 2 (typically 16 or 32). The mesh charge density, $\rho_s^q(\mathbf{r}_s)$, is defined at each mesh point $\mathbf{r}_s = \ell(s_x, s_y, s_z)$, where $s_\alpha = 0, \ldots s_c - 1$, and is

$$\rho_s^q(\mathbf{r}_s) = \frac{1}{\ell^3} \int_{V_{\text{box}}} d\mathbf{r}\, W(\mathbf{r}_s - \mathbf{r})\rho^q(\mathbf{r}) = \frac{1}{\ell^3} \sum_{i=1}^{N} q_i W(\mathbf{r}_s - \mathbf{r}_i). \tag{6.27}$$

The charge assignment function, W, is chosen to ensure conservation of the charge, that is, the fractional charges distributed to various mesh points must sum up to give the original charge q_i. It is useful to choose a function that involves a distribution on to a small number of 'supporting' mesh points and the function should also be smooth and easy to calculate. Hockney and Eastwood (1988) suggest a class of P-order assignment functions, which move a given charge onto its P nearest supporting mesh points along each dimension. The Fourier transform of the charge assignment function, $W^{(P)}$, is

$$\hat{W}^{(P)}(k_x) = \ell \left(\frac{\sin(k_x\ell/2)}{k_x\ell/2} \right)^P. \tag{6.28}$$

In real space, $W^{(P)}(x)$ is the convolution of a series of top-hat (or boxcar) functions. $W^{(P)}(x - x_s)$ is the fraction of the charge at x assigned to the mesh point at x_s. The lowest-order cases, $P = 1, 2, 3$, are termed 'nearest grid point', 'cloud in cell', and 'triangular-shaped cloud', respectively. The functional forms for $W^{(P)}(x), P = 1 \ldots 7$ are given in Appendix E of Deserno and Holm (1998). In the limit of large P, the function tends to a centred Gaussian set to zero beyond $\pm P\ell/2$.

To provide a concrete example, we consider distributing a charge across its three nearest mesh points in one dimension ($P = 3$). In this case the relevant function is the triangular-shaped cloud

$$W^{(3)}(x) = \begin{cases} \frac{3}{4} - (x/\ell)^2 & |x| \leq \frac{1}{2}\ell \\ \frac{1}{2}\left(\frac{3}{2} - |x|/\ell\right)^2 & \frac{1}{2}\ell \leq |x| \leq \frac{3}{2}\ell \\ 0 & \text{otherwise} \end{cases}$$

where x is the distance between the charge and any mesh point (Pollock and Glosli, 1996, Appendix A). Suppose that the *nearest* mesh point is s, and the distance from it to the charge, in units of ℓ, is $x' = (x - x_s)/\ell$, as shown in Fig. 6.3. Then the three non-zero weights, at the mesh points $s, s \pm 1$, may all be written in terms of x'

$$W^{(3)}(x') = \begin{cases} \frac{1}{2}(\frac{1}{2} + x')^2 & \text{at } s + 1 \\ \frac{3}{4} - x'^2 & \text{at } s \\ \frac{1}{2}(\frac{1}{2} - x')^2 & \text{at } s - 1. \end{cases} \tag{6.29}$$

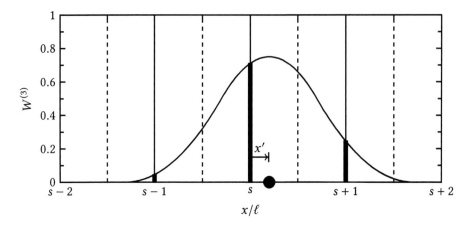

Fig. 6.3 The assignment of a charge to its three nearest mesh points in one dimension, using the triangular-shaped cloud weighting function. The vertical dashed lines are the cell boundaries, the vertical solid lines are the mesh points at the centre of each cell. ℓ is the length of the cell and $x' = (x - x_s)/\ell$ is the distance of the charge from its nearest mesh point in units of ℓ. The non-zero weights, indicated by vertical bars, are given by $W^{(3)}(x - x_s)$, $W^{(3)}(x - x_{s+1})$ and $W^{(3)}(x - x_{s-1})$.

Code 6.2　Assignment of charge to a uniform mesh

This file is provided online. It contains a program to assign charges to a cubic mesh using the 'triangular-shaped cloud' assignment of Hockney and Eastwood (1988).

```
! mesh.f90
! Assignment of charges to a 3-d mesh
PROGRAM mesh
```

In three dimensions, the 27 nearest mesh points are associated with an overall weight given by

$$W^{(3)}(\mathbf{r}_s) = W^{(3)}(x')\, W^{(3)}(y')\, W^{(3)}(z').\tag{6.30}$$

The assignment of charges to a cubic mesh is illustrated in Code 6.2.

Once the charge distribution has been assigned to the mesh, its discrete Fourier transform is

$$\hat{\rho}_s^q(\mathbf{k}) = \ell^3 \sum_{s_x=0}^{S_c-1}\sum_{s_y=0}^{S_c-1}\sum_{s_z=0}^{S_c-1} \rho_s^q(\mathbf{r}_s)\exp(-i\mathbf{k}\cdot\mathbf{r}_s) = \ell^3 \sum_{\mathbf{r}_s}\rho_s^q(\mathbf{r}_s)\exp(-i\mathbf{k}\cdot\mathbf{r}_s)\tag{6.31}$$

where $\hat{\rho}_s^q(\mathbf{k})$ is defined for the finite set of wavevectors, $\mathbf{k} = (2\pi/L)(n_x, n_y, n_z)$, with $|n_\alpha| \le (s_c - 1)/2$. Eqn (6.31) is in a form that can be used by an FFT algorithm as described in Appendix D.3. $\rho_s^q(\mathbf{r}_s)$ can be stored as an array of complex numbers, with the complex parts set to zero. The transform can be achieved with codes such as four3 for the complex-to-complex FFT in three dimensions, available in Press et al. (2007, Section 12.5), or similar codes in other standard libraries such as FFTW (Frigo and Johnson, 2005) (see Appendix D.4).

Once $\hat{\rho}_s^q$ has been calculated, the long-range part of the configurational energy, eqn (6.6), is

$$\mathcal{V}_{\text{long}}^{qq} = \frac{1}{2V} \sum_{\mathbf{k} \neq 0} \hat{G}(\mathbf{k}) |\hat{\rho}_s^q(\mathbf{k})|^2. \tag{6.32}$$

Here $\hat{G}$ is the influence function. In the ordinary Ewald method, the influence function is the product of the Green's function for the Coulomb potential and the smoothing function,

$$\hat{G}(k) = (4\pi/k^2) \exp(-k^2/4\kappa^2). \tag{6.33}$$

For the mesh approach, $\hat{G}(\mathbf{k})$ has to be optimized, because of the distortion of the isotropic charge distribution as it is moved onto the cubic lattice, and we will return to this point shortly.

$\mathcal{V}_{\text{long}}^{qq}$ can also be calculated in real space on the mesh.

$$\mathcal{V}_{\text{long}}^{qq} \approx \tfrac{1}{2}\ell^3 \sum_{\mathbf{r}_s} \rho_s^q(\mathbf{r}_s) \left(\rho_s^q \star G\right)(\mathbf{r}_s) \tag{6.34}$$

where the convolution is $\rho_s^q \star G = \mathfrak{F}^{-1}\left[\mathfrak{F}[\rho_s^q]\,\mathfrak{F}[G]\right]$ (see Appendix D).

The force on the charge, that is, the derivative of the potential, can be calculated in three different ways, by:

(a) using a finite-difference scheme in real space applied to the potential at neighbouring lattice points (Hockney and Eastwood, 1988);
(b) analytically differentiating eqn (6.34) in real space to give

$$\mathbf{f}_i = -\ell^3 \sum_{\mathbf{r}_s} \frac{\partial \rho_s^q(\mathbf{r}_s)}{\partial \mathbf{r}_i} \left(\rho_s^q \star G\right)(\mathbf{r}_s) \tag{6.35}$$

which requires the gradient of the assignment function, $\nabla W(r)$ (Deserno and Holm, 1998);
(c) differentiating in reciprocal space to obtain the electric field by multiplying the potential by ik,

$$\mathcal{E}(\mathbf{r}_s) = -\mathfrak{F}^{-1}\left[i\mathbf{k}\,\hat{\rho}_s^q(\mathbf{k})\,\hat{G}(\mathbf{k})\right]. \tag{6.36}$$

The third method, normally referred to as the ik-differentiation scheme, is the most accurate. Eqn (6.36) involves three inverse Fourier transforms; one for each component of the vector field.

In the final step of the algorithm, the field at the mesh points is used to calculate the force on the charges through the assignment function

$$\mathbf{f}_i = q_i \sum_{\mathbf{r}_s} \mathcal{E}(\mathbf{r}_s) W(\mathbf{r}_i - \mathbf{r}_s). \tag{6.37}$$

The same assignment function is used to distribute charges to the mesh and to assign forces to the charges. This is a necessary condition for the dynamics to obey Newton's third law (Hockney and Eastwood, 1988).

In the PPPM method, the optimal influence function, sometimes called the lattice Green's function, is chosen to make the results of the mesh function calculation correspond as closely as possible to those from the continuum charge density. In other words, $\hat{G}_{\text{opt}}$ can be determined by the condition that it leads to the smallest possible errors in the energy or the forces, on average, for an uncorrelated random charge distribution. Clearly, the precise form of $\hat{G}_{\text{opt}}$ depends on whether the optimization is applied to the energy or the force calculation and, in the case of the force, on which of the three differentiation methods is used (Ballenegger et al., 2012; Stern and Calkins, 2008).

For example, for the ik-differentiation for the force,

$$\hat{G}_{\text{opt}}(\mathbf{k}) = \frac{\sum\limits_{\mathbf{k}'} \left(\mathbf{k} \cdot (\mathbf{k} + s_c\mathbf{k}') \right) \hat{U}^2(\mathbf{k} + s_c\mathbf{k}') \, \hat{G}(\mathbf{k} + s_c\mathbf{k}')}{|\mathbf{k}|^2 \left[\sum\limits_{\mathbf{k}'} \hat{U}^2(\mathbf{k} + s_c\mathbf{k}') \right]^2} \tag{6.38}$$

where $\mathbf{k}' = 2\pi\mathbf{n}'/L$ and $\hat{U}(\mathbf{k})$ is the normalized charge assignment function. For example, for $P = 3$,

$$\hat{U}(\mathbf{k}) = \frac{\hat{W}^{(3)}(\mathbf{k})}{\ell^3} = \prod_\alpha \left(\frac{\sin k_\alpha \ell/2}{k_\alpha \ell/2} \right)^3 \tag{6.39}$$

where the product is over the three Cartesian coordinates. The sum over $\mathbf{k}'$ in eqn (6.38) extends the k-space beyond the first Brillouin zone. In the numerator extending the sum in this way improves the truncation error for a particular $\hat{G}_{\text{opt}}(\mathbf{k})$ and in the denominator it addresses the aliasing errors when copies of higher-order Brillouin zones are folded into the first zone (Press et al., 2007). It transpires that the term in square brackets in the denominator of eqn (6.38) can be evaluated analytically for all P. For $P = 3$

$$\sum_{\mathbf{k}'} \hat{U}^2(\mathbf{k} + s_c\mathbf{k}') = \prod_\alpha \left(1 - \sin^2(k_\alpha \ell/2) + \tfrac{2}{15} \sin^4(k_\alpha \ell/2) \right). \tag{6.40}$$

Although eqn (6.38) looks formidable, it is independent of the ion positions and can be evaluated once and for all at the beginning of the simulation (at least for simulations at constant volume). The sum over the alias vector $\mathbf{k}'$ is rapidly convergent and can be truncated at $|\mathbf{n}'| < 2$. The optimal influence function $\hat{G}_{\text{opt}}(\mathbf{k})$ replaces $\hat{G}$ in eqn (6.36) when calculating the field. The appropriate $\hat{G}_{\text{opt}}(\mathbf{k})$ for the forces calculated by analytical differentiation and the finite-difference method are given in Ballenegger et al. (2012) and Hockney and Eastwood (1988) respectively.

The situation for the calculation of the energy is simpler. Ballenegger et al. (2012) have shown that for a rapidly decaying $\hat{G}(\mathbf{k})$, the sum in the numerator can be truncated at $\mathbf{n}' = 0$ without loss of accuracy, providing a completely closed form for the optimal influence function

$$\hat{G}_{\text{opt}}(\mathbf{k}) = \frac{\hat{U}^2(\mathbf{k}) \, \hat{G}(\mathbf{k})}{\left[\sum\limits_{\mathbf{k}'} \hat{U}^2(\mathbf{k} + s_c\mathbf{k}') \right]^2}. \tag{6.41}$$

Equation (6.41) is used in eqn (6.34) for the calculation of the reciprocal-space part of the energy.

There are a number of different mesh algorithms based on the PPPM idea. These involve changes in the type and order of the assignment function W, the choice of the influence function G, and the approach to the gradient. Two of the main lines are the particle mesh Ewald (PME) method (Darden et al., 1993) and the smooth particle mesh Ewald (SPME) method (Essmann et al., 1995).

The PME method is the simplest approach since it assumes that eqn (6.33) can also be used without any alteration when the charges are assigned to a mesh. The only difference between this and the Ewald method is that the charges densities are transformed using the FFT. The use of a Lagrange interpolation scheme for the charge assignment leads to a cancellation of some of the discretization errors in this approach. The appropriate polynomials for the Lagrange interpolation have been tabulated by Petersen (1995).

Equation (6.24) suggests that an alternative approach, involving an approximation to $\exp(\mathrm{i}\mathbf{k} \cdot \mathbf{r}_i)$ might be used to assign charges to the supporting mesh points. This is the basis of the SPME method, which uses a cardinal B-spline (Massopust, 2010) at a particular order, P, to perform the assignment. The method relies on the following approximation which is valid for even P (Essmann et al., 1995; Griebel et al., 2007)

$$\exp(\mathrm{i}k_x x_i) \approx \sum_{m_x} \sum_{s_x=0}^{S_c-1} b(k_x) \exp(\mathrm{i}s_x k_x \ell) \, M_P\left(\frac{x_i - \ell s_x - m_x L}{\ell}\right) \tag{6.42}$$

where

$$b(k_x) = \frac{\exp\left(\mathrm{i}(P-1)k_x \ell\right)}{\sum_{q=0}^{P-2} \exp(\mathrm{i}k_x \ell q) M_P(q+1)}. \tag{6.43}$$

The cardinal B-splines are defined by the recurrence relationship

$$M_1(x') = \begin{cases} 1 & 0 \le x' < 1 \\ 0 & \text{otherwise} \end{cases}$$

$$M_P(x') = \frac{x'}{P-1} M_{P-1}(x') + \frac{P-x'}{P-1} M_{P-1}(x'-1) \qquad P \ge 2,\ x' \in [0, P]. \tag{6.44}$$

In eqn (6.42), the sum over the integer s_x is over the equally spaced mesh points in the x direction separated by the distance ℓ, and the sum over the integer m_x is over the periodic images. The sum over m_x is finite since $M_P(x')$ has finite support over $[0, P]$. Similar approximations in the three Cartesian coordinates are multiplied to obtain $\exp(\mathrm{i}\mathbf{k} \cdot \mathbf{r}_i)$.

This approximation for an exponential with a complex argument can be used to assign the charge density to a lattice

$$\rho_s^q(\mathbf{r}_s) = \frac{1}{\ell^3} \sum_{\mathbf{m}} \sum_{i=1}^{N} q_i \prod_{\alpha} M_P\left(\frac{r_{i\alpha} - s_\alpha \ell - m_\alpha L}{\ell}\right) \tag{6.45}$$

where the product is over the three Cartesian coordinates. Equation (6.45) can be used with an FFT to calculate $\hat{\rho}_s^q(\mathbf{k})$ and the reciprocal-space energy is

$$
\begin{aligned}
\mathcal{V}_{\text{long}}^{qq} &= \frac{1}{2V} \sum_{\mathbf{k}\neq 0} \hat{G}(\mathbf{k}) \left|\hat{b}(\mathbf{k})\right|^2 \left|\hat{\rho}_s^q(\mathbf{k})\right|^2 \\
&= \frac{1}{2V} \sum_{\mathbf{k}\neq 0} \hat{G}_{\text{opt}}(\mathbf{k}) \left|\hat{\rho}_s^q(\mathbf{k})\right|^2
\end{aligned}
\tag{6.46}
$$

where

$$
\hat{b}(\mathbf{k}) = b(k_x)b(k_y)b(k_z).
\tag{6.47}
$$

The use of the B-splines as compared to the Lagrangian interpolation in PME improves the accuracy of the calculation of both the energy and forces. The derivatives required for the forces can be readily obtained from eqn (6.35) in real space, since the cardinal B-splines can be differentiated directly $(M_P'(x) = M_{P-1}(x) - M_{P-1}(x-1))$. Finally, the b coefficients can be very accurately approximated by

$$
\left|\hat{b}(\mathbf{k})\right|^2 = \left(\sum_{\mathbf{k}'} \hat{U}(\mathbf{k} + s_c\mathbf{k}')\right)^{-2}
\tag{6.48}
$$

and therefore, the lattice Green function for the SPME has a closed form (Ballenegger et al., 2012)

$$
\hat{G}_{\text{opt}}(\mathbf{k}) = \frac{\hat{G}(\mathbf{k})}{\left[\sum_{\mathbf{k}'} \hat{U}(\mathbf{k} + s_c\mathbf{k}')\right]^2}.
\tag{6.49}
$$

A detailed implementation of the SPME with supporting code is given in Griebel et al. (2007, Chapter 7). It is worth pointing out that meshing with splines can also be used in the PPPM method as an alternative to the charge assignment function, W. The only difference then would be the choice of the optimal influence function (eqn (6.38) rather than eqn (6.49)). To simplify the notation in this section, we have considered cubic cells with mesh points equally spaced in the three coordinate directions. These three methods can be used with grids that are different sizes in each of the three coordinate directions and with cells that are triclinic (Ballenegger et al., 2012; Essmann et al., 1995).

Deserno and Holm (1998) have compared the Ewald, PPPM, PME, and SPME methods as a function of the charge assignment order (P), mesh size (S), and the differentiation scheme employed. The standard Ewald summation is unsurpassed for very high accuracy and is used as the benchmark for the other lattice methods. In comparing the mesh methods, it is essential to find the optimal value of κ for each technique and to consider, as a measure of the accuracy, the root-mean-squared deviation of the force from the mesh method as compared to that of a well-converged Ewald method, at that particular value of κ. With this criterion, the PPPM method is slightly more accurate than the SPME and both approaches can be used to replace the Ewald method. The PPPM method can be implemented with a charge assignment of $P = 3$ or 5 and a mesh size of $M = 32^3$. The most accurate results for the forces are obtained with the ik-differentiation method. The SPME method has been used with a real-space cutoff $r_c = 9$ Å, a mesh size $M = 16^3$ and spline order of $P = 6$ to give a root-mean-square deviation on the force of ca. $10^{-5} e^2/\text{Å}^2$ for a simulation of the simple point charge (SPC) model of water (Ballenegger et al., 2012). We recall that the optimal influence function is available in a closed form for the SPME approach.

The computer time for all of the mesh methods discussed in this section should scale with the number of charges to $O(N \log N)$ (the theoretical complexity of the method). This scaling factor is deduced from the behaviour of the discrete Fourier transform which scales $O(S \log S)$ where $S \propto N$ is the number of mesh points. In practice, linear scaling is observed for the mesh methods up to $N \approx 10^7$ (Arnold et al., 2013) since the FFT accounts for between 3 % and 10 % of the total time.

A replicated data, parallel version of the mesh methods (Pollock and Glosli, 1996) involves the distribution of the charges across the processors and the calculation of the real-space charge density on a particular processor, p. The total charge density is calculated as the sum over all the processors,

$$\rho_s^q(\mathbf{r}_s) = \frac{1}{\ell^3} \sum_p \sum_{i \in p} q_i W(\mathbf{r}_s - \mathbf{r}_i). \qquad (6.50)$$

At this point, copies of the complete ρ_s^q are distributed across all processors. This density is then partitioned for use with a distributed data FFT, without further communication between processors. The total energy is calculated by first summing the distributed Fourier components of the effective density and the potential on each processor and then summing over processors. The global communication involved in the first step of this approach can be avoided by using a domain decomposition of the mesh across processors. For a P-order method, only information from the $P - 1$ layers closest to the edge needs to be replicated and passed to the neighbouring processor. A suitable parallel FFT can be applied to the distributed data (Griebel et al., 2007).

6.4   Spherical truncation

The direct spherical truncation of the potential in an ionic fluid can lead to a charged sphere which would create artificial effects at its boundary, $r = r_c$. This can be countered by distributing a new charge over the surface of the sphere, equal in magnitude and opposite in sign to the net charge of the sphere, so as to guarantee local electroneutrality. The charged neutralized energy of the sphere can be obtained by shifting the Coulomb force to zero at the cutoff as described in Section 5.2.3. As early as 1983, Adams showed that the results from this approach are system size-dependent, but that for a system of $N = 512$ ions, they compare well with those obtained from the Ewald sum (Adams, 1983b).

There has been a resurgence in the use of the simple, cutoff-based techniques for long-range forces (Koehl, 2006; Fukuda and Nakamura, 2012). For example, the method of Wolf et al. (1999) replaces the Coulomb energy with a damped energy of the form

$$v_W(r_{ij}) = q_i q_j r_{ij}^{-1} \, \text{erfc}(\kappa r_{ij}). \qquad (6.51)$$

The electrostatic energy is

$$\mathcal{V}^{qq} = \frac{1}{2} \sum_{i=1}^{N} \sum_{\substack{j \neq i \\ r_{ij} < r_c}} \left[v_W(r_{ij}) - v_W(r_c) \right] - \left[\frac{\mathrm{erfc}(\kappa r_c)}{2r_c} + \frac{\kappa}{\sqrt{\pi}} \right] \sum_{i=1}^{N} q_i^2 \qquad (6.52)$$

where the first term is the shifted, damped potential and the second term is a correction for the self-energy. For every charge, i, in the cutoff sphere, an image charge of the opposite sign exists on the surface of the cutoff sphere which interacts only with i. Wolf's method has been tested against the Ewald sum for simulations of crystalline and liquid MgO and produces accurate energies for $r_c = 2.71a$, $\kappa a = 1.0$ where $a = 4.2771$ Å is the zero-temperature lattice spacing. The force associated with the potential is discontinuous at r_c and a smoothing function is applied so that the method can be used in MD simulations (Fukuda and Nakamura, 2012). This method has been extended to remove any net dipoles and higher moments in the cutoff sphere (the zero dipole method) (Fukuda et al., 2011; Fukuda, 2013). Spherical truncation is a highly efficient way of handling long-range interactions, but there is evidence that these methods introduce artefacts for highly charged and polar fluids (Koehl, 2006; Izvekov et al., 2008). A significant improvement would be to combine the spherical cutoff with a long-range contribution which depends on the configuration of the charges inside the cutoff sphere.

One such approach is the isotropic periodic sum (IPS) method (Wu and Brooks, 2005). In this method the electrostatic energy is written

$$\mathcal{V}^{qq} = \frac{1}{2} \sum_{i=1}^{N} \sum_{\substack{j \neq i \\ r_{ij} < r_c}} \left[\frac{q_i q_j}{r_{ij}} + q_i \phi_{\mathrm{IPS}}(r_{ij}) \right]. \qquad (6.53)$$

As illustrated in Fig. 6.4, a charge, 1, at the origin interacts directly with charges 2, 3, and 4 in the truncation sphere; there are other charges in the basic simulation cell, which is cubic, but these do not interact with 1. The central sphere is surrounded by an infinite number of shells (m) at distances $2mr_c$ from the origin. Images of the central sphere (image-spheres) are distributed with their centres on these image-shells. The image-spheres are simple translations of the central sphere without rotation. In a homogeneous 3D system at a fixed density, there are $n(m) = 24m^2 + 2$ image-spheres on shell m. These image-spheres are notional, and on a particular shell they can overlap or coincide with one another. The image-spheres serve only to create the long-range field.

For a pair of charges, such as 1 and 2 in Fig. 6.4, the image distribution should be symmetric around the line connecting the charges. The IPS method tackles this by defining two kinds of image-spheres: axial and random. The axial image-spheres are located at the points where the extension of the line 1–2 intersects the image-shells (that is at 1′ and 1″ in Fig. 6.4). The method places ξ image-spheres at each of the double-crossing points on each shell (where ξ is to be determined) and the remaining $n(m) - 2\xi$ image-spheres are considered to be randomly distributed on each shell. The interaction of charge 1 with all the images of charge 2, or more generally charge j, is written

$$v_{1j}(r_{1j}) = q_1 \phi_{\mathrm{IPS}}(r_{1j}), \quad \text{where} \quad \phi_{\mathrm{IPS}}(r_{1j}) = \phi_{\mathrm{axial}}(r_{1j}) + \phi_{\mathrm{random}}(r_{1j}), \qquad (6.54)$$

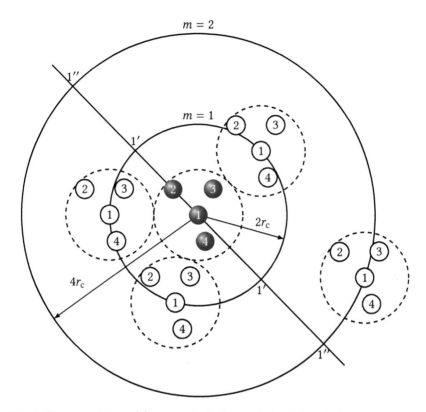

Fig. 6.4 A 2D representation of the IPS method. Charges 2, 3, and 4 are in the truncation sphere of charge 1, at the origin (all shaded). Images of the truncation sphere are located on the image-shells $m = 1$ and 2, etc. Axial images are positioned at the intersection of the line i–j with the image-shells (e.g. for 1–2 at 1′ and 1″). The axial images are not shown. Other image-spheres are distributed at random over the surface of the image-shells.

and r_{1j} is the separation of charge 1 and j in the central truncation sphere. The contribution of the axial images to the field at charge 1 is

$$\phi_{\text{axial}}(r_{1j}) = \xi q_j \sum_{m=1}^{\infty} \left[\frac{1}{2mr_c - r_{1j}} + \frac{1}{2mr_c + r_{1j}} \right]. \tag{6.55}$$

The contribution of the random image-spheres is

$$\phi_{\text{random}}(r_{1j}) = \sum_{m=1}^{\infty} \left(n(m) - 2\xi \right) \phi_{\text{shell}}(r_{1j}, m) \tag{6.56}$$

where

$$\phi_{\text{shell}}(r_{1j}, m) = \frac{q_j}{2} \int_0^{\pi} d\theta \, \sin \theta \left(\sqrt{r_{1j}^2 + (2mr_c)^2 - 4mr_c r_{1j} \cos \theta} \right)^{-1}. \tag{6.57}$$

We note that, for the electrostatic potential, eqns (6.54)–(6.57) do not converge and it is necessary to deal with the difference $[\phi_{\text{IPS}}(r_{1j}) - \phi_{\text{IPS}}(0)]$. This technical device does not affect the calculation of the energy and forces for neutral systems. The distribution parameter, ξ, can be determined by ensuring that the radial force at the boundary r_c, from the central charge and all images, is zero

$$\frac{\partial}{\partial r_{1j}}\left(q_j r_{1j}^{-1} + \phi_{\mathrm{IPS}}(r_{1j}, \xi)\right)\bigg|_{r_{1j}=r_c} = 0. \tag{6.58}$$

For the charge–charge interaction, $\xi = 1$, and Wu and Brooks (2005) show that

$$\phi_{\mathrm{IPS}}(r_{1j}) = -\frac{q_j}{2r_c}\left[2\gamma + \psi\left(1 - \frac{r_{1j}}{2r_c}\right) + \psi\left(1 + \frac{r_{1j}}{2r_c}\right)\right] \tag{6.59}$$

where $\gamma = 0.577216$ is Euler's constant and $\psi(z)$ is the digamma function

$$\psi(z) = \Gamma'(z)/\Gamma(z), \quad \text{where} \quad \Gamma(z) = \int_0^\infty dt\, t^{z-1}\exp(-t).$$

The technique can be readily extended to obtain analytical formulae for the energy and forces for potentials of the form r^{-n} in three dimensions. It can also be used to consider 3D systems that are periodic in one and two directions. Where analytical expressions are not available, for 1D and 2D systems, the IPS potentials can be conveniently expressed as numerical functions (Wu and Brooks, 2005). Comparison of the IPS method and the Ewald sum (Takahashi et al., 2010) shows that IPS is an accurate approach for estimating transport coefficients and the liquid structure of water in a homogeneous system at cutoff distances greater than 2.2 nm. The method has been extended to consider image-spheres with a radius, r_f, greater than the cutoff distance (Wu and Brooks, 2008). The region between r_f and r_c is treated using a k-space sum through the PME approach. This IPS/FFT method is highly accurate for bulk fluids, liquid–liquid and liquid–vapour interfaces, and lipid bilayers and monolayers (Venable et al., 2009).

A quite different approach to choosing a truncated potential is to seek the one that most accurately mimics the true long-range potential. For example, we might consider that the most accurate simulation of water is achieved using a large system and the Ewald method with the parameters κ, r_c, and $k_{\max}$ chosen to minimize the error in the forces. If this simulation generates the 'true' or reference trajectories, then the force-matching (FM) method (Izvekov et al., 2008) can be used to find the 'best' effective short-range electrostatic potentials that can reproduce these reference trajectories. The method is discussed in the context of coarse graining in Section 12.7, but briefly, an effective (truncated) force, $\mathbf{f}_i^{\mathrm{FM}}(\mathbf{r}^{(N)})$, is compared to the 'true' force, $\mathbf{f}_i(\mathbf{r}^{(N)})$, by calculating the residual

$$\chi^2 = \frac{1}{3N}\left\langle \sum_{i=1}^N \left|\mathbf{f}_i^{\mathrm{FM}}(\mathbf{r}^{(N)}) - \mathbf{f}_i(\mathbf{r}^{(N)})\right|^2 \right\rangle \tag{6.60}$$

where the average is over configurations generated with the 'true' force in, say, the canonical ensemble. For a spherical truncation, the effective force is of the form

$$\mathbf{f}_i^{\mathrm{FM}} = \sum_{\substack{j\neq i \\ r_{ij}<r_c}} f_{ij}^{\mathrm{FM}}(r_{ij})\hat{\mathbf{r}}_{ij}, \tag{6.61}$$

involving just the neighbours within the cutoff sphere. The trial force, f_{ij}^{FM}, in eqn (6.61), could be represented as a k-order polynomial or set of cubic spline functions with k points on a mesh $\{r_k\}$ up to r_c. Then minimizing eqn (6.60) produces a set of linear equations for the coefficients of the fitting function, resulting in the FM solution. The large number of coupled equations resulting from each step in the simulation can be reduced to a manageable number by averaging over blocks of configurations or timesteps (Noid et al., 2008b). The 'best' potential can be determined by integrating the force and setting $v_{ij}^{\mathrm{FM}}(r_c) = 0$. This method will be most useful if the effective spherical potential calculated from a reference water simulation at one density and temperature can be applied at other state points in the phase diagram. It would also be powerful if the truncated electrostatic interaction developed for the O–H interaction within water could also be applied to the O–O interaction (with an adjustment of the charges), or the potential developed with one well-known model of water, say SPC/E, could be used with another model such as TIP3P, or in the simulation of more complicated systems such as solvated ions. All this sounds like a tall order, but Izvekov et al. (2008) have demonstrated that the FM charge–charge potential is transferable across all of these cases. Even more encouraging is that the FM potential calculated between charges in a water simulation can also be used to good effect in the simulation of molten NaCl.

6.5   Reaction field

Beyond a simple spherical truncation, it is possible to treat the volume of the fluid outside the truncation sphere as a dielectric continuum. This approach is particularly valuable in the simulation of large solvated molecules such as proteins. In these systems it is possible to include some of the surrounding water molecules explicitly, but the length and size of the simulations suggests that most of the solvent will need to be included implicitly through a surrounding continuum. The most straightforward approach, the reaction field method, assumes a continuum of fixed relative permittivity, say $\epsilon_s = 80$ for water.

We will describe the method for a pure dipolar fluid. The field on a dipole at the centre of the truncation sphere consists of two parts (see Fig. 6.5): the first is a short-range contribution from molecules situated within a cutoff sphere or 'cavity' $\mathcal{R}$, and the second arises from molecules outside which form the dielectric continuum (Onsager, 1936). The size of the reaction field acting on molecule i is proportional to the moment of the cavity surrounding i,

$$\mathcal{E}_i = \frac{2(\epsilon_s - 1)}{2\epsilon_s + 1} \frac{1}{r_c^3} \sum_{j \in \mathcal{R}} \mu_j \qquad (6.62)$$

where the summation extends over the molecules in the cavity, including i, and r_c is the radius of the cavity. The contribution to the energy from the reaction field is $-\frac{1}{2}\mu_i \cdot \mathcal{E}_i$. The torque on molecule i from the reaction field is $\mu_i \times \mathcal{E}_i$. Barker and Watts (1973) first used the reaction field in a simulation of water, and there are useful discussions by Friedman (1975) and Barker (1980).

Whenever a molecule enters or leaves the cavity surrounding another, a discontinuous jump occurs in the energy due to direct interactions within the cavity and in the reaction field contribution. These changes do not exactly cancel, and the result is poor energy conservation in MD. In addition, spurious features appear in the radial distribution function at $r_{ij} = r_c$. These problems may be avoided by tapering the interactions at the cavity surface (Adams et al., 1979): the explicit interactions between molecules i and j are weighted by a

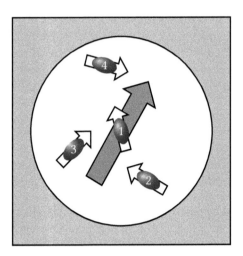

Fig. 6.5 A cavity and reaction field. Each molecule carries a permanent dipole moment (white arrows). Molecules 2, 3, and 4 interact directly with molecule 1. The continuum polarized by the molecules in the cavity produces a reaction field at 1 (shaded arrow).

factor $f(r_{ij})$, which approaches zero continuously at $r_{ij} = r_c$. For example, linear tapering may be used:

$$f(r_{ij}) = \begin{cases} 1.0 & r_{ij} < r_t \\ (r_c - r_{ij})/(r_c - r_t) & r_t \leq r_{ij} \leq r_c \\ 0.0 & r_c < r_{ij} \end{cases} \tag{6.63}$$

where an appropriate value of r_t is $r_t = 0.95 r_c$. The contribution of the molecular dipoles to the cavity dipole, and hence the reaction field, are correspondingly weighted. Investigations of tapering methods (Adams et al., 1979; Andrea et al., 1983; Berens et al., 1983) suggest that it may be rewarding to adopt more sophisticated formulae than the linear one given here.

When partial charges are used to represent a dipole, or a portion of a protein, or an ion, it is best to define a number of sets of charge groups. For example a charge group might contain the three partial charges in the spc/e model of water. The cutoff is then defined in terms of the separation of the centres of the charge groups, R_{IJ}, rather than the separation of the individual charges (Hünenberger and van Gunsteren, 1998). In this case the total energy in the reaction field geometry is

$$\mathcal{V}^{qq} = \frac{1}{2} \sum_{i=1}^{N} \sum_{\substack{j, I \neq J \\ R_{IJ} < r_c}}^{N} q_i q_j \left[\frac{1}{r_{ij}} + \left(\frac{\epsilon_s - 1}{2\epsilon_s + 1} \frac{r_{ij}^2}{r_c^3} + C \right) \right] + \mathcal{V}_{\text{Born}} + \mathcal{V}_{\text{self}} \tag{6.64}$$

where

$$\mathcal{V}_{\text{Born}} = -\frac{1}{2}\left(\frac{\epsilon_s - 1}{\epsilon_s}\right)\frac{1}{r_c}\sum_{i=1}^{N} q_i \sum_{\substack{j \\ R_{IJ}<r_c}}^{N} q_j \tag{6.65a}$$

$$\mathcal{V}_{\text{self}} = \frac{1}{2}\sum_{i=1}^{N}\sum_{j\in I}^{N} q_i q_j \left(\frac{\epsilon_s - 1}{2\epsilon_s + 1}\frac{r_{ij}^2}{r_c^3} + C\right) \tag{6.65b}$$

and $\quad C = -\frac{1}{r_c}\left[1 + \frac{\epsilon_s - 1}{2\epsilon_s + 1} - \frac{\epsilon_s - 1}{\epsilon_s}\right].$ \hfill (6.65c)

The Born energy arises from reversibly charging the atoms in a particular charge group as if they were at the centre of their own cutoff sphere. The constant C arises from a homogeneous background charge, which neutralizes any charge groups in the sphere. When all the charge groups are neutral, C vanishes and $\mathcal{V}_{\text{Born}}$ is zero. $\mathcal{V}_{\text{self}}$ is the self energy of a charge group in its own dipolar reaction field; this term is constant if the distances within a charge group remain fixed. Hünenberger and van Gunsteren (1998) discuss some inconsistencies which arise from charge groups that straddle the cutoff boundary when $r_{ij} > r_c$. These can be avoided by increasing the radius of the reaction field sphere (r_f) so that $r_f > r_c$. The appropriate extensions to eqns (6.64)–(6.65) for this case are given by Hünenberger and van Gunsteren (1998).

An alternative approach to reaction fields developed by Friedman (1975) considers the interaction between a charge inside the cutoff sphere and the infinite number of discrete image charges, along the extension of the radial line in the continuum. Accurate and efficient methods for calculating the position and size of these image charges have been developed by Lin et al. (2009). One of the difficulties of using this approach in a simulation is that as a charge approaches the cutoff, the image charge reaction field diverges. Lin et al. (2009) have suggested using a spherical reaction field of radius $r_f > r_c$. The buffer layer is filled with explicit solvent molecules, avoiding the creation of a vacuum layer between the truncation sphere and the reaction field sphere. The method has been tested on the TIP3P model of water in a truncated octahedral simulation box as a function of the width of the boundary layer and the number of image charges included along the radial line. The results for the structure, diffusion coefficient and dielectric constant obtained with two images per charge and a boundary layer of 6 Å compare well with those obtained from the PME lattice sum.

Returning to the pure dipolar fluid, Neumann and Steinhauser (1980) and Neumann (1983) have considered the nature of the simulations when a reaction field is applied. The appropriate formula for calculating the dielectric constant is eqn (6.19) where $g(\epsilon_s)$ is calculated from the fluctuation in the total moment of the complete central simulation box (Patey et al., 1982) as in eqn (6.20). The static reaction field is straightforward to calculate in a conventional MD or MC simulation, and it involves only a modest increase in execution time. A potential difficulty with the reaction field method is the need for an *a priori* knowledge of the external dielectric constant (ϵ_s). Fortunately, the thermodynamic properties of a dipolar fluid are reasonably insensitive to the choice of ϵ_s, and the dielectric constant can be calculated using eqn (6.19).

At the next level of sophistication, we can consider a continuum that contains a number of ions at a fixed concentration. The electric potential in the region outside the truncation sphere, with a dielectric continuum ϵ_s at a constant ionic strength I, obeys the Poisson–Boltzmann (PB) equation where the charge distribution is modelled using the Debye–Hückel theory

$$\rho^q(\mathbf{r}) = F \sum_{k=1}^{K} c_k z_k \exp(-z_k F \phi(\mathbf{r})/RT). \tag{6.66}$$

F is Faraday's constant ($96\,485.332\,89\,\mathrm{C\,mol^{-1}}$), R is the ideal gas constant, and c_k and z_k are the concentration and charge number (valence) for each species $k = 1, \ldots, K$ of ions in the continuum. The substitution of eqn (6.66) into eqn (6.25) defines the PB equation for $\phi(\mathbf{r})$. (We note eqn (6.25) requires an additional factor of ϵ_s on the left-hand side to account for the dielectric constant of the continuum.)

Tironi et al. (1995) have developed a generalized reaction field approach that includes the ions in the continuum by applying the linearized version of eqn (6.66) to the region outside the truncation sphere. In this case the potentials inside and outside the truncation sphere satisfy

$$\nabla^2 \phi_{\text{in}}(\mathbf{r}) = -4\pi \sum_{i=1}^{N} q_i \delta(\mathbf{r} - \mathbf{r}_i) \quad r \leq r_c$$

$$\nabla^2 \phi_{\text{out}}(\mathbf{r}) = 4\pi \kappa^2 \phi_{\text{out}}(\mathbf{r}) \quad\quad r > r_c, \tag{6.67}$$

where κ is the inverse Debye screening length

$$\kappa^2 = \frac{F^2}{\epsilon_s RT} \sum_{k=1}^{K} c_k z_k^2. \tag{6.68}$$

These two equations can be solved by matching the electric potentials and their derivatives at r_c. The potential energy for this system is

$$\mathcal{V}^{qq} = \frac{1}{2} \sum_{i=1}^{N} \sum_{\substack{j \neq i \\ r_{ij} < r_c}} q_i q_j \left[\frac{1}{r_{ij}} - \frac{(1 + B_1) r_{ij}^2}{2 r_c^3} \right] \tag{6.69}$$

where

$$B_1 = \frac{(1 - 4\epsilon_s)(1 + \kappa r_c) - 2\epsilon_s (\kappa r_c)^2}{(1 + 2\epsilon_s)(1 + \kappa r_c) + \epsilon_s (\kappa r_c)^2}. \tag{6.70}$$

As $\kappa \to 0$, eqn (6.69) reduces to the normal reaction field equation for the energy (the second term in eqn (6.64) with $C = 0$). At large κ, $B_1 \to -2$ and eqn (6.69) reaches the limit for a conducting boundary around the truncation sphere. Molecular dynamics simulations of 1 molar NaCl ($\epsilon_s = 80$, $\kappa = 3.25\,\mathrm{nm^{-1}}$, $r_c = 1.5\,\mathrm{nm}$) using the generalized reaction field method (Tironi et al., 1995) are in reasonable agreement with full Ewald simulations of the same system. The use of the PB equation avoids the clustering of ions at the cutoff, which occurs with a simple spherical cutoff. This is a stiff test of the approach as the linearized PB equation is only accurate in the limit of low ionic concentrations. The full PB equation can be solved using a finite-difference approach (Koehl, 2006; Feig et al., 2004). It has been widely used in the simulation of solvated proteins (Prabhu et al., 2004) where it is an order of magnitude faster than simulations using only explicit water molecules.

6.6   Fast multipole methods

For very large systems of charges, it possible to calculate the potential at a particular point using a fast multipole method (FMM) (Kurzak and Pettitt, 2006). The method scales $O(N)$ for large N and involves the extensive use of multipole and local expansions of the charge distribution in a particular region. It is useful to begin with a reminder of these expansion techniques.

Consider N charges of strength $q_i, i = 1 \ldots N$. The charges are located at $\mathbf{r}_i$, within a sphere $|\mathbf{r}_i| < a$, centred at the origin. The potential due to these charges at a point with polar coordinates $\mathbf{r} = (r, \theta, \varphi)$, which is outside the sphere ($|\mathbf{r}| > a$), is given by a multipole expansion

$$\phi(\mathbf{r}) = \sum_{\ell=0}^{\infty} \sum_{m=-\ell}^{m=\ell} M_{\ell,m} G_{\ell,m}(\mathbf{r}) \tag{6.71}$$

where $G_{\ell,m}$ is the irregular solid harmonic (Wang and LeSar, 1996)

$$G_{\ell,m}(\mathbf{r}) = \frac{(-1)^{\ell-m}(\ell - m)!}{r^{\ell+1}} \exp(im\varphi)P_{\ell,m}(\cos\theta). \tag{6.72}$$

The expansion coefficients are

$$M_{\ell,m} = \sum_{i=1}^{N} q_i F^*_{\ell,m}(-\mathbf{r}_i), \tag{6.73}$$

where $F^*_{\ell,m}$ is the complex conjugate of the regular solid harmonic

$$F_{\ell,m}(\mathbf{r}) = \frac{(-1)^{\ell-m} r^\ell}{(\ell + m)!} \exp(im\varphi)P_{\ell,m}(\cos\theta). \tag{6.74}$$

(Note that the moments $M_{\ell,m}$ are the spherical tensor equivalents of the Cartesian multipole moments introduced in Chapter 1 (Stone, 2013).) In practice, the upper limit of the sum over ℓ is replaced by a finite integer, p, giving rise to an error

$$\epsilon \leq \frac{A}{r-a}\left(\frac{a}{r}\right)^{p+1}, \quad \text{where} \quad A = \sum_{i=1}^{N} |q_i|. \tag{6.75}$$

One elegant feature of the multipole expansion is that it is straightforward to change the origin of the expansion using a linear multipole-to-multipole (M2M) transformation applied to the coefficients. In Fig. 6.6(a), we consider N charges $q_i, i = 1 \ldots N$ in a sphere of radius a centred at Q. Using this point as the origin of the expansion, the potential $\phi(P)$ at another point P is written in terms of a set of coefficients $M_{\ell,m}$. The centre of the

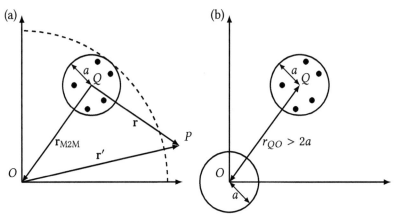

Fig. 6.6 (a) A shift in the origin of the multipole expansion. The multipole expansion of charges around the point Q in a sphere of radius a. The potential at P can be written in terms of an expansion around the origin O if $r' \geq a + |\mathbf{r}_{M2M}|$ (dashed arc). (b) The region of convergence of the local expansion. The local expansion is valid inside the sphere of radius a centred at the origin from a set of charges inside a sphere of radius a centred at Q, with $r_{QO} > 2a$.

multipole expansion can be changed from Q to O, as long as P is outside the sphere of radius $|\mathbf{r}_{M2M}| + a$. The coefficients in the new expansion, $M_{\ell', m'}$ are

$$M_{\ell', m'} = \sum_{\ell=0}^{p} \sum_{m=-\ell}^{\ell} M_{\ell, m} F^*_{\ell'-\ell, m'-m}(\mathbf{r}_{M2M}) \qquad (6.76)$$

where $\mathbf{r}_{M2M}$ is the vector from Q to O.

Turning now to the local expansion, we consider the same set of N charges in a sphere of radius a centred at Q as shown in Fig. 6.6(b). Q is at a distance greater than $2a$ from the origin. The local expansion of the potential due to these charges converges in a sphere of radius a centred at the origin. The potential at $\mathbf{r}$, inside this sphere, is

$$\phi(\mathbf{r}) = \sum_{\ell=0}^{p} \sum_{m=-\ell}^{\ell} L_{\ell, m}\, G_{\ell, m}(\mathbf{r}) \qquad (6.77)$$

where the coefficients in the local expansion are

$$L_{\ell, m} = \sum_{i=1}^{N} q_i G_{\ell, m}(-\mathbf{r}_i). \qquad (6.78)$$

The origin of a local expansion can be moved by a vector displacement $\mathbf{r}_{L2L}$ using the local-to-local (L2L) transformation

$$L_{\ell', m'} = \sum_{\ell=0}^{p} \sum_{m=-\ell}^{\ell} L_{\ell, m} F^*_{\ell-\ell', m-m'}(\mathbf{r}_{L2L}) \qquad (6.79)$$

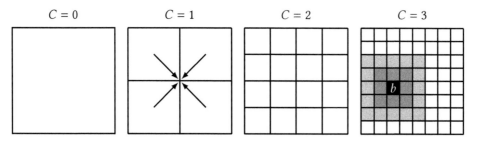

Fig. 6.7 The division of the computational box in the FMM for a two-dimensional square box. $C = 0$ corresponds to the complete square. $C = 1, 2, 3$ show the children at each level of the division. At $C = 1$ the arrows represent the shift in the origin of a multipole expansion at the centre of each box to its corner. The shifted fields are added to create the expansion at the centre of the parent box (at $C = 0$). At $C = 3$, the nearest neighbours and next-nearest neighbours of box b are shaded.

and a multipole expansion can be moved by a vector $\mathbf{r}_{\text{M2L}}$ and converted to a local expansion using the multipole-to-local (M2L) transformation

$$L_{\ell',m'} = \sum_{\ell=0}^{p} \sum_{m=-\ell}^{\ell} M_{\ell,m} G_{\ell'+\ell,m'+m}(\mathbf{r}_{\text{M2L}}).$$ (6.80)

The efficient evaluation of these three transformations (M2M, M2L, L2L) is a critical component of the FMM. The potentials are expanded in solid harmonics, rather than the more usual spherical harmonics, since this approach gives rise to computationally simpler forms of the transformations. In addition the symmetry of the solid harmonics means that one may consider only positive values of m in storing and manipulating the coefficients.

The algorithm begins by dividing the central simulation box into octants. The partition of space continues and the mesh at level $C + 1$ is obtained by dividing the mesh at C into octants. A tree structure is imposed on this mesh, so that when a particular box at level C is divided, the resulting eight boxes at level $C + 1$ are its *children* and the box itself is described as a *parent*. The number of distinct boxes at each mesh level is 8^C in three dimensions. For simplicity, this division is illustrated in two dimensions for a square box in Fig. 6.7 for $C = 0 \ldots 3$. (Of course, in our two-dimensional representation, the box is divided into quarters and there are 4^C boxes at level C.)

For a box b at level C, a nearest-neighbour box shares a boundary point with b. A second-nearest neighbour box shares a boundary point with one of the nearest-neighbour boxes of b. The use of a local expansion is only valid if the external charges contributing to the expansion and those inside the target box are well separated. In other words, the local expansion must arise from charges beyond the second-nearest neighbour boxes if the field in b is to be accurately represented.

The algorithm uses the following definitions.

(a) $\phi_{C,b}$ is the multipole expansion of the potential created by the charges contained inside box b at level C. (All expansions are truncated at the same order, p.)

(b) $\psi_{C,b}$ is the local expansion about the centre of box b at level C. This describes the potential due to all charges outside the box, its nearest-neighbour boxes and second-nearest neighbour boxes.

(c) $\tilde{\psi}_{C,b}$ is the local expansion about the centre of box b at level C from all the charges outside b's parent's box and the parent's neighbours and the parent's second-nearest neighbours.

(d) *the interaction list* for box b at level C is the set of boxes which are children of the nearest-neighbour and second-nearest neighbour of b's parents and which are not nearest-neighbours and second-nearest neighbours of b.

Suppose that at a particular level $C-1$ of the algorithm, the local expansion, $\psi_{C-1,b}$, has been obtained for all the boxes. The origin for the local expansion for a box at this level can be shifted so that the expansion acts at the centre of each of that box's children. For a particular child, j at level C, this shift will provide a representation of the potential due to all the charges outside j's parent's neighbours and next-nearest neighbours, that is $\tilde{\psi}_{C,j}$. The interaction list details the boxes whose contribution to the potential must be added to $\tilde{\psi}_{C,j}$ to produce $\psi_{C,j}$. The FMM is built around an upward pass through the mesh from the largest value of $C \approx \log_8(N)$ to $C = 0$, followed by conversion of the resulting multipole expansion to a local expansion, and a downward pass of the local expansion to the finest mesh level.

In the upward pass, the multipole expansion of the potential about the centre of each box is formed at the finest mesh level. The origin of the multipole expansion for each child is shifted to the centre of the parent box (as shown at $C = 1$ in Fig. 6.7) and the total potential is obtained at this point by addition. This process is repeated at all mesh levels up to $C = 0$.

In the downward pass, $\tilde{\psi}_{C,1} \ldots \tilde{\psi}_{C,8^C}$ at mesh level C are obtained from the local fields of their parents, $\psi_{C-1,1} \ldots \psi_{C-1,8^{(C-1)}}$, using an L2L transformation. $\tilde{\psi}_{C,b}$ is then corrected by considering the boxes on its interaction list and transforming the multipole field at the centre of these boxes to a local expansion at the centre of box b, using an M2L transformation. These corrections are added to $\tilde{\psi}_{C,b}$ to create the full local potential, $\psi_{C,b}$. This process is repeated at all mesh levels. Once the local expansions at the finest mesh level are available, they can be used to generate the potential and force on all the charges from beyond the second-nearest neighbour boxes at this level. The potential and forces from the nearest-neighbour and second-nearest neighbour boxes at this mesh level are calculated directly and added.

For an isolated cube, $\tilde{\psi}_{1,1} \ldots \tilde{\psi}_{1,8}$ are set to zero and the interaction lists are empty for $C < 3$. For a periodic system the multipole expansion for the entire cell, $\phi_{0,1}$, is available at the end of the upward pass. This is also the expansion for each of the image boxes around their centres. Now, image boxes which are not nearest-neighbour or second-nearest neighbour images are well separated from the central box and it is straightforward to form a local expansion from these images, $\psi_{0,1}$, to act at the centre of the central box (Ambrosiano et al., 1988). The precise form of this operation in three dimensions is given in Berman and Greengard (1994, eqns (34), (35)). Christiansen et al. (1993) have shown that the order of the addition of the image contributions to this sum is important. The calculation of $\psi_{0,1}$ can be thought of as the first step of the downward pass. In later versions of the algorithm (Carrier et al., 1988) different levels of mesh are used in different parts of the box depending on the local number density (the adaptive algorithm). The principles of the method are precisely the same as for the simpler uniform partition of space described here.

The important advantage of this algorithm is that the running time is of $O(N)$. Greengard and Rokhlin (1988) have shown that the required computational time goes as $N(e(\log_2 \epsilon)^2 + f)$ where e and f are constants which depend on the computer system and implementation and ϵ is the precision of the calculation of the potential. The precision fixes the order of the multipole and local expansions, that is, $p \approx -\log_2 \epsilon$. The storage requirements of the algorithm also scale with N. Similar estimates apply to the adaptive algorithm which is faster and requires less storage. In a simulation of a neutral system of 30 000 point charges embedded in Lennard-Jones disks, the FMM runs at twice the speed of the Ewald method (Sølvason et al., 1995) in a serial implementation of the code with $p = 6$.

Although the fast multipole method is capable of producing highly accurate results at an acceptable cost in two dimensions, there have been some difficulties in finding cost-efficient approaches in three dimensions. In 3D the linear translation operators are expensive to apply and this is particularly true of the operator M2L, which is used 189 times more often than M2M or L2L in a uniform system.

The transformation M2L, eqn (6.80), is a convolution, which can be evaluated as a matrix–vector multiplication in k-space. A significant gain in speed can be achieved by placing both the transformation matrix and the multipole moment on a grid and using an FFT technique to calculate the $L_{\ell',m'}$ coefficients (Elliott and Board, 1996). M2L can also be adapted to use a plane-wave representation of the expansion of $1/r$ in place of the multipole expansion, with an increase in efficiency. Additionally, an M2L translation along the z-direction is much simpler than translation along an arbitrary vector (see Cheng et al., 1999, eqns (33–35)). It is cost-effective to rotate the coefficients of the expansions into a frame with the z-axis in the direction of $\mathbf{r}_{M2L}$, then to perform the simplified translation and finally rotate the coefficients back into the original frame. These methods for reducing the computational complexity of the M2L transformation are discussed in more detail by Kurzak and Pettitt (2006).

Apart from these important efforts to optimize the serial version of the FMM, considerable effort has gone into parallelizing the algorithm. The first efforts of Greengard and Gropp (1990) have been extended to produce algorithms for both distributed-memory and shared-memory architectures (see Kurzak and Pettitt, 2005, and references therein). The FMM decomposes the simulation box into hierarchical cells organized in an octal tree. Cells at all levels can be distributed across processors and advanced graph theoretical techniques have been used to determine efficient communications and load balancing arrangements. In such a decomposition, it is inevitable that some parents and children will live on different processors and the majority of the communications occur in the M2L part of the calculation. The communications part of this particular transformation can be efficiently overlapped with the computation in the upward and downward passes (Kurzak and Pettitt, 2005).

6.7   The multilevel summation method

The multilevel summation method (MSM) (Skeel et al., 2002) is an interesting extension of the lattice mesh methods discussed in Section 6.3 combined with a tree approach, such as the FMM of the previous section. Whereas in the FMM individual charge interactions are separated into near and far pairs, the MSM separates the short-range part of the potential

---

Example 6.1 Crowded living cells

An example of the power of FMM can be seen in a recent study of molecular crowding in cellular environments using an all-atom molecular dynamics simulation (Ohno et al., 2014). The living cell is crowded with proteins comprising about 30 % of its mass. It resembles a liquid of globular proteins in an aqueous environment. The properties of proteins in this environment cannot be studied by simulating an isolated protein in a water droplet.

Using the K computer at the RIKEN Advanced Institute for Computational Science in Kobe, Japan, Ohno et al. (2014) performed calculations on 520 million atoms with a cutoff of 28 Å and electrostatic interactions evaluated by the FMM. The TIP3P rigid model of water was used with the AMBER99SB force field for the proteins: a heavy metal binding protein TTHA1718 and a crowding agent ovalbumin (as found in egg whites). The system also contained free potassium and chloride ions.

The calculations used 79 872 nodes ($P = 638\,976$ cores) in parallel, and achieved a sustained performance of 4.4 petaflops (1 petaflops is 10^{15} floating-point operations per second). The use of an FFT in, say, the PPPM method would prevent this simulation from scaling to hundreds of thousands of cores, whereas the FMM scales to the full size of the largest supercomputers currently available. This difference arises from the nature of the inter-node communications in the two methods. The communications complexity in the FFT is $O(\sqrt{P})$, whereas for FMM it is $O(\ln P)$.

As the authors point out, the total spatial extent of these simulations is approximately 0.2 μm compared to a real cell which has a minimum size of 1.0 μm. As we approach exaflops (10^{18} flops) performance, larger simulations of the whole cell will become possible. However, since the electric field propagates at 0.3 μm fs^{-1}, it will be necessary to consider relativistic corrections to classical MD at these length scales and this may have profound effects on the complexity of the algorithms for calculating the long-range forces.

---

from the long-range or smooth part at different spatial scales. The separation of scales means that the MSM will also scale linearly with the number of particles for sufficiently large N. The method begins by separating the Coulomb potential into two terms

$$\frac{1}{r} = \left(\frac{1}{r} - f_a(r)\right) + f_a(r), \tag{6.81}$$

where f_a is a softening function, chosen so that the first term is effectively zero for $r > a$. The long-range part of the potential energy can be written as

$$V_{\text{long}}^{qq} = \frac{1}{2} \sum_{i=1}^{N} \sum_{j=1}^{N} q_i q_j f_a\left(|\mathbf{r}_i - \mathbf{r}_j|\right) \tag{6.82}$$

where the excluded terms, $j = i$, are to be subtracted directly from the final results for the potential energy and forces.

The method employs a hierarchical set of cubic meshes with mesh spacing $2^{\lambda-1}\ell$ at each level $\lambda = 1 \ldots \lambda_{max}$. At a particular level the mesh point s is at $\mathbf{r}_{s,\lambda}$. At the finest mesh level, $\lambda = 1$, the softened potential is approximated as a function of the source position, $\mathbf{r}'$,

$$f_{a,1}\big(|\mathbf{r} - \mathbf{r}'|\big) \approx \sum_{s} f_{a,1}\big(|\mathbf{r} - \mathbf{r}_{s,1}|\big)\psi_{s,1}(\mathbf{r}') \tag{6.83}$$

where the basis functions, $\psi_{s,\lambda}(\mathbf{r}')$, are defined at each node point and each mesh level. $\psi_{s,1}$ is chosen to be zero on all but a small number of nodes close to s. The coefficients in the expansion (6.83) are approximated as functions of the destination position, $\mathbf{r}$,

$$f_{a,1}\big(|\mathbf{r} - \mathbf{r}_{s,1}|\big) \approx \sum_{t} f_{a,1}\big(|\mathbf{r}_{t,1} - \mathbf{r}_{s,1}|\big)\psi_{t,1}(\mathbf{r}). \tag{6.84}$$

Eqns (6.83) and (6.84) are combined, and the sums rearranged, to give

$$\mathcal{V}^{qq}_{long} = \frac{1}{2}\sum_{s}\sum_{t} f_{a,1}\big(|\mathbf{r}_{t,1} - \mathbf{r}_{s,1}|\big)\underbrace{\left(\sum_{i} q_i \psi_{t,1}(\mathbf{r}_i)\right)}_{q_{t,1}}\underbrace{\left(\sum_{j} q_j \psi_{s,1}(\mathbf{r}_j)\right)}_{q_{s,1}}. \tag{6.85}$$

The sum over charge pairs (i, j) has been replaced by a sum over charges $q_{s,1}$, $q_{t,1}$ at the lattice points (s, t). The two parts of the electrostatic potential at a grid point t are

$$\phi^{short}_{t,1} = \sum_{s}\left[|\mathbf{r}_{t,1} - \mathbf{r}_{s,1}|^{-1} - f_{a,1}\big(|\mathbf{r}_{t,1} - \mathbf{r}_{s,1}|\big)\right]q_{s,1} \tag{6.86a}$$

$$\phi^{long}_{t,1} = \sum_{s} f_{a,1}\big(|\mathbf{r}_{t,1} - \mathbf{r}_{s,1}|\big)q_{s,1}. \tag{6.86b}$$

The short-range potential, eqn (6.86a), can be evaluated using an $O(N)$ approach such as the linked-list method. The long-range range part of the potential, eqn (6.86b), is passed up the grid to the next level.

At $\lambda = 2$ (lattice spacing 2ℓ), the softening function itself is split into a short- and long-range part

$$f_{a,1}(r) = \big[f_{a,1}(r) - f_{a,2}(r)\big] + f_{a,2}(r). \tag{6.87}$$

The charges on the $\lambda = 2$ grid are

$$q_{t,2} = \sum_{s}\psi_{t,2}(\mathbf{r}_{s,1})q_{s,1} \tag{6.88}$$

and the short- and the long-range parts of the electric potential are calculated using the analogues of eqns (6.86a) and (6.86b) at $\lambda = 2$. The short-range part of the potential is retained for use in the descent phase of the algorithm.

The ascent of the levels continues in a similar manner, until $\lambda = \lambda_{max}$ where the largest grid spacing is $2^{\lambda_{max}-1}\ell$. At this level the total potential is not split into a short-

and long-range part. The total potential, $\phi = \phi^{\text{short}}$, is calculated by summing over all lattice points without the use of a cutoff,

$$\phi_{t,\lambda_{\max}} = \sum_s f_{a,\lambda_{\max}}\left(|\mathbf{r}_{t,\lambda_{\max}} - \mathbf{r}_{s,\lambda_{\max}}|\right) q_{s,\lambda_{\max}}. \tag{6.89}$$

The descent of the multilevel grid begins. At level λ, the total potential at $\lambda + 1$ is used to calculate the long-range potential at level λ

$$\phi_{t,\lambda}^{\text{long}} = \sum_s \psi_{s,\lambda+1}(\mathbf{r}_{t,\lambda})\phi_{s,\lambda+1}. \tag{6.90}$$

The total potential at level λ is calculated by adding the short-range piece (as calculated in the ascent phase). The descent continues until $\lambda = 1$ when the long-range potential at the charges is calculated by interpolation from the finest mesh points

$$\phi_i^{\text{long}} = \sum_t \psi_{t,1}(\mathbf{r}_i)\phi_{t,1}. \tag{6.91}$$

The total potential energy is

$$\mathcal{V}^{qq} = \frac{1}{2}\sum_{i=1}^{N}\sum_{j\neq i}^{N} q_i q_j \left[\frac{1}{r_{ij}} - f_{a,1}(r_{ij})\right] + \frac{1}{2}\sum_{i=1}^{N} q_i \phi_i^{\text{long}} - \frac{f_{a,1}(0)}{2}\sum_{i=1}^{N} q_i^2 \tag{6.92}$$

where the final term removes the $i = j$ interaction from the long-range part of $\mathcal{V}^{qq}$. The force on charge i can be calculated analytically by differentiating (6.92) term by term. Thus the long-range contribution to the force is

$$\mathbf{f}_i = -q_i \sum_t \phi_{t,1}\nabla_{\mathbf{r}_i}\psi_{t,1}(\mathbf{r}_i). \tag{6.93}$$

Hardy et al. (2015) suggest the following form for the softening functions

$$f_{a,\lambda}(r) = \left[\frac{1}{2^{\lambda-1}a}\right] F\left(\frac{r}{2^{\lambda-1}a}\right) \tag{6.94}$$

where

$$F(r') = \begin{cases} 1 - \frac{1}{2}(r'^2 - 1) + \frac{3}{8}(r'^2 - 1)^2 & r' \leq 1 \\ 1/r' & r' > 1 \end{cases} \tag{6.95}$$

and the function F, and its first derivative, are continuous at $r' = 1$.

The appropriate basis function is the product of one-dimensional functions for each of the Cartesian coordinates α

$$\psi_{m,\lambda}(\mathbf{r}) = \prod_\alpha \Psi\left(\frac{r_\alpha - (\mathbf{r}_{m,\lambda})_\alpha}{2^{\lambda-1}\ell}\right) \tag{6.96}$$

where for cubic interpolation

$$\Psi(x) = \begin{cases} \left(1 - |x|\right)\left(1 + |x| - \frac{3}{2}x^2\right) & 0 \leq |x| \leq 1 \\ -\frac{1}{2}\left(|x| - 1\right)\left(2 - |x|\right)^2 & 1 \leq |x| \leq 2 \\ 0 & \text{otherwise.} \end{cases} \tag{6.97}$$

The MSM can be implemented with periodic boundary conditions. In the short-range part of the calculation, the minimum image convention with a spherical cutoff is employed. In the long-range part of the calculation, the periodicity is included by wrapping around the edges of the grid as illustrated in Code 6.2. The precise number of boundary layers to be considered will depend on the range of the basis functions at each level. For a periodic system, the number of grid points must be a power of 2, which fixes the grid spacing. In a charge-neutral, periodic system, the top level is a single grid point with a charge of zero and the potential in eqn (6.89) is set to zero. A typical calculation might involve one particle level and five grid levels (Moore and Crozier, 2014). The method can be readily applied to systems that are only partially periodic and to non-periodic droplets (Hardy et al., 2015). The method has been used to simulate water and to calculate a number of properties including the dielectric constant and surface tension. Although it is not as accurate as the PPPM method, it is sufficiently accurate to produce the same thermodynamic and structural properties of TIP3P water as the PPPM method (Moore and Crozier, 2014; Hardy et al., 2015).

6.8   Maxwell equation molecular dynamics

Maxwell equation molecular dynamics (MEMD) is an alternative approach for including long-range forces whereby the electric field due to the charges is propagated by solving Maxwell's equations (Maggs and Rossetto, 2002; Rottler and Maggs, 2004). This field can then be used to calculate the force on each of the charges in a molecular dynamics calculation of their trajectories. In this case the instantaneous Coulomb potential is replaced by a retarded interaction propagating at the speed of light. The coupling of this propagated field and the charges is local and the algorithm should scale as $O(N)$. For a given configuration the charge density, ρ^q, and the current, $\mathbf{j}$, are

$$\rho^q = \sum_i q_i \, \delta(\mathbf{r} - \mathbf{r}_i), \qquad \mathbf{j} = \sum_i q_i \mathbf{v}_i \, \delta(\mathbf{r} - \mathbf{r}_i) \tag{6.98}$$

and the Maxwell equations, in standard SI units, are

$$\nabla \cdot \boldsymbol{\mathcal{E}} = \rho^q / \epsilon_0 \tag{6.99a}$$

$$\nabla \times \boldsymbol{\mathcal{E}} = -\dot{\mathbf{H}} / (\epsilon_0 c^2) \tag{6.99b}$$

$$\nabla \cdot \mathbf{H} = 0 \tag{6.99c}$$

$$\nabla \times \mathbf{H} = \epsilon_0 \dot{\boldsymbol{\mathcal{E}}} + \mathbf{j} \tag{6.99d}$$

where c is the speed of light and $\mathbf{H}$ is the free magnetic field (in this case $\mathbf{B} = \mu_0 \mathbf{H}$, see Appendix B).

It is not possible to solve the combination of eqns (6.99) for the fields with the equations of motion of charges using the real value of c, because of the enormous difference in the timescales of these motions. We need to avoid the static limit, $c \to \infty$, but to use a c that is sufficiently small to solve the coupled equations of the field and the charges. Rottler and Maggs (2004) have suggested a much lower, notional, value of c, such that $\bar{v}/c \approx 0.3$ where $\bar{v}$ is the mean velocity of the charges. This would allow the solution of the combined equations and still maintain a sufficiently large separation in the timescales so that the electric field follows the motion of the charges adiabatically. (We note that a similar approach is used in the solution of the electronic and nuclear degrees of freedom in the *ab initio* molecular dynamics method discussed in Chapter 13.)

The equations of motion can be established by considering Gauss's law, eqn (6.99a), to be a constraint that fixes a surface in the electric field space. Then all the fields on the constraint surface obey

$$\mathcal{E} = \mathcal{E}_0 + \nabla \times \Theta \qquad (6.100)$$

where $\mathcal{E}_0$ is a particular field on that surface, and Θ turns out to be related to **H**. Following Pasichnyk and Dünweg (2004), the overall Lagrangian for the system can be written as

$$\mathcal{L} = \sum_i \frac{|\mathbf{p}_i|^2}{2m_i} - \mathcal{V} + \frac{\epsilon_0}{2c^2} \int d\mathbf{r} \, |\dot{\Theta}|^2 - \frac{\epsilon_0}{2} \int d\mathbf{r} \, |\mathcal{E}|^2 + \int d\mathbf{r} \, \mathbf{A} \cdot \left(\epsilon_0 \dot{\mathcal{E}} + \mathbf{j} - \epsilon_0 \nabla \times \dot{\Theta}\right) \quad (6.101)$$

where the field **A** is a Lagrange multiplier that constrains the dynamics to the constraint surface satisfying the fourth Maxwell equation, eqn (6.99d), when we set $\mathbf{H} = \epsilon_0 \dot{\Theta}$.

The Lagrangian can be used in the normal way to define the coupled equations of motion of the system

$$\dot{\mathbf{r}}_i = \mathbf{p}_i / m \qquad (6.102a)$$
$$\dot{\mathbf{p}}_i = -\nabla_{\mathbf{r}_i} \mathcal{V} + q_i \mathcal{E} \qquad (6.102b)$$
$$\dot{\mathbf{A}} = -\mathcal{E} \qquad (6.102c)$$
$$\dot{\mathcal{E}} = c^2 \nabla \times (\nabla \times \mathbf{A}) - \mathbf{j}/\epsilon_0 \qquad (6.102d)$$

where $\mathbf{p}_i$ are the momenta of the charges. The vector field **A** is related to the free magnetic field

$$\mathbf{H} = \epsilon_0 c^2 \nabla \times \mathbf{A} \qquad (6.103)$$

so that eqn (6.102d) is simply the fourth Maxwell equation. We note that the magnetic force on the charges, the Lorentz force, which one might have expected to appear in eqn (6.102b), can and should be set to zero. The dynamics is no longer Hamiltonian but the energy

$$\mathcal{H} = \sum_i \frac{|\mathbf{p}_i|^2}{2m} + \mathcal{V} + \frac{\epsilon_0}{2} \int d\mathbf{r} \, |\mathcal{E}|^2 + \frac{1}{2\epsilon_0 c^2} \int d\mathbf{r} \, |\mathbf{H}|^2 \qquad (6.104)$$

is still conserved.

These equations are most readily solved on a cubic lattice, see Section 6.4. The grid assignments for Maxwell's equations are well known (Yee, 1966) and have been set out in detail for MEMD by Pasichnyk and Dünweg (2004). The charges, q_i are assigned to the lattice points using a charge assignment algorithm (such as the triangular-shaped cloud, or the cardinal B-splines, seen earlier). The electric field, $\mathcal{E}$, with **A** and **j**, act along the links, the edges of the cube as shown in Fig. 6.8. The divergences of the vector fields act at the lattice points (taking the differences from the six links associated with a point). The curls of these fields, e.g. **H** and Θ, are located on the cube faces or plaquettes (taking differences from the four surrounding links). The curls of the plaquette fields act along the links (taking the differences from the four plaquettes around a link). Finally, the

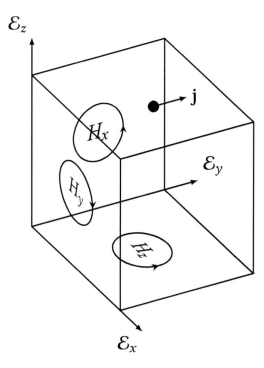

Fig. 6.8 The lattice for the MEMD algorithm. The charges in the simulation are assigned to the eight corners of the cube; the fields $(\mathcal{E}, \mathbf{j}, \mathbf{A})$ lie along the edges of the cube; the vector curls $(\Theta, \mathbf{H})$ are established on the plaquettes and are perpendicular to the surfaces of the cube (Arnold et al., 2013).

divergence of a plaquette field is placed at the centre of a cube (taking differences from the six surrounding plaquettes).

The equations are solved by using a variation of the Verlet method (Rottler and Maggs, 2004). Both the electric field and the charge positions are moved forward in two half timesteps:

(a) advance the charge momenta by half a timestep;
(b) advance the **A** field by half a timestep;
(c) looping over the three coordinate directions, x followed by y followed by z
  - advance the charge positions in the α-direction by half a timestep,
  - advance the electric field in the α-direction by half a timestep;
(d) looping over the three coordinate directions z followed by y followed by x
  - advance the charge positions in the α-direction by half a timestep,
  - advance the electric field in the α-direction by half a timestep;
(e) advance the **A** field by half a timestep;
(f) advance the charge momenta by half a timestep.

This approach preserves time reversibility and conserves the corresponding phase space volume. In addition the system remains on the constraint surface associated with the accurate solution of Gauss' equation.

6.9   Long-range potentials in slab geometry

So far, we have considered how to model electrostatic interactions in a bulk three-dimensional system. However, confined or inhomogeneous systems are also of great interest, and naturally the geometry has an effect.

Consider a pore composed of two flat interfaces represented by a static external field. The interfaces are located on the top and bottom of the simulation box containing the fluid. The box is then two-dimensionally periodic in the x and y directions. In this section, we will consider the simulations of long-range potentials in this geometry. Two cases can be considered. First, we examine a slab geometry where the charges in the fluid are distributed across the whole gap between the solid surfaces, and the extent of the fluid in the z-dimension would be comparable with the x and y dimensions of the cell. Second, we imagine a thin liquid layer, strongly adsorbed to one or both of the surfaces; the extent of the liquid in the z-direction is much less than the x and y dimensions of the cell.

In the first case, we might employ the full two-dimensional Ewald sum to calculate the energy and the forces on the charges in the fluid. This result was derived by Grzybowski et al. (2000), although there are a number of important related papers (see Lindbo and Tornberg, 2012). We consider a set of N charges confined in one dimension, $-L/2 < z_i < L/2$, but replicated in the xy plane, with $L_x = L_y = L$. The potential energy is

$$
\mathcal{V}^{qq} = \frac{1}{2} \sum_{i=1}^{N} \sum_{j=1}^{N} q_i q_j \left(\sum_{|\mathbf{m}|=0}^{\infty}{}' \frac{\mathrm{erfc}\big(\kappa |\mathbf{r}_{ij} + \mathbf{m}L|\big)}{|\mathbf{r}_{ij} + \mathbf{m}L|} \right.
$$

$$
+ \frac{\pi}{L^2} \sum_{\mathbf{k} \neq 0} \frac{\exp\big(i\mathbf{k} \cdot \mathbf{s}_{ij}\big)}{k} \left[\exp\big(kz_{ij}\big) \mathrm{erfc}\Big(\frac{k}{2\kappa} + \kappa z_{ij}\Big) + \exp\big(-kz_{ij}\big) \mathrm{erfc}\Big(\frac{k}{2\kappa} - \kappa z_{ij}\Big) \right]
$$

$$
\left. - \frac{2\sqrt{\pi}}{L^2} \left(\frac{\exp\big(-\kappa^2 z_{ij}^2\big)}{\kappa} + \sqrt{\pi} z_{ij}\, \mathrm{erf}(\kappa z_{ij}) \right) \right) - (\kappa/\pi^{1/2}) \sum_{i=1}^{N} q_i^2. \quad (6.105)
$$

In this geometry, $\mathbf{m} = (m_x, m_y, 0)$, $\mathbf{s}_{ij} = (x_{ij}, y_{ij})$ is the in-plane separation of the two ions i and j, and $\mathbf{k} = 2\pi(n_x, n_y)/L$ is the 2D reciprocal lattice vector with $k = |\mathbf{k}|$. The dash on the real-space sum indicates that the $i = j$ term is not included for $|\mathbf{m}| = 0$, and the surface term has been set to zero, corresponding to $\epsilon_s = \infty$. This result is similar in structure to eqn (6.4) for the fully three-dimensional system. However, the k-space term is more complicated and cannot be reduced from a triple to a double sum as in the three-dimensional case. This makes eqn (6.105) difficult to use in practice. Even with the use of pre-computed tables for the potential and forces, the two-dimensional Ewald sum is ten times slower to implement than the corresponding three-dimensional version (Yeh and Berkowitz, 1999).

A pragmatic solution to this problem is to use the fully three-dimensional Ewald sum by extending the unit cell in the z-direction. This geometry is shown in Fig. 6.9. The construction creates a vacuum between the sheets of charged atoms, which are confined by the walls. If the empty space in the z direction is greater than or equal to the x or y dimensions of the original simulation cell, then the forces from the image charges on the central cell should be zero due to overall charge neutrality. Spohr (1997)

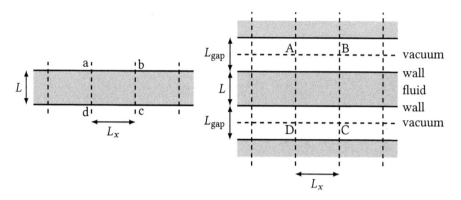

Fig. 6.9 The extended cell for use in the three-dimensional Ewald sum in a slab geometry. The walls are indicated by solid lines. The original box abcd, periodic in two dimensions, is shown on the left. The extended cell ABCD, which is periodic in three dimensions, is shown on the right.

has shown that results from eqn (6.4) converge to those from eqn (6.105) slowly as the vacuum gap between the fluid regions is increased. In making this comparison, the tinfoil boundary condition, $\epsilon_s = \infty$, for the three-dimensional Ewald sum was used, so that the final term in eqn (6.4) was omitted. Properties such as pair correlation functions, atom density profiles, charge densities, dipole densities, and dipole distributions, were not found to be sensitive to the size of the vacuum gap once $L_{gap} \geq \max(L_x, L_y)$. However, some integrated properties such as the electric potential were sensitive to the gap even when it was large compared with L_x, L_y. This presents a real difficulty, because the three-dimensional Ewald method becomes less efficient as the gap increases and more k-vectors are required in the z direction.

Yeh and Berkowitz (1999) have extended this work by inclusion of the vacuum boundary condition for the lattice of extended cells. Following Smith (1981), they note that when the infinite array shown in Fig. 6.9 is surrounded by a vacuum, there is a correction term dependent on the total z-component of the dipole moment of the simulation cell

$$\mathcal{V}^{qq}_{\text{correction}} = 2\pi V \left(\sum_i q_i z_i \right)^2 \tag{6.106}$$

and a corresponding correction for the force in the z direction. Once the correction of eqn (6.106) has been applied to eqn (6.105), the extended three-dimensional Ewald sum and the two-dimensional Ewald sum are in good agreement even for sensitive properties such as the electric field across the pore.

A completely different method for calculating the forces between ions in a slab geometry is due to Lekner (1989; 1991). The force on atom i from all of its neighbours j, and all of its images in the xy plane, is given by

$$\mathbf{f}_{ij} = q_i q_j \sum_{\mathbf{m}} \frac{\mathbf{r}_i - \mathbf{r}_j - \mathbf{m}L}{|\mathbf{r}_i - \mathbf{r}_j - \mathbf{m}L|^3}. \tag{6.107}$$

In this case, the x-component of the force on atom i can be written as a factor of $q_i q_j / L^2$ times the following dimensionless function

$$F_x = \sum_{m_x=-\infty}^{\infty} \sum_{m_y=-\infty}^{\infty} \frac{x + m_x}{\left[(x + m_x)^2 + (y + m_y)^2 + z^2\right]^{3/2}} \tag{6.108}$$

where $x_i - x_j = xL$, and y and z are defined in the same way. There are corresponding expressions for the y and z components of the force. Using the Euler transformation, $x^{-\nu} = (1/\Gamma(\nu)) \int_0^\infty dt\, t^{\nu-1} \exp(-xt)$, gives

$$\left[(x + m_x)^2 + (y + m_y)^2 + z^2\right]^{-3/2} =$$
$$\frac{2}{\sqrt{\pi}} \int_0^\infty dt\, t^{1/2} \exp\left(-t\left[(x + m_x)^2 + (y + m_y)^2 + z^2\right]\right), \tag{6.109}$$

and, after applying the Poisson–Jacobi identity,

$$\sum_{m_x=-\infty}^{\infty} \exp\left[(x + m_x)^2 t\right] = \left(\frac{\pi}{t}\right)^{1/2} \sum_{m_x=-\infty}^{\infty} \exp(-\pi^2 m_x^2/t) \cos(2\pi m_x x) \tag{6.110}$$

the following convergent series are obtained

$$F_z = \frac{2\pi \sinh 2\pi z}{\cosh 2\pi z - \cos 2\pi y} +$$
$$8\pi z \sum_{m_x=1}^{\infty} m_x \cos \pi m_x x \sum_{m_y=-\infty}^{\infty} \frac{K_1\left(2\pi m_x \left[(y + m_y)^2 + z^2\right]^{1/2}\right)}{\left[(y + m_y)^2 + z^2\right]^{1/2}}$$
$$F_x = 8\pi \sum_{m_x=1}^{\infty} m_x \sin 2\pi m_x x \sum_{m_y=-\infty}^{\infty} K_0\left(2\pi m_x \left[(y + m_y)^2 + z^2\right]^{1/2}\right) \tag{6.111}$$

where K_ν is a Bessel function of the second kind. Note that F_y need not be calculated independently since $F_y(x, y, z) = F_x(y, x, z)$. The potential energy can be obtained by integrating the force. Thus $v_{ij}^{qq} = q_i q_j \mathcal{V} / L$ where

$$\mathcal{V}(x, y, z) = 4 \sum_{m_x=1}^{\infty} \cos 2\pi m_x x \sum_{m_y=-\infty}^{\infty} K_0\left(2\pi m_x \left[(y + m_y)^2 + z^2\right]^{1/2}\right)$$
$$- \log\left[\cosh 2\pi z - \cos 2\pi y\right] + C \tag{6.112}$$

where C, the constant of integration, is 3.207 11. Typically in computing these sums, the m_x summation might be truncated at $(m_x)_{max} = 10$. For $m_x \leq 3$, m_y is taken from $-m_x$ to $+m_x$ and we take only $m_y = 0$ for other m_x values. The appropriate limits of the sum need to be checked for some typical configurations of a given system. However, there are some configurations of ions in which the sum in eqn (6.112) is slowly convergent, and these

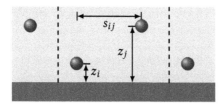

Fig. 6.10 A side view of the geometry of a thin liquid layer of ions adsorbed onto a solid surface.

will almost certainly occur among millions of moves or timesteps in a given simulation. This is due to the behaviour of the modified Bessel function K_0 as its argument tends to zero. When this sum converges slowly, one of the cyclic alternatives to eqn (6.112) can be applied. For this Lekner-cyclic technique, methods of identifying slowly convergent configurations have been reviewed by Mazars (2005). In that paper, good agreement was obtained between the two-dimensional Ewald sum and the Lekner-cyclic method for MC simulations of a Wigner crystal bilayer.

Of the three methods discussed for the slab geometry, the extended three-dimensional Ewald sum with the dipole correction is probably the best choice for simulations with long-range forces because of its simplicity and efficiency. There is considerable interest in improving the efficiency of the two-dimensional Ewald sum (Lindbo and Tornberg, 2012), and this remains an active area of research.

In the limit of a thin layer adsorbed onto a surface, such as in the physisorption of a monolayer of CO on graphite, or a Langmuir–Blodgett film, an alternative approach due to Hautman and Klein (1992) should be considered. This geometry is shown in Fig. 6.10 and we can take z_{ij}/s_{ij} to be a small parameter. This method involves the identification of the in-plane contribution to the potential between charges using the identity

$$\frac{1}{r_{ij}} = \left[\frac{1}{r_{ij}} + \sum_{n=0}^{P} \frac{a_n z_{ij}^{2n}}{s_{ij}^{2n+1}}\right] - \left[\sum_{n=0}^{P} \frac{a_n z_{ij}^{2n}}{s_{ij}^{2n+1}}\right] \tag{6.113}$$

with $a_0 = 1$, $a_1 = -\frac{1}{2}$, $a_2 = \frac{3}{8}$ as coefficients in the binomial expansion of $(1 + (z_{ij}/s_{ij})^2)^{-1/2}$. This is an identity if both series are taken to the same order P. It is the second summation in eqn (6.113) which is treated using the Ewald approach. If the charges are confined to a narrow region in z then the choice $P = 1$ allows the real-space term to be truncated at $|\mathbf{m}| = 0$. In this case, the potential is the sum of short- and long-range contributions $\mathcal{V} = \mathcal{V}_{\text{short}} + \mathcal{V}_{\text{long}}$. The short-range part is evaluated in real space

$$\mathcal{V}_{\text{short}} = \sum_{i<j} q_i q_j \left[\frac{1}{r_{ij}} - \frac{h_0(\kappa s_{ij})}{s_{ij}} + \frac{z_{ij}^2 h_1(\kappa s_{ij})}{2 s_{ij}^3}\right] \tag{6.114}$$

where κ is the normal convergence factor and

$$h_0(t) = \text{erf}(t), \qquad h_1(t) = \text{erf}(t) - \frac{2t}{\pi^{1/2}}(1 + 2t^2)\exp(-t^2). \tag{6.115}$$

Table 6.1 A comparison of the Hautman and Klein, and Lekner, methods for two opposite charges $q = \pm 1$ separated by $\Delta \mathbf{r} = (0.1, 0.1, 0.1)$ in a slab geometry with box length $L = 1$. In the Lekner method, for values of $m_x \leq 3$, $-m_x \leq m_y \leq +m_x$, and for $m_x > 3$ only $m_y = 0$ is considered. In the Hautman–Klein method, $|\mathbf{m}_{max}| = 0$, $|\mathbf{k}_{max}| L / 2\pi = 15$, and the convergence parameter $\kappa L = 6.25$.

Method	$(m_x)_{max}$	$-\mathcal{V}_{ion}$
Lekner	5	5.780 38
	10	5.772 03
	15	5.772 10
Hautman–Klein		5.771 91

The long-range, reciprocal-space contribution is

$$\mathcal{V}_{long} = \frac{\pi}{\mathcal{A}} \sum_{k \neq 0} \frac{\text{erfc}(k/2\kappa)}{k} A_k A_k^* + \frac{\pi}{\mathcal{A}} \sum_{k \neq 0} k \, \text{erfc}(k/2\kappa) \left[\frac{C_k A_k^* + C_k^* A_k}{2} - B_k B_k^* \right]$$
$$- \left(\frac{\kappa}{\pi^{1/2}} \right) \sum_i q_i^2 \tag{6.116}$$

where $\mathcal{A}$ is the cross-sectional area, $\mathbf{k}$ is the two-dimensional reciprocal lattice vector corresponding to the real-space lattice in the surface plane and

$$A_k = \sum_i q_i \exp(-i\mathbf{k} \cdot \mathbf{s}_i), \quad B_k = \sum_i q_i z_i \exp(-i\mathbf{k} \cdot \mathbf{s}_i), \quad C_k = \sum_i q_i z_i^2 \exp(-i\mathbf{k} \cdot \mathbf{s}_i).$$

By truncating the expansion in z_{ij}/s_{ij} we have again reduced the triple sum of the two-dimensional Ewald system to a double sum.

The Lekner method can also be readily applied in this geometry. As a simple comparison of these two techniques, consider two unit charges $+1$ at $(0, 0, 0)$ and -1 at $(0.1, 0.1, 0.1)$ in a periodic array of cells, with $L = 1$ in a slab geometry. The results are shown in Table 6.1. To obtain accurate values for the force between two ions in the cell, the Hautman–Klein method needs to be extended to include the additional real-space interactions from the eight next-nearest-neighbour cells for $P = 1$ and for simulations in this geometry the Lekner method is a good choice.

6.10   Which scheme to use?

With so much choice available, it is important to address the question of which of these schemes should be used when considering the simulation of long-range forces.

In the case of ionic liquids, the first thing to say is that something must be done if we want to calculate the thermodynamic, structural, and dielectric properties at the normal levels of accuracy that we expect from a simulation. Ignoring the problem with the use of a spherical cutoff or the minimum image method will not work. In terms of accuracy and ease of implementation, the straightforward Ewald sum is the gold standard. Machine precision, in calculating the energy and forces, can be reached by tuning the parameters, r_c, $\mathbf{k}_{max}$ and κ, without any additional programming effort. A number of authors (Valleau and Whittington, 1977a; Valleau, 1980; Fukuda and Nakamura, 2012) have suggested that the Ewald sum may impose an anisotropic periodic structure on an isotropic fluid, fail to damp dipolar fluctuations in the central box, and distort solvated protein structures. However, all the evidence seems to be that, with a sufficiently large number of k-vectors, the Ewald method accurately models the structural (Adams, 1983b) and time-dependent (Villarreal and Montich, 2005) properties of an isotropic fluid. Probably, the compelling reason for abandoning the Ewald sum as N increases is the computational expense.

For larger system sizes, the Ewald sum can be replaced by one of the mesh methods involving the FFT. The precise crossover point, in terms of the number of charges, for these two algorithms depends strongly on the level of optimization of both the mesh method and the Ewald sum, the required accuracy, and the computer architecture. Pollock and Glosli (1996) suggest that it is cost-effective to switch from Ewald to PPPM by $N = 516$ charges. For the SPME, Essmann et al. (1995) suggest a crossover at between $N = 600$ and $N = 900$ depending on the machine architecture. In contrast, Petersen (1995) suggest a crossover of $N = 10\,000$ between the Ewald and PME methods. These lower values of the crossover for PPPM and SPME are probably due to the efforts to optimize the FFT in these studies (Essmann et al., 1995). In all cases, it would be sensible to perform a detailed comparison of the timings of a mesh method with the standard Ewald method for systems containing more than ca. 500 charges. Both the PPPM method with $P = 3$ or 5 and ik differentiation for the forces, and the SPME method with $P = 6$ and analytical differentiation for the forces, provide sufficient accuracy when used with the appropriate optimal influence function.

For the simulation of charged or neutral biological systems in an aqueous environment, the spherical truncation methods (Section 6.4) should be considered (Fukuda and Nakamura, 2012). This is a less demanding problem, in that the partial charges that constitute the water molecule are grouped together so that the overall potential falls off as r^{-3} rather than r^{-1}. The methods that we have outlined in Sections 6.4 and 6.5 will normally do a good job in predicting many of the simple thermodynamic and structural properties of water when compared with the Ewald approach (Fennell and Gezelter, 2006). Of the methods discussed, the FM potential and the damped force-shifted (DFS) potential, which is a simple extension of the Wolf potential given in eqn (6.51) (Fennell and Gezelter, 2006), offer a good compromise between simplicity and accuracy. This is illustrated in Table 6.2 which shows a number of results from a simulation of a single Na^+ ion in water (Izvekov et al., 2008). The Ewald method is compared with a force shifted (FS) potential (where the Coulombic force is shifted to zero at the cutoff), the DFS potential, and the FM potential derived from a simulation of pure water. The agreement between the Ewald method (the standard) and the FM approach is excellent for the thermodynamic and structural properties as well as for the ion diffusion coefficient. The DFS approach is not quite as consistent as the FM approach producing a rather strange diffusion coefficient which is four times the size of the other methods, and slightly poorer thermodynamic properties.

These spherical truncation approaches also work well for sensitive properties such as the relative permittivity, ϵ_s, and the higher-order orientational correlations in the

Table 6.2 Properties of one Na^+ ion in 512 TIP3P water molecules calculated in a constant-NPT molecular dynamics simulation at 298 K and 1 bar. The interaction between the charge and the partial charges is calculated using the Ewald method, and three different approaches to spherical truncation with $r_c = 1.0$ nm: FS, DFS, and FM. We tabulate the height of the first maximum $g_{Na^+O}^{max}$ and first minimum $g_{Na^+O}^{min}$ in the ion–oxygen pair distribution function; the average potential energy $\langle V/N \rangle$; the volume per particle $\langle V/N \rangle$ and its standard deviation $\sigma(V/N)$; and the ion diffusion coefficient D. Estimated errors in last reported digits in parentheses. Adapted with permission from S. Izvekov, J. M. J. Swanson, and G. A. Voth, *J. Phys. Chem. B*, **112**, 4711–4724 (2008). Copyright (2008) American Chemical Society.

Property	Ewald	FS	DFS	FM
$g_{Na^+O}^{max}$	7.31	7.24	7.30	7.31
$g_{Na^+O}^{min}$	0.145	0.152	0.146	0.145
$\langle V/N \rangle$ (kJ mol^{-1})	−41.82(28)	−40.15(28)	−40.40(28)	−41.65(28)
$\langle V/N \rangle$ (10^{-2}nm^3)	2.946	2.960	2.953	2.943
$\sigma(V/N)$ (10^{-4}nm^3)	3.51	3.60	3.54	3.52
D (10^{-9}m^2 s^{-1})	1.07(2)	1.04(2)	4.28(2)	1.08(2)

fluid (such as the order parameters that measure the tetrahedral structure of the water hydrogen-bonded network). For example, for the SPC/E model of water, ϵ_s calculated from eqn (6.21) is 72.6, 71.1, and 71.5 for Ewald, FS and FM respectively (Izvekov et al., 2008).

The FMM and MEMD are designed to be used with much larger systems of charges. Interestingly, in a detailed comparison of these two methods with the mesh methods, Arnold et al. (2013) showed that there was no crossover between PPPM and FMM for simulations with up to 5×10^7 charges. In other words, in terms of the wall clock time per charge on a single core, the PPPM approach is faster than FMM for homogeneous fluids below this system size. This supports an earlier extrapolated estimate of the crossover at around 6×10^7 charges (Pollock and Glosli, 1996). One of the reasons for this unexpected result is that the observed complexity of the mesh methods is $O(N)$ rather than the theoretical complexity of $O(N^{3/2})$. The FMM does exhibit excellent scaling behaviour with the number of cores in a parallel implementation, and was the fastest method down to 10^3 charges per core on a well-interconnected high-performance machine such as the IBM Blue Gene/P (BG/P) (Arnold et al., 2013). The high accuracy of the FMM makes it suitable for calculations requiring high precision, and it also has the lowest memory requirement of the methods discussed. Eventually, as we approach 10^8 charges in the largest biological simulations currently performed, the FMM will be the method of choice.

The MSM has a number of appealing features. The separation of the short-range and smooth part of the potential at each level of the grid calculation means that the method should scale as $O(N)$ for large N. However, it avoids the complicated apparatus of the multipole and local expansions at the heart of the FMM and is straightforward to program. It does not use a k-space method to solve Poisson's equation and so it avoids the FFT which can pose a serious communications bottleneck for large systems on highly parallel machines. The MSM has already been efficiently parallelized (see Chapter 7) and scales better than the PPPM method. The MSM is still being developed and the principal problem of accuracy could be solved, if it were possible to apply higher-order interpolation methods without a significant loss of speed. This is an area of active research (Hardy et al., 2015).

Recent studies of the MEMD method indicate perfect linear scaling of the method up to $N \approx 10^7$ charges (Arnold et al., 2013). The accuracy of the MEMD method, as measured by the root-mean-square deviation of the potential compared with that of a fully converged Ewald sum, is at least two orders of magnitude lower than for the FMM and PPPM methods. This accuracy depends on the discretization of the charges and the problem is particularly marked for systems where the charge density is low. In the future this problem may be resolved with the use of finer grids (Arnold et al., 2013). One of the major advantages of MEMD is that it allows for an arbitrary change in the dielectric constant in different parts of the system. This feature has recently been used to simulate the electrical double layer around charged particles in aqueous solution of variable permittivity (Fahrenberger and Holm, 2014; Fahrenberger et al., 2014).

In terms of simulating slab geometries, perhaps the most effective approach is to create a fully periodic system with the use of two vacuum layers and then to apply the 3D Ewald method or one of the mesh extensions. This avoids the use of the expensive 2D Ewald approach. We have also reviewed two rather specialized slab approaches which are effective for simulating thin, physically or chemically adsorbed layers. Interestingly, the spherical cutoff methods such as the IPS and the Wolf methods have been extended to slabs, but require a large cutoff in these geometries. It is straightforward to extend the MSM to non-periodic systems such as lipid bilayers and this may become the method of choice for large simulations once the accuracy of the method is improved.

7

Parallel simulation

7.1 Introduction

Almost all high performance computing (HPC) facilities are based on parallel computers. Parallelism comes in many forms, and computer architectures are continually evolving. It is convenient to highlight two classes: *shared-memory* machines, in which many processors (often called cores) access the same memory and address space, and *distributed-memory* machines, in which the processors (often called nodes) each have their own memory, with no access to that of the others. Many HPC machines combine both approaches, consisting of a large number of independent nodes, each of which contains its own memory, shared by a number of cores. In recent years, processors based on GPUs have been shown to offer very cost-effective computing: in a crude way, these can be thought of as extremely fine-grained shared-memory devices. Many factors influence the development of architectures: the reducing scale of semiconductor chip fabrication is chiefly responsible for a general increase in processor speed (Moore's Law, briefly mentioned in Section 1.1 and Appendix A) but also important are communication speeds within and between nodes, the whole area of memory management, and the related issues of energy consumption and heat production.

With care, MD and MC algorithms may be adapted to make very effective use of parallel architectures. In this chapter we will describe some of the general features of parallel simulation algorithms. Details, however, may be very machine-dependent, and the landscape changes quickly with time, so we cannot be too specific. The optimal situation is that the program will speed up linearly with the number of processors P, for fixed system size N. (Super-linear speedup is seen occasionally, due to improved memory handling, since additional processors usually come with additional memory.) In practice, the overheads associated with parallelization (inter-process communications, memory access bottlenecks) lead to a sub-linear speedup in almost all cases: a higher degree of parallelism implies lower efficiency, where efficiency is defined as speed divided by N. Therefore, for a given number of particles, there will be a maximum number of processors on which it makes sense to run the program; this maximum may be determined by benchmarking. Usually, larger systems parallelize more efficiently than small ones, and for some cases the efficiency is roughly proportional to N/P. In implementing a program, the number of processes may not necessarily be equal to the number of processors, but for simplicity we shall assume that this is the case in the examples that follow.

Computer Simulation of Liquids. Second Edition. M. P. Allen and D. J. Tildesley.
© M. P. Allen and D. J. Tildesley 2017. Published in 2017 by Oxford University Press.

For shared-memory machines, the concept of dividing the work between programming *threads* is useful: each thread runs on a particular core and, in many cases, there are many more threads than cores, so each core handles several threads, in turn. The chief concern is to avoid different threads overwriting the same areas of memory, which would give incorrect results. Secondary concerns are to avoid, or minimize the effects of, *dependencies* between threads, which might result in delays in accessing the desired memory locations, and the whole issue of *load balancing* between threads, to make best use of the resources. In practice, it is usually most effective to parallelize loops within programs, dividing the work according to the loop index. In MC and MD programs, loops over particles are the most obvious target, and we turn to this in Section 7.2. This type of parallelization can sometimes be achieved by the compiler, given appropriate directives or indications within the language which help to identify loops which can be considered independently. The open multi-processing (openMP) framework of compiler directives, library routines, and environment variables is intended to help this process. In many cases, it is not too difficult to parallelize a serial code using this approach.

Distributed-memory machines, typically work in the single program multiple data (SPMD) paradigm: the same program is running, independently, on each node, working on its own local data. This does not prevent a 'master–slave' approach being adopted at certain times in the program, since one process can branch into the appropriate part of the code (e.g. to distribute data) while the others take the alternative branch (e.g. to receive the data). There is no direct link between the activities on each node, but the programs can send messages to each other which can include instructions to wait for the other nodes to 'catch up' and hence become synchronized. Such messages are also used to pass data between nodes. MPI is the most common way of doing this. This requires some fairly low-level consideration of the way messages are sent and received. The chief concern is to avoid deadlock: the situation where two processes are both waiting to send/receive a message to/from each other. Subsidiary concerns are the *latency* and *bandwidth* of the connecting network. Latency is a time overhead associated with sending a message, irrespective of the amount of data, while bandwidth is the rate of data transfer. Therefore, the time taken to send a message can be roughly estimated as

$$\text{time} = \text{latency} + \frac{\text{amount of data}}{\text{bandwidth}}.$$

For this reason, it is usual to gather data together into *buffers* before sending it, so as to avoid the latency costs of many small messages. One might imagine that the quantities in this equation would depend on the topology of the network between the nodes, and on details of the physical connections. For most HPC installations, at present, this is not a serious issue: the communications may, to a first approximation, be assumed to be independent of location, and the only question is whether the processing units are on the same node or on different ones. In some cases, widely separated nodes are used as the component parts of a highly distributed, and sometimes heterogeneous, computer: in these cases, communication bandwidth and latency may be critical.

Introducing MPI to a program is a non-trivial task, and often such programs must be designed and written from scratch: the differences from a serial code are too great to be introduced in an incremental manner. One useful exception is the replica-exchange or parallel tempering approach, in which each replica is executed on a different node, with comparatively loose coupling between nodes: we discuss this in Section 7.3. The domain-decomposition approach to MD and MC is well-suited to distributed-memory machines using message-passing, and we discuss this in Section 7.4. In fact, the hybrid structure of multi-core nodes may be exploited quite well in MD, using loop parallelization within each node.

Programming a GPU-based computer is somewhat more involved. First, the nodes themselves are highly parallel, and must be programmed in a dedicated language such as compute unified device architecture (CUDA) or open computing language (OPENCL). Second, data transfer rates into and out of the GPU may be a serious bottleneck. We shall not consider the details of how to do this, but note that packages such as highly object oriented molecular dynamics (HOOMD) (Glaser et al., 2015), LAMMPS, and Roskilde University molecular dynamics (RUMD) are available to conduct MD on such architectures.

Our intention in this chapter is to provide only very simple, low-level, examples of how parallel programming may be used in simulation programs. It is hoped that these can be followed without needing to study OPENMP and MPI in detail; however, for any serious work in this area, considerable preparatory study is essential. Several books provide an excellent introduction to parallel programming using OPENMP and MPI (Chandra et al., 2000; Pacheco, 1996; Quinn, 2003) and for GPUs using CUDA (Kirk and Hu, 2012).

7.2   Parallel loops

On a shared-memory machine, it is useful to approach parallelization through the concept of program *threads* which are (largely) independent; there is assumed to be a many-to-one or one-to-one correspondence between the threads and the cores on which the program is executing. A thread may correspond to one iteration of a loop. For nested loops, it is usually most efficient to parallelize the outer loop. Although the all-pairs double loop, used to calculate forces in MD, is likely to be replaced by something more efficient for a large system (see Chapter 5), it provides a good example of the salient points. Some more detailed discussions of parallelizing an MD code using OPENMP may be found elsewhere (Couturier and Chipot, 2000; Tarmyshov and Müller-Plathe, 2005).

We start with the simple Lennard-Jones double loop of Code 1.2. A parallelized version using OPENMP directives is given in Code 7.1. In Fortran, the OPENMP directives appear as comment statements; the syntax is very similar in C, using a #pragma. The first line of the directive instructs the compiler to establish a parallel region of code and share the work involved in the outer loop amongst the available threads. Several of the variables can be shared between the threads since they are never altered: n, the potential parameter sigma_sq, and the coordinate array r are in this category. The default OPENMP behaviour is to assume that variables are shared, unless otherwise stated. However, clearly, the loop index itself, i, must be private to the particular thread handling each iteration. It is not necessary, but does no harm, to declare this explicitly, and this is done on the second (continuation) line of the directive, along with several other variables, such as rij, rij_sq, which must not be shared with other threads, to avoid being overwritten. Finally, the variable v is a special case: it is updated by each iteration of the loop. This is a so-called *reduction* operation. In effect, each thread requires its own private version of the variable to be incremented during the iterations for which that thread is responsible, and then the totals for each thread should be combined at the end. This is such a common situation

Code 7.1   Parallelized double-loop, shared memory

The following code is quite standard, apart from the openMP directive indicated by
!$omp. We use the Lennard-Jones potential in this example.

```
sigma_sq = sigma ** 2
v = 0.0
!$omp   parallel do &
!$omp&  private(i,j,rij,rij_sq,sr2,sr6,sr12) &
!$omp&  reduction(+:v)
DO i = 1, n-1
   DO j = i+1, n
      rij(:) = r(:,i) - r(:,j)
      rij_sq = SUM ( rij ** 2 )
      sr2    = sigma_sq / rij_sq
      sr6    = sr2 ** 3
      sr12   = sr6 ** 2
      v      = v + sr12 - sr6
   END DO
END DO
v = 4.0 * epslj * v
```

To improve load balancing, a scheduling clause should be added to the directive, of the
form schedule(static,1) or schedule(dynamic,chunk) where chunk is the desired
chunk size, as explained in the text.

that there is a dedicated openMP clause, which specifies the variable name and the type
of arithmetic operation to be performed.

   To achieve a reasonable speedup, it is necessary to have some control over the way in
which the work is shared between threads. Quite commonly in openMP, for N particles,
that is, $N - 1$ iterations, and P threads, a static (i.e. fixed) schedule is adopted by default,
assigning the first set of iterations $i = 1 \ldots C$ to the first thread, the second set $i =
C + 1 \ldots 2C$ to the second thread, and so on, where the 'chunk size' is $C \approx (N - 1)/P$. We
can see that this will be very unbalanced because of the range of the inner loop: much
less work is involved in the later iterations than in the earlier ones. To share the work
more evenly, a scheduling clause specifying a chunk size of 1 may be used, as shown in
Code 7.1: then the first P iterations will be distributed to the P separate threads, followed
by the next P iterations, and so on. Alternatively, a dynamic schedule may be specified,
where chunks (of a desired size) are assigned to threads when they become available,
although this may carry some performance overhead.

   What happens when we wish to calculate forces f, for use in MD? Now, some care
must be taken because we typically use Newton's third law to update both f_i and f_j in
the inner loop. The first of these is not a problem because different threads will always be
handling different values of the index i. However, updating f_j is a problem because any
thread may wish to do it at any time, raising the danger of overwriting the same memory

Code 7.2   Parallelized force routine, shared memory

This file is provided online. md_lj_omp_module.f90 contains a Lennard-Jones force routine parallelized using openMP directives. This acts as a drop-in replacement for md_lj_module.f90 (Code 3.4). It can be combined with, for example, md_nve_lj.f90 from Code 3.4, and the utility modules described in Appendix A, to make a molecular dynamics program, as illustrated in the supplied SConstruct file.

```
! md_lj_omp_module.f90
! Force routine for MD simulation, LJ atoms, OpenMP
MODULE md_module
```

locations. The simple but crude solution is to abandon Newton's third law, allow the inner loop to range over $j = 1 \ldots N$ (skipping $j = i$), and increment only $\mathbf{f}_i$ within the inner loop. In this case, the force array f can be shared amongst all threads without danger. However, this involves twice as much work! The alternative is to keep the loops as they are normally written, but add the force array to the list of variables in the reduction clause. An example is given in Code 7.2.

Applying openMP to a more general MD code follows the same general principles. The approach adopted by Tarmyshov and Müller-Plathe (2005) is to parallelize only the most time-consuming parts of the code (constructing a neighbour list, computing non-bonded forces, possibly some of the bonded forces, and applying constraints) and leave the rest (data input and output, computation of most average properties) to a single thread operating in serial mode. It is possible to modify a serial program incrementally, starting with the time-critical loops, and testing at each stage to ensure that no errors have been introduced.

7.3   Parallel replica exchange

Replica exchange, or parallel tempering, was described in Section 4.9. However, we postponed until now a discussion of its most useful feature: that it can be implemented efficiently on a parallel computer. Take, for example, a distributed-memory machine, and consider the simple case of a ladder of temperatures, with each separate system simulated using standard MC in the canonical ensemble, in its own process (and, hopefully, on its own processor). We assume that the processes are identified by consecutive integer variables taking the values $m = 0, 1, \ldots, p - 1$: we call m the *rank* of the process. The temperatures are also in order $T_0 < T_1 < \ldots T_{p-1}$, and we assume that every process m stores not only its own temperature T_m, but also those of its neighbours $T_{m\pm1}$. At preset intervals (a specified number of steps or sweeps), replica exchanges are considered, as shown in Fig. 7.1. An MPI implementation is given in Code 7.3. The processes can be paired up in two different ways: $0 \leftrightarrow 1, 2 \leftrightarrow 3, \ldots$ etc., or $1 \leftrightarrow 2, 3 \leftrightarrow 4, \ldots$, and typically each of these pairings is considered in turn: in one case, any given process m is looking 'up' in temperature at $m + 1$, and in the other case, 'down' at $m - 1$. Of course, it is essential that the other process in the pair is looking in the complementary direction: so, the program is arranged to ensure that odd-m processes will look 'up' while even-m processes look

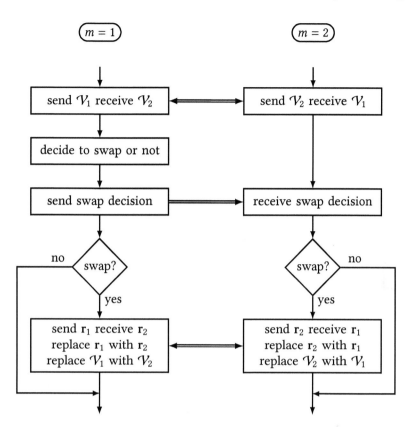

Fig. 7.1 Replica exchange, by message-passing. Systems $m = 1$ and $m = 2$ are possible exchange partners. The horizontal double arrows represent inter-process communications. At the same time, other distinct pairs of systems are exchanging in the same way. The pairing scheme changes each time, as explained in the text.

'down', and vice versa. Each pair must compare potential energies and decide whether to exchange configurations, using eqn (4.67). An important point is that this decision must be taken by *one* member of the pair, not both: it involves a floating-point calculation, and comparison with a random number, so it is vital to avoid an ambiguous verdict! In the example of Fig. 7.1, the $m = 1$ process takes the decision, so process $m = 2$ must send its own potential energy $\mathcal{V}_2$ to $m = 1$ (in the figure we assume that it is convenient to send $\mathcal{V}_1$ the other way at the same time). Process $m = 1$ executes the statements corresponding to eqn (4.67), storing the result (to swap or not) in a logical variable. This variable is sent to process $m = 2$ which has simply waited to receive the decision: this procedure guarantees that the value of the swap variable is identical on both processes. The same applies to all the other pairs in the ladder. Then, if the decision is to swap, the configurations are exchanged. It is important to realize that each process is executing the same program, but is taking a different route through the control statements. By keeping the alternating sequence of pairings in step across all processes, it is possible (and essential) to match the statements which send and receive data; also, there will be times at which processes

Code 7.3   Replica exchange, by message-passing

This file is provided online. It assumes that an MPI library has been installed, compatible with the Fortran compiler. This would normally provide a module mpi containing predefined parameters, and commands mpif90 or mpifortran to compile the code, and mpirun to run it, using a specified number of processes. The file mc_nvt_lj_re.f90 runs a set of MC simulations of systems at different temperatures, and is very closely based on mc_nvt_lj.f90 of Code 4.3; like that program, it makes use of mc_lj_module.f90 (Code 4.3) as well as the utility modules described in Appendix A. Note: Fortran and MPI data types must match, and so the variable declarations in the example code may need modifying, on any given computer.

```
! mc_nvt_lj_re.f90
! Monte Carlo, NVT ensemble, replica exchange
PROGRAM mc_nvt_lj_re
```

at the ends of the ladder of temperatures have no exchange partner and hence must do nothing while the other pairs go through their exchanges. Code 7.3 shows how this is done. One important technical point (see also Appendix E) is that the random number sequences on different processes must be independent of each other.

It is not too difficult to adapt this scheme to handle the case of exchanges between non-neighbouring temperatures. The only requirement is that an unambiguous pairing scheme be set up, and shared between all the processes, before the exchanges are attempted. One process (the master) can generate a random permutation of the numbers $0 \ldots p-1$, define the pairs as successive elements in this list, and send them to the other processes (the slaves). The modifications to handle MD rather than MC are also straightforward: velocities must be exchanged as well as coordinates, and the acceptance criterion for the move is different, as discussed in Section 4.9.

The additional MPI statements needed to implement replica exchange are not very extensive, and the amount of communication is relatively small: mainly the actual exchange of coordinates. Even this can be reduced, by exchanging the values of the temperatures rather than the configurations. This requires some additional book-keeping and more use of arrays to hold variables such as simulation run averages for each temperature, while it is running on each process. These averages would then be gathered together, by temperature, at the end. Many implementations of replica exchange actually use an external script to carry out the exchanges, rather than incorporating them into the program. This gives greater flexibility but will be less efficient in general. On the other hand, exchanging the temperatures (or other parameters), rather than configurations, fits more easily into a script-based scheme, and is a fairly common approach in packages. Earl and Deem (2004) have discussed how to allocate replicas to processes in order to optimize the use of CPU time.

A potential disadvantage of replica exchange when implemented on extremely large, possibly heterogeneous, parallel computing environments is the need to synchronize processes quite frequently. To tackle this, there have been some efforts to develop asynchronous versions of the method (Thota et al., 2011; Xia et al., 2015).

7.4   Parallel domain decomposition

The natural approach to MD on a distributed-memory machine is to divide the system into physical domains, each of which is handled by one process (again, ideally, on its own processor). A grid of cubic domains is the simplest case. To compute the forces due to short-ranged interactions, it is only necessary to know the coordinates of particles on the same process, plus those within interaction range on 'neighbouring' processes. These can be sent, at the start of the MD step, in the manner illustrated in Fig. 7.2. Coordinates (usually together with particle identifiers, needed to look up interaction potential parameters) are successively exchanged in the x, y, and z directions: the MPI routine MPI_Sendrecv provides a low-level way of doing this. Routines are also provided in MPI for easy identification of neighbouring nodes in a Cartesian grid. As shown in the figure, two sets of exchanges (in x and y) are sufficient to build up a shell, surrounding each domain, containing nearby particles from all eight neighbours (in 2D); extending this to a third exchange (in z) gives a shell surrounding the cubic domain, with nearby particles from all 26 neighbour domains in 3D (Pinches et al., 1991; Plimpton, 1995).

Given this information, forces can be calculated on each process. Within the domain, any of the standard approaches to improve efficiency (such as the cell-structure, linked-list method) may be used. The coordinates and velocities of the 'core' particles are updated by one timestep according to the MD algorithm; the coordinates of the 'shell' particles can be discarded. There then needs to be a second communication step in which particles which have moved outside their domain boundaries are sent to the process appropriate for their new domain. If, as is common at liquid densities, the particles have travelled a distance less than the shell thickness during the step, the communication pattern will be similar to that of Fig. 7.2. If this assumption cannot be made, then a slightly modified scheme is needed, to send particles to the correct domain. In either case, there are two key differences from the first exchange: velocities as well as coordinates need to be sent to the new process, and they need to be deleted from the old one.

It is clear that this algorithm will be most efficient when the communication load is relatively small: this will happen when the domains are large compared with the surrounding shell regions. Ideally, we would expect the efficiency to be the same if one increases the number of processors P in proportion to the number of particles N.

A modification of this scheme is usually more efficient (see Fig. 7.3). Coordinates near the domain boundaries are transmitted (one way) to just half of the neighbouring domains (at each stage) rather than exchanging information with every neighbour. The forces on the 'shell' particles can be calculated and transmitted back and added to those coming from particles on their home process. The advantage is that each pair force (assuming pair additivity) is computed only once, rather than twice. Note that the trick of performing exchanges successively in Cartesian axis directions does not achieve the aim of sending all the necessary information here: it is best to send directly to the desired neighbours in this case, including the diagonal ones. The method is easily implemented in 2D or 3D. Other communication patterns may be more efficient, depending on the circumstances (Liem et al., 1991; Bowers et al., 2005; 2007; Larsson et al., 2011).

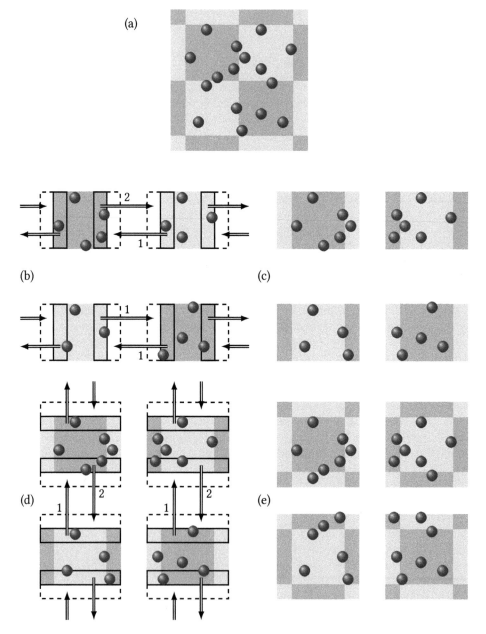

Fig. 7.2 Message-passing in parallel domain decomposition MD (2D). (a) Domains identified by checkerboard shading. For clarity we consider only particles within the central four domains. (b) First exchange, in the x-direction. The four central domains are now shown separately. Arrows indicate the number of particles being transmitted to neighbours. (c) Result of first exchange; shading indicates added regions. (d) Second exchange, in the y-direction. Note that this includes some particles which arrived in the first exchange. (e) Result of second exchange.

(a)   (b)

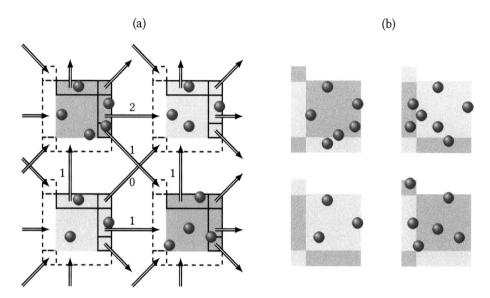

Fig. 7.3 Half-shell message-passing in parallel domain decomposition MD. We show the same 2D system as in Fig. 7.2. (a) Transmission of particle data to half the neighbouring domains. Arrows indicate the number of particles being transmitted to neighbours in two Cartesian directions (from the side rectangles), and two diagonal directions (from the corner squares). (b) Result showing half-shell surrounding each domain.

For inhomogeneous systems, a regular cubic domain decomposition may not give the best load balancing, whereby each process handles approximately the same number of particles. Common packages such as GROMACS and LAMMPS allow a cuboidal or triclinic array of domains as a compromise between simplicity and efficiency (Hess et al., 2008). A more flexible, adaptive scheme involving domains constructed from tetrahedra has been proposed recently (Begau and Sutmann, 2015).

How can we handle long-range forces on a parallel computer? Several of the methods discussed in Chapter 6, such as PPPM, PME, and SPME, rely on the fast Fourier transform (FFT), which does not scale well on parallel machines: it requires a global many-to-many communication. The FMM, on the other hand, does not rely on FFT. The MEMD method uses local data to solve the relevant equations, which makes it more suitable for parallelization. Arnold et al. (2013) have compared several of these methods, and identified both FMM and the FFT methods as being competitive, from the scalability viewpoint: however, much depends on particular physical systems, computer hardware, and on optimizing parameters of the methods. In view of this, Moore and Crozier (2014) have considered the MSM method (Stone et al., 2007; Hardy et al., 2009) as a promising candidate for computing long-range forces in a scalable way for large systems, although their first comparisons indicate that it is outperformed by PPPM.

MC simulations may also be speeded up by domain decomposition on a distributed-memory computer. There are two main points that need careful consideration. First, MC is intrinsically serial in nature: the states in the Markov chain must follow one another and,

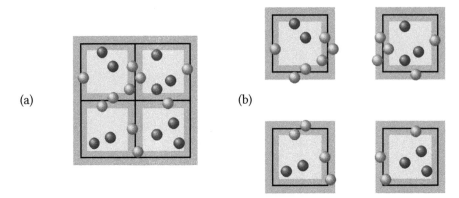

Fig. 7.4 Domain decomposition with boundary regions for parallel MC (2D example). (a) Boundaries (dark shading) containing fixed particles (light coloured), and internal regions (light shading) containing particles which can be updated in an MC sweep (dark coloured). (b) Separated domains. Note that the same boundary particles appear in more than one domain, necessitating some inter-processor communication.

in principle, each move depends on the result of the previous one. This problem can be partially overcome by constructing a 'super-move' out of a set of completely independent moves involving non-interacting sets of atoms: the result is independent of the order in which the constituent moves are carried out. The template for this is the checkerboard decomposition used for the nearest-neighbour Ising model (Pawley et al., 1985): the cubic lattice is divided into 'black' and 'white' sublattices. Spins on the black sublattice only interact with nearest neighbours on the white sublattice (and vice versa); therefore all the black spins may be updated by independent MC moves, keeping the white spins fixed, and this can be done in parallel. Then, the white spins can be updated, keeping the black spins fixed. A checkerboard scheme applied to an atomistic system (using the pattern of Fig. 7.2, for example) cannot work in exactly the same way. Even if the domains are equal to or larger than the cutoff distance, the interactions are not just restricted to nearest neighbour domains in the Cartesian axis directions. Uhlherr et al. (2002) describe a parallel MC algorithm based on dividing the system into domains separated by boundary regions which are at least r_c in width. (They discuss a striped arrangement, but in principle a cubic domain decomposition would also work.) One way of doing this is shown in Fig. 7.4. Then, fixing the boundary particles, the ones within each boundary may all be moved independently of other domains: particles in the 'active' regions are selected randomly, or sequentially, and moved in the usual way. (We assume that the maximum displacement is smaller than the boundary thickness.) Hence, each domain can be handled in parallel with the others.

The second main point that needs some care is to ensure detailed balance. What happens when a proposed particle move would take it outside the mobile part of the domain (the light regions in Fig. 7.4) and into the boundary region? In that case, the reverse move cannot happen, because a particle in the boundary region would never be selected for a move. Uhlherr et al. (2002) show that simply rejecting such moves is sufficient to ensure microscopic reversibility. Therefore, for the duration of a series of MC moves, all the domains behave independently, and are effectively separated by impenetrable walls: no particles are allowed to move outside the active part of their domain.

In this approach, the domains and boundary regions must be redefined from time to time (e.g. at the start of every sweep) to avoid violation of ergodicity. This means redistributing the particles. This can be done in a manner similar to that used in MD, described earlier, or by a master–slave mechanism (Uhlherr et al., 2002), collecting the particles centrally and sending them out again to the various processors handling each domain. Uhlherr et al. (2002) give further details of the efficiency of this approach, and the way in which it may be combined with a variety of moves for long polymer chains.

Several refinements have been proposed. O'Keeffe and Orkoulas (2009) suggest a striped domain scheme in which, again, each domain is divided into central and boundary regions. The boundaries are of width $r_c + 2\delta r_{max}$, where δr_{max} is the maximum displacement parameter. Each processor also holds the nearest boundary region of its neighbouring processors. By tackling each part of a domain in turn (all left boundaries first, then the central regions, then all right boundaries), and organizing the communications appropriately, a reasonably efficient algorithm is obtained. In a different vein, a parallelizable cluster MC algorithm has been proposed by Almarza and Lomba (2007). Here the aim is to change the positions of a large fraction of the particles simultaneously.

7.5 Parallel constraints

We close this chapter with some comments about the implementation of constraints on a parallel machine. This is particularly important for macromolecules (polymers and proteins) which may reasonably be expected to cross several domains in a domain decomposition. The iterative nature of the SHAKE and RATTLE algorithms makes them difficult to parallelize (Debolt and Kollman, 1993; Brown et al., 1994). The main issue is that the refinement of a constraint depends on the positions of the associated atoms, which may themselves be moved by refining other constraints. However one distributes the list of constraints amongst a set of processes, there are bound to be interdependencies. Since the procedure involves many iterations to converge, there will be many communication steps between processors. In practice, it becomes difficult to consider constraints across boundaries in the domain-decomposition approach.

Weinbach and Elber (2005) have suggested replacing the iterative solution of the linear equations that lie at the heart of SHAKE, with a conjugate gradient minimization based on the original matrix formulation (Ryckaert et al., 1977; Barth et al., 1995). They explain in detail how to implement the method on a distributed-memory computer. (The matrix inversion method had already been considered for shared-memory machines by Mertz et al. (1991).) Each processor handles a fixed subset of constraints. The algorithm needs to parallelize numerical tasks such as the calculation of an inner product, summation of vectors, and a matrix–vector multiplication; the last task is the most complex, in general. Weinbach and Elber (2005) describe in detail the approach to be used when all the atom coordinates are available on every processor, but also discuss briefly how it could be implemented in the context of spatial domain decomposition. In a similar spirit, the LINCS method has been parallelized (Hess, 2008) and incorporated into the GROMACS package.

Example 7.1 Massively parallel molecular dynamics

Molecular dynamics can provide a detailed mechanistic understanding of the formation of methane hydrates and the growth of ice crystals in aqueous solution (English, 2013; English and Tse, 2015). Properties such as the vibrational spectrum and the rate of break-up of the nano-clusters depend strongly on the extent of the surrounding liquid phase. These, and many other similar problems, must be modelled using large system sizes to avoid artefacts arising from the periodic boundary conditions (English et al., 2005). For these two particular problems, massively parallel MD has been used to study fluids of up to 8.6×10^6 molecules on a range of IBM Blue Gene computers: the BG/L, BG/P, and BG/Q (Mullen-Schultz, 2005; Gilge, 2012).

Two community MD codes, LAMMPS and NAMD, are employed in these studies. The LAMMPS package (Plimpton, 1995) is parallelized using a spatial decomposition and either pure MPI or hybrid MPI/openMP communications. The NAMD package (Phillips et al., 2005) uses a mixed force/spatial decomposition and pure MPI communications. Simulations were performed on both the uncharged, coarse-grained, mW model of water (see Section 1.3.4) and the atomistic, partially charged TIP4P model.

The BG/L uses a 3D toroidal network for peer-to-peer communications. The BG/P uses the same network topologies as the BG/L but at twice the bandwidth, and the BG/Q implements chip-to-chip communications in a 5D toroidal configuration. Across the series the peak performance increases from 56 gigaflops per node on BG/L to 204.8 gigaflops per chip for BG/Q. MPI versions of LAMMPS and NAMD were run on the BG/L and BG/P computers with each processor on a particular node assigned to one MPI task. The BG/Q allows up to 64 threads per 16-processor node. In the studies of all the coarse-grained and atomistic models using both packages, there is a trend to better overall performance and better scaling in moving from the BG/L to the BG/Q machine. The BG/Q delivers an almost linear parallel scaling between 1024 and 8192 threads for 1.77×10^6 molecules using LAMMPS applied to the coarse-grained model of water. This results in a speed of approximately 40 timesteps per second with 8192 threads. For the atomistic model, long-range forces were handled using the SPME method and the NAMD code demonstrated better relative scaling than LAMMPS for the 3D FFT. The faster interconnect and lower communications latency makes LAMMPS more competitive for long-range potentials on the BG/Q than on the BG/P, but it is still not as efficient as the NAMD code. These simulations have provided new insights into early stages of crystal nucleation and the first studies of the interaction between growing crystallites seeded at random positions in the fluid (English, 2014).

Community codes such as LAMMPS and NAMD continue to evolve as new computer architectures and communication protocols become available. For example, LAMMPS has been extended to run on hybrid computers with nodes containing both GPU and CPU processors using openCL and CUDA (Brown et al., 2012).

8
How to analyse the results

8.1 Introduction

It is probably true that the moment the simulation run is finished, you will remember something important that you have forgotten to calculate. Certainly, as a series of simulations unfolds and new phenomena become apparent, you may wish to reanalyse the configurations to calculate appropriate averages and correlation functions. To this end, configurations generated in the run are often stored on disk, which is then used for subsequent analysis (see Section 5.7).

In an MC simulation it would be inappropriate to store configurations after every attempted move, since successive configurations are identical or highly correlated. Typically, the configuration at the end of every fifth or tenth cycle is stored (one cycle = N attempted moves). Each stored configuration will contain vectors describing the positions of the atoms, and in the case of a molecular fluid, each orientation. It is also convenient to store the instantaneous values of the energy, virial, and any other property of interest. Although these properties can be reconstructed from the positions of the particles this is often an expensive calculation.

Equally, in an MD simulation, successive timesteps are correlated and do not contain significantly new information. In this case it is sufficient to store every fifth or tenth timestep on the disk for subsequent analysis. An MD simulation produces a significant amount of useful information, and it is normal to store vectors of positions (orientations), velocities (angular velocities), and forces (torques) for each molecule, as well as the instantaneous values of all the calculated properties. The information stored in an MD simulation is time ordered, and can be used to calculate the time correlation functions discussed in Chapter 2. The molecular positions that are stored from the MD simulation may be for particles in the central box which have been subjected to periodic boundary conditions. It is also useful to store trajectories which have not been adjusted in this way but which represent the actual movement of a molecule in space. These trajectories are particularly useful in calculating self diffusion coefficients. It is possible to convert from the central-box representation to the 'unfolded' one, on the assumption that molecules do not naturally move distances of the order of half a box length in the interval between stored timesteps. A routine for doing this is given as part of Code 8.3.

In this chapter, we discuss how to analyse a succession of configurations, or trajectory file, so as to produce structural distribution functions and time correlation functions. We then proceed to the important question of assessing statistical errors in the simulation results. Finally, we outline some techniques used to correct, extend, or smooth the raw data.

8.2   Liquid structure

We have assumed that the analysis of liquid structure will take place after a simulation is complete. As mentioned in Chapter 5, it is possible to do this during the simulation run itself, using methods very similar to those described here.

The pair distribution function $g(r)$ is formally defined by eqn (2.101), but is more simply thought of as the number of atoms a distance r from a given atom divided by the number at the same distance in an ideal gas at the same density.

We calculate $g(r)$ as follows. Configurations are analysed in turn and the minimum-image separations r_{ij} of all the pairs of atoms are calculated. These separations are stored in a histogram, $h(k)$, where each bin k has a width δr and extends from r to $r + \delta r$. A typical piece of Fortran code for sorting N atoms is given in Code 8.1.

When all the configurations have been processed, the histogram must be normalized to calculate $g(r)$. Suppose that the histogram bins $h(k)$ recording pair separations have been accumulated over n_{step} steps. Then the average number of atoms whose distance from a given atom in the fluid lies in this interval, is

$$n(k) = h(k)/(N \times n_{step}). \tag{8.1}$$

The average number of atoms in the same interval in an ideal gas at the same number density ρ is

$$n^{id}(k) = \frac{4\pi\rho}{3}\left[(r + \delta r)^3 - r^3\right]. \tag{8.2}$$

By definition the radial distribution function, in the limit of small δr, is

$$g(r + \tfrac{1}{2}\delta r) = n(k)/n^{id}(k) \tag{8.3}$$

and the code for normalizing the histogram is given in Code 8.1. The appropriate distance for a particular element of our $g(r)$ histogram is at the centre of the interval $(r, r + \delta r)$. We mention in passing that finite-size effects, and other sources of imprecision, have been considered by Kolafa et al. (2002). In the canonical ensemble, $g(r) \to 1 - O(1/N)$ as $r \to \infty$ (Gray and Gubbins, 1984, Chapter 3) and the small correction is occasionally important.

The double-loop code for sorting separations is quite expensive but cannot be vectorized because the histogram array is not accessed sequentially. Fincham (1983) has discussed a method for calculating $g(r)$ by sorting over the histogram bins rather than the molecules which is suitable for use on pipeline and parallel processors. Our code involves taking a square root for each pair in every configuration. It is also possible to sort the squared distances directly into a histogram and to calculate $g(r^2)$. A disadvantage of this is that the resulting $g(r)$ is obtained at uneven intervals in r with a larger spacing at small r, which is just the region in which the function is required with the highest resolution. Extrapolation and interpolation is difficult at small r because the function is rapidly varying (see Section 8.5.2).

Code 8.1 Calculating the pair distribution function

The first snippet of code increments the histogram of pair separations. Here, k is an INTEGER variable, and dr stores the value of δr; nk is the size of the h array, chosen such that nk*dr is the maximum required distance r, less than half the box length. This array is set to zero initially, and then the following loop is carried out for nstep configurations, each stored in the array r(3,n).

```
DO i = 1, n-1
  DO j = i+1, n
    rij(:) = r(:,i) - r(:,j)
    rij(:) = rij(:) - ANINT ( rij(:) )
    rij_sq = SUM ( rij**2 )
    k      = FLOOR ( SQRT ( rij_sq ) / dr ) + 1
    IF ( k <= nk ) h(k) = h(k) + 2
  END DO
END DO
```

The *ij* and *ji* separations are handled simultaneously, which is why h(k) is incremented by 2, not 1. The second snippet of code normalizes the results at the end.

```
const = 4.0 * pi * rho / 3.0
DO k = 1, nk
  g(k) = REAL ( h(k) ) / REAL ( n * nstep ) ! average number
  r_lo = REAL ( k - 1 ) * dr
  r_hi = r_lo + dr
  h_id = const * ( r_hi ** 3 - r_lo ** 3 ) ! ideal number
  g(k) = g(k) / h_id
END DO
```

This code is also provided online, in a program pair_distribution.f90.

```
! pair_distribution.f90
! Calculates pair distribution function g(r)
PROGRAM pair_distribution
```

An identical sorting technique can be applied to the site–site pair distribution functions mentioned in Section 2.6, and to the spherical harmonic coefficients defined in eqn (2.106). In the latter case, we average in a shell as follows (Streett and Tildesley, 1976; Gray and Gubbins, 1984)

$$g_{\ell\ell'm}(r_{ij}) = 4\pi g_{000}(r_{ij}) \langle Y^*_{\ell m}(\Omega_i) Y^*_{\ell'\bar{m}}(\Omega_j) \rangle_{\text{shell}} \tag{8.4}$$

where $\bar{m} = -m$. In this equation $\langle \ldots \rangle_{\text{shell}}$ has the following interpretation. For each pair ij, a particular bin of the $g_{000}(r_{ij})$ histogram, corresponding to a molecular centre–centre separation r_{ij}, is incremented by 2, just as in the atomic case. At the same time, the corresponding bin of each $g_{\ell\ell'm}$ histogram should have

$$Y_{\ell m}^*(\Omega_i)Y_{\ell'\bar{m}}^*(\Omega_j) + Y_{\ell m}^*(\Omega_j)Y_{\ell'\bar{m}}^*(\Omega_i)$$

added to it. At the end of the calculation, each $g_{\ell\ell'm}$ histogram bin is divided by the corresponding element of the g_{000} histogram. The result is the shell average in eqn (8.4). The function $g_{000}(r_{ij})$ is then calculated from the histogram in the usual way, and used in eqn (8.4) to give the other $g_{\ell\ell'm}$ functions.

In Example 8.1 we discuss how the careful analysis of spatial correlation functions has helped clarify the long-standing problem of two-dimensional melting.

8.3   Time correlation functions

In this section, we consider the calculation of time correlation functions from a file that contains positions, velocities, and accelerations stored at regular intervals during a molecular dynamics simulation. Bearing in mind that a wide variety of correlation functions may be of interest, analysis of a file is logistically simpler than the alternative of calculating the time correlation functions during the simulation run itself. However, it is possible to do some analysis of this kind during a simulation, and we will return to this briefly later.

8.3.1   The direct approach

The direct approach to the calculation of the time correlation functions is based on the definition eqn (2.112). Suppose that we are interested in a mechanical property, $\mathcal{A}(t)$, which may be expressed as a function of particle positions and velocities. $\mathcal{A}(t)$ might be a component of the velocity of a particle, or of the microscopic pressure tensor, or a spatial Fourier component of the particle density, for example. From the data in the file, $\mathcal{A}(t)$ will be available at equal intervals of time δt; typically, δt will be a small multiple of the timestep used in the simulation. We use τ to label successive timesteps in the file, i.e. $t = \tau \delta t$. The definition of the time average, in a discretized form, allows us to write the non-normalized autocorrelation function of $\mathcal{A}(t)$ as

$$C_{\mathcal{A}\mathcal{A}}(\tau) = \left\langle \mathcal{A}(\tau)\mathcal{A}(0) \right\rangle = \frac{1}{\tau_{\max}} \sum_{\tau_0=1}^{\tau_{\max}} \mathcal{A}(\tau_0)\mathcal{A}(\tau_0 + \tau). \tag{8.5}$$

In words, we average over $\tau_{\max}$ time origins the product of $\mathcal{A}$ at time $\tau_0\delta t$ and $\mathcal{A}$ at a time $\tau \delta t$ later. For each value of τ, the value of $\tau_0 + \tau$ must never exceed the number of values of $\mathcal{A}$, τ_{run}, stored in the file. Thus, the short-time correlations, with τ small, may be determined with slightly greater statistical precision because the number of terms in the average, $\tau_{\max}$, may be larger. We return to this in Section 8.4. Again, as written, eqn (8.5) assumes that each successive data point is used as a time origin. This is not necessary, and indeed may be inefficient, since successive origins will be highly correlated. A faster calculation will result from summation over every fifth or tenth point as time origin (with a corresponding change in the normalizing factor $1/\tau_{\max}$) and with little degradation of the statistics.

The calculations may be repeated for different values of τ, and the result will be a correlation function evaluated at equally spaced intervals of time δt apart, from zero to as

Example 8.1 Two-dimensional melting

In the so-called KTHNY theory (Kosterlitz and Thouless, 1973; Halperin and Nelson, 1978; Nelson and Halperin, 1979; Young, 1979) melting of two-dimensional crystals is predicted to occur via two continuous transitions, with the solid and liquid phases separated by a hexatic phase characterized by short-range (exponentially decaying) positional order and quasi-long-range (algebraically decaying) orientational order. The orientations are those of the vectors between neighbouring atoms. Early simulations (see e.g. Frenkel and McTague, 1979; Tobochnik and Chester, 1980; 1982) were restricted to a few thousand particles, and were somewhat inconclusive (Strandburg, 1988). However, the early work established several important quantities to analyse. The dislocation and disclination defects thought to drive the transitions were identified by a Voronoi construction (McTague et al., 1980; Weber and Stillinger, 1981), which defines each particle's neighbours in a self-consistent way; the average coordination number is six, and defects are associated with disks having five or seven neighbours. The orientational correlation functions were calculated in terms of a local order parameter

$$\psi_i = \frac{1}{n_i} \sum_{j=1}^{n_i} \exp(\mathrm{i}6\theta_{ij})$$

where the sum is over the n_i neighbours of each disk i, and θ_{ij} is the angle of the vector $\mathbf{r}_{ij}$ in a fixed coordinate system. This can be used to calculate a spatial correlation function, or a spatially averaged order parameter whose finite-size scaling behaviour can be examined (Weber et al., 1995). The elastic constants of the solid, on the approach to the transition, are also of interest (Binder et al., 2002).

A recent study (Bernard and Krauth, 2011) using $N \sim 10^6$ hard disks, and a simulation technique involving collective moves, called 'event-chain' MC, suggests that the hexatic phase exists, but that the liquid–hexatic transition is first-order. This was confirmed (Engel et al., 2013) using conventional MC and MD techniques. The analysis required accurate estimation of the pressure P, by careful extrapolation of $g(r)$ to contact (in the standard MC), or directly (in the other techniques). This showed the expected, system-size-dependent, loop in the $P(V)$ curve. The local order parameter ψ_i was used to provide visual evidence of two-phase coexistence, and positional correlation functions such as $g(r)$ confirmed that the coexisting phase is not a solid. A subsequent study of soft-disk systems (Kapfer and Krauth, 2015) suggests that this scenario changes as the softness increases, and that the first-order hexatic–liquid transition is replaced by a continuous one, in accord with the KTHNY predictions.

In passing, we note that the idea of characterizing a phase, by examining the local structure around atoms, persists in the 'topological cluster classification' for three-dimensional systems (Malins et al., 2013). A modified Voronoi construction identifies neighbours of each atom, and then every cluster of near neighbours belonging to a predefined set of typical low-energy clusters, is found. This analysis is potentially useful in discussing glass formation and dynamical arrest (Royall and Williams, 2015).

Code 8.2 Calculating a time correlation function

In this code snippet, we assume that all the data to be correlated are in the array a.

```
INTEGER                      :: t_run, t_cor, t0, t1
REAL , DIMENSION(t_run)    :: a
REAL , DIMENSION(0:t_cor) :: c, norm


c(:)     = 0.0
norm(:) = 0.0

DO t0 = 1, t_run
   t1                = MIN ( t_run, t0 + t_cor )
   c(0:t1-t0)      = c(0:t1-t0)      + a(t0) * a(t0:t1)
   norm(0:t1-t0) = norm(0:t1-t0) + 1.0
END DO
c(:) = c(:) / norm(:)
```

To select origins t0 less frequently, the loop statement may be replaced by something like DO t0 = 1, t_run, t_gap, with t_gap equal to 5 or 10, for example. We have included the foolproof counter norm(t); it should be equal to REAL(t_run-t) in our example. The central loop of this routine should be slightly more efficient than the more obvious alternative

```
DO t = 0, t_cor
   t_max = t_run - t
   c(t) = SUM ( a(1:tmax) * a(1+t:t_max+t) ) / REAL ( t_max )
END DO
```

See also Codes 8.3 and 8.4 for online programs that calculate time correlation functions.

high a value as required. In principle, $\tau \delta t$ could be extended to the entire time spanned by the data, but the statistics for this longest time would be poor, there being just one term in the summation of eqn (8.5) (the product of the first and last values of $\mathcal{A}$). In practice, $C_{\mathcal{A}\mathcal{A}}(t)$ should decay to zero in a time which is short compared with the complete run time, and it may be that only a few hundred values of τ are of interest.

A simple calculation of this kind is given in Code 8.2. The modification to deal with cross-correlations $\langle \mathcal{A}(t)\mathcal{B}(0) \rangle$ is straightforward. The procedure is almost unchanged if one wishes to calculate mean-squared displacements (see later).

In the previous example we have assumed for simplicity that all the values of $\mathcal{A}(t)$ can be stored in memory at once. On modern machines, memory is quite plentiful, and so this may be true. Nonetheless, if one is interested in single-particle properties, for large systems, this may not be practical. An alternative is to keep the data on disk, and use direct access I/O statements. However, it is generally more efficient to make a single pass through the data, assuming that enough memory is available to store all the desired elements of the autocorrelation function (rather than the data itself). In this method, τ_{cor}

Code 8.3 Program to compute diffusion coefficients

This file is provided online. `diffusion.f90` contains a program to read in a trajectory of atomic positions and velocities, and calculate the time-dependent velocity autocorrelation function, mean-squared displacement, and velocity–displacement cross-correlation, from any of which the diffusion coefficient may be obtained. The program 'unfolds' positions (i.e. removes the effects of periodic boundary conditions) on the assumption that atoms do not travel too far between successive times.

```
! diffusion.f90
! Calculates vacf and msd
PROGRAM diffusion
```

timesteps are read into memory, where $(\tau_{cor} - 1)\delta t$ is the maximum time for which the correlation function is required. As each step is read from the file, the correlations with all previous steps are computed. In the example given in Fig. 8.1(a), step 4 is correlated with the first three steps. When τ_{cor} steps have been read (Fig. 8.1(b)), the information in step 1 is no longer needed, and step $\tau_{cor} + 1$ is read into this location. This is correlated with all the other steps (Fig. 8.1(c)). The next step is read into location 2, and the correlation proceeds (Fig. 8.1(d)).

This approach can be made more efficient if time origins are chosen at intervals (e.g. every five or ten steps) in which case the a array need only store those origins. As each new data value is read in it is correlated with all previous time origins and then, only if needed, is stored (in a) for future use as a time origin. In this way, the method is economical in storage as well as being fairly efficient. Because of the way that a is cyclically overwritten with new data, it is essential to store the timestep index in an accompanying array. A sample program is given in Code 8.3, which calculates the velocity autocorrelation function, and the mean-square displacements, in order to obtain the diffusion coefficient. Essentially this same method can be used to calculate correlation functions while the run is in progress, avoiding all use of disk storage. We note in passing an almost equivalent approach (Rapaport, 2004; Dubbeldam et al., 2009) which uses buffers to store the contributions to the correlation function arising from each time origin, only adding them to the final function once all the timesteps that must be correlated with that origin have been processed.

This simple approach is still not satisfactory if very long-time correlation functions, or mean-squared displacements, are of interest, as the computational effort scales with t^2. Frenkel and Smit (2002) recognized this and devised a hierarchical scheme based on a range of sampling frequencies. Here we describe a very simple scheme, based on the conventional approach, which is easy to formulate (Dubbeldam et al., 2009). Suppose (for illustration) that, in addition to sampling every step, we also sample at intervals of 25 steps, $25^2 = 625$ steps, etc. Each 'level' of sampling has, say, $\tau_{cor} = 25$, stores its own time origins a and correlation function c, and is processed in a similar way. Then the most fine-grained analysis will give $C(\tau)$ at a resolution of one timestep, up to $\tau = 24$. The second level will cover $\tau = 25$–600, but only at intervals of 25 timesteps. The third level

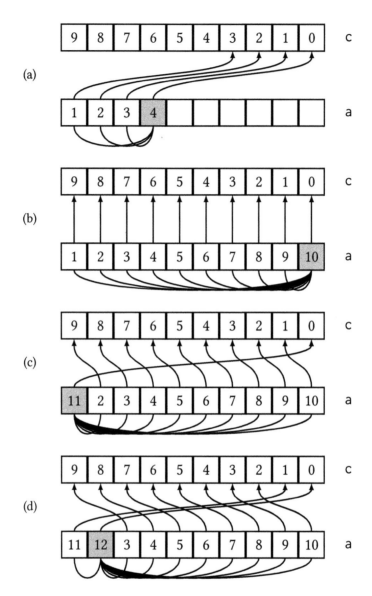

Fig. 8.1 Calculating the time correlation function in a single sweep. The data array a is correlated with itself to give the correlation function c. In this example $\tau_{cor} = 10$, correlations are computed up to nine steps, and for simplicity we assume that every step is used as a time origin. The numbers in the a array refer to the timesteps of the data stored in it. In the early stages these coincide with the array indices, but later the locations are overwritten to save space. The c array is drawn in reverse order purely for visual clarity; the numbers in this array denote the time lag between correlated values. The arrowed lines connect the latest value of a to be read in (shaded) with each of the previously stored time origins, and hence to the corresponding value to be incremented in c.

will calculate $C(\tau)$ for $\tau = 625$–15 000 at intervals of 625 steps, and so on. Five or six levels are easily sufficient for most purposes. If the sudden change of resolution at $\tau = 25$ is undesirable, τ_{cor} can be made longer; this will imply an overlap in the time ranges covered at each level, but calculation of the smaller values of τ may be omitted for the higher levels, if they have already been calculated at the lower levels. The overall decrease in time resolution at longer τ is not usually a problem since usually one assumes that $C(\tau)$ varies much more slowly than at short times; for instance, in calculating mean-squared displacements, we are often interested in a log–log plot.

The main drawback of this approach is that the number of time origins used to calculate the longer time values also decreases if the data are sampled in this way. It may be possible to mitigate this by using block-averages rather than instantaneous values as one goes from one level to the next in the hierarchical scheme: in a sense one is trading off statistical errors against systematic errors by this 'coarse graining' (Frenkel and Smit, 2002; Ramírez et al., 2010).

8.3.2 The fast Fourier transform method

It is possible to improve the speed of calculating the time correlation function by taking advantage of the very rapid algorithms available for computing discrete Fourier transforms. This particular application of the FFT was first proposed by Futrelle and McGinty (1971) and some details are given by Kestemont and Van Craen (1976) and Smith (1982a,b). The method is an application of the convolution/correlation theorem given in Appendix D. Apart from the normalizing factor τ_{max} (which may be incorporated later) the discrete correlation function, eqn (8.5), may be written as

$$C'_{\mathcal{A}\mathcal{A}}(\tau) = \sum_{\tau_0=1}^{2\tau_{run}} \mathcal{A}(\tau_0)\mathcal{A}(\tau_0 + \tau) \qquad 0 \le \tau < 2\tau_{run}. \tag{8.6a}$$

The prime reminds us of the dropped normalization. The sum runs over twice the actual number of data points: in this equation it is assumed that we have appended a set of τ_{run} zeroes to the end of our MD set. This allows us to treat the data as being cyclic in time, that is, $\mathcal{A}(2\tau_{run} + 1) = \mathcal{A}(1)$, without introducing any spurious correlations. Physically, we are only interested in $\tau = 0, \ldots, \tau_{run} - 1$. This is the easiest way of avoiding spurious correlations that would otherwise arise in the FFT method (Futrelle and McGinty, 1971; Kestemont and Van Craen, 1976). It is convenient for this purpose to renumber the time origins starting from 0 instead of 1

$$C'_{\mathcal{A}\mathcal{A}}(\tau) = \sum_{\tau_0=0}^{2\tau_{run}-1} \mathcal{A}(\tau_0)\mathcal{A}(\tau_0 + \tau) \qquad 0 \le \tau < 2\tau_{run}. \tag{8.6b}$$

Equation (8.6b) is exactly equivalent to eqn (8.5) with the normalization omitted, and the upper limit τ_{max} given by $\tau_{run} - \tau$. The equations in Appendix D give

$$\hat{C}'_{\mathcal{A}\mathcal{A}}(v) = \hat{\mathcal{A}}^*(v)\,\hat{\mathcal{A}}(v) = |\hat{\mathcal{A}}(v)|^2, \quad v = 0, 1, \ldots, 2\tau_{run} - 1 \tag{8.7}$$

Code 8.4 Calculating time correlation functions

This file is provided online. corfun.f90 computes the time correlation function of model data using the direct method and the FFT approach. The data are generated as a stochastic time series, with prescribed correlation time.

```
! corfun.f90
! Time correlation function, directly and by FFT
PROGRAM corfun
```

where v is the discrete frequency index, and $C'_{\mathcal{A}\mathcal{A}}(\tau)$ may be recovered from

$$C'_{\mathcal{A}\mathcal{A}}(\tau) = \frac{1}{2\tau_{\text{run}}} \sum_{v=0}^{2\tau_{\text{run}}-1} \left|\hat{\mathcal{A}}(v)\right|^2 \exp(2\pi i v \tau / 2\tau_{\text{run}}). \tag{8.8}$$

The steps involved in calculating the correlation function are:

(a) double the amount of data to be treated by adding τ_{run} zeroes to the end of it, storing the data in COMPLEX variables;

(b) transform the data $\mathcal{A}(\tau) \rightarrow \hat{\mathcal{A}}(v)$ using an FFT routine;

(c) calculate the square modulus $|\hat{\mathcal{A}}(v)|^2 = \hat{C}'_{\mathcal{A}\mathcal{A}}(v)$;

(d) inverse transform the result $\hat{C}'_{\mathcal{A}\mathcal{A}}(v) \rightarrow C'_{\mathcal{A}\mathcal{A}}(\tau)$ using an inverse FFT routine;

(e) apply the normalization $(\tau_{\text{run}} - \tau)^{-1}$ needed to convert $C'_{\mathcal{A}\mathcal{A}}(\tau) \rightarrow C_{\mathcal{A}\mathcal{A}}(\tau)$.

This seems a roundabout route to $C_{\mathcal{A}\mathcal{A}}(\tau)$ but each stage of the process may be carried out very speedily on a computer. For large values of τ_{run}, the full FFT takes a time proportional to $\tau_{\text{run}} \log_2 \tau_{\text{run}}$, while the direct evaluation of the full correlation function takes a time proportional to τ_{run}^2.

It is worth emphasizing that these equations are exact and may be verified using the expressions given in Appendix D. Therefore, correlation functions calculated directly and via the FFT should be identical, subject to the limitations imposed by numerical imprecision. A program comparing the two methods is provided in Code 8.4. It should be noted that the correlation function obtained is real given that the initial data are real; the imaginary part of $C_{\mathcal{A}\mathcal{A}}(\tau)$ is wasted. The way in which two correlation functions can be calculated at once, using both the real and imaginary values, has been discussed by Kestemont and Van Craen (1976).

When should we use the direct calculation and when FFT? The FFT method requires the entire set of data $\mathcal{A}(\tau)$ and an equal number of zeroes be stored in COMPLEX variables, all at once, which may cause a storage problem. Second, it produces the 'complete' correlation function over times up to the entire simulation run time. As mentioned earlier, such long-time information is normally not required and is statistically not significant because of the poor averaging; when comparing speeds it should be remembered that the conventional methods gain by not computing unwanted information, taking a time proportional to τ_{run} (not τ_{run}^2) at large τ_{run}. Third, as pointed out by Smith (1982b), the direct method may gain from vectorization on a pipeline machine when many correlation functions are required at once; the FFT method must simply compute them one at a time. Having said this, in situations where a large amount of data must be processed, if it can all be stored in memory at once, the raw speed of the FFT method should make it the preferred choice.

8.3.3   Windowing functions

Often we wish to transform a time correlation function into the frequency domain to calculate a spectrum that can be compared with experiment. The truncation of $C(t)$ after a finite time, and the presence of random statistical errors, can make the evaluation of the Fourier transform difficult. Spurious features in $\hat{C}_{\text{run}}(\omega)$ which are obtained by transforming a truncated $C_{\text{run}}(t)$, can obscure features present in the complete spectrum, $\hat{C}(\omega)$. In particular, the truncation causes spectral leakage, which often results in rapidly varying side lobes around a peak, and loss of resolution.

Windowing functions are weighting functions applied to the raw $C_{\text{run}}(t)$ to reduce the order of the discontinuity at the truncation point (t_{max}). Press et al. (2007, Chapter 13) discuss a variety of useful windowing functions. The Fourier transform of the windowing function, $\hat{W}(\omega)$, is convoluted with $\hat{C}_{\text{run}}(\omega)$ to produce the windowed spectrum, $\hat{C}_W(\omega)$:

$$\hat{C}_W(\omega) = \int_{-\infty}^{+\infty} \frac{d\omega'}{2\pi} \hat{C}_{\text{run}}(\omega')\hat{W}(\omega - \omega'). \tag{8.9}$$

The coefficients in the windowing function are chosen so that $\hat{W}$ is sharply peaked, which leads to a good resolution in the windowed spectrum. Berens and Wilson (1981) use a four-term Blackman–Harris window in computing the spectrum of liquid CO in a CO/Ar mixture by simulation. They note that multiplying $\hat{C}_W(\omega)$ by the inverse sum of the squares of the windowing function makes it possible to correct the spectral band areas for the scaling effects of the windowing function.

8.4   Estimating errors

Computer simulation is an experimental science in so far as the results may be subject to systematic and statistical errors. Sources of systematic error include size-dependence, the effects of poor random number generators, insufficient equilibration, etc. These should, of course, be estimated and eliminated where possible. It is also essential to obtain estimates of the statistical significance of the results. Simulation averages are taken over runs of finite length, and this is the main cause of statistical imprecision in the calculation of averages.

Let us begin by thinking about an MD simulation in which some property $\mathcal{A}$ is calculated at discrete values of the time, that is, $\mathcal{A}(t_1), \mathcal{A}(t_2)\ldots$. It is often possible to analyse statistical errors in quantities such as $\langle \mathcal{A} \rangle$, $\langle \delta \mathcal{A}^2 \rangle$ by assuming that the distribution of individual estimates of $\mathcal{A}(t)$ around their mean value is Gaussian. In this situation, all the moments of $\mathcal{A}$ are determined by the first two moments, the mean and the variance. Specifically,

$$\langle \delta\mathcal{A}(t_1)\,\delta\mathcal{A}(t_2)\ldots\delta\mathcal{A}(t_n)\rangle = \begin{cases} \displaystyle\sum_{\text{pairs}} \langle\delta\mathcal{A}(t_i)\,\delta\mathcal{A}(t_j)\rangle\langle\delta\mathcal{A}(t_k)\,\delta\mathcal{A}(t_\ell)\rangle\ldots & n \text{ even,} \\[2ex] 0 & n \text{ odd,} \end{cases} \tag{8.10}$$

where the sum extends over all distinct pairings of the times t_i, t_j, etc. at which the function is evaluated and $\delta \mathcal{A}(t_i) = \mathcal{A}(t_i) - \langle \mathcal{A} \rangle$. The same kind of formula applies to a discrete process such as the evolution of states in the MC method, and so much of the same analysis can be used to describe both MC and MD simulations. For Gaussian processes, our estimates of errors in $\langle \mathcal{A} \rangle$, $\langle \delta \mathcal{A}^2 \rangle$, etc. will all be traced back to the variance.

The central limit theorem of probability tells us that as the number of random estimates of a quantity increases, the distribution around the mean will become Gaussian. Thus, a simulation run average may be thought of as being sampled from some limiting Gaussian distribution function about the true mean because it is a sum over many steps. The same applies to an average taken over, say, one-tenth of a run: a so-called block-average. Any property, such as the energy, the virial, etc. is a sum of contributions from different parts of the fluid. This at least is true when the potential is not long-ranged. We expect such a property to obey properties that are approximately Gaussian. Of course, in the case of single-particle velocities and angular velocities taken at equal times, the distribution is exactly Gaussian.

Our problem, then, is to estimate the variance in a long (but finite) simulation run average. We consider this for simple averages, including structural distribution functions, for fluctuations, and for time-dependent correlation functions, in the following sections.

8.4.1   Errors in equilibrium averages

Suppose that we analyse the output file of a simulation that contains a total of τ_{run} timesteps or configurations. The run average of some property $\mathcal{A}$ is

$$\langle \mathcal{A} \rangle_{run} = \frac{1}{\tau_{run}} \sum_{\tau=1}^{\tau_{run}} \mathcal{A}(\tau). \tag{8.11}$$

If we were to assume that each quantity $\mathcal{A}(\tau)$ were statistically independent of the others, then the estimated variance in the mean would simply be given by

$$\sigma^2(\langle \mathcal{A} \rangle_{run}) = \sigma^2(\mathcal{A})/\tau_{run} \tag{8.12}$$

where $\sigma^2(\mathcal{A})$ is the bias-corrected sample variance

$$\sigma^2(\mathcal{A}) = \langle \delta \mathcal{A}^2 \rangle_{run} = \frac{1}{\tau_{run} - 1} \sum_{\tau=1}^{\tau_{run}} \left(\mathcal{A}(\tau) - \langle \mathcal{A} \rangle_{run} \right)^2 \tag{8.13}$$

(see eqn (2.48)). Note that in most statistics texts, the symbol s^2 is used for sample variance, instead of σ^2 here. The estimated error in the mean is given by $\sigma(\langle \mathcal{A} \rangle_{run})$. Of course, the data points are usually not independent: we normally store configurations sufficiently frequently that they are highly correlated with each other. The number of steps for which this correlation persists must be built into eqn (8.12).

To extract this information, the sequence of steps in the file is broken up into blocks of length τ_{blk}. Let there be n_{blk} blocks, so that $n_{blk}\tau_{blk} = \tau_{run}$. The mean value of $\mathcal{A}$ is calculated for each block

$$\langle \mathcal{A} \rangle_b = \frac{1}{\tau_{blk}} \sum_{\tau \in b} \mathcal{A}(\tau) \tag{8.14}$$

where the sum is over the configurations in block b, that is, $(b-1)\tau_{\text{blk}} + 1 \le \tau \le b\tau_{\text{blk}}$. The mean values for all blocks of this kind may be used to estimate their own variance, which we we call $\sigma^2(\langle \mathcal{A} \rangle_{\text{blk}})$

$$\sigma^2(\langle \mathcal{A} \rangle_{\text{blk}}) = \frac{1}{n_{\text{blk}} - 1} \sum_{b=1}^{n_{\text{blk}}} (\langle \mathcal{A} \rangle_b - \langle \mathcal{A} \rangle_{\text{run}})^2. \tag{8.15}$$

As the blocks become larger, the block-averages will become statistically uncorrelated and we expect the variance $\sigma^2(\langle \mathcal{A} \rangle_{\text{blk}})$ to be inversely proportional to τ_{blk} at large τ_{blk}. Our aim is to discover the constant of proportionality, which will allow us to estimate $\sigma^2(\langle \mathcal{A} \rangle_{\text{blk}})$ for the single large block that constitutes the entire run. Following Friedberg and Cameron (1970), we define the statistical inefficiency, s, as

$$s = \lim_{\tau_{\text{blk}} \to \infty} s(\tau_{\text{blk}}), \quad \text{where} \quad s(\tau_{\text{blk}}) = \frac{\tau_{\text{blk}}\, \sigma^2(\langle \mathcal{A} \rangle_{\text{blk}})}{\sigma^2(\mathcal{A})}. \tag{8.16}$$

It is the limiting ratio of the observed variance of an average to the limit expected on the assumption of uncorrelated Gaussian statistics. Having estimated s, we can write

$$\sigma^2(\langle \mathcal{A} \rangle_{\text{run}}) = s \frac{\sigma^2(\mathcal{A})}{\tau_{\text{run}}}. \tag{8.17}$$

Let us follow through an example calculation to make things clear. We have conducted a constant-NVE simulation of the Lennard-Jones fluid at a state point in the fluid region, $\rho^* = 0.78$, $T^* \approx 0.85$, using $N = 108$ particles. We measure the average kinetic temperature $\langle \mathcal{T}_k \rangle$, and its fluctuations $\sigma^2(\mathcal{T}_k) = \langle \mathcal{T}_k^2 \rangle - \langle \mathcal{T}_k \rangle^2 \approx 0.02$. Suppose our run is of length $\tau_{\text{run}} = 1000$ steps. If each step were independent of all the others, the estimated error in $\langle \mathcal{T}_k \rangle$ would be $\sqrt{\sigma^2(\mathcal{T}_k)/\tau_{\text{run}}} = \sqrt{0.02/1000} \approx 0.0045$. However, this is not the case. Figure 8.2 shows a plot of $\tau_{\text{blk}}\, \sigma^2(\langle \mathcal{T}_k \rangle_{\text{blk}})/\sigma^2(\mathcal{T}_k)$ against $1/\tau_{\text{blk}}$ (the reason for this particular choice of plot is explained shortly). An intercept value of $s_k \approx 12$ is obtained: this means that only about one configuration in every 12 steps contributes completely new information to the average. The corrected estimate of the error in $\langle \mathcal{T}_k \rangle$ is therefore $\sqrt{\sigma^2(\mathcal{T}_k) \times s_k/\tau_{\text{run}}} = \sqrt{0.02 \times 12/1000} \approx 0.015$.

Figure 8.2 shows a similar calculation for the configurational temperature $\mathcal{T}_c$, giving the (slightly worse) result $s_c \approx 15$. However, for this example, the fluctuations in $\mathcal{T}_c$ are much smaller than those in $\mathcal{T}_k$: $\sigma^2(\mathcal{T}_c) \approx 0.002$. So, the estimated error in $\langle \mathcal{T}_c \rangle$ from a 1000-step run would be $\sqrt{\sigma^2(\mathcal{T}_c) \times s_c/\tau_{\text{run}}} = \sqrt{0.002 \times 15/1000} \approx 0.005$.

The method of analysis just outlined applies to any stored set of simulation results. It is instructive to consider the particular case of time averages as estimated by MD. For an average

$$\langle \mathcal{A} \rangle_t = \frac{1}{t} \int_0^t \mathcal{A}(t')\,dt' \tag{8.18}$$

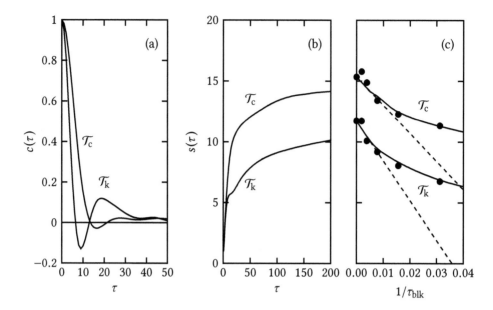

Fig. 8.2 Calculating the statistical inefficiency. Results of simulations of $N = 108$ LJ particles, $r_{cut} = 2.5\sigma$, in the liquid region, $\rho^* = 0.78$, $T^* = 0.85$. τ is measured in timesteps, $\delta t^* = 0.005$. (a) Normalized time correlation functions $c(\tau)$ of kinetic temperature $\mathcal{T}_k$ and configurational temperature $\mathcal{T}_c$. (b) s calculated by integrating $c(\tau)$ according to eqn (8.22). Further integration is needed to reach the true plateau values. (c) s calculated from block-averages using eqn (8.16) (points) plotted against inverse block size, $1/\tau_{blk}$. Solid lines are the same data shown in (b), obtained by integrating the correlation functions, using eqn (8.22). Dashed lines indicate the linear asymptotic behaviour at large τ_{blk}.

the standard result for the variance is related to the correlation function of $\mathcal{A}$ (Papoulis, 1965, Chapter 9)

$$\sigma^2(\langle \mathcal{A} \rangle_t) = \frac{2}{t} \int_0^t (1 - t'/t) \langle \delta \mathcal{A}(t')\delta \mathcal{A} \rangle dt',$$

or $\quad \dfrac{t\,\sigma^2(\langle \mathcal{A} \rangle_t)}{\sigma^2(\mathcal{A})} = 2 \displaystyle\int_0^t (1 - t'/t)\, c_{\mathcal{A}\mathcal{A}}(t')dt',$ $\qquad\qquad$ (8.19)

where $c_{\mathcal{A}\mathcal{A}}(t) = \langle \delta \mathcal{A}(t)\delta \mathcal{A} \rangle / \sigma^2(\mathcal{A})$ is the normalized correlation function. It is useful to consider times t relative to the correlation time

$$t_{\mathcal{A}} = \int_0^\infty dt'\, c_{\mathcal{A}\mathcal{A}}(t'),$$ $\qquad\qquad$ (8.20)

which is the area under the $c_{\mathcal{A}\mathcal{A}}(t')$ curve. Averaging over very short times gives

$$\sigma^2(\langle \mathcal{A} \rangle_t) = \sigma^2(\mathcal{A}), \quad \text{for } t \ll t_{\mathcal{A}}.$$ $\qquad\qquad$ (8.21)

Note that this is independent of t: the variance of short-time averages (e.g. a few timesteps) is essentially the same as that of the instantaneously sampled values. Averaging over

Code 8.5   Calculating statistical inefficiency and errors

This file is provided online. `error_calc.f90` computes the statistical inefficiency and the error in the mean of model data via block-averages and by the method of Flyvbjerg and Petersen (1989). The data are generated as a stochastic time series, with prescribed correlation time. This information can be used in plots similar to those shown in Fig. 8.2 to estimate s and the error in the mean.

```
! error_calc.f90
! Estimated error in correlated data
PROGRAM error_calc
```

times much longer than $t_{\mathcal{A}}$ gives

$$\frac{t\,\sigma^2(\langle\mathcal{A}\rangle_t)}{\sigma^2(\mathcal{A})} = 2\int_0^\infty c_{\mathcal{A}\mathcal{A}}(t')\mathrm{d}t' - \frac{2}{t}\int_0^\infty t'\,c_{\mathcal{A}\mathcal{A}}(t')\mathrm{d}t'$$

$$= 2t_{\mathcal{A}} - \frac{2}{t}\int_0^\infty t'\,c_{\mathcal{A}\mathcal{A}}(t')\mathrm{d}t', \quad \text{for } t \gg t_{\mathcal{A}}. \tag{8.22}$$

The leading term dominates as $t \to \infty$ and we may write

$$\lim_{t\to\infty} \frac{t\,\sigma^2(\langle\mathcal{A}\rangle_t)}{\sigma^2(\mathcal{A})} = 2t_{\mathcal{A}}. \tag{8.23a}$$

$$\text{or} \quad \lim_{t\to\infty} \sigma^2(\langle\mathcal{A}\rangle_t) = 2t_{\mathcal{A}}\frac{\sigma^2(\mathcal{A})}{t}. \tag{8.23b}$$

Comparing this with eqns (8.16), (8.17), we see that the statistical inefficiency is just twice the correlation time $t_{\mathcal{A}}$ divided by the time interval δt between stored configurations. Equation (8.22) also shows that the next highest term is proportional to $1/t$ at long time. This suggests that it is most sensible to plot $t\sigma^2(\langle\mathcal{A}\rangle_t)/\sigma^2(\mathcal{A})$ against $1/t$, or in general, $\tau_{\mathrm{blk}}\sigma^2(\langle\mathcal{A}\rangle_{\mathrm{blk}})/\sigma^2(\mathcal{A})$ against $1/\tau_{\mathrm{blk}}$, when a linear form at low values will be obtained (Jacucci and Rahman, 1984). Such plots appear in Fig. 8.2(c) for $\mathcal{T}_{\mathrm{k}}$ and $\mathcal{T}_{\mathrm{c}}$. A program illustrating the estimation of errors is provided in Code 8.5.

   Of course it would be possible to evaluate $t_{\mathcal{A}}$ or $\tau_{\mathcal{A}}$ by integrating the time correlation function $\langle\delta\mathcal{A}(t)\delta\mathcal{A}\rangle$ in the usual fashion and using it to estimate s (Müller-Krumbhaar and Binder, 1973; Swope et al., 1982). We can see this in Fig. 8.2(b), but note that the integration must be continued until a plateau is reached. Alternatively, if we can guess $t_{\mathcal{A}}$ in some other way, we can estimate the statistical inefficiency without carrying out a full analysis as described earlier. Smith and Wells (1984) have analysed block-averages in their MC simulations, and find an exponential decay (i.e. obeying a geometric law) of the 'correlation function' of consecutive block-averages. In the language of time-series analysis, the process is termed 'first-order autoregressive', that is, Markov (Chatfield, 1984). If such behaviour is assumed, then $\tau_{\mathcal{A}}$ may be estimated from the initial correlations $\langle\delta\mathcal{A}(\tau = 1)\delta\mathcal{A}(\tau = 0)\rangle$. In general, it is best to carry out a full analysis to establish the form of the decay of the correlation with τ; once this has been done, for a given system, it may be safe to extend the results to neighbouring state points, and here the approach of Smith and Wells might save on some effort.

Any technique that reduces s (and hence the correlation time) will help us to calculate more accurate simulation averages. As an example, we consider the calculation of the chemical potential in a molecular liquid, by Fincham et al. (1986). These authors estimated μ with a statistical inefficiency $s \approx 20$ by inserting a test-particle lattice where the orientations of the molecules were fixed throughout the simulation. By randomizing the orientations of the test molecules on the lattice at each insertion, s was reduced to $s \approx 10$. Both methods are valid, but randomizing the orientations of the test molecules allows insertions every tenth step to gain significantly new information. Inserting every tenth step in the case of a fixed lattice orientation is not a significant improvement over inserting every twentieth: twice as long a run is still required to calculate μ to a given accuracy. In a similar way it is s which has been used to compare the efficiency of different MC algorithms (see Section 4.3). As we have seen earlier in the case of $\mathcal{T}_k$ and $\mathcal{T}_c$, the error bars may also be reduced by choosing to average a variable with smaller intrinsic fluctuations.

Flyvbjerg and Petersen (1989) have given a more detailed analysis of error estimation, and propose a simple approach based on successively increasing the block size by a factor of 2. A set of data, $\mathcal{A}(1) \ldots \mathcal{A}(\tau_{\text{run}})$ is transformed to $\mathcal{A}'(1) \ldots \mathcal{A}'(\frac{1}{2}\tau_{\text{run}})$ where

$$\mathcal{A}'(\tau) = \tfrac{1}{2}\big(\mathcal{A}(2\tau - 1) + \mathcal{A}(2\tau)\big), \quad \tau = 1 \ldots \tfrac{1}{2}\tau_{\text{run}}. \tag{8.24}$$

Initially, τ is the step number, but after this 'blocking transformation' it plays the role of b in our earlier discussion, labelling blocks of length $\tau_{\text{blk}} = 2$, of which there are $n_{\text{blk}} = \frac{1}{2}\tau_{\text{run}}$. Then, as the process is repeated again and again, the blocks double each time in length (and halve in number). The mean value $\langle \mathcal{A} \rangle_{\text{run}}$, and its variance, from the underlying statistical distribution, are both invariant under this transformation. Our *estimate* of the variance, from a single data set, however, is *not* invariant. The final result of Flyvbjerg and Petersen (1989), in our notation, is

$$\sigma^2(\langle \mathcal{A} \rangle_{\text{run}}) \geq \frac{1}{n_{\text{blk}}(n_{\text{blk}} - 1)} \sum_{b=1}^{n_{\text{blk}}} (\langle \mathcal{A} \rangle_b - \langle \mathcal{A} \rangle_{\text{run}})^2 = \frac{\sigma^2(\langle \mathcal{A} \rangle_{\text{blk}})}{n_{\text{blk}}}, \tag{8.25}$$

which we can see is equivalent to eqns (8.15), (8.16) and (8.17). Using renormalization group ideas, Flyvbjerg and Petersen (1989) show that the right-hand side is a lower bound on $\sigma^2(\langle \mathcal{A} \rangle_{\text{run}})$, and tends to a 'fixed point' of the transformation; in other words, it is expected to rise towards a plateau value as τ_{blk} increases and n_{blk} falls. They also show that the relative error in the estimate of $\sigma^2(\langle \mathcal{A} \rangle_{\text{run}})$, or equivalently in s, is $\pm\sqrt{2/(n_{\text{blk}} - 1)}$. The blocking transformation provides a convenient way of doing this calculation: at each stage, the data points are combined in pairs and averaged, n_{blk} is halved (if an odd number of blocks occurs, then one of them is discarded) and the right-hand side of eqn (8.25) is recalculated, until $n_{\text{blk}} = 2$. A plot of the estimated $\sigma^2(\langle \mathcal{A} \rangle_{\text{run}})$ as a function of the number of blocking transformations rises to a plateau after, say, six to eight transformations and is constant within an increasing error bar as n_{blk} approaches 2. If this behaviour is not observed, it is a clear indication that significantly longer runs are needed. This approach is essentially equivalent to determining the statistical inefficiency from block-averages as in Fig. 8.2(c), but plotting as a function of $\log_2 \tau_{\text{blk}}$ rather than vs $1/\tau_{\text{blk}}$; the limiting result should be the same. The method is illustrated in Code 8.5.

8.4.2 Errors in fluctuations

Errors in our estimate of fluctuation averages of the type $\langle \delta \mathcal{A}^2 \rangle$ may be estimated simply on the assumption that the process $\mathcal{A}(t)$ obeys Gaussian statistics. The resulting formula is very much like eqn (8.23b)

$$\sigma^2(\langle \delta \mathcal{A}^2 \rangle_{\text{run}}) = 2t'_{\mathcal{A}} \langle \delta \mathcal{A}^2 \rangle^2 / t_{\text{run}} \tag{8.26}$$

where a slightly different correlation time appears

$$t'_{\mathcal{A}} = 2 \int_0^\infty dt \, \langle \delta \mathcal{A}(t) \delta \mathcal{A} \rangle^2 / \langle \delta \mathcal{A}^2 \rangle^2. \tag{8.27}$$

For an exponentially decaying correlation function, $t'_{\mathcal{A}} = t_{\mathcal{A}}$, the usual correlation time; it may be reasonable to assume that this is generally true, in which case the analysis of Section 8.4.1 which yields $t_{\mathcal{A}}$ leads also to an estimate of the errors in the fluctuations through eqn (8.26).

8.4.3 Errors in structural quantities

Errors in a quantity such as $g(r)$ may be estimated by considering the histogram bins that are used in its calculation. Strictly speaking, the sum which is accumulated in a histogram bin (Section 8.2) will not obey Gaussian statistics, but provided the number of counts is large, the central limit theorem of probability applies once more, and the Gaussian approximation becomes quite good. In this case, the techniques described in Section 8.4.1 may be used to estimate the standard error in any histogram bin average. When this quantity is normalized to give a particular value of $g(r)$, the standard error is divided by exactly the same normalizing factor. Carrying out a full block-average analysis for each point in $g(r)$ would be very time-consuming, and not essential. It would be sufficient in most cases to select a few points, near the first and second peaks and in the intervening minimum for example, and estimate the statistics there. A further estimate should be made at large distances: remember that statistics should be much improved as r increases, due to the increasing volume of spherical shells.

8.4.4 Errors in time correlation functions

The time correlation functions calculated in MD simulations are subject to the same kind of random errors as described for static quantities and fluctuations in the previous sections. We denote the run average by

$$C_{\mathcal{A}\mathcal{A}}^{\text{run}}(t) = \langle \mathcal{A}(t)\mathcal{A}(0) \rangle_{\text{run}} = \frac{1}{t_{\text{run}}} \int_0^{t_{\text{run}}} dt' \, \mathcal{A}(t')\mathcal{A}(t' + t) \tag{8.28}$$

where we have assumed for simplicity that $\langle \mathcal{A} \rangle$ vanishes. The error that we wish to estimate is that in

$$
\begin{aligned}
\delta C(t) &= C^{\text{run}}_{\mathcal{A}\mathcal{A}}(t) - C_{\mathcal{A}\mathcal{A}}(t) \\
&= \langle \mathcal{A}(t)\mathcal{A}(0) \rangle_{\text{run}} - \langle \mathcal{A}(t)\mathcal{A}(0) \rangle \\
&= \frac{1}{t_{\text{run}}} \int_0^{t_{\text{run}}} dt' \Big(\mathcal{A}(t')\mathcal{A}(t+t') - \langle \mathcal{A}(t')\mathcal{A}(t'+t) \rangle \Big)
\end{aligned}
\tag{8.29}
$$

where $\langle \ldots \rangle$ denotes the true, infinite time or ensemble average. The mean value $\langle \delta C(t) \rangle$ should vanish of course, but the variance of the mean is given by (Zwanzig, 1969; Frenkel, 1980)

$$
\sigma^2 \big(\langle \mathcal{A}(t)\mathcal{A} \rangle_{\text{run}} \big) = \frac{1}{t_{\text{run}}^2} \int_0^{t_{\text{run}}} \int_0^{t_{\text{run}}} dt' \, dt''
$$

$$
\times \Big(\big\langle \mathcal{A}(t')\mathcal{A}(t'+t)\mathcal{A}(t'')\mathcal{A}(t''+t) \big\rangle - \langle \mathcal{A}(t)\mathcal{A}(0) \rangle^2 \Big).
\tag{8.30}
$$

The four-variable correlation function in this equation may be simplified if we make the assumption that $\mathcal{A}(t)$ obeys Gaussian statistics, using eqn (8.10). After some straightforward manipulations described in detail by Frenkel (1980) the variance reduces to

$$
\sigma^2 \big(\langle \mathcal{A}(t)\mathcal{A} \rangle_{\text{run}} \big) \approx 2t'_{\mathcal{A}} C_{\mathcal{A}\mathcal{A}}(0)^2 / t_{\text{run}}
\tag{8.31}
$$

where $t'_{\mathcal{A}}$ is the correlation time defined by eqn (8.27). The standard error in the normalized correlation function is thus independent of time and is given by

$$
\sigma \big(\langle \mathcal{A}(t)\mathcal{A} \rangle_{\text{run}} \big) / \langle \mathcal{A}^2 \rangle \approx (2t'_{\mathcal{A}} / t_{\text{run}})^{1/2}
\tag{8.32}
$$

which has the usual appearance. As an example, for $t'_{\mathcal{A}}$ of the order of ten timesteps, it would be necessary to conduct a run of 10^5 steps in order to obtain a relative precision of $\sim 1\%$ in $C_{\mathcal{A}\mathcal{A}}(t)$. If we use the simulation average $C^{\text{run}}_{\mathcal{A}\mathcal{A}} = \langle \mathcal{A}^2 \rangle_{\text{run}}$ instead of the exact ensemble average in eqn (8.32), then the error at short times is reduced due to cancellation in the random fluctuations (Zwanzig and Ailawadi, 1969)

$$
\sigma \big(\langle \mathcal{A}(t)\mathcal{A} \rangle_{\text{run}} \big) / \langle \mathcal{A}^2 \rangle_{\text{run}} \approx (2t'_{\mathcal{A}} / t_{\text{run}})^{1/2} \big(1 - c_{\mathcal{A}\mathcal{A}}(t) \big)
\tag{8.33}
$$

where $c_{\mathcal{A}\mathcal{A}}(t) = \langle \mathcal{A}(t)\mathcal{A}(0) \rangle / \langle \mathcal{A}^2 \rangle$. Thus the error is zero at $t = 0$, but it tends to $(2t'_{\mathcal{A}} / t_{\text{run}})^{1/2}$ at long times.

This looks rather depressing, but the gloom is lightened when we turn to the calculation of single-particle correlation functions, such as the velocity autocorrelation function. The final result is then an average over N separate functions for each axis direction

$$
C_{vv}(t) = \frac{1}{N} \sum_{i=1}^{N} \langle v_{i\alpha}(t) v_{i\alpha}(0) \rangle
\tag{8.34}
$$

(and in this case a further average over equivalent axes could be carried out). The analysis of this situation follows the earlier pattern, and the estimated error is eventually found to be $\approx (2t_{\mathcal{A}}/Nt_{\mathrm{run}})^{1/2}$ at long times. The extra factor of $N^{1/2}$ in the denominator suggests that a 1 % accuracy in the velocity autocorrelation function might be achieved with 10^4 timesteps for a 100-particle system. This argument is simplistic, since the velocities of neighbouring particles at different times are not statistically independent, but single-particle correlation functions are still generally found to be less noisy than their collective counterparts. The precision with which a particular time correlation function may be estimated depends upon the range of the spatial correlations in the fluid; the size of statistically independent regions may depend upon the range of the potential and on the state point. Some of these ideas are discussed by Frenkel (1980).

In principle, a block analysis of time correlation functions could be carried out in much the same way as that applied to static averages. However, the block lengths would have to be substantial to make a reasonably accurate estimate of the errors.

We have not included in this analysis the point raised in Section 8.3, namely that the number of time origins available for the averaging of long-time correlations may be significantly less than the number of origins for short-time correlations. This limitation is imposed by the finite run length, and it means that t_{run} in the previous discussion should be replaced by $t_{\mathrm{run}} - t$ for correlations $\langle \mathcal{A}(t)\mathcal{A} \rangle$. Thus, an additional time-dependence, leading to slightly poorer statistics for longer times, enters into the formulae.

One possible source of systematic error in time correlation functions should be mentioned. The usual periodic boundary conditions mean that any disturbance, such as a sound wave, may propagate through the box, leaving through one side and re-entering through the other, so as to arrive back at its starting point. This would happen in a time of order L/v_s where L is the box length and v_s the speed of sound. With typical values of $L = 2\,\mathrm{nm}$ and $v_s = 1000\,\mathrm{m\,s^{-1}}$, this 'recurrence time' is about 2 ps, which is certainly well within the range of correlation times of interest. It is sensible, and has become the recommended practice, to inspect correlation functions for anomalous behaviour, possibly increased noise levels, at times greater than this. It is doubtful that a periodic system would correctly reproduce the correlations arising in a macroscopic liquid sample at such long times. The phenomenon was originally reported by Alder and Wainwright (1970) and also by Schoen et al. (1984). The latter workers found it hard to reproduce their results for the Lennard-Jones liquid. We would expect to see much more significant effects in solids, where sound waves are well developed, whereas phonons are more strongly damped in liquids. Nonetheless, it is obviously a good idea to keep the possibility of correlation recurrence effects in mind, particularly if 'long-time tail' behaviour is under study.

8.5   Correcting the results

When the results of a simulation have been calculated, and the errors estimated, they may still not be in the form most suitable for interpretation. The run averages may not correspond to exactly the desired state point, the structural or time-dependent properties may require extrapolation or smoothing, and it may be necessary to do some time integration or Fourier transformation to obtain the desired quantities. In this section, we will discuss all these points.

8.5.1   Correcting thermodynamic averages

In constant-NVE molecular dynamics, the kinetic temperature fluctuates around its mean value. Without using a thermostat, it is difficult to preset a desired value of T in a simulation and this is inconvenient for the comparison of results with other simulations, real experiments and theory. The determination of isotherms is useful, for example, in the calculation of a coexistence curve. Powles et al. (1982) have suggested a useful method for the correction of thermodynamic results to the desired temperature. For a particular property $\mathcal{A}$, obtained in a simulation at a mean temperature $T_{\text{run}} = \langle \mathcal{T} \rangle_{\text{run}}$, the results can be corrected to the desired temperature T using

$$\mathcal{A}(T) = \mathcal{A}(T_{\text{run}}) + (T - T_{\text{run}}) \left(\frac{\partial \mathcal{A}}{\partial T} \right)_{\rho} + \dots . \tag{8.35}$$

If the temperature difference is small, the Taylor series can be truncated at the first term. For the energy, E, the appropriate thermodynamic derivative is of course C_V. In the case of the chemical potential and the pressure, convenient expressions for the derivatives are

$$\left(\frac{\partial P}{\partial T} \right)_{\rho} = \left(P - \rho^2 \left(\frac{\partial (E/N)}{\partial \rho} \right)_T \right) \bigg/ T \tag{8.36}$$

$$\left(\frac{\partial \mu}{\partial T} \right)_{\rho} = - \left(\rho \left(\frac{\partial (E/N)}{\partial \rho} \right)_T + (E/N) - \mu \right) \bigg/ T \tag{8.37}$$

where E/N is the total energy per molecule, which is known exactly in the simulation. A series of simulation runs is carried out by varying the density, while the mean temperature of each run is kept as close to the desired value T as possible. This is achieved by using one of the thermostats described in Chapter 3 during the equilibration phase. E/N is almost a linear function of ρ, and the derivative $\partial(E/N)/\partial\rho$ is easily calculated from this series of runs. Strictly speaking, we require the derivative at fixed T (the desired temperature). In practice, the errors in the derivative arising from small temperature differences are small and can be ignored. Thus, by using eqn (8.35), values of E, P, and μ along an isotherm may be calculated, from a set of constant-NVE simulations. The technique is easily extended to other thermodynamic quantities.

8.5.2   Extrapolating $g(r)$ to contact

For a fluid with smooth repulsive interactions (such as the Lennard-Jones fluid), $g(r)$ has a maximum which corresponds to the minimum in the potential. At lower values of r, $g(r)$ falls rapidly to zero. For a hard-core fluid (such as a fluid of hard spheres or hard dumbbells), $g(r)$, or more generally $g_{ab}(r_{ab})$, is discontinuous at $r = \sigma_{ab}$, and is zero inside the core. The value of g_{ab} at contact, $g_{ab}(\sigma_{ab}^+)$, is directly related to the pressure and the other thermodynamic properties of the hard-core fluid.

For a site–site hard-core fluid the Boltzmann factor associated with the potential between two molecules i and j can be written in terms of the unit step function, $\Theta(x)$,

$$\exp\left(-\beta v(\mathbf{r}_{ij}, \Omega_i, \Omega_j)\right) = \prod_{a,b} \exp\left(-\beta v_{ab}(r_{ab})\right) = \prod_{a,b} \Theta(r_{ab} - \sigma_{ab}). \tag{8.38}$$

The product is over independent site–site distances r_{ab} between the pair of molecules. Differentiating eqn (8.38) gives the virial for the fluid,

$$w(\mathbf{r}_{ij}, \Omega_i, \Omega_j) = -\beta^{-1} r_{ij} \sum_a \sum_b \exp\left(\beta v_{ab}(r_{ab})\right) \delta(r_{ab} - \sigma_{ab}) \left(\frac{\partial r_{ab}}{\partial r_{ij}}\right)_{\Omega_i, \Omega_j}. \tag{8.39}$$

This virial can be used in eqns (2.60), (2.67) to obtain the pressure (Nezbeda, 1977; Aviram et al., 1977)

$$\frac{P}{\rho k_B T} = 1 + \frac{2\pi\rho}{3} \sum_a \sum_b \tau_{ab}(\sigma_{ab}^+) \, \sigma_{ab}^2 \, g_{ab}(\sigma_{ab}^+) \tag{8.40}$$

where

$$\tau_{ab}(r_{ab}) = \left\langle (\mathbf{r}_{ab} \cdot \mathbf{r}_{ij})/r_{ab} \right\rangle_{\text{shell}} \tag{8.41}$$

and the average is for a shell centred at r_{ab}. For a hard-sphere fluid $\tau_{ab}(r_{ab}) = r_{ab}$ and there is only one term in the sum in eqn (8.40),

$$\frac{P}{\rho k_B T} = 1 + \tfrac{2}{3}\pi\rho\sigma^3 g(\sigma^+). \tag{8.42}$$

The product $\tau_{ab}(r_{ab}) g_{ab}(r_{ab})$ for $r_{ab} = \sigma_{ab}^+$ cannot be calculated directly in a standard constant-NVT MC simulation, and has to be extrapolated from values close to contact. This extrapolation requires some care since $g_{ab}(r_{ab})$ can rise or fall rapidly close to contact. In particular there is normally a half-shell of thickness $\delta r/2$ centred at $\sigma + \delta r/4$. An accurate estimate of the function in this thin shell is required to obtain an accurate extrapolation. As a check, the extrapolated contact value should be independent of the shell thickness and it is useful to try some values in the range $\delta r = 0.025\sigma$ to $\delta r = 0.01\sigma$.

A trick, which is sometimes useful in calculating the contact value, is to extrapolate $(r_{ab}/\sigma_{ab})^\nu f(r_{ab})$, where ν is an integer, and $f(r)$ is the function of interest, to $r_{ab} = \sigma_{ab}$. This extrapolation produces $f(\sigma_{ab}^+)$ regardless of the value of ν. If the function is steeply varying, an appropriate choice of ν can facilitate this extrapolation. Freasier (1980) has reported a suggestion due to D. J. Evans, that such an extrapolation procedure be employed during the simulation run itself. The pressure itself, for hard molecular systems, is more straightforwardly estimated by a box-scaling procedure in constant-volume simulations (see Section 5.5), or from collisional impulses in event-driven MD; alternatively, of course, it is easily specified in constant-NPT MC simulations. In Example 8.1, the consistency between pressures calculated for a hard-particle system by both MC and MD, was an important issue.

8.5.3   Smoothing and extending $g(r)$

The radial distribution function and any of the angular correlation functions, such as the spherical harmonic coefficients, are subject to statistical noise. For the purposes of comparing with theoretical approximations or in order to calculate accurate Fourier transforms (see Appendix D), it is sometimes useful to smooth these data. Smoothing can be achieved by fitting a least-squares polynomial in r. However, it is difficult to find appropriate functional forms to fit a variety of correlation functions, over a wide range of temperature and density. A useful compromise is to use a smoothing formula to replace each tabulated value by a least-squares polynomial which fits a sub-range of points. For example in a five-point smoothing method, the smoothed value at point n will depend linearly on the raw data values at points $n - 2, \ldots n + 2$, with prescribed coefficients; the prescription is suitably adjusted for points lying near either end of the range. Several sophisticated smoothing schemes, such as the Savitzky–Golay filter that conserve higher-order moments of the function as it is smoothed, can be applied with good effect to simulation output (Press et al., 2007, Chapter 14).

In principle, the long-range behaviour of $g(r)$ may be deduced from its behaviour at short distances. This idea is embodied in the Ornstein–Zernike equation (Hansen and McDonald, 2013)

$$h(r) = c(r) + \rho \int \mathrm{d}\mathbf{r}' h(|\mathbf{r} - \mathbf{r}'|) c(|\mathbf{r}'|).$$
(8.43)

Eqn (8.43) just defines the direct correlation function $c(r)$ in terms of the total correlation function $h(r) = g(r) - 1$. While $h(r)$ is long-range in normal liquids, $c(r)$ has approximately the same range as the potential. The Weiner–Hopf factorization (Baxter, 1970) of the Ornstein–Zernike equation with some model for $c(r)$ such as the Percus–Yevick approximation (Hansen and McDonald, 2013) can be used to produce $c(r)$ over its complete range from a knowledge of $h(r)$ on the range $0 \leq r \leq r_c$. A knowledge of $c(r)$ over its complete range can then be used to calculate $h(r)$ over its complete range, thus extending the simulations results beyond half the length of the box, and allowing for an accurate calculation of the structure factor

$$S(k) = 1 + \rho \hat{h}(k) = \left(1 - \rho \hat{c}(k)\right)^{-1}$$
(8.44)

(see eqn (2.110)). The details of this iterative approach are provided by Jolly et al. (1976). Dixon and Hutchinson (1977) describe an alternative use of Baxter's factorization to extend $g(r)$, which avoids any explicit use of a model such as the Percus–Yevick approximation. The reader is referred to the original paper for the computational details, but the general scheme is very similar to that used by Jolly et al.

The dramatic increases in the available memory and the speed of computers, since these schemes were proposed, means that in most cases it is simply possible to increase the number of atoms N at a fixed density so that the size of the box increases and the correlation functions can be calculated directly out to much larger values of $r \leq L/2$. This is usually the preferred approach, since if long-range correlation functions are of interest, it is prudent to check directly for finite system size effects.

8.5.4   Calculating transport coefficients

The numerical integration of time correlation functions to obtain transport coefficients and correlation times is formally a straightforward exercise, given data at regularly spaced times. Simpson's rule, for example, would be quite satisfactory. However, there are a number of pitfalls to be avoided. First, there are several correlation functions that are believed to decay to zero only slowly, having a limiting algebraic dependence $t^{-\nu}$ with exponent $\nu = 3/2$ for example (Alder and Wainwright, 1969; 1970; Alder, 1986). Such a tail may extend significantly beyond the range of times for which $C(t)$ has been computed, and, as has been mentioned, statistical errors will become more severe as t increases. The integral under such a tail may nonetheless make a significant contribution to the total integral, and so the tail cannot be completely ignored. In estimating the tail it becomes necessary to attempt some kind of fit to the long-time behaviour of the correlation function, and then to use this to extrapolate to $t \to \infty$ and estimate a long-time tail correction. The importance of this correction is illustrated by the estimation, by MD, of the bulk and shear viscosities of the Lennard-Jones fluid near the triple point (Levesque et al., 1973; Hoover et al., 1980b). In all cases, the long-time behaviour of a correlation function should be examined closely before an attempt is made to calculate the time integral.

As discussed in Chapter 2, the Einstein relation provides an alternative route to the transport coefficients, which is formally equivalent to the integration of a time correlation function. This relies on the identities, for stationary processes,

$$\gamma(t) = \int_0^t dt' \langle \dot{\mathcal{A}}(t_0 + t') \dot{\mathcal{A}}(t_0) \rangle, \tag{8.45a}$$

$$= \langle \dot{\mathcal{A}}(t_0) \big(\mathcal{A}(t_0 + t) - \mathcal{A}(t_0) \big) \rangle, \tag{8.45b}$$

$$= \frac{1}{2} \frac{d}{dt} \langle \big(\mathcal{A}(t_0 + t) - \mathcal{A}(t_0) \big)^2 \rangle, \tag{8.45c}$$

$$\approx \frac{1}{2t} \langle \big(\mathcal{A}(t_0 + t) - \mathcal{A}(t_0) \big)^2 \rangle. \tag{8.45d}$$

The approximate equality becomes exact at long times. Here the angle brackets include an average over time origins t_0. We are interested in $\lim_{t \to \infty} \gamma(t)$. Thus, the diffusion coefficient may be estimated by:

(a) time-integrating the velocity autocorrelation function $\langle \mathbf{v}_i(0) \cdot \mathbf{v}_i(t) \rangle$;

(b) cross-correlating the initial velocity with the displacement $\langle \mathbf{v}_i(0) \cdot \Delta \mathbf{r}_i(t) \rangle$;

(c) numerically time-differentiating the mean-squared displacement $\langle |\Delta \mathbf{r}_i(t)|^2 \rangle$;

(d) dividing the mean-squared displacement $\langle |\Delta \mathbf{r}_i(t)|^2 \rangle$ by the time.

Alder et al. (1970) have pointed out that the transport coefficients may be more readily calculated from the mean-squared displacement by computing the gradient, eqn (8.45c), rather than by dividing by the time, eqn (8.45d). For variables with an exponentially decaying correlation function $\langle \dot{\mathcal{A}}(t) \dot{\mathcal{A}} \rangle = \langle \dot{\mathcal{A}}^2 \rangle \exp(-t/t_{\mathcal{A}})$ we have

$$\frac{1}{2} \frac{d}{dt} \langle \big(\mathcal{A}(t) - \mathcal{A}(0) \big)^2 \rangle = \int_0^t dt' \langle \dot{\mathcal{A}}(t') \dot{\mathcal{A}} \rangle = \langle \dot{\mathcal{A}}^2 \rangle t_{\mathcal{A}} \big(1 - \exp(-t/t_{\mathcal{A}}) \big) \tag{8.46}$$

but

$$\frac{1}{2t} \langle \big(\mathcal{A}(t) - \mathcal{A}(0) \big)^2 \rangle = \langle \dot{\mathcal{A}}^2 \rangle t_{\mathcal{A}} \big(1 - t_{\mathcal{A}}/t + (t_{\mathcal{A}}/t) \exp(-t/t_{\mathcal{A}}) \big). \tag{8.47}$$

In the first case, the correct result $\langle \dot{\mathcal{A}}^2 \rangle t_{\mathcal{A}}$ is approached exponentially quickly as t increases, but the second equation has a slower inverse-t dependence. In considering these equations, it is also worth noting that the limiting gradient may be calculated equally well from eqns (8.45b) and (8.45c), provided the data points are stored sufficiently close together to avoid numerical errors in the time differentiation, and the correlation function at any time t may be recovered from $\langle (\mathcal{A}(t) - \mathcal{A}(0))^2 \rangle$ by numerical differentiation, with the same caveat.

This leads us to ask when the route via the Einstein relation might be preferred to the calculation of a correlation function. The latter method is by far the most common, possibly because of the interest in the correlation functions themselves. However, there is much to be said for the Einstein relation route. In integrating the equations of motion we use (at least) the known first and second derivatives of molecular positions and orientations: this order of numerical accuracy is 'built in' to computed mean-square displacements and the like. When we numerically integrate a time correlation function using, say, Simpson's rule, especially if we have only stored the data every five or ten timesteps, we are introducing additional sources of inaccuracy. In addition, there is the tendency to stop calculating and integrating time correlation functions when the signal seems to have disappeared into the noise. This is dangerous because of the possibility of missing contributions from the small, but systematic, long-time correlations.

This is illustrated in Fig. 8.3. The diffusion coefficients of a nematic liquid crystal may be estimated by integrating the velocity autocorrelation function: there are two distinct components, respectively parallel and perpendicular, to the director. These are shown in Fig. 8.3(a). They are clearly different, as expected for this phase, but both seem to have decayed to zero at $t \approx 1$ (in reduced units) which corresponds to ≈ 500 timesteps in this case. It might be tempting to integrate the functions up to this time, or a little longer, to obtain estimates of $D_\parallel$ and $D_\perp$. However, Fig. 8.3(b) shows the time-integrated function, plotted on a logarithmic timescale out to $t = 100$, corresponding to 50 000 timesteps. This can be calculated directly, as $\langle v_\alpha(0) \, \Delta r_\alpha(t) \rangle$ for each molecule, and each Cartesian component α, or by numerically time-differentiating the mean-squared displacement $\langle \Delta r_\alpha(t)^2 \rangle$ (and dividing by two). The parallel component, especially, exhibits a residual long-time contribution, which might easily be missed if attention is focused on the velocity correlation function alone: the resulting $D_\parallel$ would be systematically underestimated by about 5 % in this case.

Alder et al. (1970) have described in detail one situation in which the Einstein expression is more convenient even when the correlation function itself is of direct interest, namely the molecular dynamics of hard systems. For some dynamic quantities there will be a 'potential' contribution involving intermolecular forces, which for hard systems act instantaneously only at collisions. Thus such contributions will be entirely absent from data stored in a file at regular intervals for correlation function analysis. The problem is exactly analogous to that of estimating the pressure in a hard system (see Section 5.5), and occurs when we wish to calculate shear or bulk viscosities from stress (pressure) tensor correlations, and thermal conductivities from local energy fluctuations. The collisional contributions to these dynamical quantities must be taken into account during the simulation run itself. Moreover, because the forces act impulsively, the appropriate dynamical quantities $\dot{\mathcal{A}}(t)$ will contain delta functions, which would make the usual correlation function analysis rather awkward.

The Einstein relation variables $\mathcal{A}(t)$ are easier to handle: they merely change discontinuously at collisions. Following Alder et al. (1970) we take as our example the calculation of the shear viscosity η via off-diagonal elements of the pressure tensor. The dynamical variable is (assuming equal masses)

$$Q_{xy} = \frac{1}{V} \sum_i m x_i \dot{y}_i. \tag{8.48}$$

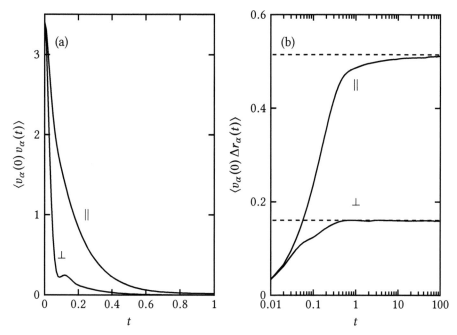

Fig. 8.3 Calculating the diffusion coefficient. Results are shown for a simulation of a nematic liquid crystal using the Gay–Berne potential (eqns (1.30)–(1.32) and Appendix C), with parameters $\mu = 1$, $\nu = 3$, $\kappa = 3$, $\kappa' = 5$ (Berardi et al., 1993), cutoff $r_c^* = 5.0$, at reduced temperature $T^* = 3.4$, density $\rho^* = 0.3$, and system size $N = 512\,000$. The timestep was $\delta t^* = 0.002$, particle mass was taken as unity, and the moment of inertia corresponded to a uniform mass distribution. Run lengths of order 10^6 steps were used to estimate the longer-time correlations. (a) Velocity autocorrelation function for components α parallel (‖) and perpendicular (⊥) to the nematic director. (b) Time-integrated velocity autocorrelation function for the same components, with a logarithmic time scale. Estimates of the plateau values are indicated by dashed lines.

For systems undergoing free flight between collisions (say at times t_1 and t_2), the change in Q_{xy} is just

$$Q_{xy}(t_2) - Q_{xy}(t_1) = \frac{1}{V}\left[\sum_i m\dot{x}_i \dot{y}_i\right](t_2 - t_1). \qquad (8.49)$$

After a collision the term in square brackets changes, but this change is easy to compute, involving just the velocities of the colliding molecules. At a collision there is also a change in Q_{xy}:

$$Q_{xy}(t^+) - Q_{xy}(t^-) = \frac{1}{V}mx_{ij}\delta\dot{y}_i \qquad (8.50)$$

where i and j are the colliding pair, $x_{ij} = x_i - x_j$, and $\delta\dot{y}_i$ is the collisional change in the velocity of i $(= -\delta\dot{y}_j)$. Thus the total change in Q_{xy} over any period of time is obtained by summing terms of the type shown in eqn (8.49), for all inter-collisional intervals in that period (including the times before the first and after the last collision) and adding in all the terms of the type shown in eqn (8.50) for the collisions occurring in that interval.

These values of $Q_{xy}(t) - Q_{xy}(0)$ may then be used in the Einstein expressions. Finally, the correlation function is recovered by numerical differentiation.

It should be noted that purely 'kinetic' correlation functions, such as the velocity autocorrelation function and correlation functions involving molecular positions and orientations, not potential 'terms', can be calculated in the normal way even for hard systems, and this is the preferred method where possible.

9

Advanced Monte Carlo methods

9.1 Introduction

The MC methods described in Chapter 4 may not be the most efficient ways of estimating certain statistical averages. The Metropolis prescription, eqn (4.21), for example, generates simulation trajectories that are naturally weighted to favour thermally populated states of the system, that is, with Boltzmann-like weights. There are a number of important properties, such as free energies, that are difficult to calculate using this approach (see Sections 2.4, 5.5). Consider the configurational partition function Q_{NVT}^{ex} defined in eqn (2.25), in terms of which the excess free energy is $A^{\mathrm{ex}} = -k_{\mathrm{B}}T \ln Q_{NVT}^{\mathrm{ex}}$. This may be written as a configurational average

$$Q_{NVT}^{\mathrm{ex}} = 1/\langle \exp(\beta \mathcal{V}) \rangle. \tag{9.1}$$

In principle, the denominator in eqn (9.1) can be calculated in a conventional simulation. Unfortunately, Metropolis Monte Carlo is designed to sample regions in which the potential energy is negative, or small and positive (not enormous compared with $k_{\mathrm{B}}T$). These regions make little contribution to the average in eqn (9.1), and this route to A is impractical. An indirect route to A, thermodynamic integration, based on conventional sampling of the Boltzmann distribution, is briefly summarized in Section 9.2.1. Direct calculation of the free energy really requires more substantial sampling over higher-energy configurations. In such non-Boltzmann sampling, ρ_n/ρ_m is no longer simply $\exp(-\beta \delta \mathcal{V}_{nm})$ but the two states n and m are additionally weighted by a suitable function. This weighting function is designed to encourage the system to explore regions of phase space not frequently sampled by the Metropolis method. The weighted averages may be estimated more accurately than in conventional Monte Carlo, and are then corrected, giving the desired ensemble averages, at the end of the simulation. We describe this technique in Section 9.2. Also in this section we describe methods to refine the weighting function iteratively so as to give a flat distribution with respect to the energy or some order parameter of interest. This provides, in principle, a route to the density of states in a fluid, and the entropy or the free energy.

A second extension involves changing the underlying stochastic matrix $\boldsymbol{\alpha}$ to make Monte Carlo 'smarter' at choosing its trial moves. In the conventional method, $\boldsymbol{\alpha}$ is symmetric and trial moves are selected randomly according to eqn (4.27). However, it is possible to choose an α that is unsymmetric and that still satisfies the condition of microscopic reversibility. This may be used in the Monte Carlo method to sample preferentially in the vicinity of a solute molecule or a cavity in the fluid, to move particles preferentially in the direction of the forces acting on them, and to move polymer chains by taking account of their local configurational energy landscape. These techniques are described in Section 9.3.

These methods introduce some of the character of MD into MC simulations. MD is totally deterministic, and intrinsically many-body in nature. By contrast, Metropolis MC is entirely stochastic and usually entails single-particle moves. Unfavourable energy configurations are avoided simply by rejecting them and standing still. In the smarter MC methods, biases are introduced to guide the system in its search for favourable configurations, which may be advantageous, particularly if collective motions are important in avoiding 'barriers' in phase space.

In Section 9.4 we describe two MC methods which have been developed to make a direct attack on the prediction of phase diagrams. The Gibbs ensemble MC method uses two coexisting simulation boxes to predict liquid–vapour or liquid–liquid coexistence lines without a direct calculation of statistical properties such as the free energy or chemical potential. The Gibbs–Duhem technique can be used to trace a coexistence line starting from a known point on the line, by estimating its slope directly in the simulation. It can be readily applied to solid–liquid equilibria. Finally in Section 9.5, we show how a reactive canonical Monte Carlo method can be developed to study chemical reactions at equilibrium by introducing the reactions as balanced forward and backward moves in the simulation.

9.2   Estimation of the free energy

9.2.1   Introduction

In this section we discuss the ways in which conventional simulations can be extended to facilitate the calculation of free energies. Grand canonical MC is a direct method for calculating the free energy, as discussed in Section 4.6, but this can become inefficient in dense fluids. The methods of thermodynamic integration and direct particle insertion have been introduced in Section 2.4, and some technical details of the particle insertion method are given in Section 5.5. Thermodynamic integration is, in fact, more generally applicable than eqns (2.72) and (2.73) might suggest. Let us begin with a less taxing problem than that of estimating A, specifically the calculation of a free-energy difference. Consider two fluids characterized by potentials $\mathcal{V}(\mathbf{r})$ and $\mathcal{V}_0(\mathbf{r})$, for which the free energy of the reference fluid, A_0, is known. If we introduce a parameter λ into the potential $\mathcal{V}(\mathbf{r}; \lambda)$ such that $\mathcal{V}_0(\mathbf{r}) \equiv \mathcal{V}(\mathbf{r}; 0)$ and $\mathcal{V}(\mathbf{r}) \equiv \mathcal{V}(\mathbf{r}; 1)$, and define a corresponding (excess) partition function $Q_{NVT}^{\text{ex}}(\lambda)$, then it easy to show

$$\left(\frac{\partial A(\lambda)}{\partial \lambda} \right)_{NVT} = -k_{\text{B}} T \frac{\partial Q_{NVT}^{\text{ex}}(\lambda)/\partial \lambda}{Q_{NVT}^{\text{ex}}(\lambda)} = \left\langle \frac{\partial \mathcal{V}(\mathbf{r}; \lambda)}{\partial \lambda} \right\rangle_{NVT;\lambda}$$

where the notation indicates that the canonical ensemble average is calculated using the potential-energy function $\mathcal{V}(\mathbf{r}; \lambda)$. The free energy of the fluid of interest, A, can be determined by integration

$$A = A_0 + \int_0^1 d\lambda \left\langle \frac{\partial \mathcal{V}(\mathbf{r}; \lambda)}{\partial \lambda} \right\rangle_{NVT;\lambda}. \tag{9.2}$$

If the parameterization is a simple linear formula, $\mathcal{V}(\mathbf{r}; \lambda) = \lambda \mathcal{V}(\mathbf{r}) + (1-\lambda)\mathcal{V}_0(\mathbf{r})$, then the quantity to be averaged is just $\Delta \mathcal{V}(\mathbf{r}) = \mathcal{V}(\mathbf{r}) - \mathcal{V}_0(\mathbf{r})$. An accurate numerical estimate of this integral (by Simpson's rule or some other quadrature scheme) requires many simulations at different values of λ, to evaluate $\langle \partial \mathcal{V}(\mathbf{r}; \lambda)/\partial \lambda \rangle_{NVT;\lambda}$. It is also important to check that the integrand does not suffer any discontinuities; that is, just as in the use of eqns (2.72), (2.73), that the path of integration does not cross any phase boundaries. With these precautions, the thermodynamic integration route can be quite reliable, but also time-consuming. It is, however, well suited to problems in which we are interested in the free-energy change associated with 'mutating' one molecular species into another.

Now we take the opportunity to go further into the free-energy problem. Considerable effort has been expended on developing novel MC methods which allow determination of the 'statistical' properties (e.g. A and S) of fluids. We present a summary of the available methods and comment on their usefulness in Section 9.2.7. First, though, we take a closer look at the nature of the problem.

9.2.2   Energy distributions

Once again, rather than tackle the taxing problem of estimating an absolute free energy, we consider the calculation of a free-energy difference. Given two fluids characterized by potentials $\mathcal{V}(\mathbf{r})$ and $\mathcal{V}_0(\mathbf{r})$, the free energy of the fluid of interest, A, relative to that of the reference fluid, A_0, can be determined from

$$\Delta A = A - A_0 = -k_B T \ln(Q/Q_0) = -k_B T \ln\left\langle \exp(-\beta \Delta \mathcal{V}) \right\rangle_0 \tag{9.3}$$

where $\Delta \mathcal{V}(\mathbf{r}) = \mathcal{V}(\mathbf{r}) - \mathcal{V}_0(\mathbf{r})$ and the ensemble average $\langle \cdots \rangle_0$ is taken in the reference system $\mathcal{V}_0$. Unless the two fluids are very similar, and $\beta \Delta \mathcal{V}$ is small for all the important configurations in this ensemble, the average in eqn (9.3) is difficult to calculate accurately. The reason for this becomes clear if we rewrite the configurational density function $\rho_0(\mathbf{r})$ as a function, $\rho_0(\Delta \mathcal{V})$, of the energy difference. Then

$$\exp(-\beta \Delta A) = Q/Q_0 = \int_{-\infty}^{\infty} d(\Delta \mathcal{V}) \exp(-\beta \Delta \mathcal{V}) \rho_0(\Delta \mathcal{V}). \tag{9.4}$$

$\rho_0(\Delta \mathcal{V})$ is the density (per unit $\Delta \mathcal{V}$) of configurations $\mathbf{r}$ in the reference ensemble which satisfy $\mathcal{V}(\mathbf{r}) = \mathcal{V}_0(\mathbf{r}) + \Delta \mathcal{V}$ for the specified $\Delta \mathcal{V}$. ρ_0 contains the Boltzmann factor $\exp(-\beta \mathcal{V}_0)$ and a factor associated with the change from $3N$ variables $(\mathbf{r})$ to one $(\Delta \mathcal{V})$. It may be sampled directly by constructing a histogram during a simulation of the reference system.

To illustrate this, we use a very simple system suggested by Wu and Kofke (2004), consisting of N independent harmonic oscillators. The reference and perturbed potential energies are, respectively,

$$\mathcal{V}_0 = \sum_{i=1}^{N} c_0 x^2, \quad \text{and} \quad \mathcal{V} = \sum_{i=1}^{N} c x^2, \tag{9.5}$$

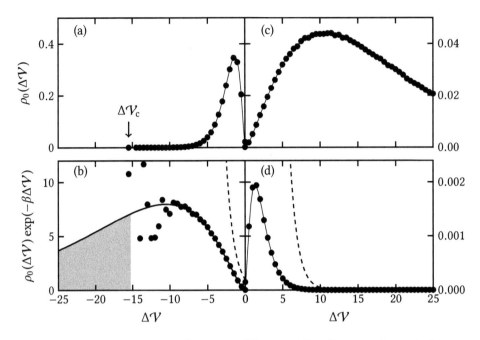

Fig. 9.1 The problem in estimating free-energy differences. The reference and perturbed systems are defined by eqn (9.5) with $N = 5$, and we take $k_B T = 1$. MC simulations directly generate the coordinates x_i, from a normal distribution: 10^6 configurations are used. (a) Potential parameters $c_0 = 8$ (reference), $c = 1$ (perturbed). The function $\rho_0(\Delta\mathcal{V})$, where $\Delta\mathcal{V} = \mathcal{V} - \mathcal{V}_0$: exact (line) and MC simulation results (points). (b) The integrand of eqn (9.4) corresponding to (a): exact (line) and MC simulation results (points). The function $\exp(-\beta\Delta\mathcal{V})$ is shown as a dashed line. The shaded region corresponding to $\Delta\mathcal{V} < \Delta\mathcal{V}_c$ is not sampled at all in the MC simulation. The simulation estimate in this case, from eqn (9.3), is $\Delta A \approx -4.52$, compared with the exact value $\Delta A = -5.1986$. (c) and (d) As for (a) and (b), but with the roles of reference and perturbed system interchanged. Now the simulation estimate of ΔA is within 0.1 % of the exact value $\Delta A = +5.1986$.

with chosen parameters c_0 and c. The free-energy difference is $\Delta A = \frac{1}{2} N k_B T \ln(c/c_0)$, and the distribution $\rho_0(\Delta\mathcal{V})$ may be calculated exactly (Wu and Kofke, 2004). This is shown in Fig. 9.1(a) for the particular choice $N = 5$, $c_0 = 8$, $c = 1$, along with the results of an MC simulation which, at first sight, seems to reproduce it very well. However, $\rho_0(\Delta\mathcal{V})$ decreases rapidly away from the mean value. In a simulation run of finite length, very low values of $\Delta\mathcal{V}$ are not sampled accurately. Indeed, in a histogram recording the potential energies which arise in such a simulation, there will be no entries at all for $\Delta\mathcal{V}$ less than some value $\Delta\mathcal{V}_c$ (see Fig. 9.1(a)). The true distribution (i.e. that obtained from an infinite run) would be small but non-zero below $\Delta\mathcal{V}_c$. For estimating most properties, this would not matter. However, when multiplied by the rapidly growing value of $\exp(-\beta\Delta\mathcal{V})$, these low-energy points should make a substantial contribution to the integral in eqn (9.4). This contribution is the shaded area in Fig. 9.1(b), which in the finite-length simulation is incorrectly reckoned to be zero. The resulting simulation estimate of ΔA from eqn (9.3) is therefore seriously in error.

If one interchanges the parameters, making $c_0 = 1$, $c = 8$, the situation is different, as shown in Fig. 9.1(c) and (d). Now the low-energy tail of the distribution is well sampled, so that, even when multiplied by the rapidly varying $\exp(-\beta\Delta V)$ function, all the important contributions to the integral of eqn (9.4) are also sampled satisfactorily. In this case, an accurate estimate of ΔA (which now has the opposite sign, because the identities of the reference and perturbed systems have been switched) is obtained.

Kofke (2004; 2005; 2006) has considered this problem from the viewpoint of the relationship between the important regions of phase space for the two systems. There will usually be one 'direction' of perturbation, that is, one choice of reference system, which is better than the other. This roughly corresponds to a simulation which samples a larger region of phase space, enveloping the regions which are important for the perturbed system. However, there is no guarantee that either direction will give a satisfactory result, particularly if the important regions of phase space do not overlap much. Moreover, unlike many simulation averages, the results are systematically *biased* rather than being randomly distributed: they may be quite reproducible (but wrong). Kofke (2004; 2005; 2006) has summarized various situations: sometimes it is possible to estimate the size of the bias (the shaded area in Fig. 9.1(b)). When the simple approach of eqn (9.3) is unreliable (in either direction), it may be possible to remedy the situation by a different sampling scheme, and we turn to this now.

9.2.3   Non-Boltzmann sampling

The solution to this problem is to sample on a non-Boltzmann distribution which favours configurations with large negative values of ΔV. This bias must be introduced in such a way that it can subsequently be removed. Torrie and Valleau (1974; 1977b) sample from a general density function

$$\rho_W(\mathbf{r}) = W(\mathbf{r})\exp\left(-\beta V_0(\mathbf{r})\right) \Big/ \int d\mathbf{r}\, W(\mathbf{r})\exp\left(-\beta V_0(\mathbf{r})\right). \tag{9.6}$$

Here $W(\mathbf{r}) = W(\Delta V(\mathbf{r}))$ is a positive-valued weighting function which is specified at the beginning of a simulation run. The method described in Chapter 4 is used to produce a Markov chain of states with a limiting distribution given by eqn (9.6). Specifically, a trial move, from state m to state n, is accepted with a probability given by $\min\{1, (W_n/W_m)\exp[-\beta(\delta V_0)_{nm}]\}$. The average of any property in the reference ensemble, $\langle \mathcal{A}\rangle_0$, can be related to averages taken over MC trials, that is, in the weighted ensemble, using

$$\langle\mathcal{A}\rangle_0 = \frac{\langle\mathcal{A}/W\rangle_{\text{trials}}}{\langle 1/W\rangle_{\text{trials}}} = \frac{\langle\mathcal{A}/W\rangle_W}{\langle 1/W\rangle_W} \tag{9.7}$$

where the notation $\langle\cdots\rangle_W$ reminds us of the weighting. This means that the ratio $\mathcal{A}(\mathbf{r})/W(\mathbf{r})$ is calculated for each step of the simulation, and averaged over the run; the average of $1/W(\mathbf{r})$ is also required in order to obtain the final result. The densities $\rho_0(\Delta V)$ and $\rho_W(\Delta V)$ are related by

$$\rho_0(\Delta V) = \frac{\rho_W(\Delta V)/W(\Delta V)}{\left\langle 1/W(\Delta V)\right\rangle_W}. \tag{9.8}$$

Thus, the density function ρ_0 itself may be calculated by building up a histogram during the simulation. An appropriate choice of $W(\Delta V)$ with an accurate estimate of the denominator in eqn (9.8) gives ρ_0 over a much wider range of $\Delta\mathcal{V}$ than is possible in a conventional simulation. The improved ρ_0 can be used in eqn (9.4) to calculate the required free-energy difference. Equivalently, eqn (9.7) can be used with $\mathcal{A} = \exp(-\beta\Delta\mathcal{V})$.

One of the difficulties with this method is that there is no *a priori* recipe for $W(\Delta\mathcal{V})$. It is often adjusted by trial and error until ρ_W is as wide and uniform as possible, forming an 'umbrella' over the two systems $\mathcal{V}$ and $\mathcal{V}_0$. A useful rule of thumb is that it should extend the range of energies sampled in a conventional MC simulation by a factor of three or more, allowing accurate calculation of much smaller ρ_0 values (Torrie and Valleau, 1977b). A popular choice is $W(\Delta\mathcal{V}) = \exp(-\beta\Delta\mathcal{V}/2)$ (see e.g. Lee and Scott, 1980), in other words, the weighted sampling uses a potential equal to $\frac{1}{2}[\mathcal{V}(\mathbf{r}) + \mathcal{V}_0(\mathbf{r})]$. However Kofke (2004) has pointed out that this (or any linear combination of the potentials) may be a bad choice, from the viewpoint of sampling the relevant areas of configuration space. A better approach seems to be to construct the overall sampling weight function as a linear combination of Boltzmann distributions corresponding to the systems of interest and, possibly, intermediate systems (Valleau, 1999, section II.A). A certain amount of iteration is required, since the coefficients in such a linear combination, which would give equal sampling weights to the various systems, are the inverse partition functions, which are related to the free energies of interest themselves.

A limitation of the method is that, in practice, unlike particle insertion or grand canonical MC, it only gives free-energy differences between quite similar systems. The calculation of absolute free energies requires an accurate knowledge of the reference system value. Umbrella sampling is normally performed on small systems. The larger the system, the smaller the relative fluctuations, and the more sharply varying the density functions. This reduces the overlap between the distributions and makes the accurate calculation of the free-energy differences more difficult. Fortunately, the N-dependence of relative free energies is thought to be small, and such simulations are economical. Exploring the fluctuations in larger systems requires a more sophisticated algorithm such as the Wang–Landau method described in Section 9.2.5.

If the two systems $\mathcal{V}$ and $\mathcal{V}_0$ are very different from one another, it may be necessary to introduce an intermediate stage, or many intermediate stages. In this case, eqn (9.3) can be generalized to

$$\exp\left(-\beta(A - A_0)\right) = \left\langle\exp\left(-\beta(\mathcal{V} - \mathcal{V}_n)\right)\right\rangle_n \left\langle\exp\left(-\beta(\mathcal{V}_n - \mathcal{V}_{n-1})\right)\right\rangle_{n-1} \times \cdots$$
$$\cdots \times \left\langle\exp\left(-\beta(\mathcal{V}_2 - \mathcal{V}_1)\right)\right\rangle_1 \left\langle\exp\left(-\beta(\mathcal{V}_1 - \mathcal{V}_0)\right)\right\rangle_0 \quad (9.9)$$

where systems $\mathcal{V}_1 \ldots \mathcal{V}_n$ have been introduced with properties intermediate between those of $\mathcal{V}$ and $\mathcal{V}_0$. This multistage sampling (Valleau and Card, 1972) has been employed directly to calculate the free-energy difference between hard spheres and Coulombic hard spheres. Each of the separate averages in eqn (9.9) can be evaluated with the help of umbrella sampling, which reduces the number of intermediate stages required (Torrie and Valleau, 1977b). As an illustration of the umbrella sampling technique, Torrie and Valleau (1977b) have related the free energy of the Lennard-Jones fluid to that of the inverse twelfth-power fluid, and these same authors also found it useful in the study of liquid mixtures (Torrie and Valleau, 1977a).

Shing and Gubbins (1981; 1982) used umbrella sampling in conjunction with test particle insertion to calculate the chemical potential. We describe the second of their two methods, which is the more generally applicable. A single test particle is inserted in the fluid, at intervals during the normal simulation. It moves through the fluid, keeping the real particles fixed, using a non-Boltzmann sampling algorithm which favours configurations of high $\exp(-\beta \Delta V_{test})$ (see Section 2.4, eqn (2.75)). Each configuration is weighted by a factor $W(V_{test})$. One particularly simple form for W is

$$W(V_{test}) = \begin{cases} 1 & V_{test} \leq V_{max} \\ 0 & V_{test} > V_{max}. \end{cases} \tag{9.10}$$

In the simulation of a Lennard-Jones fluid, V_{max} was taken to be $200\,\epsilon$, and the weighting function rejected all moves which led to a significant overlap. The test particle has no real interaction with the atoms in the fluid, and, in general, a test particle move from position $\mathbf{r}_{test}^m$ to $\mathbf{r}_{test}^n$ is accepted if

$$W(\mathbf{r}_{test}^n)/W(\mathbf{r}_{test}^m) \geq \xi \tag{9.11}$$

where ξ is a random number in the range $(0, 1)$. After 100–200 moves (the first few of which are discarded) the test particle is removed and the regular simulation resumed. During the run the distribution of test particle energies, $\rho_W(V_{test})$, is calculated. The distribution is proportional to the unweighted distribution, $\rho_0(V_{test})$, for $V_{test} \leq V_{max}$ (see eqn (9.8)). The constant of proportionality is most easily obtained in this case by performing a parallel set of unweighted test particle insertions, and comparing the two distributions in the region where they are both well known. Once $\rho_0(V_{test})$ is known accurately over its complete range, then the chemical potential can be calculated from

$$\mu^{ex} = -k_B T \ln \left(\int_{-\infty}^{\infty} dV_{test}\, \rho_0(V_{test}) \exp(-\beta V_{test}) \right). \tag{9.12}$$

The usual problem with the insertion method, namely the high probability of finding overlaps at high densities, is controlled by the weighted sampling. Shing and Gubbins (1982) have also proposed a method which concentrates the sampling on the configurations that exhibit suitable 'holes' for the insertion. A useful modification of the particle insertion method has been to 'turn on' the test particle interaction gradually (Mon and Griffiths, 1985). The idea of using a variable coupling parameter has been used to estimate solubilities (Swope and Andersen, 1984). Kofke and Cummings (1998) have studied the calculation of chemical potentials by staged insertion, using a reduced-radius test particle as the intermediate stage.

All of the preceding discussion has focused on distributions of the potential energy, and the calculation of the Helmholtz free energy. However, much of it can be applied, unchanged, to distributions of an arbitrary order parameter q, and hence the determination of the Landau free energy $\mathcal{F}$ defined in Section 2.11. Sometimes our interest lies in determining the full curve $\mathcal{F}(q)$ (or, more generally, a surface depending on two or more order parameters); sometimes the main aim is to calculate a free-energy difference between two low-lying states, which are separated by a barrier. The methods of umbrella sampling may be applied to this problem, to generate a broad distribution of q which covers all the states of interest, as well as the more sophisticated 'flat histogram' sampling methods described in Section 9.2.5.

Here, however, we turn to a method which, superficially, has the opposite intention: each simulation is restricted to a narrow range of q, or window. The idea is to build up the free-energy curve by combining the results of all these simulations. The term 'umbrella sampling' is widely used in biomolecular simulation to refer to this method of multistage sampling for the calculation of a free-energy barrier (Kaestner, 2011). Like the methods discussed previously, it involves the application of an additional potential to the Hamiltonian, the so-called umbrella potential. To illustrate this technique let us consider a small peptide in water (Mu and Stock, 2002). The object is to calculate the free energy as a function of one of the dihedral angles, ϕ, in the peptide (see Fig. 9.2(a)). This can be determined from the probability density $\rho(\phi)$ in an unrestrained simulation (see Section 2.11). The umbrella potential is of the form

$$v_W(\phi, \phi_k) = \tfrac{1}{2}c(\phi - \phi_k)^2 \equiv v_{W,k}(\phi) \tag{9.13}$$

and it is defined for a set of reference angles $\{\phi_k\}$, $k = 1, \dots n$, covering the range of interest. This is added to the normal potential for the peptide in solution $\mathcal{V}_0(\mathbf{r})$. Mc simulations are performed with the potential $\mathcal{V}_0(\mathbf{r}) + v_{W,k}(\phi)$ for each value of k. The umbrella potential restrains the peptide to sample states in the harmonic well around the value ϕ_k. If the restraint potential is removed then the system will simply not sample the higher-energy configurations away from the minima in the torsional potential.

For a particular k, the biased probability distribution function is

$$\rho_{W,k}(\phi) = \exp(\beta A_k) \int d\mathbf{r} \, \exp\left(-\beta \mathcal{V}_0(\mathbf{r})\right) \exp\left[-\beta v_{W,k}\left(\phi(\mathbf{r})\right)\right] \delta\left(\phi(\mathbf{r}) - \phi\right) \tag{9.14}$$

where A_k is proportional to the free energy of the system with the biased potential, and $\mathbf{r}$, as usual, represents the complete set of coordinates of peptide and solvent. $\rho_{W,k}(\phi)$ can be calculated in the simulation as

$$\rho_{W,k}(\phi) = \frac{\left\langle H\left(\phi(\mathbf{r}) - \phi, \Delta\phi\right)\right\rangle_k}{\Delta\phi \, M_k}, \quad H = \begin{cases} 1 & |\phi(\mathbf{r}) - \phi| < \tfrac{1}{2}\Delta\phi \\ 0 & \text{otherwise.} \end{cases} \tag{9.15}$$

In eqn (9.15), the top-hat function, H, sorts the dihedral angle, $\phi(\mathbf{r})$, into bins of width $\Delta\phi$ around ϕ and M_k is the total number of histogram entries for simulation k. The probability function for the unbiased simulation can be recovered from the biased probability using

$$\rho_k(\phi) = \exp\left(-\beta(A_k - A_0)\right) \exp\left(\beta v_{W,k}(\phi)\right) \rho_{W,k}(\phi) \tag{9.16}$$

where A_0 is the free energy of the unbiased system. On the left, the index k simply indicates which simulation gave rise to the function: in principle they are all the same, and $\rho(\phi)$ can be obtained from any one of them, but in practice an individual $\rho_{W,k}(\phi)$ distribution only contains information around ϕ_k, and the same will be true of the function $\rho_k(\phi)$ obtained from it via eqn (9.16). All of the $\rho_{W,k}(\phi)$ distributions should be combined to produce the full unbiased distribution. This process is illustrated in the inset of Fig. 9.2(b): each portion of $\mathcal{F}(\phi)$, $-k_B T \ln \rho_k(\phi)$, is determined up to an unknown additive constant,

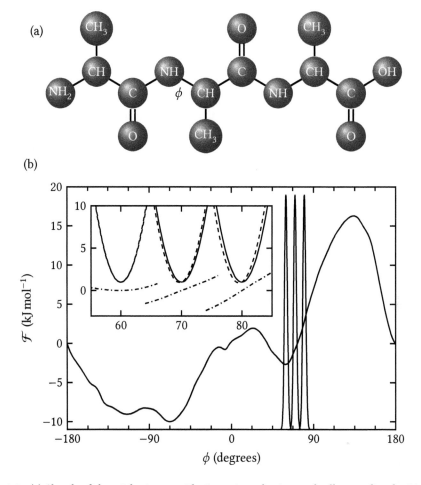

Fig. 9.2 (a) Sketch of the trialanine peptide, investigated using umbrella sampling by Mu and Stock (2002), with the central ϕ angle indicated. (b) Approximate representation of the Landau free energy for rotation around ϕ. Also shown are three successive distributions $\rho_{W,k}(\phi)$, centred at 10° intervals of ϕ_k, as might be determined by umbrella sampling. The inset shows the corresponding biasing potentials $v_{W,k}(\phi)$ (solid curves), the sampled distributions in the form $-k_B T \ln \rho_{W,k}(\phi)$ (dashed curves), and their difference, the reconstructed portions of free energy $\mathcal{F}(\phi)$ (dash-dotted curves). All the data is idealized, and we make no attempt to represent the statistical noise.

and subject to statistical errors (not shown) which depend on ϕ. This can be accomplished using the weighted histogram analysis method (WHAM) (Kumar et al., 1992).

In the WHAM procedure, we consider a linear combination of the unbiased distributions, which can also be expressed in terms of the biased ones

$$\rho(\phi) = \sum_{k=1}^{n} C_k(\phi)\rho_k(\phi) = \sum_{k=1}^{n} C_k(\phi) \exp\left(-\beta(A_k - A_0)\right) \exp\left(\beta v_{W,k}(\phi)\right)\rho_{W,k}(\phi) \quad (9.17)$$

where the coefficients must satisfy the constraints

$$\sum_{k=1}^{n} C_k(\phi) = 1, \qquad \forall \phi. \tag{9.18}$$

Using the method of Lagrange multipliers, it is possible to minimize the variance in our estimate of $\rho(\phi)$. After some straightforward algebra (Frenkel and Smit, 2002; Tuckerman, 2010) we can show that the optimum distribution is given by the coefficients

$$C_k(\phi) = \frac{M_k \, \exp(\beta A_k) \exp\big(-\beta v_{W,k}(\phi)\big)}{\sum_{k'=1}^{n} M_{k'} \, \exp(\beta A_{k'}) \exp\big(-\beta v_{W,k'}(\phi)\big)}$$

and so $\rho(\phi)$ is conveniently written in terms of the biased distributions

$$\rho(\phi) = \frac{\sum_{k=1}^{n} M_k \, \rho_{W,k}(\phi)}{\sum_{k=1}^{n} M_k \, \exp\big(\beta(A_k - A_0)\big) \exp\big(-\beta v_{W,k}(\phi)\big)} \tag{9.19}$$

where

$$\exp\big(-\beta(A_k - A_0)\big) = \int d\phi \, \rho(\phi) \exp\big(-\beta v_{W,k}(\phi)\big). \tag{9.20}$$

Equations (9.18)–(9.20) can be solved iteratively to convergence given an initial guess for the $(A_k - A_0)$ parameters. The result is illustrated, as a Landau free energy $\mathcal{F}(\phi) = -k_B T \ln \rho(\phi)$, in Fig. 9.2(b), for the peptide example. (Note that Mu and Stock (2002) analysed their data using an alternative formula to WHAM.) The procedure relies on the assumption that there is the same quality of sampling in each of the n simulations performed. It is clear that the sampling can be performed using either an MC or MD technique, as is most convenient.

We have chosen to illustrate the method with a simple order parameter, ϕ, but our umbrella potential could describe any reaction coordinate, $q(\mathbf{r})$, or be a function of many such coordinates. This could describe the pathway from reactants, across a barrier, to the products in a chemical reaction. For a simple gas-phase dissociation of a dimer, the appropriate reaction coordinate might be the separation of the two constituent atoms. However, in the case of the dissociation of (say) water into hydroxyl and hydroxonium ions, the reaction pathway is a complicated function of the coordinates of the many water solvent molecules surrounding the reacting pair (Dellago et al., 2002; Bolhuis et al., 2002). In fact, one of the major difficulties in applying these methods and some of the corresponding molecular dynamics techniques discussed in Chapter 10 is the choice of an appropriate reaction coordinate or order parameter.

WHAM is also an extremely useful tool for combining the results of REMC (parallel tempering) simulations at a range of different state points (see Section 4.9) and the necessary modifications to the prescription are discussed by Chodera et al. (2007).

Virnau and Müller (2004) have proposed a method which they term 'successive umbrella sampling' in which the order parameter space is divided into rather narrow, overlapping, windows. In their original paper, the order parameter is discrete (it is the number of particles, in a grand canonical simulation), and we adopt the same description here, but the method may be equally well applied to a system with a continuous order parameter. Suppose that the width of each window is $w + 1$, so that window 0 corresponds to $0 \leq N \leq w$, window 1 to $w \leq N \leq 2w$, and in general window k to $kw \leq N \leq (k + 1)w$. Any GCMC move that would take the system outside its window is rejected, and the current state counted again in the usual way. An occupation histogram $H_k(N)$ of states N is constructed within each window. After each simulation has run for a pre-determined number of GCMC moves, the histograms are combined together, matching up the boundary values for successive windows to obtain an overall probability distribution

$$\frac{P(N)}{P(0)} = \frac{H_0(w)}{H_0(0)} \frac{H_1(2w)}{H_1(w)} \frac{H_2(3w)}{H_2(2w)} \cdots \frac{H_k(N)}{H_k(kw)} \tag{9.21}$$

where N lies in the k-th window. From this the free energy can be obtained as a function of N.

Virnau and Müller (2004) discuss various practical details. As described earlier, the simulations are conducted independently within each window, and hence can be done in parallel if desired. However, this involves the preparation of starting configurations for each window, and for this reason it may be more practically convenient to sample the windows successively, taking advantage of the overlapping states to provide starting configurations for each window from ones sampled in the previous one. There is no need to apply any weight function within the window, as long as the windows are narrow enough to avoid large changes in $H_k(N)$ between the low-N and high-N window boundaries. However, the efficiency might be improved by including a weight function, to make the sampling more uniform within each window, and if the windows are tackled successively, this weight can be extrapolated from the results of the previous window. Virnau and Müller (2004) also discuss the estimation of errors, and argue that the optimum window size will be system (and system-size) specific: a wider window will be less prone to kinetic trapping, but may be less efficient at sampling, especially in the absence of an accurate weight function.

In the next section, we shall meet another method for combining the results of two or more simulations, which has some similarities to WHAM. Finally, it is worth remembering that, in the limit of narrow windows, the umbrella-sampling method becomes very similar to conventional thermodynamic integration. The advantage of umbrella sampling lies in the overlap between successive windows, which helps to stitch together the separate contributions to the free-energy curve (using WHAM or eqn (9.21)). Without this overlap, each window can just give an estimate of free-energy *derivatives*, of which the first derivative will be the most accurately determined, and the problem of generating the full free-energy curve becomes very similar to simple thermodynamic integration.

9.2.4 Acceptance ratio methods

An interesting extension of the ideas introduced in the previous section is the work of Bennett (1976). In the canonical ensemble, the ratio of the partition functions of two fluids is given in terms of an arbitrary weighting function $W(\mathbf{r})$

$$\frac{Q_1}{Q_0} = \frac{Q_1 \int d\mathbf{r} \, W(\mathbf{r}) \exp(-\beta(\mathcal{V}_1 + \mathcal{V}_0))}{Q_0 \int d\mathbf{r} \, W(\mathbf{r}) \exp(-\beta(\mathcal{V}_1 + \mathcal{V}_0))} = \frac{\langle W \exp(-\beta\mathcal{V}_1)\rangle_0}{\langle W \exp(-\beta\mathcal{V}_0)\rangle_1}. \tag{9.22}$$

The choice $W = \exp(\beta\mathcal{V}_0)$ or $W = \exp(\beta\mathcal{V}_1)$ leads to eqn (9.3), but, as we have seen, this is likely to be impractical. Bennett shows that a particular choice of W will minimize the variance in the estimation of Q_1/Q_0. The best choice is

$$W = \text{constant} \times \left(\frac{Q_0}{(\tau_0/s_0)} \exp(-\beta\mathcal{V}_1) + \frac{Q_1}{(\tau_1/s_1)} \exp(-\beta\mathcal{V}_0) \right)^{-1} \tag{9.23}$$

where (τ_0/s_0) and (τ_1/s_1) are the number of statistically independent configurations generated in each of the two Markov chains (see eqn (8.16)). Substitution of eqn (9.23) into eqn (9.22) gives

$$\frac{Q_1}{Q_0} = \frac{\langle f(+\beta\Delta\mathcal{V} + C)\rangle_0}{\langle f(-\beta\Delta\mathcal{V} - C)\rangle_1} \exp(C) \tag{9.24}$$

where $C = \ln(Q_1 s_1 \tau_0 / Q_0 s_0 \tau_1)$, $\Delta\mathcal{V} = (\mathcal{V}_1 - \mathcal{V}_0)$ and f is the Fermi function

$$f(x) = \left(1 + \exp(x)\right)^{-1}. \tag{9.25}$$

Writing eqn (9.24) in terms of energy distributions, we obtain

$$\frac{Q_1}{Q_0} = \frac{\int d(\Delta\mathcal{V}) \, f(+\beta\Delta\mathcal{V} + C)\rho_0(\Delta\mathcal{V})}{\int d(\Delta\mathcal{V}) \, f(-\beta\Delta\mathcal{V} - C)\rho_1(\Delta\mathcal{V})} \exp(C). \tag{9.26}$$

The constant C acts as a shift in potential, so as to bring the two systems into as close a correspondence as possible. The method works as follows. A simulation of each fluid is performed, and the density functions ρ_1 and ρ_0 calculated by constructing histograms as functions of $\Delta\mathcal{V}$. A value of C is guessed, and the ratio Q_1/Q_0 calculated from eqn (9.26). C is recalculated from

$$C \approx \ln(Q_1/Q_0). \tag{9.27}$$

In eqn (9.27) we have assumed that $\tau_1/s_1 \approx \tau_0/s_0$. This can be checked by direct calculation using the methods described in Section 8.4. An iterative solution of eqns (9.26) and eqn (9.27) gives a value for C and Q_1/Q_0. Bennett (1976) presents an alternative graphical solution. The method works well if there is any overlap between the functions ρ_1 and ρ_0. The overlap can be improved using umbrella or multi-stage sampling. Equation (9.26) has been used to calculate the free energy of a model of liquid nitrogen from that of the hard dumb-bell fluid (Jacucci and Quirke, 1980a).

Equations (9.26) and (9.27) minimize the error in the estimate of the free-energy difference between two systems, 1 and 0, whereas eqns (9.18) and (9.19), associated with the restraint potentials in umbrella sampling, minimize the error in the estimate of the distribution function, ρ. Interestingly for the special case of two simulations of the same length ($M_1 = M_0$) and with an umbrella potential $\Delta\mathcal{V} = \mathcal{V}_1 - \mathcal{V}_0$, the two minimizations produce exactly the same result.

A slightly modified form of eqn (9.24) is applicable in the constant-NVE ensemble (Frenkel, 1986)

$$\frac{Q_1}{Q_0} = \frac{\langle f[+(\Delta\mathcal{V} - \Delta E)/k_B \mathcal{T}_0 + C]\rangle_0}{\langle f[-(\Delta\mathcal{V} - \Delta E)/k_B \mathcal{T}_1 - C]\rangle_1} \exp(C) \tag{9.28}$$

where $C = \ln(Q_1 s_1 \tau_0 / Q_0 s_0 \tau_1)$, Q means Q_{NVE}, and $\Delta E = E_1 - E_0$ is the difference between the total energies in the two simulations. $\mathcal{T}_0$ and $\mathcal{T}_1$ are the instantaneous values of the temperature and eqn (9.28) assumes that $\langle \mathcal{T}_1 \rangle = \langle \mathcal{T}_0 \rangle$. The microcanonical partition function can be related to the entropy through eqn (2.18). In the MD simulations the two density functions $\rho_1[(\Delta \mathcal{V} - \Delta E)/k_B \mathcal{T}_1]$ and $\rho_0[(\Delta \mathcal{V} - \Delta E)/k_B \mathcal{T}_0]$ are calculated and eqn (9.28) is solved iteratively for C.

Bennett's acceptance ratio method can be readily extended to incorporate data from many states to estimate free-energy differences and thermodynamic averages for an arbitrary thermodynamic state (Shirts and Chodera, 2008). The multi-state Bennett acceptance ratio (MBAR) method combines n independent simulations. To illustrate the method, we will consider n systems at constant-NVT but with different potential-energy functions, $\mathcal{V}_I, I = 1 \ldots n$. The method can be readily extended to other ensembles or to combine the data from systems at different temperatures or chemical potentials. For a particular potential $\mathcal{V}_I$, τ_I independent configurations are sampled from the probability distribution $\rho_I = Q_I^{-1} \exp(-\beta \mathcal{V}_I(\mathbf{r}))$. These configurations are labelled $\{\mathbf{r}_{I1}, \mathbf{r}_{I2} \ldots \mathbf{r}_{I\tau_I}\}$; they are chosen from the trajectory with a statistical inefficiency $s_I = 1$. The average of some configurational property $\mathcal{A}$ is

$$\langle \mathcal{A} \rangle_I = \frac{1}{\tau_I} \sum_{i=1}^{\tau_I} \mathcal{A}(\mathbf{r}_{Ii}). \tag{9.29}$$

The Helmholtz free-energy difference between two states I and J is

$$A_J - A_I = -k_B T \ln(Q_J/Q_I). \tag{9.30}$$

For an arbitrary function, W_{IJ}, we note the identity

$$Q_I \left\langle W_{IJ}(\mathbf{r}) \exp\left(-\beta \mathcal{V}_J(\mathbf{r})\right) \right\rangle_I = Q_J \left\langle W_{IJ}(\mathbf{r}) \exp\left(-\beta \mathcal{V}_I(\mathbf{r})\right) \right\rangle_J \tag{9.31}$$

which is essentially the same as eqn (9.22) in our extended notation. Using eqn (9.29) in eqn (9.31) and summing over systems J, we obtain

$$\sum_{J=1}^{n} \frac{Q_I}{\tau_I} \sum_{i=1}^{\tau_I} W_{IJ}(\mathbf{r}_{Ii}) \exp\left(-\beta \mathcal{V}_J(\mathbf{r}_{Ii})\right) = \sum_{J=1}^{n} \frac{Q_J}{\tau_J} \sum_{j=1}^{\tau_J} W_{IJ}(\mathbf{r}_{Jj}) \exp\left(-\beta \mathcal{V}_I(\mathbf{r}_{Jj})\right) \tag{9.32}$$

for each system I. Eqn (9.32) is nothing more than a set of n coupled equations that can be solved for all of the Q_I, given a set of configurations for each system. The solution will be unique up to a multiplicative constant. As in the original method of Bennett, W_{IJ} (the extended bridge sampling estimator) is chosen to minimize the variance in Q_J/Q_I and the optimal form suggested by maximum-likelihood methods (Kong et al., 2003) is

$$W_{IJ}(\mathbf{r}) = \frac{\tau_J Q_J^{-1}}{\sum_{K=1}^{n} \tau_K Q_K^{-1} \exp\left(-\beta \mathcal{V}_K(\mathbf{r})\right)}. \tag{9.33}$$

Substitution of eqn (9.33) into eqn (9.32) gives an expression for the free energy

$$(A_I - A_0) = -k_B T \ln\left[\sum_{J=1}^{n} \sum_{j=1}^{\tau_J} \frac{\exp\left(-\beta \mathcal{V}_I(\mathbf{r}_{Jj})\right)}{\sum_{K=1}^{n} \tau_K \exp\left[\beta\left((A_K - A_0) - \mathcal{V}_K(\mathbf{r}_{Jj})\right)\right]} \right]. \tag{9.34}$$

These equations can be solved iteratively for A_I up to an additive constant, A_0. The similarities with the WHAM method, eqns (9.19) and (9.20), are evident. In fact, WHAM and

MBAR are equivalent in the limit that the histogram bin width in WHAM is reduced to zero. MBAR reduces to the original Bennett method in the limit of two systems. Unlike WHAM, MBAR also provides a direct estimate of the uncertainties in the calculation of free-energy differences (Kong et al., 2003). The free-energy differences can also be calculated for states that are not directly sampled. The set of n states is increased by including a new state, $\mathcal{V}_{\text{new}}$, with $\tau_{\text{new}} = 0$ in eqn (9.34) and without further need to check for self-consistency. A general configurational property, $\langle \mathcal{A} \rangle$, can be estimated by including a new state $[\mathcal{V}(\mathbf{r}) - k_{\text{B}}T \ln \mathcal{A}(\mathbf{r})]_{\text{new}}$ with $\tau_{\text{new}} = 0$ into eqn (9.32).

Recently, Paliwal and Shirts (2013) have used MBAR to explore the effect of the Coulomb and Lennard-Jones cutoffs on the results of free-energy calculations, while Aimoli et al. (2014) have used the method, in the isothermal–isobaric ensemble, to make extensive predictions of the properties of supercritical CO_2 and CH_4 over a wide range of temperatures and pressures. A Python implementation of the MBAR algorithm has been made available by Shirts and Chodera (2008).

9.2.5 Wang–Landau methods

As we have established already, the probability distribution function $\rho(q)$ of some variable $q(\mathbf{r})$, measured in, for example, the canonical ensemble, is related to the Landau free energy $\mathcal{F}(q)$ (Section 2.11) by

$$\mathcal{F}(q) = -k_{\text{B}}T \ln \rho(q) + \text{constant} \quad \Rightarrow \quad \rho(q) \propto \exp(-\mathcal{F}(q)/k_{\text{B}}T).$$

It follows that introducing a weight function of the form $W(q) = \exp(+\mathcal{F}(q)/k_{\text{B}}T)$ into the Metropolis acceptance/rejection criterion will result in a flat distribution $\rho_W(q)$. Several simulation methods are based on turning this idea around, and devising an iterative method of refining an initial estimate of $\mathcal{F}(q)$ in such a way as to generate, eventually, such a flat distribution. For reasons to become clear shortly, these are commonly referred to as 'density of states' methods; recent reviews have been provided by Singh et al. (2012) and Janke and Paul (2016).

Early versions of this method are called *multicanonical* or *entropic* sampling (Berg and Celik, 1992; Lee, 1993; Berg et al., 1995). Assume that a histogram is used to accumulate $\rho_W(q)$, and that a table stores the current estimate of $\mathcal{F}(q)$ at corresponding values of q. The simulation begins with all the entries set to zero. A typical sequence of runs is:

1. Conduct a simulation, of a pre-determined length, sampling with effective potential
   $\mathcal{V}_{\text{eff}}(\mathbf{r}) = \mathcal{V}(\mathbf{r}) - \mathcal{F}(q(\mathbf{r}))$.
2. Determine $\rho_W(q)$ as a histogram, averaged over this simulation run.
3. Wherever $\rho_W(q) > 0$, reset $\mathcal{F}(q) \to \mathcal{F}(q) - k_{\text{B}}T \ln \rho_W(q)$.

Then the process returns to step 1. The last step sets the scene for the next simulation: it effectively discriminates against the more popular states, and in favour of the less popular ones (amongst those visited so far). The steps are repeated until some convergence criterion is met. The first (unweighted) simulation samples just the low-lying free-energy region of q, but as the weight function builds up, higher values of $\mathcal{F}(q)$ are visited more frequently, and eventually $\rho_W(q) \to \text{constant}$. When this is achieved, the table of free energies is the desired Landau free energy.

This approach has been refined by Wang and Landau (2001a,b), and applied to the case of determining the density of states. This corresponds to the case where $q = E$ (the total energy) or $q = \mathcal{V}$ (the potential energy). The corresponding density $\rho(E)$ or $\rho(\mathcal{V})$ is just the microcanonical ensemble partition function; we shall use Ω instead of ρ for this, as it is the more usual notation. The 'Landau free energy' $\mathcal{F}(q)/k_{\mathrm{B}}T$ is actually the negative of the entropy $-S(E)/k_{\mathrm{B}}$ or its configurational part. When this technique is successful, therefore, it achieves the objective that we had previously suggested was very difficult or impossible to achieve: the determination of the entropy or free energy in a single simulation.

Wang and Landau (2001a,b) proposed essentially two refinements of the scheme just described. First, the updating of the 'free energy' or 'entropy' term, and hence the weighting that appears in the acceptance/rejection criterion, is performed at every Monte Carlo step. Second, the amount by which this weight is updated is set at the start, and is progressively adjusted downwards as the measured histogram becomes more and more flat. In outline, the scheme looks like this. Initially $S(\mathcal{V}) = 0$ for all the tabulated values of $\mathcal{V}$, and the adjustment factor is set to $\delta S = k_{\mathrm{B}}$.

1. Set all $\Omega_W(\mathcal{V}) = 0$.
2. Conduct MC moves using $S(\mathcal{V})$. Suppose after each move the new potential energy is $\mathcal{V}$, then reset $S(\mathcal{V}) \rightarrow S(\mathcal{V}) + \delta S$ and increment $\Omega_W(\mathcal{V}) \rightarrow \Omega_W(\mathcal{V}) + 1$.
3. Continue the simulation until the sampled density of states $\Omega_W(\mathcal{V})$ is 'sufficiently flat'.
4. Reduce the entropy increment $\delta S \rightarrow \frac{1}{2}\delta S$.

Then the process returns to step 1, resetting the sampled density $\Omega_W(\mathcal{V})$ to zero, and restarting the accumulation of this histogram. The whole sequence, of determining a 'flat' histogram and then adjusting δS downwards, continues until δS is 'sufficiently small'. At this point, S should have converged to the (configurational part of the) microcanonical entropy $S(\mathcal{V})$, and the true density of states is given by $\Omega(\mathcal{V}) \propto \exp\left(S(\mathcal{V})/k_{\mathrm{B}}\right)$ (compare the usual Boltzmann expression $S = k_{\mathrm{B}} \ln \Omega$). An example of a Wang–Landau program is given in Code 9.1.

Several details of this scheme need to be explained. First, in step 2, the Monte Carlo scheme uses a standard Metropolis-like acceptance/rejection formula, but this is based completely on the function $S(\mathcal{V})$:

$$P_{\mathrm{acc}} = \min\left(1, \exp(-\Delta S/k_{\mathrm{B}})\right), \quad \Delta S = S\left(\mathcal{V}(r_{\mathrm{new}})\right) - S\left(\mathcal{V}(r_{\mathrm{old}})\right).$$

Notice that temperature plays no part in this algorithm. In terms of the statistical weights (numbers of states) at each energy, $\Omega(\mathcal{V})$, this updating scheme will sample states with a probability proportional to $\exp(-S/k_{\mathrm{B}})$, that is, *inversely proportional* to Ω. This results in a flat distribution of sampled energies (typically, the number of accessible states $\Omega(\mathcal{V})$ increases very rapidly with $\mathcal{V}$, but this is exactly cancelled by the weighting, which goes inversely with Ω).

Second, we need to consider what flatness criterion would be appropriate in step 3. Typically this condition is expressed as a relation between the minimum value in the histogram, and the mean value:

$$\frac{\min \Omega_W(\mathcal{V})}{\overline{\Omega_W(\mathcal{V})}} > f \tag{9.35}$$

Code 9.1 Wang–Landau simulation of chain molecule

These files are provided online. The program mc_chain_wl_sw.f90 conducts an MC simulation for a single chain molecule composed of hard square-well atoms. This allows
the chain to explore the entire energy landscape, and the aim is to sample the potential
energy distribution uniformly. The chain exhibits a collapse transition. A variety of
moves, such as reptation and pivot, are provided in mc_chain_sw_module.f90. For
comparison, a simple constant-NVT program, using the same moves, is given in
mc_chain_nvt_sw.f90. As usual, the programs use the utility modules of Appendix A;
the program initialize.f90 (Code 5.6) may be employed to prepare initial configurations.

```
! mc_chain_wl_sw.f90
! Monte Carlo, single chain, Wang-Landau, square wells
PROGRAM mc_chain_wl_sw

! mc_chain_sw_module.f90
! Monte Carlo, single chain, square wells
MODULE mc_module

! mc_chain_nvt_sw.f90
! Monte Carlo, single chain, NVT, square wells
PROGRAM mc_chain_nvt_sw
```

where f is typically chosen to be 0.8–0.95. Higher values give a more stringent requirement for flatness, which is generally preferable, but more expensive.

Third, when should we decide that the scheme has converged? Usually, δS is reduced
so that $\exp(-\delta S/k_B)$ is extremely small, perhaps even comparable with the limit imposed
by the built-in discrete representation of floating-point numbers. This is important, because
the density of states $\Omega(\mathcal{V})$, which is effectively being cancelled by the weight function
to give a flat sampled distribution $\Omega_W(\mathcal{V})$, may easily cover many orders of magnitude.
Therefore, a very accurately determined $\Omega(\mathcal{V})$ is essential.

There are some practical issues with the scheme. The ideal situation is where all
accessible energies are visited in the first run, and subsequent runs simply refine the
weights at each energy, giving a smoother and smoother function $S(\mathcal{V})$. This is seldom the
case in practice: states having new energies are often discovered later on in the sequence.
The flatness calculation of eqn (9.35) can only be based on the energy bins in the histogram
that have actually been visited, so this gives a false impression during the early stages.
Also, when a new energy bin is discovered, the simulation typically gets stuck in it, until
enough increments have been added to the weight function to escape from the bin. This
may take a substantial amount of time, if δS has been adjusted down many times since
the start. A second problem is that the entropy adjustment factor δS is reduced 'too fast'
by the standard scheme to cope with the roughness of the landscape. This gives a false
impression of convergence: the algorithm reaches a halt before all the peaks and troughs

in $S(V)$ have been cancelled out. In principle, this can be tackled by changing to a method which, at long MC 'times' t, reduces δS in a way proportional to $1/t$ (Belardinelli and Pereyra, 2007a,b). There is one final issue that should be mentioned. The ultimate output of this scheme is the $S(V)$ which generates a flat distribution. From this, a wide range of canonical ensemble averages may be generated, at any desired temperature, by including the weight factor, along with the appropriate Boltzmann factor, in an integral over V. However, technically the weight has not been determined from an equilibrium simulation: even though the final increment δS is very small, the weighting function is changing throughout the process. To be sure, a final, long, simulation should be undertaken, with a weight that is no longer being updated, and this run should be used to calculate results.

We should spend a moment thinking about the practicalities of calculating averages at a chosen temperature. Suppose that we are interested in a variable $\mathcal{A}$, and that we have stored the Wang–Landau simulation averages of this quantity, as a function of V, in a histogram: $\langle \mathcal{A} \rangle_V$. The formula for the canonical ensemble average is

$$\langle \mathcal{A} \rangle = \frac{\int dV\, \Omega(V)\exp(-\beta V)\,\langle \mathcal{A} \rangle_V}{\int dV\, \Omega(V)\exp(-\beta V)} = \frac{\int dV\, \exp(-\beta V + S(V)/k_B)\,\langle \mathcal{A} \rangle_V}{\int dV\, \exp(-\beta V + S(V)/k_B)}$$

$$= \frac{\int dV\, \exp(-\beta \mathcal{F}(V))\,\langle \mathcal{A} \rangle_V}{\int dV\, \exp(-\beta \mathcal{F}(V))},$$

where we introduce a 'Landau free energy' $\mathcal{F}(V)$ for convenience. The histogram values may be used to compute both the integrals here. However, we must pay attention to the enormous range of values covered by the integrands: the function $\exp[-\beta \mathcal{F}(V)]$ will be very sharply peaked near some value of V which depends on the temperature. The solution is to find this maximum value, and concentrate the integration on the important region nearby. Equivalently, find the minimum value of $\mathcal{F}(V)$, call it $\mathcal{F}_{min}$, and subtract it from $\mathcal{F}(V)$ in both numerator and denominator:

$$\langle \mathcal{A} \rangle = \frac{\int dV\, \exp\left[-\beta\big(\mathcal{F}(V) - \mathcal{F}_{min}\big)\right]\langle \mathcal{A} \rangle_V}{\int dV\, \exp\left[-\beta\big(\mathcal{F}(V) - \mathcal{F}_{min}\big)\right]}.$$

This will guard against extremely large value of the integrands, which might otherwise cause overflow, and the range of integration can be truncated when the integrands fall to extremely low values, so as to avoid underflow.

The Wang–Landau method is currently regarded as the first choice for determining densities of states, and hence simulation results (including the free energy) across a wide range of state points, out of a single set of runs. However, it is not guaranteed to produce accurate results: the effectiveness of the sampling scheme must be system-dependent. The method is easily generalized to the case of variables other than the energy, and to handle more than a single variable, although it works less well as the dimensionality of the space to be explored increases.

9.2.6 Nested sampling

Nested sampling (NS) has the same aims as the Wang–Landau approach (estimation of the density of states, and hence the partition function and related quantities) and is, similarly, an iterative method. It arose from an idea due to Skilling (2006), and was applied to molecular simulation by Pártay et al. (2010). The method then seems to have been rediscovered and applied to condensed phase systems of $N = 300$ water molecules (Do et al., 2011) and the solid state (Do and Wheatley, 2013). The key elements are: (a) uniform sampling of configuration space without using a Boltzmann weight; (b) progressive reduction of the accessible configuration space by removing the high-energy regions. Uniform sampling means choosing coordinates at random; as explained in Chapter 4, this means that the system will spend most of its time in very high-energy states in which there are many overlaps between particles, because the configurational density of states is a very rapidly increasing function of potential energy. This is not enough, in itself, but it gives an estimate of the density of states $\rho(V)$ at high potential energy. It is then possible to divide the set of sampled energies in two, such that a specified fraction f of them lie in the lower part, and a fraction $1 - f$ in the upper part. The next stage is restricted to simulating the lower-energy portion of the density of states in a similar way, building up more information about $\rho(V)$ in this region. At each stage i of the NS process, then, we do as follows.

1. Sample configurations randomly, subject to $V < V_i$, constructing an energy distribution $\rho(V)$. This can be done by Markov chain MC, rejecting moves that would violate the condition.
2. Determine the energy V_{i+1} at a fixed fraction f of $\rho(V)$ (a typical value would be $f = 1/2$, in which case V_{i+1} is the median value).
3. Increment i and repeat until some convergence criterion is met.

In the first stage, the upper energy limit is set to a very high value. Going from one stage to the next, it is necessary to generate the starting condition in some way consistent with the (reduced) upper limit on potential energy. One way of doing this is to use many independent 'walkers' at each stage, and at the start of the next stage prune the ones that lie above the new threshhold, and duplicate the ones that lie below it. This approach also helps reduce the danger of becoming trapped in a single low-lying minimum. We expect $\rho(V)$ at each stage to increase sharply with V up to the imposed limit; each successive stage samples a low-energy subset of the configurations sampled at the previous stage. This does not mean that the energy is dramatically reduced at each stage; indeed, it may only require a small lowering of the 'ceiling' to reduce the configuration-space volume by half. Afterwards, an estimate of the excess partition function at any chosen inverse temperature $\beta = 1/k_BT$, follows from

$$Q_{ex} \approx \frac{1}{W_f} \sum_i (f^{i-1} - f^i) \exp\left(-\tfrac{1}{2}\beta(V_{i-1} + V_i)\right), \quad \text{where} \quad W_f = \sum_i (f^{i-1} - f^i).$$

The term $(f^{i-1} - f^i)$ represents the fraction of configuration space that is explored between energies V_{i-1} and V_i, and the Boltzmann factor uses the mean of these energies as an estimate of the typical energy of that region. Nested sampling is not restricted to 'energy' space: it may be used to compute the isothermal–isobaric partition function (Pártay et al., 2014; Wilson et al., 2015).

A drawback of NS is that the random sampling method makes it difficult to simulate outside the framework of MC. Recently, it has been adapted so that conventional canonical ensemble simulations may be used (Nielsen, 2013) opening up the possibility of wider application.

9.2.7 Summary

Statistical properties, such as entropy and free energy, can be calculated directly by simulating in the grand canonical ensemble (see Sections 3.10, 4.6). Such simulations are not useful at high density without some biased sampling trick to improve the probability of a successful particle insertion.

Free-energy differences may be calculated by averaging the Boltzmann factor of the energy difference between the two systems. This is easy to incorporate as a black-box procedure, but is fraught with danger: whichever system is chosen as the reference in which to perform the simulation, there may be a poor overlap of the important regions of phase space with the other system. As emphasized by Lu and Kofke (2001a,b), usually one direction of perturbation (in which the entropy of the target is less than the entropy of the reference) is preferable to the other. It is almost never a good idea to simply average the results obtained by the forward and backward routes (Lu et al., 2003; Pohorille et al., 2010). Usually it is advisable to examine the underlying data, such as energy distributions, provided by forward and backward routes, as illustrated in Section 9.2.2. Bennett (1976) recommends calculating

$$g_0(\Delta \mathcal{V}) = \ln \rho_0(\Delta \mathcal{V}) - \tfrac{1}{2}\beta \Delta \mathcal{V}, \tag{9.36a}$$

$$g_1(\Delta \mathcal{V}) = \ln \rho_1(\Delta \mathcal{V}) + \tfrac{1}{2}\beta \Delta \mathcal{V}, \tag{9.36b}$$

$$g_1(\Delta \mathcal{V}) - g_0(\Delta \mathcal{V}) = \beta \Delta A \tag{9.36c}$$

and plotting g_0, g_1, and their difference against $\Delta \mathcal{V}$, which should be a constant in the region of overlap. This may or may not give a good estimate of ΔA, but at least it should be consistent with the estimate obtained, for instance, from the acceptance-ratio method, otherwise some reconsideration is needed (Pohorille et al., 2010).

The umbrella sampling method does give a useful route to free-energy differences. However, it cannot give absolute free energies, and there is always a subjective element in choosing the appropriate weighting function. Two systems that are quite different can only be linked by performing several intermediate simulations, even with the use of umbrella sampling at each stage. If there is any overlap between the distributions of configurational energy of a set of systems, then the MBAR method is a powerful route to the free-energy differences. It also provides estimates of the free-energy difference and configurational properties of any new states within the envelope of the systems studied.

Perhaps the most direct attack is to calculate the chemical potential by the particle insertion method in any ensemble (using the appropriate formula). This method is easy to program and fits neatly into an existing code. The additional time required for the calculation is approximately 20 % of the normal run time. This method may also fail at densities close to the triple point, although there is some disagreement about its precise range of validity. A useful check (as per eqn (9.36)) is to calculate the distribution of test particle energies and real molecule energies during a run. When the logarithm of the ratio of these distributions is plotted against $\beta \mathcal{V}_{\text{test}}$ it should be a straight line of slope one, and the intercept should be $-\beta \mu^{\text{ex}}$ (Powles et al., 1982). If this method gives a different

Example 9.1 Computational alchemy

The efficacy of a medicinal drug is determined, in part, by its ability to bind to the active site of a protein. The binding free energy of a molecule is the difference between the free energy of the drug in the active site and in aqueous solution. Jorgensen and Ravimohan (1985) first showed that the difference between the binding free energies of two molecules B and A, $\Delta\Delta G^{\text{binding}}$, could be calculated using a thermodynamic cycle, such that

$$\Delta\Delta G^{\text{binding}} = \Delta G^{\text{bound}}_{\text{BA}} - \Delta G^{\text{aq}}_{\text{BA}}$$

where ΔG_{BA} is the change in the Gibbs free energy when molecule A changes to molecule B in either the bound or aqueous environment. The Gibbs free energies of 'mutation' can be estimated using the perturbation approach of eqn (9.3) with the averages calculated in the constant-NPT ensemble using either a Monte Carlo method with 11 windows of overlap sampling (Jorgensen and Thomas, 2008) or molecular dynamics using the REST approach (Cole et al., 2015).

Inhibitors of the HIV-1 reverse transcriptase enzyme are important in anti-HIV therapy. A particularly potent inhibitor of the wild-type enzyme is the catechol diether (1) with X = Cl or F. However, there are concerns about the cyano-vinyl group in the structure and the possibility that electrophilic addition could lead to covalent modification of proteins and nucleic acids. For this reason Lee et al. (2013) used free-energy simulations to consider bicyclic (2) replacements of the cyano-vinyl group. A pyrrole ring with the N at position W is used as the reference compound. 18 perturbations of the five-membered ring using C, N, O at the W, X, Y, Z positions are considered. The calculation of $\Delta\Delta G^{\text{binding}}$ suggests important candidates for synthesis and for further studies of the crystal structure of the bound inhibitors. The indolizine (R, R' = H, R'' = CN, W, Y, Z = C, and X = N) is found to be particularly potent as measured by its EC_{50} value: the dose required to obtain 50 % protection of the infected cells.

result from the straightforward average of the Boltzmann factor of the test particle energy, then there is a problem with convergence. In this case the particle insertion should be enhanced by umbrella sampling (Shing and Gubbins, 1982).

The internal energy can be accurately calculated by simulation in the canonical ensemble, and the temperature can be accurately calculated in the microcanonical ensemble.

This makes the thermodynamic integration of eqn (2.72) an accurate route to free-energy differences. One possible disadvantage is that a large number of simulations may be required to span the integration range. This is not a problem if the aim of the simulation is an extensive exploration of the phase diagram, and one short cut is to plan simulations at appropriate temperatures along the integration range to enable you to perform a Gauss–Legendre quadrature of eqn (2.72) without the need for interpolation (Frenkel, 1986). One other possible difficulty is the requirement of finding a reversible path between the state of interest and some reference state. Ingenious attempts have been made to integrate along a thermodynamic path linking the liquid with the ideal gas (Hansen and Weis, 1969) or with the harmonic lattice (Hoover and Ree, 1968) without encountering the irreversibility associated with the intervening phase transitions. In the solid state, it may be necessary to apply an external field to reach the Einstein crystal (Frenkel and Ladd, 1984) and a similar technique may be used to calculate the free energy of a liquid crystal (Frenkel and Mulder, 1985).

In Section 11.6 we return to the calculation of free energies, by nonequilibrium work measurement. The direct approach involving averaging of Boltzmann factors, and the thermodynamic integration method, may be considered as extreme examples of this in the limits, respectively, of fast and slow perturbations.

9.3 Smarter Monte Carlo

In the conventional MC method, all the molecules are moved with equal probability, in directions chosen at random. This may not be the most efficient way to proceed: we might wish to attempt moves for some molecules more often than others, or to bias the moves in preferred directions. This preferential sampling can be accomplished using an extension of the Metropolis solution (eqn (4.21)) of the following form:

$$\pi_{mn} = \alpha_{mn} \qquad\qquad \alpha_{nm}\rho_n \geq \alpha_{mn}\rho_m \qquad m \neq n$$

$$\pi_{mn} = \alpha_{mn}\left(\frac{\alpha_{nm}\rho_n}{\alpha_{mn}\rho_m}\right) \qquad \alpha_{nm}\rho_n < \alpha_{mn}\rho_m \qquad m \neq n$$

$$\pi_{mm} = 1 - \sum_{n \neq m} \pi_{mn}. \tag{9.37}$$

We recall that π_{mn} is the one-step transition probability of going from state m to state n. In this case it is easy to show that microscopic reversibility is satisfied, even if $\alpha_{mn} \neq \alpha_{nm}$. The Markov chain can be easily generated by making random trial moves from state m to state n according to α_{mn}. The trial move is accepted with a probability given by $\min(1, \alpha_{nm}\rho_n/\alpha_{mn}\rho_m)$. The details of this type of procedure are given in Section 4.4. We make use of the prescription of eqn (9.37) in the following.

9.3.1 Preferential sampling

In a dilute solution of an ion in water, for example, the most important interactions are often those between solute and solvent, and between solvent molecules in the primary solvation shell. The solvent molecules further from the ion do not play such an important role. It is sensible to move the molecules in the first solvation shell more frequently than the more remote molecules. Let us define a region $\mathcal{R}_{sol}$ around the solute molecule, solvent molecules within the region being designated 'in', and the remainder being 'out'. A parameter p defines how often we wish to move the 'out' molecules relative to the 'in' ones: p lies between 0 and 1, smaller values corresponding to more frequent moves of the 'in' molecules. A move consists of the following steps (Owicki and Scheraga, 1977b).

(a) Choose a molecule at random.
(b) If it is 'in', make a trial move.
(c) If it is 'out', generate a random number uniformly on $(0, 1)$. If p is greater than the random number then make a trial move. If not, then return to step (a).

In step (c), if it is decided not to make a trial move, we return to step (a) immediately, and select a new molecule, without accumulating any averages. This procedure will attempt 'out' molecule moves with probability p relative to 'in' molecule moves. Trial moves are accepted with a probability $\min(1, \alpha_{nm}\rho_n/\alpha_{mn}\rho_m)$ and the problem is to calculate the ratio α_{nm}/α_{mn} for this scheme. Consider a configuration m with N_{in} 'in' molecules and N_{out} 'out' molecules. The chance of selecting an 'in' molecule is

$$p_{in} = \frac{N_{in}}{N} + (1-p)\frac{N_{out}}{N}\frac{N_{in}}{N} + \left((1-p)\frac{N_{out}}{N}\right)^2\frac{N_{in}}{N} + \ldots = \frac{N_{in}}{N'} \tag{9.38}$$

where $N' = pN + (1-p)N_{in}$. Note how, in eqn (9.38) we count all the times that we look at 'out' molecules, decide not to try moving them, and return to step (a), eventually selecting an 'in' molecule.

Once we have decided to attempt a move, there are four distinct cases, corresponding to the moving molecule in states m and n being 'in' or 'out' respectively. Let us consider the case in which we attempt to move a molecule which was initially 'in' the region $\mathcal{R}_{sol}$ to a position outside that region (see Fig. 9.3). Suppose that trial moves may occur to any of $N_{\mathcal{R}}$ positions within a cube $\mathcal{R}$ centred on the initial position of the molecule. Then α_{mn} is the probability of choosing a specific 'in' molecule, and attempting to move it to one of these sites as shown in Fig. 9.3:

$$\alpha_{mn} = \frac{1}{N_{in}}\frac{N_{in}}{N'}\frac{1}{N_{\mathcal{R}}} = \frac{1}{N'N_{\mathcal{R}}}. \tag{9.39}$$

The chance of attempting the reverse move, from a state containing $N_{out} + 1 = N - N_{in} + 1$ 'out' molecules and $N_{in} - 1$ 'in' molecules, is

$$\alpha_{nm} = \frac{1}{N - N_{in} + 1}\left(1 - \frac{N_{in} - 1}{pN + (1-p)(N_{in} - 1)}\right)\frac{1}{N_{\mathcal{R}}} = \frac{p}{N'N_{\mathcal{R}}}\left(1 - \frac{1-p}{N'}\right)^{-1} \tag{9.40}$$

and the desired ratio can be obtained. Summarizing for all four cases we have

$m \to n$	α_{nm}/α_{mn}	
in $\to$ out	$p\left(1 - (1-p)/N'\right)^{-1}$	(9.41a)
out $\to$ out	1	(9.41b)
in $\to$ in	1	(9.41c)
out $\to$ in	$\left[p\left(1 + (1-p)/N'\right)\right]^{-1}$	(9.41d)

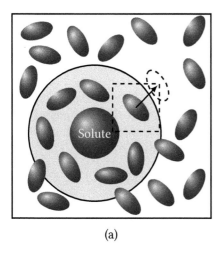 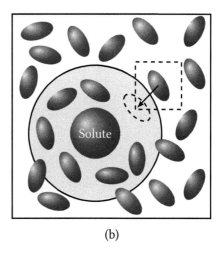

(a)   (b)

Fig. 9.3 (a) Attempting to move a solvent molecule out of the region $\mathcal{R}_{\text{sol}}$ around the solute (shaded). The cube of possible new positions $\mathcal{R}$ centred on the initial position of the molecule is indicated by dashed lines. (b) Attempting the reverse move.

where N' is calculated for N_{in} molecules in state m. In the simulation, p is chosen so that the initial probability of attempting an 'in' molecule move is typically 0.5. In an unweighted simulation, the probability of moving 'in' molecules obviously depends on the system size, but would be much lower, say $10\,\%$–$20\,\%$.

Owicki (1978) has suggested an alternative method for preferential sampling which has been used in the simulation of aqueous solutions (Mehrotra et al., 1983). In this method, the probability of choosing a solvent molecule decays monotonically with its distance from the solute. We define a weighting function, which typically takes the form

$$W'(r_{i0}) = r_{i0}^{-\nu} \tag{9.42}$$

where ν is an integer. Here r_{i0} is the distance of molecule i from the solute, which we label 0. At any instant, a properly normalized weight may be formed

$$W(r_{i0}) = W'(r_{i0}) \Big/ \sum_{j} W'(r_{j0}) \tag{9.43}$$

and used to define a probability distribution for the current configuration. A molecule i is chosen from this distribution using a rejection technique as described in Appendix E. An attempted move is then made to any of the $N_{\mathcal{R}}$ neighbouring positions. Denoting $W(r_{i0})$ in the initial and final states simply by W_m and W_n respectively, the required ratio of underlying transition matrix elements is

$$\alpha_{nm}/\alpha_{mn} = W_n/W_m. \tag{9.44}$$

We have tacitly assumed that the solute molecule is fixed. This is permissible, but relaxation of the first neighbour shell will be enhanced if the solute is allowed to move as well. This will, of course, change all the interactions with solvent molecules. In the scheme just described, the solute may be moved as often as desired, with $\alpha_{nm}/\alpha_{mn} = 1$, without any additional modifications.

A useful example of preferential sampling is the cavity-biased GCMC method (Mezei, 1980). GCMC becomes less useful at high densities because of the difficulty of making successful creation and destruction attempts. In the cavity-biased method insertion is only allowed at points where a cavity of a suitable radius, r_c, exists. The probabilities of accepting a creation or destruction attempt (eqns (4.41) and (4.42)) are modified by an additional factor p_N, the probability of finding a cavity of radius r_c, or larger, in a fluid of N molecules. A creation attempt is accepted with a probability given by

$$\min\left(1, \exp\left[-\beta\delta\mathcal{V}_{nm} + \ln\left(\frac{zVp_N}{N+1}\right)\right]\right) \tag{9.45a}$$

and a destruction attempt is accepted with a probability given by

$$\min\left(1, \exp\left[-\beta\delta\mathcal{V}_{nm} + \ln\left(\frac{N}{zVp_{N-1}}\right)\right]\right). \tag{9.45b}$$

The simulation is realized by distributing a number of test sites uniformly throughout the fluid. During the run each site is tested to see whether it is at the centre of a suitable cavity or not. In this way p_N is calculated with a steadily improving reliability and at the same time it is possible to locate points in the fluid suitable for an attempted creation. In the event that no cavity is available, we can continue with the next move or use a scheme which mixes the cavity sampling with the more conventional GCMC method. Details of the mixed scheme, which requires complicated book-keeping to ensure microscopic reversibility, are given in the original paper. Mezei reports an eightfold increase in the efficiency of creation/destruction attempts in a simulation of a supercritical Lennard-Jones fluid. Cavity bias has been included in the set of moves used by the MC module of the CHARMM package (Hu et al., 2006). It can be considered a precursor to configurational-bias Monte Carlo (CBMC) (see Section 9.3.4) and an extension of the method has been used to speed up reaction ensemble MC (see Section 9.5). Loeffler et al. (2015) have adapted the grand canonical Monte Carlo algorithm by introducing an energy-bias method for trial insertions and deletions. This approach allows molecules to be inserted into and removed from regions of high (least negative) energy in the fluid (e.g. at the surface of a cluster). This approach is combined with either the aggregation-volume-bias (AVB) MC method (Chen and Siepmann, 2000) or the unbonding–bonding (UB) MC method (Wierzchowski and Kofke, 2001). The two algorithms preferentially insert or remove molecules in a particular sub-region of fluid (e.g. the bound region directly around another molecule). The combined targetting of particular regions and energies can deliver a significant increase in insertion/removal efficiency and an accelerated rate of convergence for the thermodynamic properties of the system (Loeffler et al., 2015).

9.3.2 Force-bias Monte Carlo

In real liquids, the movement of a molecule is biased in the direction of the forces acting on it. It is possible to build this bias into the underlying stochastic matrix α of the Markov chain. The reason for adopting a force-bias scheme is to improve convergence to the limiting distribution, and steer the system more efficiently around the bottlenecks of phase space (see Section 4.3). Pangali et al. (1978) adopt the following prescription for the underlying Markov chain:

$$\alpha_{mn} = \begin{cases} \exp\left(+\lambda\beta(\mathbf{f}_i^m \cdot \delta\mathbf{r}_i^{nm})\right)/C(\mathbf{f}_i^m, \lambda, \delta r_{\max}) & n \in \mathcal{R} \\ 0 & n \notin \mathcal{R}. \end{cases} \tag{9.46}$$

Here we have assumed that just one atom i is to be moved, $\mathbf{f}_i^m$ is the force on this atom in state m, $\delta\mathbf{r}_i^{nm} = \mathbf{r}_i^n - \mathbf{r}_i^m$ is the displacement vector in a trial move to state n, λ is a constant and C is a normalizing factor. Typically, λ lies between 0 and 1. When $\lambda = 0$, eqn (9.46) reduces to eqn (4.27) for the conventional transition probability. As usual, $\mathcal{R}$ is a cube of side $2\delta r_{\max}$ centred on the initial position $\mathbf{r}_i^m$ (see Fig. 4.2). A little manipulation shows that

$$C(\mathbf{f}_i^m, \lambda, \delta r_{\max}) = \prod_{\alpha=x,y,z} \frac{2\sinh(\lambda\beta\delta r_{\max} f_{i\alpha}^m)}{\lambda\beta f_{i\alpha}^m}. \tag{9.47}$$

It is clear from eqn (9.46) that this prescription biases $\delta\mathbf{r}_i^{nm}$ in the direction of the force on the atom.

The force bias (FB) method is implemented as follows. An atom i is chosen at random and given a trial random displacement $\delta\mathbf{r}_i^{nm}$ selected using a rejection technique (see Appendix E) from the probability distribution determined by eqn (9.46). The trial move is accepted with a probability given by $\min(1, \alpha_{nm}\rho_n/\alpha_{mn}\rho_m)$. The ratio appearing here is given by

$$\frac{\alpha_{nm}\rho_n}{\alpha_{mn}\rho_m} = \exp\left[-\beta\left(\delta\mathcal{V}_{nm} + \lambda\delta\mathbf{r}_i^{nm} \cdot (\mathbf{f}_i^m + \mathbf{f}_i^n) + \delta W^{\mathrm{FB}}\right)\right] \tag{9.48}$$

where

$$\delta W^{\mathrm{FB}} = -k_{\mathrm{B}}T \ln\left(\frac{C(\mathbf{f}_i^m, \lambda, \delta r_{\max})}{C(\mathbf{f}_i^n, \lambda, \delta r_{\max})}\right) \tag{9.49}$$

can be calculated using eqn (9.47). For small values of $\delta r_{\max}$

$$\delta W^{\mathrm{FB}} \approx \tfrac{1}{6}\lambda^2\beta\delta r_{\max}^2\left(|\delta\mathbf{f}_i^{nm}|^2 + 2\delta\mathbf{f}_i^{nm} \cdot \mathbf{f}_i^m\right) \tag{9.50}$$

where $\delta\mathbf{f}_i^{nm} = \mathbf{f}_i^n - \mathbf{f}_i^m$. The two parameters in the method, λ and $\delta r_{\max}$, can be adjusted to maximize the root-mean-square displacement of the system through phase space: a simple, though not unique, measure of efficiency (Rao et al., 1979; D'Evelyn and Rice, 1981). The FB method is particularly powerful when dealing with hydrogen-bonded liquids such as water, which are susceptible to bottlenecks in phase space. Molecular translation is handled as just described, and the extension to include torque-biased rotational moves is straightforward. The analogous equation to eqn (9.46) is

$$\alpha_{mn} = C^{-1}\exp\left(+\lambda\beta\mathbf{f}_i^m \cdot \delta\mathbf{r}_i^{nm} + \lambda\beta\boldsymbol{\tau}_i^m \cdot \delta\boldsymbol{\phi}_i^{nm}\right) \qquad n \in \mathcal{R} \tag{9.51}$$

where $\boldsymbol{\tau}_i^m$ is the torque on molecule i in state m, and $\delta\boldsymbol{\phi}_i^{nm}$ is the trial angular displacement, that is, $\delta\boldsymbol{\phi}_i^{nm} = \delta\phi^{mn}\mathbf{e}$ where $\mathbf{e}$ is the axis of rotation. A study of a model of water using force bias (Rao et al., 1979) demonstrated clear advantages over conventional MC methods, and better agreement with MD results for this system. A further study (Mehrotra et al., 1983) showed an improvement in convergence by a factor 2 to 3 over conventional MC.

9.3.3 Smart Monte Carlo

Force-bias Monte Carlo involves a combination of stochastic and systematic effects on the choice of trial moves. A similar situation applies to the motion of a Brownian molecule in a fluid: it moves around under the influence of random forces (from surrounding solvent molecules) and systematic forces (from other nearby Brownian molecules). We will turn to the simulation of Brownian motion in Chapter 12 when we describe a range of coarse-grained methods, and simply describe here the smart Monte Carlo (SMC) scheme devised by Rossky et al. (1978), which is derived from it. The trial displacement of a molecule i from state m to state n may be written

$$\delta \mathbf{r}_i^{nm} = \beta A \mathbf{f}_i^m + \delta \mathbf{r}_i^{G}. \tag{9.52}$$

$\delta \mathbf{r}_i^{G}$ is a random displacement whose components are chosen from a Gaussian distribution with zero mean and variance $\langle (\delta r_{i\alpha}^{G})^2 \rangle = 2A$, $\alpha = x, y, z$. The quantity A is an adjustable parameter (equal to the diffusion coefficient multiplied by the timestep in a Brownian dynamics simulation). The underlying stochastic matrix for this procedure is

$$\alpha_{mn} = (4A\pi)^{-3/2} \exp\left(-\left|\delta \mathbf{r}_i^{nm} - \beta A \mathbf{f}_i^m\right|^2 / 4A\right). \tag{9.53}$$

In practice a trial move consists of selecting a random vector $\delta \mathbf{r}_i^{G}$ from a Gaussian distribution as described in Appendix E, and using it to displace a molecule chosen at random according to eqn (9.52). The move is accepted with probability $\min(1, \alpha_{nm}\rho_n / \alpha_{mn}\rho_m)$ (see eqn (9.37)) and the required ratio is

$$\frac{\alpha_{nm}\rho_n}{\alpha_{mn}\rho_m} = \exp\left[-\beta\left(\delta \mathcal{V}_{nm} + \tfrac{1}{2}(\mathbf{f}_i^n + \mathbf{f}_i^m) \cdot \delta \mathbf{r}_i^{nm} + \delta W^{\mathrm{SMC}}\right)\right] \tag{9.54}$$

where

$$\delta W^{\mathrm{SMC}} = \frac{\beta A}{4}\left(|\delta \mathbf{f}_i^{nm}|^2 + 2\delta \mathbf{f}_i^{nm} \cdot \mathbf{f}_i^m\right) \tag{9.55}$$

and the notation is the same as for eqn (9.50). Rossky et al. (1978) tested the method by simulating ion clusters, and Northrup and McCammon (1980) have used SMC to study protein structure fluctuations. There are clear similarities, and slight differences, between the FB and SMC methods. One difference is that eqn (9.53) puts no upper limit on the displacement of a given molecule at any step, using a Gaussian probability distribution instead of a cubic trial displacement region. However, if we write eqn (9.53) in the form

$$\alpha_{mn} = (4A\pi)^{-3/2} \exp\left(-\frac{\beta^2 A}{4}|\mathbf{f}_i^m|^2 - \frac{1}{4A}|\delta \mathbf{r}_i^{nm}|^2\right) \exp\left(+\tfrac{1}{2}\beta \mathbf{f}_i^m \cdot \delta \mathbf{r}_i^{nm}\right) \tag{9.56}$$

and compare with eqn (9.46), we note that the distributions are particularly similar for $\lambda = \tfrac{1}{2}$. For this choice, the two ratios governing acceptance of a move are identical if (Rao et al., 1979)

$$\delta W^{\mathrm{SMC}} = \delta W^{\mathrm{FB}} \tag{9.57}$$

and for small step sizes, this holds for $A = \tfrac{1}{6}\delta r_{\mathrm{max}}^2$. Comparisons between the two techniques are probably quite system-dependent. Both offer a substantial improvement over conventional MC on a step-by-step basis in many cases, but they are comparable with molecular dynamics in complexity and expense since they involve calculation of forces and torques. Both methods improve the acceptance rate of moves. The most efficient method, in this sense, would make $\alpha_{nm}/\alpha_{mn} = \rho_m/\rho_n$ when every move would be accepted, but of course ρ_n is not known before a move is tried. SMC, and FB with $\lambda = \frac{1}{2}$, both approach 100 % acceptance rates quadratically as the step size is reduced. This makes multi-molecule moves more feasible. In fact, SMC simulations with N-molecule moves and small step sizes are almost identical with the Brownian dynamics (BD) or Schmoluchowski equation simulations of Chapter 12. The extra flexibility of SMC and FB methods lies in the possibility of using larger steps (and rejecting some moves) and also in being able to move any number of molecules from 1 to N. An example program appears in Code 12.2.

FB and SMC simulation are not limited to the canonical ensemble. Mezei (1983) has introduced a virial-bias volume move for simulations in the constant-NPT ensemble and a similar technique can be used in an SMC simulation (Mezei, 1983). SMC simulations have also been used to simulate crystal growth from Lennard-Jones solutions (Huitema et al., 1999) and polymer systems in the melt and lamellar microphases (Müller and Daoulas, 2008). More recently, Moucka et al. (2013) have compared various FB and SMC algorithms for the simulation of polarizable models of water and aqueous electrolytes.

9.3.4 Configurational-bias Monte Carlo

In the basic Metropolis method, an atom is moved to a new trial position by sampling the space around its old position randomly and uniformly (see Fig. 4.2). However, it would be more efficient to move the atom to a point in the surrounding space where its potential energy is large and negative, and the probability density of the trial configuration is correspondingly high. To accomplish this, we would need a probability map of the space surrounding the atom in advance of the trial move. This is precisely the technique used in the CBMC method. A detailed probability map of the space surrounding the atom is constructed and a trial move is made by sampling from this probability distribution (i.e. biasing the move to positions of high probability). The underlying stochastic matrix, α_{mn}, is then known for the forward move. Since we also need the underlying stochastic matrix α_{nm} for the move back, a similar mapping strategy is applied in the direction from the chosen trial state to the original atom position.

To describe the method in detail we will consider the MC simulation of a polymer chain in a solvent. In Chapter 4 we described the use of a number of Monte Carlo methods to study the properties of polymer melts and polymer solutions. The problem with these methods is that it is difficult to move the polymer to a new trial position without creating a significant overlap with a solvent molecule or with another polymer chain. It might be possible to place the first monomer in a chain at a new trial position of low energy by randomly sampling the space around its current position, but by the time four or five monomer beads have been added in this way the chance of a significant overlap and a rejection are very high.

One way to avoid this problem is to sample the space around a given bead in the polymer chain in a biased way, so that one attempts to add the next bead in a favourable position of low energy and high probability (Siepmann and Frenkel, 1992). CBMC requires the creation of a map of the energy around a given monomer as the chain is grown and the

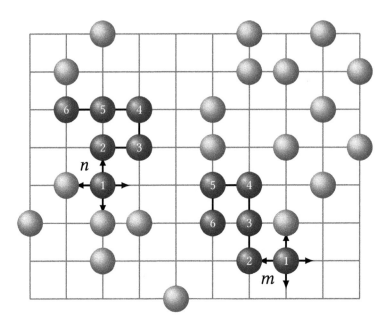

Fig. 9.4 A six-bead polymer on a square lattice. m denotes the old chain and n the new one. The arrows point to candidate positions for bead 2. The light-grey beads represent solvent atoms.

calculation of the so-called Rosenbluth weights, associated with this biasing (Rosenbluth and Rosenbluth, 1955).

This approach can be illustrated by considering a polymer (of say, six monomers) plus a set of solvent beads, on a 2D periodic square lattice, as shown in Fig. 9.4. Each site may only be occupied by one bead (solvent or monomer), or can be vacant. For simplicity, we assume nearest-neighbour interactions between solvent beads, between solvent and monomer beads, and between monomer beads which are not directly bonded to each other. Moves of the solvent atoms are handled by standard methods; here we consider moves of the polymer chain. At the beginning of a trial move, the polymer in state m, occupying positions $\mathbf{r}_1^m \cdots \mathbf{r}_6^m$, is removed from the lattice. To construct the new trial position for the polymer, state n, we choose a random position, $\mathbf{r}_1^n$, for the first monomer. The weight of this monomer is defined as $w_n(1) = k \exp(-\beta \mathcal{V}_1^n(\mathbf{r}_1^n))$ where $k = 4$ is the coordination number of the lattice and $\mathcal{V}_1^n(\mathbf{r}_1^n)$ is the interaction energy between this monomer and solvent atoms on the lattice. To place the second monomer, we consider the four available lattice points, surrounding $\mathbf{r}_1^n$, labelling them $j = 1, \ldots, 4$ (arrowed in Fig. 9.4). We calculate the potential energy $\mathcal{V}_2^n(\mathbf{r}_{2,j}^n)$ of monomer 2 at position j with the solvent atoms on surrounding lattice sites (excluding the intramolecular interaction with monomer 1, because it is bonded to 2). From these energies we construct the probability of placing the monomer at each point j:

$$p(\mathbf{r}_{2,j}^n) = \frac{\exp\left(-\beta \mathcal{V}_2^n(\mathbf{r}_{2,j}^n)\right)}{w_n(2)}, \quad \text{where} \quad w_n(2) = \sum_{j=1}^{k} \exp\left(-\beta \mathcal{V}_2^n(\mathbf{r}_{2,j}^n)\right); \qquad (9.58)$$

$w_n(2)$ is the Rosenbluth weight of monomer 2. We now select one of the four possible positions $\mathbf{r}_{2,j}^n$ by sampling randomly from the distribution of eqn (9.58); let $\mathbf{r}_2^n$ denote the chosen position and write the energy contribution as $\mathcal{V}_2^n(\mathbf{r}_2^n)$. The contribution of this part of the move to the underlying stochastic matrix α_{mn} is $\exp(-\beta\mathcal{V}_2^n(\mathbf{r}_2^n))/w_n(2)$. We continue to build the complete chain in this way for each of the six monomers i. Note that k is effectively reduced from four to three for $i > 2$, since one of the neighbouring positions $(\mathbf{r}_{i-1}^n)$ is already occupied. In general, the potential-energy term $\mathcal{V}_i^n(\mathbf{r}_{i,j}^n)$ will include interactions with non-bonded previously placed monomer beads (e.g. in Fig. 9.4, monomer 5 interacts with monomer 2, and this term would appear in $\mathcal{V}_5^n(\mathbf{r}_5^n)$), as well as with solvent atoms. The overall underlying matrix for the forward move is a product of the probabilities for each stage (each of which is conditional on the result of the previous one):

$$\alpha_{mn} = \prod_{i=1}^{6} \frac{\exp\left(-\beta\mathcal{V}_i^n(\mathbf{r}_i^n)\right)}{w_n(i)} = \frac{\exp\left(-\beta\sum_{i=1}^{6}\mathcal{V}_i^n(\mathbf{r}_i^n)\right)}{\prod_{i=1}^{6}w_n(i)} = \frac{\exp(-\beta\mathcal{V}_n)}{W_n} \tag{9.59}$$

where W_n is the overall Rosenbluth weight of the new configuration

$$W_n = \prod_{i=1}^{6} w_n(i). \tag{9.60}$$

Notice how, in eqn (9.59), the partial energies $\mathcal{V}_i^n(\mathbf{r}_i^n)$ calculated at each stage sum up to the total potential energy $\mathcal{V}_n$ associated with the insertion of the chain: each contribution (monomer–solvent and monomer–monomer) is counted once, as it should be. At any stage in the process, one of the candidate positions may coincide with a solvent bead, or an already-placed monomer bead (e.g. two of the four candidates for $\mathbf{r}_2^n$ in Fig. 9.4 are already occupied): in this case, the potential energy would be infinite and the contribution to the Rosenbluth weight would be zero. In the event that *all* of the candidate sites are occupied, the Rosenbluth weight for that stage will be zero, and the whole move can be rejected without any further calculation.

The stochastic matrix for the reverse move α_{nm} can be constructed by considering the Rosenbluth weights for the conformation of the chain in state m. The procedure is similar to the one just outlined, with the simplifying feature that the actual monomer positions $\mathbf{r}_i^m$ are already known, and there is no need to select them from a distribution such as (9.58). We first calculate the weight $w_m(1) = k\exp(-\beta\mathcal{V}_1^m(\mathbf{r}_1^m))$ for the first monomer in its original position. In the next step we determine the energy of monomer 2 in its original position $\mathbf{r}_2^m$ and at the other three positions $\mathbf{r}_{2,j}^m$, $j = 1, \ldots, k-1$ at which it could have been placed around monomer 1, if we were actually conducting the reverse move. The Rosenbluth weight is

$$w_m(2) = \exp\left(-\beta\mathcal{V}_2^m(\mathbf{r}_2^m)\right) + \sum_{j=1}^{k-1}\exp\left(-\beta\mathcal{V}_2^m(\mathbf{r}_{2,j}^m)\right). \tag{9.61}$$

As before, $\mathcal{V}_2^m$ contains only interactions between the monomer and nearby solvent beads. The process is continued for each monomer i in sequence: the Rosenbluth weight is calculated from the actual position $\mathbf{r}_i^m$ and the $k-1$ other positions $\mathbf{r}_{i,j}^m$ surrounding monomer $i - 1$ which might have been selected in a reverse move, and in general $\mathcal{V}_i^m$ will include interactions between monomer i and previously placed non-bonded monomers (with indices $< i$), as well as solvent. After the whole chain m has been retraced, the stochastic matrix for the reverse move may be written

$$\alpha_{nm} = \prod_{i=1}^{6} \frac{\exp\left(-\beta \mathcal{V}_i^m(\mathbf{r}_i^m)\right)}{w_m(i)} = \frac{\exp\left(-\beta \sum_{i=1}^{6} \mathcal{V}_i^n(\mathbf{r}_i^m)\right)}{\prod_{i=1}^{6} w_m(i)} = \frac{\exp(-\beta \mathcal{V}_m)}{W_m} \qquad (9.62)$$

where

$$W_m = \prod_{i=1}^{6} w_m(i). \qquad (9.63)$$

Microscopic reversibility in the configurational-bias algorithm is achieved if we select trial states with a probability given by

$$\min\left(1, \frac{\alpha_{nm}\rho_n}{\alpha_{mn}\rho_m}\right) = \min\left(1, \frac{W_n}{W_m}\right) \qquad (9.64)$$

where we have substituted eqns (9.59) and (9.62) for α_{mn} and α_{nm}. Thus, in CBMC, all of the configurational weighting is included in the underlying matrix used to create the trial moves.

The extension of this method to off-lattice fluids can be illustrated by considering a chain of Lennard-Jones atoms joined by harmonic springs

$$v^s(d) = \tfrac{1}{2}\kappa(d - d_{eq})^2$$

where d is the distance between successive atoms, d_{eq} is the equilibrium bond length and κ is the spring constant. In this example the spring is the only intramolecular bonding potential in the problem; there are no bond bending or dihedral potentials controlling the conformation of the chain. Lennard-Jones interactions exist between non-bonded atoms in the chain, as well as between monomers and surrounding solvent atoms.

The method is essentially the same as that applied to the lattice problem with two additional considerations. First, we can no longer calculate the Rosenbluth weights of the growing chain exactly (since we do not have a finite number of lattice directions). In this case we can choose k random locations for each monomer (where k is typically 30–50) to sample the energy landscape around the growing chain. One can show, formally, that the convergence of the Markov chain to its correct limiting probability is unaffected by the choice of k. Second, the intramolecular potentials need to be included. This can be achieved by sampling the bond vectors directly as the new chain is constructed. For example, the possible k positions for atom 2 are

$$\mathbf{r}_{2,j}^n = \mathbf{r}_1^n + d_j \mathbf{e}_j, \qquad j = 1 \dots k, \qquad (9.65)$$

where the bond length d_j is chosen from the distribution $d^2 \exp[-\beta v^s(d)]$ (the factor d^2 comes from the volume element in spherical coordinates) and its direction $\mathbf{e}_j$ is chosen randomly on the unit sphere (see E.3 and E.4 for details). The additivity of the spring and Lennard-Jones potentials means that the corresponding probabilities can be simply multiplied, that is, in constructing α_{mn}, the Boltzmann factor $\exp[-\beta v^s(d)]$ has already been implicitly included through this sampling method.

The general procedure is exactly as for the lattice model. The first step in a trial move of the polymer from state m to n is to remove the chain in state m and to randomly place the first bead of the new chain at the trial position $\mathbf{r}_1^n$. The corresponding Rosenbluth weight is $w_n(1) = k \exp[-\beta \mathcal{V}_1^n(\mathbf{r}_1^n)]$, where $\mathcal{V}_1^n$ consists of Lennard-Jones interactions of the monomer with solvent atoms. The next step is to select k candidate positions for atom 2, as described earlier. The corresponding potential energies $\mathcal{V}_2^n(\mathbf{r}_{2,j}^n)$ again consist just of Lennard-Jones interactions with the solvent; the intramolecular spring potential is *not* included, as just explained. These k potential-energy values are used to construct a Rosenbluth weight, and to select one of the k positions from a probability distribution, defined by eqn (9.58); this position becomes $\mathbf{r}_2^n$. This procedure is repeated to build the trial positions for successive atoms in the chain. Note that, as the chain grows, it will be necessary to include the Lennard-Jones interaction of the trial chain atom i with the non-bonded atoms in the same chain that have already been placed, as well as with solvent atoms. Once the entire chain has been constructed, the Rosenbluth factor, W_n, is calculated from an equation like (9.60).

The Rosenbluth weight of the old state, m, is calculated in a similar manner to the lattice case. First, $w_m(1) = k \exp[-\beta \mathcal{V}_1^m(\mathbf{r}_1^m)]$ is calculated. Then, for atom 2, the original position, $\mathbf{r}_2^m$, supplemented by $(k-1)$ further positions $\mathbf{r}_{2,j}^m$ around $\mathbf{r}_1^m$, chosen by the random sampling method just described, are used to calculate the Rosenbluth weight $w_m(2)$ through eqn (9.61). This procedure continues until the whole length of the old chain has been retraced. At each stage, just as for the forward move, Lennard-Jones interactions of each monomer $\mathbf{r}_i^m$ or $\mathbf{r}_{i,j}^m$, $j = 1 \dots k-1$ with solvent atoms, as well as with previously considered monomers $\mathbf{r}_{i'}^m$ whose index is $i' < i$, are included in the potential energies used to calculate $w_m(i)$, but the spring potentials are omitted, since they are used implicitly in the sampling of positions. The overall Rosenbluth weight for state m is given by an equation like (9.63). The trial state n is accepted with a probability given by $\min(1, W_n/W_m)$ as before. An example of this kind of calculation is given in Code 9.2.

The method has been described for a simulation of one polymer chain in a solvent of atoms but can be straightforwardly extended to a fluid of chains (a polymer melt). In this case, one of the polymers in the old state m is selected at random and moved in the way just described. As always, if the move is rejected, the trial polymer is removed, the polymer in the old configuration is restored, and the old state m is recounted in the Markov chain.

The method can be readily extended to more realistic polymer potentials, where the chain is described by not only bond-stretching potentials, but also bond-angle potentials, $v^{\text{bend}}(\theta)$, and torsional potentials, $v^{\text{torsion}}(\phi)$. Here θ is the angle between adjacent bonds and ϕ is the dihedral angle describing the relative orientation of four adjacent monomers (see Appendix C.2). In many cases, it is still possible to include these extra terms directly in the sampling of monomer positions, and hence avoid incorporating them explicitly in the Rosenbluth weights. The extra $v^{\text{bend}}(\theta)$ terms may arise, when choosing $\mathbf{r}_{i,j}^n$ after $\mathbf{r}_{i-1}^n$ and $\mathbf{r}_{i-2}^n$ since θ is defined in terms of those positions; similarly a $v^{\text{torsion}}(\phi)$ term may involve $\mathbf{r}_{i,j}^n$ and the positions of the three preceding monomers. The angle θ is sampled from the distribution $\sin\theta \exp[-\beta v^{\text{bend}}(\theta)]$ (again, the factor $\sin\theta$ comes from the volume

Code 9.2   Configuration-biased simulation of chain molecule

This file is provided online. mc_chain_nvt_cbmc_lj.f90 carries out a CBMC simulation of a single Lennard-Jones chain, illustrating the calculation of Rosenbluth weights from the non-bonded interactions. The program also illustrates the direct sampling of intramolecular bond lengths, dictated by a harmonic spring potential. Routines to calculate energies and carry out moves are contained in mc_chain_lj_module.f90 and, as usual, the utility modules described in Appendix A handle various functions. The program initialize.f90 (Code 5.6) may be employed to prepare initial configurations.

```
! mc_chain_nvt_cbmc_lj.f90
! Monte Carlo, single chain, NVT, CBMC
PROGRAM mc_chain_nvt_cbmc_lj


! mc_chain_lj_module.f90
! Monte Carlo, single chain, LJ atoms
MODULE mc_module
```

element in polar coordinates) and ϕ from the distribution $\exp[-\beta v^{\text{torsion}}(\phi)]$. Then, the unit vector $\mathbf{e}_i$ in eqn (9.65) is expressed in terms of θ, ϕ, and the positions of the preceding monomers.

CBMC has now been widely applied in the simulation of polymer solutions and melts. It has been used to model fluids of normal alkanes with as many as 70 carbon atoms (de Pablo et al., 1993) and biologically important systems of linear and cyclic peptides (Deem and Bader, 1996). The method has also been used to make trial insertions of polymer chains into fluids in grand canonical simulations and it provides a route to the free energy in such systems. Using this approach, Smit (1995) has used constant-μ, V, T simulations to calculate the adsorption isotherms of butane and hexane in the zeolite silicate.

CBMC can be extended to study branched polymer systems (Dijkstra, 1997). For a simple twofold branch (e.g. 3 goes to 4 and 4′) it is necessary to generate two random vectors ($\mathbf{r}_{34}$ and $\mathbf{r}_{34'}$) on the surface of a sphere with probabilities proportional to the Boltzmann factors associated with all of the angle potentials $v^{\text{bend}}(\theta_{234})$, $v^{\text{bend}}(\theta_{234'})$, and $v^{\text{bend}}(\theta_{434'})$. More complicated branching geometries have been successfully tackled using a two-stage biased insertion scheme (Martin and Siepmann, 1999).

CBMC can be combined with some of the other techniques discussed later in this chapter for the efficient calculation of the phase diagrams of polymer-containing systems. For example, CBMC has been used with the Gibbs ensemble Monte Carlo method to model the selective adsorption of linear and branched alkanes from methanol onto a carbon slit pore (Bai and Siepmann, 2013), and Maerzke et al. (2012) have combined CBMC with the Wang–Landau algorithm to simulate the density of states of complex chemical systems.

Finally, apart from the obvious advantages when modelling polymer systems, CBMC can be used to good effect for any systems where there is a strong association of the monomers. Recently, McGrath et al. (2010) have used the technique to perform first

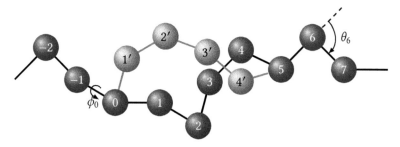

Fig. 9.5 A concerted rotation driven by a change in the driver angle ϕ_0. The light-grey beads, atoms $1', 2', 3'$ and $4'$, move during the rotation.

principles Monte Carlo calculations of clusters of ten HF molecules. In this case, CBMC is used to increase the sampling efficiency of the cluster formation and destruction.

9.3.5 Concerted rotations

Theodorou (2010) and his co-workers have developed a series of new moves for the equilibration of dense polymer systems. Unlike the CBMC method, which works from the end of the polymer by threading segments through the surrounding dense phase, the concerted rotation algorithms use trial moves that rearrange the conformation of interior segments.

The simplest form of concerted rotation algorithm (Dodd et al., 1993) involves changing seven consecutive torsional angles in the chain. In the example shown in Fig. 9.5, the concerted rotation begins by changing the driver angle ϕ_0 by some randomly chosen amount between $-\pi$ and $+\pi$. The six torsions $\{\phi_1 \ldots \phi_6\}$ will change so that r_5 (the position of atom 5), e_6 (the unit vector along bond 5), and γ_6 (the third Euler angle of the triad $5, 6, 7$ in the space-fixed frame), remain unchanged. Atoms 1, 2, 3, and 4 will move, while all other atoms in the chain are fixed. We consider the case when the bond lengths d_i and bond angles θ_i are constrained, noting that this condition can be readily relaxed.

This geometrical problem is solved in the frame of bond 1 after the change in the angle ϕ_0 has been applied. In that frame the constraining equations are

$$r_5^{(1)} = d_1 + T_1\left(d_2 + T_2\left(d_3 + T_3(d_4 + T_4 d_5)\right)\right) \tag{9.66}$$

and

$$e_6^{(1)} = T_1 T_2 T_3 T_4 T_5 e_1 \tag{9.67}$$

where $r_5^{(1)}$ and $e_6^{(1)}$ are the column vector representations of the position of atom 5 and the unit bond vector of bond 6 in the frame of bond 1. $d_i = (d_i, 0, 0)^T = d_i e_1$ where $e_1 = (1, 0, 0)^T$ and the superscript T denotes the transpose. The matrix T_i transforms a vector in the frame of reference of bond $i + 1$ into its representation in the frame of bond i:

$$T_i = \begin{bmatrix} \cos\theta_i & \sin\theta_i & 0 \\ \sin\theta_i\cos\phi_i & -\cos\theta_i\cos\phi_i & \sin\phi_i \\ \sin\theta_i\sin\phi_i & -\cos\theta_i\sin\phi_i & -\cos\phi_i \end{bmatrix} \tag{9.68}$$

where θ_i is the bond angle with apex at atom i. Equation (9.67) can be expressed as an algebraic equation in the four unknowns $\{\phi_1, \phi_2, \phi_3, \phi_4\}$

$$[e_6^{(1)}]^{\mathrm{T}} T_1 T_2 T_3 T_4 - \cos \theta_5 = 0. \tag{9.69}$$

Note that γ_6, which is a constraint of the geometrical problem, does not explicitly appear in this solution. The three equations, eqn (9.66), can be used to express ϕ_2, ϕ_3, and ϕ_4 in terms of ϕ_1, and we can rewrite eqn (9.69) as a non-linear equation in ϕ_1 and the constraints

$$f(\phi_1; \phi_0, r_5, e_6, \gamma_6) = 0. \tag{9.70}$$

Once $\phi_1 \ldots \phi_4$ are determined, ϕ_5 can be extracted from the y and z components of eqn (9.66). ϕ_6 follows from simple geometrical considerations. Thus, eqns (9.66) and (9.69) for a fixed ϕ_0 can produce many solutions $\{\phi_1 \ldots \phi_6\}$ which enable the chain to rejoin the main skeleton at atom 5. These solutions are the possible concerted rotations for a particular ϕ_0.

Although this sounds straightforward, the actual solution of eqns (9.66) and (9.69) is a *tour de force* of numerical analysis. Equation (9.66) has two possible values of ϕ_2 for a given ϕ_1 and two possible values of ϕ_4 for a given ϕ_2, so that eqn (9.69) has four branches. For each branch, eqn (9.69) may not exist, may have no real solutions, or may have multiple real solutions. Every solution along each branch has to be identified, so that intervals have to be searched exhaustively using a weighted bisection method. The full details are provided in Dodd et al. (1993, Appendix B). Finally, we note, that our description applies to concerted rotations in the middle of the chain and a slightly different algorithm applies to rotations where the driver angle is closer than eight skeletal bonds from the chain ends, in which case we would need a concerted rotation involving fewer torsional angles (Dodd et al., 1993).

This solution of the concerted rotation problem involves a change from the torsional angles, $\{\phi_1 \ldots \phi_6\}$ to the positional and orientational variables, $r_5, e_6^{(1)}, \gamma_6$, involved in the constraint. The Jacobian for the transformation, J, is given by

$$J = \left| \frac{1}{\det(B)} \right| \tag{9.71}$$

where B is the 5×5 matrix

$$B = \begin{pmatrix} (e_1 \times r_{51}) & (e_2 \times r_{52}) & (e_3 \times r_{53}) & (e_4 \times r_{54}) & 0 \\ (e_1 \times e_6)_x & (e_2 \times e_6)_x & (e_3 \times e_6)_x & (e_4 \times e_6)_x & (e_5 \times e_6)_x \\ (e_1 \times e_6)_y & (e_2 \times e_6)_y & (e_3 \times e_6)_y & (e_4 \times e_6)_y & (e_5 \times e_6)_y \end{pmatrix} \tag{9.72}$$

and $r_{ij} = r_i - r_j$ and $0 = (0, 0, 0)^{\mathrm{T}}$. At the end of the trial move the $\{\phi_1 \ldots \phi_6\}$ values are used to calculate $\{r_1 \ldots r_4\}$ and hence $\det(B)$. If the final two rows of eqn (9.72) are linearly dependent, the z component is used in place of the x or y component. Incorporating the Jacobian in the MC move is essential to guarantee microscopic reversibility.

To perform a move, a torsional angle is chosen randomly to be ϕ_0, from all of the chains in the liquid, and a propagation direction for the concerted rotation, towards one end of the chain or the other, is chosen randomly. For simplicity, we only consider the case that ϕ_0 is far from the ends; other cases are considered in detail in Dodd et al. (1993).

(1) Randomly select a $\delta\phi_0$ on $(-\delta\phi_0^{\max}, +\delta\phi_0^{\max})$ and calculate the trial value of the driver angle $\phi_0^n = \phi_0^m + \delta\phi_0$.

(2) Find all the ϕ_1 on $-\pi$ to π such that $f(\phi_1; \phi_0^n, \mathbf{r}_5, \mathbf{e}_6, \gamma_6) = 0$.

    (a) If f does not exist or has no real solution, then reject the attempted move.

    (b) Otherwise note the number of real roots, N_n, found and uniformly select one of these roots to be the trial state $\{\phi_0^n, \phi_1^n, \ldots, \phi_6^n\}$.

(3) Calculate the Jacobians of the trial state, J^n and the original state J^m using eqns (9.71) and (9.72).

(4) Find all the solutions for the reverse concerted rotation, $f(\phi_1; \phi_0^m, \mathbf{r}_5, \mathbf{e}_6, \gamma_6) = 0$. One of these roots will be ϕ_1^m. Note the number of roots, N_m, produced by the reverse move.

(5) Calculate the change in potential energy, $\delta\mathcal{V}_{nm}$, including all interactions within the polymer and with other surrounding polymers.

(6) Accept the trial move with a probability given by

$$\min\left[1, \frac{N_m J_n}{N_n J_m} \exp(-\beta\delta\mathcal{V}_{nm})\right]. \tag{9.73}$$

The acceptance criterion is based on eqn (9.37) with the ratio of underlying stochastic matrix elements $\alpha_{nm}/\alpha_{mn} = N_m/N_n$, and with the Jacobians taken into account.

The concerted rotation algorithm has been included in the set of moves used by the MC module of the CHARMM package (Hu et al., 2006) and in the MCCCS Towhee package (Martin, 2013). It is also used in the MCPRO package to generate trial moves for protein backbones (Jorgensen and Tirado-Rives, 2005; Ulmschneider and Jorgensen, 2003).

A number of extensions of the original concerted rotation method have been developed. The intramolecular double rebridging (IDR) (Pant and Theodorou, 1995) is initiated by selecting, at random, an internal trimer in the polymer; for example the three beads behind bead i, connecting it to j, as shown in Fig. 9.6(a). This trimer is removed. A second trimer, the three beads behind k, connecting it to ℓ, is also excised. Beads i–k and j–ℓ are now joined with trimers constructed using the nine geometrical constraints that are in play for each trimer. There may be many different ways of achieving this rebridging. If j and k are separated by at least three beads, a second rearrangement can be considered by excising the three beads ahead of i and k and reconstructing the chain as before. Each possible rebridging in the forward and backward move needs to be identified and the corresponding Jacobians calculated for the particular move that is chosen. The move is accepted with a probability given by eqn (9.73). Double rebridging can also be used for two trimers in different molecules (Karayiannis et al., 2002).

The end-bridging algorithm (Mavrantzas et al., 1999) randomly identifies a trimer within the first polymer, for example the three atoms between i and j in Fig. 9.6(b). The algorithm uses a list of polymer ends such as k that are within the all-trans bridging distance $4d\cos(\theta_0/2)$ of j, where d is the carbon–carbon bond length and θ_0 is the equilibrium bond angle. One of these polymer ends is chosen at random and a trimer is inserted between j and k, subject to the normal constraints, to create a new polymer. In this method a polydisperse mixture of polymers of different lengths is being created and these simulations should be conducted in the semi-grand ensemble at a fixed chemical potential difference between the components (Peristeras et al., 2007).

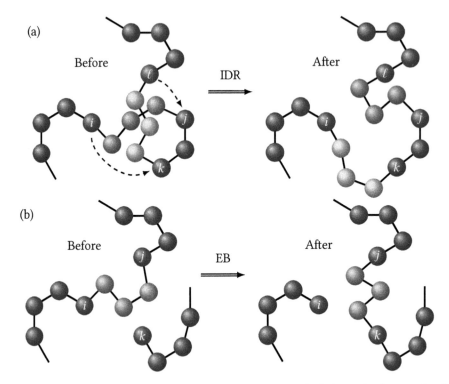

Fig. 9.6 Further concerted rotation moves. (a) The IDR move, in which the triplets between beads i and j and between k and ℓ are exchanged using a concerted-rotation algorithm. (b) The EB move, in which the triplet between beads i and j is extracted from one polymer and attached to the k end of a second polymer to create new polymers of different lengths.

Ulmschneider and Jorgensen (2004) have extended the concerted rotation algorithm. The use of a single driver angle to initiate the concerted rotation has been extended to allow the modification of many bond lengths, bond angles and torsional angles in the moving section of the chain (the open chain). This pre-rotation phase of the move is followed by an examination of the different ways of linking the open chain back to the fixed chain; this is the post-rotation part of the move. Bottaro et al. (2012) have shown that it is possible to make choices in the pre-rotation part of the move by selecting changes in the degrees of freedom from a correlated Gaussian distribution and that in this case the post-rotation, or chain relinking, can be solved analytically. Zamuner et al. (2015) have extended these ideas by considering the manifold of all the chain states compatible with the fixed portion of the chain. The chain is opened in the pre-rotation phase in the space that is tangent to the manifold and then closed in the orthogonal space using a root-finding algorithm. This method neatly avoids the explicit calculation of the Jacobians of the forward and backward moves. The reader is referred to the original papers for details of these refinements.

9.4 Simulation of phase equilibria

9.4.1 Gibbs ensemble Monte Carlo

The Gibbs ensemble Monte Carlo method (Panagiotopoulos, 1987) allows for the direct simulation of two coexisting fluid phases, for example a liquid and a gas. The method uses two independent, periodic, simulation boxes at the same temperature: we can imagine box I to be deep in the coexisting liquid phase and box II to be deep in the coexisting gas phase. Box I contains N^{I} atoms in volume V^{I} and box II contains N^{II} in a volume V^{II}. Each box is surrounded by periodic images in three dimensions and there are no real interfaces in this method. The combined boxes form a representative system from a canonical ensemble at constant $N = N^{\mathrm{I}} + N^{\mathrm{II}}$, constant $V = V^{\mathrm{I}} + V^{\mathrm{II}}$ and constant T (Smit et al., 1989). Since the two boxes are at equilibrium, then $P^{\mathrm{I}} = P^{\mathrm{II}} = P$ and $\mu^{\mathrm{I}} = \mu^{\mathrm{II}} = \mu$, and the algorithm is designed to ensure that this is the case.

The simulation proceeds with three different types of move: an atom displacement; a volume rearrangement; and an atom exchange. For an atom-displacement move, a cycle of trial displacements is attempted in box I using the normal Metropolis criterion for accepting or rejecting such moves (see Section 4.5). A corresponding cycle of moves is attempted in box II.

For a volume rearrangement, equal and opposite changes in the volume are made in boxes I and II at the same overall pressure. For a trial attempted volume change, δV in box I, the ratio of the probabilities for the old and new state is (eqn (4.33))

$$\left(\frac{\rho_n}{\rho_m}\right)_{\mathrm{I}} = \exp\left[-\beta\delta V_{nm}^{\mathrm{I}} - \beta P\delta V + N^{\mathrm{I}}\ln(V^{\mathrm{I}} + \delta V) - N^{\mathrm{I}}\ln V^{\mathrm{I}}\right]. \tag{9.74a}$$

At the same time, a corresponding volume change of $-\delta V$ is made in box II

$$\left(\frac{\rho_n}{\rho_m}\right)_{\mathrm{II}} = \exp\left[-\beta\delta V_{nm}^{\mathrm{II}} + \beta P\delta V + N^{\mathrm{II}}\ln(V^{\mathrm{II}} - \delta V) - N^{\mathrm{II}}\ln V^{\mathrm{II}}\right]. \tag{9.74b}$$

The ratio of probabilities for the combined moves in the two independent boxes is obtained by multiplying these ratios together

$$\left(\frac{\rho_n}{\rho_m}\right)_{\mathrm{vol}} = \exp\left[-\beta\left(\delta V_{nm}^{\mathrm{I}} + \delta V_{nm}^{\mathrm{II}}\right) + N^{\mathrm{I}}\ln\left(1 + \frac{\delta V}{V^{\mathrm{I}}}\right) + N^{\mathrm{II}}\ln\left(1 - \frac{\delta V}{V^{\mathrm{II}}}\right)\right]. \tag{9.75}$$

The combined volume move is accepted with probability $\min[1, (\rho_n/\rho_m)_{\mathrm{vol}}]$. Note that the unknown coexistence pressure, P, is not required for the acceptance test.

For an atom exchange move, one of the two boxes is chosen with equal probability, say box I. The first part of the trial move consists of creating an atom in box I at a constant chemical potential μ. The ratio of the probabilities of the new and old states is (eqn (4.41))

$$\left(\frac{\rho_n}{\rho_m}\right)_{\mathrm{I}} = \exp\left[-\beta\delta V_{nm}^{\mathrm{I}} + \ln\left(\frac{zV^{\mathrm{I}}}{N^{\mathrm{I}} + 1}\right)\right]. \tag{9.76a}$$

At the same time, a corresponding trial atom destruction is made in box II

$$\left(\frac{\rho_n}{\rho_m}\right)_{\mathrm{II}} = \exp\left[-\beta\delta V_{nm}^{\mathrm{II}} + \ln\left(\frac{N^{\mathrm{II}}}{zV^{\mathrm{II}}}\right)\right]. \tag{9.76b}$$

Code 9.3   Gibbs ensemble simulation

This file is provided online. The program mc_gibbs_lj.f90 controls the simulation, reads in the run parameters, selects moves, and writes out the results. It uses the routines in mc_gibbs_lj_module.f90 and the utility routines of Appendix A. The program initialize.f90 (Code 5.6) may be employed to generate a pair of initial configurations.

```
! mc_gibbs_lj.f90
! Monte Carlo, Gibbs ensemble
PROGRAM mc_gibbs_lj


! mc_gibbs_lj_module.f90
! Energy and move routines for Gibbs MC, LJ potential
MODULE mc_module
```

The ratio of probabilities for the combined exchange move is the product of these

$$\left(\frac{\rho_n}{\rho_m}\right)_{\text{ex}} = \exp\left[-\beta\left(\delta \mathcal{V}^{\text{I}}_{nm} + \delta \mathcal{V}^{\text{II}}_{nm}\right) + \ln\left(\frac{V^{\text{I}}N^{\text{II}}}{V^{\text{II}}(N^{\text{I}}+1)}\right)\right]. \tag{9.77}$$

The combined exchange move is accepted with probability $\min[1, (\rho_n/\rho_m)_{\text{ex}}]$. Note that the unknown coexistence chemical potential, μ, is not required for the acceptance test. An example Gibbs ensemble simulation program is given in Code 9.3.

During the course of a simulation using the Gibbs method, the number of atoms in each box and the volume of each box will evolve until the individual boxes reach the coexisting densities, ρ_ℓ and ρ_g, at the specified temperature. This evolution should be monitored by plotting the individual density of each box as a function of the number of MC cycles. Individual simulations, at different temperatures, will provide pairs of densities along the binodal or coexistence curve. It is straightforward to check that equilibrium has been achieved by calculating the virial pressure and chemical potential directly in each box and making sure that they are equivalent. In the Gibbs method, the appropriate form of the Widom particle insertion equation for the total chemical potential in box I is

$$\mu^{\text{I}} = -k_{\text{B}}T \ln \frac{1}{\Lambda^3}\left\langle\left(\frac{V^{\text{I}}}{N^{\text{I}}+1}\right)\exp(-\beta \mathcal{V}^{\text{I}}_{\text{test}})\right\rangle_{\text{I}} \tag{9.78}$$

where $\mathcal{V}^{\text{I}}_{\text{test}}$ is the potential energy of inserting a ghost particle into box I.

The method will only work at temperatures below the critical temperature of the model fluid. As this critical temperature is approached from below, the method becomes less reliable. As ρ_ℓ approaches ρ_g, the identity of the two boxes will change frequently during the course of the simulation and it may be impossible to identify a 'liquid' and a 'gas' box. This problem will be evident in the plot of densities as a function of the number of cycles. Although it is not possible to use the Gibbs method up to the critical point, it is possible to use the results of simulations below T_c to obtain a first estimate of T_c by fitting the data to the law of rectilinear diameters

$$\frac{\rho_\ell + \rho_g}{2} = \rho_c + A\left|T - T_c\right| \tag{9.79}$$

and the scaling law

$$\rho_\ell - \rho_g = B\left|T - T_c\right|^\beta. \tag{9.80}$$

A and B are constants and here β is the scaling exponent ($\beta = 0.32$ in three dimensions and $\beta = 0.125$ in two dimensions).

One of the strengths of the Gibbs method is that it is easily extended to mixtures (Panagiotopoulos et al., 1988). The only important new consideration is in the atom exchange move. N_I and N_{II} in eqn (9.77) now refer to the number of atoms of one particular species in boxes I and II respectively. There may be a number of distinct species in both boxes. For the condition of microscopic reversibility to hold there are two steps:

(a) choose either box for the trial creation with equal probability;
(b) select, with a fixed (but otherwise arbitrary) probability, which of the species is to be interchanged.

We note that during the simulation, one of the regions can empty of atoms of one or more species. This is likely to happen if the species are present in low concentrations at coexistence. The simulation can continue in the normal way under these circumstances. When a box is empty, an attempt to transfer an atom from that box is immediately rejected.

For mixtures, the Gibbs method can be performed with the total system in the constant-NPT ensemble. Equation (9.75) is modified to allow for independent and different volume moves in boxes I and II. In this case, the two independent volume changes can occur simultaneously, but faster convergence is achieved by changing the volume of only one region at a time (using eqn (4.33)). Of course, if we attempted this kind of simulation on a single component system we would end up with both boxes in a liquid phase or gas phase, depending on whether the imposed value of the pressure, P, was above or below the actual coexistence pressure. The Gibbs method can be readily extended to osmotic equilibria (Panagiotopoulos et al., 1988). In this case, the two regions are imagined to be separated by a semipermeable membrane which restricts the exchange of certain species but allows the exchange of the membrane-permeable species. Only atoms of the latter species are interchanged between the boxes. The volume rearrangement criterion is modified to take into account the osmotic pressure difference imposed across the membrane, $\Pi = P^I - P^{II}$. The volume rearrangement move is accepted with a probability given by $\min[1, (\rho_n/\rho_m)_{osm}]$ where

$$\left(\frac{\rho_n}{\rho_m}\right)_{osm} = \exp\left[-\beta(\delta V_{nm}^I + \delta V_{nm}^{II}) + N^I \ln\left(1 + \frac{\delta V}{V^I}\right) + N^{II} \ln\left(1 - \frac{\delta V}{V^{II}}\right) - \beta \Pi \, \delta V\right]. \tag{9.81}$$

A check of the validity of this method is to compare the applied osmotic pressure, Π, with the calculated average pressure difference between the two boxes.

The Gibbs ensemble method can be used to consider the vapour–liquid equilibria of simple molecular fluids (Stapleton et al., 1989), and extended to more complicated binary molecular mixtures (de Pablo et al., 1992; Do et al., 2010). With the addition of CBMC, the method can be used to predict the phase equilibria of fluids containing long-chain hydrocarbons (Bai and Siepmann, 2013). Vorholz et al. (2004) have used it to study the solubility of carbon dioxide in aqueous solutions of sodium chloride where the so-called salting-out effect, the reduction in the solubility of CO_2 on addition of the ions, is predicted by the simulation. The method can also be used to consider the equilibria of quantum fluids such as neon using quantum wave packet methods where a Lennard-Jones potential ($\epsilon/k_B = 35.68$ K, $\sigma = 2.761$ Å) with quantum parameter $\hbar/(\sigma\sqrt{m\epsilon}) = 0.0940$ gives a good fit to the experimental liquid–vapour curve (Georgescu et al., 2013). Finally the method can be extended to study multiphase equilibria by increasing the number of boxes that are used concurrently in the simulation (Lopes and Tildesley, 1997).

9.4.2   The Gibbs–Duhem method

Once a single point (P, T) on the coexistence curve has been accurately established, it is possible to use the Gibbs–Duhem integration method (Kofke, 1993a,b) to calculate other points at coexistence. The method starts from the Clapeyron equation (itself derived from the Gibbs–Duhem equation)

$$\left(\frac{d \ln P}{d\beta}\right)_\sigma = -\frac{\Delta H}{\beta P \Delta V} = f(\beta, P),\tag{9.82}$$

where the derivative is taken along the coexistence line. $\Delta H = H_\alpha - H_\beta$ is the enthalpy difference between the two coexisting phases α and β, and ΔV is the corresponding volume difference (both are extensive quantities). Equation (9.82) is a smooth, first-order differential equation where the right-hand side, f, depends on both the independent and dependent variables.

   This equation can be solved using a predictor–corrector method (Press et al., 2007, section 17.6) which requires a number of efficient evaluations of f. The Gibbs–Duhem method uses two independent simulations of the coexisting phases to estimate $f(\beta, P)$ at each step of the process. The two coexisting phases must be simulated by constant-NPT MC (see Section 4.5) or MD (Section 3.9). The strength of the method is that it avoids any attempted particle insertions, and can therefore be used to model solid–liquid coexistence, where the grand canonical and Gibbs Monte Carlo methods would fail.

   Consider the vapour–liquid equilibrium of a single-component Lennard-Jones fluid. We require a starting point on the coexistence curve, (β_0, P_0). This could come, for example, from a Gibbs simulation or by thermodynamic integration to calculate the free energy. Once this point is established we have a first estimate of $f(\beta_0, P_0) = f_0$. We can now attempt to move up or down the coexistence curve by a small interval of inverse temperature, $\delta\beta$. Using the simplest trapezoidal predictor, we can estimate the coexisting pressure at $\beta_1 = \beta_0 + \delta\beta$,

$$\ln P_1 = \ln P_0 + \delta\beta f_0.\tag{9.83}$$

The two separate simulations of the gas and liquid phase can now be performed at (β_1, P_1) and from the new estimates of ΔH and ΔV we calculate f_1. These can now be used in the corresponding corrector formula to obtain a better estimate of P_1

$$\ln P_1 = \ln P_0 + \frac{\delta\beta}{2}\left(f_1 + f_0\right).\tag{9.84}$$

A short run is now performed at the new P_1 (left side of eqn (9.84)) to obtain a better estimate of f_1. This cycle is repeated until the corrector step has converged and the final (β_1, P_1) is the new point on the coexistence curve.

Simulations of this type can be continued along any coexistence curve in small steps of β. Once we have determined two points on the curve, f_0 and f_1, it is possible to use the more accurate mid-point predictor–corrector scheme and when three points are known, we can employ the Adams predictor–corrector (Press et al., 2007).

One immediate question is why, or whether, the method is thermodynamically stable. At various points, we will be simulating two boxes at a value of (T, P) that is not exactly on the coexistence curve. Consequently, one of the two phases will be unstable with respect to the other. There is a possibility that the gas phase will condense or the liquid phase evaporate. This not likely to happen at low temperatures where the free-energy barrier between the phases is large. The significant metastable range of the two phases saves the day. However, as we approach the critical point in a liquid–gas equilibrium this unwanted phase change can occur and the method will cease to converge. Kofke (1993a) demonstrates that this problem can be mitigated by coupling the volume changes in the two simulation boxes. This introduces a Jacobian into the coupled partition function which modifies the Hamiltonian and the sampling distribution.

The Gibbs–Duhem method, although originally applied to the coexistence line in the (P, T) projection of the phase diagram for a single-component fluid, can be extended to other projections for liquid mixtures. The Clapeyron-like equations for the liquid-vapour coexistence of binary mixtures (for use in the semi-grand ensemble) are

$$\left(\frac{\partial \beta}{\partial \xi_2}\right)_{\sigma, P} = \frac{\Delta x_2}{\xi_2(1 - \xi_2)\Delta h}$$

$$\left(\frac{\partial \ln P}{\partial \xi_2}\right)_{\sigma, \beta} = \frac{\Delta x_2}{\xi_2(1 - \xi_2)(Z_\ell - Z_g)} \qquad (9.85)$$

where $\Delta h = \Delta H / N$ is the difference in enthalpy per particle between liquid and gas, ξ_2 is the fugacity fraction ($\xi_2 = f_2 / (f_1 + f_2)$ where $f_i = \exp[\beta(\mu_i - \mu_i^\ominus)]$ is the fugacity of component i), $\Delta x_2 = x_{2,\ell} - x_{2,g}$ (x_2 is the mole fraction of component 2 in a coexisting phase), and $Z = PV / Nk_\mathrm{B}T$ is the compressibility factor in each coexisting phase. Equation (9.85) determines how the temperature and pressure change with fugacity fraction along the coexistence curve. Mehta and Kofke (1994) have calculated pressure–composition projections of the mixture phase diagram using Gibbs–Duhem integration. Simulations of the two coexisting phases are accomplished using either a semi-grand approach in which molecules attempt to change their species identity, keeping the total number of molecules fixed, or an osmotic approach involving insertions or deletions of one of the species, while the number of molecules of the other species remain fixed. (Note, the Clapeyron-like equations for the osmotic approach are given in terms of f_2 rather than ξ_2 (Mehta and Kofke, 1994).) Galbraith and Hall (2007) have used the Gibbs–Duhem method to study the solid–liquid coexistence of CO_2, C_2H_6, and F_2 mixtures modelled using the diatomic Lennard-Jones potential. Moucka et al. (2013) have used the method to model aqueous NaCl solutions at ambient conditions, using the standard simple point charge/extended (SPC/E) force field for water and the Joung–Cheatham force field for the electrolyte. The water chemical potential is calculated using the osmotic ensemble Monte Carlo algorithm by varying the number of water molecules at a constant amount of solute.

The method can be readily extended to consider phase equilibria for polydisperse fluids. For example, in a fluid of polydisperse hard-spheres, the pressure and activity distribution in both the solid and liquid phases are equal. The activity ratio (or fugacity ratio) distribution

$$\frac{f(\sigma)}{f(\sigma_0)} = \exp\left[\beta\left(\mu(\sigma) - \mu(\sigma_0)\right)\right] = \exp\left[\beta\Delta\mu(\sigma)\right]$$

is related to the chemical potential difference relative to a particular diameter $\sigma = \sigma_0$. Bolhuis and Kofke (1996) have modelled $f(\sigma)$ as a Gaussian distribution peaked at σ_0 with variance v. In this case, the Gibbs–Duhem method traces the locus of the solid–liquid transition in the (P, v) plane. Monte Carlo simulations are performed in the isobaric semi-grand ensemble. The appropriate Gibbs–Duhem-like equation is

$$\frac{dP}{dv} = \frac{\Delta m_2}{2v^2\beta\Delta v}, \tag{9.86}$$

where m_2 is the second moment of the size distribution of the hard spheres (measured with respect to the origin, $m_1 = 0$), v is the volume per hard sphere, and Δ indicates the difference between the coexisting phases. Integration of eqn (9.86) from the monodisperse limit ($v \to 0$) is used to trace the coexistence pressure as a function of the variance of the imposed activity distribution. A 'terminal' polydispersity is observed above which there can be no fluid–solid coexistence. It is possible to extend this technique to use a more general form of the activity distribution (Kofke and Bolhuis, 1999).

9.5 Reactive Monte Carlo

A Monte Carlo simulation technique for modelling chemical reactions at equilibrium was developed independently by Smith and Triska (1994) and Johnson et al. (1994). Reactive canonical Monte Carlo (also known as the reaction ensemble method) considers a general chemical reaction

$$\sum_{i=1}^{C} v_i A_i = 0 \tag{9.87}$$

where A_i is the chemical symbol for species i and v_i is its stochiometric coefficient (positive for products, negative for reactants, and zero for species that do not change in the reaction). The sum is over all of the C species. For a chemical reaction at equilibrium, the chemical potentials satisfy

$$\sum_{i=1}^{C} v_i \mu_i = 0. \tag{9.88}$$

Consider a state of the system m before the chemical reaction occurs. Using the approximation of separable molecular internal degrees of freedom (Münster, 1969; Gray and Gubbins, 1984), the probability of finding the system in state m is

$$\rho_m = \frac{1}{Q_{\mu VT}} \exp\left[\sum_{i=1}^{C} \left(\beta N_i \mu_i - \ln(N_i!) + N_i \ln q_i\right) - \beta \mathcal{V}_m\right], \tag{9.89}$$

where N_i is the number of molecules of species i, q_i is the partition function for an isolated molecule of species i, and $\mathcal{V}_m$ is the intermolecular potential energy for all of the molecules in the configurational state m. Generally, $q_i = q_{t,i} q_{r,i} q_{v,i} q_{el,i}$, is the product of the translational, rotational, vibrational, and electronic contributions to the single molecule partition function and $\mathcal{V}_m$ is assumed to be independent of the internal quantum state of the molecule. For a single-component fluid containing N atoms, eqn (9.89) reduces to eqn (4.39). A single reaction proceeds in the forward direction to a new state n, changing the number of molecules of each species by ν_i. The probability of the new state is

$$p_n = \frac{1}{Q_{\mu VT}} \exp\left[\sum_{i=1}^{C} \left(\beta(N_i + \nu_i)\mu_i - \ln[(N_i + \nu_i)!] + (N_i + \nu_i) \ln q_i \right) - \beta \mathcal{V}_n \right]. \quad (9.90)$$

The ratio of the probabilities for the single reaction is

$$\frac{p_n}{p_m} = \exp(-\beta \delta \mathcal{V}_{nm}) \left(\prod_{i=1}^{C} q_i^{\nu_i} \right) \left(\prod_{i=1}^{C} \frac{N_i!}{(N_i + \nu_i)!} \right) \quad (9.91)$$

where we have used eqn (9.88) to eliminate the chemical potentials. The partition functions for the isolated molecules can be calculated from first principles. They can also be expressed in terms of the ideal gas equilibrium constant (Hill, 1956)

$$\prod_{i=1}^{C} q_i^{\nu_i} = K_{id}(T) \prod_{i=1}^{C} (\beta p^\ominus V)^{\nu_i} = K_{id}(T)(\beta p^\ominus V)^{\bar\nu} \quad (9.92)$$

where $p^\ominus$ is the standard state pressure and $\bar\nu$ is the sum of the stochiometric coefficients. In this way eqn (9.91) can be written as

$$\frac{p_n}{p_m} = \exp(-\beta \delta V_{nm})(\beta p^\ominus V)^{\bar\nu} \exp(-\Delta G^\ominus / RT) \prod_{i=1}^{C} \frac{N_i!}{(N_i + \nu_i)!} \quad (9.93)$$

where $\Delta G^\ominus$ is the standard molar Gibbs free energy of the reaction, which can be estimated from thermodynamic tables, and R is the gas constant.

The ratio of the probabilities for the reverse reaction can be obtained by substituting $-\nu$ for ν in eqn (9.91) and in eqn (9.93) additionally changing the sign of $\Delta G^\ominus$ and $\bar\nu$. It is possible to develop a Monte Carlo scheme in which the forward and backward reactions occur with equal probability. The transition probability is independent of the chemical potential and it is possible to fix the temperature and work at constant density (in the constant-NVT ensemble) or at constant pressure (in the constant-NPT ensemble) where N is now the total number of atoms rather than molecules.

As a simple example, consider the dimerization reaction (Johnson et al., 1994)

$$2\,NO_2 \rightleftharpoons N_2O_4. \quad (9.94)$$

For the forward reaction

$$\frac{p_n}{p_m} = \exp(-\beta \delta V_{nm}) \frac{N_{NO_2}\left(N_{NO_2} - 1\right) q_{N_2O_4}}{\left(N_{N_2O_4} + 1\right) q_{NO_2}^2} \quad (9.95)$$

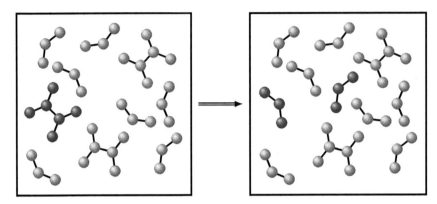

Fig. 9.7 The backward trial move for a reactive canonical Monte Carlo simulation of the NO_2 dimerization reaction. Unaffected molecules are a lighter shade. An N_2O_4 molecule is replaced by an NO_2 molecule (randomly oriented) plus a second NO_2 molecule inserted at a random position.

and for the back reaction

$$\frac{\rho_n}{\rho_m} = \exp(-\beta\delta V_{nm})\frac{N_{N_2O_4}\, q_{NO_2}^2}{\left(N_{NO_2} + 2\right)\left(N_{NO_2} + 1\right)q_{N_2O_4}}. \tag{9.96}$$

The canonical version of the algorithm would proceed as follows (see Fig. 9.7). One of the three following possible trial moves is attempted with a fixed probability.

(1) Choose a molecule at random and attempt a change in its position (and orientation).
(2) Attempt a forward reaction step.
    (a) Choose an NO_2 molecule at random.
    (b) Change this molecule to an N_2O_4 molecule, picking a random orientation.
    (c) Choose another NO_2 molecule at random and delete it.
    (d) Accept the move with a probability $\min(1, \rho_n/\rho_m)$ where ρ_n/ρ_m is given by eqn (9.95).
(3) Attempt the reverse reaction step.
    (a) Choose an N_2O_4 molecule at random.
    (b) Change the molecule to an NO_2 molecule, picking the orientation at random.
    (c) Randomly insert another NO_2 molecule into the fluid.
    (d) Accept the move with a probability $\min(1, \rho_n/\rho_m)$ where ρ_n/ρ_m is given by eqn (9.96).

Steps (2) and (3) must have equal probability to ensure microscopic reversibility. To perform simulations at constant-NPT, it is also necessary to perform a trial volume change move as discussed in Section 4.8.

The method can be readily generalized to a set of independent linear chemical reactions. It does not require an *a priori* knowledge of the chemical potentials. The method can be straightforwardly extended for use with the Gibbs ensemble Monte Carlo method to study combined physical and chemical equilibria. In this case only one of the reacting species needs to be transferred between the simulation boxes, since the chemical reaction step will establish equilibrium in each of the coexisting phases. The method will not be successful at high densities when it is unlikely that a product molecule can be successfully inserted into the fluid. The method is also complicated to implement for systems with many chemical reactions since each reaction would need to be considered many times to achieve equilibrium.

The reaction ensemble Monte Carlo method has been widely applied to reactions confined in porous solids or near solid surfaces, reactions at high temperature and high pressure, reactions in solution and at phase boundaries (Turner et al., 2008). For example, Malani et al. (2010) have developed a model for silica polymerization at ambient temperatures and low densities which focuses on SiO_4 coordination. The model explicitly includes the energetics of hydrolysis and condensation reactions. Nangia and Garrison (2010) have used the method to study dissolution at mineral–water interfaces with specific emphasis on silicates. The equilibrium properties of the system are explored using the combined reactive Monte Carlo and configurational-bias Monte Carlo methods. The full three-dimensional structure of the mineral with explicit water molecules can be tackled with this combination. The reaction ensemble method has also been combined with dissipative particle dynamics (see Chapter 12) to study both static and dynamical properties of a simple polydisperse homopolymer system (Lisal et al., 2006). The technique has been used to predict the effect of solvents, additives, temperature, pressure, shear, and confinement on the polydispersity of the polymer.

10

Rare event simulation

10.1 Introduction

The simulation of 'rare events' poses a particular challenge. Broadly speaking, a rare event consists of a transition from one region of phase space to another. The system typically resides for a long period in either region alone: long enough sensibly to define distinct time-averaged properties. At equilibrium, one state may be significantly more stable than the other, in which case we say that the other state is thermodynamically metastable. Typically, we wish to study the rate of conversion of one state into the other, and possibly the 'path' that the system takes (in some suitable set of reaction coordinates). There are several common examples. The orientational switching of a liquid crystal, following the application of an electric field, may fall into this category. Conformational interconversion in polymers and biomolecules, the nucleation of a crystal from the melt, and chemical reactions are other examples.

The problem for the simulator is that the residence periods may be much longer than the maximum practical length of a simulation. The transition process itself may occur on a microscopic timescale, suitable for direct simulation by molecular dynamics or Brownian dynamics; however, the transition rate is not simply given by the speed of this part of the process.

In this chapter, we shall discuss some of the simulation techniques developed to study the dynamics of rare events. We begin in Section 10.2 with the simplest situation, when the assumptions of transition state theory (TST) apply, and discuss how this leads to an exact result in Section 10.3. This leads, in Section 10.4, to the question of identifying suitable reaction coordinates and paths. Then we go to the other extreme, and consider in Section 10.5 how to measure a rate of transition while making the fewest possible assumptions: the so-called transition path sampling (TPS) approach. This method is formally exact but can be quite expensive to implement. In Section 10.6, we discuss methods such as forward flux sampling (FFS) and transition interface sampling (TIS) which attempt to combine features of both approaches, to give rigorous results while remaining computationally efficient. Finally, we give a summary in Section 10.7.

In the present chapter we cannot cover all the details of this field. Fortunately, the background theory, and many of the practical details, are described in several excellent books and review articles (Chandler, 1987; Dellago et al., 2002; Moroni, 2005; Bolhuis and Dellago, 2010; van Erp, 2012).

Computer Simulation of Liquids. Second Edition. M. P. Allen and D. J. Tildesley.
© M. P. Allen and D. J. Tildesley 2017. Published in 2017 by Oxford University Press.

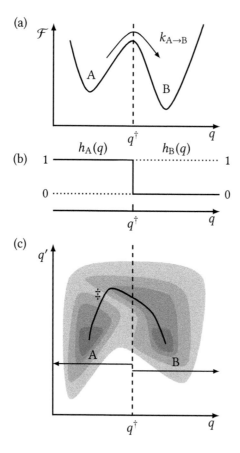

Fig. 10.1 Reaction coordinates and free energies (schematic). (a) Free energy as a function of a single reaction coordinate $\mathcal{F}(q)$, for a system with two minima, A and B, separated by a maximum at $q^\dagger$ (vertical dashed line). (b) The indicator functions $h_A(q)$ (solid line) and $h_B(q)$ (dotted line). (c) Free energy (dark for low values, light for high values) as a function of two reaction coordinates $\mathcal{F}(q, q')$, for a system with two basins A and B, with a saddle point at ‡, and a typical 'transition path' (solid line). The vertical dashed line marks the 'transition state' that would be deduced by considering the single reaction coordinate q.

10.2   Transition state approximation

In some circumstances it is physically reasonable to divide phase space into two regions, each associated with one of the two states, which we call A and B (see Fig. 10.1). It may be possible to identify a reaction coordinate q, a function of coordinates and momenta, for which a Landau free energy

$$\mathcal{F}(q) = -k_B T \ln\langle \delta(q - q(\mathbf{r})) \rangle + C$$

may be defined (see Section 2.11); in the following, q is assumed to be a function of coordinates only, $q = q(\mathbf{r})$, but this may not be the case in general. The dividing line between A and B is located at the maximum of this function, which we call the *transition state* $q^\dagger$. We assume that the probability density near the maximum is much lower than in the basins on either side. Let us take A to be the region $q < q^\dagger$, and B to correspond to $q > q^\dagger$, and define two *indicator functions*

$$h_A(q) = \Theta(q^\dagger - q), \quad h_B(q) = \Theta(q - q^\dagger), \quad \text{where} \quad \Theta(x) = \begin{cases} 0 & x < 0 \\ 1 & x > 0 \end{cases} \quad (10.1)$$

which take values of one or zero in the appropriate regions. Note that $h_A + h_B = 1$, and that the ensemble averages

$$\langle h_A \rangle = \frac{\int_{r \in A} dr \exp(-\beta \mathcal{V})}{\int dr \exp(-\beta \mathcal{V})} = \frac{\int_{-\infty}^{q^\dagger} dq \exp[-\beta \mathcal{F}(q)]}{\int_{-\infty}^{\infty} dq \exp[-\beta \mathcal{F}(q)]} \equiv \frac{Q_A}{Q}, \quad (10.2a)$$

$$\langle h_B \rangle = \frac{\int_{r \in A} dr \exp(-\beta \mathcal{V})}{\int dr \exp(-\beta \mathcal{V})} = \frac{\int_{q^\dagger}^{\infty} dq \exp[-\beta \mathcal{F}(q)]}{\int_{-\infty}^{\infty} dq \exp[-\beta \mathcal{F}(q)]} \equiv \frac{Q_B}{Q}, \quad (10.2b)$$

where $Q = Q_A + Q_B$, give the probability of the system being in region A or region B, respectively. The forward rate is expressed as $k_{A \to B} \langle h_A \rangle$ where $\langle h_A \rangle$ represents the (normalized) population of reactants and $k_{A \to B}$ is the rate constant. A similar expression applies to the reverse reaction and, at equilibrium, $\langle h_B \rangle k_{B \to A} = \langle h_A \rangle k_{A \to B}$.

All of the foregoing discussion may seem straightforward, given q, but there is no general, unique way of defining a reaction coordinate, and different choices of q will lead to different free-energy curves. More seriously, Fig. 10.1(c) reminds us that it may be appropriate to discuss the transition in terms of several reaction coordinates, and that the 'obvious' location of a transition state (a saddle point in the multidimensional case) may be unrelated to the location in the one-dimensional case.

In the TST approximation, a trajectory crossing the plane $q = q^\dagger$ is assumed to have negligible probability of returning. In this case, the rate constant for the transition may be evaluated from purely static ensemble averages

$$k_{A \to B}^{\text{TST}} = \frac{\langle \dot{q}\, \Theta(\dot{q})\, \delta(q - q^\dagger) \rangle}{\langle h_A \rangle} \quad (10.3)$$

with a similar expression for $k_{B \to A}^{\text{TST}}$. The factor $\Theta(\dot{q})$ is included to ensure that only positive-going fluxes are counted; sometimes this is dropped, and $\dot{q}$ is replaced by $\frac{1}{2}|\dot{q}|$, to give the same result. This may be written

$$k_{A \to B}^{\text{TST}} = \frac{\langle \dot{q}\Theta(\dot{q})\, \delta(q - q^\dagger) \rangle}{\langle \delta(q - q^\dagger) \rangle} \frac{\langle \delta(q - q^\dagger) \rangle}{\langle h_A \rangle} = \frac{\langle \dot{q}\Theta(\dot{q})\, \delta(q - q^\dagger) \rangle}{\langle \delta(q - q^\dagger) \rangle} \frac{\exp[-\beta \mathcal{F}(q^\dagger)]}{\int_{-\infty}^{q^\dagger} dq \exp[-\beta \mathcal{F}(q)]}.$$

The first, kinetic, term is essentially the average of $\dot{q}\, \Theta(\dot{q})$ for states at $q^\dagger$, and may often be evaluated exactly without the need of a simulation: it will be proportional to $\sqrt{k_B T / M}$, where M is a mass associated with the generalized coordinate q. The second term is the probability density at the transition state; this, or equivalently the free energy $\mathcal{F}(q^\dagger)$ relative to A, may be evaluated by the methods discussed in Section 9.2. TST is discussed in more detail in the reviews cited earlier and by Vanden-Eijnden and Tal (2005).

10.3 Bennett–Chandler approach

Bennett (1977) and Chandler (1978) have extended the TST approach to give an expression for the rate constant, which is (in principle) exact. Consider the correlation function

$$C_{A \to B}(t) = \frac{\langle h_A(0) h_B(t) \rangle}{\langle h_A \rangle}. \tag{10.4}$$

At long times, $t > t_{mol}$, where t_{mol} is a typical molecular relaxation time, $C_{A \to B}(t)$ approaches its plateau value $\langle h_B \rangle$ with an extremely slow exponential growth (Chandler, 1987; Moroni, 2005) determined by the sum of the rate constants

$$C_{A \to B}(t) \approx \langle h_B \rangle \left(1 - \exp(-kt) \right),$$
$$\dot{C}_{A \to B}(t) \approx \langle h_B \rangle k \exp(-kt)$$

where $k = k_{A \to B} + k_{B \to A}$. Using the expressions $\langle h_B \rangle k_{B \to A} = \langle h_A \rangle k_{A \to B}$ and $h_A + h_B = 1$, this may be written as

$$\dot{C}_{A \to B}(t) \approx k_{A \to B} \exp(-kt),$$

and hence

$$k_{A \to B} \approx \dot{C}_{A \to B}(t) \quad \text{for } t_{mol} < t \ll k^{-1}. \tag{10.5}$$

This expression is only true for $t > t_{mol}$. It is possible to show that the TST expression may be re-written as

$$k_{A \to B}^{TST} = \dot{C}_{A \to B}(t = 0_+) \tag{10.6}$$

that is, TST makes the assumption that the exponential behaviour holds even when $t \to 0$, which is consistent with the underlying assumption that reactive trajectories only cross $q = q^\dagger$ once. This results in an overestimate of the true rate, and the correction factor is usually called the *transmission coefficient* κ

$$k_{A \to B} = \kappa \, k_{A \to B}^{TST}, \quad 0 \le \kappa \le 1 \tag{10.7}$$

where $\kappa = \kappa(t)$ is approximately constant for $t_{mol} < t \ll k^{-1}$, and

$$\kappa(t) = \frac{\left\langle \dot{q}(0) h_B \big(q(t) \big) \right\rangle_{q^\dagger}}{\left\langle \dot{q}(0) h_B \big(q(0_+) \big) \right\rangle_{q^\dagger}} = \frac{\left\langle \dot{q}(0) h_B \big(q(t) \big) \right\rangle_{q^\dagger}}{\left\langle \dot{q}(0) \Theta \big(\dot{q}(0) \big) \right\rangle_{q^\dagger}}. \tag{10.8}$$

Both averages are taken with the constraint $q(0) = q^\dagger$ applied. This is easily done in molecular dynamics by the methods described in Section 3.4, remembering to include corrections for the metric tensor; the resulting ensemble is sometimes called the 'blue-moon ensemble' (Carter et al., 1989; Ciccotti and Ferrario, 2004). Alternatively, the constraint may be applied approximately, by umbrella sampling, in either MD or MC. To obtain the numerator of eqn (10.8), it is necessary to use the configurations sampled in the constrained ensemble as starting points for unconstrained trajectories, which are used to correlate the flux crossing the transition state at $t = 0$ with the population of state B at later times. This takes account of all the recrossings of the transition state, which are neglected in TST. However, many independent trajectories may need to be used in order to calculate κ with reasonable statistical uncertainty.

The scheme just described proceeds from the choice of reaction coordinate q, through the calculation of the free-energy curve $\mathcal{F}(q)$, to the identification of a transition state $q^\dagger$ at its maximum, and finally the calculation of the TST result and the transmission coefficient κ. In principle, the same answer should be obtained from different choices of q; in practice, bad choices will result in very inefficient, and imprecise, calculations of κ. In fact, it may be difficult to find a good choice. A small value of κ indicates that many recrossings of the $q = q^\dagger$ surface are occurring. The fact that positive-going and negative-going trajectories all contribute to the numerator of eqn (10.8), with opposite signs, is a problem. Various modifications of the scheme, to counter these deficiencies, have been proposed (Bergsma et al., 1986; White et al., 2000).

It is also possible to improve the definition of the transition state (Hummer, 2004). Because of the variational inequality $k_{A \to B} \leq k_{A \to B}^{TST}$, eqn (10.7), we may define the best dividing surface, and in principle the best reaction coordinate, so as to minimize $k_{A \to B}^{TST}$. As pointed out by Chandler (1978), this is equivalent to maximizing $\mathcal{F}(q^\dagger)$. This gives rise to a variety of practical approaches (Truhlar et al., 1996; see also Vanden-Eijnden and Tal, 2005), broadly termed variational TST; in practice, these methods are limited by the expense of varying the definition of q, and thereby parameterizing the reaction path in terms of a single reaction coordinate. Variational TST does not attempt to calculate the exact result, although it might well provide the optimum starting point for dynamical trajectories which could be used to efficiently compute the transmission coefficient.

10.4   Identifying reaction coordinates and paths

Figure 10.1(c) gives the traditional cartoon of a reaction path, linking two basins (local minima) A and B in a low-dimensional free-energy landscape, and passing through a saddle point ‡ which acts as a transition state. In the previous section, we have discussed the even simpler picture, Fig. 10.1(a), of a single reaction coordinate, with a transition state † located at the maximum, and not (in general) coinciding in any sense with ‡. It is important to realize that the real situation is much more complicated: the system actually explores a $3N$-dimensional potential-energy landscape (or even a $6N$-dimensional manifold, if momenta are included) which is intrinsically 'rough' and contains large numbers of high-dimensional local minima and saddle points (Wales, 2004). The basins A and B will be linked by an *ensemble* of paths, representing all thermally accessible routes through this landscape, and a proper description of the transition will require us to sample this ensemble. Even if we adopt a reduced description in terms of a few reaction coordinates, and use them to define a free energy, it is still an ensemble of paths, not a single path; by the same token, it will be misleading to think that there is a single transition state. Before turning to methods aimed at rigorously sampling this ensemble, in Section 10.5, we mention methods which attempt to deduce the relevant reaction coordinate(s) and/or to identify a single representative path, or a small number of such paths. The hope is that the path ensemble may be characterized by small fluctuations away from the representative paths, and that perhaps a coarse-grained dynamics, or even TST, involving just the reduced variables will be satisfactory. For an excellent recent review, see Rohrdanz et al. (2013). Examples of these approaches are the zero-temperature and finite-temperature string method (E and Vanden-Eijnden, 2010), diffusion maps (Rohrdanz et al., 2011; Zheng et al., 2011), diffusion-map-directed molecular dynamics (Preto and Clementi, 2014), and sketch maps (Ceriotti et al., 2011; Tribello et al., 2012).

If one can identify a small set of candidate reaction coordinates, then it may be possible to accelerate the sampling of those variables, and possibly select from them the most suitable one to describe the reaction. Metadynamics (Laio and Parrinello, 2002) is a method that combines molecular dynamics with an idea that is common to several accelerated simulation techniques: progressively flattening the free-energy landscape, so as to sample uniformly in a chosen reduced set of coordinates. It is reviewed in detail elsewhere (Laio and Gervasio, 2008), and is widely implemented in simulation packages. The usual approach is to add to the potential-energy function a set of Gaussian functions of variables such as q, q' in Fig. 10.1(c), which are centred on configurations that have already been visited. The effect is to bias the dynamics against visiting those same regions again and, as the trajectory proceeds, it explores higher and higher free energies. The same approach is used in Wang–Landau sampling, discussed in Section 9.2.5. The effective potential evolves in time; ideally it will eventually provide an estimate of the desired free-energy landscape. The convergence to this limit, and the range of exploration of different basins, may be controlled in a scheme known as well-tempered metadynamics (Barducci et al., 2008). An advantage of the method is its flexibility and ease of implementation for a wide choice of reduced variables. In the current context, the dynamical nature of the evolution seems to enhance the likelihood of discovering the saddle-points lying between basins, although (of course) the dynamics is biased by the added terms in the potential. In so-called reconnaissance metadynamics (Tribello et al., 2010) the method is extended to allow the use of a very large number of generalized coordinates, and this opens up the possibility of using it to identify the relevant reaction coordinates and pathways.

Various other simulation methods, based on the TST assumption, involve accelerating the dynamics within basins, while attempting to estimate the transition rates between basins. Examples are hyperdynamics (Kim and Falk, 2014) and temperature-accelerated dynamics (Sorensen and Voter, 2000); they are most commonly applied to dynamics in the solid state but have also been used to study protein folding problems. Similar ideas appear in the conformational flooding approach of Grubmüller (1995). We do not discuss these further but refer the reader to a recent review (Perez et al., 2009b).

10.5 Transition path sampling

In this section we modify the definitions of the regions A and B so that they only refer to the basins around the reactant and product states. We once again define functions $h_A(\mathbf{r})$ and $h_B(\mathbf{r})$ taking values $h = 1$ around the states $\mathbf{r} \in A$ and $\mathbf{r} \in B$ respectively, $h = 0$ otherwise, but we allow the possibility of a substantial region of configuration space in which *both* are zero. Hence we no longer require $h_A(\mathbf{r}) + h_B(\mathbf{r}) = 1$. However, we anticipate that it will be possible to define these regions such that the probability of configurations where $h_A(\mathbf{r}) = h_B(\mathbf{r}) = 0$ is low, that is, the system spends most of its time in either A or B. Hence $\langle h_A \rangle + \langle h_B \rangle \approx 1$. This removes attention from a critical 'transition region' separating A and B, and focuses it on the *paths* which link A and B. By the same reasoning, it is hoped that the results are not critically dependent on the definitions of the boundaries of the two regions, and there is no need to define a reaction coordinate.

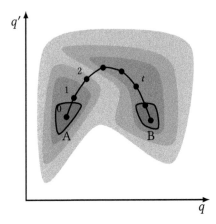

Fig. 10.2   A transition path joining two states A and B in a two-dimensional free-energy landscape. The boundaries of the states are indicated by solid lines. The coordinates q, q', are not needed to define A, B, or the path.

With these definitions in place, it is possible to define an equilibrium ensemble of *transition paths*, that is, those trajectories beginning in A and ending in B. A path (see Fig. 10.2) may be thought of as a sequence of points in phase space $(\mathbf{r}(t), \mathbf{p}(t))$, separated by regular intervals of time δt (or some other parameter), and linked by the normal time evolution (which could be deterministic, as in MD, or stochastic, as in Brownian or Langevin dynamics). The probability of a given path is then the product of all the conditional probabilities linking successive phase space points, $P(\mathbf{r}(t), \mathbf{p}(t) \to \mathbf{r}(t + \delta t), \mathbf{p}(t + \delta t))$ multiplied by the initial probability $P(\mathbf{r}(0), \mathbf{p}(0))$. The probability of the subset of transition paths is then obtained by including factors of $h_A(\mathbf{r}(0), \mathbf{p}(0))$ and $h_B(\mathbf{r}(t_{\mathrm{path}}), \mathbf{p}(t_{\mathrm{path}}))$.

Transition path sampling (TPS) involves generating a succession of such reactive trajectories by a standard importance-sampling Monte Carlo method (Bolhuis et al., 2002). The methods only differ slightly depending on the type of dynamics being used; here we assume standard MD. As usual in importance sampling, there are two stages: proposal of a new trajectory, constructed in some way from an existing one, and acceptance or rejection of this proposal in a way that obeys detailed balance. There are two common methods of proposing new trajectories: *shifting* and *shooting*. The shifting move deletes a section of the trajectory from one end, and adds a new section, of equal time, at the other end: hence the path is shifted forwards or backwards in time. In the simplest case of microscopically reversible dynamics, and an equilibrium distribution of initial conditions, the acceptance–rejection criterion reduces to simply applying the conditions that the start point still lies within A, and the end point still lies within B. The shooting move is illustrated in Fig. 10.3. It entails the selection of a timestep, at random, somewhere along the trajectory, and the perturbation of (typically) particle momenta to give new values, followed by forward and backward time integration until a new trajectory is complete. In the simple case of dynamics which preserve the stationary equilibrium distribution, such as the canonical one, and assuming that the starting points are selected in a symmetric way, the acceptance–rejection criterion is once again fairly simple: it includes factors to

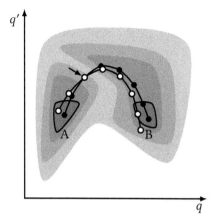

Fig. 10.3 Generating a new path (open circles) from an existing one (filled circles) by a shooting move initiated from the arrowed position.

enforce the requirement that the trajectory go from A to B, multiplied by a Metropolis-like factor involving ratios (new/old) of products of ensemble weights for the configurations along the new and old trajectories. Different variations of these moves for different kinds of simulation are discussed elsewhere (Bolhuis et al., 2002). Shifting and shooting moves are complementary in nature. The shifting move is relatively cheap to implement and may have a high acceptance rate because much of the new trajectory is the same as the old one; on the other hand, it does not achieve a dramatic exploration of the landscape. The shooting move often produces a radically different path, which must be followed in its entirety, and may have a lower acceptance rate; however, when successful, the move is very effective. Typically the perturbations applied in shooting moves are small, although 'aimless shooting' (Mullen et al., 2015), in which all momenta are reselected from the Maxwell–Boltzmann distribution, combined with flexibility in the trajectory lengths, can be useful.

Although TPS may seem complicated, the dynamics involved in the computation of the trajectories is quite standard, and the additional features of generating new starting points, and accepting or rejecting the new trajectories, can be handled by a surrounding script or shell program. The method can therefore be added to established simulation packages without too much effort.

In terms of averages over those paths, denoted $\langle\ldots\rangle_{A\to B}$, the correlation function $C_{A\to B}(t)$, and through eqn (10.5) the rate constant $k_{A\to B}$, may be calculated (Dellago et al., 2002). In a sense, it is necessary to connect the weights associated with the ensemble of paths to those of a conventional ensemble. This follows by recognizing eqn (10.4), the definition of $C_{A\to B}(t)$, as a ratio of two partition functions: $Q_{A\to B}(t)/Q_A$. $Q_{A\to B}(t)$ counts the paths that start in A and end in B at time t. Q_A counts all the paths that start in A, irrespective of their end point, and hence is proportional to the probability of the system being in region A. This ratio may be regarded as the exponential of a free-energy difference, and computed by umbrella sampling, using methods akin to those described in Chapter 9. However, at first sight, this needs doing, complete with the sampling of many independent paths, over a range of t in order to establish the plateau behaviour characteristic of intermediate times $t_{mol} < t \ll k^{-1}$. Further consideration leads to a more effective method. The quantity $\langle h_B(t)\rangle_{A\to B}$ can be determined by averaging over paths. Then, the ratio

$$\frac{\langle h_B(t)\rangle_{A\to B}}{\langle h_B(t')\rangle_{A\to B}} = \frac{Q_{A\to B}(t)}{Q_{A\to B}(t')}$$

can be calculated for arbitrary t and t', without doing any umbrella sampling. Next, the value of $C_{A\to B}(t')$ may be computed by umbrella sampling for just one time t'. Finally, the equation

$$C_{A\to B}(t) = \frac{Q_{A\to B}(t)}{Q_A} = \frac{\langle h_B(t)\rangle_{A\to B}}{\langle h_B(t')\rangle_{A\to B}} C_{A\to B}(t')$$

allows us to obtain $C_{A\to B}(t)$ at arbitrary times. It is expected that there will be a range of values of t for which $C_{A\to B}(t)$ increases linearly with t, so that its time derivative has the necessary plateau, and in fact this can be checked from the behaviour of $\langle h_B(t)\rangle_{A\to B}$ in the first stage. Then eqn (10.5) gives $k_{A\to B}$.

Once the transition path ensemble is properly sampled, it should be possible to compute not only the rate but also some features of the paths themselves, perhaps even identifying a suitable reaction coordinate and transition state, if one exists. This can be done by analysing so-called committor surfaces, which connect configurations having the same likelihood of arriving in B rather than A (Best and Hummer, 2005; Peters and Trout, 2006).

Whatever method is used, the TPS approach is quite expensive, involving an extended dynamical simulation covering the timescale of the process of interest, before accepting or rejecting each new trajectory as a member of the desired ensemble; the whole procedure then needs to be repeated enough times to give statistically significant results. Of course, this expense must be compared with the alternatives: brute-force simulation of rare events is never likely to be competitive, while the techniques described in the previous sections involve serious approximations, assumptions, or biasing factors. A recent review (Bolhuis and Dellago, 2015) covers a variety of practical pitfalls and misconceptions regarding the method, including some of the attempts made to improve efficiency, discussed in the next section.

10.6   Forward flux and transition interface sampling

The approaches described in this section rely on the definition of a single coordinate q, to describe the progress of the reaction, but without needing to identify a transition state value $q^\dagger$ as in sections 10.2 and 10.3. Instead, in the spirit of the previous section, it is assumed that region A may be defined by $q < q_A \equiv q_0$, and region B by $q > q_B \equiv q_n$, in a sensible way. The assumption is that trajectories entering region A or B reach equilibrium, and lose memory, before leaving again. Then $n - 1$ intermediate values of q are inserted between these boundaries, as illustrated in Fig. 10.4. The forward rate constant may be written (van Erp et al., 2003)

$$k_{A\to B} = \frac{\langle \Phi_0\rangle}{\langle h_A\rangle} P(q_0 \to q_n) = \frac{\langle \Phi_0\rangle}{\langle h_A\rangle} \prod_{i=0}^{n-1} P(q_i \to q_{i+1}). \tag{10.9}$$

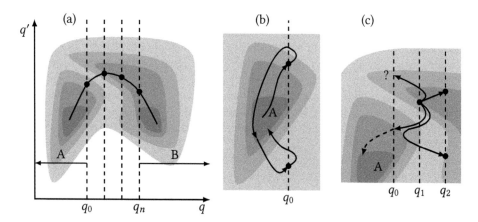

Fig. 10.4 States and interfaces defined in terms of a reaction coordinate q. (a) Schematic free-energy landscape, and definition of interfaces. An example reactive trajectory is shown. (b) Calculating the outgoing flux at q_0 using an equilibrium trajectory. Two outward crossings of the interface (black dots) are counted, and stored for future use. (c) Calculating the conditional probability $P(q_1 \rightarrow q_2)$ from stored configurations at q_1. Two partial paths are shown which reach q_2 without first returning to A (black dots), and hence contribute to the transition path ensemble. One path returns to A, and does not contribute. A fourth path, indicated by a question mark, is discussed in the text.

The first term is the flux of states from basin A to the boundary at q_0. The second term is the conditional probability that a configuration starting at $q = q_0$, heading out from A, will arrive at a later time at interface $q = q_n$, entering region B, rather than returning to A. In the second part of eqn (10.9), this conditional probability is factorized into a product of terms. Each term represents a conditional probability of reaching interface q_{i+1} from q_i, rather than returning to A; included in this condition is the requirement that the trajectory started in A, and crossed q_i for the first time before proceeding onwards. The aim of the techniques described in this section is the same as in Section 10.5: to generate trajectories sampled from the transition path ensemble linking A and B. The hope is that it can be done in a more efficient way.

Here we explain how this is tackled in FFS (Allen et al., 2006a,b). The flux term $\langle \Phi_0 \rangle / \langle h_A \rangle$ is typically measured in an equilibrium molecular dynamics simulation, by simply dividing the number of outgoing crossings of the boundary surface by the simulation time. Each time this occurs, the configuration $(\mathbf{r}, \mathbf{p})$ is saved for the next stage. The more saved configurations, the better: numbers of order 10^3 have been used in tests (Allen et al., 2006a,b). Therefore, the position of the boundary q_0 needs to be chosen such that equilibrium trajectories in the A basin reach $q = q_0$ infrequently (because only an insignificant part of the basin should lie on the $q > q_0$ side), but not too infrequently. This is illustrated in Fig. 10.4(b). On the rare occasions in which a trajectory escapes from A and makes the transition to B spontaneously, the procedure is restarted with a new configuration, equilibrated in A. In subsequent stages, the probability $P(q_i \rightarrow q_{i+1})$ is evaluated for each successive pair of interfaces, by picking a starting configuration randomly from the stored set at $q = q_i$, and following the dynamics until it either reaches

$q = q_{i+1}$ or returns all the way to A. The configurations which do reach q_{i+1} are themselves stored, to be used as starting points for the next section. This is shown diagrammatically in Fig. 10.4(c). The dynamics, therefore, needs to have a stochastic element to it, in order that different trajectories are generated by the same starting configuration. It is tempting, but wrong in principle, to achieve this effect simply by randomizing the velocities at the start of each $q_i \rightarrow q_{i+1}$ trajectory. In a similar vein, it is important to realize that the method does not make any Markovian assumptions about the successive conditional probabilities appearing in eqn (10.9): they are all linked by the conditions applied to the ensemble of starting points.

The outcome from these successive calculations is a set of conditional probabilities, which are multiplied together, as well as the initial outgoing flux, according to eqn (10.9). The method relies on obtaining good statistical estimates of each of these quantities, which in turn requires a suitable choice of the number and location of interfaces. Allen et al. (2006a) have described ways to estimate the efficiency of this, and related, methods; Borrero and Escobedo (2008) and Kratzer et al. (2013) have described ways to optimize the approach. As a practical point, it is not essential to make q_0 coincide with the criterion used to define the basin A (Allen et al., 2006b). Indeed, it may be desirable to have a 'tighter' boundary around the minimum. Some partial trajectories might cross the $q = q_0$ boundary at locations q' which lie within the basin of attraction of B, and hence not actually be on the way back to A. Such a path is indicated with a question mark in Fig. 10.4(c). To make sure that trajectories are correctly identified as not belonging to the transition path ensemble, it may be useful to define the initial state separately from the interface q_0, for example as in Fig. 10.3: all reactive transition paths would originate within this region, and a non-reactive partial path would be identified by following it all the way back to this region. This highlights one possible inefficiency of the method: the need to follow some trajectories for a very long time, until their fate becomes clear. The FFS method has been reviewed by Allen et al. (2009). A harness to implement FFS, and a generalization of the approach called stochastic process rare event sampling (SPRES) (Berryman and Schilling, 2010), has been provided for a range of MD packages (Kratzer et al., 2014).

A potential drawback of these methods, which follow trajectories forwards from the reactant basin A to the product basin B, is the effect of attrition. In other words, only a certain fraction of the starting configurations at each interface will succeed in reaching the next interface, and this effect is cumulative. As a result, if one traces the trajectories back from B to A, it is likely that they originate from just a few starting points, or perhaps even just one point, on the interface q_0. This may mean that the true ensemble of transition paths is poorly sampled, as discussed by van Erp (2012). For this reason, although FFS is conceptually simple and fairly easy to implement, this is a reason to prefer the somewhat more complicated TIS approach. However, FFS has one significant advantage, namely that the rate equation is equally valid for an out-of-equilibrium, driven system.

The original TIS method (van Erp et al., 2003; Moroni et al., 2004a; Moroni, 2005; van Erp and Bolhuis, 2005) works within the same framework of interfaces defined by a coordinate q, but uses a more sophisticated way of building the ensemble of transition paths. From a given path, which crosses the q_i interface, the 'shooting' method of the previous section is used to generate perturbed trajectories, which are typically followed

Example 10.1 Homogeneous crystal nucleation

The formation of a crystal nucleus, in an undercooled liquid, is an excellent example of a rare event for which the simulation techniques of this chapter may be useful (Yi and Rutledge, 2012; Turci et al., 2014). A typical experimental nucleation rate of 10^6 cm^{-3} s^{-1} translates into a single nucleation event every 10^{12} s in a simulation box of 10^4 molecules. Major questions to be addressed are whether or not classical nucleation theory applies, whether the structure of the critical nucleus is the same as that of the final product crystal, what is the free-energy barrier, and, of course, what determines the rate? Early, brute-force, simulations of Lennard-Jones atoms used temperatures of order 50 % of the equilibrium freezing temperature, and system sizes $N \approx 10^6$ (Swope and Andersen, 1990) to suggest that both body centred cubic (BCC) and FCC nuclei formed, but only the latter would go on to form crystals. Later work, using umbrella sampling based on bond-orientational order parameters (ten Wolde et al., 1995) employed systems $N \approx 10^5$, but also a much more modest degree of undercooling, 20 %, to suggest that the critical nuclei had a core–shell structure. The same authors made a first attempt to calculate the rate through the Bennett–Chandler method (ten Wolde et al., 1996). Lennard-Jones freezing has been subsequently examined by metadynamics (Trudu et al., 2006) and TIS (Moroni et al., 2005b). These simulations showed that there is a range of 'critical' nucleus sizes and degrees of crystallinity, with shape and internal structure playing a role.

A particularly nice example of these techniques is the study of a soft-core model of colloidal suspensions by Lechner et al. (2011). These authors obtained a reweighted path ensemble from replica-exchange TIS simulations, which was then analysed in detail to obtain information about the nature of the reaction coordinate. They showed that, as well as the structure of the nucleus, the size of a prestructured cloud of particles surrounding the nucleus could also be a significant order parameter in the crystallization mechanism.

---

forward in time, to interface q_{i+1}, and backward in time, to check that they originated in the A basin. There are acceptance–rejection criteria for the new path, relative to the old one, based on reaching the correct end points, and on the path length. Moroni et al. (2005a) have shown how the free-energy profile of the process may be obtained at no additional computational cost, while Borrero et al. (2011) have discussed how to optimize the method.

If an assumption is made that the trajectory loses memory over two interfaces, then full transition paths may be constructed from partial paths in a more efficient way (Moroni et al., 2004b). This is likely to be true if the dynamics is diffusive in nature, that is, heavily overdamped due to large frictional and random force effects (see Section 12.2). A similar assumption, of memory loss during the transition from one interface to the next, underpins the milestoning approach (Faradjian and Elber, 2004).

10.7   Conclusions

Sampling of rare events remains a significant challenge for simulation, and there is no single technique that will work in every case. There are a few regularly used small-scale testbeds, such as the alanine dipeptide system, and the 38-atom Lennard-Jones cluster, which may be used to calibrate and compare different methods. However, systems of interest are often much larger, of high dimensionality, and contain very many metastable minima and saddle points. The area of crystal nucleation provides a good example of the way in which improved simulation techniques have shed more and more light on a difficult problem, and this is discussed in Example 10.1.

It may be that the ensemble of transition paths cannot be thought of in terms of small variations from a 'representative' trajectory, but instead includes many contributing 'channels', which are hard to convert into one another using the methods described in this chapter. It is possible to combine TPS and its relations with techniques such as parallel tempering/replica exchange (Section 4.9) (Vlugt and Smit, 2001), or Wang–Landau sampling (Section 9.2.5) (Borrero and Dellago, 2010), to improve efficiency and calculate rates at a range of temperatures simultaneously. Other tricks of this kind are possible within the TIS scheme (van Erp and Bolhuis, 2005; van Erp, 2007; Bolhuis, 2008). This area remains highly active, and the reader is recommended to consult the recent literature for the latest developments.

11

Nonequilibrium molecular dynamics

11.1 Introduction

So far in this book, we have considered the computer simulation of systems at equilibrium. Even the introduction, into the molecular dynamics equations, of terms representing the coupling to external systems (constant-temperature reservoirs, pistons, etc.) have preserved equilibrium: the changes do not induce any thermodynamic fluxes. In this chapter, we examine adaptations of MD that sample nonequilibrium ensembles: nonequilibrium molecular dynamics (NEMD). One motivation for this is to improve the efficiency with which transport coefficients are calculated, the route via linear response theory and time correlation functions (eqns (2.116)–(2.130)) being subject to significant statistical error. This has been discussed in part in Chapter 8. Another is to examine directly the response of a system to a large perturbation lying outside the regime of linear response theory, as occurs in a shock wave. Finally, it proves possible to calculate free-energy differences by using nonequilibrium techniques.

One problem with time correlation functions is that they represent the average response to the naturally occurring (and hence fairly small) fluctuations in the system properties. The signal-to-noise ratio is particularly unfavourable at long times, where there may be a significant contribution to the integral defining a transport coefficient. Moreover, the finite system size imposes a limit on the maximum time for which reliable correlations can be calculated. The idea behind nonequilibrium methods is that a much larger fluctuation may be induced artificially, and the signal-to-noise level of the measured response improved dramatically. By measuring the steady-state response to such a perturbation, problems with long-time behaviour of correlation functions are avoided. NEMD measurements are made in much the same way as that used to estimate simple equilibrium averages such as the pressure and temperature. These methods correspond much more closely to the procedure adopted in experiments: shear and bulk viscosities, and thermal conductivities, are measured by creating a flow (of momentum, energy, etc.) in the material under study. The growing interest in understanding liquid flow in porous media, microfluidics, and nanofluidics, have all stimulated the development of reliable methods of simulating such systems directly.

Computer Simulation of Liquids. Second Edition. M. P. Allen and D. J. Tildesley.
© M. P. Allen and D. J. Tildesley 2017. Published in 2017 by Oxford University Press.

Early attempts to induce momentum or energy flow in a molecular dynamics simulation have been reviewed by Hoover and Ashurst (1975). One possibility is to introduce boundaries, or boundary regions, where particles are made to interact with external momentum or energy reservoirs (Ashurst and Hoover, 1972; 1973; 1975; Hoover and Ashurst, 1975; Tenenbaum et al., 1982). These methods, however, are incompatible with periodic boundary conditions, and so they introduce surface effects into the simulation. Most of the approaches we shall describe avoid this by being designed for consistency with the usual periodic boundaries, or by modifying these boundaries in a homogeneous way, preserving translational invariance and periodicity. However, in some cases, the flow against an interface is itself of interest, and we shall give some examples later.

Nonequilibrium molecular dynamics simulations have grown in popularity over the last few years, and several excellent reviews (Ciccotti et al., 1979; Hoover, 1983a,b; Evans and Morriss, 1984a) and books (Evans and Morriss, 2008; Tuckerman, 2010) may be consulted for further details.

The methods we are about to describe all involve perturbing the usual equations of motion in some way. Such a perturbation may be switched on at time $t = 0$, remaining constant thereafter, in which case the measured responses will be proportional to time-integrated correlation functions. The long-time steady-state responses (the infinite time integrals) may then yield transport coefficients. Alternatively, the perturbation may be applied as a delta function pulse at time $t = 0$ with subsequent time evolution occurring normally. In this case, the responses are typically proportional to the correlation functions themselves: they must be measured at each timestep following the perturbation, and integrated numerically to give transport coefficients. Finally, an oscillating perturbation $\propto \sin \omega t$ may be applied. After an initial transient period, the measured responses will be proportional to the real and imaginary parts of the Fourier–Laplace transformed correlation functions, at the applied frequency ω. To obtain transport coefficients, several experiments at different frequencies must be carried out, and the results extrapolated to zero frequency. The advantages and disadvantages of these different techniques will be discussed following a description of some of the perturbations applied.

The perturbations typically appear in the equations of motion as follows (Evans and Morriss, 1984a)

$$\dot{q} = p/m + \mathcal{A}_p \cdot \mathcal{F}(t) \tag{11.1a}$$

$$\dot{p} = f - \mathcal{A}_q \cdot \mathcal{F}(t). \tag{11.1b}$$

The condensed notation disguises the complexity of these equations in general. Here, $\mathcal{F}(t)$ is a $3N$-component vector representing a time-dependent applied field. It can be thought of as applying to each molecule, in each coordinate direction, separately. The quantities $\mathcal{A}_q(q, p)$ and $\mathcal{A}_p(q, p)$ are functions of particle positions and momenta. They describe the way in which the field couples to the molecules, perhaps through a term in the system Hamiltonian. Each can be a $3N \times 3N$ matrix in the general case, but usually many of the components vanish. The perturbation can often be thought of as coupling separately to some property of each molecule (e.g. its momentum), in which case $\mathcal{A}_q$ and $\mathcal{A}_p$ become very simple indeed. However, some properties (e.g. the energy density, the pressure tensor), while being formally broken down into molecule-by-molecule contributions, actually depend on inter-molecular interactions, and so $\mathcal{A}_q$ and $\mathcal{A}_p$ must be functions of all particle positions and momenta in the general case.

In standard linear response theory, the perturbation is represented as an additional term in the system Hamiltonian

$$\mathcal{H}^{ne} = \mathcal{H} + \mathcal{A}(q, p) \cdot \mathcal{F}(t) = \mathcal{H} + \sum_i \mathcal{A}_i(q, p) \cdot \mathcal{F}_i(t) \tag{11.2}$$

in which case we simply have

$$\mathcal{A}_q = \nabla_q \mathcal{A}(q, p), \qquad \mathcal{A}_p = \nabla_p \mathcal{A}(q, p). \tag{11.3}$$

The average value $\langle \mathcal{B} \rangle_{ne}$ of any phase function $\mathcal{B}(q, p)$ in the nonequilibrium ensemble generated by the perturbation is given by

$$\langle \mathcal{B}(t) \rangle_{ne} = -\frac{1}{k_B T} \int_0^t dt' \langle \mathcal{B}(t - t') \dot{\mathcal{A}}(0) \rangle \cdot \mathcal{F}(t') \tag{11.4}$$

assuming that the equilibrium ensemble average $\langle \mathcal{B} \rangle$ vanishes and that the perturbation is switched on at time $t = 0$. However, it has long been recognized that the perturbation need not be derived from a Hamiltonian (Jackson and Mazur, 1964). Provided that

$$\nabla_q \cdot \dot{q} + \nabla_p \cdot \dot{p} = \left(\nabla_q \cdot \mathcal{A}_p - \nabla_p \cdot \mathcal{A}_q \right) \cdot \mathcal{F}(t) = 0 \tag{11.5}$$

the incompressibility of phase space still holds, and eqn (11.4) may still be derived. In this case, however, $\dot{\mathcal{A}}$ cannot be regarded as the time derivative of a variable $\mathcal{A}$. Rather, it is simply a function of q and p, defined by the rate of change of internal energy

$$\dot{\mathcal{H}} = -\left((p/m) \cdot \mathcal{A}_q + f \cdot \mathcal{A}_p \right) \cdot \mathcal{F}(t) \equiv -\dot{\mathcal{A}} \cdot \mathcal{F}(t). \tag{11.6}$$

Thus, $\mathcal{A}_q$ and $\mathcal{A}_p$ are sufficient to define $\dot{\mathcal{A}}$ in eqn (11.4). These equations have been developed and extended by Evans and co-workers (Evans and Morriss, 1984a; 2008).

When a perturbation is applied in molecular dynamics, typically the system heats up. This heating may be controlled by techniques analogous to those employed in constant-temperature MD, as discussed in Chapter 3. The thermostatting method is important in defining the nonequilibrium ensemble being simulated (Evans and Morriss, 2008). In the following sections, for simplicity, we shall omit the extra terms in the equations of motion which serve this purpose. This choice corresponds to a perturbation which is applied adiabatically, that is, the work done on the system exactly matches the increase in the internal energy. We shall return to this in Section 11.7. Moreover, in most of this chapter, the perturbations are assumed to apply to an atomic system, or to the centres of mass of a system of molecules. Accordingly, we shall revert to the notation r, p rather than q, p.

11.2 Spatially oscillating perturbations

Some of the earliest nonequilibrium simulations attempted to measure the shear viscosity of an atomic Lennard-Jones fluid. One technique, which maintains conventional cubic periodic boundary conditions, is to use a spatially periodic perturbation to generate

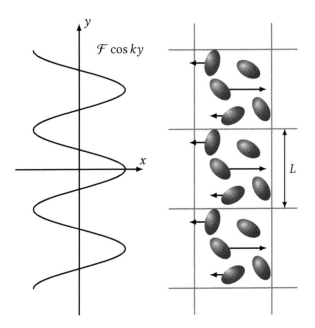

Fig. 11.1 Spatially oscillating shear flow perturbation. Three periodic boxes are shown. The force function $\mathcal{F} \cos ky$ is plotted as a function of y on the left. The arrows on the right indicate the forces acting on each molecule.

an oscillatory velocity profile (Gosling et al., 1973; Ciccotti et al., 1976b; 1979). At each timestep in an otherwise conventional MD simulation, an external force in the x-direction is applied to each molecule. The magnitude of the force depends upon the molecule's y-coordinate as follows:

$$f_{ix}^{\text{ext}} = \mathcal{F} \cos(2\pi n y_i / L) = \mathcal{F} \cos k y_i \tag{11.7}$$

where $\mathcal{F}$ is a constant, and the wavevector $\mathbf{k} = (0, k, 0) = (0, 2\pi n/L, 0)$, with n an integer, is commensurate with the side L of the simulation box. This force field is illustrated in Fig. 11.1 for the lowest wavelength case, $n = 1$. On applying this perturbation and waiting, a spatially periodic velocity profile develops. Specifically, at a given y-coordinate, the mean x-velocity of a molecule should be

$$\langle v_x(y) \rangle_{\text{ne}} \approx \frac{\rho}{k^2 \eta} \mathcal{F} \cos ky. \tag{11.8}$$

By fitting their results to this equation, Gosling et al. (1973) were able to estimate the shear viscosity η with significantly less computational effort than that required using equilibrium methods.

It is worth examining carefully the origins of eqn (11.8). The perturbation of eqn (11.7) is a non-Hamiltonian one, but falls into the general scheme of the last section. Writing in a slightly more general form

$$f_{ix}^{\text{ext}}(t) = \mathcal{F}(t) \exp(-iky_i) \tag{11.9}$$

we can show that the response in any k-dependent quantity $\mathcal{B}(k)$ is related to a correlation function involving the transverse current $j_x^\perp(k,t)$ or the transverse momentum $p_x^\perp(k,t)$:

$$j_x^\perp(k,t) = \frac{1}{V} \sum_{i=1}^{N} v_{ix} \exp\left(iky_i(t)\right) \tag{11.10a}$$

$$p_x^\perp(k,t) = \frac{1}{V} \sum_{i=1}^{N} p_{ix} \exp\left(iky_i(t)\right). \tag{11.10b}$$

These equations are analogous to eqns (2.137b), (2.137c), except that we take the k-vector in the y direction and make the x-component of the velocity explicit. Specifically, the perturbation of eqn (11.9) appears in eqn (11.1) in the following way. For all i, we have $\mathcal{A}_{qix} = -\exp(-iky_i)$, while all the remaining $\mathcal{A}_q$ and $\mathcal{A}_p$ terms vanish. We therefore have (eqn (11.6)) $\dot{\mathcal{A}} = \sum_i (p_{ix}/m)\mathcal{A}_{qix} = -Vj_x^\perp(-k)$. Thus

$$\langle \mathcal{B}(k,t)\rangle_{\mathrm{ne}} = \frac{V}{k_{\mathrm{B}}T} \int_0^t dt' \left\langle \mathcal{B}(k,t-t')j_x^\perp(-k,0)\right\rangle \mathcal{F}(t') \tag{11.11}$$

and for the response in the current itself

$$\langle j_x^\perp(k,t)\rangle_{\mathrm{ne}} = \frac{V}{k_{\mathrm{B}}T} \int_0^t dt' \left\langle j_x^\perp(k,t-t')j_x^\perp(-k,0)\right\rangle \mathcal{F}(t'). \tag{11.12}$$

The equilibrium time correlation function is real and is linked to the shear viscosity through the Fourier–Laplace transform, valid at low k,ω (Hansen and McDonald, 2013):

$$\frac{V}{k_{\mathrm{B}}T}\left\langle j_x^\perp(k,\omega)j_x^\perp(-k)\right\rangle \approx \frac{\rho/m}{i\omega + k^2\eta(k,\omega)/\rho m}. \tag{11.13}$$

This equation may be used to define a k,ω-dependent shear viscosity $\eta(k,\omega)$, which goes over to the transport coefficient η on taking the limit $\omega \to 0$ followed by the limit $k \to 0$. Taking the zero-frequency limit here means time-integrating from $t = 0$ to $t = \infty$, so

$$\frac{V}{k_{\mathrm{B}}T} \int_0^\infty dt \left\langle j_x^\perp(k,t)j_x^\perp(-k)\right\rangle = \frac{\rho^2}{k^2\eta(k)}. \tag{11.14}$$

If the perturbation remains constant, $\mathcal{F}(t) = \mathcal{F}$ from $t = 0$ onwards, this is essentially the quantity appearing on the right of eqn (11.12) as the integration limit goes to infinity and the steady state is obtained. Thus

$$\langle j_x^\perp(k,t \to \infty)\rangle_{\mathrm{ne}} = \frac{\mathcal{F}\rho^2}{k^2\eta(k)}. \tag{11.15}$$

Apart from a factor of ρ linking the current and velocity profile, and the explicit appearance of the $\cos ky$ term reflecting the fact that a real perturbation (the real part of eqn (11.9)) yields a real, in-phase response, this is eqn (11.8). Note how the method relies on a k-dependent viscosity going smoothly to η as $k \to 0$. This means that in a real application, several different k-vectors should be chosen, all orthogonal to the x-direction, and an extrapolation to zero k should be undertaken (Ciccotti et al., 1976b).

The same sinusoidal perturbation may be applied as a delta function in the time $\mathcal{F}(t) \propto \delta(t)$, which superimposes a sinusoidal initial velocity profile on the equilibrium distribution. The subsequent time decay may then be observed and the viscosity extracted. Averaging over several repetitions of the experiment is required to obtain good statistics. This approach has been termed 'transient molecular dynamics' (Thomas and Rowley, 2007).

A variant of this approach, in which the applied force has the form

$$f_{ix}^{\text{ext}} = \text{sgn}(y_i)\mathcal{F}(t) = \begin{cases} \mathcal{F}(t) & 0 \le y_i < L/2 \\ -\mathcal{F}(t) & -L/2 \le y_i < 0 \end{cases} \quad \text{with} \quad \mathcal{F}(t) = \begin{cases} \mathcal{F} & t > 0 \\ 0 & t \le 0 \end{cases} \quad (11.16)$$

has been proposed by Backer et al. (2005). This corresponds to a constant force, with equal and opposite values depending on whether the particle lies in the upper or lower half of the box. Decomposing this square-wave form in a Fourier series, and applying the same analysis as before, gives the velocity profile

$$\langle v_x(y) \rangle_{\text{ne}} \approx \frac{\rho}{2\eta}\mathcal{F} y(\tfrac{1}{2}L - |y|). \quad (11.17)$$

Extracting the viscosity then amounts to fitting to a piecewise parabolic, rather than sinusoidal, velocity profile, similar to the classical Poiseuille flow experiment between planar walls. A disadvantage of this method is the discontinuous nature of the flow at $y = 0$ and $y = L/2$, which will cause significant heating, and possibly density variation, in the vicinity. Although the perturbation is periodic, this version of the method is more comparable with the inhomogeneous, boundary-driven methods discussed in Section 11.4.

There is plenty of scope to extend NEMD methods to study quantities other than the transport coefficients of hydrodynamics and their associated correlation functions. In an atomic fluid, an example is the direct measurement of the dynamic structure factor $S(k, \omega)$ (Evans and Ojeda, 1992). Here, a spatially periodic perturbation is applied which couples to the number density. The method yields a response function that may be converted directly to $S(k, \omega)$ in a manner preserving the quantum mechanical detailed balance condition, eqn (2.153), whereas conventional methods of obtaining $S(k, \omega)$ (through eqn (2.136)) yield the symmetrical, classical function.

The pitfalls inherent in using finite-k perturbations are well illustrated by another attempt to measure η via coupling to the transverse momentum density (Ciccotti et al., 1979). In this case, the perturbation is of Hamiltonian form, eqn (11.2), with

$$\mathcal{A} \cdot \mathcal{F}(t) = V p_x^{\perp}(-k, t)\mathcal{F}(t) \quad (11.18)$$

and $\mathbf{k} = (0, k, 0)$ as before. The responses in this case are given by

$$\langle \mathcal{B}(k, t) \rangle_{\text{ne}} = \frac{V}{k_{\text{B}}T} \int_0^t \mathrm{d}t' \left\langle \mathcal{B}(k, t - t')\mathcal{P}_{yx}(-k, 0) \right\rangle ik\mathcal{F}(t') \quad (11.19)$$

where

$$\mathcal{P}_{yx}(k, t) = \frac{1}{V}\sum_i m v_{ix} v_{iy} \exp(iky_i) + \frac{1}{V}\sum_i \sum_{j>i} y_{ij} f_{ijx} \left(\frac{\exp(iky_i) - \exp(iky_j)}{iky_{ij}} \right) \quad (11.20)$$

is defined so that $\dot{p}_x^{\perp}(k,t) = ik\mathcal{P}_{yx}(k,t)$ (compare eqn (2.122)). Specifically

$$\langle \mathcal{P}_{yx}(k,t)\rangle_{\text{ne}} = \frac{V}{k_{\mathrm{B}}T}\int_0^t dt'\left\langle \mathcal{P}_{yx}(k,t-t')\mathcal{P}_{yx}(-k,0)\right\rangle ik\mathcal{F}(t'). \tag{11.21}$$

Now the infinite time integral of $\left\langle \mathcal{P}_{yx}(k,t)\mathcal{P}_{yx}(-k)\right\rangle$ can be calculated by applying a steady perturbation and measuring the long-time response of the out-of-phase component of $\mathcal{P}_{yx}(k)$ (i.e. the $\sin ky$ component if a real $\cos ky$ field is applied, because of the ik factor in eqn (11.21)). Unfortunately, this quantity vanishes identically for non-zero k. This is because in the equation

$$\eta = \lim_{\omega\to 0}\lim_{k\to 0}\frac{V}{k_{\mathrm{B}}T}\int_0^{\infty} dt\, \exp(-i\omega t)\left\langle \mathcal{P}_{yx}(k,t)\mathcal{P}_{yx}(-k)\right\rangle \tag{11.22}$$

the $k\to 0$ limit must be taken first (yielding eqn (2.119)). It is not possible to define a quantity $\eta(k,\omega)$ which has sensible behaviour at low ω simply by omitting the limiting operations in eqn (11.22). Of course, the same difficulty applies in any attempt to measure η via an equilibrium MD calculation of eqn (11.22). A way around this problem has been given by Evans (1981a), but in either case (equilibrium or nonequilibrium calculations) the wavevector and frequency-dependent correlation function in eqn (11.22) is required before the extrapolation to give η can be undertaken.

In a similar vein, it is straightforward to introduce a Hamiltonian perturbation which couples to a Fourier component of the energy density and which can yield the energy current autocorrelation function (Ciccotti et al., 1978; 1979). The link with the transport coefficient, however, suffers from the same drawback: the limit $k \to 0$ must be taken before $\omega \to 0$ (i.e. before a steady-state time integration) or the function vanishes.

11.3   Spatially homogeneous perturbations

If a zero-wavevector transport coefficient is required, then a zero-wavevector technique is preferred. Generally this requires both a modification of the equations of motion, and a compatible modification of the periodic boundary conditions. In this section we consider several examples, mostly targetted at the transport coefficients defined in Chapter 2.

11.3.1   Shear flow

A useful review of the literature, combined with an explanation of many of the nonequilibrium algorithms used to study viscous flow of liquids, has been provided by Todd and Daivis (2007). The simplest approach is to simulate planar Couette flow, in which a uniform shear is applied to the system. A set of suitably modified periodic boundaries was proposed by Lees and Edwards (1972) and is illustrated in Fig. 11.2. In essence, the infinite periodic system is subjected to a uniform shear in the xy plane. The simulation box and its images centred at $(x, y) = (\pm L, 0), (\pm 2L, 0)$, etc. (for example, A and E in Fig. 11.2) are taken to be stationary. Boxes in the layer above, $(x, y) = (0, L), (\pm L, L), (\pm 2L, L)$, etc. (e.g. B, C, D) are moving at a speed $(dv_x/dy)L$ in the positive x direction (dv_x/dy is the shear rate, or strain rate, and we will use the symbol $\mathcal{F}(t)$ for it). Boxes in the layer below, $(x, y) = (0, -L), (\pm L, -L), (\pm 2L, -L)$, etc. (e.g. F, G, H) move at a speed $\mathcal{F}L$ in the negative x direction. In the more remote layers, boxes are moving proportionally faster relative

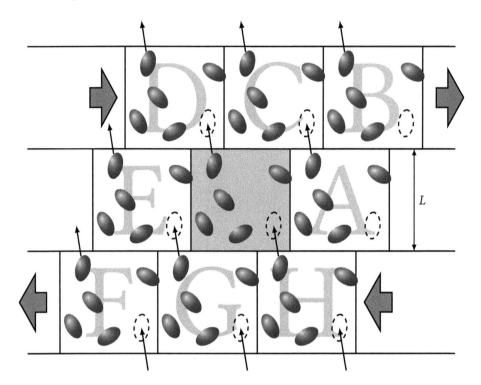

Fig. 11.2 Homogeneous shear boundary conditions. Each row is moving relative to the ones above and below, as indicated by the large arrows.

to the central one. Box B, for example, starts off adjacent to A but, if $\mathcal{F}$ is a constant, it will move away in the positive x-direction throughout the simulation, possibly ending up hundreds of box lengths away. This corresponds to a simple shear in one direction (see Fig. 11.2). It is most convenient to represent this using shifted cubic boxes, rather than by deforming the box, since a steady shear can then be maintained without the box angles becoming extremely acute (compare Fig. 9.4). Of course it is also possible to make $\mathcal{F}$ vary sinusoidally in time, in which case the box B will oscillate about its initial position. The periodic minimum image convention must be modified in this case. The upper layer (BCD in Fig. 11.2) is displaced relative to the central box by an amount δx which is equal to the box length L multiplied by the strain (the time-integrated strain rate):

$$\delta x = \text{strain} \times L = \left(\int_0^t dt' \mathcal{F}(t') \right) L = \mathcal{F} t L \quad \text{for constant } \mathcal{F}.$$

Suppose, as usual, that we work in a cubic box of unit length. Then the minimum image correction becomes

```
rij(1) = rij(1) - ANINT ( rij(2) ) * strain
rij(:) = rij(:) - ANINT ( rij(:) )
```

where `strain` stores the current value of the strain. Thus, an extra correction is applied to the x component, which depends upon the number of boxes separating the two molecules

Code 11.1 Molecular dynamics using Lees–Edwards boundaries

These files are provided online. The program `md_nvt_lj_le.f90` contains an MD program which, together with `md_lj_le_module.f90` and the utility modules of Appendix A, implements molecular dynamics of shear flow using Lees–Edwards boundaries. An isokinetic version of the SLLOD algorithm is used (Pan et al., 2005).

```
! md_nvt_lj_le.f90
! MD, NVT ensemble, Lees-Edwards boundaries
PROGRAM md_nvt_lj_le
```

```
! md_lj_le_module.f90
! Force routine for MD, LJ atoms, Lees-Edwards boundaries
MODULE md_module
```

in the y direction. Note that adding or subtracting whole box lengths to or from the horizontal displacement (i.e. whole units to or from the strain) makes no difference to this correction and it is convenient to take the strain to lie in the range $(-\frac{1}{2}, \frac{1}{2})$. It is advisable to keep replacing molecules in the central box as they cross the boundaries, especially if a steady-state shear is imposed, to prevent the build-up of substantial differences in the x coordinates. When this is done, the x velocity of a molecule must be changed as it crosses the box boundary in the y-direction, for consistency with the applied velocity gradient. The difference in box velocities between adjacent layers is $\mathcal{F}L$, or, if the velocity is being handled in units where $L = 1$, simply $\mathcal{F}$. Suppose this is stored in the variable `strain_rate`. Then periodic boundary crossing is handled as follows.

```
r(1,:) = r(1,:) - ANINT ( r(2,:) ) * strain
v(1,:) = v(1,:) - ANINT ( r(2,:) ) * strain_rate
r(:,:) = r(:,:) - ANINT ( r(:,:) )
```

However, as we shall see shortly, most algorithms for shear flow in Lees–Edwards boundary conditions actually store and use the *peculiar velocities*, that is, molecular velocities relative to the local streaming velocity created by the applied velocity gradient. These peculiar velocities are also the ones to which a thermostat should be applied. They should *not* be subjected to the periodic boundary correction, as they are the same for every periodic image. An example of a program using Lees–Edwards boundaries may be found in Code 11.1.

For large simulations, it may be desirable to speed up the calculation by using a neighbour list. If so, it makes more sense to use a cell-structure, linked-list method rather than the simple Verlet approach, because of the steadily changing box geometry (Evans and Morriss, 1984a). There is a subtlety here, since the shifting layers of boxes may necessitate searching more cells than is the case in a conventional simulation. The way this is done is shown in Code 11.2.

The Lees–Edwards boundary conditions alone can be used to set up and maintain a steady linear velocity profile, with gradient dv_x/dy. The shear viscosity is then estimated

Code 11.2 Cell structure and linked lists in sheared boundaries

This file is provided online. This module, together with `link_list_module.f90` (Code 5.3), is intended as a drop-in replacement for `md_lj_le_module.f90` (Code 11.1). It may be built with `md_nvt_lj_le.f90` of Code 11.1 and the utility modules of Appendix A, to give a molecular dynamics program for shear flow using Lees–Edwards boundaries.

```
! md_lj_llle_module.f90
! Force routine for MD, LJ, Lees-Edwards, using linked lists
MODULE md_module
```

from the steady-state nonequilibrium average of $\mathcal{P}_{yx}(k = 0)$,

$$\langle \mathcal{P}_{yx}(t \to \infty)\rangle_{\text{ne}} = -\eta(dv_x/dy). \tag{11.23}$$

This technique was used by Naitoh and Ono (1976; 1979) and Evans (1979c,a), and by many others subsequently. It is a satisfactory way to mimic steady Couette flow occurring in real systems (Schlichting, 1979).

However, the modified boundaries alone are not sufficient to drive the most general time-dependent perturbations. Now we wish to apply a shear perturbation to each molecule, instead of just relying on the modified boundaries. It is possible to devise a perturbation term of Hamiltonian form, eqn (11.2), to do this (Hoover et al., 1980b):

$$\mathcal{A} \cdot \mathcal{F}(t) = \left(\sum_i y_i p_{ix}\right)\mathcal{F}(t). \tag{11.24}$$

$\mathcal{F}$ is the instantaneous rate of strain, that is, $\mathcal{F} = dv_x/dy$. This gives equations of motion of the form (eqns (11.1), (11.3))

$$\begin{aligned}
\dot{x}_i &= p_{ix}/m + y_i\mathcal{F}(t) & \dot{p}_{ix} &= f_{ix} \\
\dot{y}_i &= p_{iy}/m & \dot{p}_{iy} &= f_{iy} - p_{ix}\mathcal{F}(t) \\
\dot{z}_i &= p_{iz}/m & \dot{p}_{iz} &= f_{iz}.
\end{aligned} \tag{11.25}$$

These equations are implemented in conjunction with the periodic boundary conditions of Lees and Edwards (consider replacing y_i with $y_i \pm L$ in eqn (11.25)). We will show that they are a consistent low-k limit of eqns (11.18), (11.19). If the perturbation in eqn (11.18) is divided by $-ik$ to give instead

$$\mathcal{A} \cdot \mathcal{F}(t) = V p_x^{\perp}(-k, t)\mathcal{F}(t)/(-ik) = \sum_i \frac{\exp(-iky_i)}{-ik}p_{ix}\mathcal{F}(t), \tag{11.26}$$

the exponential may be expanded and the first term dropped because of momentum conservation. Taking the limit $k \to 0$ then gives eqn (11.24). $\mathcal{F}(t)$ is the instantaneous rate of strain. The analogue of eqn (11.19) is then

$$\langle\mathcal{B}(t)\rangle_{\text{ne}} = -\frac{V}{k_B T}\int_0^t dt'\langle\mathcal{B}(t - t')\mathcal{P}_{yx}\rangle\mathcal{F}(t') \tag{11.27}$$

and eqn (11.21) becomes

$$\langle \mathcal{P}_{yx}(t)\rangle_{\text{ne}} = -\frac{V}{k_{\text{B}}T} \int_0^t dt' \langle \mathcal{P}_{yx}(t-t')\mathcal{P}_{yx}\rangle \mathcal{F}(t') \tag{11.28}$$

where zero-k values are implied throughout. Historically, this approach is termed the DOLLS tensor algorithm.

In fact, rather than eqns (11.25), a different set of equations is preferred for the simulation of shear flow (Evans and Morriss, 1984a,b; Ladd, 1984), and these are usually referred to as the SLLOD equations:

$$\begin{aligned}
\dot{x}_i &= p_{ix}/m + y_i\mathcal{F}(t) & \dot{p}_{ix} &= f_{ix} - p_{iy}\mathcal{F}(t) \\
\dot{y}_i &= p_{iy}/m & \dot{p}_{iy} &= f_{iy} \\
\dot{z}_i &= p_{iz}/m & \dot{p}_{iz} &= f_{iz}.
\end{aligned} \tag{11.29}$$

These equations are non-Hamiltonian but generate the same linear responses as eqns (11.25)–(11.28). They are preferred because they are believed to give correct non-linear properties (Evans and Morriss, 1984b) and the correct distribution in the dilute gas (Ladd, 1984) where eqn (11.25) fails (see also Daivis and Todd, 2006). They also generate trajectories identical to the straightforward Lees–Edwards boundary conditions when $\mathcal{F}(t)$ is a constant. This follows from an elimination of momenta

$$\ddot{x}_i = f_{ix}/m + y_i\dot{\mathcal{F}}(t), \qquad \ddot{y}_i = f_{iy}/m, \qquad \ddot{z}_i = f_{iz}/m. \tag{11.30}$$

If a step function perturbation is applied, that is, $\mathcal{F}(t) = \text{constant}$ for $t > 0$, eqns (11.30) are integrated over an infinitesimal time interval at $t = 0$ (this sets up the correct initial velocity gradient) and evolution thereafter occurs with normal Newtonian mechanics ($\dot{\mathcal{F}} = 0$) plus the modified boundaries. For the more general case (e.g. $\mathcal{F}(t)$ oscillating in time) step-by-step integration of eqns (11.29) or (11.30) is needed.

11.3.2 Extensional flow

Couette flow, as simulated through the preceding equations, contains some rotational character. The shear viscosity may, alternatively, be investigated by combining an expansion in the x-direction with a compression in the y-direction. This generates irrotational, planar extensional flow. The appropriate SLLOD equations for this case are

$$\begin{aligned}
\dot{x}_i &= p_{ix}/m + x_i\mathcal{F}(t) & \dot{p}_{ix} &= f_{ix} - p_{ix}\mathcal{F}(t) \\
\dot{y}_i &= p_{iy}/m - y_i\mathcal{F}(t) & \dot{p}_{iy} &= f_{iy} + p_{iy}\mathcal{F}(t) \\
\dot{z}_i &= p_{iz}/m & \dot{p}_{iz} &= f_{iz}.
\end{aligned} \tag{11.31}$$

(We note that there has been some disagreement in the literature as to the correct equations to use for extensional flow (Daivis and Todd, 2006; Edwards et al., 2006) which seems, as yet, unresolved.) At first sight, it seems impossible to apply these perturbations continuously and indefinitely in time, because the simulation box becomes extremely elongated and narrow. However, as shown by Kraynik and Reinelt (1992) and implemented in MD by Todd and Daivis (1998; 1999) and Baranyai and Cummings (1999), it is possible

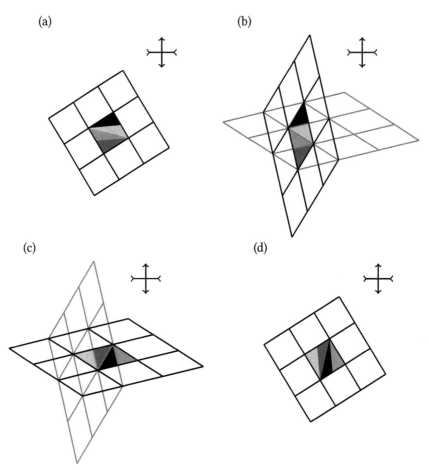

Fig. 11.3 Operation of the Kraynik and Reinelt (1992) periodic boundaries for planar elongational flow. Expansion occurs vertically, and contraction horizontally. (a) Initially cubic periodic boundary system, tilted with respect to the flow axes, highlighting four triangular regions of the reference cell. (b) At time $\frac{1}{2}\Delta t$, before remapping of periodic boundary conditions. (c) At time $\frac{1}{2}\Delta t$, after remapping. Attention has been switched to different periodic images of particles, in three of the four shaded regions of the box. Those in the lightest-shaded triangle are unchanged. (d) At time Δt, giving cubic periodic boundaries once more. The simulation will continue for another$\frac{1}{2}\Delta t$ before the next remapping.

to simulate planar extensional flow using a box which is oriented at a particular angle with respect to the extension/compression axes, and is periodically remapped onto itself. We denote this time period as Δt: it will be significantly longer than the MD timestep δt. The way this happens is very similar to the famous Arnold 'cat map' (Hunt and Todd, 2003; Todd and Daivis, 2007). The deformation, and remapping, of the simulation box is illustrated in Fig. 11.3. The initially cubic periodic boundary system (a) is tilted at an angle $\theta = \tan^{-1}((\sqrt{5}-1)/2) \approx 31.72°$ to the elongation direction. Four different regions of the original reference cell are highlighted. Expansion and contraction occur in the xy

plane (the z direction is unperturbed), so the initially square cross-section deforms into a parallelogram shape. At time $\frac{1}{2}\Delta t$, illustrated in (b), it is possible to choose a different combination of the periodic images of particles lying in the four regions to create a new parallelogram, at a perpendicular orientation, as shown in (c). This simply amounts to reselecting the set of particles defining the basic simulation cell: there is no physical change to the system. After another $\frac{1}{2}\Delta t$ of elongation, the cross-section becomes a square again, as seen in (d). The simulation then proceeds as before.

Several technical issues arise in this method. For a given choice of boundaries, with a prescribed rotation angle of the box axes relative to the flow directions, there is an exactly calculable strain, and hence a well-defined time interval Δt, between remappings. It makes sense to choose the timestep δt, and the rate of deformation, such that Δt consists of an integer number of steps. During that period, the principal axes of the periodic system will evolve in a well-defined way, but to implement the minimum image and periodic boundary conventions as the particles move, it is necessary to transform back and forth between coordinate systems. This needs to be done in a numerically robust way, avoiding the build-up of large numbers as the strain increases with time. The finite precision available on a computer can cause a drift in total momentum, which needs to be corrected. Several of these points are discussed by Todd and Daivis (2007). Finally, it would also be possible to start with a cubic box, and wait until a time Δt before remapping; this would mean that a highly elongated box would be transformed back into the original cubic shape. However, the efficiency of the program will be highest if the maximum elongation of the periodic box is kept as small as possible (or, more specifically, the minimum distance between periodic images is not allowed to become too small). Switching periodic boundary systems at the $\frac{1}{2}\Delta t$ mark relative to the cubic shape, as we describe here, rather than at Δt, works better (Daivis, 2014).

Similar considerations have been applied to the simulation of mixed shear and elongational flow (Hunt et al., 2010). Long-time uniaxial extensional flow, in which expansion along x is matched by compression along both y and z, can in principle be tackled using modified boundaries (Hunt, 2015), although a remapping of the kind discussed earlier is not possible.

11.3.3   Expansion and contraction

To estimate the bulk viscosity directly, a homogeneous NEMD technique based on the zero-k expression, eqn (2.124), very closely related to those used for shear viscosity, has been developed by Hoover et al. (1980a,b). The modified equations of motion are

$$\dot{\mathbf{r}}_i = \mathbf{p}_i/m + \mathbf{r}_i \mathcal{F}(t), \qquad \dot{\mathbf{p}}_i = \mathbf{f}_i - \mathbf{p}_i \mathcal{F}(t) \qquad (11.32)$$

and they correspond to a homogeneous dilation or contraction of the system. They are combined with uniformly expanding or contracting periodic boundary conditions, of the kind described for the constant-pressure dynamics of Andersen (1980) in Section 3.9. In fact, there is a close connection between the two methods, the difference lying in the quantities held fixed and those allowed to vary. In practice, the time dependence is oscillatory, and an extrapolation to zero frequency is required to estimate the bulk viscosity.

11.3.4   Heat flow

The development of a nonequilibrium method of determining the thermal conductivity has been a non-trivial exercise. Essentially identical homogeneous algorithms, compatible with periodic boundaries, were developed independently by Evans (1982), Evans and Morriss (1984a), and Gillan and Dixon (1983). The modified equations are

$$\dot{\mathbf{r}}_i = \mathbf{p}_i/m \tag{11.33a}$$

$$\dot{\mathbf{p}}_i = \mathbf{f}_i + \delta\epsilon_i \mathcal{F}(t) + \frac{1}{2}\sum_j \mathbf{f}_{ij}\left(\mathbf{r}_{ij} \cdot \mathcal{F}(t)\right) - \frac{1}{2N}\sum_j\sum_k \mathbf{f}_{jk}\left(\mathbf{r}_{jk} \cdot \mathcal{F}(t)\right). \tag{11.33b}$$

Here, $\mathcal{F}(t)$ is a three-component vector chosen to lie (say) in the x-direction: $\mathcal{F} = (\mathcal{F}, 0, 0)$. The term $\delta\epsilon_i = \epsilon_i - \langle\epsilon_i\rangle$ is the deviation of the 'single-particle energy' from its average value (see eqn (2.129)). The last term in eqn (11.33b) ensures that momentum is conserved (it redistributes non-conservative terms, with a negative sign, equally amongst all the particles). These equations are non-Hamiltonian, but satisfy the condition laid down in eqn (11.5), and so allow linear response theory to be applied in the usual way. The responses are related to correlations with the zero-k energy flux j_x^ϵ:

$$\langle\mathcal{B}(t)\rangle_{\text{ne}} = \frac{V}{k_{\text{B}}T}\int_0^t dt'\left\langle\mathcal{B}(t - t')j_x^\epsilon\right\rangle\mathcal{F}(t') \tag{11.34}$$

where

$$j_x^\epsilon = \frac{1}{V}\left(\sum_i \delta\epsilon_i\dot{x}_i + \sum_i\sum_{j>i}\left(\mathbf{v}_i \cdot \mathbf{f}_{ij}\right)x_{ij}\right). \tag{11.35}$$

In particular,

$$\langle j_x^\epsilon(t)\rangle_{\text{ne}} = \frac{V}{k_{\text{B}}T}\int_0^t dt'\left\langle j_x^\epsilon(t - t')j_x^\epsilon\right\rangle\mathcal{F}(t') \tag{11.36}$$

so (compare eqn (2.127)) the thermal conductivity is given by a steady-state experiment, with $\mathcal{F}(t') = \mathcal{F}$ after $t = 0$,

$$\lambda_T T = \langle j_x^\epsilon(t \to \infty)\rangle_{\text{ne}}/\mathcal{F}. \tag{11.37}$$

The method induces an energy flux, without requiring a temperature gradient which would not be compatible with periodic boundaries. Note that in mixtures, the formulae are more complicated, and involve the heat flux rather than the energy flux; these two are identical if all the molecules have the same mass (Hansen and McDonald, 2013).

11.3.5   Diffusion

Nonequilibrium methods to measure the diffusion coefficient or mobility are most closely connected with the original derivations based on linear-response theory (Kubo, 1957; 1966; Luttinger, 1964; Zwanzig, 1965). The mobility of a single molecule in a simulation may be measured by applying an additional force to that molecule and measuring its drift velocity at steady state (Ciccotti and Jacucci, 1975). This is a useful approach when a single solute molecule is present in a solvent. The generalization of this approach to measure mutual diffusion in a binary mixture was considered by Ciccotti et al. (1979), and it is simplest to consider in the context of measuring the electrical conductivity in a binary electrolyte. A Hamiltonian perturbation (eqn (11.2)) is applied with

$$\mathcal{A} \cdot \mathcal{F}(t) = -\sum_i q_i x_i \mathcal{F}(t) \tag{11.38}$$

so the equations of motion are conventional except for

$$\dot{p}_{ix} = f_{ix} + q_i \mathcal{F}(t). \tag{11.39}$$

Here, $q_i = \pm 1$ (say) is the charge on each ion. Responses are then related to correlations with the charge current

$$j_x^q(t) = \frac{1}{V} \sum_i q_i \dot{x}_i \tag{11.40}$$

and in particular

$$\langle j_x^q(t) \rangle_{\text{ne}} = \frac{V}{k_B T} \int_0^t dt' \langle j_x^q(t - t') j_x^q \rangle \mathcal{F}(t'). \tag{11.41}$$

Applying a steady-state field for $t > 0$, and measuring the steady-state induced current, gives the electrical conductivity. Ciccotti et al. (1979) made it clear, however, that it is not necessary for the particles to be charged; the quantities q_i simply label different species. In the case of a neutral 1:1 binary mixture, the steady-state response in j_x^q is simply related to the mutual diffusion coefficient D_m (Jacucci and McDonald, 1975):

$$D_m = \frac{V}{\rho} \int_0^\infty dt \langle j_x^q(t) j_x^q(0) \rangle \tag{11.42}$$

where ρ is the total number density. Hence

$$D_m = \frac{k_B T}{\rho \mathcal{F}} \langle j_x^q(t \to \infty) \rangle_{\text{ne}} \tag{11.43}$$

if we apply $\mathcal{F}$ at $t = 0$.

This approach has been taken to its natural conclusion, when the two components become identical (Evans et al., 1983; Evans and Morriss, 1984a). Now the q_i are simply labels without physical meaning, in a one-component system: half the particles are labelled $+1$ and half labelled -1 at random. When the perturbation of eqn (11.38) is applied, eqn (11.43) yields the self-diffusion coefficient. Evans et al. (1983) compare the Hamiltonian algorithm described here with one derived from Gauss's principle of least constraint, and find the latter to be more efficient in establishing the desired steady state. For a one-component fluid, of course, there is less motivation to develop nonequilibrium methods of measuring the diffusion coefficient, since equilibrium simulations give this quantity with reasonable accuracy compared with the other transport coefficients. Before leaving this section, we should mention that k-dependent perturbations which induce charge currents may also be applied, much as for the other cases considered previously (Ciccotti et al., 1979).

11.3.6   Other perturbations

When it comes to molecular fluids, many more quantities are of interest. All the methods described previously can be applied, but there is the choice of applying the perturbations to the centres of mass of the molecules, or to other positions such as the atomic sites (if any) (Allen, 1984; Ladd, 1984; Allen and Maréchal, 1986). This choice does not affect the values of hydrodynamic transport coefficients, but differences can be seen at non-zero wavevector and frequency. Shear–orientational coupling can be measured like this (Evans, 1981b; Allen and Kivelson, 1981) as can the way in which internal motions of chain molecules respond to flows (Brown and Clarke, 1983).

In addition, totally new NEMD techniques may be applied to molecular fluids. Evans and Powles (1982) have investigated the dielectric properties of polar liquids by applying a suitable electric field and observing the response in the dipole moment. Normal periodic boundaries are employed in this case. Evans (1979b) has described a method of coupling to the antisymmetric modes of a molecular liquid via an imposed 'sprain rate'. No modification of periodic boundaries is necessary, merely the uniform adjustment of molecular angular velocities. The transport coefficient measured in this way is the vortex viscosity: its measurement by equilibrium simulations is fraught with danger (Evans and Streett, 1978; Evans and Hanley, 1982). Evans and Gaylor (1983) have proposed a method for coupling to second-rank molecular orientation variables. This is useful for determining transport coefficients which appear in the theory of Rayleigh light scattering and flow birefringence experiments. Indeed, predicted Rayleigh and Raman spectra may be generated directly by this method. Director reorientation in field-induced alignment of a nematic liquid crystal has also been studied by NEMD (Luckhurst and Satoh, 2010).

11.4   Inhomogeneous systems

Nonequilibrium simulations of inhomogeneous systems arise in two contexts: when a boundary-driven perturbation is preferred to one of the homogeneous methods described in the previous section, and when dynamics, such as fluid flow, in the vicinity of an interface is of intrinsic interest.

As mentioned at the start of the chapter, boundary-driven flow (of momentum, energy, etc.) was one of the earliest approaches in NEMD (Ashurst and Hoover, 1972; 1973; 1975; Hoover and Ashurst, 1975; Tenenbaum et al., 1982). The disadvantage of this technique is that measurements of bulk transport coefficients must be made in the region far from the boundaries, requiring a simulation box which is large enough to give a substantial bulk region. However, if this requirement is met, then the method has the advantage that the bulk region can be simulated using unperturbed equations of motion, with all external effects (including the thermostat) confined to the boundaries. This approach is closer to what is done in most real experiments, although we must not lose sight of the fact that the applied fields and measured flows in a simulation may be several orders of magnitude larger than in real life.

An elegant technique due to Holian (reported by Hoover and Ashurst (1975) and by Erpenbeck and Wood (1977)) provides a starting point to discuss so-called reverse NEMD methods. The technique of Holian measures the diffusion coefficient by 'colouring' or 'labelling' particles in a conventional MD simulation. A very similar method has recently been published by Dong and Cao (2012), The idea is to establish a steady-state flux of particles of two types, in opposite directions, by a suitable relabelling scheme which is localized to specific regions of the box. In Holian's method, the relabelling occurs when particles cross the periodic box boundary wall in particular directions. In the method of Dong and Cao (2012), two regions are identified, half a box length apart, and particle labels are swapped between them. The essential points are that the relabelling has no effect on the physical dynamics (because all the atoms are identical, except for the labels) and that, away from these highly localized regions, the labels are conserved quantities, so diffusion determines the relation between labelled particle flux and concentration gradient. Measuring the ratio of these two quantities gives the diffusion coefficient. (A slightly different approach, but based on a similar idea of studying the density and flux of particles initially labelled by position, has also appeared (Powles et al., 1994).) Just to be clear, in all the methods of this paragraph, the system remains (physically) homogeneous: the treatment of diffusion serves as an introduction to the methods about to be described, which do cause some variation of properties with position.

This basic idea of reverse NEMD has been proposed by Müller-Plathe, and applied to the calculation of thermal conductivity (Müller-Plathe, 1997; Bedrov and Smith, 2000; Zhang et al., 2005), shear viscosity (Müller-Plathe, 1999; Bordat and Müller-Plathe, 2002), and the Soret coefficient (Müller-Plathe and Reith, 1999; Polyakov et al., 2008). In these cases, atomic velocities in the different regions of the box are exchanged so as to generate a heat flow or shear flow. For example, the 'hottest' particle in one region is exchanged with the 'coldest' particle in the other region. In contrast to the diffusion method just described, this exchange does affect the dynamics, so we can expect a non-Boltzmann, nonequilibrium, distribution of positions and velocities within the exchange regions, and close to them. However, in the bulk region, once more, unperturbed Newtonian dynamics will apply. An advantage of the method is that the exchanges conserve overall energy and momentum, so despite the nonequilibrium nature of the simulation, thermostatting is, at first sight, not required. The desired transport coefficient is calculated as the ratio of the flux (of energy or momentum) to the observed gradient (of temperature or velocity, respectively). In principle, the former is known exactly from the imposed exchange rate. A key parameter is the frequency with which exchanges are attempted, which governs the strength of the perturbation: exchanging too rarely will produce a small response compared with the thermal noise, while doing it too often will take the system out of the linear response regime. Tenney and Maginn (2010) point out that, for example in the shear viscosity case, gradients in fluid temperature and density may arise, causing nonlinear velocity profiles and erroneous results. For molecular fluids, it is simplest to exchange centre-of-mass velocities (Bedrov and Smith, 2000; Bordat and Müller-Plathe, 2002) to avoid a sudden perturbation of internal degrees of freedom (which may include constrained bond lengths, angles, etc.) resulting from an exchange of atomic velocities not accompanied by a change in molecular orientations and internal coordinates.

11.5   Flow in confined geometry

We turn now to the nonequilibrium simulation of flow in the vicinity of surfaces, where the aim is to measure intrinsic dynamical properties of the fluid–surface interfacial region itself. The systems of interest may include planar or structured solid surfaces, surfaces modified by the attachment of molecules, as in the case of polymer brushes, defouling

agents, etc., or simply fluid–fluid interfaces. The flow of molecular fluids through microscopic channels is of great interest, thanks to potential applications such as desalination.
In nanofluidics, surface effects are of critical importance.

A common geometry consists of a slab formed of parallel walls, separated by a distance
in the z direction, with periodic boundary conditions in the x and y directions. This will
be further discussed in Chapter 14. We assume that the walls are not perfectly flat and
smooth, that is, that the potential energy changes if they are moved in the xy-plane,
relative to the fluid confined between them. Keeping the walls stationary, but applying a
constant force in the x direction to all the fluid molecules, will generate planar Poiseuille
flow, at least for the situation where the walls are far enough apart. Moving the two
walls in the $+x$ and $-x$ directions, respectively, without any other external forces, will
produce the analogue of boundary-driven planar Couette flow. Of course, depending
on the physics of the system, the actual response to these perturbations may be more
complicated. At the smallest wall separations, the problem is better regarded as one of
surface friction and lubrication, rather than fluid hydrodynamics.

In the hydrodynamic context, for the simplest situation in which the transverse xy
structure of the wall is taken to act in an averaged way, we model the flow velocity as
$\mathbf{v}(z) = (v_x(z), v_y(z), v_z(z))$. The effect of each boundary can be characterized by a single
parameter, a friction coefficient ζ between the fluid and the wall, supplemented by an
estimate of the position, z_w, at which the friction is assumed to act:

$$v'_x(z_w) = \frac{1}{\ell} v_x(z_w) \qquad v'_y(z_w) = \frac{1}{\ell} v_y(z_w) \qquad v_z(z_w) = 0 \qquad (11.44)$$

where the prime denotes the derivative with respect to z, $\ell = \eta/\zeta$ is the slip length, and
η is the shear viscosity in the bulk. In all these equations, the velocities are measured
in the frame of reference of the wall. The x and y conditions represent the equality of
the transverse stresses (forces per unit area) acting on an infinitesimal boundary layer at
$z = z_w$, namely a viscous stress coming from the bulk fluid, and a frictional stress which
acts between the boundary and fluid:

$$\mathcal{P}_{zx} = \eta v'_x(z_w) = \zeta v_x(z_w), \qquad \mathcal{P}_{zy} = \eta v'_y(z_w) = \zeta v_y(z_w). \qquad (11.45)$$

The no-slip boundary condition corresponds to $\zeta \to \infty$, $\ell \to 0$; if this is assumed, then
the position of the boundary corresponds to the vanishing of the transverse velocity field
(relative to the wall) under any flow conditions. On the other hand, if the boundary
position z_w is assumed known, then ℓ can be deduced from the velocity and its gradient
at this position: geometrically it is the extrapolation length of the velocity profile (i.e.
the distance beyond z_w at which the linearly extrapolated velocity vanishes). With an
increasing interest in nanofluidics, in which channel sizes may approach molecular
dimensions, and smooth or superhydrophobic surfaces, which may exhibit very low
friction (Kannam et al., 2011; 2012b) determining both these parameters by molecular
simulation may be of interest.

On the face of it, performing NEMD simulations and measuring the velocity profile gives
a route to the surface friction coefficient ζ, and possibly the position of the hydrodynamic
boundary z_w. Typically, a single Couette flow or Poiseuille flow simulation is not sufficient
to calculate them. Either results from simulations of both kinds must be combined, or a

systematic study of varying slab width must be conducted (or, of course, both). On top of this, the apparent friction coefficient may vary with flow rate, reflecting changes in the surface structure or other nonlinear effects. Some case studies, illustrating how these problems are approached in practice, are presented briefly in Example 11.1.

Kannam et al. (2012b) have argued that the NEMD approach, involving a fit to the velocity profile, is not practical for determining the slip length, especially in the interesting regime when it becomes large, because of the sensitivity of the method to variations in the profile. Before discussing NEMD further, it is appropriate to ask whether an equilibrium time correlation function exists for the friction coefficient ζ, and whether this provides a practical route to determining it.

Bocquet and Barrat (1994) provided a derivation of equilibrium time correlation functions for the wall parameters. The relevant expressions are:

$$\zeta = \frac{1}{\mathcal{A}k_BT} \int_0^\infty dt \, \langle F_x^W(t)F_x^W(0)\rangle \tag{11.46a}$$

$$z_w = \frac{\int_0^\infty dt \, \langle F_x^W(t)\mathcal{P}_{zx}(0)\rangle}{\int_0^\infty dt \, \langle F_x^W(t)F_x^W(0)\rangle} \tag{11.46b}$$

where $\mathcal{A}$ is the wall area, and

$$\mathcal{P}_{zx} = \sum_i p_{ix}p_{iz}/m_i + \left(f_{ix}^f + f_{ix}^W\right)z_i \tag{11.47a}$$

$$F_x^W = \sum_i f_{ix}^W. \tag{11.47b}$$

f_{ix}^f is the force on particle i due to all other particles *in the fluid* and f_{ix}^W is the force on i due to the wall. The pressure tensor component $\mathcal{P}_{zx}$ is calculated using these forces multiplied by the z_i coordinate (contrast the usual case of periodic boundary conditions which use an explicitly pairwise form of the virial term). The correlation functions also involve the total wall–fluid force F_x^W. ζ is the wall friction coefficient. Note that a shift in the origin of z coordinates simply shifts z_w by the same amount and leaves ℓ invariant.

Eqn (11.46a) for ζ has been contested (Petravic and Harrowell, 2007), on the grounds that it seems to depend on the wall separation, and is therefore not a true single-surface property. It is clear that there are some subtleties associated with taking the limits in the correct order (Bocquet and Barrat, 2013). A careful analysis in terms of the finite-slab hydrodynamic modes (Chen et al., 2015) has suggested the origin of the discrepancy: the formulae of Bocquet and Barrat (1994), eqns (11.46), appear to be valid only in the limit of large wall separation.

Hansen et al. (2011) have proposed an alternative equilibrium MD method. This relies on defining a fluid slab of width Δ near one of the walls, which is significantly smaller than the gap L_z between the walls, but larger than the thickness of the first fluid layer near the wall. Correlation functions involving the slab velocity and the wall–slab force are now calculated

$$C_{vF}(t) = \left\langle v_x^{slab}(0)F_x^W(t)\right\rangle, \qquad C_{vv}(t) = \left\langle v_x^{slab}(0)v_x^{slab}(t)\right\rangle \tag{11.48}$$

Example 11.1 Measuring the slip length

Different groups have approached the practical problem of quantifying the friction coefficient ζ, or slip length ℓ, of a fluid near a solid wall, in different ways.

Barrat and Bocquet (1999b) use the method of simultaneously fitting Couette and Poiseuille flow profiles to study slip lengths at a non-wetting interface: they use Lennard-Jones interactions, with the solid represented by a fixed FCC lattice, and with the attractive r^{-6} term adjusted so that the fluid–solid potential is less attractive than the fluid–fluid potential. The wall separation is $L_z/\sigma \approx 18$, and the observed slip lengths are in the range $2 \le \ell/\sigma \le 40$, so $0.1 \le \ell/L_z \le 2$.

Duong-Hong et al. (2004) fit Poiseuille and Couette flow profiles for a DPD model with each wall composed of a double-layer of DPD beads, supplemented by a bounce-back boundary condition. They assume that the hydrodynamic boundary coincides with the physical wall position, and fit their results with the no-slip boundary condition: since they observe smooth profiles and measure very low flow velocities near the walls, this is probably quite an accurate assumption.

Pastorino et al. (2006) and Müller and Pastorino (2008) examine Poiseuille and Couette flow of a polymer melt confined between smooth walls with grafted polymer brushes. They illustrate the consequences of taking the hydrodynamic boundary condition to coincide with the physical wall position, on the apparent slip length. With no brush, $\ell \to \infty$, but as the grafting density increases, ℓ decreases to a very small value, and even apparently goes negative. This corresponds to the vanishing of the extrapolated velocity profile within the brush, and is clearly associated with partial penetration of the brush by the melt. They also observe a layer, near the wall, in which the flow is in the 'wrong' direction, due to cyclic motion of the grafted chain ends.

Varnik and Binder (2002) consider Poiseuille flow of polymer chains. They explicitly consider the hydrodynamic boundary position, and estimate it via the local viscosity which is calculated from z-resolved stress and velocity profiles. The hydrodynamic boundary is taken to be the place where the local viscosity diverges. This position is also observed to be where the velocity gradient vanishes. The hydrodynamic boundary is observed to be of order one monomer length inside the physical wall, and to move away from it as T is lowered, that is, when fluid–wall attraction is effectively increased. Slip lengths of order 10σ are illustrated.

Priezjev and Troian (2004) study the slip length in FENE polymeric systems (see eqn (1.33)) between double-layer FCC walls with Couette flow, for chain lengths up to 16. Again, it is assumed that the hydrodynamic boundary and the physical wall position coincide. They choose $L/\sigma \approx 25$ and observe $2 \le \ell/\sigma \le 35$ at low shear rates. They also show a close proportionality with the slip length as estimated from the shear viscosity in bulk, and the structure factor and diffusion coefficient in the first fluid layer, using the expression of Barrat and Bocquet (1999a). Similar assumptions are used in their study of fluid flow near patterned surfaces (Priezjev et al., 2005).

where now F_x^{w} is the total force on the *slab* due to the wall. Hansen et al. (2011) then write a relation between the Laplace transforms of these correlation functions, and a Laplace transformed friction kernel $\tilde{\zeta}(s)$:

$$\tilde{\zeta}(s) = -\frac{1}{\mathcal{A}}\frac{\tilde{C}_{vF}(s)}{\tilde{C}_{vv}(s)} \quad \Rightarrow \quad \zeta = -\frac{1}{\mathcal{A}}\frac{\int_0^\infty dt\, C_{vF}(t)}{\int_0^\infty dt\, C_{vv}(t)} \qquad (11.49)$$

and they determine ζ by fitting the functions $\tilde{C}_{vF}(s)$ and $\tilde{C}_{vv}(s)$.

Mundy et al. (1996; 1997) and Tuckerman et al. (1997) presented a nonequilibrium method to calculate these coefficients. Neglecting the thermostat, their method is based on a form of the SLLOD equations

$$\dot{x}_i = p_{ix}/m_i + \gamma(z_i - Z_0), \qquad \dot{p}_{ix} = f_{ix} - \gamma p_{iz} \qquad (11.50)$$

with conventional expressions for the other components. The perturbation is defined to be zero at a chosen coordinate Z_0. When implemented with appropriate thermostatting this generates a linear velocity profile with gradient γ, having a zero value at Z_0. It should be noted that, in this method, *the boundaries do not move*. Linear response theory based on these equations and the analysis of Bocquet and Barrat (1994) then gives

$$\langle F_x^{\mathrm{w}} \rangle_{\mathrm{NE}} = \frac{\gamma}{k_{\mathrm{B}}T} \int_0^\infty dt\, \langle F_x^{\mathrm{w}}(t)\dot{Q}(0)\rangle$$

where $\dot{Q} = \mathcal{P}_{zx} - Z_0 F_x^{\mathrm{w}}$ and $\langle \cdots \rangle_{\mathrm{NE}}$ denotes the steady-state (long-time) nonequilibrium average. Comparison with eqns (11.46) gives

$$\frac{\langle F_x^{\mathrm{w}} \rangle_{\mathrm{NE}}}{\mathcal{A}} = \zeta\gamma(z_{\mathrm{w}} - Z_0)$$

which has the same physical interpretation as eqn (11.45). Hence, simulations at two chosen values of Z_0, measuring $\langle F_x^{\mathrm{w}} \rangle_{\mathrm{NE}}$, will give enough information to determine the two boundary parameters. In principle, these could come from a single simulation using the response at both walls, but it is quite likely that the flow at one of them will be so high as to lie outside the linear response regime. Therefore, care should be taken. In practice, thermostatting needs to be applied to all the degrees of freedom, as discussed in the papers (Mundy et al., 1996; 1997; Tuckerman et al., 1997). We note in passing that the relation between the SLLOD equations, Lees–Edwards boundaries, and wall-driven Couette flow has been discussed by Petravic (2007). Finally, at the risk of repetition, the reservations of Kannam et al. (2012b) about NEMD methods may still apply to this approach, especially when the slip length is high, and the underlying equations (Bocquet and Barrat, 1994) seem to be inapplicable when the wall separation is small.

When the fluid flow occurs in highly confined and/or non-planar geometries, notably nanotubes, additional effects come into play. The curvature of the surface itself may modify the friction coefficient. Surface structuring of the liquid may extend through most of the channel, so that the shear viscosity of the bulk fluid is not a useful parameter. For the smallest nanotubes, the motion may be effectively restricted to a single file of atoms, when the conventional equations of diffusion and viscous flow no longer apply.

The flow of water in carbon nanotubes has been an especially challenging problem, with reported slip lengths spanning several orders of magnitude, in both simulation and experiment (for a summary see Kannam et al., 2013). In addition to all the technical simulation issues, the results in highly confined geometries can be sensitive to the water model employed: for example, whether it is flexible or rigid, whether hydrogen-bond networks are easily formed and broken, and even by subtle electronic structure effects (see e.g. Cicero et al., 2008). Although NEMD methods are quite commonly applied to this problem, great care must be taken in interpreting the results, and it may be advisable to compare with equilibrium MD: Kannam et al. (2012a) have extended the approach of Hansen et al. (2011) to handle cylindrical geometry. For sufficiently narrow channels, the standard hydrodynamic description of fluid flow must incorporate effects such as rotation–translation coupling. Increasingly, molecular simulations are used to supplement experiments in this area, and cast light on the origins of interesting flow effects (Hansen et al., 2009; Bonthuis et al., 2009; Hansen et al., 2010; Bonthuis et al., 2011).

Thermostatting the system, as always, is a sensitive issue (Bernardi et al., 2010). One approach, when the walls are physically constructed of atoms which are allowed to vibrate, is to apply a thermostat to the wall particles only, and allow a thermal gradient to develop in the fluid. This has the merit of corresponding to the physical situation in which heat is dissipated at the walls. A second approach is to re-thermalize the particle velocities when they collide with the wall. However, Bernardi et al. (2010) discourage the use of 'frozen' (i.e. rigid) walls, and fluid atom thermostatting, because of consequent introduction of artificial effects.

11.6 Nonequilibrium free-energy measurements

In this section we turn to the use of nonequilibrium simulations to measure free-energy differences. Already, in Chapter 9, we have encountered methods in which some parameter, λ, is varied, so as to convert one system, A, into another, B. In principle, if this is done quasi-statically (i.e. infinitesimally slowly), the work done $\mathcal{W}$ may be expressed as the integral of a thermodynamic force, $-\partial A/\partial\lambda$, itself the average of a mechanical force $\langle \partial \mathcal{H}/\partial\lambda \rangle$. Hence, by numerical integration, the free energy $\Delta A = \mathcal{W}$ may be obtained. (One may consider repeating the experiment many times and averaging the work, but for quasi-static changes we do not distinguish between $\mathcal{W}$ and $\langle \mathcal{W} \rangle$.) Several variants of this technique are described in Chapter 9.

What happens if the variation of λ is done rapidly? Examples from real life include stretching a biomolecule (Liphardt et al., 2002; Collin et al., 2005), or pulling a colloidal particle through water using some laser tweezers (Wang et al., 2002; Carberry et al., 2004). In a simulation, this would correspond to a finite-time τ, and repeating the simulation many times will give a distribution of results. Each simulation gives

$$\mathcal{W} = \int_0^\tau dt \left(\frac{\partial \mathcal{H}(\lambda_t; \mathbf{r}_t, \mathbf{p}_t)}{\partial \lambda_t} \right) \dot{\lambda}_t = \mathcal{H}(\lambda_\tau; \mathbf{r}_\tau, \mathbf{p}_\tau) - \mathcal{H}(\lambda_0, \mathbf{r}_0, \mathbf{p}_0). \tag{11.51}$$

The system coordinates and momenta will follow a path $(\mathbf{r}_t, \mathbf{p}_t)$, $t = 0 \ldots \tau$, and the work $\mathcal{W}$ will depend on this whole path; hence $\mathcal{W}$ depends on the initial configuration $(\mathbf{r}_0, \mathbf{p}_0)$, and we will need to average over this with (let us say) a canonical ensemble.

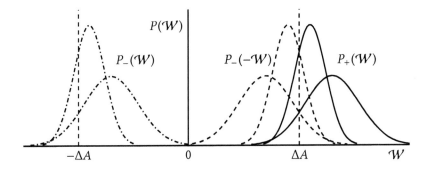

Fig. 11.4 Schematic plot of probability distributions for work $P_+(W)$ involved in the forward process (full lines), and negative work $P_-(-W)$ of the reverse process (dashed lines). We also show the distribution of work $P_-(W)$ of the reverse process (dash-dotted lines), of which $P_-(-W)$ is a reflection. Narrower peaks are for low speeds, and broader curves for high speeds.

Let us call A $\rightarrow$ B the 'forward' process, denoted '+', and B $\rightarrow$ A the 'reverse' process, denoted '−'. Then the Second Law of thermodynamics tells us that

$$\langle W \rangle_+ \geq \Delta A, \quad \text{and} \quad \langle W \rangle_- \geq -\Delta A, \tag{11.52}$$

with the equalities only holding for reversible (infinitesimally slow) processes. Following from the work of Evans and Searles (1994; 2002), Jarzynski (1997), and Crooks (1999), one can write down much more precise and useful equations. The Crooks fluctuation theorem states that

$$\frac{P_+(W)}{P_-(-W)} = \exp\left(\frac{W - \Delta A}{k_B T}\right) \tag{11.53}$$

relating the probability distributions for the work done in the forward and reverse processes. Figure 11.4 illustrates what is meant by the two probability functions in this expression. The Jarzynski relation

$$\langle \exp(-W/k_B T) \rangle_+ = \exp(-\Delta A/k_B T) \tag{11.54}$$

relates the work of the forward process to the corresponding free-energy change. It is very surprising at first sight, because the quantity on the left is the average of a nonequilibrium quantity apparently depending on the speed of the transformation, while the right-hand side is purely based on equilibrium thermodynamics. Nonetheless, it is an exact equality, easily derived from eqn (11.53), and the more familiar inequality $\langle W \rangle_+ \geq \Delta A$ is perfectly consistent with it.

These two equations, (11.53) and (11.54), and variations of them, underpin several techniques for measuring free-energy differences in simulations. The distributions of Fig. 11.4 show several features that influence practical concerns. Just as for some of the techniques discussed in Chapter 9, it is usually safer to compare probability distributions than simply calculate a single number, so a method that uses eqn (11.53) is likely to be more reliable than a 'black box' implementation of eqn (11.54). Any pairs of forward and reverse trajectories that are being compared must be the opposite of each other, that is, the control parameter λ must change at the same speed, in opposite directions. All the corresponding forward and reverse distributions cross at the same value of $\mathcal{W}$, which according to eqn (11.53) is ΔA. Moreover, where $P_+(\mathcal{W})$ and $P_-(-\mathcal{W})$ overlap, it is possible to plot $\ln(P_+(\mathcal{W})/P_-(-\mathcal{W}))$ vs $\beta \mathcal{W}$, giving a straight line of gradient 1 passing through zero at $\mathcal{W} = \Delta A$. Wu and Kofke (2005) and Kofke (2006) have given some heuristic guidelines to help assess the reliability of the forward and reverse processes.

A significant simplification occurs if the probability distribution of the work $\mathcal{W}$ done in a nonequilibrium process is Gaussian, because then it is completely defined by its first two moments: $\langle \mathcal{W} \rangle$ and $\sigma^2 = \langle \mathcal{W}^2 \rangle - \langle \mathcal{W} \rangle^2$. The properties of Gaussian distributions imply that

$$\left\langle \exp(-\beta\mathcal{W}) \right\rangle_+ = \exp\left(-\beta\langle\mathcal{W}\rangle_+\right) \exp(\sigma^2\beta^2/2) = \exp(-\beta\Delta A)$$

where the last equation is Jarzynski's. Hence, taking logs,

$$\langle \mathcal{W} \rangle_+ - \frac{\sigma^2}{2k_\mathrm{B}T} = \Delta A, \quad \text{or} \quad (\beta\sigma)^2 = 2\beta(\langle\mathcal{W}\rangle_+ - \Delta A).$$

This makes it clear that $\langle \mathcal{W} \rangle_+ \geq \Delta A$, and also that the average amount of work dissipated is related to the fluctuations in the work done. The variance of the dissipative work is exactly twice the distance of the mean value from ΔA, when measured in units of $k_\mathrm{B}T$. Figure 11.4 has been sketched, roughly, with this in mind, that is, the further from equilibrium, the broader the distribution. Of course, there is no guarantee that the actual distribution of work in a nonequilibrium simulation will be Gaussian, but it gives a guide to the expected behaviour. The faster the process is performed, the further the peak of the distribution will be from the desired free energy.

It is important to understand the nature of the averaging going on in such a process. Although the *equilibrium* free-energy difference $\Delta A = A_\mathrm{B} - A_\mathrm{A}$ appears in both eqn (11.53) and (11.54), we only ever need to assume that the *starting* configurations for these trajectories (forward or reverse) are at equilibrium! After the driving forces are switched off at $t = \tau$, the final configurations will almost certainly *not* be at equilibrium: they will subsequently relax to equilibrium (at the new state point) without any further work being done; this does not affect the validity of the equations, which have been verified under a range of assumptions about the type of dynamics involved in the time evolution, and the applied thermostatting.

The rate of switching is one of the parameters that can be chosen to optimize the method. At one extreme, very slow switching, we recover the thermodynamic integration route mentioned at the start of this section. At the other extreme, instantaneous switching, there is no dynamical evolution at all and we get eqn (9.3). As well as this choice, one can imagine introducing a bias into the sampling, much as was done for static free-energy methods described in Chapter 9. A useful summary of optimization approaches has been published by Dellago and Hummer (2014).

Frequently the free energy is desired as a function $\mathcal{F}(q)$ of some quantity, $q(\mathbf{r})$, a combination of atomic coordinates. Common examples of this are the effective free energy (including averaged solvent effects) between two biomolecules, as a function of their separation, or the barrier to transfer of a molecule through a lipid bilayer, as a function of centre-of-mass position. The Landau free energy is written, as in (2.172), in terms of a probability distribution function $\mathcal{F}(q) = -k_B T \ln\langle \delta(q - q(\mathbf{r}))\rangle$ calculated in the canonical ensemble.

By accumulating and analysing the work done at intermediate stages, one can calculate a free-energy profile, $A(\lambda)$, which is a function of the control parameter λ. If λ and q are the same thing, then this is identical with $\mathcal{F}(q)$. However, it is quite common for λ and q to be different: λ is typically a control parameter in the Hamiltonian, with q being a generalized coordinate to which it is coupled. The Hamiltonian might take the form

$$\mathcal{H}(\lambda_t) = \mathcal{H}(\lambda_0) + \Delta\mathcal{V}(\lambda_t, q), \quad \text{with} \quad \mathcal{H}(\lambda_0) = \mathcal{H}_A, \quad \text{and} \quad \mathcal{H}(\lambda_\tau) = \mathcal{H}_B. \quad (11.55)$$

Hummer and Szabo (2001; 2005) approached this problem using the same kind of histogram reweighting scheme described in Chapter 9 (Ferrenberg and Swendsen, 1989) and writing the desired distribution as a linear combination of distributions in q obtained at different times in the nonequilibrium process. The weights appearing in this combination are intended to emphasize the contributions from the trajectory which sample best the region around each value of q. Oberhofer and Dellago (2009) have reconsidered the selection of weights, and propose an optimal way of selecting them, but conclude that the original choice of Hummer and Szabo performs well. Amongst several other approaches to improving the method, Vaikuntanathan and Jarzynski (2011) have proposed a way of reducing dissipation in the calculated work, by introducing artificial 'escorted' trajectories.

11.7   Practical points

In almost all of the perturbations described earlier, work is done on the system, and it will heat up if no thermostat is applied. So far, we have neglected the additional terms in the equation of motion that are necessary to thermostat the system. However, the thermostatting method is very important, and an incorrect approach to temperature control can invalidate the results of a nonequilibrium measurement.

Usually, the state point is controlled at each step of the simulation. This can be done by the methods described in Chapter 9, to give either isothermal or adiabatic equations of motion. A term $-\xi(\mathbf{r}, \mathbf{p})\mathbf{p}$ is added to the momentum equations given in the previous sections, with ξ chosen by a Lagrange multiplier technique so as to constrain the kinetic temperature, or the energy, to the desired value.

The most important question is 'what to thermostat?' In fluid flow, the particle velocity consists of two parts. The first of these is the 'streaming velocity', which is free of thermal fluctuations and should correspond (at least over length scales and time scales which are large compared with molecular ones) to solutions of the appropriate hydrodynamic equations (for example, the Navier–Stokes equation). The second component, the 'thermal' or 'peculiar' velocity, is measured with respect to the streaming velocity. This is subject to all the thermal fluctuations that are neglected in continuum hydrodynamics. Nearby molecules in a simulation will have the same streaming velocity (or very nearly so) but different thermal velocities, in general. Only the latter should be thermostatted: the temperature corresponds to the mean-squared deviations of the molecular velocities with respect to the local flow. However, in the equations of motion, the thermostatting term $-\xi(\mathbf{r}, \mathbf{p})\mathbf{p}$ involves momenta calculated in an external, stationary, frame of reference. This raises a problem, because the streaming velocity is one of the 'outputs' from a simulation, not one of the 'inputs'. There is no simple way of separating the two parts of the overall velocity. In the simplest geometry, using the Lees–Edwards boundaries and the SLLOD equations to simulate Couette flow, it might seem reasonable to assume a linear velocity profile determined by the applied velocity gradient. However, this is not foolproof: in an inhomogeneous system (for example, containing a fluid–fluid interface) this linear profile will be incorrect (Padilla et al., 1995), and deviations from such a profile arise spontaneously in fluids which exhibit shear banding.

One approach, the profile unbiased thermostat (PUT), involves averaging the velocity, as a function of the spatial coordinates, over a period of time, and using this as the streaming velocity (Evans and Morriss, 1986). This is then subtracted from each molecule's velocity prior to thermostatting, and added back again afterwards. The averaging requires the accumulation of a histogram, on some chosen spatial resolution. For a simple planar flow (Couette or Poiseuille, for instance), a histogram dividing the system into slabs (Padilla et al., 1995), or a Fourier decomposition in one direction (Travis et al., 1995), would be suitable. Travis et al. (1995) also point out the need to consider the orientation dependence of the streaming velocity for molecular systems. The procedure is clearly easiest if the flow can be assumed to be steady in time, although a method to handle dynamical streaming velocities has been proposed (Bagchi et al., 1996).

A second approach is to use a thermostat which is based on relative velocities, such as the pairwise Lowe–Andersen thermostat (Lowe, 1999) described in Chapter 12, or a pairwise version of the Nosé–Hoover thermostat (Allen and Schmid, 2007). An alternative is the configurational temperature thermostat of Braga and Travis (2005), eqn (3.76). This last method was used by Costa et al. (2013) in dynamic NEMD simulations of Couette flow. Daivis et al. (2012) have compared different thermostatting methods for shear flow, finding that measured shear viscosities are generally insensitive to the details, but normal stress differences are strongly dependent on the thermostat.

A second practical difficulty, especially relevant for shear and extensional flows, is the implementation of a cell neighbour list scheme to improve the efficiency of the program. For a shearing system, one can either use the 'shifting box' representation (Lees and Edwards, 1972) of Fig. 11.2, or an equivalent scheme (Evans, 1979c; Hansen and Evans, 1994) in which each box is sheared into a parallelogram (in 3D, a parallelepiped); for elongational flow, Fig. 11.3 shows how the boxes are deformed into such a shape, but it is also possible to rotate and remap the periodic system into a 'shifting box' form, when constructing lists. Whichever representation is used, the periodic system of cells may be made commensurate with the simulation box; however, attention needs to be paid to the minimum dimensions of the (possibly elongated) cells, and to the (possibly time-dependent) definition of the set of 'neighbouring cells' which are scanned for atoms within interaction range. Hansen and Evans (1994) and Bhupathiraju et al. (1996) have explained how to do this for shear flow; in the shifting-box format, it is just necessary to reset the cell connectivity when the neighbouring boxes reach an offset of half a box length, or equivalently when the shear angle reaches $\tan^{-1} 0.5 \approx 26.6°$. Matin et al. (2003) have discussed practical ways of doing this for the Kraynik–Reinelt periodic boundaries of planar elongational flow. A significant difference is that the number of cells must change with time, as the box dimensions change.

11.8 Conclusions

Should we use nonequilibrium methods? It is still an open question as to whether or not they are more efficient at estimating bulk transport coefficients than equilibrium simulations. It should be remembered that the techniques described in Chapter 8 can produce an entire range of correlation functions and transport coefficients from the output of a single equilibrium run. Nonequilibrium simulations are normally able to provide only one fundamental transport coefficient, plus the coefficients describing cross-coupling with the applied perturbation, at once. If the basic simulation is expensive (e.g. for a complicated molecular model) compared with the correlation function analysis, then equilibrium methods should be used. On the other hand, NEMD is claimed to be much more efficient when the comparison is made for individual transport coefficients. By applying a steady-state perturbation, the problem of integrating a correlation function which has noisy long-time behaviour (or similarly examining an Einstein plot) is avoided, but it is replaced with the need to extrapolate, in a possibly ill-defined manner, to zero applied perturbation. Chen et al. (2009) argue, in a short note, that the NEMD approach is rarely advantageous, provided the equilibrium correlation functions are analysed in a sensible way. The debate will, no doubt, continue.

A similar discussion surrounds the use of nonequilibrium methods for the calculation of free-energy differences: are they really more economical (for a given accuracy) than conventional methods? Oberhofer et al. (2005) consider this question, and conclude that, provided similar consideration is given to biasing the sampling in each case, fast switching is generally inferior to slow switching. They make the point that the time evolution implicit in the nonequilibrium approach is really not critical, in the sense that it is used simply to generate a mapping between ensembles. Dellago and Hummer (2014) essentially reinforce this conclusion, highlighting the fact that the exponential work average of the Jarzynski equation, for strong driving, tends to be dominated by a small number of large contributions, giving high statistical errors, and a risk of bias in the result. They suggest that fast switching may be more successful when several distinct pathways connect the states of interest. They also point out that, as this is a relatively newly developed field, there is still the potential for more sophisticated variants to be devised. However, of course, it is essential to make a fair comparison between standard and newly developed techniques before drawing conclusions.

12

Mesoscale methods

12.1 Introduction

The simulation techniques we shall describe in this chapter are motivated by the desire to bridge the time and/or length scales between a detailed description of the molecular system of interest and the behaviour that it exhibits. This may involve approximating the interactions between molecular sub-units; an extension of the united-atom approach which is often called 'coarse graining'. It may also mean dividing the system into (at least) two parts, and treating them at different levels of detail. For static properties this is sometimes described as 'integrating out' some of the degrees of freedom; if dynamical properties are of interest, it involves adopting a 'reduced' equation of motion, in which the time-dependent effects of the omitted variables are approximately accounted for.

Consider the simulation of a large molecule (e.g. a protein) or a collection of suspended particles (a colloid) in a solvent such as water. Even though the motion of the solvent molecules is of little intrinsic interest, they will be present in large numbers, and they will make a full molecular simulation very expensive. In such a case, it may be reasonable to adopt an approximate approach: the solvent particles are omitted from the simulation, and their effects upon the solute represented by a combination of random forces and frictional terms. Newton's equations of motion are thus replaced by some kind of Langevin equation, and this will be the subject of the first few sections.

Such a simple description makes many approximations, and neglects one possibly important effect on the solute motion: the hydrodynamic flow of the solvent. Several methods have been proposed to put this ingredient back into the simulation, and we shall describe three of these: dissipative particle dynamics, the multiparticle collision method (originally called stochastic rotation dynamics), and the lattice Boltzmann method. These methods have been particularly successful in the study of complex fluids, for example amphiphilic molecules which form a variety of phases including micelles, vesicles, and bilayers. In the terminology of fluid dynamics, these methods tackle problems at intermediate values of the Knudsen number Kn, the ratio between the mean free path of the particles and the characteristic length scale of inhomogeneities in the system (Raabe, 2004); molecular dynamics would apply for Kn $\gtrsim$ 1, while the solution of the continuum Navier–Stokes equations would be suitable for Kn $\ll 10^{-1}$.

These techniques allow us to tackle problems at the mesoscale, rather than the microscale (or atomistic) level. There are other aspects to this: the systematic approach to coarse graining the interactions, and the relation between dynamical algorithms and Monte Carlo methods, to name but two. We discuss these at the end of the chapter.

12.2   Langevin and Brownian dynamics

The Langevin equation is a stochastic differential equation, describing the Brownian motion of particles in a liquid, as well as a number of other physical systems (Chandrasekhar, 1943; Snook, 2007; Coffey and Kalmykov, 2012). Formally, it may be derived by applying projection operator methods to the equations of motion for the phase space distribution function (Zwanzig, 1960; 1961a,b) or the dynamical variables themselves (Mori, 1965a,b); an elegant unified treatment has been presented by Nordholm and Zwanzig (1975). The essential physical idea behind the derivation is *time scale separation*: the variables that are retained are assumed to vary much more slowly than those that are represented by stochastic terms. More details may be found elsewhere (Berne and Pecora, 1976; McQuarrie, 1976; Hansen and McDonald, 2013).

Our starting point is the Langevin equation in the following form:

$$\dot{\mathbf{r}} = \mathbf{v} = \mathbf{p}/m, \quad \dot{\mathbf{p}} = \mathbf{f} - \xi\mathbf{v} + \sigma\dot{\mathbf{w}} = \mathbf{f} - \gamma\mathbf{p} + \sigma\dot{\mathbf{w}}. \tag{12.1}$$

As usual, each vector represents the complete set of N particles. The last three terms are, respectively, the effects of the systematic forces of interaction between the particles, the frictional forces on them due to the solvent, and the so-called random forces, which are represented by the time derivative of a ($3N$-dimensional) Wiener process $\mathbf{w}$. Here the friction coefficient ξ (or equivalently the damping constant γ) is related by the equation

$$\xi = m\gamma = \frac{k_BT}{D}$$

to the diffusion coefficient D of the particles *in the absence of any interactions*, that is, if we set $\mathbf{f} = 0$. The coefficient σ governs the strength of the random forces. It is related to ξ through the *fluctuation–dissipation theorem*

$$\sigma = \sqrt{2\xi k_BT} = \sqrt{2\gamma mk_BT}.$$

Given this equation, and the properties of $\mathbf{w}$ discussed later, it can be shown that eqn (12.1) generates a trajectory that samples states from the canonical ensemble at temperature T. The key relations between ξ, σ, and D essentially go back to Einstein's analysis of Brownian motion; details can be found elsewhere (Chandrasekhar, 1943; Kubo, 1966; Snook, 2007; Marconi et al., 2008; Coffey and Kalmykov, 2012). Simulation methods which use eqn (12.1), as well as the version without inertia which will be given shortly, are generally referred to as Brownian dynamics (BD) techniques, or sometimes as Langevin dynamics.

We should spend a moment discussing the physics, and the mathematics, implicit in eqn (12.1), especially the random force term. Each of the $3N$ components $w_{i\alpha}$ ($i = 1\ldots N$, $\alpha = x, y, z$), is assumed to be independent. The definition of a Wiener process $w_{i\alpha}$ is that its change over a differentially small time interval dt is a random variable, normally distributed, with variance equal to dt. In other words, we may write

$$dw = w(t + dt) - w(t) = \sqrt{dt}\, G \tag{12.2}$$

where each component of G, $G_{i\alpha}$, is an independent Gaussian random variable, with zero mean $\langle G_{i\alpha} \rangle = 0$ and unit variance, $\langle G_{i\alpha}G_{j\beta} \rangle = \delta_{ij}\delta_{\alpha\beta}$. So, $\dot{w}$ is not properly defined, since dw is not proportional to dt in the limit $dt \to 0$. The $\sqrt{dt}$-dependence means that the momentum part of eqn (12.1) should really be written

$$dp = f\, dt - \gamma p\, dt + \sigma\, dw$$

and this form, together with eqn (12.2), translates straightforwardly into a numerical integration algorithm with a nonvanishing timestep δt, as we shall see. Before discarding $\dot{w}$, however, we note that its time correlation function is proportional to a Dirac delta function $\langle \dot{w}_{i\alpha}(0)\dot{w}_{i\alpha}(t) \rangle \propto \delta(t)$. This is consistent with the picture of rapid, random, buffeting of a particle by the surrounding solvent. It is also the defining characteristic of 'white noise', which is a term sometimes used to describe the random force term.

We should note two more important features of the stochastic term: it is *uncorrelated* with any of the dynamical variables (positions and velocities) at earlier times, and it is an *additive* term, which avoids some complications of stochastic calculus. This simplifies the construction of a numerical algorithm. The earliest approach (Ermak and Buckholz, 1980) had the disadvantage of not reducing to a stable molecular dynamics method in the limit of low friction $\xi \to 0$. Later suggestions (Allen, 1980; 1982; van Gunsteren and Berendsen, 1982; 1988) were designed to give Verlet-equivalent algorithms in this limit, and the particular suggestion of Brünger et al. (1984) has been widely adopted.

The fact that the stochastic terms are independent of the coordinates and momenta means that the equations in the absence of forces are exactly soluble, and this allows the construction of a symplectic integration algorithm, using an approach similar to the one that yields velocity Verlet (see Section 3.2.2). An operator-splitting method may be devised in various different ways (see e.g. Cotter and Reich, 2006; Bussi and Parrinello, 2007; Melchionna, 2007). These have been analysed in some detail (Leimkuhler and Matthews, 2013a,b). The optimal method, which they call 'BAOAB', involves inserting, into the middle of the usual 'kick–drift–kick' sequence, the exact solution of the force-free momentum equation

$$dp = -\gamma p\, dt + \sqrt{2\gamma m k_B T}\, dw$$

which is

$$p(t + \delta t) = \exp(-\gamma \delta t)p(t) + \sqrt{1 - \exp(-2\gamma \delta t)}\sqrt{m k_B T}\, G.$$

Code 12.1 Brownian dynamics program

This file is provided online. The BD programme bd_nvt_lj.f90, combined with the standard Lennard-Jones module md_lj_module.f90 (Code 3.4) for the forces, and the utility modules of Appendix A, carries out Langevin equation dynamics using the BAOAB algorithm of Leimkuhler and Matthews (2013a).

```
! bd_nvt_lj.f90
! Brownian dynamics , NVT ensemble
PROGRAM bd_nvt_lj
```

This gives the following algorithm

$$p(t + \tfrac{1}{2}\delta t) = p(t) + \tfrac{1}{2}\delta t f(t) \tag{12.3a}$$

$$r(t + \tfrac{1}{2}\delta t) = r(t) + \tfrac{1}{2}\delta t p(t + \tfrac{1}{2}\delta t)/m \tag{12.3b}$$

$$p'(t + \tfrac{1}{2}\delta t) = \exp(-\gamma\delta t)p(t + \tfrac{1}{2}\delta t) + \sqrt{1 - \exp(-2\gamma\delta t)}\sqrt{mk_BT}\,G \tag{12.3c}$$

$$r(t + \delta t) = r(t + \tfrac{1}{2}\delta t) + \tfrac{1}{2}\delta t p'(t + \tfrac{1}{2}\delta t)/m \tag{12.3d}$$

$$p(t + \delta t) = p'(t + \tfrac{1}{2}\delta t) + \tfrac{1}{2}\delta t f(t + \delta t). \tag{12.3e}$$

Here, eqns (12.3a) and (12.3e) are the (completely standard) half-step kicks. Eqns (12.3b) and (12.3d) are the usual drift equations, but for half a step at a time. The frictional and random force terms together appear in the middle step (12.3c). Leimkuhler and Matthews have shown that this algorithm performs well at both high and low values of the friction, using test systems such as a one-dimensional oscillator, Lennard-Jones and Morse potential atomic clusters, and alanine dipeptide (solvated and unsolvated). An example BD program is given in Code 12.1.

At high friction, the relaxation of the momenta can be assumed to occur instantaneously. Setting $\dot{p} = 0$ in eqn (12.1) and substituting into the equation for $\dot{r}$ gives the Brownian dynamics equation of motion (i.e. the Langevin equation without inertia)

$$\dot{r} = \xi^{-1}\left(f + \sigma\dot{w}\right) = \frac{D}{k_BT}\left(f + \sigma\dot{w}\right). \tag{12.4}$$

A simple algorithm for this (Ermak and Yeh, 1974; Ermak, 1975) is

$$r(t + \delta t) = r(t) + \frac{D}{k_BT}f(t)\delta t + \sqrt{2D\delta t}\,G. \tag{12.5}$$

Again, the diffusion coefficient enters as a parameter of the method, and determines the variance of the last term, the random Gaussian displacements. In the absence of forces, we would see displacements $\delta r_{i\alpha} = r_{i\alpha}(t + \delta t) - r_{i\alpha}(t)$ satisfying $\langle \delta r_{i\alpha}^2 \rangle = 2D\delta t$ as expected. This algorithm can also be interpreted as a Monte Carlo method of the kind discussed in Section 9.3, and we return to this in Section 12.3. The formalism may easily be extended from atomic systems to include rigid and non-rigid molecules, and the incorporation of constraints is straightforward (van Gunsteren and Berendsen, 1982) although the usual care should be taken in their application (van Gunsteren, 1980; van Gunsteren and Karplus, 1982).

The previous equations all ignore memory effects in the random force as well as indirect interactions between the atoms, mediated by the solvent. In principle, the inclusion of a specified memory function into the Langevin equation is straightforward (Ciccotti et al., 1976a; Doll and Dion, 1976; Adelman, 1979; Ermak and Buckholz, 1980; Ciccotti and Ryckaert, 1980). Essentially, this corresponds to the inclusion of additional derivatives of the momentum in the equations of motion, or to the extension of the time correlation function of the random force from a delta function to, for example, a decaying exponential.

The simplest effect of the surrounding solvent is to replace the bare interaction between solute particles by a potential of mean force. We shall discuss this in Section 12.7. The second effect, neglected in the simple Langevin and Brownian dynamics equations, is the effect that solvent flow, induced by one molecule, has on the surrounding molecules. If this hydrodynamic effect is not tackled directly (as in Sections 12.4–12.6), it can be introduced approximately into the Brownian dynamics algorithm via a configuration-dependent diffusion coefficient. The way this appears in the equation of motion depends on the convention adopted for stochastic differentials; however, the integration algorithm is unambiguously written (Ermak and McCammon, 1978)

$$\mathbf{r}(t + \delta t) = \mathbf{r}(t) + \frac{\mathbf{D}(t)}{k_{\mathrm{B}}T} \cdot \mathbf{f}(t)\delta t + \boldsymbol{\nabla} \cdot \mathbf{D}(t)\delta t + \mathbf{R}. \tag{12.6}$$

Here, as usual, the vectors contain $3N$ components; $\mathbf{D}$ is a $3N \times 3N$ diffusion tensor or matrix, whose components depend on molecular positions. The random part of the displacement, $\mathbf{R}$, is selected from the $3N$-variate Gaussian distribution with zero means and covariance matrix

$$\langle \mathbf{RR} \rangle = 2\mathbf{D}\delta t. \tag{12.7}$$

As a consequence, the components of $\mathbf{R}$ are *correlated* with each other. Sampling these random variables is a comparatively time-consuming exercise, depending on some expensive manipulations of the matrix $\mathbf{D}$. Two forms of $\mathbf{D}$ are commonly adopted. The simplest, suggested by the equations of macroscopic hydrodynamics, is the Oseen tensor. Writing $\mathbf{D}$ as an $N \times N$ set of 3×3 matrices $\mathbf{D}_{ij}$ for each pair of molecules,

$$\mathbf{D} = \begin{pmatrix} \mathbf{D}_{11} & \mathbf{D}_{12} & \cdots & \mathbf{D}_{1N} \\ \mathbf{D}_{21} & \mathbf{D}_{22} & \cdots & \mathbf{D}_{2N} \\ \vdots & \vdots & \ddots & \vdots \\ \mathbf{D}_{N1} & \mathbf{D}_{N2} & \cdots & \mathbf{D}_{NN} \end{pmatrix}$$

this takes the form

$$\mathbf{D}_{ij} = \begin{cases} \dfrac{k_{\mathrm{B}}T}{6\pi\eta a}\mathbf{1} & i = j \\ \dfrac{k_{\mathrm{B}}T}{8\pi\eta r_{ij}}\left(\mathbf{1} + \hat{\mathbf{r}}_{ij}\hat{\mathbf{r}}_{ij}\right) & i \neq j \end{cases} \tag{12.8}$$

where η is the viscosity, a an estimate of the hydrodynamic radius (not diameter!), and $\mathbf{r}_{ij}$ the vector between the molecules, with $\hat{\mathbf{r}}_{ij} = \mathbf{r}_{ij}/r_{ij}$ as usual. This tensor has the property that $\mathbf{\nabla} \cdot \mathbf{D} = 0$, so this term may be dropped from eqn (12.6). The standard approach to generating the random displacements is to factorize $\mathbf{D}$ by the Cholesky square root method (Press et al., 2007), that is, determine the (unique) lower triangular real matrix $\mathbf{L}$ such that $\mathbf{D} = \mathbf{L} \cdot \mathbf{L}^T$ (where $\mathbf{L}^T$ is the transpose of $\mathbf{L}$). Then, given *independently* sampled normal (unit-variance) variables $\mathbf{G}$,

$$\mathbf{R} = \sqrt{2\Delta t}\, \mathbf{L} \cdot \mathbf{G} \tag{12.9}$$

will be the desired set of *correlated* Gaussian displacements (Ermak and McCammon, 1978) (see also Appendix E). However, the decomposition is only valid for positive definite matrices, and the Oseen tensor does not always fulfil this condition.

For this reason, the Oseen tensor is commonly replaced by the Rotne–Prager–Yamakawa tensor (Rotne and Prager, 1969; Yamakawa, 1970), which has the identical form for $i = j$, but which for different particles $i \neq j$ is

$$\mathbf{D}_{ij} = \begin{cases} \left(\dfrac{k_{\mathrm{B}}T}{8\pi\eta r_{ij}}\right)\left[\left(1 + \hat{\mathbf{r}}_{ij}\hat{\mathbf{r}}_{ij}\right) + \dfrac{2a^2}{r_{ij}^2}\left(\tfrac{1}{3}1 - \hat{\mathbf{r}}_{ij}\hat{\mathbf{r}}_{ij}\right)\right] & \text{for } r_{ij} \geq 2a, \\[3mm] \left(\dfrac{k_{\mathrm{B}}T}{6\pi\eta a}\right)\left[\left(1 - \dfrac{9}{32}\dfrac{r_{ij}}{a}\right)1 + \dfrac{3}{32}\dfrac{r_{ij}}{a}\hat{\mathbf{r}}_{ij}\hat{\mathbf{r}}_{ij}\right] & \text{for } r_{ij} < 2a. \end{cases} \tag{12.10}$$

This also has the property of zero divergence, and is positive definite for all r_{ij}. We should bear in mind that all these tensors are only leading approximations (Felderhof, 1977; Schmitz and Felderhof, 1982), and what is more they assume pairwise additivity in what is really a many-body problem in hydrodynamics (Mazur, 1982; Mazur and van Saarloos, 1982; van Saarloos and Mazur, 1983).

The solution of these equations of motion may be speeded up by a variety of techniques. The Cholesky method has computational cost $O(N^3)$ which rapidly becomes expensive for large N. Fixman (1986) has proposed a method for improving this to $O(N^{2.25})$, using an expansion in Chebyshev polynomials. Beenakker (1986) has suggested handling the long-range hydrodynamic interactions by Ewald sum; there are some subtleties associated with this (Smith, 1987; Smith et al., 1987). Banchio and Brady (2003) apply a fast Fourier transform method to compute many-body long-range hydrodynamic interactions, and divide the Brownian forces into near-field and far-field components, resulting in a method that is $O(N^{1.5}\log N)$ for sufficiently large N. Jain et al. (2012) have optimized both the Chebyshev and Ewald elements of the calculation, for semidilute polymer solutions, giving an overall cost $O(N^{1.8})$. More recently, a method called the truncated expansion approximation (Geyer and Winter, 2009) was proposed to calculate the correlated random displacements more rapidly, and a technique using Krylov subspaces (Ando et al., 2012) seems very promising.

12.3 Brownian dynamics, molecular dynamics, and Monte Carlo

The BD method may be related to MD, and MC, techniques (Rossky et al., 1978; Horowitz, 1991; Akhmatskaya et al., 2009; Allen and Quigley, 2013). Starting with eqn (12.5), and making the change of variables

$$D = \frac{k_{\mathrm{B}}T}{2m}\delta t$$

we obtain

$$r(t + \delta t) = r(t) + \tfrac{1}{2}(\delta t^2 / m) f(t) + \sqrt{\frac{k_B T}{m}} \delta t\, G. \tag{12.11}$$

This is recognizable as the velocity Verlet MD algorithm, with all the initial velocities $v(t)$ selected randomly from the Maxwell–Boltzmann distribution, as per the thermostat of Andersen (1980). This link suggests that the timescales of MD (inertial dynamics) and BD (diffusional dynamics) simulations may be related, effectively replacing the inverse mass by the diffusion coefficient, and that neither technique is, in principle, preferable to the other, in terms of the choice of timestep for a given force field.

Suppose that we wish to sample configurations r from the canonical ensemble. Then, as discussed in Chapter 9, the acceptance probability for proposed moves $r_m \rightarrow r_n$ may be written

$$\min\left(1, \exp\left(-\beta \delta \mathcal{V}_{nm}\right) \frac{\alpha_{nm}}{\alpha_{mn}}\right) \tag{12.12}$$

for $n \neq m$, where α_{mn} measures the probability of attempting the move, α_{nm} that of the reverse move, and $\delta \mathcal{V}_{nm}$ is the potential-energy change (Metropolis et al., 1953; Hastings, 1970). Even if the calculation of dynamical properties is not of interest, it may be convenient to introduce the conjugate momenta p, to assist in the configurational sampling. Rossky et al. (1978) noted that BD is equivalent to the SMC method of Chapter 9, where N-particle moves are proposed with displacements chosen from a Gaussian distribution, biased in the direction of the forces. If the momentum part of the velocity Verlet scheme is included with the position update of eqn (12.11), then the move selection may be written

$$p_m = \sqrt{m k_B T}\, G$$
$$r_n = r_m + \tfrac{1}{2}(\delta t^2 / m) f_m + p_m (\delta t / m)$$
$$p_n = p_m + \tfrac{1}{2} \delta t (f_m + f_n).$$

The new momenta p_n are not needed for the next step, because they are randomly resampled, but they greatly simplify the SMC move acceptance probability, which becomes

$$\min\left[1, \exp\left(-\beta \delta \mathcal{V}_{nm}\right) \exp\left(-\beta \delta \mathcal{K}_{nm}\right)\right] = \min\left[1, \exp\left(-\beta \delta \mathcal{H}_{nm}\right)\right] \tag{12.13}$$

where $\delta \mathcal{K}_{nm} = \mathcal{K}_n - \mathcal{K}_m$ is the change in kinetic energy, calculated from old and new momenta. This same conclusion was reached by Duane et al. (1987), and they termed the algorithm hybrid Monte Carlo (HMC). The acceptance probability is not unity, because the velocity Verlet algorithm does not exactly conserve the Hamiltonian. However, as Duane et al. (1987) made clear, the *symplectic* property of the algorithm, in particular the conservation of phase space volume, is crucial. This is because the ratio of move selection probabilities appearing in eqn (12.12) includes, as well as the factor $\exp(-\beta \delta \mathcal{K}_{nm})$ appearing in eqn (12.13), the ratio of phase space volumes $dr\,dp$ for the initial and final states. This ratio will be unity for a symplectic algorithm. A sample program is given in Code 12.2.

Viewed as a Monte Carlo method which happens to use a particular way of selecting the new configuration, HMC has plenty of flexibility. The dynamical step may employ a different Hamiltonian (perhaps a cheaper one to evaluate) from the one used in the

Code 12.2 Smart Monte Carlo simulation

These files are provided online. The programme `smc_nvt_lj.f90`, combined with `smc_lj_module.f90` for the energies and forces, and the utility modules of Appendix A, carries out SMC, using a notation similar to that of BD and HMC, as described in the text. The user may select the number of particles to move.

```
! smc_nvt_lj.f90
! Smart Monte Carlo, NVT ensemble
PROGRAM smc_nvt_lj
```

```
! smc_lj_module.f90
! Energy, force, and move routines for SMC, LJ potential
MODULE smc_module
```

acceptance–rejection criterion. Of course, the acceptance ratio for moves will be low if the two Hamiltonians do not give similar results for most configurations. In principle, the acceptance rate may be increased arbitrarily by reducing the timestep, and eventually a purely dynamical scheme with no move rejection will be recovered. An attraction of the method is the ability to increase the timestep, and correct for the discretization error through the acceptance criterion. Unfortunately, since the energy is an extensive quantity, as system size increases, the acceptance rate will decrease if the other parameters are held constant.

Viewed as a molecular dynamics method with thermostatting and (occasional) move rejections, this approach is also quite flexible. There is a tendency, in MD, to increase the timestep to as large a value as possible without introducing noticeable errors. The danger is that systematic errors may still occur, due to an over-large timestep, even if they are not noticed! The approach of 'Metropolizing' an MD simulation, by accepting or rejecting steps on the basis of eqn (12.13), should guarantee that the correct equilibrium ensemble is sampled, at the expense of perturbing any measured dynamical properties. Just as in the original Andersen thermostat, one can devise schemes which randomize only a fraction of the momenta at the start of the step, preserving the others as is usually done in MD (Akhmatskaya and Reich, 2008; Akhmatskaya et al., 2009).

It is perhaps also useful to point out that a distance-dependent diffusion tensor in BD, such as eqn (12.10), translates into a distance-dependent mass tensor in MD. Adjusting the masses, to improve the efficiency of MD, is an idea that goes back to Bennett (1975), and there has been a recent revival of interest in this approach in the biomolecular simulation community (Lin and Tuckerman, 2010; Kunz et al., 2011; Michielssens et al., 2012; Wright and Walsh, 2013). A reversible, symplectic, algorithm for solving the equations of motion derived from the corresponding Hamiltonian has been given (Akhmatskaya and Reich, 2008). The correct sampling of correlated momenta in the Andersen thermostatting step is non-trivial, in general, but can be tackled in the same way as the random displacement in Brownian dynamics with hydrodynamic interactions.

12.4   Dissipative particle dynamics

An alternative kind of stochastic dynamics, which incorporates hydrodynamic effects directly in the fluid simulation is DPD (Hoogerbrugge and Koelman, 1992). The method has been reviewed recently (Liu et al., 2014). The DPD equations formally resemble eqn (12.1)

$$\dot{\mathbf{r}} = \mathbf{v} = \mathbf{p}/m \tag{12.14a}$$

$$\dot{\mathbf{p}} = \mathbf{f}(\mathbf{r}) - \xi \mathbf{V}(\mathbf{r}, \mathbf{p}) + \sigma \dot{\mathbf{W}}(\mathbf{r}, \mathbf{p}). \tag{12.14b}$$

As before, ξ is a friction coefficient, $\sigma = \sqrt{2\xi k_B T}$, and $\mathbf{f}$ are forces acting between the particles, which are derived from a potential-energy function (often called 'conservative' forces). $\xi \mathbf{V}(\mathbf{r}, \mathbf{p})$ and $\sigma \dot{\mathbf{W}}(\mathbf{r}, \mathbf{p})$ are respectively dissipative (frictional) forces and random forces, but $\mathbf{V}$ and $\dot{\mathbf{W}}$ are more complicated functions than their Langevin equation counterparts.

The particles themselves represent *regions* of fluid, not individual atoms or molecules. The potential from which the conservative forces are derived is not an attempt to represent realistic interparticle interactions, except in the most approximate fashion. Roughly speaking, it is fitted to the liquid compressibility (see later). The force usually takes the form of a linear function of separation, characterized by a single 'repulsion parameter' a in the following way:

$$\mathbf{f}_{ij} = a\varphi_{ij}(r_{ij}/r_c)\hat{\mathbf{r}}_{ij}, \qquad \hat{\mathbf{r}}_{ij} = \mathbf{r}_{ij}/r_{ij} \tag{12.15}$$

where typically $\varphi(x) = 1 - x$ for $x < 1$, $\varphi(x) = 0$ for $x > 1$. This is much softer than, say, the Lennard-Jones force, and, in turn, means that much longer timesteps (e.g. of order 10ps) may be used than in conventional molecular dynamics simulations. This is the main attraction of the method.

The dissipative and random terms are also pairwise additive:

$$\mathbf{V}_i = \sum_{j \neq i} \mathbf{V}_{ij}, \qquad \dot{\mathbf{W}}_i = \sum_{j \neq i} \dot{\mathbf{W}}_{ij} \tag{12.16}$$

where

$$\mathbf{V}_{ij} = \varphi(r_{ij}/r_c)\left(\mathbf{v}_{ij} \cdot \hat{\mathbf{r}}_{ij}\right)\hat{\mathbf{r}}_{ij}, \qquad \dot{\mathbf{W}}_{ij} = \sqrt{\varphi(r_{ij}/r_c)}\, \dot{w}_{ij}\, \hat{\mathbf{r}}_{ij}. \tag{12.17}$$

Here, $\mathbf{v}_{ij} = \mathbf{v}_i - \mathbf{v}_j$ is the relative velocity of the pair, while the $w_{ij}(t) = w_{ji}(t)$ are independent Wiener processes. The weight functions are chosen to satisfy the fluctuation–dissipation theorem, as shown by Español and Warren (1995). There is no fundamental reason for using the same function $\varphi(x)$ in eqn (12.17) as in eqn (12.15), but it is convenient and usual. The key feature of these definitions is that the dissipative and random terms, like the forces $\mathbf{f}_{ij}$, conserve momentum. This gives rise to hydrodynamic flow. Another way of stating the same thing is that the dynamics is Galilean-invariant. This is not true of the Langevin equation, eqn (12.1), since the frictional term is proportional to the particle velocity in a fixed reference frame.

DPD resembles a particle-based method, initially developed for solving fluid dynamics problems in astrophysics, smoothed particle hydrodynamics (SPH) (Gingold and Monaghan, 1977; Lucy, 1977), as well as a general technique for continuum mechanics problems, smoothed particle applied mechanics (SPAM) (Hoover and Hoover, 2001). The parameters in the equations of motion should be chosen to reproduce the desired properties of the fluid. A very useful paper (Groot and Warren, 1997) sets out a way of choosing these parameters. By taking the cutoff distance to define the unit of length, matching the compressibility to that of water under ambient conditions, and adjusting the friction coefficient to give the desired viscosity, Groot and Warren arrived at a consistent set of values: number density $\rho = 3$, stochastic force strength $\sigma = 3$, and temperature $k_\mathrm{B}T = 1$.

In soft matter, the DPD method has been used very successfully to model amphiphilic systems: for example, to study microphase separation of block copolymers (Groot et al., 1999), flow effects on polymer brushes (Goujon et al., 2004; 2005), and also in modelling lipid bilayers (Kranenburg et al., 2003; Kranenburg and Smit, 2004; Venturoli et al., 2005; 2006; Shillcock and Lipowsky, 2006), and membrane fusion processes (Grafmüller et al., 2007). This kind of application uses different DPD beads for water, and for the hydrophilic and hydrophobic parts of the relevant molecules. The species are distinguished by the pairwise repulsion parameters acting between them (a in eqn (12.15)). The beads are taken to represent many atoms, in a highly coarse-grained fashion. The connections between the beads in a lipid are usually taken to be simple harmonic spring potentials.

The implementation of an accurate algorithm for DPD is not as straightforward as for Langevin dynamics, because the frictional forces depend on both separation and relative velocity. Various schemes have been put forward (Pagonabarraga et al., 1998; Besold et al., 2000; den Otter and Clarke, 2001; Vattulainen et al., 2002; Nikunen et al., 2003; Peters, 2004) many of them based on modifying the velocity Verlet approach. Groot and Warren (1997) recommended a timestep $0.04 \le \Delta t \le 0.06$ (for water simulations using the typical parameters and units just mentioned) based on requiring the simulation temperature to lie within 3 % of the value provided as an input parameter. However, it soon became apparent that a drift in temperature would occur even in the absence of conservative forces (Marsh and Yeomans, 1997). On top of this, the thermostatting implicit in the method tended to obscure the consequences of using timesteps that were too long for compatibility with the conservative forces (something which could be tested separately by standard MD), and various artefacts were reported (Hafskjold et al., 2004; Jakobsen et al., 2005; Allen, 2006b). Some of these problems may be alleviated by a multiple timestep approach similar to that described in Section 3.5 (Jakobsen et al., 2006). Algorithms derived by an operator-splitting approach similar to that used for standard MD have been proposed by Shardlow (2003), and the consensus seems to be that these are the ones which perform best (Lisal et al., 2011). An example DPD program is given in Code 12.3.

An alternative approach is to implement an MD method, employing soft conservative DPD potentials and a momentum-conserving thermostat. A 'pairwise' stochastic thermostat constructed in the spirit of Andersen (1980), was proposed by Lowe (1999), and is usually referred to as the Lowe–Andersen thermostat (see Section 3.8.1). No frictional forces are used: instead, a standard MD integrator is applied to the conservative forces, and at intervals the thermostat is applied, or not, with a predefined probability, to the pairs which lie in range. The range over which the thermostat acts, and the frequency of randomization, are parameters of the method. In addition, the randomization probability may be a constant, or it may include a weight function based on pair separation (similar to the one used in DPD). An example of this thermostat is also given in Code 12.3.

Leimkuhler and Shang (2015) have reviewed several DPD algorithms and proposed a further one which performs very well at low friction: a version of the pairwise Nosé–

Code 12.3 Dissipative particle dynamics

These files are provided online. The DPD programme dpd.f90, combined with a small module in the file dpd_module.f90 and the utility modules of Appendix A, carries out dynamics of beads interacting through the soft DPD potential using the operator-splitting algorithm of Shardlow (2003) or, as an alternative, the pairwise Lowe–Andersen thermostat (Lowe, 1999).

```
! dpd.f90
! Dissipative particle dynamics
PROGRAM dpd


! dpd_module.f90
! Dissipative particle dynamics module
MODULE dpd_module
```

Hoover thermostat of Allen and Schmid (2007), with an additional Langevin equation acting on the dynamical friction coefficient. However, it is not possible to tune the viscosity over a wide range using this method alone: combination with the Lowe–Andersen approach, or something similar, would be required for this.

12.5   Multiparticle collision dynamics

Now we turn to methods in which the solvent is modelled explicitly, albeit approximately. The idea is to conserve momentum, and be able to prescribe (at least) the solvent viscosity, but avoid the expense of a microscopically realistic representation of the solvent particles. This is then combined with a suitable model for solvent–solute interactions. Most of our interest is then likely to focus on the motion of the solute particles. First, we consider the solvent model alone.

The multiparticle collision dynamics (MPCD) method, originally termed the stochastic rotation method, was proposed by Malevanets and Kapral (1999). The solvent particles do not interact with each other through a potential, so there are no systematic forces, and the solvent obeys an ideal gas equation of state. In the absence of solute, the solvent particles undergo free flight in between discrete collisions, which occur at regular intervals δt:

$$\mathbf{r}(t + \delta t) = \mathbf{r}(t) + \mathbf{v}(t)\,\delta t.$$

The collisions act to redistribute momenta between particles, and one can imagine several ways of doing this in a physically reasonable manner: momentum and energy should be conserved, and the effects should be 'local' in the sense of having a restricted spatial range. The Lowe–Andersen thermostat discussed in the previous section is constructed with similar constraints in mind. The usual approach is to divide the simulation box into cuboidal cells, of a specified side length a, and redefine the momenta of all the particles within each cell, independently of the other cells. Taking all the particle masses to be equal for simplicity, this is accomplished by:

1. calculating the centre-of-mass velocity of particles in each cell;
2. subtracting this from the velocities of all the particles in that cell;
3. rotating all particle velocities by a random amount about a random axis;
4. adding back the cell centre-of-mass velocity.

Denoting the times before a collision by t_-, and after by t_+, this may be expressed as

$$\mathbf{v}_i(t_+) = \mathbf{v}_{cm} + \mathbf{A} \cdot \left(\mathbf{v}_i(t_-) - \mathbf{v}_{cm}\right) \qquad \forall i \in \text{cell}$$

$$\mathbf{v}_{cm} = \frac{1}{N_{cell}} \sum_{i \in cell} \mathbf{v}_i(t_-) = \frac{1}{N_{cell}} \sum_{i \in cell} \mathbf{v}_i(t_+)$$

where N_{cell} is the number of particles in the cell. The matrix $\mathbf{A}$ is a 3×3 rotation matrix of the kind seen in Chapter 3: it is defined by a unit vector, randomly chosen on the unit sphere, which specifies the axis of rotation, and an angle of rotation ϕ.

Evidently this algorithm conserves both momentum and kinetic energy, as can be seen most simply in the centre-of-mass frame. It does not locally conserve angular momentum, in the manner of the hard-sphere collisions discussed in Chapter 3, which act along the line of centres between atomic pairs, but this is not generally considered to be a serious deficiency. However, the use of a cubic grid of cells to define the collisions violates the Galilean invariance (including the rotational isotropy) of a fluid. To avoid this, the absolute position of the grid is changed randomly before the collisions are implemented (Ihle and Kroll, 2001; 2003a). In practice, this may be implemented by shifting all the particles by a common, randomly selected, vector before the collisions, including any effects of periodic boundary conditions, and applying the reverse transformation afterwards.

The time δt between collisions, the size a of the cells, the magnitude of rotation ϕ, and the fluid particle density, are the key parameters determining the transport properties of the fluid. In fact, so simple is the kinetic scheme, that these can be exactly calculated (Ihle and Kroll, 2003b; Tüzel et al., 2003; Kikuchi et al., 2003; Pooley and Yeomans, 2005).

Padding and Louis (2006) have given a useful summary of the MPCD method, together with a discussion of how to match up the parameters with the desired properties of the fluid, such as the Reynolds, Peclet, and Schmidt numbers.

When solute molecules or particles are added to the system, the dynamical scheme just described needs to be modified. There are two typical ways of modelling the interaction between solute and solvent. The simplest approach, from the technical point of view, is to include the solute particles in the collision step. A standard MD algorithm, such as velocity Verlet, is used to advance the positions and velocities of the solute particles between collisions, while the solvent undergoes free flight. Of course, there is no particular need for the solute and solvent timesteps to be made equal: the MD timestep will naturally reflect the intermolecular interactions (between the particles in a colloidal suspension or the monomers in a polymer solution, for instance). The interval between collisions may be some multiple of this. This method has been used to study the influence of hydrodynamics on polymer collapse (Kikuchi et al., 2005). A slightly unrealistic feature is that it allows the solvent to penetrate freely into the solute particles: there is no 'excluded volume' effect.

In the second approach, there is an explicit potential of interaction between solute and solvent particles (Malevanets and Kapral, 2000), giving rise to forces in the velocity

Verlet algorithm, which applies to all the particles. Therefore, the solvent particles may be excluded from the regions of space occupied by the solute. The solvent motion between collisions is no longer 'free flight', but is still quite inexpensive since there are no solvent–solvent interactions. This method has been used to study sedimenting colloids (Hecht et al., 2005; Padding and Louis, 2006), and these two papers also include many technical details and alternative choices in the implementation. Batôt et al. (2013) have compared both the solvent–solute interaction schemes, described earlier, with Brownian dynamics.

The MPCD method has been reviewed by Gompper et al. (2009), and Winkler et al. (2013) have illustrated its application to a range of particle suspensions. An interesting example of both MPCD and DPD in action is the modelling of blood flow, and this is discussed in Example 12.1.

It is important to realize that the particles of the solvent do not truly represent clusters of molecules in a real system. The MPCD scheme aims to solve the Navier–Stokes equation, including some effects of fluctuations: nothing more. In the next section we shall discuss an even more idealized approach to the same problem.

12.6   The lattice-Boltzmann method

We only give a brief summary of the lattice-Boltzmann (LB) method here; more comprehensive treatments may be found in several review articles (Benzi et al., 1992; Chen and Doolen, 1998; Raabe, 2004; Dünweg and Ladd, 2009; Aidun and Clausen, 2010) and books (Wolf-Gladrow, 2000; Succi, 2013). In common with the multiparticle collision dynamics method described in Section 12.5, an LB simulation code proceeds in an alternating sequence of 'streaming' steps and 'collision' steps, and space is subdivided into a lattice of cells. However, the fluid model is further simplified, in that the particle velocities are restricted to values which, over the course of one timestep, would take each particle precisely along a lattice vector between nearby cells. As a consequence, it is only necessary to keep track of the *numbers* of particles having each allowed velocity, and in fact the LB method simulates the time evolution of the probability distribution function of velocities in each cell: it is a numerical method to solve the so-called Boltzmann equation. This equation is essentially a molecular-scale analogue of the continuum Navier–Stokes equation of hydrodynamics.

Let us denote the velocities by c_i where i simply labels the allowed values. The single-particle, time-dependent, probability distribution function $f(\mathbf{r}, \mathbf{c}_i, t)$ is abbreviated $f_i(\mathbf{r}, t)$, where the positions $\mathbf{r}$ lie on a lattice. A shorthand is used to identify different LB geometries, based on the dimensionality and the number of velocities considered (Qian et al., 1992). D2Q9, for instance, corresponds to a two-dimensional square lattice, in which the velocities may take a particle to any of the four adjacent squares, any of the four diagonal neighbours, or remain stationary. D3Q19 would be a three-dimensional cubic lattice, in which any of the 27 cells in the $3 \times 3 \times 3$ surrounding cube are accessible, except for the eight that lie in the directions $(\pm 1, \pm 1, \pm 1)$. These two examples are illustrated in Fig. 12.1. The choice of allowed velocities affects the isotropy of the resultant flows. The combined collision and streaming step is then simply

$$f_i(\mathbf{r} + \mathbf{c}_i \delta t, t + \delta t) = f_i^{\dagger}(\mathbf{r}, t) = f_i(\mathbf{r}, t) + \Omega_i\big(\{f_i(\mathbf{r}, t)\}\big),$$

Example 12.1 Blood flow

Modelling the flow of blood is a high-profile example of the application of mesoscale simulation techniques. The fluid contains a high concentration of red blood cells, each of which is a soft, deformable, body. The distortion of these cells is an intrinsic part of the problem; some diseases change their stiffness and hence may affect the flow. Often, one is interested in how the blood flows in the highly confined environment of capillaries. The coupling between flow and cell deformation is a critical issue. For all these reasons, a method that includes both hydrodynamics and a particle-based description is highly desirable. The field has been reviewed by Fedosov et al. (2013). An early example (Noguchi and Gompper, 2005) used multiparticle collision dynamics to represent the hydrodynamics of the solvent. The blood cell membrane consisted of a set of beads, having mass and interacting via a repulsive potential to give them an effective excluded volume. In addition, bonding or tethering potentials were used to triangulate the assembly of beads into a closed surface, and the curvature elasticity was reproduced through a discretized version of the Helfrich free energy, eqn (2.190). The fluidity of the membrane was facilitated by allowing tethers to flip, at intervals, between the two possible diagonals of adjacent triangles. Finally, two global potential-energy terms were added to control the enclosed volume and surface area. By a suitable choice of parameters, the equilibrium shape of each cell could be made to reproduce that seen in real life: a disk-like form, with the two opposite faces deformed inward (biconcave). This approach allowed Noguchi and Gompper (2005) to study capillary flow at low Reynolds number. As the flow rate increased, cell deformation into an elongated shape, and a parachute shape, were observed. Both transitions reduced the flow resistance.

In subsequent papers, the aggregation of red blood cells into stacks, known as 'rouleaux', has been studied using a different mesoscale model, based on DPD (Fedosov et al., 2011). The triangulation of the cell network was also fixed, rather than changing dynamically; the network spring parameters were precalculated so as to minimize stress, and the springs included some dissipative effects to mimic the internal viscosity (Fedosov et al., 2010). The attractive forces responsible for aggregation were represented with a Morse potential. Highly non-Newtonian behaviour resulted from the interplay of shear flow, aggregation, cell deformation, and reorientation, and the predictive power of such simulations in the medical sphere was considered.

---

that is, the population f_i of each velocity in each cell $\mathbf{r}$ is subjected to the effects of a *collision operator* Ω_i, which produces the post-collisional value $f_i^{\dagger}(\mathbf{r}, t)$, and this is then translated by the appropriate amount to a nearby cell.

The collision operator represents the averaged effect of particle interactions on the velocity distribution, and is, of necessity, an approximation. Even making the assumption that it is a local effect, in principle it depends on *all* the velocity component populations f_i, as indicated in the previous equation, and it may, in general, be a nonlinear function of the f_i. It must, of course, satisfy the laws of conservation of mass and momentum.

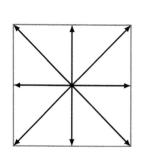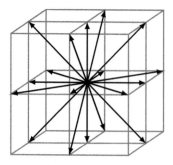

Fig. 12.1 Velocities in the D2Q9 and D3Q19 lattice-Boltzmann models.

The simplest version, the so-called BGK approximation (Bhatnagar et al., 1954; Qian et al., 1992), is a linear equation

$$\Omega_i\big(\{f_i(\mathbf{r}, t)\}\big) = -\Lambda\big(f_i(\mathbf{r}, t) - f_i^{\text{eq}}(\mathbf{r}, t)\big), \quad \text{where} \quad \Lambda = \frac{\delta t}{\tau}.$$

This corresponds to a single relaxation time τ, for each velocity component i independently, towards its equilibrium distribution f_i^{eq}. In practice, a more general (but still linear) collision operator, which mixes up the different velocity component distributions, is usually adopted:

$$\Omega_i\big(\{f_i(\mathbf{r}, t)\}\big) = -\sum_j \Lambda_{ij}\big(f_j(\mathbf{r}, t) - f_j^{\text{eq}}(\mathbf{r}, t)\big).$$

The eigenvalues of the collision matrix Λ_{ij} represent inverse relaxation times, and this is generally called a multiple relaxation time model (d'Humières et al., 2002). As well as being a more stable numerical method, the flexibility of this approach (in being able to adjust the eigenvalues) allows it to be applied to a wider range of flow problems.

   Actually, these equations are better termed *quasi-linear*, because the equilibrium distribution function $f_i^{\text{eq}}(\mathbf{r}, t)$ depends on position and time through the local fluid density ρ and local fluid velocity $\mathbf{u}$, which are themselves defined via

$$\rho(\mathbf{r}, t) = \sum_i f_i(\mathbf{r}, t)$$

$$\rho(\mathbf{r}, t)\mathbf{u}(\mathbf{r}, t) = \sum_i f_i(\mathbf{r}, t)\mathbf{c}_i$$

and the form of f_i^{eq} may be obtained by considering a moment expansion.

   The LB equations just described are completely deterministic. Although they may reproduce hydrodynamics at large length and time scales, they do not incorporate the fluctuations that characterize behaviour at the mesoscale: the number of particles within a cell, for instance, would be expected to fluctuate significantly, unless the cells become macroscopically large. To account for this, a stochastic term is typically added to the collision operator, giving an equation resembling the Langevin equation of Section 12.2. This approach was pioneered by Ladd (1994a,b) (see also Ladd and Verberg, 2001) in a way that reproduced the appropriate fluctuation–dissipation relation at long wavelength, and then improved by Adhikari et al. (2005) so as to treat fluctuations correctly at all wavelengths. Thermal fluctuations affect the relaxation of the stress tensor, and also the other (non-conserved) modes.

The boundary conditions used in molecular dynamics to simulate shear flow (Lees and Edwards, 1972) can be introduced into LB simulations (Wagner and Yeomans, 1999; Wagner and Pagonabarraga, 2002). If it is desired to model fluids in the vicinity of solid surfaces, some physical boundary conditions need to be introduced. The simplest is the no-slip or bounce-back condition, which may be applied along the links between neighbouring cells: the corresponding component f_i is reflected back to the originating node, and the velocity reversed. This leads in turn to the modelling of particles embedded in a lattice-Boltzmann fluid: if the particles are allowed to move, one needs a consistent way of transferring momentum from the fluid to the particles and vice versa. Ahlrichs and Dünweg (1999) have shown how to combine LB and MD methods to simulate a polymer chain in solution; the application to particle–fluid suspensions has been reviewed by Ladd and Verberg (2001). Both types of system are addressed in Dünweg and Ladd (2009), together with examples such as colloid sedimentation and polymer migration in a confined geometry.

Amongst several methods for modelling multiphase systems, and phase separation, we shall concentrate on the approach of Swift et al. (1995), Orlandini et al. (1995), and Swift et al. (1996). Here, in the spirit of the Cahn and Hilliard (1958) approach to nonequilibrium dynamics, a non-ideal, equilibrium, free-energy term is introduced in the form of a functional of the particle density: for example, a squared-gradient van der Waals form. The pressure tensor derived from this functional then enters into the LB equations. The approach may be extended to encompass multi-component fluids, and, through the introduction of a free energy depending on a tensor order parameter, liquid crystals (Denniston et al., 2000; 2001; Henrich et al., 2010).

A key practical issue in applying the lattice-Boltzmann approach to a physical system is the matching of the parameters of the model to the physical fluid of interest. The basic units of LB are the cell size, the timestep, and the fluid density. These may be used to define dimensionless quantities (times, distances, velocities) in terms of which the numbers that characterize fluid flow (especially the Reynolds number) may be calculated. Comparison with the physical system may then be made via these numbers.

LB simulations are intrinsically well-suited to implementation on a parallel computer, since the cell structure of the system is regular, constant in time, and involves only local transfer of information. The scalability of LB calculations has been discussed (Stratford and Pagonabarraga, 2008; Bernaschi et al., 2009; Clausen et al., 2010).

12.7 Developing coarse-grained potentials

All of the approaches described earlier in this chapter make some dramatic approximations regarding the interactions between the molecules, and how to replace them with interactions between larger units. The development of systematic ways of doing this is a long-standing, but still very active, area of research. Several reviews are available (Saunders and Voth, 2013; Chu et al., 2006; 2007).

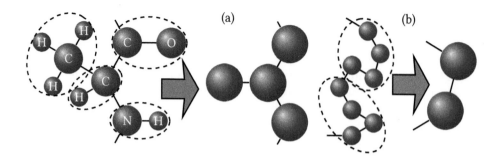

Fig. 12.2 Schematic illustration of coarse graining of (a) an amino acid within a peptide or protein, and (b) a polymer chain.

Superficially, the process of 'coarse graining' consists of replacing groups of atoms by single beads, as illustrated schematically in Fig. 12.2. The coarse-grained (CG) model of an amino acid, shown in Fig. 12.2(a), has been proposed by Bereau and Deserno (2009) to represent protein or peptide chains. Four beads are used per monomer: three to represent the $-NH-CH-CO-$ groups of the backbone, and one to represent the side chain (the example in the figure is CH_3 corresponding to the amino acid alanine). In Fig. 12.2(b) we represent a more generic coarse graining of a polymer, in which several monomers are replaced by a single CG bead.

For simplicity, we shall assume that all the atoms j in the original model, at positions $\mathbf{r}_j$, are divided into N_{CG} mutually exclusive groups which we label J. Moreover, we will represent each group by a position vector $\mathbf{R}_J$, given by some function $\mathbf{R}_J = C_J(\{\mathbf{r}_j\}_{j \in J})$ of the coordinates of the atoms belonging to it. Typically, $\mathbf{R}_J$ will be the centre of mass of the atoms comprising J, in which case this is a linear function

$$\mathbf{R}_J = \sum_{j \in J} C_{Jj} \mathbf{r}_j, \quad \text{where} \quad C_{Jj} = m_j \bigg/ \sum_{j' \in J} m_{j'},$$

or, for short, $\mathbf{R} = \mathbf{C} \cdot \mathbf{r}$. Here $\mathbf{C}$ is a $3N_{CG} \times 3N$ matrix. This amounts to replacing each group of atoms J by a single spherical bead. Including the internal coordinates of J, or orientational degrees of freedom, would be possible, but we do not consider this here. These beads are the particles in the CG model.

However, a geometrical mapping between the two descriptions is by no means the end of the story: the interactions must be chosen in such a way as to make simulation averages of selected properties of the two systems match as closely as possible. Having done that, further questions present themselves. Is there a unique way of doing this, or at least an optimal one? If the properties match at one state point, how well will they match at another? Is it possible to relate the dynamics of the CG system to those of the original one? The answer to most of these questions, unfortunately, is in the negative. Here we consider two common approaches.

12.7.1 Force matching

Originally based on a method to extract classical effective forces from *ab initio* simulations (Ercolessi and Adams, 1994), the force-matching approach (Izvekov and Voth, 2005a,b;

Noid et al., 2008a) aims to make the forces acting on the beads, derived from a CG interaction potential, as close as possible to the ensemble-averaged forces, acting on the same entities, in a fully atomistic simulation. Again, for simplicity, let us consider the statistical mechanics of the two models in the canonical ensemble. The probability distribution function for **R**, determined by a simulation of the fully atomistic model, will be given by

$$P(\mathbf{R}) = \left\langle \delta\big(\mathbf{R} - \mathbf{C} \cdot \mathbf{r}\big) \right\rangle \propto \int d\mathbf{r}\, \delta\big(\mathbf{R} - \mathbf{C} \cdot \mathbf{r}\big)\, \exp\big(-\beta V(\mathbf{r})\big).$$

The same distribution, and hence the same ensemble-averaged functions of **R**, may be generated by a canonical ensemble simulation using the potential $V^{\text{CG}}(\mathbf{R})$ defined by

$$P(\mathbf{R}) \propto \exp\big(-\beta V^{\text{CG}}(\mathbf{R})\big) \tag{12.18a}$$

$$\Rightarrow \quad V^{\text{CG}}(\mathbf{R}) = -k_{\text{B}}T \ln\left[\int d\mathbf{r}\, \delta\big(\mathbf{R} - \mathbf{C} \cdot \mathbf{r}\big)\, \exp\big(-\beta V(\mathbf{r})\big) \right] + \text{constant} \tag{12.18b}$$

where an arbitrary constant has been introduced. This is the same as the definition of the Landau free energy, eqn (2.172). Many degrees of freedom have been 'integrated out' of the problem; it is important to remember that this makes the resulting $V^{\text{CG}}(\mathbf{R})$ depend on density, temperature, and other factors determining the state point such as solvent composition or, in general, the concentrations of other species. It is, therefore, not a 'potential energy' in the usual sense. Assuming that the CG beads have been defined in the way described earlier, and neglecting complicating issues such as constraints acting between the atoms on different beads, it is possible to derive from this the relation

$$\mathbf{F}_J^{\text{CG}} = -\nabla_{\mathbf{R}_J} V^{\text{CG}}(\mathbf{R}) = \frac{\int d\mathbf{r}\, \left(\sum_{j \in J} \mathbf{f}_j\right) \delta\big(\mathbf{R} - \mathbf{C} \cdot \mathbf{r}\big)\, \exp\big(-\beta V(\mathbf{r})\big)}{\int d\mathbf{r}\, \delta\big(\mathbf{R} - \mathbf{C} \cdot \mathbf{r}\big)\, \exp\big(-\beta V(\mathbf{r})\big)} = \left\langle \sum_{j \in J} \mathbf{f}_j \right\rangle_{\mathbf{R}},$$

with the total force acting on all the atoms belonging to bead J appearing, and being averaged, on the right. The form of this equation leads to the terminology 'potential of mean force' for $V^{\text{CG}}(\mathbf{R})$. However, this simple result does depend on the aforementioned assumptions, and the formal derivation leading to a more general formula in the case where some of those assumptions are relaxed may be found in Noid et al. (2008a). The angle brackets $\langle \cdots \rangle_{\mathbf{R}}$ denote an ensemble average over the atomic coordinates **r**, subject to the constraint $\mathbf{C} \cdot \mathbf{r} = \mathbf{R}$. One might imagine accumulating multidimensional histograms of these forces, on each bead J, as functions of all the coarse-grained coordinates $\mathbf{R}_J$, during a standard MD or MC simulation; alternatively, one could apply constraints on the $\mathbf{R}_J$ in MD, at systematically chosen values, in which case account should be taken of the metric tensor discussed in Section 2.10.

The original implementation of Izvekov and Voth (2005a,b) involves variationally minimizing the quantity

$$\chi^2 = \left\langle \frac{1}{3N_{\text{CG}}} \sum_{J=1}^{N_{\text{CG}}} \left| \left(\sum_{j \in J} \mathbf{f}_j(\mathbf{r})\right) - \mathbf{F}_J^{\text{CG}}(\mathbf{C} \cdot \mathbf{r}) \right|^2 \right\rangle. \tag{12.19}$$

The angle brackets represent a simulation average of the atomistic system, each snapshot of which consists of a set of coordinates **r**. There is considerable flexibility in the way in which the functions $F_J^{CG}(R)$ are parameterized. Early applications used a pairwise decomposition of the non-bonded interactions between CG sites, with each interaction represented as a set of splines. More generally, the CG forces may be derived from a linear combination of force functions, which act as an incomplete basis set in the vector space of all possible such forces. The coefficients are varied so as to minimize χ^2 in eqn (12.19), and the fact that they appear linearly makes it possible to use efficient, standard, algorithms for the minimization, even when the number of parameters is quite large. Further details, and discussions that go beyond the simple case presented here, may be found elsewhere (Noid et al., 2008a,b).

Even though the true potential $\mathcal{V}_{CG}(R)$ is unlikely to be a simple sum of pairwise functions of the CG coordinates R_J, this approach has been quite successful in modelling biomolecules (Izvekov and Voth, 2005a; Shi et al., 2006) and ionic liquids (Wang et al., 2006). An extension to incorporate three-body CG interactions (Larini et al., 2010) has been shown to improve the modelling of water.

12.7.2   Structure matching

A slightly different approach to deriving a CG model is based on reproducing structural information, such as a set of pair distribution functions. These might be calculated in an atomistic simulation, or indeed determined by X-ray or neutron diffraction, and so these methods have a direct application to the interpretation of experimental results. Underpinning these methods is a theorem due to Henderson (1974), establishing the uniqueness of a pair potential which gives rise to a given pair distribution function (including, if relevant, angular variables). This has been extended to more general forms of the potential (Rudzinski and Noid, 2011). It has also been put in the context of density functional theory (Chayes et al., 1984; Chayes and Chayes, 1984), and shown to be related to the relative entropy of atomistic and coarse-grained systems (Shell, 2008).

This is not the same as eqn (12.18), which does not imply that any specified pair correlation functions will be reproduced. There is no formal proof of *existence* of such a CG potential; nonetheless, its uniqueness (if it exists) encourages the development of methods to optimize approximations to it.

An early scheme, based on the work of the experimental community (Schommers, 1983; Soper, 1996), called iterative Boltzmann inversion (IBI), proceeds as follows (Reith et al., 2003). Consider a CG model similar to the one defined before: a set of beads (molecules) centred at R_J, each of which consists of atoms whose coordinates are r_j. Suppose that the pair distribution function $g(R) = g(|R_J - R_K|)$ has been calculated from an atomistic simulation. Then, we expect

$$v_0(R) = -k_B T \ln g(R) \tag{12.20}$$

to be a 'zeroth' approximation to the CG pair potential. For anything other than an extremely *dilute* system, performing a simulation with this pair potential will produce a pair distribution function $g_0(R)$ which differs from $g(R)$. This is used as the start of an iterative scheme, based on successive simulations of the CG system, and refinements of the CG potential, according to the equation

$$v_{\kappa+1}(R) = v_\kappa(R) - k_B T \ln\left(\frac{g(R)}{g_\kappa(R)}\right). \tag{12.21}$$

When $g_\kappa(R) > g(R)$, the correction term will be positive, increasing the pair potential at that separation in the next iteration, and (hopefully) making $g_{\kappa+1}(R) < g_\kappa(R)$. When $g_\kappa(R) < g(R)$, the opposite effect should be seen. Often the method is applied to polymeric systems, for which a CG version of some intramolecular degrees of freedom may be needed. A simple approach is to make the approximation that the separate distribution functions (of intramolecular bond-stretching coordinates, or bend and twist angles) are independent of each other, and that the CG potential may be represented as a sum of independent contributions. One might hope to apply a formula similar to eqn (12.21) separately to each part. In fact, early attempts to model polymers (Tschöp et al., 1998) used direct Boltzmann inversion formulae resembling eqn (12.20), without further refinement, to obtain the intramolecular potentials from the corresponding distribution functions. For example for a torsion angle, $v(\phi) \approx -k_BT \ln P(\phi)$, and for a bend angle $v(\theta) \approx -k_BT \ln P(\theta)/\sin\theta$, where we highlight the typical (Jacobian) scaling factor needed to convert a raw histogram of probabilities into a probability density. Non-bonded effective interactions are typically treated in a more sophisticated way. However, in practice, the assumption of independent probability distributions is often found to be inaccurate, so cross-correlations may invalidate, or reduce the efficiency of, the approach.

The so-called inverse Monte Carlo (IMC) or inverse Newton methods were proposed by Lyubartsev and Laaksonen (1995) and Lyubartsev et al. (2010). In the IMC method, the CG potential is expressed as a linear combination of terms

$$\mathcal{V}(\mathbf{R}) = \sum_k C_k \mathcal{V}^{(k)}(\mathbf{R}) \tag{12.22}$$

where the $\mathcal{V}^{(k)}(\mathbf{R})$ constitute a basis set in the space of potentials, and the coefficients C_k determine a particular choice of potential. As in the previous section, we may regard the $\mathbf{R}$ as a subset of the full set of atomic coordinates, or as a suitable linear combination such as molecular centres of mass. The idea, as in IBI, is to match structural quantities which are functions of the $\mathbf{R}$, as measured in atomistic and CG simulations. To assist this process, averages of the $\mathcal{V}^{(k)}$ terms are calculated, together with *derivatives* of these quantities with respect to the coefficients $\{C_k\}$. Near a particular set of values $\{C_k\}$, the change in an average $\langle \mathcal{V}^{(k)} \rangle$ resulting from changes $\{\Delta C_k\}$ is given by

$$\Delta\langle \mathcal{V}^{(k)} \rangle = \langle \mathcal{V}^{(k)} \rangle_R - \langle \mathcal{V}^{(k)} \rangle = \sum_\ell \frac{\partial \langle \mathcal{V}^{(k)} \rangle}{\partial C_\ell} \Delta C_\ell + \text{higher-order terms} \tag{12.23a}$$

with the standard fluctuation expression for the derivative

$$\frac{\partial \langle \mathcal{V}^{(k)} \rangle}{\partial C_\ell} = -\beta \left(\langle \mathcal{V}^{(k)} \mathcal{V}^{(\ell)} \rangle_R - \langle \mathcal{V}^{(k)} \rangle_R \langle \mathcal{V}^{(\ell)} \rangle_R \right). \tag{12.23b}$$

Using an initial guess $\{C_k\}$, a CG simulation is carried out, measuring the values of $\langle \mathcal{V}^{(k)} \rangle_R$ and $\langle \mathcal{V}^{(k)} \mathcal{V}^{(\ell)} \rangle_R$. The differences between the values calculated in the CG simulations, $\langle \mathcal{V}^{(k)} \rangle_R$, and those obtained in the atomistic simulations, $\langle \mathcal{V}^{(k)} \rangle$, are used in eqn (12.23a) to calculate $\Delta\langle \mathcal{V}^{(k)} \rangle$. Given the measured fluctuations, and hence the derivatives $\partial \langle \mathcal{V}^{(k)} \rangle / \partial C_\ell$, this equation may be solved for the changes ΔC_ℓ which should, to lowest order, bring the CG measurements into line with the atomistic ones. A new CG simulation is carried out with the new coefficients, and the process continued to convergence. The method is similar to the solution of a system of algebraic equations by the Newton–Raphson method. Lyubartsev and Laaksonen (1995) point out that a tabulated CG potential can be regarded as a particular case of eqn (12.22), in which case the quantities being matched are essentially the pair distribution functions. In a later paper, Lyubartsev et al. (2010) generalize these equations to the case when the properties to be matched are not the same as the terms in the potential energy, and they call this the inverse Newton method.

12.7.3   Top-down methods

The coarse-graining methods discussed in the previous sections adopt a 'bottom-up' strategy: the interaction potential is essentially based on a microscopic model, then a computer simulation is used to predict the equation of state, and other macroscopic properties. This information can then be used to refine the potential, in an iterative way. In the chemical engineering community, an alternative approach has been gaining popularity: the potential parameters are deduced directly from the equation of state, which has been fitted to available experimental data. This is, in practice, much faster than the bottom-up method, and may produce force fields that apply over a wide range of state points. The approach relies on an accurate theory connecting the equation of state with the potential, valid for particular kinds of molecular model. A very widespread theory of this kind originates in a series of papers due to Wertheim (1984a,b; 1986a,b) for fluids of atoms having a repulsive core plus one or more short-range attractive sites giving rise to directional ordering. This theory was re-expressed in a form more suitable for engineering applications by Jackson et al. (1988) and Chapman et al. (1988; 1989; 1990), resulting in the general approach known as statistical associating fluid theory (SAFT).

Typically, each molecule is taken to be built from monomeric units, for example hard spheres. The starting point is an expression for the excess Helmholtz free energy of the monomer system, A^{mono}, representing the excluded volume and any dispersion effects: the Carnahan and Starling (1969) equation would be suitable for hard spheres, and perturbation expressions based on it would apply to other potentials (see e.g. Hansen and McDonald, 2013). Association into chains, through bonding between infinitely strongly attractive sites on the surface of the monomers, is handled by the first-order perturbation approach of Wertheim: this results in an additive term A^{chain} which depends on an estimate of the cavity distribution function $y^{\mathrm{mono}}(\ell)$ between monomers at bond length ℓ. An essential feature is that the attraction sites only act between pairs of monomers, and this restricts their interaction range. The same theory is applied to any association sites on the monomers, acting between chains: these attractions are not infinitely strong, and can be thought of as short-range square-well potentials; they must, again, only act between pairs. The resulting free energy A^{assoc} depends on the fraction of these sites that are bonded. All the aforementioned free energy terms add together and can lead to an accurate equation of state for liquids and liquid mixtures consisting of small molecules, or chains, formed from tangent spheres, which may optionally associate together. This simple approach is less successful when the monomers overlap with each other (i.e. ℓ is significantly smaller than the diameter), and no account is taken of bond bending potentials or intramolecular attraction within the chains. The reviews of Müller and Gubbins (2001),

Economou (2002), and McCabe and Galindo (2010) give many details of the method, and some of the measures that may be taken to address these shortcomings. The monomers need not be hard spheres: they may be modelled using, for example, Lennard-Jones potentials, square wells of variable range, or Yukawa potentials. The aforementioned reviews describe the application of SAFT to polar and nonpolar liquids, ionic liquids, inhomogeneous systems, and even liquid crystals and solid phases.

Müller and Jackson (2014) describe the rationale behind using this theoretical framework to construct a coarse-grained force field, already fitted to the experimental equation of state, for use in simulations. They concentrate in particular on a version of the theory based on the Mie $n-m$ potential, eqn (1.34), for the monomers. This conveniently allows the adjustment of the potential softness and range (through the exponents n and m) as well as the core diameter σ and well depth ϵ. Performing simulations with this potential allows one to predict structural and dynamical properties, and interfacial properties, that are not directly accessible through SAFT itself. In Section 1.3.4 we already mentioned a CG single-bead model of water using the Mie 8–6 potential, which is quite successful in the modelling of aqueous mixtures. When it comes to modelling hydrocarbon chains, several CH_2 units may be represented as a single monomer. Müller and Jackson (2014) point out that it may be necessary to add an intramolecular bond-bending potential to the SAFT-derived model, to ensure a physically reasonable chain rigidity, as this feature is missing from the theory. The parameters in this part of the force field must be derived by conventional methods. Lobanova et al. (2016) give an example involving alkane–H_2O–CO_2 ternary mixtures. An encouraging feature of the SAFT approach is that the parameters describing different groups in a molecule are, to first order at least, transferable. Therefore, large molecules may be built up from monomers representing different functional groups, whose parameters have been determined from the equations of state of smaller molecules. Further details may be found in Müller and Jackson (2014).

12.7.4   Comments

As will be clear from the foregoing sections, there are strong incentives to coarse grain the interactions in molecular simulation, both from the viewpoint of improving efficiency, and also to obtain some insight into the basic scientific phenomena, by simplifying the description and dropping the (hopefully less important) details. Although coarse graining is not completely routine, various packages have been provided to help. Several coarse-graining schemes have been incorporated into the VOTCA suite (Rühle et al., 2009). The IBI method has been implemented in the software package IBISCO (Karimi-Varzaneh et al., 2011). Both the IBI and IMC methods have been combined in the software package MAGIC (Mirzoev and Lyubartsev, 2013).

A key drawback of coarse graining is the impossibility of reproducing all the properties of a system at once. From the earliest force-matching studies it was recognized that the CG simulation pressure could not be made to match the CG value (or, as a consequence, the system density would be incorrect) due to missing intramolecular terms in the virial expression. This problem would typically be tackled by adding a constraint. Actually, this problem comes in two flavours: *representability* (lack of self-consistency in properties for a given system and state point) and *transferability* (inability to use the same CG model in

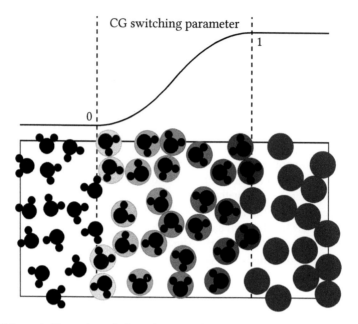

Fig. 12.3 Schematic illustration of AdRess (Praprotnik et al., 2005; 2008). In the interaction region (centre), a parameter smoothly switches the interactions between an atomistic model (on the left) and a coarse-grained model (on the right).

different circumstances or at different state points). This has been illustrated for a range of exemplary systems (Louis, 2002; Johnson et al., 2007).

When it comes to dynamical properties, a CG model may be highly inaccurate. The softer, smoother, interaction potentials that result from coarse graining may result in faster dynamics (higher diffusion coefficients, and lower viscosities) than in the atomistic case. This is actually beneficial, in that the exploration of configuration space is accelerated. However, the prediction of dynamical properties is problematic. One approach is to establish a rough scaling factor for the timescales by comparing diffusion coefficients, and then assume that other properties scale in a similar way. Another is to add frictional and random forces in the way described in Section 12.2, adjusting their strength so as to match some of the transport coefficients.

Finally, we should mention the attempts that have been made to simulate systems with different degrees of coarse graining, in contact with each other through an interface. The motivation here is to concentrate the most detailed simulation effort on the important regions, perhaps the active area of a biomolecule with its surrounding water molecules, while treating the surroundings, such as the more remote solvent, in an approximate way. Such ideas have been commonplace in specific contexts for many years. The quantum mechanics/molecular mechanics (QM/MM) approach (Warshel and Levitt, 1976) treats part of the system using quantum mechanics, and the rest by classical molecular mechanics (see Section 13.3); a celebrated simulation of crack propagation combined a continuum description of the solid far from the crack with an atomistic model (Broughton et al., 1999). Also, we mentioned in Section 12.6 the combination of a lattice-Boltzmann solvent with a particle description of a solute. The implementation of schemes like this is simplified when there is little transfer of matter between the different length scales. Liquid state systems, however, require some consideration of the fluxes of density, momentum, and energy, across the respective boundaries. Matching these fluxes, as well as the bulk thermodynamic properties of the different components, is an almost insoluble problem.

The adaptive resolution scheme (AdRess) (Praprotnik et al., 2005; 2008) compromises by switching smoothly (as a function of position) between the different force fields (atomistic and CG) across an interaction region, and applying an additional 'thermodynamic' force (Poblete et al., 2010; Fritsch et al., 2012) which is determined iteratively. This is illustrated schematically in Fig. 12.3. A disadvantage is that the method lacks a properly defined Hamiltonian; therefore the equilibrium distribution function is not known and it is not possible, for example, to construct a Monte Carlo algorithm for the same system. More recently the problem has been revisited by constructing a global Hamiltonian which includes a switching function (Potestio et al., 2013a); the resulting H-AdRess method conserves energy, and permits a Monte Carlo scheme to be used (Potestio et al., 2013b). However, the introduction of an extra term, to compensate for the difference in thermodynamic properties between the two sub-systems, is unavoidable.

13

Quantum simulations

13.1 Introduction

The dynamics of a quantum system is described by the time-dependent Schrödinger equation

$$i\hbar \frac{\partial \Phi(\mathbf{r}^{(n)}, \mathbf{R}^{(N)}; t)}{\partial t} = \mathcal{H}\Phi(\mathbf{r}^{(n)}, \mathbf{R}^{(N)}; t) \tag{13.1}$$

where Φ is the complete wave function for the nuclei and the electrons. In this chapter, we use the notation $\mathbf{r}^{(n)}$ for the complete set of positions of all of the n electrons, and $\mathbf{R}^{(N)}$ for the positions of the N nuclei, to avoid any confusion with the position $\mathbf{r}$ in space. The Hamiltonian for the system is

$$\mathcal{H} = -\sum_{I=1}^{N} \frac{\hbar^2}{2M_I} \nabla_I^2 - \sum_{i=1}^{n} \frac{\hbar^2}{2m_e} \nabla_i^2 + \mathcal{V}_{\text{n-e}}\left(\mathbf{r}^{(n)}, \mathbf{R}^{(N)}\right)$$

$$= -\sum_{I} \frac{\hbar^2}{2M_I} \nabla_I^2 + \mathcal{H}_e\left(\mathbf{r}^{(n)}, \mathbf{R}^{(N)}\right) \tag{13.2}$$

where $\mathcal{V}_{\text{n-e}}$ is the sum of all the Coulombic interactions (nuclei–nuclei, electrons–electrons, and nuclei–electrons). m_e is the mass of the electron, M_I the mass of nucleus I, and $\mathcal{H}_e$ is the Hamiltonian for the electronic sub-system.

In Section 13.2, we consider approximations to the solution of eqn (13.1) using the *ab initio* molecular dynamics method. In this case the total wave function, Φ, is factored into a part depending on the electronic degrees of freedom, Ψ, and a part corresponding to the nuclei, χ. Ψ is calculated for the clamped positions of the nuclei using a static electronic structure calculation. The corresponding forces on the nuclei from the electrons can be calculated and the equations of motion of the nuclei are then solved classically. In this case it is the electrons associated with the classical nuclei that are being treated quantum mechanically.

In Section 13.3, we introduce the QM/MM approach. In these methods, a small part of the system must be studied using a quantum mechanical technique (e.g. *ab initio* molecular dynamics) while the remainder of the system can be adequately simulated using a classical approach (e.g. classical molecular dynamics). The interesting problem is how to couple these two regions consistently.

An alternative approach to quantum mechanical systems is through the non-normalized quantum-mechanical density operator

$$\varrho = \exp(-\beta\mathcal{H}) = 1 - \beta\mathcal{H} + \frac{\beta^2}{2}\mathcal{H}\mathcal{H} + \dots \tag{13.3}$$

which satisfies the Bloch equation

$$\partial\varrho/\partial\beta = -\mathcal{H}\varrho. \tag{13.4}$$

In the coordinate representation of quantum mechanics, we may define the density matrix as

$$\varrho\left(\mathbf{R}^{(N)}, \mathbf{R}'^{(N)}; \beta\right) = \left\langle\mathbf{R}^{(N)}\left|\varrho\right|\mathbf{R}'^{(N)}\right\rangle = \left\langle\mathbf{R}^{(N)}\right| \exp(-\beta\mathcal{H}) \left|\mathbf{R}'^{(N)}\right\rangle \tag{13.5}$$
$$= \sum_s \chi_s(\mathbf{R}^{(N)})\, \varrho\, \chi_s^*(\mathbf{R}'^{(N)})$$

where s is a quantum state of the system with a nuclear wave function, χ_s, and we assume the electrons associated with each nucleus remain in their ground state.

A formal solution of the Schrödinger equation, eqn (13.1), suggests that, for time-independent Hamiltonians, the quantum-mechanical propagator from time 0 to time t is

$$U(t) = \exp(\mathcal{H}t/i\hbar). \tag{13.6}$$

This propagator converts $\Phi(0)$ to $\Phi(t)$. Thus we see an analogy between the propagator of eqn (13.6) and the definition of ϱ, eqn (13.3), or similarly between eqn (13.1) and the Bloch equation (13.4). This isomorphism is achieved with the transformation $\beta \rightarrow it/\hbar$. Some of the techniques discussed in this chapter use high-temperature or short-time approximations to the quantum-mechanical density matrix or propagator. Nonetheless, these techniques are often useful in low-temperature simulations where the $\hbar^2$-expansions might fail.

One successful approach treating the nuclei quantum mechanically has been to use the path-integral formulation of quantum mechanics (Feynman and Hibbs, 1965). By taking the trace of $\varrho(\mathbf{R}^{(N)}, \mathbf{R}'^{(N)}; \beta)$, that is, by setting $\mathbf{R}^{(N)} = \mathbf{R}'^{(N)}$ and then integrating over $\mathbf{R}^{(N)}$, we obtain the quantum partition function.

$$Q_{NVT} = \int d\mathbf{R}^{(N)} \left\langle\mathbf{R}^{(N)}\right| \exp(-\beta\mathcal{H}) \left|\mathbf{R}^{(N)}\right\rangle = \int d\mathbf{R}^{(N)} \varrho\left(\mathbf{R}^{(N)}, \mathbf{R}^{(N)}; \beta\right). \tag{13.7}$$

Using this relationship, thermodynamic properties, static structural properties, and dynamic properties may, in some circumstances, be estimated by the temperature–time analogy already discussed. In Section 13.4 we describe a simulation algorithm which has arisen directly out of this formalism. This approach is often used as a semiclassical finite-temperature technique, which will provide a measure of improvement over quantum-corrected classical results (e.g. for liquid neon). In this section, we concentrate on the application of path-integral methods to the nuclei.

Path-integral techniques can also be applied in the strongly quantum-mechanical low-temperature limit (e.g. for liquid helium, Morales et al., 2014), but special techniques have also been developed for these problems. As an example we consider a random-walk estimation of the electronic ground state in Section 13.5.

13.2   *Ab-initio* molecular dynamics

13.2.1   Approximate quantum dynamics

We consider the time evolution of a quantum mechanical systems of nuclei and electrons. At a particular time t, the eigenfunctions of the electrons can be obtained from the solution of the time-independent electronic Schrödinger equation

$$\mathcal{H}_e\left(\mathbf{r}^{(n)}, \mathbf{R}^{(N)}\right)\Psi_k\left(\mathbf{r}^{(n)}, \mathbf{R}^{(N)}\right) = E_k\left(\mathbf{R}^{(N)}\right)\Psi_k\left(\mathbf{r}^{(n)}, \mathbf{R}^{(N)}\right) \qquad (13.8)$$

for a fixed configuration $\mathbf{R}^{(N)}$ of the nuclei. $\mathcal{H}_e$ is defined in eqn (13.2) and the eigenfunctions Ψ_k are orthonormal. The eigenvalues E_k are the energies of the electronic sub-system in the fixed field of the nuclei (the so-called adiabatic energies).

Using the Born–Huang ansatz (Kutzelnigg, 1997), we can expand the total wave function

$$\Phi\left(\mathbf{r}^{(n)}, \mathbf{R}^{(N)}; t\right) = \sum_{\ell=0}^{\infty} \Psi_\ell\left(\mathbf{r}^{(n)}, \mathbf{R}^{(N)}\right)\chi_\ell\left(\mathbf{R}^{(N)}; t\right), \qquad (13.9)$$

where the functions χ_ℓ are time-dependent coefficients in the expansion. Substitution of eqn (13.9) into eqn (13.1), followed by an integration over $\mathbf{r}^{(n)}$ and a neglect of the non-adiabatic coupling operators leads to the Born–Oppenheimer (BO) approximation

$$\left[-\sum_I \frac{\hbar^2}{2M_I}\nabla_I^2 + E_k\left(\mathbf{R}^{(N)}\right)\right]\chi_k\left(\mathbf{R}^{(N)}; t\right) = i\hbar\frac{\partial \chi_k\left(\mathbf{R}^{(N)}; t\right)}{\partial t}, \qquad (13.10)$$

where we can now identify χ as the set of nuclear wave functions for a selected electronic state k. Eqn (13.10) is the quantum-mechanical equation of motion of the nuclei in the BO approximation. The corresponding equation for the electronic degrees of freedom is

$$\mathcal{H}_e\Psi\left(\mathbf{r}^{(n)}, \mathbf{R}^{(N)}; t\right) = i\hbar\frac{\partial \Psi\left(\mathbf{r}^{(n)}, \mathbf{R}^{(N)}; t\right)}{\partial t}, \qquad (13.11)$$

where Ψ, the electronic wave function, can be expressed in the basis of the electronic states

$$\Psi\left(\mathbf{r}^{(n)}, \mathbf{R}^{(N)}; t\right) = \sum_{\ell=0}^{\infty} c_\ell(t)\Psi_\ell\left(\mathbf{r}^{(n)}, \mathbf{R}^{(N)}; t\right), \qquad (13.12)$$

and the coefficients, $c_\ell(t)$ are the occupation numbers of the states.

Equations (13.10) and (13.11) can be solved, in principle, to give the quantum dynamics of the nuclei and electrons each moving in a time-dependent effective potential defined by the other part of the system. However, even for modest-sized condensed phase problems, this is not a pragmatic approach, and there are three different ways in which the problem can be simplified to make it more tractable. All of these can be termed *ab initio* molecular dynamics (AIMD). The first of these, Ehrenfest dynamics (Marx and Hutter, 2012), solves the quantum mechanical motion of the electrons at every step using eqn (13.11) and then uses the wave function to compute the force on each nucleus I, propagating the nuclei forward in time using classical mechanics:

$$M_I \ddot{\mathbf{R}}_I = -\nabla_I \int d\mathbf{r}^{(n)} \, \Psi^* \mathcal{H}_e \Psi = -\nabla_I \langle \mathcal{H}_e \rangle. \tag{13.13}$$

Ehrenfest dynamics is a mean-field approach that includes non-adiabatic transitions between Ψ_k and Ψ_ℓ. One particular advantage of the Ehrenfest approach is that during the motion the wave functions, Ψ_k, remain normalized and orthogonal to one another and this orthonormality does not need to be imposed in the dynamics using constraints.

In the second approach, known as Born–Oppenheimer (BO) dynamics, the electronic wave function Ψ is restricted to the ground-state adiabatic wave function, Ψ_0 at all times, corresponding to a single term in eqn (13.12). The equations of motion are

$$M_I \ddot{\mathbf{R}}_I(t) = -\nabla_I \min_{\Psi_0} \left\{ \left\langle \Psi_0 \middle| \mathcal{H}_e \middle| \Psi_0 \right\rangle \right\}$$

$$E_0 \Psi_0 = \mathcal{H}_e \Psi_0, \tag{13.14}$$

where the nuclei move classically on the BO potential-energy surface and E_0 is obtained from the solution of the time-independent Schrödinger equation for the ground state. In contrast to Ehrenfest dynamics, the ground state has to be established at each step in the dynamics by diagonalizing the Hamiltonian, $\mathcal{H}_e$.

In the third approach, developed by Car and Parrinello (1985), $\langle \Psi_0 | \mathcal{H}_e | \Psi_0 \rangle$ is considered to be a functional of the set of orbitals, $\{\psi_i\}$, that form the basis of the wave function. It is now possible to construct a Lagrangian in which the functional derivative with respect to these orbitals leads to the force that will drive the orbitals forward in time. This Lagrangian can be represented as

$$\mathcal{L} = \sum_I \tfrac{1}{2} M_I \dot{\mathbf{R}}_I^2 + \sum_i \mu \left\langle \dot{\psi}_i \middle| \dot{\psi}_i \right\rangle - \left\langle \Psi_0 \middle| \mathcal{H}_e \middle| \Psi_0 \right\rangle + \{\text{constraints}\}, \tag{13.15}$$

where μ is a fictitious mass controlling the dynamics of the orbitals and we have included a set of constraints in the Lagrangian that will keep the underlying orbitals orthogonal and normalized during the dynamics. The equations of motion resulting from $\mathcal{L}$ are

$$M_I \ddot{\mathbf{R}}_I(t) = -\nabla_I \left\langle \Psi_0 \middle| \mathcal{H}_e \middle| \Psi_0 \right\rangle + \nabla_I \{\text{constraints}\},$$

$$\mu \ddot{\psi}_i(t) = -\frac{\delta \langle \Psi_0 | \mathcal{H}_e | \Psi_0 \rangle}{\delta \psi_i^*} + \frac{\delta}{\delta \psi_i^*} \{\text{constraints}\}, \tag{13.16}$$

where $\delta/\delta\psi_i^*$ indicates a functional derivative. (Note that in taking the functional derivative of the orbital kinetic energy, $\dot{\psi}_i^*$ and $\dot{\psi}_i$ are treated as independent functions.) For both BO and CP dynamics the force on the nuclei can, in many cases, be evaluated efficiently using the Hellmann–Feynman theorem

$$-\nabla_I \left\langle \Psi_0 \middle| \mathcal{H}_e \middle| \Psi_0 \right\rangle \approx -\left\langle \Psi_0 \middle| \nabla_I \mathcal{H}_e \middle| \Psi_0 \right\rangle. \tag{13.17}$$

In these equations of motion there is a real temperature associated with the kinetic energy of the nuclei and a fictitious temperature associated with the electronic degrees of freedom arising from the kinetic energy, $\sum_i \mu \langle \dot{\psi}_i | \dot{\psi}_i \rangle$. If this electronic temperature is kept low, with a suitable choice of μ then the electronic degrees of freedom will evolve on the BO surface and we should recover the Born–Oppenheimer dynamics of eqn (13.14). We will return to the expressions for the constraint terms in eqn (13.16) when we have considered the details of the electronic wave function.

13.2.2 Density functional theory and the Kohn–Sham method

The use of the Born–Oppenheimer or Car–Parrinello dynamics requires a knowledge of the electronic wave function for the ground state of the system. Since the electrons are fermions, Ψ_0 must be antisymmetric to the exchange of two electrons. This symmetry can be included by writing the full wave function as a Slater determinant of the individual wave functions of each of the n electrons

$$\Psi_0 \left(\mathbf{r}^{(n)} \right) = \frac{1}{\sqrt{n!}} \begin{vmatrix} \psi_1(\mathbf{r}_1) & \psi_2(\mathbf{r}_1) & \cdots & \psi_n(\mathbf{r}_1) \\ \psi_1(\mathbf{r}_2) & \psi_2(\mathbf{r}_2) & \cdots & \psi_n(\mathbf{r}_2) \\ \vdots & \vdots & \ddots & \vdots \\ \psi_1(\mathbf{r}_n) & \psi_2(\mathbf{r}_n) & \cdots & \psi_n(\mathbf{r}_n) \end{vmatrix}. \tag{13.18}$$

Traditionally, Ψ_0 is then obtained using the Hartree–Fock (HF) approximation (Schaefer, 1972). This approach scales as $O(n^4)$ and more sophisticated version of the theory such as the configuration-interaction methods are significantly more expensive, with the MP4 method scaling as $O(n^7)$.

The breakthrough in electronic structure calculations for condensed phase systems has been the development of density functional theory (DFT) (Hohenberg and Kohn, 1964). In DFT, the expectation value

$$F_{\mathrm{HK}}[\rho_0] = \left\langle \Psi_0[\rho_0] \middle| \mathcal{H}'_e \middle| \Psi_0[\rho_0] \right\rangle \tag{13.19}$$

defines a functional, F_{HK}, of the ground-state electron density, $\rho_0(\mathbf{r})$, that is completely independent of the environment of the electrons and only depends on the spatial coordinate, $\mathbf{r}$. (Note that the electron density, ρ, should not be confused with the density operator, ϱ, used in other sections of this chapter.)

In eqn (13.19)

$$\mathcal{H}'_e = \mathcal{K}[\rho] + \mathcal{V}_{ee}[\rho] \tag{13.20}$$

where $\mathcal{K}$ is the quantum mechanical kinetic-energy operator of the electrons and $\mathcal{V}_{ee}$ is the potential energy between the electrons. The corresponding energy functional,

$$E[\rho] = F_{\mathrm{HK}}[\rho] + \int d\mathbf{r} \, \mathcal{V}_{ext}(\mathbf{r}) \, \rho \left(\mathbf{r}, \mathbf{R}^{(N)} \right), \tag{13.21}$$

depends additionally on the external field due to the positively charged nuclei acting on the electrons. The first Hohenberg–Kohn theorem states that the full many-particle ground state is a unique functional of $\rho(\mathbf{r})$. The second H–K theorem states that the energy functional, $E[\rho]$, is a minimum when $\rho = \rho_0$, the true ground-state density.

The next step is to find the universal functional F_{HK}. The approach of Kohn and Sham (1965) is to replace the system of n interacting electrons by an auxiliary system of n

independent electrons. We denote this non-interacting system by the subscript s, with the total energy $E_s = \mathcal{K}_s + \int d\mathbf{r} \, \mathcal{V}_s \rho_s$. There will be a particular, local potential, $\mathcal{V}_s(\mathbf{r})$, such that the exact ground-state density of the interacting system, ρ_0, is equal to ρ_s. If the ground state of the non-interacting systems is singly degenerate then

$$\rho_0(\mathbf{r}) = \sum_{i=1}^{\text{occ}} f_i |\psi_i(\mathbf{r})|^2 \tag{13.22}$$

where the sum is over the occupied orbitals, f_i is the occupancy number of the orbital (0, 1, or 2), and the single-particle orbitals are obtained from the Schrödinger equation

$$\mathcal{H}_{KS}\psi_i = \left[-\frac{\hbar^2}{2m_e}\nabla^2 + \mathcal{V}_s \right]\psi_i = \epsilon_i \psi_i. \tag{13.23}$$

Thus the total energy of the interacting system is

$$E_{KS}[\rho_0] = \mathcal{K}_s[\rho_0] + \int d\mathbf{r} \, \mathcal{V}_{\text{ext}}(\mathbf{r})\rho_0(\mathbf{r}) + \frac{1}{2}\iint d\mathbf{r} \, d\mathbf{r}' \frac{\rho_0(\mathbf{r})\,\rho_0(\mathbf{r}')}{|\mathbf{r} - \mathbf{r}'|} + E_{xc}[\rho_0]. \tag{13.24}$$

The energy contains four terms: the first is the kinetic energy for the non-interacting system,

$$\mathcal{K}_s[\rho_0] = -\frac{\hbar^2}{2m_e}\sum_{i=1}^{\text{occ}} f_i \langle \psi_i | \nabla^2 | \psi_i \rangle; \tag{13.25}$$

the second is the energy of the electrons in the static field of the nuclei ($\mathcal{V}_{\text{ext}}$ also includes the interaction between the nuclei themselves, $\sum_{I>J} q_I q_J / |\mathbf{R}_I - \mathbf{R}_J|$); the third is the classical Coulomb energy between the electrons; and the fourth is the exchange–correlation energy. Note that for simplicity, we continue to set $4\pi\epsilon_0 = 1$ in the Coulomb terms. E_{xc} is simply the sum of the error made in using a non-interacting kinetic energy and the error made in treating the electron–electron interaction classically. From eqn (13.24), the single-particle potential is given by

$$\mathcal{V}_s(\mathbf{r}) = \mathcal{V}_{\text{ext}}(\mathbf{r}) + \int d\mathbf{r}' \frac{\rho_0(\mathbf{r}')}{|\mathbf{r} - \mathbf{r}'|} + \frac{\delta E_{xc}}{\delta \rho(\mathbf{r})}. \tag{13.26}$$

This approach would be exact if we knew E_{xc} exactly. However, this is not the case, and to proceed we will need a reasonable approximation for the exchange–correlation energy. In the first instance, it is often represented using the local density approximation (LDA),

$$E_{xc}[\rho_0(\mathbf{r})] \approx \int d\mathbf{r} \, \rho_0(\mathbf{r})\varepsilon_{xc}[\rho_0(\mathbf{r})] \tag{13.27}$$

where $\varepsilon_{xc}[\rho_0(\mathbf{r})]$ is the exchange–correlation functional for a uniform electron gas, per electron, at a particular density, $\rho_0(\mathbf{r})$. $\varepsilon_{xc}[\rho_0(\mathbf{r})]$ is the sum of an exchange term, which is known exactly for the homogeneous electron gas, and a correlation term that can be accurately calculated by quantum Monte Carlo (Ceperley and Alder, 1980), and readily parameterized (Perdew and Zunger, 1981). The LDA approach is surprisingly accurate and this is probably due to a cancellation of errors in the exchange and correlation terms (Harrison, 2003).

A significant improvement in E_{xc} can be obtained by using the generalized gradient approximations (GGA). These approximations are still local but take into account the gradient of the density at the same coordinate (Harrison, 2003). The typical form for a GGA functional is;

$$E_{xc}^{GGA}[\rho_0(\mathbf{r})] \approx \int d\mathbf{r} \, f_{xc}[\rho_0(\mathbf{r}), \nabla\rho_0(\mathbf{r})].$$ (13.28)

Considerable ingenuity has produced accurate, locally based functionals, f_{xc}, such as the PW91 functional of Perdew et al. (1992) which contains an accurate description of the Fermi holes resulting from the exclusion principle. There are also a number of important semi-empirical approaches; the most widely used of these, Becke–Lee–Yang–Parr (BLYP), uses the exchange functional of Becke (1988) combined with the correlation functional of Lee et al. (1988). There is also an important class of hybrid functionals, such as Becke, 3-parameter, Lee–Yang–Parr (B3LYP), in which the Becke exchange functional is mixed with the energy from Hartree–Fock theory. These hybrid functionals have been implemented in MD calculations (Gaiduk et al., 2014), but are expensive to employ. The development of the simple GGA-based functionals has allowed the application of DFT to real problems in chemical bonding and they are now widely used in *ab initio* MD methods.

Now all the elements of eqn (13.26) are in place, a simple approach to calculating the ground-state energy for a given nuclear configuration would be as follows:

(a) starting from a guess for $\rho_0(\mathbf{r})$, calculate the single particle potential, $\mathcal{V}_s(\mathbf{r})$, using eqn (13.26);

(b) solve the single orbital Schrödinger equations, eqn (13.23), by diagonalizing the Kohn–Sham (KS) Hamiltonian;

(c) calculate a new estimate for the ground-state density from eqn (13.22);

(d) iterate until convergence and calculate the ground-state energy from eqn (13.24).

The time for this calculation is determined by the diagonalization step which scales as $O(n^3)$.

Before leaving this section, we must address two further, important, considerations: representation of the single-particle orbitals, ψ_i, in the KS approach, and the use of pseudopotentials to represent the tightly bound electrons. The single-electron orbitals are often expanded in Gaussians or plane waves that fit neatly into the periodic boundary conditions of the simulation. For plane waves

$$\psi_i^g(\mathbf{r}) = \exp(i\mathbf{g} \cdot \mathbf{r}) \sum_{\mathbf{k}} c_i^g(\mathbf{k}) \exp(i\mathbf{k} \cdot \mathbf{r})$$ (13.29)

where $\mathbf{k} = 2\pi\mathbf{n}/L$ is a reciprocal lattice of the MD cell, the wavevector $\mathbf{g}$ is in the first Brillouin zone of the reciprocal lattice of the MD cell, and the coefficient c_i^g is a complex number. For ordered systems, we need to consider wave functions corresponding to many different points in the Brillouin zone, but for disordered or liquid-like systems, in a large MD cell, it is possible to work at the zone centre, $\mathbf{g} = 0$. In this case, the orbitals are simply

$$\psi_i(\mathbf{r}) = \sum_{\mathbf{k}} c_i(\mathbf{k}) \exp(i\mathbf{k} \cdot \mathbf{r})$$ (13.30)

and $c_i(-\mathbf{k}) = c_i^*(\mathbf{k})$. The KS potential converges rapidly with increasing $\mathbf{k}$ and in practical calculations it is necessary to choose a cutoff energy, E_{cut} such that

$$\frac{\hbar^2}{2m_e}\left|\mathbf{k}_{max}\right|^2 \le E_{cut}. \tag{13.31}$$

The precision of the density functional calculation is determined by the choice of E_{cut}.

There are many other orthonormal, localized functions that can be used to expand ψ_i; these include Gaussians and wavelets. At present there is considerable interest in the use of Wannier functions, the Fourier transforms of the Bloch eigenstates, as a basis set (Martin, 2008, Chapter 21). In principle, these alternative basis sets might be more efficient at describing highly inhomogeneous charge distributions. However, the plane-wave basis is versatile and accurate and is still widely used. For example, for plane waves, the kinetic-energy term in E_{KS} can be written simply in terms of the coefficients, $c_i(\mathbf{k})$,

$$\mathcal{K}_s[\rho_0] = \frac{\hbar^2}{2m_e} \sum_{i=1}^{occ} \sum_{\mathbf{k}} f_i(\mathbf{k}) \left|\mathbf{k}\right|^2 c_i(\mathbf{k}) c_i^*(\mathbf{k}) \tag{13.32}$$

and the Pulay force (the correction to the Hellmann–Feynman approximation) vanishes exactly. A Gaussian basis also provides an analytical expression for $\mathcal{K}$ and the Coulomb potential for an isolated system.

The electrons around the nucleus can be divided into two types: core electrons that are strongly bound to a particular nucleus (e.g. the 1s electrons in Cl); and the valence electrons that are polarizable and take part in the formation of new bonds (e.g. the $2p_z$ electron in Cl). In AIMD, it is efficient to concentrate on the evolution of the valence electrons and to combine the core electrons with the nucleus by defining a pseudopotential. This potential, $v_{PP}(\mathbf{r})$, is the potential acting on the valence electrons from the nucleus and the core electrons. In our calculations, it will replace the potential of the bare nuclear charge when constructing the $\mathcal{V}_{ext}$ as used in eqns (13.24) and (13.26). In the rest of this section, when we discuss the dynamics of the ions, we mean the dynamics of the positively charged nucleus and the core electrons.

A useful pseudopotential for a particular atom is calculated by comparing the solutions of the Schrödinger equation for the all-electron system and for the corresponding system with a trial pseudopotential. An accurate and transferable $v_{PP}(\mathbf{r})$ is designed so that, at large distance from the nucleus (outside a core region, $r > R_c$), the pseudo-wavefunction matches the all-electron wave function and that, in the core region, the pseudo-wavefunction produces the same charge as the all-electron wave function (i.e. it is norm-conserving). In addition, the pseudopotential is chosen to be as smooth as possible to minimize the number of plane waves required to describe it accurately.

The overall pseudopotential around a nucleus is often constructed separately for each angular momentum component, ℓ, m of the wave function

$$v_{PP}(\mathbf{r}) = \sum_{\ell=0}^{\infty} \sum_{m=-\ell}^{+\ell} v_{\ell,m}(r)\mathcal{P}_{\ell,m} = \sum_{L=0}^{\infty} v_L(r)\mathcal{P}_L, \qquad r < R_c \tag{13.33}$$

where L is a combined index $\{\ell, m\}$, $v_L(r)$ is the pseudopotential for a particular angular momentum channel and $\mathcal{P}_L = |L\rangle\langle L|$ is the projection operator for the angular momentum which picks out a particular state, L. Note that $v_L(r)$ is equal to the L-independent, all-electron potential outside the core and $v_L(r) \to -q_{ion}/r$ as $r \to \infty$, where q_{ion} is the net charge on the ion. The ionic pseudopotential can be split into a local L-independent part, which contains all the effects of the long-range Coulomb potential, and a non-local part, $\Delta v_L(r)$

$$v_{PP}(\mathbf{r}) = \sum_{L=0}^{\infty} v_{local}(r)\mathcal{P}_L + \sum_{L=0}^{\infty} \left[v_L(r) - v_{local}(r)\right]\mathcal{P}_L$$

$$= v_{local}(r) + \sum_{L=0}^{L_{max}-1} \Delta v_L(r)\mathcal{P}_L \tag{13.34}$$

where $\Delta v_L(r) = 0$ for $r > R_c$ and $L \geq L_{max}$. Normally L_{max} is set to 1 for first-row atoms and 2 or 3 for heavier atoms. $v_{local}(r)$ can be readily included in the KS energy

$$E_{local} = \int d\mathbf{r}\, v_{local}(\mathbf{r})\rho(\mathbf{r}). \tag{13.35}$$

The non-local contribution to the pseudopotential can be calculated by projecting the operator onto a local basis set, using the Kleinman–Bylander projection (Kleinman and Bylander, 1982). The energies and forces associated with the non-local part of the pseudopotential are readily evaluated in k-space. Using these approaches it is straightforward to generate pseudopotentials fitted to simple analytical forms such as a combination of polynomials and Gaussians, and examples of such calculations are available for many different atom types (Gonze et al., 1991; Goedecker et al., 1996).

This short introduction to pseudopotentials has allowed us to establish the vocabulary and principles associated with their use in *ab initio* MD. Fuller accounts of their properties can be found in Marx and Hutter (2012, Chapter 4) and Martin (2008, Chapter 11).

13.2.3   Car–Parrinello dynamics revisited

As we have seen in the last section, the electronic structure problem can be efficiently solved for a given position of the ions using DFT. The development of the Car–Parrinello (CP) method, in 1985, was a breakthrough that avoided the explicit diagonalization of the KS Hamiltonian. There are a number of excellent reviews (Remler and Madden, 1990; Galli and Parrinello, 1991; Galli and Pasquarello, 1993) and a book (Marx and Hutter, 2012) covering the CP method and we provide a short reprise here.

For a plane-wave basis set, the Lagrangian, eqn (13.15) can be written as

$$\mathcal{L} = \mu \sum_i \sum_{\mathbf{k}} \dot{c}_i^*(\mathbf{k})\dot{c}_i(\mathbf{k}) + \tfrac{1}{2}\sum_I M_I \dot{\mathbf{R}}_I^2 - E_{KS}[\rho] + \sum_{ij} \Lambda_{ij}\left(\sum_{\mathbf{k}} c_i^*(\mathbf{k})c_j(\mathbf{k}) - \delta_{ij}\right) \tag{13.36}$$

where Λ_{ij} is the Lagrange multiplier that keeps the orbitals, ψ_i and ψ_j, orthonormal. The resulting equations of motion are

$$\mu\ddot{c}_i(\mathbf{k}) = -\frac{\partial E_{KS}}{\partial c_i^*(\mathbf{k})} + \sum_j \Lambda_{ij}c_j(\mathbf{k}) \tag{13.37a}$$

$$M_I\ddot{\mathbf{R}}_I = -\boldsymbol{\nabla}_I E_{KS}. \tag{13.37b}$$

Equation (13.37a) describes the motion of the Fourier components of the plane-wave expansion in time and eqn (13.37b) the motion of the ions. These coupled equations can be solved using the Verlet or velocity Verlet algorithm and the constraints can be applied using a standard technique such as SHAKE or RATTLE (Tuckerman and Parrinello, 1994). These algorithms are discussed in Chapter 3.

A number of parameters need to be fixed at the beginning of a Car–Parrinello MD simulation, and we use a number of studies of liquid water to illustrate some of these choices. The first consideration would be the basis set used to describe the electronic structure. For example, Laasonen et al. (1993) and Zhang et al. (2011a) both used a plane-wave basis with the simplicity and advantages already discussed. In contrast, Lee and Tuckerman (2006) have employed a discrete variable representation (DVR) of the basis. This approach combines some of the localization of the density that would be typical of a Gaussian basis set with the orthogonality offered by the plane-wave functions. This choice improves energy conservation for a particular E_{cut} and grid size, but avoids problems with the Pulay forces and the basis set superposition errors associated with an atom-centred basis set. Second, the exchange interaction for the calculations needs to be chosen. The LDA is not appropriate for liquid water but Laasonen et al. (1993) used the GGA method of Becke, while Lee and Tuckerman (2006) used the BLYP. Zhang et al. (2011b) have employed the hybrid functional PBE0 with a Hartree–Fock correlation energy and a more advanced functional designed to improve the modelling of the O–O dispersion interactions. The time required for the simulations increases significantly as the exchange functional becomes more complicated and accurate, so that the simulations using more advanced functionals are often performed on as few as 32 H_2O molecules. Water simulations are typically performed at the zone centre, $g = 0$, with between 32 and 100 molecules, and take into account explicitly the four electronic states per molecule corresponding to the valence electrons of the oxygen atom. The hydrogen atoms (nucleus and electron) and the oxygen nucleus and core electrons can be represented by pseudopotentials of the type developed by Troullier and Martins (1991) and others (Vanderbilt, 1985; Goedecker et al., 1996). For H, the wave function corresponding to its pseudopotential is smooth, avoiding the kink at the nuclear position (associated with the divergence of the Coulomb potential) and it can be represented with fewer plane waves. The accuracy of the calculation is controlled by the choice of E_{cut} or the number of wavevectors representing each electronic state. E_{cut} might range from 35 Ha to 150 Ha. (Note that, in much of the literature, the cutoff energy is given in either hartree or rydberg units, where $1\,\text{Ha} = 2\,\text{Ry} = 4.3597 \times 10^{-18}\,\text{kJ mol}^{-1}$.) From eqn (13.31), the number of **k** vectors corresponding to a particular energy cutoff is given by

$$N_{PW} = \left(\frac{\sqrt{2}}{3\pi^2}\right) V (E_{cut})^{3/2} \tag{13.38}$$

where, in this equation, V is the volume corresponding to a single molecule of water, and V and E_{cut} are in atomic units ($(bohr)^3$ and hartrees respectively). For a cutoff of 35 Ha, each orbital requires 1000 wavevectors for each of the electronic states associated with each of the water molecules in the simulation. Note that this number takes into account the symmetry in the coefficients at the zone centre. The density,

$$\rho_0(\mathbf{r}) = \sum_i f_i \sum_{\mathbf{k}} \sum_{\mathbf{k}'} c_i^*(\mathbf{k}') c_i(\mathbf{k}) \exp(i(\mathbf{k} - \mathbf{k}') \cdot \mathbf{r}) \tag{13.39}$$

and the corresponding single-particle potential $\mathcal{V}_s(\mathbf{r})$ vary twice as fast as the wave function and would require 8000 wavevectors per electronic state.

In a plane-wave calculation, the real-space grid spacing is (Lee and Tuckerman, 2006)

$$\ell = \left(\frac{\pi^2}{8E_{cut}} \right)^{1/2} \tag{13.40}$$

where ℓ is in bohr and E_{cut} in hartrees. So, for example, a single water molecule in a small cubic box at a density of $997\ \mathrm{kg\,m^{-3}}$ with an orbital cutoff of 35 Ha would require $(31)^3$ grid points (the number of grid points scales linearly with the number of water molecules in the simulation). The DVR basis can use a much coarser grid for the real-space density than the plane-wave basis (Lee and Tuckerman, 2006) at the same level of accuracy as judged by the energy conservation in the simulation.

Finally, the timestep for the CP simulation used by Lee and Tuckerman (2007) is 0.05 fs (or 2.07 a.u.). This is used with a fictitious mass $\mu = 500$ a.u. (see eqn (13.37a)). The earlier work of Laasonen et al. (1993) used a timestep of 0.169 fs (6.99 a.u.). In order to achieve a separation of the ionic and electronic kinetic energies with this long timestep, they used a fictitious mass $\mu = 1100$ a.u. and doubled the mass of the protons to simulate D_2O rather than water.

The starting configuration for the CP simulation comprises the ionic positions and velocities and the initial values of all the $c_i(\mathbf{k})$ and their first time derivatives at $t = 0$. The choice of the ionic positions and velocities follows the ideas developed in Sections 5.6.1 and 5.6.2. The coefficients for the wave functions can be chosen at random. It is possible to reflect the importance of the different plane waves by sampling from a Gaussian distribution in $|\mathbf{k}|$. Alternatively one can superimpose the electron densities of the underlying atomic configuration to produce an estimate of the overall density and diagonalize the corresponding KS matrix in any reasonable basis to give the starting coefficients. Marx and Hutter (2012) suggest using the pseudo-atom densities and the pseudo-atomic wave functions, already used in the calculation of the pseudopotential, in this context. The velocities of the coefficients at $t = 0$ can be set consistently by moving forward and backward one timestep, calculating the $c_i(\mathbf{k}, -\delta t)$ and $c_i(\mathbf{k}, +\delta t)$ from $\mathbf{R}^{(N)}(-\delta t)$ and $\mathbf{R}^{(N)}(+\delta t)$ and using these coefficients to estimate $\dot{c}_i(\mathbf{k}, 0)$ (Remler and Madden, 1990).

Once the simulation parameters and the starting configuration are established the simulation itself proceeds like a classical MD calculation. The heart of a typical CP MD code is as follows. At $t = 0$ an initial guess is made for $\psi_i = \sum_{\mathbf{k}} c_i(\mathbf{k}) \exp(i\mathbf{k} \cdot \mathbf{r})$. Then, at each subsequent timestep the following operations are carried out.

1. Calculate $\rho_0(\mathbf{r})$, $\mathcal{V}_s[\rho_0(\mathbf{r})]$, $E_{KS}[\rho_0(\mathbf{r})]$, $\nabla_I E_{KS}$.
2. For each electronic state i and for each plane wave $\mathbf{k}$:
    (a) calculate $\partial E_{KS}/\partial c_i^*(\mathbf{k})$;
    (b) integrate $\mu \ddot{c}_i(\mathbf{k}) = -\partial E_{KS}/\partial c_i^*(\mathbf{k})$.
3. Orthogonalize wave functions.
4. For each ion, integrate $M_I \ddot{\mathbf{R}}_I = -\nabla_I E_{KS}$.
5. Compute averages.

At a particular timestep, the $c_i(\mathbf{k})$ are used to calculate $\hat{\rho}_0(\mathbf{k})$, which is transformed to produce $\rho_0(\mathbf{r})$ at every point on a real-space lattice and hence E_{KS} (eqn (13.24)). The force on the ions is calculated from the derivatives of the energy with respect to the ionic positions. This force is determined by the derivatives of the local and non-local pseudopotentials and the electrostatic energy with respect to R_I; the kinetic energy and exchange energy make no contribution. We recall that the second and third terms in E_{KS}, eqn (13.24), are long-ranged and the energies and forces resulting from these electrostatic interactions are calculated using the Ewald summation method, or one of the other techniques described in Chapter 6.

Looping over all of the electronic states i and wavevectors $\mathbf{k}$, the force on the wave function coefficients, $-\partial E_{KS}/\partial c_i^*(\mathbf{k})$, is calculated from all the terms in eqn (13.24). For example, for a plane-wave basis, the force from the kinetic energy,

$$\frac{\partial \mathcal{K}}{\partial c_i^*(\mathbf{k})} = \frac{\hbar^2 f_i}{2m_e}|\mathbf{k}|^2 c_i(\mathbf{k}) \tag{13.41}$$

is evaluated in k-space and the same is true of the electronic force associated with the non-local pseudopotential. In contrast the electronic forces involving the local pseudopotential, the Coulomb and the exchange energies are convolutions (double sums in reciprocal space). They are most efficiently evaluated by multiplying the corresponding real-space functions and then Fourier transforming to obtain the force. For the precise pathway, through real and reciprocal space to $\partial E_{KS}/\partial c_i^*(\mathbf{k})$, see Payne et al. (1989) and Galli and Pasquarello (1993, Fig. 7). The ability to move backwards and forwards between real and reciprocal space using fast Fourier transforms is at the heart of these molecular dynamics calculations and can often be achieved efficiently using standard scientific libraries tuned for a particular machine.

Once the forces on the coefficients are calculated, the coefficients are moved forwards in time without constraints using, say, the velocity Verlet algorithm, and the orthogonalization is applied using the RATTLE algorithm. Finally, the forces on the ions are used to advance the ionic positions.

The hallmark of a properly functioning MD simulation is that the energy associated with the Lagrangian, eqn (13.36) should be conserved. That is

$$E_{total} = \mu \sum_i \sum_\mathbf{k} \dot{c}_i(\mathbf{k})\dot{c}_i^*(\mathbf{k}) + \tfrac{1}{2}\sum_I M_I|\dot{\mathbf{R}}_I|^2 + E_{elec}, \tag{13.42}$$

where E_{elec} is the sum of the last three terms in eqn (13.24) including the potential between the ions. E_{total} should not drift and should remain constant to ca. ±0.000 01 Ha over the course of the simulation. Remler and Madden (1990) have pointed out, for a well-defined trajectory on the BO surface, it is

$$E_{real} = \tfrac{1}{2}\sum_I M_I|\dot{\mathbf{R}}_I|^2 + E_{elec}, \tag{13.43}$$

that should be conserved, and that this can only be achieved if the fictitious kinetic energy is several orders of magnitude smaller than the variation in E_{real}. For example, Fig. 2 of Lee and Tuckerman (2007) shows the clear separation of the kinetic energies associated with the ionic and electronic degrees of freedom and excellent overall conservation of energy over the 60 ps of a simulation of water. The separation of the characteristic frequencies of the ionic and electronic degrees of freedom prevents energy exchange between these systems and is at the root of the success of the CP method.

The CP method just described samples states in the microcanonical ensemble with E_{total} constant. It is also possible to simulate at constant NVT (Tuckerman and Parrinello, 1994) using the Nosé–Hoover chain (Martyna et al., 1992) described in Section 3.8.2, where the heat bath is coupled to the ionic degrees of freedom. Normally, one would not apply a separate thermostat to the electrons since this can disturb the correlation between the electronic and ionic motion, resulting in an additional friction on the ions with a corresponding flow of energy from the ionic system to the electrons. When the gap between the electronic and nuclear degrees of freedom is very small, separate heat baths with different chain lengths can be coupled to the ions and the electrons to ensure that the system remains on the BO surface.

CP simulations in the constant-NPT ensemble can be performed with isotropic changes in a cubic box (Ghiringhelli and Meijer, 2005), or with varying box shape (Focher et al., 1994). In a constant-NPT simulation the volume of the cell changes, and therefore the plane-wave basis ($2\pi n/L$) changes. This creates a Pulay force arising from the Hellmann–Feynman theorem. This problem can be avoided by imposing an effective energy cutoff, $E_{cut}^{eff} \ll E_{cut}$ (eqn (13.31)) and smoothly suppressing the contribution of all plane waves above E_{cut}^{eff} (Marx and Hutter, 2012, p.187).

13.2.4   Summary

The CP method is a powerful technique in the simulation of liquids and solids. For example, it can be used to calculate equilibrium and structural properties of water (Pan et al., 2013; Alfe et al., 2014) and the time correlation functions and their corresponding condensed phase spectra (Zhang et al., 2010). It can be readily extended to consider fluids in confined geometries and pores (Donadio et al., 2009) and to study mixture such as ions in aqueous solution (Kulik et al., 2012).

The development of the CP method in 1985 was the historical breakthrough in AIMD. However, once that Rubicon had been crossed, a reexamination of BO dynamics confirmed the power of this direct approach, and BO dynamics is the most widely used method at the present time. Bo dynamics, eqn (13.14), implies a full diagonalization of the KS matrix for a fixed position of the ions. This can be recast as a constrained minimization with respect to the orbitals

$$\min_{\{\psi_i\}}\left[\langle\Psi_0|\mathcal{H}^{KS}|\Psi_0\rangle\right]\Big|_{\langle\psi_i|\psi_j\rangle=\delta_{ij}} \tag{13.44}$$

which can be efficiently solved using a conjugate gradient method (Teter et al., 1989; Arias et al., 1992). The speed of these minimizations depends on the quality of the initial guess for the orbitals for a given ionic configuration $\mathbf{R}_I^{(N)}$ and this can be improved by efficiently extrapolating the electronic configuration or the density matrix forward in time (Payne et al., 1992; VandeVondele et al., 2005).

Which of these methods is most efficient in terms of the computer time required to solve a particular problem? This issue is discussed in some detail by Marx and Hutter (2012, section 2.6) and we simply highlight a number of observations from their discussion. We start from the premise that the method which allows us to use the longest timestep will allow us to calculate more accurate properties for a particular computing budget.

Example 13.1 *Ab initio* ionic liquids

A number of independent simulations of the ionic liquid, dimethyl imidazolium chloride, $[DMIM]^+Cl^-$, have been performed using *ab initio* molecular dynamics.

Del Popolo et al. (2005) performed BO dynamics using the SIESTA package, with the electronic orbitals expanded in a non-orthogonal basis set of atom-centred functions. Buhl et al. (2005) and Bhargava and Balasubramanian (2006) carried out CP dynamics in the CPMD code with a plane-wave basis set. These simulations, although all quite different in terms of system size, run length, and exchange functional, produced a consistent picture of the structure of the ionic liquid at 425 K.

Each imidazolium cation is surrounded by 6 Cl^- anions out to a distance of 0.65 nm. The neutron-weighted radial distribution functions obtained from the *ab initio* simulations are in good agreement with the corresponding experimental functions (Hardacre et al., 2003). However, the scattering, from deuterated samples, is dominated by the intramolecular contribution from the H and D atoms in the cation (Del Popolo et al., 2007), and it is difficult to extract information on the intermolecular structure from these experiments. The vibrational spectra derived from the *ab initio* simulations are also in good agreement with experiment (Bhargava and Balasubramanian, 2006).

There are interesting differences between the intermolecular structures obtained from these *ab initio* simulations and from classical molecular dynamics carried out under the same conditions, using the best force fields available for these ionic liquids (Lopes et al., 2004). The main difference occurs in the site–site radial distribution function between the Cl^- anion and the H_a atom, the unique, acidic hydrogen atom indicated in the figure. The first maximum in $g_{H_aCl}(r)$ shifts from 0.28 nm (classical) to 0.23 nm (*ab initio*) indicating the formation of a hydrogen bond between H_a and the counter ion. Additionally the classical model predicts a hole in the density of the Cl^- anions directly above and below the H_a atom. In the *ab-initio* simulations the density hole is filled. This is again indicative of H-bond formation and is observed in both the simulations of Del Popolo et al. (2005) and Bhargava and Balasubramanian (2006), and inferred from the experimental neutron scattering results (Hardacre et al., 2003). The *ab initio* simulations suggest that it is necessary to refine the classical force fields for $[DMIM]^+Cl^-$ to reproduce the hydrogen bond.

CP dynamics can be conducted with a timestep approximately one order of magnitude larger than that used with Ehrenfest dynamics (eqn (13.13)). The timestep advantages of BO dynamics over CP dynamics are more pronounced for heavier nuclei: in this case there is an approximate order of magnitude advantage in favour of the BO approach. Model studies of Si employed a timestep at least five times larger than for the CP method with only a small loss in the quality of the energy conservation. However, in the important case of water, the longer timestep that can be employed in the BO dynamics is outweighed by the number of iterations required to achieve convergence of the minimizations, and the CP dynamics is more efficient by a factor of between 2 and 4, at comparable levels of energy conservation. For water, the choice of CP over BO depends on how tightly one wants to control the energy conservation for a given expenditure of computer time. Those from a classical simulation background will probably feel most comfortable with the higher level of energy conservation provided by the CP dynamics. As one would expect both methods produce the same results within the estimated error for the static and dynamic properties of water (Kuo et al., 2006). The issue of the choice of timestep to control the energy conservation is particularly important when one wants to accurately sample in a particular ensemble or to calculate time-dependent properties. Both BO and CP dynamics are also frequently used as simulated annealing techniques to reach a minimum energy state particularly in the study of solids. In this case the BO approach offers some clear advantages. When comparing different techniques, the implementation of multiple timestep methods may also tip the balance one way or the other (Luehr et al., 2014).

This section has highlighted some of the complexities in developing AIMD codes and it is unlikely that a single researcher is going to create a new AIMD code from scratch. Fortunately, there are many packages available that can perform BO, CP, or Ehrenfest dynamics, with a variety of pseudopotentials and basis sets. We have listed a number of these in Table 13.1. There are web references to the software in the Table which will allow the interested reader to access the software under the terms of its distribution.

Throughout this discussion, we have adopted the common '*ab initio*' nomenclature for this class of molecular dynamics. It is worth reflecting on how *ab initio* these methods really are. There is considerable freedom in choosing the exchange functional and in choosing the combination of a particular exchange functional with a pseudopotential and a basis set. In many recent papers one observes some ingenuity in adjusting these combinations to give the best possible fit to a particular experimental observable. This in itself is no mean feat because the simulations are often on small system sizes for quite short times. It is difficult to ascribe a disagreement with experiment unambiguously to either a limitation in the methodology or a particular choice in constructing the KS energy. Unquestionably, these methods have opened up important new areas such as aqueous electrochemistry, ionic liquids, and liquid metals to the power of accurate simulation, but there is still some way to go before we have full control of the model required to study a new system without recourse to adjustment against experiment.

Finally, we note that the techniques developed for the *ab initio* dynamics of quantum systems can be applied with good effect to more classical systems, containing the type of induced interactions described by eqn (1.36). In such cases (see Section 3.11), we can perform a molecular dynamics on the ions and allow the induced dipole moments to follow this motion on the appropriate Born–Oppenheimer surface (Salanne et al., 2012).

13.3   Combining quantum and classical force-field simulations

The techniques described in Section 13.2 are computationally demanding and are normally applied to small systems. In modelling a large system such as a solvated protein, it

Table 13.1 A number of codes available for performing density functional calculations and AIMD. This list is not complete and is merely representative of the range of packages available. A number of these are free for academic user in particular countries and some are also available through commercial suppliers. The precise terms and conditions for use are available at the referenced websites.

Code	Functionality	Source
ABINIT	A DFT electronic structure code, with pseudopotentials, plane waves and wavelets; relaxation by BO dynamics, TDDFT.	www.abinit.org/
CASTEP	A DFT electronic structure code with pseudopotential, plane waves, direct minimization of the KS energy functional; a wide range of spectroscopic features.	www.castep.org/
VASP	A DFT electronic structure code with BO dynamics; plane waves, pseudopotentials, hybrid functionals plane-wave basis; robust code for non-experts.	www.vasp.at/
SIESTA	A DFT electronic structure code, with relaxation and BO dynamics; linear scaling; based on numerical atomic orbitals.	departments.icmab.es/leem/siesta/
CPMD	AIMD code with DFT; CP dynamics, BO dynamics, Ehrenfest dynamics; plane-wave basis, PIMD, QM/MM.	www.cpmd.org/
CP2K	AIMD; a mixed Gaussian and plane-wave approach, BO dynamics.	www.cp2k.org/
Quantum Espresso	CP dynamics, BO dynamics, plane-wave basis; quantum transport; normal DFT features.	www.quantum-espresso.org/

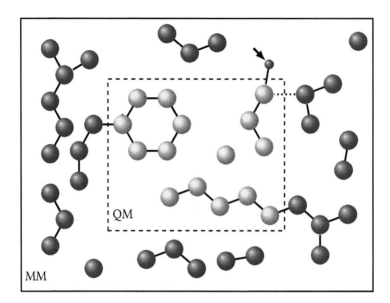

Fig. 13.1 A QM system (dashed box) embedded in a classical molecular mechanics (MM) system. The atoms that are treated quantum mechanically are light grey, the classical atoms are dark grey. The arrowed dark atom is a capping H atom used to preserve the valence of the quantum fragment at the quantum–classical boundary. The nearby horizontal dotted line indicates a monovalent pseudopotential discussed in the text.

is possible to treat a particular region of interest using a quantum mechanical (QM) approach and to model the environment around this region classically. An example is the transformation of an unconjugated α-keto acid to its conjugated isomers by enzyme catalysis in aqueous solution (Siegbahn and Himo, 2009). It is essential to model the acid molecule and the four important residues of the enzyme, 4-oxalocrotonate tautomerase, in a full QM way to capture the making and breaking of the bonds. However it would not be efficient or necessary to include other parts of the enzyme and the solvent in such detail and these could be usefully represented by a classical force field.

The idea of embedding a quantum mechanical sub-system in a larger classical system was first developed for static calculations by Warshel and Levitt (1976). For this reason it is often described as the QM/MM method (Groenhof, 2013). In the approach, the energy of the QM part of the system can be calculated using post-HF methods, DFT, or semi-empirical quantum methods such as AM1 and CNDO. The dynamics in the quantum region can be explored using the BO or CP methods described in Section 13.2. The classical region can be treated using a standard force field and an MD calculation. In this section, we will focus on the use of DFT methods in the quantum region. Fig. 13.1 shows a QM system embedded in a cluster of classical atoms and molecules. The choice of the QM sub-system is always arbitrary and its definition often requires an initial insight into the problem at hand. The sub-system can be defined either by labelling atoms or by defining a geometrical region to be treated quantum mechanically. For the moment, we will assume a fixed QM region with no movement of molecules across the QM/MM boundary. If the molecules in the two regions are distinct, the forces across the boundary will normally consist of dispersion interactions and electrostatic interactions. As is often the case, the boundary cuts a covalent bond and then there are also bond stretching, bond angle, and torsional forces between the regions.

There are two ways to calculate the energy of the combined system. In the simplest subtractive scheme (Maseras and Morokuma, 1995), the energy of the whole system is calculated using an MM force field, the energy of the QM system is added, and finally the MM energy of the QM sub-system is subtracted:

$$E = E^{QM}(QM) + E^{MM}(QM + MM) - E^{MM}(QM). \tag{13.45}$$

All the calculations are performed with a particular QM or MM method and there is no specific consideration of the interface. The interaction between the two regions is only included at the MM level. A more accurate, additive, approach is widely used in practice (Röthlisberger and Carloni, 2006). In this case the energy is composed of three terms

$$E = E(QM) + E(MM) + E(QM/MM) \tag{13.46}$$

where $E(QM)$ is the QM energy in the QM region, $E(MM)$ is the MM energy in the MM region, and the interaction energy between the regions is (Ippoliti et al., 2012)

$$
\begin{aligned}
E(QM/MM) &= \sum_{I'=1}^{MM} \left[\int dr \, \frac{q_{I'} \rho(\mathbf{r})}{|\mathbf{R}_{I'} - \mathbf{r}|} + \sum_{I=1}^{QM} \frac{q_{I'} q_I}{|\mathbf{R}_{I'} - \mathbf{R}_I|} \right] \\
&+ \sum_{\substack{\text{non-bonded} \\ \text{pairs}}} \left(\frac{A_{II'}}{R_{II'}^{12}} - \frac{B_{II'}}{R_{II'}^{6}} \right) + \sum_{\text{bonds}} k_r (R_{II'} - r_0)^2 \\
&+ \sum_{\text{angles}} k_\theta (\theta - \theta_0)^2 + \sum_{\text{torsions}} \sum_n k_{\phi,n} [\cos(n\phi + \delta_n) + 1]. \tag{13.47}
\end{aligned}
$$

The index I labels nuclei (or cores) in the QM region, while I' labels the MM atoms. The first two terms represent the interaction between the total charge density (due to electrons and cores) in the QM region and the classical charges in the MM region. The third term represents the dispersion interactions across the QM/MM boundary, and the fourth term consists of all the covalent bond stretching potentials that cross this boundary. The final two terms account for the energy across the boundary due to covalent bond angle bending and torsional potentials; here it is understood that at least one of the atoms involved in the angles θ and ϕ is a QM atom, with the others being MM atoms.

The terms $E(MM) + E(QM/MM)$ are included in E_{KS} in the Lagrangian, eqn (13.36), and a plane-wave basis set is used to solve the CP equations of motion for the QM system. The equations of motion of the MM charges are solved using a standard MD technique including all the classical forces and the forces from the QM region calculated from the gradient of eqn (13.47).

For a system of 10^6 grid points for the electron density in the QM region and 10^4 MM atoms, the full evaluation of the charge–charge term in $E(QM/MM)$, eqn (13.47), remains prohibitively expensive. This problem can be mitigated by apportioning the MM charges into three concentric spheres around the QM region. In the innermost sphere, the classical MM charges interact with the QM density as described by eqn (13.47).

The MM charges in the next, intermediate, region interact with a set of constructed charges in the QM region. These constructed charges are associated with each of the QM nuclei and are calculated using the RESP approach, to mimic the electron density, $\rho(\mathbf{r})$, as described in Section 1.4.2. This dynamical-RESP approach allows the fluctuating charges on the QM nuclei to be calculated on-the-fly during the course of the simulation. It also includes a restraining term to prevent these charges from fluctuating too strongly away from the underlying values of the Hirshfeld charges (Laio et al., 2002b). (Note: the Hirshfeld charges are calculated by sharing the molecular charge density at each point between the atoms, in proportion to their free-atom densities at the distances from the nuclei (Hirshfeld, 1977).)

In the third, outermost region, the classical charges interact with the multipole moments of the quantum charge distribution (Laio et al., 2004). The calculation of the charge–charge interaction using different, successively more approximate, methods in the three shells enables a very significant reduction in computational cost without a significant loss of accuracy (Laio et al., 2004). Finally, the behaviour of the dynamical-RESP charges provides a useful method for monitoring the chemical state of the QM atoms (Laio et al., 2002b).

When the QM–MM boundary intersects a covalent bond, an artificial dangling bond is created in the QM system. This is known as the link-atom problem and it can be tackled by capping the unsaturated valence with an H or F atom (Field et al., 1990). (This capping atom, which belongs to the QM region, is indicated by an arrow in Fig. 13.1.) This is not a perfect solution, since new, unphysical atoms are introduced into the QM region and considerable care has to be taken to remove the interactions between these capping atoms and the real MM atoms; this is particularly important for any charge–charge interaction (Laio et al., 2004). A more sophisticated approach compensates for the dangling bond by the addition of a localized orbital centred on the terminal QM atom before the boundary (Assfeld and Rivail, 1996). These localized orbitals have to be determined by calculations on small fragments of the real system and they remain frozen throughout the simulation without responding to changes in the electronic structure of the QM system. The most flexible approach is to introduce a monovalent pseudopotential at the first MM atom beyond the boundary; this is along the dotted line in Fig. 13.1. This is constructed so that the electrons in the QM region are scattered correctly by the classical region (von Lilienfeld et al., 2005). Analytic, non-local pseudopotentials (Hartwigsen et al., 1998) have been employed to good effect in this approach. They do not introduce additional interactions or degrees of freedom into the QM systems and they are sufficiently flexible to change as the QM region evolves.

There are a number of important issues to consider when implementing the QM/MM method. First, the method facilitates the simulation of larger systems, which include a QM core, but the maximum timestep that is applied to the whole system is controlled by the CP or BO dynamics. The short timestep required would normally limit the simulation to a few picoseconds, unless the classical dynamics is included using a multiple-timestep method as described by Nam (2014). Second, the method does not include the Pauli repulsion between the electrons in the QM region and the electrons associated with the classical atoms in the MM region. The result is that electrons from the QM region can be strongly and incorrectly associated with positively charged atoms close to the boundary in the MM region. This problem, known as electron spill-out, can be avoided by replacing the MM atoms and any corresponding point charges in the region close to the boundary by ionic pseudopotentials with screened electrostatic interactions (Laio et al., 2002a).

Finally, it is useful to be able to lift the restriction that all the atoms that start in the QM region remain in that region throughout the simulation. This is particularly important for liquid-like environments where solvent atoms (water) might diffuse around the active site of a protein. One possibility is to label atoms as either QM or MM at the start and to stick with these labels throughout. Unfortunately, in this case, solvent atoms that are explicitly involved in the reaction could be treated as MM atoms while solvent atoms which have moved far away from the reaction could be treated as QM atoms. An improvement is to define the QM region geometrically, as (say) a sphere of radius R_{QM}, and to change the nature of the atoms that cross the sphere from QM to MM and vice-versa. This will lead to significant discontinuities in the energy as atoms change their type, which can create instabilities in the dynamics and drifts in the temperature (Bulo et al., 2009). This problem has been addressed by defining an additional transition region between the QM and MM regions. The outer boundary of the transition region is set to $R_{MM} > R_{QM}$. The N adaptive atoms in the transition region, at any point in time, can be either QM or MM atoms. For example, if $N = 3$, there are $2^N = 8$ partitions between the two types: {QM, QM, QM}, {QM, QM, MM} ... {MM, MM, MM}. The potential energy and force from the adaptive atoms is given by a weighted sum over all the possible partitions, p:

$$\mathcal{V}^{ad} = \sum_{p=1}^{2^N} w_p \mathcal{V}_p\left(\mathbf{R}^{(N)}\right) \tag{13.48}$$

where w_p is a weight and $\mathcal{V}_p\left(\mathbf{R}^{(N)}\right)$ is the potential energy of a particular partition. Each of the N atoms is given a λ-value which determines its degree of MM character:

$$\lambda(R) = \begin{cases} 0 & R < R_{QM} \\ \dfrac{(R - R_{QM})^2(3R_{MM} - R_{QM} - 2R)}{(R_{MM} - R_{QM})^3} & R_{QM} \leq R \leq R_{MM} \\ 1 & R > R_{MM}. \end{cases} \tag{13.49}$$

λ depends on the distance R of an atom from the centre of the QM region. In a particular partition, the set of QM atoms has λ values denoted by $\{\lambda\}_p^{QM}$ while the set of MM atoms is described by $\{\lambda\}_p^{MM}$. The weight given to a particular partition is

$$w_p = \max\left[\min\left(\{\lambda\}_p^{MM}\right) - \max\left(\{\lambda\}_p^{QM}\right), 0\right]. \tag{13.50}$$

A detailed rationalization of this particular choice of w_p is provided by Bulo et al. (2009). This approach produces a smooth force that changes from pure QM to pure MM across the transition region. The computational overhead for the calculation of $\mathcal{V}^{ad}$ and its derivatives appears to scale as 2^N. However, using eqn (13.50), any partition in which an MM atom is closer to the QM region than any of the QM atoms in the transition region, has a weight of zero. There are only $N + 1$ contributing partitions and this makes the calculation of the adaptive potential efficient enough for many applications (Park et al., 2012).

The QM/MM method can be used in the CPMD program, see Table 13.1. The CPMD-QM option in this code requires the definition of the topology and coordinates of the protein and solvent, the classical force field to be used (either AMBER or GROMOS), and the three radii required to define the charge–charge interaction in the calculation of $E(\text{QM/MM})$.

13.4   Path-integral simulations

In this section, we consider fluids where the nuclei of the atoms are modelled quantum mechanically. In Section 2.9 we described the way in which first-order quantum corrections can be applied to classical simulations. The corrections to the thermodynamic properties arise from the Wigner–Kirkwood expansion of the phase-space distribution function (Green, 1951; Oppenheim and Ross, 1957) which may in turn be treated as an extra term in the Hamiltonian (Singer and Singer, 1984)

$$\mathcal{H}_{\text{qu}} = \mathcal{H}_{\text{cl}} + \frac{\hbar^2 \beta}{24m}\left[-\frac{\beta}{m}\left(\sum_I \mathbf{P}_I \cdot \boldsymbol{\nabla}_I\right)^2 \mathcal{V}^{\text{cl}} + 3\sum_I \nabla_I^2 \mathcal{V}^{\text{cl}} - \beta \sum_I |\boldsymbol{\nabla}_I \mathcal{V}^{\text{cl}}|^2 \right]. \quad (13.51)$$

Here $\boldsymbol{\nabla}_I$ is short for the gradient $\boldsymbol{\nabla}_{\mathbf{R}_I}$ with respect to the position $\mathbf{R}_I$ of the atom or, more precisely, the nucleus, and $\mathbf{P}_I$ is the momentum of the nucleus. Of course, as an alternative, the quantum potential can be obtained by integrating over the momenta

$$\mathcal{V}_{\text{qu}} = \mathcal{V}_{\text{cl}} + \frac{\hbar^2 \beta}{24m}\left[2\sum_I \nabla_I^2 \mathcal{V}^{\text{cl}} - \beta \sum_I |\boldsymbol{\nabla}_I \mathcal{V}^{\text{cl}}|^2 \right]. \quad (13.52)$$

This potential may be used in a conventional Monte Carlo simulation to generate quantum-corrected configurational properties. It is the treatment of this additional term by thermo-dynamic perturbation theory that gives rise to the quantum corrections mentioned in Section 2.9. Alternatively, a molecular dynamics simulation, based on the Hamiltonian of eqn (13.51), can be employed (Singer and Singer, 1984). Apart from measures to cope with numerical instabilities resulting from derivatives of the repulsive part of the potential, the technique is quite standard. In this section, we consider techniques that go beyond the expansion in $\hbar^2$ typified by eqn (13.51).

One of the most straightforward of these simulation techniques is that based on a discretization of the path-integral form of the density matrix (Feynman and Hibbs, 1965), because the method essentially reduces to performing a classical simulation. Since the early simulation work (Fosdick and Jordan, 1966; Jordan and Fosdick, 1968) and the work of Barker (1979), the technique has become popular, in part because the full implications of the quantum–classical isomorphism have become clear (Chandler and Wolynes, 1981; Schweizer et al., 1981). This type of approach is particularly useful when we wish to consider the nucleus as a quantum particle to capture phenomena such as quantum tunnelling and zero-point motion. Consider a single neon atom, which remains in its ground electronic state and whose position is described by $\mathbf{R}_1$. Starting with eqn (13.7) we can use the Trotter factorization (Tuckerman, 2010, Appendix C) to divide the exponential into P equal parts

$$Q_{1VT}(\beta) = \int dR_1 \langle R_1 | e^{-\beta\mathcal{H}/P} \ldots e^{-\beta\mathcal{H}/P} \ldots e^{-\beta\mathcal{H}/P} | R_1 \rangle \qquad (13.53)$$

and inserting unity in the form

$$1 = \int dR \, |R\rangle\langle R| \qquad (13.54)$$

between each pair of exponentials gives

$$Q_{1VT}(\beta) = \int dR_1 dR_2 \ldots dR_P \langle R_1 | e^{-\beta\mathcal{H}/P} | R_2 \rangle \langle R_2 | e^{-\beta\mathcal{H}/P} | R_3 \rangle$$
$$\ldots \langle R_{P-1} | e^{-\beta\mathcal{H}/P} | R_P \rangle \langle R_P | e^{-\beta\mathcal{H}/P} | R_1 \rangle$$
$$= \int dR_1 dR_2 \ldots dR_P \, \varrho(R_1, R_2; \beta/P)\varrho(R_2, R_3; \beta/P) \ldots \varrho(R_P, R_1; \beta/P). \qquad (13.55)$$

We seem to have complicated the problem; instead of one integral over diagonal elements of ϱ, we now have many integrals involving off-diagonal elements. However, each term involves, effectively, a higher temperature (or a weaker Hamiltonian) than the original. At sufficiently large values of P, the following approximation becomes applicable:

$$\varrho(R_a, R_b; \beta/P) \approx \varrho_{\text{free}}(R_a, R_b; \beta/P) \exp\left(-(\beta/2P)\left[\mathcal{V}^{\text{cl}}(R_a) + \mathcal{V}^{\text{cl}}(R_b)\right]\right) \qquad (13.56)$$

where $\mathcal{V}^{\text{cl}}(R_a)$ is the classical potential energy as a function of the configurational coordinates, and where the free-particle density matrix is known exactly. For a single molecule, of mass m, it is

$$\varrho_{\text{free}}(R_a, R_b; \beta/P) = \left(\frac{Pm}{2\pi\beta\hbar^2}\right)^{3/2} \exp\left(-\frac{Pm}{2\beta\hbar^2} R_{ab}^2\right) \qquad (13.57)$$

where $R_{ab}^2 = |R_a - R_b|^2$. Now the expression for Q is

$$Q_{1VT} = \left(\frac{Pm}{2\pi\beta\hbar^2}\right)^{3P/2} \int dR_1 \ldots dR_P \exp\left(-\frac{Pm}{2\beta\hbar^2}\left(R_{12}^2 + R_{23}^2 + \ldots + R_{P1}^2\right)\right)$$
$$\times \exp\left(-(\beta/P)\left[\mathcal{V}^{\text{cl}}(R_1) + \mathcal{V}^{\text{cl}}(R_2) + \ldots + \mathcal{V}^{\text{cl}}(R_P)\right]\right). \qquad (13.58)$$

These formulae are almost unchanged when we generalize to a many-molecule system. For N atoms,

$$Q_{NVT} = \frac{1}{N!}\left(\frac{Pm}{2\pi\beta\hbar^2}\right)^{3PN/2} \int dR_1^{(N)} \ldots dR_P^{(N)}$$
$$\exp\left(-\frac{Pm}{2\beta\hbar^2}\left[|R_{12}^{(N)}|^2 + |R_{23}^{(N)}|^2 + \ldots + |R_{P1}^{(N)}|^2\right]\right)$$
$$\times \exp\left(-(\beta/P)\left[\mathcal{V}^{\text{cl}}(R_1^{(N)}) + \mathcal{V}^{\text{cl}}(R_2^{(N)}) + \ldots + \mathcal{V}^{\text{cl}}(R_P^{(N)})\right]\right). \qquad (13.59)$$

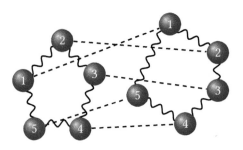

Fig. 13.2 Two ring-polymer 'molecules' ($P = 5$) representing the interaction between two atoms in a path-integral simulation. The straight dashed lines are the intermolecular potential interactions, the wavy lines represent the intramolecular spring potentials.

We must consider carefully what eqn (13.59) represents. Each vector $R_a^{(N)}$ represents a complete set of $3N$ coordinates, defining a system like our N-atom quantum system of interest. The function $\mathcal{V}^{cl}(R_a^{(N)})$ is the potential-energy function for each one of these systems, calculated in the usual way. Imagine a total of P such systems, which are more or less superimposed on each other. Each atom in system a is quite close to (but not exactly on top of) the corresponding atom in systems $b, c, \dots$ etc. Each contributes a term $\mathcal{V}^{cl}(R_a^{(N)})$ to the Boltzmann factors in eqn (13.59), but the total is divided by P to obtain, in a sense, an averaged potential. The systems interact with each other through the first exponential term in the integrand of eqn (13.59). Each vector $R_{ab}^{(N)}$ ($R_{12}^{(N)}$, $R_{23}^{(N)}$, etc.) represents the complete set of N separations between corresponding atoms of the two systems a and b. Specifically the squared terms appearing in eqn (13.59) are

$$\left|R_{ab}^{(N)}\right|^2 = \left|R_a^{(N)} - R_b^{(N)}\right|^2 = \sum_{i=1}^{N}\left|R_{ia} - R_{ib}\right|^2 \tag{13.60}$$

where R_{ia} is the position of atom i in system a. These interactions are of a harmonic form, that is, the systems are coupled by springs.

There is an alternative and very fruitful way of picturing our system of NP atoms (Chandler and Wolynes, 1981). It can be regarded as set of N molecules, each consisting of P atoms which are joined together by springs to form a classical ring polymer. This is illustrated in Fig. 13.2. We write the integral in eqn (13.59) in the form of a classical configurational integral

$$Z_{NVT} = \int \exp\left(-\beta\mathcal{V}(R^{(NP)})\right) dR_{11}\dots dR_{ia}\dots dR_{NP} \tag{13.61}$$

where $R^{(NP)}$ is the complete set of NP atomic coordinates, R_{ia} corresponding to atom a on molecule i, with the configurational energy consisting of two parts

$$\mathcal{V}(R^{(NP)}) = \mathcal{V}^{cl}(R^{(NP)}) + \mathcal{V}^{qu}(R^{(NP)}). \tag{13.62}$$

The classical part is

$$
\begin{aligned}
\mathcal{V}^{\text{cl}} &= \frac{1}{P}\left[\mathcal{V}^{\text{cl}}(\mathbf{R}_1^{(N)}) + \mathcal{V}^{\text{cl}}(\mathbf{R}_2^{(N)}) \dots + \mathcal{V}^{\text{cl}}(\mathbf{R}_P^{(N)})\right] \\
&= \frac{1}{P}\sum_{a=1}^{P}\sum_{i<j}^{N} v^{\text{cl}}\left(\left|\mathbf{R}_{ia} - \mathbf{R}_{ja}\right|\right) = \frac{1}{P}\sum_{a=1}^{P}\sum_{i<j}^{N} v^{\text{cl}}(R_{iaja}).
\end{aligned}
\tag{13.63}
$$

We have assumed pairwise additivity here and in Fig. 13.2 for simplicity, although this is not essential. The quantum part of the potential is

$$
\begin{aligned}
\mathcal{V}^{\text{qu}} &= \left(\frac{Pm}{2\beta^2\hbar^2}\right)\left(\left|\mathbf{R}_{12}^{(N)}\right|^2 + \left|\mathbf{R}_{23}^{(N)}\right|^2 \dots + \left|\mathbf{R}_{P1}^{(N)}\right|^2\right) \\
&= \left(\frac{Pm}{2\beta^2\hbar^2}\right)\sum_{a=1}^{P}\sum_{i=1}^{N}\left|\mathbf{R}_{ia} - \mathbf{R}_{ia+1}\right|^2 = \sum_{i}\sum_{a} v^{\text{qu}}(R_{iaia+1})
\end{aligned}
\tag{13.64}
$$

where we take $a + 1$ to equal 1 when $a = P$. Note how the interactions between molecules only involve correspondingly numbered atoms a (i.e. atom 1 on molecule i only sees atom 1 on molecules j, k, etc.), while the interactions within molecules just involve atoms with adjacent labels. The system is formally a set of polymer molecules, but an unusual one: the molecules cannot become entangled, because of the form of eqn (13.63), and the equilibrium atom–atom bond lengths in each molecule, according to eqn (13.64), are zero.

The term outside the integral of eqn (13.59) may be regarded as the kinetic contribution to the partition function, if the mass of the atoms in our system is chosen appropriately. Actually, this choice is not critical, if the configurational averaging is the key problem to solve. Nonetheless it proves convenient (as we shall see shortly) to use an MD-based simulation method, and De Raedt et al. (1984) recommend making each atom of mass $Pm = M$. Then the kinetic energy of the system becomes

$$
\mathcal{K} = \tfrac{1}{2}\sum_{ia}(Pm)|\mathbf{v}_{ia}|^2 = \tfrac{1}{2}\sum_{ia}|\mathbf{p}_{ia}|^2/(Pm) = \tfrac{1}{2}\sum_{ia}|\mathbf{p}_{ia}|^2/M
\tag{13.65}
$$

and the integration over momenta, in the usual quasi-classical way, yields

$$
Q_{NVT}(\beta) = \frac{1}{(NP)!}\left(\frac{M}{2\pi\beta\hbar^2}\right)^{3NP/2}\int d\mathbf{R}^{(NP)}\,\exp\left[-\beta(\mathcal{V}^{\text{cl}} + \mathcal{V}^{\text{qu}})\right].
\tag{13.66}
$$

Apart from the indistinguishability factors, which may usually be ignored as far as the calculation of averages is concerned, this is the approximate quantum partition function eqn (13.59) for our N-particle system.

Thus a Monte Carlo simulation of the classical ring polymer system with potential energy $\mathcal{V}$ given by eqn (13.62) may be used to generate averages in an ensemble whose configurational distribution function approximates that of a quantum system. Examples appear in Code 13.1.

Code 13.1 Path-integral Monte Carlo

These files are provided online. Two separate programs are supplied. qmc_pi_sho.f90 is a simple path-integral MC program to simulate a harmonic oscillator, for which the average energy may be compared with the exact result (see Fig. 13.3). qmc_pi_lj.f90 is a program to simulate a liquid of N Lennard-Jones atoms represented by $P \times N$ beads. This code is combined with qmc_pi_lj_module.f90, to define the interactions, and both programs use the utility modules of Appendix A.

```
! qmc_pi_sho.f90
! Quantum Monte Carlo, path-integral, harmonic oscillator
PROGRAM qmc_pi_sho


! qmc_pi_lj.f90
! Quantum Monte Carlo, path-integral method
PROGRAM qmc_pi_lj


! qmc_pi_lj_module.f90
! Energy and move routines for PIMC simulation, LJ potential
MODULE qmc_module
```

Equally, an MD simulation with Hamiltonian

$$\mathcal{H} = \tfrac{1}{2} \sum_{ia} |\mathbf{p}_{ia}|^2 / M + \mathcal{V}^{\text{cl}}(\mathbf{R}^{(NP)}) + \mathcal{V}^{\text{qu}}(\mathbf{R}^{(NP)}) \qquad (13.67\text{a})$$

or indeed

$$\mathcal{H} = \tfrac{1}{2} \sum_{ia} |\mathbf{p}_{ia}|^2 / m + \mathcal{V}^{\text{cl}}(\mathbf{R}^{(NP)}) + \mathcal{V}^{\text{qu}}(\mathbf{R}^{(NP)}) \qquad (13.67\text{b})$$

will achieve the same result. We have explicitly written down two versions of the Hamiltonian to remind ourselves that we are completely free to choose the fictitious mass M corresponding to the momenta whereas the m appearing in $\mathcal{V}^{\text{qu}}$ is not adjustable. The use of MD techniques to generate equilibrium states in this way is often referred to as path-integral molecular dynamics (PIMD).

These approaches can be readily extended to the simulation of an isolated quantum atom in a classical solvent bath, where the classical atoms behave like polymers contracted to a point. The method has also been used to study the behaviour of an excess electron in a classical fluid (Miller, 2008), the transfer of an electron between ions in water (Menzeleev et al., 2011), and to shed light on the long-standing problem of the anomalous mobility of the proton (Marx et al., 1999) and the hydroxyl ion (Tuckerman et al., 2002) in water. The extension of PIMD to molecular systems is possible and desirable when, in a case such as water, translational motion may be regarded as classical while rotation is quantum-mechanical (Kuharski and Rossky, 1984). There are additional complications in the case of asymmetric tops (Noya et al., 2011). These simulations of water have been extended to consider models such as SPC/E (Berendsen et al., 1987), which comprise a Lennard-Jones

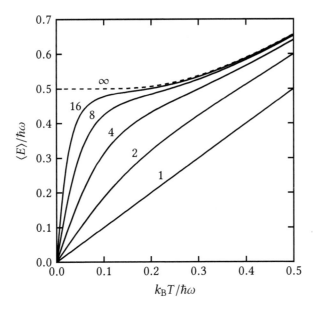

Fig. 13.3 The average energy of the path-integral approximation to the quantum harmonic os-cillator of frequency ω as a function of temperature. We give the results for various values of P, the number of 'atoms' in the ring polymer. $P = 1$ is the classical result, and $P \to \infty$ is the quantum mechanical limit.

site on the oxygen atom and partial charges on the hydrogen and oxygen atoms. Each polymer bead in the PIMD simulation is a replica of this scaffold and a technique such as SHAKE is used to constrain the internal structure of these beads during the dynamics (Miller and Manolopoulos, 2005).

As the number of particles P in our ring polymer grows we obtain a better approxima-tion to the quantum partition function; these equations become formally exact as $P \to \infty$, going over to the Feynman path-integral representation (Feynman and Hibbs, 1965). How well do we expect to do at finite P? Some idea of what we may expect can be obtained by studying the quantum harmonic oscillator for which exact solutions are available in the classical $P = 1$ case, in the quantum-mechanical limit $P \to \infty$, and for all intermediate P (Schweizer et al., 1981). The computed average energy is plotted in Fig. 13.3. It can be seen that the finite-P energy curves deviate from the true result as the temperature decreases, leaving the zero-point level ($\frac{1}{2}\hbar\omega$) and dropping to the classical value at $T = 0$, $\langle E \rangle = 0$. The region of agreement may be extended to lower temperatures by increasing P.

Now to the practical matters. The classical polymer model is easily simulated by standard techniques, such as constant-NVT MC or MD. In principle, a better approximation to the quantum system will be obtained by making P as large as possible. A P-bead system is expected to be roughly P times more expensive than the classical one. Markland and Manolopoulos (2008b,a) have shown that if the interaction potentials can be separated into short- and long-ranged contributions, including electrostatics, the latter can be evaluated more economically by constructing a contracted ring polymer with fewer beads. Therefore, some of the expense associated with increasing P may be offset. There are, however, some additional technical problems to be expected as P increases. According to eqn (13.64) the internal spring constant increases with P as well as with temperature, while the external forces felt by each individual atom decrease as P increases. In an MC simulation this might mean that separate attention would have to be given to moves which altered intramolecular distances and those involving the molecule as a whole. In some cases, a normal mode analysis of the polymer may help in choosing MC step sizes (Herman et al., 1982). More directly, one can abandon intramolecular Metropolis moves and build the polymer from scratch, by sampling from the internal, free-molecule, distribution (Jacucci and Omerti, 1983). In PIMD simulations there is the corresponding danger that the time-scale separation of the internal and external motions may become acute at high P and this will necessitate a shorter timestep. A sensible choice of the dynamic mass M helps to overcome this problem: with the choice of eqn (13.65) the stiffest internal modes of the polymer will be characterized by a frequency $k_B T/\hbar$; alternative choices of M may be made so as to match the internal frequencies with the external time scales (De Raedt et al., 1984). Nonetheless, the danger of slow energy exchange between internal and external modes, leading to non-ergodic behaviour, is a real concern. One of a number of approaches to this problem (Berne, 1986) is to use the freedom we have in changing the individual masses of the polymer beads. A judicious choice of these masses will ensure that all wavelength modes of the ring polymer are optimally sampled in the dynamics. This can be achieved by linearly transforming the internal degrees of freedom of the ring polymer

$$\mathbf{u}_{ia} = \sum_{a'=1}^{P} \mathbf{U}_{aa'}^{i}\, \mathbf{R}_{ia'} \tag{13.68}$$

into normal modes $\{\mathbf{u}_{ia}\}$ so as to diagonalize the harmonic quantum potential

$$\mathcal{V}^{\mathrm{qu}}(\mathbf{R}_{i1}\dots\mathbf{R}_{iP}) = \mathcal{V}^{\mathrm{qu}}(\mathbf{u}_{i2}\dots\mathbf{u}_{iP}) = \tfrac{1}{2}\sum_{a=2}^{P}\left(\frac{P\tilde{m}_{ia}}{\beta^2\hbar^2}\right)\left|\mathbf{u}_{ia}\right|^2. \tag{13.69}$$

$\mathbf{U}^{i}$ is a matrix of unit determinant describing the transformation for each polymer i, and $\{\tilde{m}_{ia}\}$ are the transformed masses of the beads. The first normal mode $a = 1$ corresponds to the translational motion of the centroid of the ring polymer, and does not contribute to $\mathcal{V}^{\mathrm{qu}}$.

There are two common choices for $\mathbf{U}$ used in path-integral simulations. The traditional normal-mode transformation uses a simple Fourier transform. The alternative 'staging' transformation breaks up the path-integral chain to achieve a separation of the long- and short-wavelength modes of the system. (Staging was originally developed as a method for improving sampling in path-integral Monte Carlo by creating a primary chain with the correct thermal wavelength, but composed of only a few particles, and then adding the secondary chains with adjacent particles in the primary chain as the end points (Sprik et al., 1985).) The precise form of $\mathbf{U}$ for both these transformations is described in detail in Martyna et al. (1999, Appendix A).

Comparing with eqn (13.67b), the Hamiltonian for the transformed variables is

$$\mathcal{H}' = \sum_{i=1}^{N}\sum_{a=1}^{P}\frac{\left|\mathbf{p}_{ia}\right|^2}{2\tilde{m}_{ia}} + \frac{P\tilde{m}_{ia}}{2\beta^2\hbar^2}\left|\mathbf{u}_{ia}\right|^2 + \frac{1}{P}\sum_{a=1}^{P}\sum_{i>j} v^{\mathrm{cl}}\left(\left|\mathbf{R}_{ia}(\{\mathbf{u}_i\}) - \mathbf{R}_{ja}(\{\mathbf{u}_j\})\right|\right) \tag{13.70}$$

where, in the case of the normal mode transformation, the $\tilde{m}_{ia}$ are proportional to the normal mode eigenvalues. There remains the freedom to choose a set of new masses, $\{\tilde{m}_{ia}\}$, associated with the fictitious momenta. For the normal mode transformation, all modes can be placed on the same time scale, with $\tilde{m}_{i1} = m_i = M/P$ and $\tilde{m}_{ia} \propto \tilde{m}_{ia}$, for $a = 2, \ldots P$. The dynamics associated with eqn (13.70) can then be performed to follow the development of the normal modes with time. The **u** coordinates and associated forces can be efficiently calculated using FFTs. Similar relationships can be established for the 'staging' transformation (Tuckerman et al., 1993).

In addition to the transformations, the sampling of the phase space of the stiff ring polymers can be enhanced by coupling a Nosé–Hoover chain thermostat to each of the $3NP$ degrees of freedom of the system (Martyna et al., 1999). The optimal sampling of phase space is achieved for a thermostat 'mass' of $Q_{ia} = \beta \hbar^2/P$. Alternatively, a specially designed stochastic thermostat may be applied (Ceriotti et al., 2010). Finally, multiple-timestep methods of the type described in Section 3.5 can also be applied to improve the efficiency. A short timestep can be applied to the harmonic bonds while the classical force from the external degrees of freedom can be evaluated using a significantly longer step (Tuckerman et al., 1993).

There are some subtleties in the way a path-integral simulation is used to estimate ensemble averages. Averages that are only a function of the position variables, such as the spatial distribution of atoms, can be calculated using the diagonal elements of the density matrix. Averages that are only a function of the momentum operators, such as the momentum distribution, require both diagonal and off-diagonal elements of the density matrix, and in a path-integral representation this requires the evaluation of paths that are not cyclic. In the canonical ensemble, the partition function is accessible through the evaluation of cyclic paths, and properties that depend on both position and momentum can be evaluated through the normal thermodynamic relationships (Tuckerman, 2010). For example, the energy is obtained in the usual way by forming $-Q_{NVT}^{-1}(\partial Q_{NVT}/\partial \beta)$; however, the 'quantum spring' potential is temperature-dependent, and the result is

$$\langle E \rangle = \langle \mathcal{V}^{\mathrm{cl}} \rangle + \tfrac{3}{2} NPk_{\mathrm{B}}T - \langle \mathcal{V}^{\mathrm{qu}} \rangle = \langle \mathcal{V}^{\mathrm{cl}} \rangle + \langle \mathcal{K} \rangle - \langle \mathcal{V}^{\mathrm{qu}} \rangle. \tag{13.71}$$

Note the sign of the $\mathcal{V}^{\mathrm{qu}}$ term. This might cause some confusion in an MD simulation, where the total energy $\mathcal{V}^{\mathrm{cl}} + \mathcal{V}^{\mathrm{qu}} + \mathcal{K}$ is the conserved variable (between stochastic collisions if these are applied). This quantity is not the correct estimator for the quantum energy. There is yet a further wrinkle. The 'quantum kinetic energy' part of eqn (13.71) is the difference between two quantities, $\tfrac{3}{2} NPk_{\mathrm{B}}T$ and $\langle \mathcal{V}^{\mathrm{qu}} \rangle$, both of which increase with P. This gives rise to loss of statistical precision: in fact the relative variance in this quantity increases linearly with P, making the estimate worse as the simulation becomes more accurate (Herman et al., 1982). The solution to this is to use the virial theorem for the harmonic potentials to replace eqn (13.71) by the following

$$\langle E \rangle = \left\langle \frac{1}{P} \sum_a \sum_{i<j} v^{\mathrm{cl}}(R_{iaja}) \right\rangle + \frac{3}{2} Nk_{\mathrm{B}}T + \frac{1}{2} \left\langle \frac{1}{P} \sum_a \sum_{i<j} \mathbf{R}_{iaja} \cdot \nabla_{\mathbf{R}_{iaja}} v^{\mathrm{cl}}(R_{iaja}) \right\rangle. \tag{13.72}$$

Actually, it is not clear that this will give a significant improvement. Jacucci (1984) has pointed out that the statistical inefficiency (see Section 8.4.1) may be worse for the quantity in eqn (13.72) than for that in eqn (13.71), due to the persistence of correlations, thus wiping out the advantage. Other configurational properties are estimated in a straightforward way. For example, the pressure may be obtained from the volume derivative of Q. The method is particularly well-suited to estimating the free-energy differences needed to calculate isotope effects (Ceriotti and Markland, 2013). The pair distribution function becomes essentially a site–site $g(r)$ for atoms with the same atom label (Chandler and Wolynes, 1981). It is accumulated in the normal way. The 'size' of each quantum molecule is also of interest (De Raedt et al., 1984; Parrinello and Rahman, 1984). This may be measured by the radius of gyration R_i of polymer i,

$$R_i^2 = \frac{1}{P} \sum_{a=1}^{P} |\mathbf{R}_{ia} - \mathbf{R}_i|^2 \tag{13.73}$$

where $\mathbf{R}_i = (1/P) \sum_{a=1}^{P} \mathbf{R}_{ia}$ is the centre of mass of the polymer.

Having described a set of developments that enable accurate PIMD to be performed, it is tempting to use the time evolution in these systems to measure the quantum correlations and spectra in the fluid phase, or perhaps to follow the motion of a proton or electron in a liquid. Unfortunately, Hamiltonians of the form eqn (13.67a) or eqn (13.67b) do not result in a dynamics with any physical meaning.

An indirect approach, based on the time–temperature analogy, has been advanced by Thirumalai and Berne (1983; 1984). This involves measuring the internal 'spatial' correlation functions of the polymer chain. More recently, Craig and Manolopoulos (2004; 2005a,b) have developed a ring polymer molecular dynamics (RPMD) method, in an attempt to study the dynamics of quantum systems (for a review see Habershon et al., 2013). In this method a system is prepared using a PIMD calculation with the full panoply of transformation, thermostatting and multiple timesteps to ensure a number of well-equilibrated starting configurations. All of this is then switched off and, following a suggestion by Parrinello and Rahman (1984), the fictitious mass associated with the momenta is set to m. The system exhibits a classical equilibrium density corresponding to a fluid with $3NP$ degrees of freedom at a temperature of PT. Although the trajectories developed by RPMD are not those of the quantum dynamical operator $\exp(i\mathcal{H}t/\hbar)$, Craig and Manolopoulos (2004) show that 'classical correlation' functions of the system exhibit some properties of the full quantum time correlation function. For two operators $\mathcal{A}$ and $\mathcal{B}$, the Kubo-transformed real-time correlation function (Kubo, 1957) is

$$\tilde{C}_{AB}(t) = \frac{1}{\beta Q_{NVT}} \int_0^\beta d\lambda \, \mathrm{Tr}\left[e^{-(\beta-\lambda)\mathcal{H}} \mathcal{A} e^{-\lambda\mathcal{H}} e^{i\mathcal{H}t/\hbar} \mathcal{B} e^{-i\mathcal{H}t/\hbar} \right]. \tag{13.74}$$

For these correlation functions

$$\tilde{C}_{AB}(t) = \tilde{C}_{BA}(-t), \qquad \tilde{C}_{AB}(t) = \tilde{C}_{AB}(t)^*, \qquad \tilde{C}_{AB}(t) = \tilde{C}_{AB}(-t)^*. \tag{13.75}$$

The standard correlation functions of the RPMD obey these symmetries for all P. One can also show that the RPMD produces the exact result for $\tilde{C}_{AB}(t)$, at all times, for a harmonic potential, when $\mathcal{A}$ and $\mathcal{B}$ are linear operators. In the case of weak anharmonic oscillators, RPMD agrees well, but not exactly, with the full quantum results. RPMD is exact in the high-temperature limit and in the short-time limit. It evolves the system so that it preserves the exact quantum distribution of states. Also, a classical molecular dynamics result such as

$$\frac{1}{2}\frac{d}{dt}\left\langle \left|\mathbf{R}_c(t) - \mathbf{R}_c(0)\right|^2\right\rangle_{RP} = \int_0^t dt \left\langle \mathbf{v}_c(t) \cdot \mathbf{v}_c(0)\right\rangle_{RP}, \tag{13.76}$$

where $\mathbf{R}_c$ is the centroid and $\mathbf{v}_c(t)$ is the bead-averaged or centroid velocity of a particular ring polymer, is obeyed.

Significant indirect evidence has been gathered to assess the performance of RPMD in describing quantum dynamics. For example, calculations indicate that RPMD can also be used to probe the 'deep-tunnelling' regime at low temperatures where the barrier is not parabolic. Richardson and Althorpe (2009) have rationalized this observation by establishing a connection between RPMD and semi-classical instanton theory (an established tool for calculating tunnelling rates). In studying the dynamics of an excess electron in a fluid of helium atoms, Miller (2008) calculated the exact mean-squared displacements, $\langle|\mathbf{R}_{ia}(s) - \mathbf{R}_{ia}(0)|^2\rangle$, in imaginary time, s, for the polymer beads as a function of solvent density. The mean-squared displacement can also be constructed indirectly from the analytic continuation of the velocity auto-correlation function for the solvated electron calculated by RPMD. A comparison of these two approaches shows that the RPMD becomes more accurate with increasing solvent density in the regime where the electron rattles in the solvent cage.

These results might encourage one to use RPMD to look at correlation functions and even to move beyond this formalism and the linear response regime to model processes far from equilibrium; for example, the dynamics of electron transfer between mixed-valence transition metal ions in water (Menzeleev et al., 2011) or the prediction of quantum reaction rates for bimolecular gas-phase reactions (Stecher and Althorpe, 2012).

However, great care is required with this approach; the RPMD description of the dynamics is not exact, and it is known to fail in a number of specific instances. RPMD methods overestimate the tunnelling of electrons in electron transfer reactions, particularly in the inverted regime. RPMD fails to describe the dynamics in a number of strongly coherent quantum systems (Craig and Manolopoulos, 2004). Witt et al. (2009) have used RPMD to calculate vibrational spectra of the diatomic and polyatomic molecules OH, H_2O and CH_4, using between 16 and 64 polymer beads on each atom and averages accumulated over 50 independent runs. The spectra are calculated from the Fourier transform of the quantum dipole autocorrelation function. In these simulations the intrinsic dynamics of the ring polymers can interfere with the physical frequencies of the molecules. Spurious artificial peaks can grow at the ring polymer frequencies and physical peaks can split due to resonant coupling. Similar artefacts are observed in the simulation of the far infrared spectrum of water (Habershon et al., 2008).

Progress in RPMD edges forward by comparison with more exact quantum simulation results, where possible, and by detailed comparison with experiment. It is early in its development for a widespread and general application of the method. This situation will improve as efforts continue to develop a real justification of RPMD starting from the quantum Liouville equation (Jang et al., 2014).

An important alternative approach to quantum time correlation has been the development and application of centroid molecular dynamics (CMD) (Voth, 1996), following the path centroid ideas developed by Feynman and Hibbs (1965). The equations of motion of the centroid of the ring polymer are

$$\dot{\mathbf{R}}_c = \mathbf{p}_c/m, \qquad \dot{\mathbf{p}}_c = -\nabla_{\mathbf{R}_c}\mathcal{V}_c = \mathbf{f}_c \tag{13.77}$$

where $\mathbf{p}_c$ is the momentum of the centroid and $\mathcal{V}_c$ is the centroid potential of mean force, corresponding to the density $\varrho_c(\mathbf{R}_c) = \exp[-\beta\mathcal{V}_c(\mathbf{R}_c)]$ (Cao and Voth, 1994). The force on the centroid is

$$\mathbf{f}_c(\mathbf{R}_c) = -\frac{1}{P}\left\langle \sum_{a=1}^{P}\nabla_{\mathbf{R}_a}\mathcal{V}^{cl}(\mathbf{R}_a)\,\delta\left(\frac{1}{P}\sum_{a=1}^{P}(\mathbf{R}_a - \mathbf{R}_c)\right)\right\rangle \tag{13.78}$$

where the gradient describes the total force on a bead, a, in one ring polymer from the corresponding bead a in the other ring polymers and the delta function extracts those configurations of the ring polymer where the centroid is at $\mathbf{R}_c$.

The direct use of eqn (13.78) is problematic. At first glance, the constraint of including only configurations corresponding to the centroid at $\mathbf{R}_c$ means that a full PIMD has to be run to sample the whole configuration space and the constraint applied retrospectively. However, the centroid force can be evaluated at each step by switching to the normal mode description of the ring polymer. The first normal mode, with associated mass $\tilde{m}_1 = m$, corresponds to the translation of the centroid. Solving the equations of motion of the other $P - 1$ normal modes automatically imposes the constraint in eqn (13.78).

In an alternative formulation of the problem, 'adiabatic' CMD, the masses associated with non-centroid modes are decreased (Cao and Martyna, 1996):

$$\tilde{m}_a = \lambda_a\gamma^2 m \qquad 2 \le a \le P, \tag{13.79}$$

where λ_a are the eigenvalues of the normal-mode transformation and $0 < \gamma < 1$ is the adiabaticity parameter. As γ decreases, the timestep required to solve accurately the dynamics of the $P - 1$ non-centroid modes can be reduced. A multiple-timestep approach can then be applied where the dynamics of the non-centroid modes are followed over a series of short timesteps, calculating the force on the centroid at each step. At each of the longer timesteps, the average force on the centroid is calculated from the preceding short timesteps and is used to update the position of the centroid (Pavese et al., 2000).

This 'adiabatic' CMD has been widely applied to calculate time correlation functions of liquids (Voth, 1996). There have been extensive comparisons between CMD and RPMD. Perez et al. (2009a) have shown that for para-H_2 at 14 K, the Kubo-transformed velocity autocorrelation function from both methods is essentially the same. However in the calculation of vibrational and far infrared spectra in liquids (Witt et al., 2009), the unphysical resonances that appear in the RPMD spectra do not appear in CMD. In other cases, CMD exhibits broadening and shifting of peak positions as compared with experiment. The assumptions in both the CMD and RPMD methods have been analysed in detail (Jang et al., 2014).

The path-integral method described in this section is often termed the 'primitive algorithm'; it uses the most crude approximation to the density matrix. Other improved approximations have been advocated (Schweizer et al., 1981) with the aim of reducing the number of polymer units required and possibly improving the convergence. One important improvement (Takahashi and Imada, 1984) is

$$\varrho(\mathbf{R}_a, \mathbf{R}_b; \beta/P) \approx$$

$$\varrho_{\text{free}}(\mathbf{R}_a, \mathbf{R}_b; \beta/P) \exp\left[-\frac{\beta}{P}\left(\mathcal{V}^{\text{cl}}(\mathbf{R}_a) + \frac{\hbar^2}{24m}\left(\frac{\beta}{P}\right)^2 \left|\nabla_{\mathbf{R}_a}\mathcal{V}^{\text{cl}}(\mathbf{R}_a)\right|^2\right)\right] \quad (13.80)$$

where the second term in the exponential of eqn (13.80) arises from the double commutator $[[\mathcal{V}, \mathcal{K}], \mathcal{V}]$ in the expansion of $\exp(-\beta\mathcal{H})$ (Brualla et al., 2004). This density matrix is fourth order in the action and has been used in MC simulations of liquid He and Ne at low temperatures (Brualla et al., 2004). Zillich et al. (2010) have developed a new class of propagators for path-integral Monte Carlo that are fourth, sixth and eighth order in the action, and that do not require higher-order derivatives of the potential, by using an extrapolation of the primitive second-order propagator.

13.5 Quantum random walk simulations

The methods discussed in Section 13.4 are suitable for the simulation of liquid neon, liquid water and gas-phase infrared spectra of CH_4, CH_5^+. These are systems where the quantum effects are significant but not dominant. However, where the behaviour is essentially quantum mechanical, as in liquid helium, we need to consider other techniques such as diffusion Monte Carlo, where the many-body Schrödinger equation is solved by generating a random walk in imaginary time. These simulations are normally used to calculate the ground-state wave function and the corresponding energy for systems of bosons and fermions. They are zero-temperature methods.

The adoption of an imaginary time evolution converts the Schrödinger equation into one of a diffusional kind.

$$-\frac{\partial\Psi(\mathbf{r}, s)}{\partial s} = \left(-D\nabla_{\mathbf{r}}^2 + \mathcal{V}(\mathbf{r}) - E_{\text{T}}\right)\Psi(\mathbf{r}, s) \quad (13.81)$$

where $s = it/\hbar$, $\mathcal{V}$ is the potential and E_{T} is an arbitrary zero of the energy which is useful in this problem. In this section, for notational simplicity, $\Psi(\mathbf{r}, s)$ is the wave function for all the n particles in the fluid and $\mathbf{r}$ rather than $\mathbf{r}^{(n)}$ is used to represent their $3n$ coordinates. The 'diffusion coefficient' is defined to be

$$D = \hbar^2/2m. \quad (13.82)$$

The simulation of this equation to solve the quantum many-body problem is a very old idea, possibly dating back to Fermi (Metropolis and Ulam, 1949), but it is the implementation of Anderson (1975; 1976) that interests us here.

If we interpret $\Psi(\mathbf{r}, s)$ (note: not $|\Psi|^2$!) as a probability density, then eqn (13.81) is essentially the Schmoluchowski equation for the configurational distribution (Doi and Edwards, 1988),

$$\frac{\partial}{\partial t}\rho(\mathbf{r}, t) + \frac{D}{k_{\text{B}}T}\nabla_{\mathbf{r}} \cdot \left(\mathbf{f}\,\rho(\mathbf{r}, t)\right) = D\nabla_{\mathbf{r}}^2\rho(\mathbf{r}, t) \quad (13.83)$$

with the systematic force, $\mathbf{f}$, set to 0 and $k_{\text{B}}T = 1$. The diffusive part of the Schmoluchowski equation can be simulated by using the Brownian dynamics algorithm of eqn (12.5). The additional complication is that the $(V(\mathbf{r}) - E_T)\Psi$ term in eqn (13.81) acts as a birth and death process (or a chemical reaction) which changes the weighting (probability) of configurations with time. To incorporate this in a simulation means allowing the creation and destruction of whole systems of molecules. These copies of the systems are often referred to as walkers. Simulations of many individual systems are run in parallel with one another. Although this sounds cumbersome, in practice it is a feasible route to the properties of the quantum liquid. That such a simulation may yield a ground-state stationary solution of the Schrödinger equation may be seen by the following argument. Any time-dependent wave function can be expanded in a set of stationary states $\Psi_n(\mathbf{r})$, when the time evolution becomes

$$\Psi(\mathbf{r}, s) = \sum_n c_n \exp\left[-s(E_n - E_T)\right]\Psi_n(\mathbf{r}), \text{ where } s = it/\hbar, \tag{13.84}$$

and the c_n are the initial condition coefficients. In the imaginary time formalism, the state with the lowest energy becomes the dominant term at long times. If we have chosen $E_T < E_0$, then the ground-state exponential decays the least rapidly with time, while if $E_T > E_0$, the ground-state function grows faster than any other. If we are lucky enough to choose $E_T = E_0$, then the function $\Psi(\mathbf{r}, s)$ tends to $\Psi_0(\mathbf{r})$ at long times while the other state contributions decay away. For $\Psi(\mathbf{r}, s)$ to be properly treated as a probability density, it must be everywhere positive (or negative) and this will be true for the ground state of a liquid of bosons.

The reaction part of the 'reaction-diffusion' equation is treated as usual, by integrating over a short timestep δs. Formally

$$\Psi(s + \delta s) = \Psi(s) \exp\left[-(V(\mathbf{r}) - E_T)\delta s\right]. \tag{13.85}$$

This enters into the simplest quantum Monte Carlo algorithm as follows. Begin with a large number (100–1000) of replicas of the N-body system of interest. Then:

(a) Perform a Brownian dynamics step using eqn (12.5), with $\mathbf{f}(t) = 0$ and with D given by eqn (13.82), on each system. (Note that the temperature does not enter into the random walk algorithm since there are no systematic forces.)

(b) For each system, evaluate $V(\mathbf{r})$, compute $\exp[-(V(\mathbf{r}) - E_T)\delta s] = K$ and replace the system by K identical copies (clones) of itself.

(c) Return to step (a).

Step (b), the cloning step, requires a little more explanation, since in general K will be a positive real number. The system is replaced by $\lfloor K \rfloor$ replicas of itself ($\lfloor K \rfloor$ is the largest integer less than K, as given by the Fortran FLOOR function) and a further copy is added with a probability $K - \lfloor K \rfloor$ using a random number ζ generated uniformly on the range $(0, 1)$. If $K < 1$, this is equivalent to deleting the current system from the simulation with probability $(1 - K)$, and retaining it (a single copy) with probability K. A little thought reveals that the number of copies can always be expressed as $\lfloor K + \zeta \rfloor$. If new copies are generated, they evolve independently of each other thereafter.

The simple example of a particle in a one-dimensional quadratic potential is shown in Fig. 13.4, and in Code 13.2. The seven walkers are lined up at the start with an arbitrary distribution in the x-coordinate. As the simulation progresses, walkers that stray into

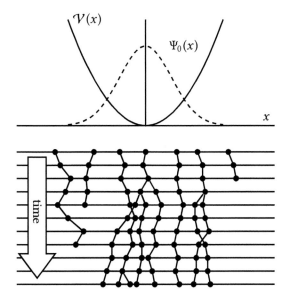

Fig. 13.4 An impression of a quantum Monte Carlo simulation of a particle in a one-dimensional quadratic potential $\mathcal{V}(x)$, after Foulkes et al. (2001). In the long-time limit, the distribution of walkers should approach the form of the ground-state wave function $\Psi_0(x)$ (dashed).

Code 13.2   Quantum Monte Carlo using a diffusion equation

This file is provided online. qmc_walk_sho.f90 is a simulation of the ground state of a particle in a quadratic potential using a diffusion simulation with birth and death events to model the Schrödinger equation. Gaussian random numbers are generated by a utility routine (see Appendix A). Several of the parameters in this program, and the scheme for updating E_T, are taken from a useful tutorial introduction (Kosztin et al., 1996). The time evolution of this very simple implementation is quite noisy and can be sensitive to the initial value of E_T.

```
! qmc_walk_sho.f90
! Quantum MC, random walk, simple harmonic oscillator
PROGRAM qmc_walk_sho
```

regions of high potential are likely to be annihilated, while those that sample the region of low potential (high probability) are likely to proceed to the finish line, and occasionally to be cloned. The distribution of walkers on the finishing line, in the limit of long simulation time and a large number of walkers, will be the Gaussian distribution corresponding to the ground-state wave function of the particle.

This scheme is fairly crude. Clearly, depending on E_T, the number of systems still under consideration may grow or fall dramatically, and the value of this parameter is continually adjusted during the simulation to keep the current number approximately constant (Anderson, 1975). Hopefully in the course of a run conducted in this way $E_T \rightarrow E_0$. The fluctuation in the number of systems is substantial and this makes the estimate of E_T subject to a large statistical error. A number of ways around this difficulty have been proposed (Anderson, 1980; Mentch and Anderson, 1981) and we shall concentrate on one approach (Kalos et al., 1974; Reynolds et al., 1982) which uses importance sampling. Suppose we multiply $\Psi(\mathbf{r}, s)$ by a specified trial wave function $\Psi_T(\mathbf{r}, s)$ and use the result

$$\Upsilon(\mathbf{r}, s) = \Psi(\mathbf{r}, s)\Psi_T(\mathbf{r}, s) \tag{13.86}$$

in the Schrödinger equation. Then we obtain

$$
\begin{aligned}
-\frac{\partial \Upsilon}{\partial s} &= -D\nabla_{\mathbf{r}}^2 \Upsilon + \left(E_T(\mathbf{r}) - E_T\right)\Upsilon + D\nabla_{\mathbf{r}} \cdot \left(\Upsilon\nabla_{\mathbf{r}} \ln |\Psi_T(\mathbf{r})|^2\right) \\
&= -D\nabla_{\mathbf{r}}^2 \Upsilon + \left(E_T(\mathbf{r}) - E_T\right)\Upsilon + D\nabla_{\mathbf{r}} \cdot \left(\Upsilon\mathbf{F}_{\mathrm{qu}}\right),
\end{aligned} \tag{13.87}
$$

where the local energy is defined by

$$E_T(\mathbf{r}) = \Psi_T^{-1}\mathcal{H}\Psi_T, \tag{13.88}$$

and should not be confused with E_T. We have also defined the quantum force $\mathbf{F}_{\mathrm{qu}}$, which is derived from the pseudopotential $v_{\mathrm{PP}}(r_{ij})$ if Ψ_T is given (as is common) by

$$\Psi_T(\mathbf{r}) = \exp\left[-\tfrac{1}{2}\sum_i \sum_{j>i} v_{\mathrm{PP}}(r_{ij})\right] = \prod_{j>i} \exp\left[-\tfrac{1}{2}v_{\mathrm{PP}}(r_{ij})\right]. \tag{13.89}$$

Eqn (13.87), often described as the 'importance-sampled' Schrödinger equation (Gillan and Towler, 2011), resembles the Schmoluchowski equation, eqn (13.83), with the force term retained and with $k_B T = 1$ throughout. It can be solved using eqn (12.5) with $\mathbf{f}(t)$ replaced by $\mathbf{F}_{\mathrm{qu}}$. All the techniques described in this section are now applied to the function Υ rather than to Ψ. The procedure for duplicating or deleting systems now depends on $(E_T(\mathbf{r}) - E_T)$, where $E_T(\mathbf{r})$ is evaluated for each system. This process is controlled more easily by a sensible choice of $\Psi_T(\mathbf{r})$ as discussed by Reynolds et al. (1982). The quantum force appears in these simulations just as the classical force appears in the smart MC method described in Chapter 9, or the Brownian dynamics of Chapter 12. This force guides the system in its search for low 'energy', that is, high Ψ_T^2. If Ψ_T is a good approximation to the ground state Ψ_0, then the energy $E_T(\mathbf{r})$ tends to E_0 independently of $\mathbf{r}$, and so is subject to little uncertainty. If E_T is adjusted so as to maintain the steady-state population of systems, then this will also tend to E_0. As Reynolds et al. (1982) point out, the average $\langle E_T(\mathbf{r})\rangle$ obtained without any system creation/destruction attempts would correspond to a variational estimate based on Ψ_T. Identical variational estimates can also be calculated using the Monte Carlo method (McMillan, 1965; Schmidt and Kalos, 1984; Morales et al., 2014). In the random walk techniques, the MC simulation is replaced by Brownian dynamics. The inclusion of destruction and creation allows $\Psi(\mathbf{r})$ to differ from Ψ_T and the simulation probes the improved $\Psi(\mathbf{r})$. Of course making Ψ_T more complicated and hence more complete, adds to the computational expense.

The particular problems of fermion systems are discussed by Reynolds et al. (1982). The essential point is that the ground-state fermion wave function must contain multidimensional nodal surfaces. Each region of configuration space bounded by these 'hypersurfaces' within which Ψ_T may be taken to have one sign throughout, may be treated separately by the random walk technique. The nodal positions themselves are essentially fixed by the predetermined form of $\Psi_T(\mathbf{r})$. This introduces a further variational element into the calculation. The fixed-node approximation, and alternative approaches for fermion systems, are described in detail by Reynolds et al. (1982).

The work of Booth et al. (2009) represents an important advance in the solution of the fermion problem. This approach switches the focus from the wave function in terms of the particle position, $\Psi(\mathbf{r})$, to the space of Slater determinants, $|D_i\rangle$. The structure of the Hartree–Fock Slater determinant is given in eqn (13.18). For a general problem of N electrons chosen from $2M$ spin-orbitals there are approximately $\binom{M}{N/2}^2$ such determinants, a number which grows factorially with both M and N. The full configuration interaction (FCI) wave function for the ground state can be represented as

$$\Psi_0^{FCI} = \sum_i c_i |D_i\rangle \qquad (13.90)$$

where c_i are the coefficients of the expansion with the appropriate sign. The beauty of this representation is that the structure of $|D_i\rangle$ ensures the overall antisymmetry of the fermionic wave function. The vector $\{sgn(c_i)\}$ is composed of a set of ± 1's and 0's. Any useful trial wave function must have the same sign vector as the true wave function, and this vector cannot be determined without a full knowledge of Ψ_0; this is a restatement of the familiar sign problem in the Slater determinant space. We note that this space, consisting of many excitations from the ground state, can be very large with i ranging from 10^6 to 10^{14} for calculations on small molecular species. Two determinants in this space are said to be connected or coupled if $\langle D_i| \mathcal{H} |D_j\rangle \neq 0$.

Substituting eqn (13.90) into the imaginary-time Schrödinger equation gives a set of coupled equations for the coefficients

$$-\frac{\partial c_i}{\partial s} = (K_{ii} - E_T)c_i + \sum_{j\neq i} K_{ij}c_j, \qquad (13.91)$$

where E_T is the familiar shift energy used to control the population size in any stochastic solution of the equation and

$$K_{ij} = \langle D_i| \mathcal{H} |D_j\rangle - E_{HF}\delta_{ij}, \qquad (13.92)$$

where E_{HF} is the Hartree–Fock energy of the uncorrelated problem. In the long-time limit $E_T \rightarrow E_{corr}$, where E_{corr} is the correlation energy, and in the steady state, the coefficients $\{c_i\}$ represent the ground state, which is the exact wave function for the basis set under consideration.

Eqn (13.91) can be solved by the type of diffusion Monte Carlo algorithm already described in this section. The full configuration interaction quantum Monte Carlo (FCIQMC) method considers only the birth and death of the community of walkers; there are no diffusive moves in this approach. There are three processes at each step:

(a) For each walker α on $|D_i\rangle$, a coupled determinant $|D_j\rangle$ is chosen with a fixed probability, and an attempt is made to spawn one or more new walkers at j with a probability proportional to $|K_{i_\alpha j}|\, \delta s$, where δs is the imaginary timestep in the simulation. The spawned walker has the same sign as its parent if $K_{i_\alpha j} < 0$ or the opposite sign to the parent otherwise (Booth et al., 2009).

(b) A particular walker α at $|D_i\rangle$ is cloned or dies at i with a probability given by $(K_{i_\alpha i_\alpha} - E_T)\delta s$. Walkers that die disappear from the space immediately.

(c) Finally, at each determinant, i, the signs of all of the spawned, cloned and surviving walkers are considered, and pairs of the opposite sign are annihilated, reducing the number of walkers by two in each case.

In the steady state each walker, α, on a particular determinant i_α will have a sign $s_\alpha = \pm 1$. The resulting coefficient or amplitude is

$$c_i \propto N_i = \sum_{\alpha=1}^{N_w} s_\alpha \delta_{i_\alpha i} \tag{13.93}$$

where $N_w = \sum_i |N_i|$. Values of $\delta s = 10^{-4}$–10^{-3} a.u. have been used in the simulations and the important details of the techniques can be found in the original literature (Booth et al., 2009).

The method has been successfully used to calculate the ground-state energy of species such as C_2, H_2O, N_2, O_2, and NaH (Booth et al., 2009). The results are as accurate as the FCI quantum calculations which require the diagonalization in the full space of Slater determinants. The FCIQMC method requires less storage and scales more efficiently than the brute-force diagonalization. Steps such as the spawning and the death of walkers only require information at a particular determinant so that the algorithm is eminently parallelizable (Gillan and Towler, 2011). The method has also been used to model the homogeneous electron gas (Shepherd et al., 2012) to obtain a finite-basis energy, which is significantly and variationally lower than any previously published work. The FCIQMC method has been extended to complex wave functions and used to study rare gas, ionic, and covalent solids (Booth et al., 2013). In FCIQMC we have a direct approach to solving the many-body fermion problem and a confluence of stochastic simulation methods and traditional quantum chemistry. It remains to be seen if these methods can be applied more widely to disordered systems and systems with very strong quantum correlations. At the very least, FCIQMC will provide detailed information on the force fields and exchange energies of classical and *ab initio* MD.

13.6   Over our horizon

We conclude our short introduction to quantum simulations by pointing to some of the many important techniques that lie beyond the scope of this book.

In considering ground-state or zero-temperature methods, Section 13.5, we have concentrated on the diffusion Monte Carlo algorithm. The simplest and most widely used approach to the problem is the variational Monte Carlo method. The total energy and its variance are calculated through a Monte Carlo evaluation of their expectation values using a trial wave function. The trial wave function is adjusted so that the energy and the variance are minimized at the ground state (Ceperley et al., 1977; Morales et al., 2014). There are a whole class of projector methods, Green's function Monte Carlo, that attempt to extract the ground-state wave function for many-body problems using stochastic algorithms (Kalos and Whitlock, 2008). The Schrödinger equation in imaginary time has an associated Green's function, $G(s)$, which drives the wave function forward in 'time' towards its ground state. Eqn (13.85) can be written as

$$\Psi(s + \delta s) = G(\delta s)\Psi(s), \quad \text{where} \quad G(s) = \exp\left[(\mathcal{H} - E_T)s\right]. \tag{13.94}$$

The imaginary time Green's function Monte Carlo method (Schmidt et al., 2005) uses a random walk technique with importance sampling to determine an accurate $G(s)$ from an initial guess. One important development in this area has been to treat segments of the Langevin trajectories produced in stochastic simulations as 'polymers'. Well-known Monte Carlo polymer algorithms, such as reptation (see Section 4.8.2), are used to develop these trajectories to produce accurate expectation values of local observables, and their static and dynamic response functions in imaginary time (Baroni and Moroni, 1999).

At non-zero temperatures, the path-integral methods, described for semi-classical systems in Section 13.4, can be readily extended to more strongly quantum systems by increasing the number of beads in the ring polymer and by using more accurate forms of the density matrix (Morales et al., 2014). Indeed path-integral approaches can be applied to electrons and other fermions by restricting paths to those that are node avoiding, as exemplified by recent simulations of the electron gas (Brown et al., 2013). Simulations of nuclei and electrons have been performed using the coupled electron–ion Monte Carlo method. In this method, classical or path-integral Monte Carlo simulations of the nuclei are conducted with the BO energy of the ground-state electrons determined by a separate variational or diffusion Monte Carlo simulation (Pierleoni and Ceperley, 2006).

So far in this chapter, we have focused on ground-state simulations. It is possible to use the AIMD methods described in Section 13.4 to tackle problems involving many-electron excited states. Ehrenfest dynamics, eqn (13.13), produces strong adiabatic coupling which mixes the excited states in the dynamics (Tully, 1998) and is therefore unsuitable for following trajectories that split onto different and distinct potential surfaces. This problem can be addressed using surface-hopping molecular dynamics (Tully, 1990). Typically, the equations of motion for the atoms are solved on the energy surface

$$\mathcal{V}_{kk}(\mathbf{R}^{(N)}) = \left\langle \Psi_k(\mathbf{r}^{(n)}, \mathbf{R}^{(N)}) \middle| \mathcal{H}_e \middle| \Psi_k(\mathbf{r}^{(n)}, \mathbf{R}^{(N)}) \right\rangle$$

where the integration is over the electronic degrees of freedom at fixed nuclear position. The density matrix elements, describing the electronic degrees of freedom, can be integrated along this surface

$$\frac{\partial \varrho_{kk}}{\partial t} = \sum_{\ell \neq k} b_{k\ell} \tag{13.95}$$

with

$$b_{k\ell} = 2\hbar^{-1} \operatorname{Im}(\varrho_{k\ell}^* \mathcal{V}_{k\ell}) - 2 \operatorname{Re}(\varrho_{k\ell}^* \mathbf{R}^{(N)} \cdot \mathbf{d}_{k\ell}) \tag{13.96}$$

where $\mathcal{V}_{k\ell}$ is the matrix element of the electronic Hamiltonian and $\mathbf{d}_{k\ell} = \langle \Psi_k | \nabla_{\mathbf{R}} \Psi_\ell \rangle$ is the non-adiabatic coupling vector (Tully, 1990). At a particular integration step, the switching probabilities, g_{kj}, from state k to all other states j, are computed in terms of the density matrix element ϱ_{kk} and the matrix $b_{k\ell}$

$$g_{kj} = \frac{\Delta t\, b_{jk}}{\varrho_{kk}}. \tag{13.97}$$

The switch to a state j is made with a probability g_{kj}. If the switch is successful, the velocity of the particles has to be adjusted and the motion proceeds on the potential surface $\mathcal{V}_{jj}$. This technique is often described as the 'fewest switches' algorithm, since it makes the minimum number of changes of state required to maintain the correct statistical distribution among the possible states. For example, the technique has been used successfully to study tunnelling and electronic excitation (Müller and Stock, 1997).

More generally, the DFT method described in Section 13.4 can be extended to excited states using the time-dependent density functional theory (TDDFT) or Runge–Gross theory (Marques and Gross, 2004; Ullrich, 2011). In the dynamical cases, the quantum mechanical action

$$S[\Psi] = \int_{t_{\text{initial}}}^{t_{\text{final}}} dt\, \left\langle \Psi(t) \left| \frac{\partial}{\partial t} - \mathcal{H}_e \right| \Psi(t) \right\rangle \tag{13.98}$$

has to be stationary to produce the time-dependent Schrödinger equation and $S[\rho]$ is stationary at the exact time-dependent density. This stationarity gives rise to the TDDFT equations of motion, which, in a similar spirit to equilibrium density functional theory, require an approximation for the time-dependent exchange correlation potential. It is possible to propagate the KS orbitals directly in real time using a number of mean-field approaches (Watanabe and Tsukada, 2002; Marx and Hutter, 2012) and excursions to excited electronic states can be achieved by imposing an external perturbation such as a strong laser field. Alternatively a version of TDDFT can be combined directly with surface-hopping to explore excited states (Craig et al., 2005).

Finally, we briefly mention methods that move us towards the real quantum dynamics of systems. One approach starts from the von Neumann equation for the time-dependent density matrix

$$\frac{\partial \varrho(t)}{\partial t} = -\frac{i}{\hbar} \left[\mathcal{H}, \varrho(t) \right], \tag{13.99}$$

where $[\mathcal{A}, \mathcal{B}]$ is the normal commutator. The density matrix can be transformed to the Wigner distribution function (Singer and Smith, 1990), which is the solution to the Wigner–Liouville equation (the quantum mechanical equivalent of the Liouville equation for movement in phase space). In this representation it may be possible to split the system into a small quantum sub-system surrounded by a bath of particles. The mass of the bath particles is increased so that they can be treated classically (Kapral and Ciccotti, 1999). Propagation of the Wigner–Liouville equation for this type of mixed quantum–classical system is a useful route to the accurate calculation of the time correlation functions of the quantum sub-system (Egorov et al., 1999; Shi and Geva, 2004).

There have been important developments in the application of path-integral methods to study dynamics of strongly quantum systems. Hernandez et al. (1995) have developed a Feynman path centroid density that offers an alternative perspective to the Wigner prescription for the evaluation of equilibrium and dynamical properties. Poulsen et al. (2004) have applied the Feynman–Kleinert linearized path-integral approximation to study quantum diffusion in liquid para-hydrogen. Coker and Bonella (2006) have also developed a new approach to calculating time correlation functions for mixed systems. The evolution of the highly quantum mechanical part of the system, the electronic sub-system, is mapped onto the evolution of a set of fictitious harmonic oscillators. At the same time the bath is treated using a path-integral representation of the nuclear part of the problem, which is approximated using a linearization of the path integrals. The linearization makes this approach tractable and it has been used to study the diffusion of an excess electron in a dilute metal–molten salt solution.

14

Inhomogeneous fluids

In this chapter we consider various aspects of the simulation of systems that contain gas–liquid, liquid–liquid, or solid–liquid interfaces. Often, we are interested in the surface properties themselves: the surface tension, the structure within the interfacial region, or any wetting/drying or layering transitions. Inhomogeneous systems require particular care in their preparation, and in the tests that are performed to ensure that they are at equilibrium, before any property measurements can take place. Additionally, there may be pitfalls in properly defining the quantities to be averaged. Long-range forces can be particularly problematic; coexistence properties of even the simple Lennard-Jones system are known to vary dramatically with the chosen pair potential cutoff (Trokhymchuk and Alejandre, 1999). The following sections aim to discuss some of these issues. For a recent review of simulations of interfacial phenomena in soft matter, see Binder (2014). Usually it is necessary to determine, ahead of time, the state point(s) and properties of the two bulk phases that are at equilibrium with each other. Vega et al. (2008) have reviewed the whole area of determination of phase diagrams by simulation, including a discussion of the modelling of coexisting phases.

14.1 The planar gas–liquid interface

14.1.1 The starting configuration

In simulating a planar gas–liquid interface of, say, the Lennard-Jones fluid, it is possible to perform an MC or MD simulation in a three-dimensional periodic system based on a slab geometry (Chapela et al., 1975; Rao and Levesque, 1976).

For reasonably low temperatures (closer to the triple point than the critical point) it is possible to prepare the system starting from a simulation of the bulk liquid using normal cubic periodic boundary conditions. A density slightly above the required coexisting liquid density, and at the required coexisting temperature, is chosen. Once the liquid has equilibrated, the central box is separated from its periodic images, and extended in the z direction by adding a number of empty boxes to each of the opposite faces. These boxes have the same dimensions as the original box. This extended system is now periodically reproduced throughout space to create slabs of the liquid in the xy plane surrounded by vacuum. The precise dimensions will depend on the system and properties of interest. For example, detailed studies of the dependence of surface tension, γ, on box size, for the Lennard-Jones potential (Chen, 1995; Orea et al., 2005; Biscay et al., 2009), show significant

Computer Simulation of Liquids. Second Edition. M. P. Allen and D. J. Tildesley.
© M. P. Allen and D. J. Tildesley 2017. Published in 2017 by Oxford University Press.

artefacts (specifically oscillations) for small values of L_x, L_y, but indicate that γ reaches a limiting value for boxes of size $11\sigma \times 11\sigma \times 60\sigma$. Other interfacial properties may be less sensitive to this choice, depending upon the interaction potential, but this at least suggests that we might surround the original box by two empty boxes on each side to create an extended system of size $\{L, L, 5L\}$ with $L \approx 11\sigma$. This would correspond to $N \approx 1100$ Lennard-Jones atoms close to the triple point of the liquid.

For temperatures closer to the critical point, the vapour-phase coexisting density is much closer to that of the liquid, and fluctuations are more important. Typically, much larger systems are required. Watanabe et al. (2012) used box shapes $\{L, L, 2L\}$, and system sizes up to $L = 128\sigma$, corresponding to $N \approx 1.5 \times 10^6$. They prepared the initial configuration by placing particles in both the high-density and low-density regions before equilibrating. As we shall see later, in this regime, some coexistence properties may be studied without explicitly simulating in slab geometry.

From this starting configuration, simulations of the slab can be performed in the canonical (NVT) or microcanonical (NVE) ensembles. The densities (and in the latter case, temperatures) of the two phases will adjust themselves automatically to maintain the coexistence condition, resulting in uniform pressure and chemical potential throughout. It is best to avoid simulating at constant-NPT (or μVT) because, unless the system is simulated at the *exact* coexistence value of P (or μ) for the model, one of the two phases will grow at the expense of the other, resulting eventually in a single phase. (As we shall see in Section 14.4.1, it may be possible to avoid this by dynamically controlling P.) During the equilibration, if the 'vapour' boxes are initially empty, atoms will migrate into them from the bulk liquid until a coexisting vapour is formed. In early simulations, the centre of mass of the liquid slab was observed to move by ca. $0.1\,\sigma$. This effect, which is small for larger system sizes, can be minimized by carefully removing any net momentum during equilibration of an MD calculation and, if necessary, re-centering the liquid slab at the centre of the box.

14.1.2 Establishing equilibrium

After the initial preparation, we must equilibrate the coexisting-phase slab system: it is important to establish thermal, mechanical and chemical equilibrium. Thermal equilibrium is established by monitoring the temperature across the simulation cell in a direction normal to the surface. We define a series of slabs of thickness Δz, parallel to the xy plane. We will calculate properties at values of z corresponding to the centres of these slabs. Then, in an MD simulation of N atoms,

$$T(z) = \frac{1}{3k_B} \left\langle \frac{\sum_{i=1}^{N} H(z_i - z, \Delta z) m_i |v_i|^2}{\sum_{i=1}^{N} H(z_i - z, \Delta z)} \right\rangle \tag{14.1}$$

where m_i is the mass of atom i, v_i its velocity and H is the top-hat function

$$H(z_i - z, \Delta z) = \begin{cases} 1 & z - \frac{1}{2}\Delta z < z_i < z + \frac{1}{2}\Delta z \\ 0 & \text{otherwise.} \end{cases} \tag{14.2}$$

The way this translates into practice is illustrated in Code 14.1, where the final profile is stored in temp(k). At equilibrium $T(z)$ should be constant as a function of z across the whole system.

Code 14.1 **Calculating the temperature and density profiles**

The number of slabs is nk: larger values give thinner slabs and better resolution, but poorer statistics in each slab. The histograms h1(nk), ht(nk), and the normalization factor norm, should be declared and set to zero at the start; dz holds the value of Δz and box(3) contains the box length L_z. Particle coordinates and velocities are in the arrays r(3,n) and v(3,n); we assume that the z-coordinates satisfy $|z_i| \leq \pm L_z/2$. For simplicity k_B and all masses are taken to be unity. Then at intervals, the following loop is executed.

```
dz = box(3) / REAL ( nk )
DO i = 1, n
   k = 1 + FLOOR ( ( r(3,i) + box(3)/2.0 ) / dz )
   k = MAX(1,k)    ! guard against roundoff
   k = MIN(nk,k)   ! guard against roundoff
   h1(k) = h1(k) + 1.0
   ht(k) = ht(k) + SUM ( v(:,i)**2 ) / 3.0
END DO
norm = norm + 1.0
```

At the end of the block or run, the temperature and density profiles can be calculated.

```
area = box(1)*box(2)
DO k = 1, nk
   IF ( h1(k) > 0.5 ) THEN
        temp(k) = ht(k) / h1(k)
        rho(k)  = h1(k) / ( norm * area * dz )
        rhot(k) = ht(k) / ( norm * area * dz )
   ELSE
        temp(k) = 0.0
        rho(k)  = 0.0
        rhot(k) = 0.0
   END IF
END DO
```

Position k in the array corresponds to $z = -\frac{1}{2}L_z + (k - \frac{1}{2})\Delta z$, at the centre of the slab.

To establish mechanical equilibrium it is necessary to evaluate the pressure tensor. Using eqn (2.182) of Chapter 2, we can write the Irving–Kirkwood (IK) definition of the pressure in a form convenient for evaluation in a simulation (Walton et al., 1983):

$$P_{\alpha\beta}(z) = \rho(z)k_B T\,\delta_{\alpha\beta} + \frac{1}{\mathcal{A}}\left\langle \sum_{i=1}^{N-1}\sum_{j>i}^{N}(\mathbf{r}_{ij})_\alpha(\mathbf{f}_{ij})_\beta\,\frac{1}{|z_{ij}|}\Theta\left(\frac{z_i - z}{z_{ij}}\right)\Theta\left(\frac{z - z_j}{z_{ij}}\right)\right\rangle, \quad (14.3)$$

where α and β are Cartesian components, $z_{ij} = z_i - z_j$, and Θ is the unit step function.

The first, ideal-gas, term has been written for a constant-T ensemble, and requires the evaluation of the single-particle density profile $\rho(z)$. This is formally defined in eqn (2.175)

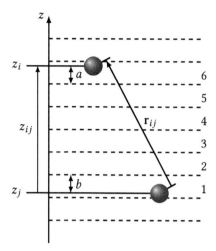

Fig. 14.1 The evaluation of the Irving–Kirkwood pressure between two particles i and j in a simulation. Successive slabs are labelled 1–6; the interparticle vector has a z component which completely spans four of the slabs, plus end contributions a and b.

and can be calculated via

$$\rho(z) = \frac{1}{\mathcal{A}\Delta z}\left\langle \sum_{i=1}^{N} H(z - z_i, \Delta z) \right\rangle \tag{14.4}$$

where $\mathcal{A}$ is the area of the interface, and H and Δz have the same meanings as in eqns (14.1), (14.2). The way this works in practice is also illustrated in Code 14.1, giving the result rho(k), which may be multiplied by $k_{\mathrm{B}}T$ to give the diagonal components of the ideal-gas contribution. In a constant-NVE MD simulation, the term should strictly be written $\rho(z)k_{\mathrm{B}}T(z)\delta_{\alpha\beta}$ (i.e. allowing for temperature variation in different slabs) and this is also given in Code 14.1 as rhot(k).

The second, excess, contribution to eqn (14.3) is evaluated as shown in Fig. 14.1. The product of Θ-functions is 1 when z lies between z_i and z_j, and 0 otherwise. The force between i and j contributes to slabs between the atoms (labelled 1 to 6, for example, in Fig. 14.1). For those slabs completely contained in the range $z_i \ldots z_j$ (slabs 2, 3, 4, and 5 in the figure), the value of $(\mathbf{r}_{ij})_\alpha (\mathbf{f}_{ij})_\beta/|z_{ij}|$ is added to the corresponding histogram bin. This value is multiplied by $a/\Delta z$ and $b/\Delta z$ for the end slabs (where a and b are shown in Fig. 14.1). In the event that z_i and z_j lie in the same slab, $(\mathbf{r}_{ij})_\alpha (\mathbf{f}_{ij})_\beta/\Delta z$ is added to the corresponding bin. The histogram is averaged over a run, divided by the area $\mathcal{A}$, and added to the ideal-gas part to give the final pressure profile.

At equilibrium, the off-diagonal elements of the tensor are zero. The normal pressure,

$$P_{\mathrm{N}} = P_{zz},$$

is constant for all z as we move through the liquid–gas interface and this is a signature of mechanical stability. The tangential component,

$$P_{\mathrm{T}} = \frac{P_{xx} + P_{yy}}{2},$$

changes with z in the vicinity of the interface, as discussed in Chapter 2, but should tend to P_N in the bulk regions. There is no unique definition of the tangential component of the pressure tensor and, for example, one alternative form has been proposed by Harasima (1958), and studied by simulation (Walton et al., 1983). Here the contour between atoms i and j is at first normal to the surface and then perpendicular. In this case

$$P_T(z) = \rho(z)k_BT + \frac{1}{2\mathcal{A}\Delta z}\left\langle \sum_{i=1}^{N-1}\sum_{j>i}^{N}\left(x_{ij}(\mathbf{f}_{ij})_x + y_{ij}(\mathbf{f}_{ij})_y\right)H(z - z_i, \Delta z)\right\rangle, \tag{14.5}$$

where once more H is defined in eqn (14.2). The normal pressure is unchanged by the choice of contour.

Todd et al. (1995) have presented an alternative definition of the pressure tensor based on the mass and continuity equations of hydrodynamics. This technique, known as the method of planes (MOP), was developed for the MD simulation of a fluid between two walls under shear, but is equally applicable to the situation of interest here. It provides the $P_{\alpha z}$ components of the tensor where z is the direction perpendicular to the interface. An atom i crosses a particular z-plane in the simulation at a set of times $t_{i,m}$ where $m = 1, 2 \ldots$ indexes all the crossings in the course of the simulation. Then

$$P_{\alpha z}(z) = \lim_{\tau\to\infty}\frac{1}{\mathcal{A}\tau}\sum_{i=1}^{N}\sum_{0<t_{i,m}<\tau} m_i(v_i)_\alpha(t_{i,m})\ \text{sgn}\left[(v_i)_z(t_{i,m})\right]$$

$$+ \frac{1}{2\mathcal{A}}\left\langle\sum_{i=1}^{N}\sum_{j\neq i}^{N}(\mathbf{f}_{ij})_\alpha\left[\Theta(z_i - z)\Theta(z - z_j) - \Theta(z_j - z)\Theta(z - z_i)\right]\right\rangle \tag{14.6}$$

where

$$\text{sgn}\,x = \begin{cases} -1 & x < 0 \\ 0 & x = 0 \\ 1 & x > 0. \end{cases}$$

The sgn function of the velocity distinguishes right to left crossings of the plane from those that are in the opposite sense. Eqn (14.6) is formally equivalent to eqn (14.3) although the MOP is more computationally efficient and gives rise to less oscillatory structure in $P_{\alpha z}(z)$ than the IK approach. Note that the values of z defining the planes may be chosen arbitrarily: a regular set of planes, spaced apart by a distance Δz, may be convenient, but is not necessary.

It is possible to calculate the local stress tensor in any volume of a fluid of arbitrary shape using the volume averaging method (Cormier et al., 2001). In this case

$$P_{\alpha\beta}(z) = \frac{1}{\Omega}\left\langle\sum_{i=1}^{N}m_i(v_i)_\alpha(v_i)_\beta\,\Lambda_i + \sum_{i=1}^{N-1}\sum_{j>i}^{N}(\mathbf{r}_{ij})_\alpha(\mathbf{f}_{ij})_\beta\,\ell_{ij}\right\rangle \tag{14.7}$$

where Ω is a volume of arbitrary shape and size which is less than or equal to the total volume, $\Lambda_i = 1$ if atom i is inside Ω and is zero otherwise, and ℓ_{ij} is the fraction of $|r_{ij}|$ which lies within Ω. Heyes et al. (2011) have shown that, for the planar interface, the

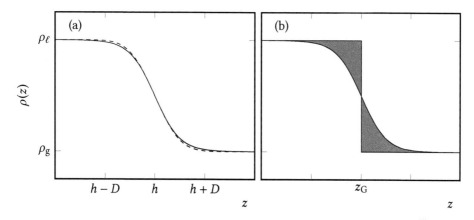

Fig. 14.2 Theoretical density profiles $\rho(z)$ in the vicinity of a liquid–gas interface. (a) Equation (14.8), hyperbolic tangent form (solid curve) and eqn (14.9), error function form (dashed curve). Parameters are chosen such that the gradients at $z = h$ are equal. (b) The Gibbs dividing surface: z_G is defined to make the shaded areas equal.

volume averaging method is exactly equivalent to the MOP and IK approaches. It has the advantage over MOP of giving all the elements of the pressure tensor. Finally we note that eqns (14.3)–(14.7) are only applicable to fluids acting through pair additive forces.

Finally chemical equilibrium might be established by test-particle insertion using eqn (2.185). However, in the gas phase, the statistics for the insertion will be poor, since there are so few atoms for the test particle to interact with at a particular value of z; and in the liquid phase, the overlaps on insertion will mean that almost all estimates of the Boltzmann factor of the test particle energy will be zero. This situation is most difficult close to the triple point of the liquid, and better estimates of $\mu(z)$ can be obtained for simulations at orthobaric densities moving towards the critical point. Accurate values of $\mu(z)$ through the particle insertion method can be obtained for softer potentials, such as the conservative potential in dissipative particle dynamics of planar liquid–liquid interfaces (Goujon et al., 2004).

14.1.3   The structure in the gas–liquid interface

For the planar interface, the single-particle density profile, $\rho(z)$, defined in eqn (2.175), may be calculated as discussed in the previous section. In practice $\rho(z)$ can often be accurately fitted by a function of the form

$$\rho(z) = \tfrac{1}{2}(\rho_\ell + \rho_g) - \tfrac{1}{2}(\rho_\ell - \rho_g)\tanh\big(2(z-h)/D\big) \tag{14.8}$$

where h is the interface position, the limiting densities are ρ_ℓ ($z \ll h$) and ρ_g ($z \gg h$), and D is a measure of the interfacial thickness (see Figs 2.4 and 14.2(a)). Eqn (14.8) can be inverted and the two empirical parameters, h and D, can be determined directly from $\rho(z)$. This is not the only possible functional form: there is simulation (Ismail et al., 2006) and experimental (Bu et al., 2014) evidence in favour of an error function profile,

$$\rho(z) = \tfrac{1}{2}(\rho_\ell + \rho_g) - \tfrac{1}{2}(\rho_\ell - \rho_g)\,\text{erf}\big(\sqrt{\pi}(z-h)/D\big), \tag{14.9}$$

also illustrated in Fig. 14.2(a), and in principle the vapour–liquid interface may exhibit asymmetry (Parry et al., 2015) due to the different correlation lengths in the two phases.

For many purposes, h can be taken equal to the position of the Gibbs dividing surface z_G which is formally defined in terms of the density profile, without reference to a model function. For a single interface (see Fig. 14.2(b))

$$\int_{-\infty}^{z_G} dz \left(\rho_\ell - \rho(z)\right) = \int_{z_G}^{+\infty} dz \left(\rho(z) - \rho_g\right), \tag{14.10}$$

where again we choose the liquid (gas) to lie at $z < 0$ ($z > 0$). In slab geometry, with two interfaces, the $\pm\infty$ limits in this equation must be replaced by positions deep in the bulk of either phase; similarly, eqn (14.8) or (14.9) should be fitted separately to the region around each interface. A smooth $\rho(z)$ that is completely symmetrical around the centre of the slab is a signature of a well-equilibrated simulation. In the calculation of $\rho(z)$ it is necessary to make sure that the centre of the slab is not drifting. This could artificially smooth any structure in the calculated $\rho(z)$ and, at very least, artificially broaden the intrinsic profile.

It is also useful to be able to calculate the pair distribution function in a simulation of the planar interface. Consider two atoms at positions $\mathbf{r}$ and $\mathbf{r}'$. Cylindrical symmetry means that we can write $\rho^{(2)}(\mathbf{r}, \mathbf{r}')$ as $\rho^{(2)}(z, z', s)$, where $s = \sqrt{(x-x')^2 + (y-y')^2}$ is their separation in the plane of the surface, or equivalently as $\rho^{(2)}(z, c, r)$, where $r = |\mathbf{r} - \mathbf{r}'|$ is their separation in 3D, $c = \cos\theta = (z-z')/r$, and θ is the polar angle of $\mathbf{r} - \mathbf{r}'$ with respect to the surface normal. The coordinates (z, c, r) are the easiest to employ in a system with periodic boundary conditions since everything is referenced to the position $\mathbf{r}$ in the central box. Starting from eqn (2.176) we can derive a form suitable for calculation in a simulation. The first step is to manipulate the δ functions to obtain

$$\rho^{(2)}(z, \mathbf{r} - \mathbf{r}') = \frac{1}{\mathcal{A}} \left\langle \sum_{i=1}^{N} \sum_{j \neq i} \delta(z - z_i) \, \delta(\mathbf{r} - \mathbf{r}' - \mathbf{r}_{ij}) \right\rangle. \tag{14.11}$$

Here we have used the result $\int d\mathbf{r} \, \delta(\mathbf{r} - \mathbf{a})\delta(\mathbf{r} - \mathbf{b}) = \delta(\mathbf{a} - \mathbf{b})$, and translational invariance in the xy-plane to integrate freely over $\mathbf{s} = (x, y)$, giving $\int d\mathbf{s} = \mathcal{A}$. Factorizing the second delta function into spherical co-ordinates and considering the cylindrical symmetry, we have

$$\rho^{(2)}(z, c, r) = \frac{1}{2\pi r^2 \mathcal{A}} \left\langle \sum_{i=1}^{N} \sum_{j \neq i} \delta(z - z_i) \, \delta(r - r_{ij}) \, \delta(c - c_{ij}) \right\rangle. \tag{14.12}$$

Converting this to a formula based on a grid with spacing Δr along the radial direction, Δc for the polar coordinate, and Δz for the direction normal to the interface, by replacing the delta functions with the top-hat functions of eqn (14.2), we obtain

$$\rho^{(2)}(z, c, r) = \frac{1}{2\pi r^2 \mathcal{A} \Delta r \Delta c \Delta z} \times$$

$$\left\langle \sum_{i=1}^{N} \sum_{j \neq i} H(z - z_i, \Delta z) \, H(c - c_{ij}, \Delta c) H(r - r_{ij}, \Delta r) \right\rangle. \tag{14.13}$$

Code 14.2 Radial distribution function in a planar interface

This file is provided online. `grint.f90` contains a program to read in a trajectory of atomic positions, and calculate the two-body density $\rho^{(2)}(z, c, r)$, and hence the radial distribution function $g^{(2)}(z, c, r)$ for a system whose single-particle density varies in the z direction.

```
! grint.f90
! g(z,c,r) in a planar interface
PROGRAM grint
```

Note that $\rho^{(2)}$ has dimensions (length)$^{-6}$ as does the right-hand side. Eqn (14.13) can be implemented in the simulation, or afterwards, as shown in Code 14.2.

The dimensionless radial distribution function is given by

$$g^{(2)}(z_1, c_{12}, r_{12}) = \frac{\rho^{(2)}(z_1, c_{12}, r_{12})}{\rho(z_1)\rho(z_1 + r_{12}c_{12})}$$

and will tend to 1 at large values of r_{12}. Once $g^{(2)}$ has been calculated as a function of the three variables it can be readily transformed to another coordinate system as required:

$$g^{(2)}(z_1, c_{12}, r_{12})\, r_{12}^2\, dz_1 dc_{12} dr_{12} = g^{(2)}(z_1, z_2, s_{12})\, s_{12}\, dz_1 dz_2 ds_{12}$$

$$= g^{(2)}(z_1, z_2, r_{12})\, r_{12}\, dz_1 dz_2 dr_{12}. \qquad (14.14)$$

$\rho(z)$ and $\rho^{(2)}(z_1, z_2, s_{12})$ are related through the first Born–Green–Yvon equation (Rowlinson and Widom, 1982)

$$k_B T \ln\left(\frac{\rho(z)}{\rho_\ell}\right) = -2\pi \int_{-\infty}^{z} dz_1 \int_{-\infty}^{\infty} dz_2\, z_{12}\, \rho(z_2) \times$$

$$\int_0^\infty ds_{12} \left(\frac{s_{12}}{r_{12}}\right) v'(r_{12})\, g^{(2)}(z_1, z_2, s_{12}), \qquad (14.15)$$

where $v'(r_{12})$ is the derivative of the intermolecular potential, $z_{12} = z_1 - z_2$, ρ_ℓ is the limiting density in the liquid phase, and $g^{(2)}(z_1, z_2, s_{12}) = \rho^{(2)}(z_1, z_2, s_{12})/\rho(z_1)\rho(z_2)$. This can be a useful check of the simulated distribution functions since they should satisfy eqn (14.15) exactly.

After our discussion of capillary-wave fluctuations in the next section, we return briefly to the description of structure in the interface in Example 14.1.

14.1.4 The surface tension

As discussed in Chapter 2, the surface tension can be obtained from the difference between the normal and tangential components of the pressure (see eqn (2.183a)), and we have explained how to measure these in Section 14.1.2. It can also be more directly calculated using the expression due to Buff (1952), eqn (2.184)

$$\gamma = \frac{1}{4\mathcal{A}} \left\langle \sum_i \sum_{j>i} \left(r_{ij} - \frac{3z_{ij}^2}{r_{ij}} \right) v'(r_{ij}) \right\rangle, \tag{14.16}$$

where it is assumed that the simulation contains two planar interfaces normal to z.

The value of the surface tension is particularly sensitive to potential truncation: the long-range correction to eqn (14.16) can be as much as 35 % of the total value of γ, in a simulation using, for example, a Lennard-Jones potential with a potential cutoff $r_c = 3.0 \, \sigma$. Blokhuis et al. (1995) show that this correction can be calculated from eqn (2.184). Taking $\rho^{(2)}(r_{12} > r_c, z_1, z_2) = \rho(z_1)\rho(z_2)$, substituting eqn (14.8) for $\rho(z)$, integrating over z_1 and introducing $c_{12} = (z_1 - z_2)/r_{12}$, we obtain

$$\gamma_{\text{LRC}} = \frac{\pi(\rho_\ell - \rho_g)^2}{2} \int_0^1 \mathrm{d}c_{12} \int_{r_c}^\infty \mathrm{d}r_{12} \, \coth\left(\frac{2r_{12}c_{12}}{D}\right) v'(r_{12}) r_{12}^4 \left(3c_{12}^3 - c_{12}\right). \tag{14.17}$$

This integral can be readily evaluated using a simple quadrature. It is normally calculated at the end of a simulation using the values of $\rho(z)$ corresponding to the truncated potential, and, as such, provides a lower bound to the accurate long-range correction to γ for the full potential.

To improve on this estimate, values of the long-range correction to the energy or force are required at each configuration or step of the simulation, so that they can be used to drive the evolution of the system. This problem was first tackled by Guo and Lu (1997) and in a simpler fashion by Janeček (2006).

The method of Janeček is based on the calculation of the energy of atom i with all the atoms in a slab z_k with a number density $\rho(z_k)$. The long-range correction to the energy associated with atom i is

$$\mathcal{V}_i^{\text{LRC}}(z_i) = \sum_{k=1}^{n_s} w\big(|z_i - z_k|\big)\rho(z_k)\Delta z, \tag{14.18}$$

where the sum is over all slabs in the simulation cell in the z-direction. The contribution $w(z)$ is calculated by assuming a uniform distribution of atoms in the slab. For the Lennard-Jones potential

$$w(z) = \begin{cases} 4\pi\epsilon\sigma^2 \left[\frac{1}{5}\left(\frac{\sigma}{r_c}\right)^{10} - \frac{1}{2}\left(\frac{\sigma}{r_c}\right)^4 \right] & z \le r_c \\[4mm] 4\pi\epsilon\sigma^2 \left[\frac{1}{5}\left(\frac{\sigma}{z}\right)^{10} - \frac{1}{2}\left(\frac{\sigma}{z}\right)^4 \right] & z > r_c. \end{cases} \tag{14.19}$$

The total long-range correction for a particular configuration of atoms is

$$\mathcal{V}^{\text{LRC}} = \tfrac{1}{2} \sum_{i=1}^N \mathcal{V}_i^{\text{LRC}}(z_i). \tag{14.20}$$

This expression can be used in an MC simulation when calculating the energy change in a trial move. Eqn (14.20) is not convenient, as written, since it requires a recalculation of the density profile for each new trial configuration. MacDowell and Blas (2009) have shown that by substituting eqn (2.175) for the density profile into eqn (14.18) and replacing the summation over slabs by an integration, the long-range correction can be written conveniently as a sum over all pairs of atoms and N self-energy terms:

$$\mathcal{V}^{\text{LRC}} = \frac{1}{\mathcal{A}} \sum_{i=1}^{N-1} \sum_{j>i} w\big(|z_i - z_j|\big) + \frac{1}{2\mathcal{A}} \sum_{i=1}^{N} w(0). \tag{14.21}$$

$\mathcal{V}^{\text{LRC}}$ can be evaluated at each trial move of an MC simulation and differentiated to give a force that can be used in an MD calculation.

The methods of Guo and Lu, and of Janeček, applied at each step of the simulation, provide the same density profile corresponding to the full potential (Goujon et al., 2015). Once the simulation is completed and the density profile is available, then eqn (14.17) can be used to calculate the long-range correction to the surface tension γ. An accurate value of γ_{LRC} can also be calculated directly at each step in the simulation without reference to the profile (Goujon et al., 2015). These various methods of calculating the long-range corrections have been evaluated and assessed for a number of atomic and molecular simulations for planar interfaces (Ghoufi et al., 2008; Míguez et al., 2013; Goujon et al., 2015).

The LRC for slab simulations with a dispersion interaction between atoms can be included by applying an Ewald sum to the extended box (López-Lemus and Alejandre, 2002). The Ewald sum for an r_{ij}^{-6} potential has been discussed in Section 6.2. The calculation of γ is independent of the precise real-space truncation of the potential. Alejandre and Chapela (2010) and Isele-Holder et al. (2012) have used the SPME and PPPM methods to simulate the vapour–liquid interface of water treating both the dispersion and the Coulombic interactions within the framework of the lattice sum. The results for the surface tension and coexistence curves using the Ewald sum are in good agreement with simulations conducted using spherical cutoffs for the dispersion interaction.

In the case of non-pairwise-additive potentials such as the Axilrod–Teller potential (described in Appendix C.4) there is an additional contribution to eqn (14.16) for the surface tension (Toxvaerd, 1972)

$$\gamma^{\text{AT}} = \frac{3}{4\mathcal{A}} \left\langle \sum_{i<j<k} \frac{\partial \mathcal{V}^{(3)}}{\partial r_{ij}} \left(\frac{r_{ij}^2 - 3z_{ij}^2}{r_{ij}} \right) \right\rangle, \tag{14.22}$$

again, assuming two planar interfaces. The derivative with respect to r_{ij} is at constant r_{ik}, r_{jk}. Note that, for the Axilrod–Teller potential, there are three identical terms for the derivative with respect to each variable, giving rise to the factor of 3 in the numerator. The long-range correction to the three-body part of γ can be estimated using the superposition approximation as discussed in Section 2.8. It can be evaluated numerically using the simulated radial distribution function, $g^{(2)}(z_1, c_{12}, r_{12})$ (Goujon et al., 2014).

An alternative and useful method of calculating the surface tension is the test area (TA) method of Gloor et al. (2005). This method uses the thermodynamic definition of the surface tension as the change in the Helmholtz free energy A with area $\mathcal{A}$

$$\gamma = \lim_{\Delta\mathcal{A}\to 0} \left(\frac{\Delta A}{\Delta\mathcal{A}} \right)_{NVT}. \tag{14.23}$$

A normal MD or MC simulation of the gas–liquid interface is performed as a reference system (0), and perturbations around this reference are performed making small changes to the area $\mathcal{A} = 2L_x L_y$ (again, assuming two interfaces) at a constant overall volume. If $\Delta\mathcal{A} = \mathcal{A}_1 - \mathcal{A}_0$ then

$$\gamma = \lim_{\Delta\mathcal{A}\to 0} -\frac{k_B T}{\Delta\mathcal{A}} \ln\langle\exp(-\beta\Delta\mathcal{V})\rangle_0 \qquad (14.24)$$

where $\Delta\mathcal{V} = \mathcal{V}_1 - \mathcal{V}_0$ and the average is calculated over the states of the reference system. The perturbations are achieved by scaling the sides of the box

$$L_{x,1} = L_{x,0}(1+\epsilon)^{1/2}, \quad L_{y,1} = L_{y,0}(1+\epsilon)^{1/2}, \quad L_{z,1} = L_{z,0}(1+\epsilon)^{-1}, \qquad (14.25)$$

where $\epsilon = \Delta\mathcal{A}/\mathcal{A}_0 \ll 1$. The particle coordinates are scaled in the same way; this happens automatically if they are stored in box-scaled form, $\mathbf{s}_i = (x_i/L_x, y_i/L_y, z_i/L_z)$. (See also the box-scaling method for pressure estimation in hard-particle simulations, discussed in Section 5.5.) These trial perturbations can be performed regularly throughout the simulation, without affecting the underlying evolution of the reference system. ϵ needs to be chosen to be sufficiently small that eqn (14.24) is an accurate representation of γ, and large enough to allow an accurate calculation of the Boltzmann factor. A value of $\epsilon = 5 \times 10^{-4}$ produces values of γ comparable to those from eqn (14.16) for a variety of atomic and molecular systems (Ghoufi et al., 2008). There is nothing unique about a positive choice for ϵ, and one useful check is to make corresponding reductions in the surface area throughout the reference simulation, and to check that these result in the same value of γ. The techniques discussed in this section for calculating the long-range correction apply to the TA method, and the technique can be used for pair-additive, non-pair-additive and molecular potentials. Errington and Kofke (2007) have used the TA method to calculate the surface tension of the square-well fluid. Their results suggest that in the case of discontinuous potentials, the configuration spaces sampled by the reference and perturbed systems differ significantly even for small $\Delta\mathcal{A}$ and this leads to inaccuracies in the estimates of γ. This problem can be avoided by the application of Bennett's method to the calculation of the free-energy difference for this model (see Section 9.2.4).

Table 14.1 shows the estimates of the liquid–vapour surface tension of methane modelled as a single Lennard-Jones atom. The TA and IK methods give results for γ that agree within the statistical uncertainties of the simulation. Either method can be used with confidence for continuous potentials. At a cutoff $r_c = 2.5\sigma$ the long-range correction for γ is 36 % of the total value reducing to 8 % at a cutoff of 8σ. The tail correction applied at the end of the simulation (Blokhuis et al., 1995) consistently underestimates the more accurate correction of Janeček (2006). If the full long-range correction is calculated at every step, then a cutoff of 3σ is sufficient for an accurate estimate of γ. However, if the correction of Blokhuis et al. is employed at the end of the simulation, then a cutoff of $r_c = 4.5\sigma$ is required for the accurate calculation of ρ_ℓ in eqn (14.17) (Goujon et al., 2015).

As an alternative to the aforementioned methods, the surface tension may be obtained by analysing the capillary-wave fluctuations. The idea is to determine the interface position or height $h(x, y)$ and fit its Fourier coefficients to eqn (2.187) or its mean-square deviation, or the interface width, to eqn (2.188). A useful summary of the different variants of this approach is provided by Werner et al. (1999), and a comparison with other methods by

Table 14.1 Simulation of the liquid–vapour surface tension $\gamma/\mathrm{mJ\,m^{-2}}$ for the Lennard-Jones methane model at 120 K as a function of cutoff radius. Results are given for the truncated potential without a long-range correction; with the LRC of Blokhuis et al. (1995); and with the LRC of Janeček (2006). Subscripts indicate the test area (TA) and Irving–Kirkwood (IK) methods respectively. The number in parentheses is the estimated error in the final quoted decimal place. Reprinted from J. M. Míguez, M. M. Piñeiro, and F. J. Blas, *J. Chem. Phys.*, **138**, 034707 (2013) with the permission of AIP Publishing.

r_c/σ	No correction		Blokhuis et al. (1995)		Janeček (2006)	
	γ_{TA}	γ_{IK}	γ_{TA}	γ_{IK}	γ_{TA}	γ_{IK}
2.5	8.67(7)	8.67(7)	12.66(6)	12.66(7)	13.61(7)	13.64(7)
3	10.41(7)	10.41(7)	13.55(1)	13.55(7)	13.9(1)	13.8(1)
4	11.75(9)	11.75(9)	13.66(9)	13.66(9)	13.8(1)	13.8(1)
5	12.74(9)	12.73(9)	14.02(9)	14.03(9)	13.8(1)	13.8(1)
8	12.7(1)	12.7(1)	13.4(1)	13.4(1)	13.7(1)	13.8(1)

Schrader et al. (2009). The simplest approach, conceptually, is to fit the density profile of the system in a box $L \times L \times L_z$ to an equation such as eqn (14.8), for several different transverse dimensions L, and observe the increase in the apparent width of the interface, $\langle D \rangle$, with L. In the key formula, eqn (2.189),

$$\langle D \rangle^2 = D_0^2 + \frac{k_B T}{4\gamma} \ln\left(\frac{L}{a}\right) = \left[D_0^2 - (k_B T/4\gamma) \ln a\right] + (k_B T/4\gamma) \ln L,$$

it is not possible to disentangle the intrinsic width D_0 and the small-scale cutoff a: both are unknown. It is therefore essential to measure D at many values of L, confirm the logarithmic behaviour predicted by this equation, and extract the coefficient of $\ln L$.

Instead of simulating many systems with different transverse dimensions L, it is possible to perform a block analysis on a single system of large L, subdividing it into $M \times M$ columns of size $\ell \times \ell$ where $M = L/\ell = 2, 3$, etc. This is illustrated in Fig. 14.3. For each column, it is possible to determine the Gibbs dividing surface and set $h_\ell = z_G$, or to fit the profile, giving h_ℓ and D_ℓ, where we indicate the dependence on column width. It is convenient to measure the height relative to the average interface position for the whole box (translations of the interface in the z direction do not change its surface area, and hence are independent of the capillary waves). From these measurements, averaging over the separate columns and over the simulation run, we can compute the average width D_ℓ, and the probability distribution of heights h_ℓ, including the mean-squared deviation $\langle \delta h_\ell^2 \rangle = \langle h_\ell^2 \rangle - \langle h_\ell \rangle^2$. The relevant capillary-wave predictions, eqns (2.189) and (2.188) are

$$\langle D_\ell \rangle^2 = D_0^2 + \frac{k_B T}{4\gamma} \ln\left(\frac{\ell}{a}\right), \tag{14.26a}$$

$$\langle \delta h_\ell^2 \rangle = \frac{k_B T}{2\pi\gamma} \ln\left(\frac{L}{\ell}\right). \tag{14.26b}$$

This makes it possible to determine γ from simulations of a single system, from the dependence of D_ℓ on the column dimension ℓ, and we expect eqn (14.26a) to hold for

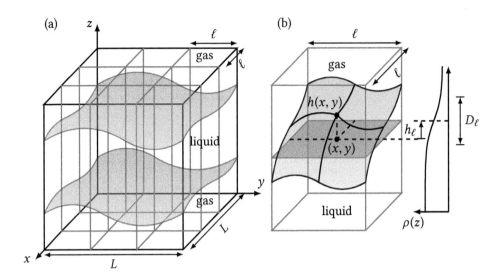

Fig. 14.3 Refinement of the gas–liquid interface. (a) Subdivision of the simulation box into $M \times M$ columns, with $M = 3$. The two interfaces are shaded; the transverse box width L and column width $\ell = L/M$ are indicated. (b) Upper half of one column, containing part of one of the interfaces. The interface height $h(x, y)$ within this column is defined with respect to the average for the whole simulation box (dark-grey plane). The density profile for this column may be used to calculate a column-averaged height h_ℓ and interface width D_ℓ.

sufficiently large ℓ. However, this does require a sizeable system, and there are still drawbacks: for instance, Ismail et al. (2006) have reported that the results depend on the choice of functional form used to fit the profiles, with eqn (14.9) giving results in better agreement with the stress tensor route to γ than eqn (14.8). In principle, eqn (14.26b) relies only on determining the distribution of interface *positions*, which can be done without invoking a specific model profile. The capillary waves of wavelength $\lambda < \ell$ contribute to the width D_ℓ within each column, while those for which $\ell < \lambda < L$ contribute to the positional fluctuations. In practice, we expect eqn (14.26b) to give good results for large ℓ, except possibly for $\ell = L/2$ when the values of h_ℓ are constrained by the (effectively fixed) box-averaged interface. An essentially equivalent method involves fitting to eqn (2.187), having determined the Fourier components $\hat{h}(\mathbf{k})$. This can be achieved by the same kind of process described here, choosing a small value of ℓ to determine $h(x, y)$ at fine resolution, followed by Fourier transformation.

This is an appropriate point to mention the 'surface-pinned' definition of the intrinsic surface (Chacón and Tarazona, 2003). Tarazona and Chacón (2004) have developed a method for calculating this in a simulation. For a particular configuration, all the atoms within the liquid slab are identified. (For example, for a Lennard-Jones fluid, atoms with less than three nearest neighbours closer than $r_{\mathrm{c}} = 1.5\sigma$ are assigned to the gas phase and any spurious overhangs or vapour cluster atoms are removed from the liquid phase.) The atoms are then assigned to $M \times M$ columns, as shown in Fig. 14.3, choosing the most external 'liquid' atom in each column to provide the initial set of M^2 pivot points. $M = 3$ provides an initial, coarse representation of $h(x, y) \equiv h(\mathbf{s})$. The initial intrinsic surface is defined as the minimal-area surface going through all the pivots, obtained by minimizing a combination of two terms. The first term is a sum of squared distances of the refined pivot points from the existing ones. The second, much smaller, contribution is the sum of Fourier terms appearing in eqn (2.187), for $|\mathbf{k}| < 2\pi/a$, a being the short-range cutoff, which essentially gives the extra surface area due to non-planarity. At the end of this process we have a refined set of M^2 pivot points. A number of new pivot points are added by including atoms that are within a fixed, predefined distance of the old pivot point. A new intrinsic surface is then computed using the extended set of pivot points, and re-minimizing the surface area. This procedure is repeated until there are no additions to the list of pivot points. The resulting $h(\mathbf{s})$, or $\hat{h}(\mathbf{k})$, may be compared with the predictions of capillary-wave theory, and used to estimate γ using the approaches described earlier. As we discuss in Example 14.1, the 'intrinsic surface' also gives us the opportunity to look more closely at the structure within the interface.

The methods described in this section can be applied to the calculation of γ for fluids where the density difference between the gas and the liquid is significant and the interface is stable. As we approach the critical temperature of the fluid, this is not the case and we need an alternative approach. If we are not *too* close to the critical point, we can take advantage of the well-understood finite-size scaling behaviour near the first-order liquid–vapour transition (Binder and Landau, 1984; Borgs and Kotecký, 1990); see also the reviews of Binder et al. (2011) and Binder (2014), and especially Binder et al. (2012). The study of probability distributions of density, or number of particles, using the method of histogram reweighting, has been applied to the calculation of γ for the Lennard-Jones system by Potoff and Panagiotopoulos (2000) and Block et al. (2010), and for simple alkanes by Virnau et al. (2004). These simulations do not involve the simulation of a prepared interface, but all the effects arise from interfacial contributions to configurations involving the two phases. We describe them here to complete our discussion of γ.

We consider the grand canonical MC simulation of a system (see Chapter 4.6) in the two-phase region of the phase diagram. Close to the critical point, the system will fluctuate between the liquid and gas phases and the probability, $\rho_{\mu VT}(N)$, of finding a certain number of atoms in the simulation will take the characteristic bimodal distribution shown in Fig. 14.4. It is assumed that μ takes the coexistence value, corresponding to equal areas under the two peaks; histogram reweighting

$$\rho_{\mu VT}(N) = \exp\left[\beta(\mu - \mu')N\right]\rho_{\mu'VT}(N)$$

may be used to convert the results of a simulation at a nearby value μ'. This is illustrated schematically in Fig. 14.4(a), and the same function is plotted on a logarithmic scale in Fig. 14.4(b). This is proportional to the Landau free energy with respect to N

$$\mathcal{F}(N) = -k_\mathrm{B}T \ln \rho_{\mu VT}(N).$$

Assuming that the dominant configurations in the region between N_g and N_ℓ consist of a slab of liquid surrounded by gas, this allows us to estimate the surface tension (for two

---

Example 14.1 Intrinsic structure

The capillary waves discussed in this section have the effect of smoothing out all the interfacial structure. The possibility of removing this effect, and measuring the 'intrinsic' structure, has been debated for many years. If one adopts the 'surface-pinned' definition of the intrinsic surface $h(\mathbf{s})$ (Chacón and Tarazona, 2003; Tarazona and Chacón, 2004), defined for a configuration snapshot, this may be used to calculate the intrinsic density profile (compare eqn (2.175))

$$\tilde{\rho}(z) = \frac{1}{\mathcal{A}_0}\left\langle \sum_{i=1}^{N} \delta\left(z - z_i + h(x_i, y_i)\right)\right\rangle, \tag{14.27}$$

where $\mathcal{A}_0 = L_x L_y$. The value of h at the coordinates (x_i, y_i) of any atom may be calculated from the Fourier representation $\hat{h}(\mathbf{k})$. The profile is then averaged over independent configurations. The intrinsic structure is richer than the normal averaged density profile which decays monotonically in moving from the liquid to the gas, as in Fig. 14.2. The intrinsic structure just inside the liquid is layered (as in the case of liquid against a planar wall).

Mecke and Dietrich (1999) have pointed out that it would be more exact to use a z-coordinate normal to the surface to calculate the profile. This will be important if the surface is particularly rough. The difficulty in this approach is in the calculation of the appropriate normalizing area. Tarazona and Chacón (2004) have described a simple interpolation scheme to calculate the area under these circumstances.

In a recent review, Tarazona et al. (2012) have shown the application of the intrinsic surface method to the gas–liquid interface of molten salts. They note the importance of the technique in helping to develop an understanding of the wavevector-dependent surface tension and the density of the outermost liquid layer, where there are currently no theories. The simulations will help to understand the differences between the area-independent profiles from density functional theory and the area-dependent profile from capillary-wave theory and reflectivity experiments (Pershan, 2009).

---

interfaces of area $\mathcal{A} = L^2$) according to

$$\Delta A = 2\gamma \mathcal{A} = k_{\mathrm{B}}T\left[\ln \rho_{\mu VT}^{\max} - \ln \rho_{\mu VT}^{\min}\right] \tag{14.28}$$

where $\rho_{\mu VT}^{\min} = \rho_{\mu VT}(N_{\min})$ is the minimum between the two peaks and

$$\rho_{\mu VT}^{\max} = \tfrac{1}{2}\left(\rho_{\mu VT}(N_\ell) + \rho_{\mu VT}(N_{\mathrm{g}})\right) \tag{14.29}$$

is the average of the maxima of the liquid and gas peak heights. (These peak heights are, in general, not equal because the equilibrium condition is specified by equal peak *areas*, while the peak widths are typically different.)

However, there is some uncertainty associated with the assumption that $\rho_{\mu VT}^{\min}$ corresponds to the desired slab geometry. If this is the case, the probability distribution of

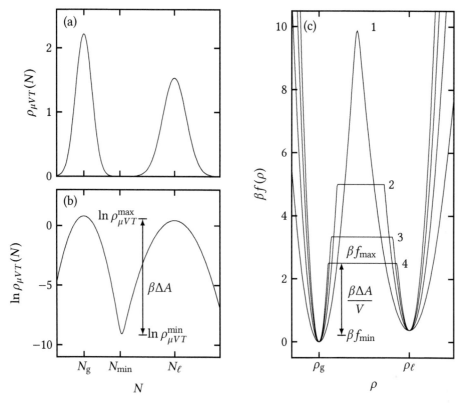

Fig. 14.4 Probability distributions and free-energy densities in the liquid–vapour coexistence region (schematic). (a) Bimodal probability distribution $\rho_{\mu VT}(N)$ of number of atoms N in a grand canonical simulation. (b) The same function, on a logarithmic scale. The indicated difference $\beta\Delta A$ should be equal to the interfacial free energy $2\gamma\mathcal{A}$. (c) The free-energy density $\beta f(\rho) = -\ln\rho_{\mu VT}(N)/V$ for different system sizes. Curve 1 is the same as illustrated in (b), box length L. The other curves represent systems of size $2L$, $3L$ and $4L$ respectively.

N should have an observable flat region in between the two single-phase peaks. This is because the two planar interfaces should be far enough apart not to influence each other, and the conversion of liquid into vapour (or vice versa) will simply change the interface positions, at no free-energy cost. This is more likely to be observed for larger system sizes, and/or when the box is chosen to be somewhat elongated with $L_x = L_y = L < L_z$. It is instructive to convert $\mathcal{F}(N)$ into a Landau free-energy density, $f = \mathcal{F}/V$ and express it as a function of number density $\rho = N/V$. This is schematically illustrated in Fig. 14.4(c), for different system sizes L. As L increases, the expected plateau region develops and becomes better defined. This makes the definition of $f_{\max}$, corresponding to $\rho_{\mu VT}^{\min}$, less ambiguous. At the same time, the free-energy difference per unit volume decreases in proportion to L^{-1}. (In the thermodynamic limit it becomes a straight line between ρ_g and ρ_ℓ.) It is arguable that one should distrust the value of γ obtained by this route, unless a plateau region has been observed.

In general, the two-phase density range will consist of macroscopic configurations of spherical or cylindrical droplets or bubbles, as well as the slabs discussed earlier; the precise mix will depend on system size and box geometry. It is possible to study these in their own right (MacDowell et al., 2006; Binder et al., 2011; 2012). Recently Wilding (2016) has introduced ways to improve the sampling by suppressing the non-slab configurations, thereby extending the range of applicability of the method.

There is still a residual system-size dependence in the measured value of γ. For a cubic box of side L, scaling arguments (Binder, 1982) predict

$$\beta \gamma_L = \beta \gamma_\infty - s \frac{\ln L}{L^2} - \frac{C}{2L^2} \tag{14.30}$$

where s is a universal exponent and C a constant. γ_L can be calculated for a series of simulations of different box sizes (Potoff and Panagiotopoulos, 2000; Hunter and Reinhardt, 1995); plots of $\beta \gamma_L$ vs $L^{-2} \ln L$ are linear and, for temperatures below the critical point, can be extrapolated to give an estimate of γ_∞. Schmitz et al. (2014) discuss the origins of these effects and also consider different box geometries.

The method as described will work straightforwardly for $T \gtrsim 0.95\,T_c$. For temperatures below this, the free-energy barrier between the two phases is sufficiently high that it is necessary to employ umbrella sampling or multicanonical sampling techniques to obtain accurate estimates of $\rho_{\mu VT}(N)$ (Errington, 2003; Virnau and Müller, 2004). For the Lennard-Jones fluid, the finite-size scaling methods can be used to calculate γ for $T \gtrsim 0.7\,T_c$ (Potoff and Panagiotopoulos, 2000). This method forms a useful complementary technique to the direct simulation of the interface for calculating γ. Note, however, that direct simulations of large-scale LJ systems near the critical point (Watanabe et al., 2012) do not completely agree with the conclusions of Potoff and Panagiotopoulos (2000).

14.2   The gas–liquid interface of a molecular fluid

The ideas discussed in the last section can be readily applied to molecular fluids. The pressure tensor of a molecular fluid is calculated from the pair-additive force on the centre of mass of molecule i from another molecule centred at j. In the case of an interaction site model, the Kirkwood–Buff equation (see eqns (2.184) and (14.16)) for the surface tension becomes

$$\gamma_{KB} = \frac{1}{4\mathcal{A}} \left\langle \sum_i \sum_{j>i} \sum_a \sum_b \left(\frac{r_{ij} \cdot r_{ab} - 3z_{ij}z_{ab}}{r_{ab}} \right) \frac{dv_{ab}(r_{ab})}{dr_{ab}} \right\rangle \tag{14.31}$$

where a and b range over the individual atoms in molecules i and j, respectively, v_{ab} is the site–site potential, and we again assume two planar interfaces in a slab geometry.

If the fluid is composed of ions in solution or small molecules such as water, where a number of partial charges have been used to model the dipole, then the interactions are long-range and we need to employ techniques such as the Ewald sum in the simulation. Since the inhomogeneous system, modelled as a slab of liquid surrounded by vapour, is periodic, the Ewald sum can be applied in a straightforward manner to calculate the total energy and pressure of the charged system, and this approach is discussed in detail in Section 6.9. For now, we will simply consider the contribution of the charge–charge interaction to the pressure tensor and the surface tension.

Table 14.2 Simulated values of the surface tension, in $\mathrm{mJ\,m^{-2}}$, for three molecular fluids. γ_{LJ} is the contribution from the repulsion–dispersion potential, γ_R, $\gamma_{K,1}$ and $\gamma_{K,2}$ are the contributions from the Ewald sum, and γ_{LRC} is the contribution from the long-range part of the repulsion–dispersion potential evaluated by the method of Guo and Lu (1997). The water model was TIP4P/2005. Reprinted from A. Ghoufi, F. Goujon, V. Lachet, and P. Malfreyt, *J. Chem. Phys.*, **128**, 154716 (2008) with the permission of AIP Publishing.

System	$T\,/\,\mathrm{K}$	γ_{LJ}	γ_R	$\gamma_{K,1}$	$\gamma_{K,2}$	γ_{LRC}	γ_{tot}
H_2O	478	−87.6	112.1	−6.4	12.0	3.5	33.6
CO_2	238	6.1	2.0	−0.8	1.7	2.5	11.5
H_2S	187	21.0	8.0	−1.5	2.8	8.0	38.3

For an inhomogeneous system, the components of the pressure tensor can be calculated using an approach described by Nosé and Klein (1983, Appendix A). Thus

$$\sum_{\gamma} P_{\alpha\gamma} V H^{-1}_{\beta\gamma} = -\frac{\partial V}{\partial H_{\alpha\beta}} \tag{14.32}$$

where the matrix $\mathbf{H} = (\mathbf{h}_1, \mathbf{h}_2, \mathbf{h}_3)$ is composed of the column vectors $\mathbf{h}_\alpha$ that are the sides of the box of volume V, and $\mathbf{H}^{-1}$ is its inverse. Then, for a cubic box,

$$
\begin{aligned}
VP_{\alpha\beta} = \Bigg\langle & \frac{1}{2}\sum_i \sum_{j\neq i}\sum_a \sum_b q_{ia}q_{jb}\left(\frac{2}{\sqrt{\pi}}\kappa r_{ab}\exp(-\kappa^2 r^2_{ab}) + \mathrm{erfc}(\kappa r_{ab})\right)\frac{(\mathbf{r}_{ij})_\alpha \cdot (\mathbf{r}_{ab})_\beta}{r^3_{ab}} \\
& + \left[\frac{2\pi}{V}\sum_{k\neq 0}Q(k)S(\mathbf{k})S(-\mathbf{k})\left(\delta_{\alpha\beta} - \frac{2k_\alpha k_\beta}{k^2} - \frac{k_\alpha k_\beta}{2\kappa^2}\right)\right] \\
& - \left[\frac{2\pi}{V}\sum_i \sum_a (d_{ia})_\beta\, q_{ia}\sum_{k\neq 0}Q(k)ik_\alpha\big(S(\mathbf{k})\exp(-i\mathbf{k}\cdot\mathbf{r}_{ia}) - S(-\mathbf{k})\exp(i\mathbf{k}\cdot\mathbf{r}_{ia})\big)\right]\Bigg\rangle.
\end{aligned}
\tag{14.33}
$$

In eqn (14.33) $\mathbf{k}$ is a reciprocal lattice vector for the box, $\mathbf{d}_{ia} = \mathbf{r}_{ia} - \mathbf{r}_i$,

$$S(\mathbf{k}) = \sum_i \sum_a q_{ia}\exp(-i\mathbf{k}\cdot\mathbf{r}_{ia}), \quad \text{and} \quad Q(k) = 4\pi^2\exp(-k^2/4\kappa^2)/k^2. \tag{14.34}$$

The elements of $\mathbf{P}$ are all that is required to calculate the ionic contribution to γ. $P_{\alpha\beta}(z)$ can be calculated using the techniques discussed in Section 14.1.1; explicit formulae are given in Ghoufi et al. (2008, appendix A). This ionic contribution to γ arising from the three terms in eqn (14.33), γ_R, $\gamma_{K,1}$, $\gamma_{K,2}$, is added to the contribution from the repulsion–dispersion potential, and its long-range correction, to obtain the total surface tension. The size of these contributions for three typical molecular fluids is shown in Table 14.2.

The reaction field method has also been used to include the charge–charge interaction in modelling the surface tension of water. From a formal point of view there is no straightforward extension of the reaction field method to inhomogeneous systems. However, it has been used with a spherical cutoff and a surrounding medium with a dielectric constant corresponding to that of bulk water. Since the long-range field does not have spherical symmetry, and the dielectric tensor is a function of z, this approximation cannot be correct. Nevertheless, the calculated values of γ appear to be in good agreement with those obtained using the full Ewald method for this system (Míguez et al., 2010). More work needs to be done to examine the reaction field method for this geometry.

14.3   The liquid–liquid interface

The methods of the previous sections may be applied, almost unchanged, to the study of liquid–liquid interfaces. Symmetrical mixtures of Lennard-Jones atoms have been used extensively as a test-bed: the two components A and B typically have the same ϵ and σ parameters, but the A–B cross-interactions are made unfavourable, so as to promote phase separation (Block et al., 2010; Martínez-Ruiz et al., 2015). Preparing a system in a slab geometry proceeds by first equilibrating the two bulk liquids (typically an A-rich phase and a B-rich phase) at the coexistence state point, or at least a reasonable estimate thereof. It is sometimes convenient to use the constant-$NP_{zz}T$ ensemble, allowing the L_z box dimensions to vary, while keeping both cross-sectional areas $\mathcal{A} = L \times L$ equal to each other. Then the boxes are combined, end to end, with small gaps in between to avoid the worst overlaps. Constant-NVT simulations then proceed, the gaps are quickly filled, and some time must be allowed for the interfaces to stabilize, and for thermodynamic equilibrium to be achieved.

Simulations of oil–water or oil–water–surfactant systems have focused on interfacial properties for many years. Zhang et al. (1995) discuss various methodological aspects; Neyt et al. (2014) give a recent account of surface tension measurements in such systems.

Further examples, using large systems, come from the field of complex fluids. Vink et al. (2005) studied capillary-wave fluctuations in the Asakura–Oosawa model of colloid–polymer mixtures using $L \times L \times L_z$ systems of size $L = 60\sigma$, $L_z = 120\sigma$, where σ is the colloid radius. They used the block analysis described earlier, in which the interface is localized within columns (see Fig. 14.3). As well as the surface tension, it proved possible to extract a bending rigidity (cf. Sections 2.13, 14.6) (Blokhuis et al., 2008; Blokhuis, 2009). The nematic–isotropic interface of rod-like particles of aspect ratio $\ell/d = 15$ was studied by Akino et al. (2001) and Wolfsheimer et al. (2006). In both cases box dimensions $L = 10\ell$, $L_z = 20\ell$ were employed ($N \approx 10^5$ molecules) and capillary-wave fluctuations again studied by block analysis. For this kind of system, nematic orientational ordering occurs parallel to the plane of the interface, and it is possible to distinguish capillary fluctuations with wavevector components parallel and perpendicular to the director. Once more, a bending rigidity term can be identified.

14.4   The solid–liquid interface

In simulating a solid–liquid interface, the atoms of the solid can be considered as part of the system interacting with each other and with atoms in the fluid through pair potentials, or as a static external field interacting only with the atoms in the fluid. This field will be z-dependent if the solid–liquid interface in the xy plane is considered flat, or can also depend on x and y if the interface is structured. The full range of MC and MD techniques can be applied to either of these two models. First, we consider how to determine the

melting point by direct simulation of solid–liquid coexistence, and second we examine
methods for calculating the surface energy. The methods of handling long-range potentials
in this geometry have been discussed in Section 6.9.

14.4.1 Determining the melting point

Zhang and Maginn (2012) have compared various methods for determining the melting
point, or more generally the melting curve in the phase diagram. Directly heating the solid
at a given pressure P, until it melts, will give an overestimate of the melting temperature,
due to hysteresis. Similarly, lowering the temperature of the liquid until it starts to freeze
will give an underestimate, due to supercooling. Separate simulations of the two phases
may be carried out, and free energies estimated by some of the methods of Section 9.2;
thermodynamic integration may then be used to establish a point on the melting curve. The
curve may then be traced out by the Gibbs–Duhem method of Section 9.4.2. Alternatively,
it is possible to connect the liquid and solid phases by a thermodynamic integration path
that avoids the phase transition (Grochola, 2004; 2005; Eike et al., 2005), or via gateway
states using a biased sampling scheme (Wilding and Bruce, 2000; Errington, 2004; McNeil-
Watson and Wilding, 2006).

A simpler method is direct simulation of the solid and liquid phases in coexistence.
The system is prepared in a manner similar to the liquid–liquid case: periodic boxes of
solid and liquid, having the same x and y dimensions, with P and T reasonably close to
the melting line are equilibrated separately. These are brought together, generating two
interfaces in the xy-plane, and a further period of equilibration is allowed for overlaps in
the interfacial regions to relax. During the early part of this period, it may be convenient
to keep the solid particles fixed, to avoid generating internal stresses, and to use the
constant-$NP_{zz}T$ ensemble (Zykova-Timan et al., 2009).

Then the two-phase system is allowed to reach thermodynamic equilibrium. It is worth
considering the best ensemble for this. As discussed in Section 14.1.1, a constant-NPT
simulation will slowly evolve to give a single phase, determined by whether the state
point (P, T) lies above or below the melting curve. Instead, it is recommended to use the
constant-NVE (Morris, 2002; Morris and Song, 2002; Yoo et al., 2004) or constant-NPH
ensemble (Wang et al., 2005; Brorsen et al., 2015). In either case, a significant two-phase
coexistence region exists, so (provided the fixed thermodynamic variables lie within this
region) the interfaces between the two phases should be preserved. Also, in either case,
the temperature will automatically adjust towards the coexistence value. If the system
is too hot, melting will occur, extracting latent heat and lowering the temperature. The
reverse will happen if the system is too cool. The constant-NPH ensemble is particularly
convenient, because as well as being able to specify the pressure, one may allow the
box dimensions to vary separately, so as to relieve stress in the solid phase. Naturally,
profiles of density, pressure, and temperature should be monitored, as a lack of equilibrium,
especially near the interfaces, is an ever-present danger.

Technically, it is not correct to carry out simulations at constant pressure in the
presence of interfaces which span the simulation box (Vega et al., 2008), precisely because
of the existence of a surface-energy term (which is often of interest in its own right, see
Section 14.4.2). However, if the system is long in the direction normal to the interface,
the surface term may have a relatively small effect. After the melting temperature and

box dimensions have been determined, it is possible to turn to simulations in the $NP_{zz}T$ ensemble, with fixed cross-sectional area but varying box length in the z direction. If the pressure component P_{zz} is not exactly at the coexistence value, the solid phase will grow or shrink; however, very small adjustments of P_{zz} may be made to reduce the velocity of the two interfaces to zero, and in this way enable the melting point (P, T) to be determined accurately (Zykova-Timan et al., 2010).

14.4.2   Solid–liquid interfacial energy

The interfacial thermodynamics in the solid–liquid case (and indeed for solid–vapour) differs from the liquid–vapour case. Denoting the excess surface free energy per unit area by γ, and the surface stress by σ, the relationship between them is (Shuttleworth, 1950; Tiller, 1991)

$$\sigma = \gamma + \mathcal{A}\frac{\mathrm{d}\gamma}{\mathrm{d}\mathcal{A}}, \quad \text{or, in general} \quad \sigma_{\alpha\beta} = \gamma\delta_{\alpha\beta} + \frac{\partial\gamma}{\partial\varepsilon_{\alpha\beta}},$$

where $\mathcal{A}$ is the cross-sectional area and $\varepsilon_{\alpha\beta}$ is a component of the strain tensor. In the case of an interface between two fluids, the second term is zero, and this allows a route to γ through the stress profile, as discussed in the previous sections. To determine the solid–liquid interfacial energy, alternative methods must be used.

The solid–liquid interfacial energy may be determined by capillary-wave oscillations, in the same way as for the liquid–vapour interface (Karma, 1993). However, the physics is a little more complicated: the fluctuations are determined by an interfacial *stiffness*, which includes a curvature term added to the surface tension, and it also depends on the particular solid surface being examined. A convenient and practical approach to this problem has been developed by Hoyt et al. (2001) and Morris and Song (2002) and applied to the crystal–melt interfaces of the Lennard-Jones system (Morris and Song, 2003), hard spheres (Davidchack et al., 2006), metals (Hoyt et al., 2001; Morris, 2002), water (Benet et al., 2014), and ionic systems (Benet et al., 2015). The simulation box is chosen to be very long in the z-direction, perpendicular to the interfaces, but rather short in the y direction. The system is therefore quasi-two-dimensional, and the interfaces are almost one-dimensional 'ribbons', as illustrated in Fig. 14.5.

With the definitions of Fourier components in Appendix D, eqn (2.187) gives the mean-square amplitude of each component, except that the surface tension γ is replaced by $\tilde{\gamma}$, the interfacial stiffness. In general this is written in tensorial form

$$\tilde{\gamma}_{\alpha\beta}(\hat{z}) = \gamma(\hat{z}) + \left.\frac{\partial^2\gamma(\hat{n})}{\partial n_\alpha \partial n_\beta}\right|_{\hat{n}=\hat{z}}$$

where $\hat{z}$ is the unit vector normal to the average interface, $\hat{n}$ is the local surface normal, and $\alpha, \beta = x, y$. Indeed, this form has been used for boxes of (nearly) square cross-section (Härtel et al., 2012; Turci and Schilling, 2014). In the thin slab geometry under consideration here, a simpler form applies:

$$\tilde{\gamma} = \gamma + \left.\frac{\mathrm{d}^2\gamma(\theta)}{\mathrm{d}\theta^2}\right|_{\theta=0} \tag{14.35}$$

where the angle θ is defined in Fig. 14.5. In fact, the interfacial free energy γ will depend on the orientation of the crystal plane which is exposed to the liquid: this dependence is

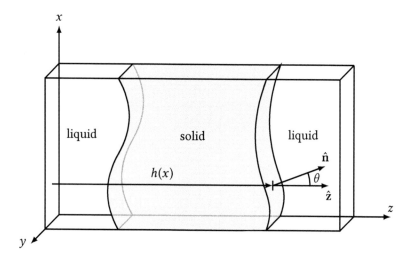

Fig. 14.5 Thin, elongated periodic box used to study crystal–melt interface fluctuations. The coordinate system is chosen with z normal to the average interface plane. Fluctuations in height as a function of transverse coordinate $h(x)$ also imply fluctuations in the angle θ between the local surface normal $\hat{n}$ and its average $\hat{z}$. The resulting curvature contribution to the surface free energy may be represented through the interfacial stiffness, eqn (14.35).

usually written as an expansion in spherical harmonic functions of the angles defining the crystal plane, and a corresponding expansion for $\tilde{\gamma}$ via eqn (14.35). The problem reduces to the determination of the coefficients in this expansion. This, therefore, requires one to conduct many simulations, measuring $\tilde{\gamma}$ for a range of crystal orientations within the box. For this geometry, the wavevectors corresponding to height fluctuations in the capillary-wave theory need only be considered in the x-direction.

We have not yet discussed how to determine the height $h(x)$ of each interface, and hence the Fourier coefficients. This relies on being able to assign an order parameter to each atom, characterizing it as belonging to the liquid or solid phase, or having an intermediate value characteristic of the interface. The most obvious candidates, such as the density, or the potential energy of each atom, are generally not robust or accurate enough for this task. Usually, some kind of bond-order parameter (based on a set of vectors connecting a particle with its nearest neighbours) is used for this purpose (Steinhardt et al., 1983; Hoyt et al., 2001; Morris, 2002; Lechner and Dellago, 2008). It may help the analysis to average the coordinates over a short time period. Discretizing the height on a grid in x, Fourier transforming, averaging the mean-squared amplitudes $\langle \hat{h}(k)^2 \rangle$, and fitting the low-$k$ data to eqn (2.187), gives $\tilde{\gamma}$ for a single crystal orientation.

A more direct route to the crystal–melt surface free energy was proposed by Broughton and Gilmer (1986), and has been refined and extended by Davidchack and Laird (2000). It is usually referred to as the *cleaving* method; it had previously been suggested for the liquid–vapour interface (Miyazaki et al., 1976) but the capillary wave and stress profile approaches are generally preferred in this case. The idea is to compute the free energy required to create the interface from the bulk phases, by thermodynamic integration. An

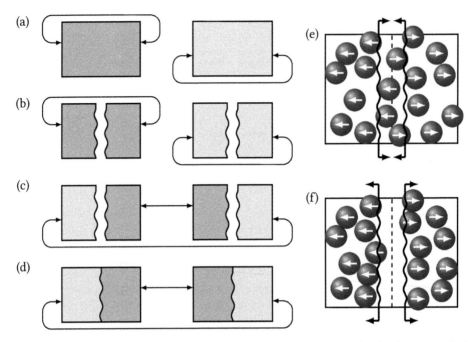

Fig. 14.6 (a)–(d) The successive steps in evaluating the surface free energy by the cleaving method. Liquid and solid phases are indicated by light and dark grey, respectively. Periodic boundaries are shown as solid lines connected by arrows. The wavy lines represent the cleaving boundaries, or walls, and their operation is indicated in (e) and (f). Atoms are first identified as lying to the left or right of the dashed line, and are accordingly labelled with arrows. The walls (wavy lines) are pushed in the direction indicated by their arrows, and act only on the similarly labelled atoms.

external potential is introduced, which effectively creates two planar interfaces separated by a region from which the atoms are excluded. A parameter in this potential is used to reversibly increase the width of the exclusion region. The free energy of creating such an interface in both the bulk liquid and solid phases is measured, by thermodynamic integration between states schematically illustrated in Fig. 14.6(a) → (b). Following this, the periodic boundaries are rearranged to give (c); for very short-range interactions, this involves no free-energy change, and even in the presence of long-range interactions, the free-energy change may be computed without too much trouble. Effectively, two separate periodic simulation boxes, each containing one bulk phase, become a single combined box, containing two interfaces. In the final stage (c) → (d), the external potential is slowly removed, resulting in two physical interfaces between the phases, and once again the free-energy change is measured by thermodynamic integration. Summing all the contributions gives an estimate of the interfacial free energy, for two surfaces, each spanning the cross-section of the periodic box.

The operation of the cleaving potential is illustrated in Fig. 14.6(e) and (f). Initially, the molecules are identified as lying to the left or right of a specified plane. An external potential is introduced, consisting of two terms, each acting as a kind of planar 'wall' interacting only with one type of molecule. To begin with, these walls are located well within the region of the other type of molecule, and so they have no effect. This is illustrated in Fig. 14.6(e). The walls are then slowly pushed towards, and through, each other, leading to the situation of Fig. 14.6(f), in which a region has been cleared of particles, and the direct interactions across this region are dramatically reduced (to zero in the case of short-ranged potentials). The free-energy change during the process is calculated by thermodynamically integrating the applied force on the walls, with respect to position.

Success depends on the cleaving process being reversible, and this turns out to be sensitive to the form of the external potential. It is desirable to perturb the system as little as possible. However, it is also necessary, when cleaving the liquid, to introduce the kind of structure which will be a good match to that at the surface of the crystal. A flat, planar wall is not a good choice for this; instead it is usual to construct the walls from atoms frozen in the ideal solid lattice positions. This approach has been applied to hard spheres (Davidchack and Laird, 2000), Lennard-Jones and soft-sphere potentials (Laird and Davidchack, 2005), metallic systems (Liu et al., 2013), and silicon, using a three-body potential (Apte and Zeng, 2008). Comparison with the capillary fluctuation method has led to further refinement (Davidchack, 2010), and although it seems possible to reconcile results obtained in both ways, the cleaving method is less sensitive to the anisotropy of the interfacial energy. There remain some potential hysteresis issues associated with the thermodynamic integration route for cleaving, especially if the solid–liquid interface drifts during the final removal of the external potential. The procedure can be adapted (Benjamin and Horbach, 2014; 2015), or alternative thermodynamic integration routes used (Espinosa et al., 2014; Schmitz and Virnau, 2015) to avoid these problems.

14.5   The liquid drop

There are many properties of small liquid drops, such as the radial dependence of the pressure, the sign of the Tolman length, and the size-dependence of the surface tension, which are of fundamental interest and which are not readily available from experiment (Malijevský and Jackson, 2012). A small drop, in equilibrium with its vapour, is an obvious candidate for computer simulation. This section discusses some of the technical problems associated with the preparation and equilibration of stable systems in the two-phase region, and highlights some of the important results that have emerged from recent studies. The main thrust of this work has been to explore fundamental properties of drops rather than to make a connection with the scant experimental results. For this reason, these simulations employ simple models such as the truncated Lennard-Jones potential, or the Stockmayer potential. In this section we will concentrate exclusively on the Lennard-Jones potential model and its truncated variations. The three earliest studies of the Lennard-Jones drop (Rusanov and Brodskaya, 1977; Powles et al., 1983a; Thompson et al., 1984) used the MD method. Unbiased MC methods can lead to bottlenecks, which have caused artificial structure in the density profile of a planar interface (Lee et al., 1974). Large system sizes and long runs are required to obtain useful results for drops. These early studies were limited by the size and speed of the available computers. They studied $N \lesssim 2000$ atoms for up to 3.5×10^5 timesteps. More recent studies (van Giessen and Blokhuis, 2009; Lee, 2012) used as many as 5×10^5 atoms for up to 5×10^7 timesteps. Runs of this length and size are required to understand some of the subtler physics of the drop.

The simulation of the drop begins by performing a normal bulk simulation of the Lennard-Jones fluid using periodic boundary conditions. The drop is excised from the bulk and placed either at the centre of a new periodic system with a larger central box (Powles et al., 1983a) or in a spherical container (Rusanov and Brodskaya, 1977; Thompson et al., 1984). The size of the central box or the spherical container must be large enough so that two periodic images of the drop, or the drop and the wall of the container, do not interfere with one another. On the other hand, if the system size is chosen to be too large, the liquid drop will evaporate to produce a uniform gas. The difficulty of choosing a suitable starting density can only be resolved by trial and error (and ideally a prior knowledge of the coexisting densities at the desired temperature). In practice, the distance between the outside of the two periodic images of the drop should be at least a drop diameter. In the case of a container, its radius should be two to three times larger than the radius of the drop.

The spherical container is best thought of as a static external field which confines the molecules to a constant volume. Thompson et al. (1984) use the repulsive Lennard-Jones potential $v^{\mathrm{RLJ}}(d)$ (eqn (1.10a)) to model this wall; d is the distance along a radius vector (from the container centre) between the molecule and the wall. Lee (2012) uses a half-harmonic potential to confine the atoms. Solving Newton's equations for this system will conserve energy and angular momentum about the sphere centre. The drop moves around inside the spherical container as atoms evaporate from the surface of the liquid and subsequently rejoin the drop. In another variant of this technique, the external field moves so that it is always centred on the centre of mass of the system. Solution of Newton's equations in a time-dependent external field does not conserve energy; in this particular instance (Thompson et al., 1984) the simulation was also performed at constant temperature using momentum scaling (see Section 3.8) and the equilibrium results were shown to be equivalent to those obtained in the more conventional microcanonical ensemble. More recent simulations fix the temperature using a Nosé–Hoover thermostat for each atom in the system (van Giessen and Blokhuis, 2009; Lee, 2012).

Figure 14.7 shows a schematic snapshot of part of a drop after equilibration. At any instant the drop is non-spherical, but on average the structure is spherical and the drop is surrounded by a uniform vapour. The radius of the drop, defined shortly in eqn (14.37), should change very little, say 1 %, during the production phase of the run. The temperature profile through the drop should be constant.

The principal structural property of the drop is the density profile, $\rho(r)$. It is defined as the average number of atoms per unit volume a distance r from the centre of the drop. Since the drop moves during the run, it is necessary to recalculate its centre of mass as an origin for $\rho(r)$ at each step. This is defined, assuming equal-mass atoms, by

$$\mathbf{r}'_{\mathrm{cm}}(t) = \frac{1}{N'} \sum_{i=1}^{N'} \mathbf{r}_i(t) \tag{14.36}$$

where N' is the number of atoms in the drop at time t. This has to be defined in some way. Powles et al. (1983b) have implemented the nearest-neighbour distance criterion of Stoddard (1978) for identifying atoms in the drop. The method makes use of a clustering algorithm. This begins by picking an atom i. All atoms j that satisfy $r_{ij} < r_{\mathrm{cl}}$, where r_{cl} is a critical atom separation, are defined to be in the same cluster as i. Each such atom j is

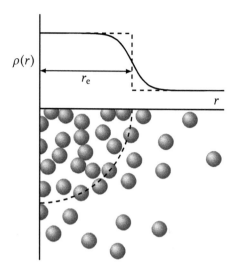

Fig. 14.7 A snapshot of a drop with its centre at the origin. For illustration, we use two dimensions and show one quadrant only. We also show the density profile and the equimolar dividing surface, r_e which defines the radius of the droplet (for details see Thompson et al., 1984).

Code 14.3 Cluster analysis

This file is provided online. `cluster.f90` reads in a configuration, using a utility module routine (Appendix A) and, for a given critical separation r_{cl}, produces a set of circular linked lists, one for each identified cluster (Stoddard, 1978).

```
! cluster.f90
! Identify atom clusters in a configuration
PROGRAM cluster
```

added to the cluster, and is subsequently used in the same way as i, to identify further members. When this first cluster is complete, an atom outside the cluster is picked, and the process repeated to generate a second cluster, and so on. The whole procedure partitions the complete set of atoms into mutually exclusive clusters. In the case of the liquid drop system, the largest cluster is the drop itself and the algorithm works most efficiently if the first atom i is near the centre of the drop. The atoms that are not in the drop cluster are defined to be in the vapour. Fowler (1984) has described in detail the implementation of the method. An efficient clustering algorithm (Stoddard, 1978) is given in Code 14.3. This method, however, does not scale well as N becomes very large, and other approaches may be preferred (Edvinsson et al., 1999).

r_{cl} has to be chosen sensibly. Studying the dependence of N' upon r_{cl} provides guidance in the choice of this parameter. Values of r_{cl} between 1.3σ and 1.9σ (Thompson et al., 1984; Powles et al., 1983b) have been used. In more recent simulations the net momentum of the drop is set to zero at every 100 timesteps. At the same time the position of every atom is shifted so that the centre of mass of the drop is at the origin. In this case the drop is defined as all atoms within a fixed distance (e.g. $r_e + 2\sigma$, where r_e is defined shortly) of the origin (van Giessen and Blokhuis, 2009), although clearly this depends on the precise system studied.

The equimolar dividing surface, r_e, provides a useful measure of the size of the liquid droplet:

$$r_e^3 = \frac{1}{(\rho_g - \rho_\ell)} \int_0^\infty \frac{d\rho(r)}{dr} r^3 dr. \tag{14.37}$$

r_e is defined so that if the limiting densities of the two phases were constant up to $r = r_e$ and changed discontinuously at $r = r_e$ ($D \to 0$ in eqn (14.8)), the system would contain the same number of molecules. (Compare eqn (14.10) and Fig. 14.2(b) defining the Gibbs dividing surface for a planar interface.) r_e is shown schematically in Fig. 14.7.

Studies show that the width of the surface increases rapidly with temperature, and that drops disintegrate at temperatures below the bulk critical temperature. The most recent simulation results suggest that the thickness of the interface, D, decreases slightly below that of a planar interface at the same temperature, and the liquid density increases above the planar limit, as the drop size decreases. For example, in the usual reduced Lennard-Jones units, at $T^* = 0.9$, van Giessen and Blokhuis (2009) obtain

r_e^*	∞	49.132	9.033
ρ_ℓ^*	0.664 743	0.667 495	0.678 697

This behaviour is also consistent with the Laplace equation, valid for large drops, where the pressure difference, $\Delta P = P_\ell - P_g$, is inversely proportional to the radius of the surface of tension, r_s

$$\Delta P = 2\gamma_s / r_s \tag{14.38}$$

where γ_s is the surface tension (the suffix reminds us that it is for a spherical drop), which is defined to act at the surface of tension. For very small drops, eqn (14.38) does not apply, the opposite trend is observed, and ρ_ℓ decreases with drop size as the attractive cohesive forces decrease.

The pressure tensor in a spherical drop can be written in terms of two independent components, normal, $P_N(r)$, and transverse, $P_T(r)$, defined as follows:

$$\mathbf{P}(r) = P_N(r)\,\hat{\mathbf{r}}\hat{\mathbf{r}} + P_T(r)\left(1 - \hat{\mathbf{r}}\hat{\mathbf{r}}\right) \tag{14.39}$$

where $\hat{\mathbf{r}}$ is the radial unit vector from the centre of the droplet. The condition for mechanical equilibrium, $\nabla \cdot \mathbf{P} = 0$, relates these components through a differential equation:

$$P_T(r) = P_N(r) + \frac{r}{2}\frac{dP_N(r)}{dr}. \tag{14.40}$$

This is in contrast to the planar interface where the components must be calculated separately. For the drop, the calculation of $P_N(r)$ is sufficient to describe $\mathbf{P}$, and to calculate γ, through a thermodynamic or a mechanical route.

The IK normal component of the pressure tensor can be calculated in a simulation by considering the intersection of the vector $\mathbf{r}_{ij}$, describing the interaction between atoms i and j, and the spherical surface of radius r centred on the drop (Thompson et al., 1984)

$$P_N(r) = \rho(r)k_B T - \frac{1}{4\pi r^3}\left\langle \sum_i \sum_{j>i} \sum_{\text{intersections}} |\mathbf{r} \cdot \mathbf{r}_{ij}| \frac{1}{r_{ij}} \frac{dv(r_{ij})}{dr_{ij}} \right\rangle. \tag{14.41}$$

Here the configurational term is the component of the ij force along the normal to the surface at the points of intersection with $\mathbf{r}$; this term is positive for a repulsive force and negative for an attractive force. For any i and j, r_{ij} may have zero, one, or two intersections with the sphere, depending on the positions of the atoms. In the case of two intersections, both contributions are equal by symmetry. Simple formulae for the number of intersections and the dot product in eqn (14.41) are available (Thompson et al., 1984, eqns (A11) and (A12)). The simulations of van Giessen and Blokhuis (2009) show that $P_N(r)$ is flat inside the droplet with a small minimum on the vapour side of r_e. $P_T(r)$ has a strong minimum in the interfacial region, with a small maximum on the vapour side of r_e. (These results do not agree with the simulations of Lee (2012) where $P_N(r)$ decreases in the liquid phase as $r \to 0$. These differences may be due to the different boundary conditions in the simulations or the different cutoff used in the potential model.) Interestingly, the pressure of the vapour can also be obtained by the normal virial calculation, eqn (2.61), in the case of a drop simulated using periodic boundary conditions. The only requirement is that the surface of the central box is always in the vapour part of the system. The virial equation, using the mean density of the sample, simply gives the average of the normal pressure over the surface of the box.

The Tolman equation, based on purely thermodynamic arguments, relates the surface tension of a droplet of radius r_s to the surface tension in the planar limit

$$\frac{\gamma_s}{\gamma_\infty} = 1 - \frac{2(r_e - r_s)}{r_s} = 1 - \frac{2\delta}{r_s} \tag{14.42}$$

where δ is the Tolman length, r_s is the radius of tension and γ_∞ is the surface tension of the fluid in the planar limit. It is likely that there are additional small terms, that are quadratic in $1/r_s$, to be added to eqn (14.42).

The surface of tension can be calculated through the rearrangement of eqns (14.38) and (14.42) to give

$$r_s = \frac{3\gamma_\infty - \left(9\gamma_\infty^2 - 4\gamma_\infty r_e \Delta P\right)^{1/2}}{\Delta P} \tag{14.43}$$

which can then be used in the thermodynamic expression, eqn (14.38) to calculate γ_s. Since P_ℓ, P_g and r_e are unambiguously determined in the simulation, γ_s can be calculated in terms of γ_∞.

A mechanical definition of the surface tension can be obtained by considering the forces on a strip cutting the surface of the droplet (Rowlinson and Widom, 1982):

$$\gamma_{s,m} = \int_0^\infty dr \left(\frac{r}{r_{s,m}}\right)^2 \left(P_N(r) - P_T(r)\right). \tag{14.44}$$

The subscript (s,m) indicates the mechanically defined surface tension calculated at the mechanically defined radius of tension

$$r_{s,m} = \int_0^\infty dr\, r\, P_N'(r) \,\bigg/\, \int_0^\infty dr\, P_N'(r), \tag{14.45}$$

where $P_N'(r) = dP_N(r)/dr$. This approach can be readily implemented in the course of a simulation (Lee, 2012) using

$$\gamma_{s,m}^3 = -\frac{(\Delta P)^2}{8} \int_0^\infty dr\, r^3 P_N'(r), \tag{14.46}$$

which can be obtained from eqn (14.44) and the Laplace equation written in terms of $r_{s,m}$. Equation (14.46) requires a knowledge of ΔP which can be determined from eqn (14.40)

$$\Delta P = 2 \int_0^\infty dr\, \frac{1}{r}\Big(P_N(r) - P_T(r)\Big) \tag{14.47}$$

where the integration is across the interface. Detailed simulation studies of the Lennard-Jones droplet (ten Wolde and Frenkel, 1998; Blokhuis and Bedeaux, 1992) show that the surfaces r_s and $r_{s,m}$ are displaced by ca. 1σ and that the mechanical and thermodynamic definitions of the surface tension acting at the respective surfaces of tension are different. These are also different from the surface tension acting at the Gibbs dividing surface, r_e. This situation is further complicated, since the moments of the pressure tensor profile (e.g. r_s) can depend on the particular definition of the microscopic stress (e.g. Irving–Kirkwood versus Harasima).

Recently, Lau et al. (2015) have extended the TA method (see eqn (14.24)) to a spherical droplet to estimate the thermodynamic surface tension γ_s. As in the case of the planar interface, the box containing the droplet is scaled, together with the atom coordinates, according to eqn (14.25). (Equivalent, independent scalings can be applied in each of the Cartesian coordinate directions due to the spherical symmetry of the droplet.) This conserves the total volume of the box. A positive value of ϵ in eqn (14.25) distorts the average spherical shape into an oblate ellipsoid, whereas a negative ϵ produces a prolate ellipsoid. Since both distortions increase the surface area, the change in free energy is of $O(\epsilon^2)$. In this case

$$\gamma_s = \frac{1}{c}\left(\langle b \rangle_0 - \frac{\beta}{2}\langle a^2 \rangle_0\right) \tag{14.48}$$

$$\text{where} \quad a = \frac{\partial \Delta V}{\partial \epsilon}, \quad b = \frac{1}{2}\frac{\partial^2 \Delta V}{\partial \epsilon^2}, \quad c = \frac{8\pi r_e^2}{5},$$

r_e is the equimolar radius of the spherical droplet, and $\Delta V = V_1 - V_0$, the change in potential energy in making the attempted perturbation. The terms involving a and b are typically of the same order of magnitude as each other, and tend to cancel, to yield a small γ_s. Hence an increased computational effort is usually required: Lau et al. (2015) note that runs of $\sim 100\,\text{ns}$ are required for each droplet compared to runs of $\sim 10\,\text{ns}$ for the application of the TA method to the planar interface.

The Tolman length can be extracted from measurements of ΔP as a function of drop size, and an accurate estimate of γ_∞. Expanding eqns (14.42) and (14.38) in powers of δ/r_e gives

$$\frac{r_e}{2}\Delta P = \gamma_\infty\left(1 - \frac{\delta}{r_e} + \cdots\right). \tag{14.49}$$

A plot of the left-hand side against $1/r_e$ gives, in the usual reduced units, $\delta^* = -0.10 \pm 0.02$ for the truncated, shifted Lennard-Jones potential at $T^* = 0.9$ (van Giessen and Blokhuis, 2009). This important result, of a negative Tolman length, is supported by the work of Block et al. (2010). Using simulations in the grand canonical ensemble and applying a finite-size scaling analysis, they were able to calculate γ_L for the droplet as a function of system size L, using the techniques described in Section 14.1.4. Although there is considerable scatter in γ_L versus $1/L$, estimates of $\delta^* = -0.11 \pm 0.06$ at $T^* = 0.68$ and $\delta^* = -0.07 \pm 0.04$ at $T^* = 0.78$ were obtained. Finally, Homman et al. (2014) have used a free-energy perturbation approach to calculate r_s and estimate a Tolman length of $\delta^* = -0.04 \pm 0.01$. The negative sign of the Tolman length is also predicted by density functional theory (Barrett, 2006).

There is still an important puzzle concerning δ. It can also be calculated from a simulation of the planar interface using a virial-like expression

$$\delta = -\frac{1}{\gamma_\infty}\int_{-\infty}^{\infty} dz\Big[(z - z_G)f_1(z) - f_2(z)\Big] \tag{14.50}$$

where z_G is the Gibbs dividing surface, and

$$f_1(z) = \frac{1}{2\mathcal{A}\Delta z}\left\langle\sum_i\sum_{j>i}\frac{dv(r_{ij})}{dr_{ij}}\left(r_{ij} - \frac{3z_{ij}^2}{r_{ij}}\right)H(z - z_i, \Delta z)\right\rangle$$

$$f_2(z) = \frac{1}{4\mathcal{A}\Delta z}\left\langle\sum_i\sum_{j>i}\frac{dv(r_{ij})}{dr_{ij}}\left(r_{ij} - \frac{3z_{ij}^2}{r_{ij}}\right)z_{ij}H(z - z_i, \Delta z)\right\rangle \tag{14.51}$$

with $z_{ij} = z_i - z_j$. This route consistently produces a positive value: $\delta^* = 0.207 \pm 0.002$ at $T^* = 0.9$ (van Giessen and Blokhuis, 2009); $\delta^* = 0.16 \pm 0.04$ at $T^* = 0.75$ to 0.50 ± 0.12 at $T^* = 0.95$ (Haye and Bruin, 1994). The reason for this discrepancy is unclear. van Giessen and Blokhuis (2009) suggest that this may be due to the capillary wave fluctuations of the droplet which are significantly greater than the Tolman length itself.

14.6   Fluid membranes

As discussed in Section 2.13, the simulation of planar fluid bilayer membranes is a very active area, with links to biological systems. The molecules comprising the bilayers are usually amphiphilic: in the biological context they may be lipids, while model systems may be formed from block copolymers. A public force-field parameter repository for lipids is available (Domański et al., 2010), and many practical details are discussed by Tieleman (2010; 2012). A review, concentrating on atomistic and coarse-grained force fields, has recently appeared (Pluhackova and Böckmann, 2015). Biomembrane simulation is a huge field which we cannot review here; instead we pick out a few issues of technical interest.

Example 14.2 Wetting and drying

Wetting and drying phenomena involve three phases: a simple example is when a fluid, which may exist as liquid (ℓ) or vapour (v), is near a solid wall (w). If the bulk fluid is in the vapour state, and the wall is sufficiently attractive, it may be covered by a thin film of liquid: this is *complete wetting*. Alternatively, droplets of liquid may sit on the wall, with the liquid–vapour interface approaching it at the so-called *contact angle* θ. This is *partial wetting*, and macroscopically balancing the forces acting on the three-phase contact line gives Young's equation $\gamma_{wv} - \gamma_{w\ell} = \gamma_{v\ell} \cos\theta$, relating θ to the three interfacial tensions. Wetting corresponds to $\theta < 90°$, $\cos\theta > 0$, while the drying regime, which is seen for walls which prefer vapour to liquid, is characterized by $\theta > 90°$, $\cos\theta < 0$. On changing the state point, or increasing the wall attraction, the contact angle may decrease, $\theta \to 0°$, $\cos\theta \to 1$, giving complete wetting: the nature of this transition is of interest. The corresponding drying transition occurs as $\theta \to 180°$, $\cos\theta \to -1$. Large-scale atomistic simulations of water (Giovambattista et al., 2016) seem to suggest that the foregoing macroscopic description is valid down to a few nanometres.

Direct measurement of contact angles from droplet profiles is possible (Ingebrigtsen and Toxvaerd, 2007; Becker et al., 2014; Svoboda et al., 2015), but there will always be some uncertainty associated with extrapolating the profiles very close to the surface, particularly if the wall has some structure. Also, the line tension may give a non-negligible contribution to the free energy (Schrader et al., 2009), and hence the profiles. If the wetting or drying transition itself is of interest, a thermodynamic route becomes preferable. Pioneering work in this direction was undertaken by Sikkenk et al. (1988) and van Swol and Henderson (1989; 1991) using conventional (but, at the time, quite demanding) simulation methods. Nowadays, an approach based on distribution functions and free energies is very revealing (Grzelak and Errington, 2010; Kumar and Errington, 2013; Evans and Wilding, 2015). The latter authors studied SPC/E water near flat walls of variable attractive strength, using GCMC. A variety of smart sampling techniques (see Chapter 9) were used to make this possible at liquid densities. By studying the density distribution, and using eqn (14.28), Evans and Wilding (2015) were able to determine both $\gamma_{v\ell}$ (in a fully periodic system) and $\gamma_{wv} - \gamma_{w\ell}$ (in slit geometry with two planar walls), assuming, as usual, that configurations with planar interfaces were dominant. Modest system sizes of a few hundred molecules were used. Their results provided evidence that the wetting transition is a first-order surface phase transition, at least for long-range surface attraction, but that drying is a continuous, critical, phase transition. This in turn sheds light on the nature of the very large density fluctuations seen in water near hydrophobic surfaces, as being due to the proximity of a surface critical point, giving rise to a long-ranged density correlation function parallel to the surface, and a dramatically enhanced local compressibility. These conclusions were supported by a more detailed study of the Lennard-Jones system, for which critical point exponents could be determined (Evans et al., 2016).

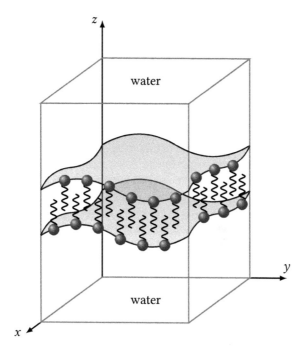

Fig. 14.8 Schematic illustration of the geometry of a bilayer membrane simulation. The head-group positions are indicated by surfaces, and a few amphiphilic molecules are shown explicitly, with spheres indicating the hydrophilic head groups and wavy lines denoting the hydrophobic tails.

The typical geometry involves a bilayer spanning the xy-dimensions of the periodic simulation box, as illustrated in Fig. 14.8. The box length in the z direction should be long enough to minimize interactions between periodic images (recall that the head groups are typically charged or polar): 30–40 water molecules per amphiphile seems quite typical (Ding et al., 2015). This arrangement is similar to that used to study interfaces, Sections 14.1 and 14.4, but there are some important differences. For practical purposes, the number of amphiphiles in the bilayer is fixed. Because of this, the cross-sectional area (or the corresponding transverse pressure) is a relevant thermodynamic variable in its own right, even though the bilayer is fluid in nature. As the area increases, the membrane comes under tension, while a decrease in area results in compression, and possible crumpling.

Specifying the surface tension requires some care. As emphasized in Section 2.13, a zero-tension ensemble should be adopted, to match the usual real-life situation, particularly if membrane fluctuations are to be dominated by curvature elasticity terms (eqn (2.190)), rather than surface tension terms, in the free energy. This may be achieved by measuring the tension via the stress profile of eqn (2.183a), as a function of cross-sectional area $\mathcal{A}$, and then choosing the value of $\mathcal{A}$ for which it vanishes (Goetz and Lipowsky, 1998). Alternatively, the simulation may be conducted in a constant-$NPT\gamma$ or constant-NP_TP_NT ensemble (Zhang et al., 1995; Venturoli and Smit, 1999; Venturoli et al., 2006; Rodgers and Smit, 2012). Ideally, the x and y box lengths should be scaled together, keeping a fixed, roughly square, cross-sectional shape; if they are allowed to vary independently, the cross-section may become very long and thin. Therefore, the situation is intermediate between the kind of isotropic box scaling typically used in liquid-state NPT simulations, and the fully anisotropic ensemble sometimes used in the solid state.

A complication, however, is that general-purpose force fields often fail to reproduce the experimentally observed area per lipid, and other membrane properties, under these conditions. To fix this, an artificial tension is sometimes applied to the membrane, but this will certainly affect some of the other properties. A more satisfactory alternative approach is to reparameterize the force field (Slingsby et al., 2015), in an effort to generate a full set of results consistent with experiment.

Within a zero-tension state, membrane height fluctuations, as a function of transverse wavevector, may be used to obtain the bending modulus κ using eqn (2.191) (Goetz et al., 1999; Lindahl and Edholm, 2000; Brandt et al., 2011). The method suffers from the need to simulate a system with a large cross-sectional area, to carry out the low-k extrapolation, and from the very slow timescales associated with the fluctuations. To attempt to address these problems, various alternative techniques have been devised using umbrella sampling (den Otter and Briels, 2003), possibly combined with field-theoretic methods (Smirnova and Müller, 2015), or by applying deformations through a guiding potential (Kawamoto et al., 2013), or by stretching a cylindrical vesicle (Harmandaris and Deserno, 2006). For the Gaussian elasticity $\bar{\kappa}$, a method has been proposed which involves the closing of disk-like lamellae into spherical vesicles (Hu et al., 2012). It has been suggested that a Fourier analysis of lipid orientation fluctuations, rather than height fluctuations, delivers more robust results without requiring such large cross-sectional areas (Watson et al., 2012; Levine et al., 2014). It may also be possible to extract elastic moduli from distributions of local lipid tilt and splay functions (Johner et al., 2016). Although the stress profile through a membrane is of some intrinsic interest (Ollila and Vattulainen, 2010), and is often used as a route to the elastic moduli via eqns (2.192) and (2.193) (see e.g. Marrink et al., 2007; Orsi et al., 2008) it is not clear that results obtained this way agree with those obtained by other methods (Hu et al., 2012).

Other areas in which simulation may make a significant contribution include the estimation of the free-energy change to insert molecules (such as proteins or cholesterol) into the bilayer from the solvent, and the calculation of single-particle diffusion coefficients, either of the amphiphiles themselves or of the inserted molecules. There is also considerable interest in phase transitions that occur within the bilayer, especially in the multi-component case. In principle, these calculations use perfectly standard methods, such as umbrella sampling (see Chapter 9) and the study of Einstein relations (see Chapter 8). However, bilayer membranes may require very long simulation times to obtain statistically significant results, and Tieleman (2010) has highlighted this as a major concern. For the free-energy calculations, there is concrete evidence that bilayer reorganization during the insertion process may result in systematic sampling errors (Neale et al., 2011). The diffusion problem is particularly interesting, being a combination of quasi-two-dimensional motion with the transfer of momentum to and from the solvent (Saffman and Delbrück, 1975; Saffman, 1976). However, Camley et al. (2015) have argued that finite-size effects in this situation lead to significant errors in estimating diffusion coefficients by simulations using periodic boundary conditions.

14.7 Liquid crystals

Finally, we turn to a class of systems that exhibit inhomogeneity in orientation space, namely liquid crystals. As discussed in Section 2.14, the simplest example, the *nematic* phase, is characterized by a director **n**. The orientational distribution is anisotropic, and both static and dynamic quantities need to be calculated in a coordinate frame based on **n**.

A key complication in all these calculations is that the director may vary with time. For large system sizes, director motion becomes extremely slow, and it is possible to define an effectively static director frame. For smaller systems, it may be necessary to constrain the director artificially (Allen et al., 1996). In principle, the orientation of the director within the simulation box is arbitrary; it is sometimes convenient to apply a small orienting field along, say, the z-direction, to align the director. The field is then removed, and the system allowed to equilibrate, before calculating simulation averages.

In smectic phases, the molecules order into layers, so that there is effectively long-range order in one direction (at least). There are many types of smectic phase: the simplest, having only short-range translational order within each layer, are smectic-A, in which the director is parallel to the layer normal, and smectic-C, in which the director is tilted. In either case, one can think of the layers as having two-dimensional fluid character, but molecules are also able to transfer between layers. If the phase is allowed to form spontaneously, the layers may adopt any orientation commensurate with the simulation box. As in the case of solid-state simulations, the box dimensions may need to vary in order to relieve any strain associated with the layers. The layer orientation may be determined by calculating the structure factor $S(\mathbf{k})$, based on the particle centre-of-mass coordinates, for a range of **k**-vectors compatible with the box. High values of $S(\mathbf{k})$ should correspond to vectors **k** perpendicular to the layers, and the layer spacing can be inferred from the corresponding values of $2\pi/|\mathbf{k}|$. This approach then allows particles to be assigned to layers. For smectic phases, once the layers are identified in this way, it is possible to compute spatial correlation functions for pairs of particles which lie in the same layer, within adjacent layers, and so on (Brown et al., 1998).

Finally, we should mention chiral phases, the simplest of which is the chiral nematic or cholesteric phase, formed by chiral molecules (which are not superimposable on their mirror image). Here, the director lies in a plane perpendicular to a certain direction in space, for example the z-axis, and rotates in a helical fashion:

$$\mathbf{n} = (\cos\phi(z), \sin\phi(z), 0), \quad \phi(z) = \phi_0 + k_{\mathrm{P}}z \qquad (14.52)$$

where $k_{\mathrm{P}} = 2\pi/\lambda_{\mathrm{P}}$, and λ_{P} is called the *pitch*. Measuring the equilibrium pitch, and relating it to molecular properties, is a key goal in simulation, theory and experiment.

The main problem in simulating such systems is ensuring compatibility with the periodic boundary conditions, since the equilibrium pitch is typically quite long compared with practical simulation box sizes. One way around this is to abandon periodic boundaries in the z-direction, and instead simulate the system between parallel flat walls, which interact with the molecules in such a way as to encourage alignment in the xy-plane, but otherwise do not influence their orientations. In these circumstances, there is no incompatibility with the change in twist angle $\Delta\phi = \phi(L_z/2) - \phi(-L_z/2) = k_{\mathrm{P}}L_z$, where L_z is the box length, which arises from the intrinsic chirality, and the pitch can be measured directly. The simplest way of doing this is to define the Q-tensor, eqn (2.194) in a succession of slabs defined by $z_i \pm \frac{1}{2}\Delta z$, where $z_i = i\Delta z$ and i is an integer, diagonalize each tensor to obtain $\mathbf{n}(z_i)$, and fit to the expected form, eqn (14.52).

However, the timescales for developing and equilibrating such a structure may be quite long. An alternative approach is to study the system in periodic boundaries, in which the twist angle will be forced to have values of $\Delta\phi = 0, \pi, 2\pi$ etc., and so $k = 0, \pm\frac{1}{2}k_0, \pm k_0, \ldots$, where $k_0 = 2\pi/L_z$. Additionally, it is possible to devise a modified periodic boundary convention (Allen, 1993; Allen and Masters, 1993) in which all molecular position and orientation vectors are rotated by $\pm\pi/2$ in the xy-plane, when one applies a translation $z \rightarrow z \pm L_z$, and this leads to $k = \pm\frac{1}{4}k_0, \pm\frac{3}{4}k_0, \ldots$. Such a system will be (orientationally) strained, since the imposed pitch will not have its equilibrium value $k \neq k_P$; measuring the (orientational) stress for various values of k can yield the desired k_P. The elastic free energy may be written

$$\mathcal{F} = \frac{1}{2}VK_2\left(k - k_P\right)^2$$

where K_2 is the appropriate elastic constant and V is the volume. It may be shown that

$$\frac{\partial\mathcal{F}}{\partial k} = VK_2\left(k - k_P\right) = \langle\Pi_{zz}\rangle$$

$$\text{where} \quad \Pi_{zz} = -\frac{1}{2}\sum_{i\neq j}z_{ij}\tau_{ij}^z = -\frac{1}{2}\sum_{i<j}z_{ij}\left(\tau_{ij}^z - \tau_{ji}^z\right),$$

τ_{ij}^z being the z-component of the torque on i due to j, and $z_{ij} = z_i - z_j$ as usual. Therefore, two (or more) simulations measuring $\langle\Pi_{zz}\rangle$ at different values of k are sufficient to determine both K_2 and k_P. This, and related, methods are discussed by Allen and Masters (2001) and Germano et al. (2002).

Appendix A
Computers and computer simulation

A.1 Computer hardware

In the early days of molecular simulation, it was necessary to have access to one of a few dedicated computing facilities, housed in national laboratories or other research institutions. Even as computer hardware became more widespread, it was still possible to identify a hierarchy of designs: microcomputers, workstations, and mainframes; high-performance computing facilities relied on special designs such as vector/pipeline machines, parallel computers with custom-built interconnects, or even special-purpose chips. This largely changed with the development of personal computers, and the Internet, since when the technology has spread through the workplace, the home, and on mobile devices. In recent years, even high-performance computers have tended to be made out of commodity building blocks, and are used for financial transactions, search engines, data storage, database management, and scientific computing, as well as in gaming platforms. The development of cloud computing has challenged the idea that the user needs to have access to a particular computer, or even know where the computer is.

Although computer architectures may have changed dramatically, many of the underlying features remain the same as in the old days. The fast processing of floating-point numbers lies at the heart of most scientific computing, and molecular simulation in particular. Moore's Law (Moore, 1965) famously gave a rule of thumb that the number of electronic components on an integrated circuit doubles every year (later revised to every 18–24 months). This has effectively translated into an exponential growth in computer speed over the subsequent 50 years as the scale of semiconductor fabrication has steadily shrunk, and this indirectly leads to the growth in scientific research performed with computers, as illustrated in Fig. 1.1. The reasons for the sustained improvement are, of course, complicated, and there is no guarantee that it will continue (Golio, 2015; Shalf and Leland, 2015).

The speed of access to memory is often a rate-limiting step, and Moore's Law has not helped this aspect of performance to keep pace with CPU speed. There is not the space in this book to discuss this in detail but the reader should be aware of a few general principles. Due to the relative costs of fast-access and slow-access memory, it is generally organized into a hierarchy of memory *caches*, the fastest and smallest of which is physically closest to the processing unit. The design relies on the principle of *locality*: if a process uses a given memory location, it is assumed that it is very likely to require other nearby memory locations in the future. Therefore, program efficiency is improved by fetching a contiguous chunk of memory at once, and storing it in a *cache line* for subsequent fast access. This is the reason why the order of loop indices, when accessing elements of multidimensional arrays consecutively, is important. An extension of the idea is the vector or pipeline processor whereby successive low-level operations are conducted on data which are passed from one processing element to the next, rather like a factory production line: this also relies on organizing the data into a contiguous chunk.

High performance nowadays cannot be achieved without using a parallel architecture in which many processors work on the same problem (see Chapter 7). For parallel computers, an important question is whether the processors share memory between them, or have the memory distributed amongst them so that each processor only accesses its own. For shared-memory machines, the processors may have to compete for access to the memory, and this will affect program efficiency. Worse than this is the possibility of different processors writing to the same memory locations, especially through the intermediate levels of cache, as this may lead to erroneous results. For distributed-memory machines, the speed of communications between processors is a key factor determining efficiency. These architectures also have to organize their communications so as to avoid 'deadlock', which is when two or more processors are each waiting for the other to complete its own activity before starting their own.

At the time of writing, most high-performance computers typically employ a hybrid architecture: an internal network connects a set of compute nodes, each carrying its own memory, which is accessed by a set of cores belonging to the node. There might be hundreds or thousands of nodes, and the number of cores per node varies from just a few, to a few tens, at present. In recent years there has been increased use of processors based on GPUs, which can be thought of as extremely parallel nodes containing thousands of cores. These usually require programming in a special-purpose language, and memory management is a critical issue.

A.2   Programming languages

With such a variety of computer architectures available it can be challenging to write a portable simulation program, especially one which takes advantage of parallel architectures. Nonetheless, several standard packages do this, as mentioned in Chapter 7, where we try to give a flavour of the kinds of algorithms that are adopted. A slightly different issue is the choice of computer language. This is influenced more by stylistic preferences, especially towards one or another programming *paradigm*: usually a choice between an object-oriented approach and a procedural approach. Most molecular simulation programs are written in one of four languages: Fortran, C, C++, and Python. The first two of these are procedural, based around subroutine and function calls which take in data and return results. In procedural languages, a modular approach to any large programming task is recommended: this isolates the effects of any routine to its own local variables, explicitly interacting with the calling routine through a list of arguments and function return values. The latter two languages are object-oriented, based on the concept of objects which encapsulate both the data structures and the procedures, usually called methods, that act on them. This allows one to build steadily more complicated objects out of simpler ones, and it is tempting to associate real-world objects (such as molecules) with digital counterparts.

Both procedural and object-oriented paradigms are intended to lead to code which is reusable and easy to maintain. In this book we have decided to give examples entirely using the procedural approach, and Fortran allows us to do this in a very simple and clear way. Actually, in its recent incarnations, Fortran also supports object-oriented programming, but we have decided not to follow that route here. For examples of object-oriented approaches to molecular simulations see Hinsen (2000), Halverson et al. (2013), and Schultz and Kofke (2015).

We also supply Python versions of some of the codes online, making extensive use of the NumPy and SciPy libraries to handle arrays and carry out numerical processing. The aim is to make these examples more accessible to those without a Fortran background.

A.3   Fortran programming considerations

Throughout this book, and in the online code

http://www.oup.co.uk/companion/allen_tildesley

we use Fortran conforming to a relatively modern standard (2008). This has some advantages: a built-in syntax for array operations, a straightforward approach to modular programming, and a basic simplicity. It is also a compiled language, which means that it is quite efficient, and widely used, so it is easy to find compilers which are optimized for different machine architectures. The common tools for parallelizing scientific codes (openMP and MPI) are compatible with Fortran as well as C. We hope that those who are used to other program languages will find little difficulty in understanding these examples (or the supplied Python alternatives); also we point out the provisions, in current Fortran standards, for interoperability with C codes. Many of the design decisions in the examples have been taken with the intention of making the code uncluttered and easy to follow. We emphasize that all the software accompanying this book is intended for illustrative purposes only, and we make no claim of fitness for any purpose, including research applications. The software is distributed without any warranty, and the authors disclaim any liability associated with its use.

For those unfamiliar with Fortran here are a few notes that may help understand the example codes. Fortran supports multidimensional arrays, with an index notation such as r(k,i) for particle positions where k might represent the Cartesian coordinate, taking values 1, 2, 3, for x, y, z, and i is the particle number, taking values between 1 and N. Variable declarations, including arrays, are given at the start of any program or subprogram, as follows:

```
REAL , DIMENSION(3,n)  ::  r
```

Note that Fortran indices begin, by default, at 1, in contrast to C where the numbering begins at 0. However, the lower limit may also be specified explicitly, so a quaternion, for instance, with components (q_0, q_1, q_2, q_3), might be declared q(0:3). Array slices may be accessed using a notation such as r(:,i:j) which selects all the Cartesian coordinates of those particles whose indices lie in the range $i \ldots j$ inclusive. An extension of the notation, r(:,i:j:2), introduces a stride of 2, that is, particles with indices $i, i + 2$, etc. are selected.

Code A.1   Utility modules

These files are provided online. They contain a range of routines falling broadly into the following categories:

1. Routines to handle the input and output of configurations, in `config_io_module.f90`.
2. Routines to handle the quantities that we wish to average in a simulation, in `averages_module.f90`.
3. Routines to generate random numbers, including the Metropolis function, other mathematical quantities, translational and orientational order parameters, in `maths_module.f90`.

```
! config_io_module.f90
! Routines for atomic/molecular configuration I/O
MODULE config_io_module


! averages_module.f90
! Calculation of run averages with output to output_unit
MODULE averages_module


! maths_module.f90
! routines for maths, random numbers, order parameters
MODULE maths_module
```

Finally, an entire array may be referenced by r(:,:) or, most simply, by r. We use this notation extensively in our example programs, in statements such as r=r+v*dt and it matches the mathematical notation used in the text, where **r** represents the entire set of coordinate vectors. Note that when an array such as v is multiplied by a scalar such as dt, every element of the array is multiplied by the same number. An element-by-element division of arrays, such as r(:,i)/box(:) is allowed, provided the dimensions match (here, they are both equal to 3).

We make extensive use of Fortran modules in our examples. A module contains entities, that is, data and procedures (functions and subroutines), and there is precise control over which entity identifiers within the module are accessible outside it (PUBLIC), and which are not (PRIVATE). In addition, the entities may be PROTECTED against being changed outside the module. The USE statement appears at the start of programs and subprograms wishing to access the contents of a module; although this is optional, we usually give an explicit list, following the word ONLY, of the data and procedures that are actually required. Modules are useful for storing reusable pieces of code, that is, effectively libraries of routines. Three 'utility' modules are described in Code A.1. Almost every online example program uses routines from some or all of these modules. We also use modules having the same name, stored in different files, as a convenient way to provide compatible alternative versions of routines: for instance, force routines with and without neighbour lists. Partly for this reason (as will be discussed shortly) each program is built in its own directory.

For simplicity, we make no use of pointers in our examples, even in Code 3.7 which handles lists of upcoming collisions, and in Codes 5.2–5.4, which illustrate force and energy routines using linked lists. It is quite possible to reformulate the algorithms so as to use standard computing data structures in which pointers play a natural role.

Most computers now are 64-bit (8 bytes), which means that floating-point variables should have sufficient precision for most of the purposes of this book. Particular areas where this is important are in molecular dynamics algorithms, and in the use of accumulators for run-averages, where loss of precision may have unintended consequences. The issue is less critical for Monte Carlo algorithms. However, some care needs to be taken in the source code. For historical reasons, the *default* precision of REAL variables in many compilers corresponds to 32 bits (4 bytes) rather than 64. For this reason, it is common to declare the KIND of variables explicitly using statements such as

```
INTEGER , PARAMETER :: sp = SELECTED_REAL_KIND(6, 37)
INTEGER , PARAMETER :: dp = SELECTED_REAL_KIND(15, 307)
```

where the first argument of SELECTED_REAL_KIND is the minimum number of digits of decimal precision and the second argument is the minimum exponent range. Alternatively, Fortran 2008 provides predefined constants, which may be used in a similar way

```
USE , INTRINSIC :: iso_fortran_env
INTEGER , PARAMETER :: sp = REAL32
INTEGER , PARAMETER :: dp = REAL64
```

These statements are typically placed in a MODULE which may be used throughout the program, to provide these KIND variables in statements such as

```
REAL(sp) :: x
REAL(dp) :: y
```

declaring x as a 'single precision' (i.e. 32-bit-equivalent) floating-point variable, and y as a 'double precision' (i.e. 64-bit-equivalent) one.

In the book, and in the online codes, for simplicity, *we do not do this*. Instead, we assume that a compiler option such as -fdefault-real-8 has been used to define the default REAL variables as corresponding to 8 bytes. This should be borne in mind when using the codes. This compiler option is selected in the supplied SConstruct file, which uses scons to compile and build the examples. The statements near the start of this file, defining these options, and the locations of various libraries are the most likely ones that will need changing to suit a given computing platform.

The scons approach is not the only way to build the codes; one could use make, or any one of innumerable alternative utilities, or simply give the compilation statement on the command line. For each individual program, there are very few dependencies and the process should be simple. The SConstruct file itemizes all the required source files in each case. It should be noted that each program is built in its own subdirectory, into which the source files are copied at the start of the process. This is the cleanest way of avoiding conflicts, especially between the names of intermediate module files which are produced by the compiler. It is also possible, in principle, to use an integrated development environment (IDE) which combines the essential functions of source code editing, building, and debugging.

The Python versions of the online examples generally have file names derived from the corresponding Fortran codes, and we have tried to keep to the same general modular structure. They do not require compiling. They do, however, rely on reasonably up-to-date installations of Python and the numeric and scientific computing libraries NumPy and SciPy, available from www.scipy.org and elsewhere. The user should be aware that these libraries are regularly updated, so there is always the risk of incompatibility with a particular version.

A full list of codes appears at the end of the book.

Appendix B
Reduced units

B.1 Reduced units

For systems consisting of just one type of molecule, it is sensible to use the mass of the molecule as a fundamental unit, that is, set $m_i = 1$. As a consequence, the particle momenta $\mathbf{p}_i$ and velocities $\mathbf{v}_i$ become numerically identical, as do the forces $\mathbf{f}_i$ and accelerations, $\mathbf{a}_i$. This approach can be extended further. If the molecules interact by pair potentials of a simple form, such as the Lennard-Jones potential (eqn (1.6)), they are completely specified by a few parameters such as ϵ and σ; then further fundamental units of energy, length, etc. may be defined. From these definitions, units of other quantities (pressure, time, momentum, etc.) follow directly. Static and dynamic properties of the Lennard-Jones system are invariably quoted in reduced units

density	$\rho^* = \rho\sigma^3$	(B.1a)
temperature	$T^* = k_B T/\epsilon$	(B.1b)
energy	$E^* = E/\epsilon$	(B.1c)
pressure	$P^* = P\sigma^3/\epsilon$	(B.1d)
time	$t^* = (\epsilon/m\sigma^2)^{1/2} t$	(B.1e)
force	$\mathbf{f}^* = \mathbf{f}\sigma/\epsilon$	(B.1f)
torque	$\tau^* = \tau/\epsilon$	(B.1g)
surface tension	$\gamma* = \gamma\sigma^2/\epsilon$	(B.1h)

and so on. The reduced thermodynamic variables determine the state point or, to be precise, a set of corresponding states with closely related properties. Quite generally, if the potential takes the form $v(r) = \epsilon f(r/\sigma)$, where f is an arbitrary function, there is a principle of corresponding states which applies to thermodynamic, structural, and dynamic properties (Helfand and Rice, 1960). Thus, the Lennard-Jones potential may be used to fit the equation of state for a large number of systems (Rahman, 1964; Mc-Donald, 1972). For the even simpler soft-sphere potential of eqn (1.9), a single reduced variable $(\rho\sigma^3)(\epsilon/k_B T)^{3/\nu}$ defines the excess (i.e. non-ideal) properties (see e.g. Hoover et al., 1970; 1971). In the limit of the hard-sphere potential (formally corresponding to $\nu \to \infty$) the temperature becomes a totally redundant variable so far as static quantities

Computer Simulation of Liquids. Second Edition. M. P. Allen and D. J. Tildesley.
© M. P. Allen and D. J. Tildesley 2017. Published in 2017 by Oxford University Press.

are concerned, and enters the dynamic properties only through the definition of a reduced time

$$t^* = (k_B T / m\sigma^2)^{1/2} t. \tag{B.2}$$

The use of reduced units avoids the possible embarrassment of conducting essentially duplicate simulations. There are also technical advantages in the use of reduced units. If parameters such as ϵ and σ have been given a value of unity, they need not appear in a computer simulation program at all; consequently some time will be saved in the calculation of potential energies, forces, etc. Of course, the program then becomes unique to the particular functional form of the chosen potential. For complicated potentials, with many adjustable parameters, or in the case of mixtures of species, there is only a slight technical advantage to be gained by choosing one particular energy parameter, one characteristic length, and one molecular mass, to be unity. In this case, it is common practice to define these units in such a way that the quantities appearing in the program take numerical values that are not too extreme (for instance, lying between 10^{-6} and 10^6).

In SI units, Coulomb's law is

$$v^{qq} = q_i q_j / 4\pi\epsilon_0 r_{ij} \tag{B.3}$$

where q_i and q_j are charges in Coulombs, r_{ij} is the separation in metres, and

$$\epsilon_0 = 8.8542 \times 10^{-12} \, C^2 \, N^{-1} \, m^{-2}$$

is the permittivity of free space. In reduced units based on the Lennard-Jones energy and length parameters, ϵ and σ respectively, the charge, dipole, and quadrupole are

$$q^* = \frac{q}{\sqrt{4\pi\epsilon_0\sigma\epsilon}}, \qquad \mu^* = \frac{\mu}{\sqrt{4\pi\epsilon_0\sigma^3\epsilon}}, \qquad Q^* = \frac{Q}{\sqrt{4\pi\epsilon_0\sigma^5\epsilon}}. \tag{B.4}$$

Many older papers give the moments in electrostatic units (e.s.u.). Useful conversion factors are

charge:	$1\,C = 2.9979 \times 10^9$ e.s.u.
dipole:	$1\,C\,m = 2.9979 \times 10^{11}$ e.s.u. cm
quadrupole:	$1\,C\,m^2 = 2.9979 \times 10^{13}$ e.s.u. cm^2.

It is convenient to use an alternative definition of the unit of charge, whether or not other reduced units are employed. In most of this book eqn (B.3) is used without the factor $4\pi\epsilon_0$. In this case the charge q is divided by $(4\pi\epsilon_0)^{1/2}$ and has units of $m\,N^{1/2}$.

Maxwell's equations describe the classical evolution of the combined electric ($\mathcal{E}$) and magnetic (**B**) fields in a microscopic system. In SI units these equations are

$$\nabla \cdot \mathcal{E} = \rho^q / \epsilon_0 \tag{B.5a}$$

$$\nabla \times \mathcal{E} = -\frac{\partial \mathbf{B}}{\partial t} \tag{B.5b}$$

$$\nabla \cdot \mathbf{B} = 0 \tag{B.5c}$$

$$\nabla \times \mathbf{B} = \mu_0 \left(\mathbf{j} + \epsilon_0 \frac{\partial \mathbf{B}}{\partial t} \right) \tag{B.5d}$$

Table B.1 Some atomic units and their conversion to SI units.

Dimension	Name	Symbol	Expression	SI equivalent
length	bohr	a_0	$4\pi\epsilon_0\hbar^2/(m_e e^2)$	$5.291\,772 \times 10^{-11}$ m
energy	hartree	Ha or E_h	$(m_e e)^4/(4\pi\epsilon_0\hbar)^2$	$4.359\,745 \times 10^{-18}$ J
time			$\hbar/\text{Ha}$	$2.418\,884 \times 10^{-17}$ s
dipole			$e a_0$	$8.478\,354 \times 10^{-30}$ C m
quadrupole			$e a_0{}^2$	$4.486\,551 \times 10^{-40}$ C m^2

where ρ^q and $\mathbf{j}$ are the charge and electrical current density, respectively, and

$$\mu_0 = 4\pi \times 10^{-7}\,\text{T}\,\text{m}\,\text{A}^{-1}$$

is the permeability of free space. These equations can also be written with $4\pi\epsilon_0 = 1$. In this case, only eqns (B.5a) and (B.5d) are modified to give

$$\nabla \cdot \mathcal{E} = 4\pi\rho^q \tag{B.6a}$$

$$\nabla \times \mathbf{B} = \frac{4\pi\mathbf{j}}{c^2} + \frac{1}{c^2}\frac{\partial \mathbf{B}}{\partial t} \tag{B.6b}$$

where we have used the identity $c^2 = 1/\mu_0\epsilon_0$. Eqn (B.6a), Poisson's equation, is used extensively in the discussion of PPPM methods in Section 6.3. For the discussion of MEMD in Section 6.8, it is more convenient to use the full SI version, eqn (B.5).

In discussing the electronic calculations in Chapter 13, and the electrostatic moments in Chapter 1, atomic units (a.u.) are used in a number of places. These units are defined in terms of the electron rest mass, m_e, the elementary charge, e, and the action, $\hbar$. The definitions of the relevant quantities and the corresponding SI values are given in Table B.1. The energy unit of a rydberg (Ry) is frequently used for the orbital or density cutoff in electronic structure calculations; $1\text{Ha} = 2\text{Ry}$.

With a mesoscale approach such as the DPD method described in Section 12.4, it is convenient to use a set of units in which the mass, m_i, of the DPD particle is set to 1.0, the distances are reduced with respect to r_c, the cutoff in the conservative force, and the energies are scaled by setting $k_B T = 1.0$. This choice leads to a simple form for the code when considering a one-component fluid (see Code 12.3). In the main, the connection between the reduced and real variables is the same as that defined in eqn (B.1) with ϵ replaced by $k_B T$ and σ by r_c. However, the connection between the DPD units and the real units is not as simple when trying to model a real system. First, there is not a one-to-one correspondence between the particles: a single DPD bead represents several molecules. Second, as discussed in Section 12.7.4, for any coarse-grained model, it is not possible to reproduce the real thermodynamics in a consistent way.

We illustrate the connection between the simulated quantities calculated in reduced units and the experimental values in Table B.2 for a DPD simulation of the gas–liquid interface of water. In this study, the model is a density-dependent generalization of the standard DPD force, where the conservative force depends on the local density (Ghoufi and Malfreyt, 2011). A DPD bead representing water contains $N_m=3$ water molecules. The simulations are performed at 298 K. A number of experimental properties of water at

Table B.2 The conversion from DPD to real units for a simple model of water. ρ_M is the mass density of liquid water, M is the molar mass, and N_m is the number of water molecules represented by a single bead. Reprinted with permission from A. Ghoufi and P. Malfreyt, *Phys. Rev. E*, **83**, 051601 (2011). Copyright (2011) by the American Physical Society.

DPD simulation		Conversion formula	Experiment
Parameter	Value		
r_c^*	1	$r_c = (\rho^* N_m v)^{1/3}$	$8.52\,\text{Å}$
ρ^*	6.88	$\rho_M = \rho^* N_m (M/N_A)/r_c^3$	$997\,\text{kg m}^{-3}$
P^*	0.1	$P = P^* k_B T/r_c^3$	$0.67\,\text{MPa}$
γ^*	12.4	$\gamma = \gamma^* k_B T/r_c^2$	$70.3\,\text{mN m}^{-1}$
β_T^{-1*}	48.0	$\beta_T^{-1} = (\rho k_B T/N_m)\beta_T^{-1*}$	$2.2\,\text{GPa}$
δt^*	0.01	$\delta t = \delta t^* (N_m D_{\text{bead}}^* r_c^2/D_{\text{water}})$	$6.8\,\text{ps}$

this temperature are required to make the conversion: the isothermal compressibility, $\beta_T = 4.55 \times 10^{-10}\,\text{Pa}^{-1}$, or its inverse $\beta_T^{-1} = 2.2\,\text{GPa}$; the volume of one water molecule, $v = 30\,\text{Å}^3 = 3 \times 10^{-29}\,\text{m}^3$, the reciprocal of which is the number density of water molecules $\rho = 3.336 \times 10^{28}\,\text{m}^{-3}$; the molar mass, $M = 18 \times 10^{-3}\,\text{kg mol}^{-1}$ and hence the molecular mass $M/N_A = 2.989 \times 10^{-26}\,\text{kg}$; and the diffusion coefficient of the water molecule, $D_{\text{water}} = 2.43 \times 10^{-9}\,\text{m}^2\,\text{s}^{-1}$. Because the numbers of particles in the simulated and real systems are different, it is necessary to match the *mass densities* ρ_M rather than the number densities.

The conversion factor for the timestep can be determined by calculating the diffusion coefficient of the water bead in the DPD simulation (Groot and Rabone, 2001). D_{bead}^* is determined from the slope of $\langle |\mathbf{r}_i(t) - \mathbf{r}_i(0)|^2 \rangle/6$ as a function of t. The conversion factor is given in the final line of Table B.2. Note, the value of the timestep scales linearly with the number of molecules in the DPD bead.

For a standard DPD fluid, it is possible to estimate the parameter a, which governs the size of the conservative force (see eqn (12.15)), by appeal to experiment. This is achieved by matching the reciprocal of the isothermal compressibility, β_T^{-1*}, calculated from the simulated equation of state (Groot and Warren, 1997) with the experimental value of β_T^{-1} for the fluid. For the density-dependent generalization of DPD, it is possible to fit the two parameters for the conservative force, a and b, by matching the simulation values of the experimental isothermal compressibility and the experimental surface tension (Ghoufi and Malfreyt, 2011).

Appendix C
Calculation of forces and torques

C.1 Introduction

The correct calculation of the forces and torques resulting from a given potential model is essential in the construction of a properly functioning molecular dynamics program. In this appendix, we consider forces and, where appropriate, torques, arising from five complicated potential models:

(a) a polymer chain with constrained bond lengths, but realistic bond angle-bending and torsional potentials;

(b) a molecular fluid of linear molecules, where the permanent electrostatic interactions are handled using a multipole expansion;

(c) a fluid of atoms with three-body interactions modelled using the Axilrod–Teller triple-dipole potential;

(d) a system of ions, where the Coulomb interactions are handled using the Ewald sum;

(e) the Gay–Berne potential for nonspherical rigid bodies.

The formulae given here will be useful to anyone constructing simulation programs containing these potential models. In addition, the methods of derivation may assist the reader in handling a range of more complicated potentials. In most cases, for a given potential, the force expressions would be derived with the aid of a symbolic algebra package, and then converted directly into computer code, usually with some purely cosmetic editing to aid readability. Then, they should be numerically checked; we return to this at the end of the appendix.

C.2 The polymer chain

We consider a simple model of a polymer, consisting of n_a atoms linked by rigid bonds. The angle between successive bonds, θ_a, and the torsional angle ϕ_a defined by three successive bonds, are both allowed to vary. The way in which the atoms and angles are labelled is shown in Fig. C.1 (see also Fig. 1.10). If the bond vector between atoms $a - 1$

Computer Simulation of Liquids. Second Edition. M. P. Allen and D. J. Tildesley.
© M. P. Allen and D. J. Tildesley 2017. Published in 2017 by Oxford University Press.

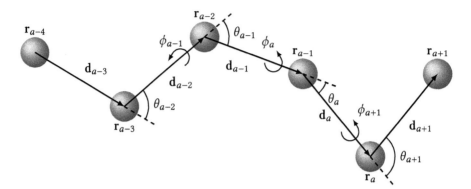

Fig. C.1 A polymer chain. The bending angle θ_a is the angle between the bond vectors $\mathbf{d}_a$ and $\mathbf{d}_{a-1}$. The torsional angle ϕ_a is the angle between the plane defined by $\mathbf{d}_a$ and $\mathbf{d}_{a-1}$ and that defined by $\mathbf{d}_{a-1}$ and $\mathbf{d}_{a-2}$.

and a is $\mathbf{d}_a = \mathbf{r}_a - \mathbf{r}_{a-1}$, then θ_a may be calculated from

$$\cos \theta_a = \frac{\mathbf{d}_a \cdot \mathbf{d}_{a-1}}{|\mathbf{d}_a| \, |\mathbf{d}_{a-1}|} \tag{C.1}$$

and ϕ_a may be obtained from

$$\cos \phi_a = -\frac{(\mathbf{d}_a \times \mathbf{d}_{a-1}) \cdot (\mathbf{d}_{a-1} \times \mathbf{d}_{a-2})}{|\mathbf{d}_a \times \mathbf{d}_{a-1}| \, |\mathbf{d}_{a-1} \times \mathbf{d}_{a-2}|}. \tag{C.2}$$

Actually, to completely determine ϕ_a (including its sign) one should also calculate $\sin \phi_a$ from the vectors appearing in eqn (C.2), or use a formula involving the ATAN2 function. In the following discussion, we assume that the value of $\cos \phi_a$ will be sufficient, but this depends on the precise form of the torsional potential.

Associated with each θ_a and ϕ_a will be potential terms of the kind appearing in eqn (1.37). The precise forms are not necessary for the following discussion, however; we simply assume that they may be written as functions $v_a^\theta (\cos \theta_a)$ and $v_a^\phi (\cos \phi_a)$ and that derivatives with respect to these cosines may be written down. The key quantities to be calculated, then, are the forces on each atom due to these potentials (as well as contributions from other sources such as non-bonded interactions). Following the approach of Pear and Weiner (1979), we use the chain rule to calculate these forces. The position coordinate of atom a will appear in the bending potentials for angles θ_a, θ_{a+1} and θ_{a+2}, and also in the torsional potentials for angles ϕ_a, ϕ_{a+1}, ϕ_{a+2} and ϕ_{a+3}. Hence there will be contributions to the force on atom a from all these sources:

$$\begin{aligned} \mathbf{f}_a &= -\sum_{c=a}^{a+2} \nabla_a v_c^\theta (\cos \theta_c) - \sum_{c=a}^{a+3} \nabla_a v_c^\phi (\cos \phi_c) \\ &= -\sum_{c=a}^{a+2} \left(\frac{\mathrm{d} v_c^\theta (\cos \theta_c)}{\mathrm{d} \cos \theta_c} \right) \nabla_a \cos \theta_c - \sum_{c=a}^{a+3} \left(\frac{\mathrm{d} v_c^\phi (\cos \phi_c)}{\mathrm{d} \cos \phi_c} \right) \nabla_a \cos \phi_c. \end{aligned} \tag{C.3}$$

Here, and henceforth, we use ∇_a as an abbreviation for $\nabla_{\mathbf{r}_a}$. Conversely, a bending term $v_a^\theta (\cos \theta_a)$ will produce forces on atoms a, $a-1$, and $a-2$, while a torsional term $v_a^\phi (\cos \phi_a)$

will produce forces on atoms a, $a-1$, $a-2$, and $a-3$, with the labelling of Fig. C.1. In the following we assume that we wish to consider the potential term by term, and calculate the gradients of the cosine functions that will help us to do it.

The formulae are simplified if we define

$$C_{ab} = C_{ba} = \mathbf{d}_a \cdot \mathbf{d}_b \tag{C.4}$$

$$D_{ab} = D_{ba} = C_{aa}C_{bb} - C_{ab}^2. \tag{C.5}$$

in terms of which the cosines may be expressed

$$\cos \theta_a = C_{aa-1} \left(C_{aa}C_{a-1a-1} \right)^{-1/2} \tag{C.6}$$

$$\cos \phi_a = -\left(C_{aa-1}C_{a-1a-2} - C_{aa-2}C_{a-1a-1} \right) \left(D_{aa-1}D_{a-1a-2} \right)^{-1/2}. \tag{C.7}$$

Using the vector identity

$$\nabla(\mathbf{A} \cdot \mathbf{B}) = (\mathbf{B} \cdot \nabla)\mathbf{A} + (\mathbf{A} \cdot \nabla)\mathbf{B} + \mathbf{B} \times (\nabla \times \mathbf{A}) + \mathbf{A} \times (\nabla \times \mathbf{B}) \tag{C.8}$$

in which the last two terms vanish for the vectors involved here, it is possible to derive a simple set of rules governing the vector differentiation of the C and D functions:

$$\nabla_a C_{aa} = 2\mathbf{d}_a$$
$$\nabla_a C_{aa+1} = \mathbf{d}_{a+1} - \mathbf{d}_a$$
$$\nabla_a C_{a+1a+1} = -2\mathbf{d}_{a+1}$$
$$\nabla_a C_{ab} = \mathbf{d}_b \qquad\qquad (b \neq a, a+1)$$
$$\nabla_a C_{a+1b} = -\mathbf{d}_b \qquad\qquad (b \neq a, a+1)$$
$$\nabla_a C_{bc} = 0 \qquad\qquad (b, c \neq a, a+1)$$

and

$$\nabla_a D_{aa+1} = 2C_{a+1a+1}\mathbf{d}_a - 2C_{aa}\mathbf{d}_{a+1} - 2C_{aa+1}\mathbf{d}_{a+1} + 2C_{aa+1}\mathbf{d}_a$$
$$\nabla_a D_{ab} = 2C_{bb}\mathbf{d}_a - 2C_{ab}\mathbf{d}_b \qquad\qquad (b \neq a, a+1)$$
$$\nabla_a D_{a+1b} = -2C_{bb}\mathbf{d}_{a+1} + 2C_{a+1b}\mathbf{d}_b \qquad\qquad (b \neq a, a+1)$$
$$\nabla_a D_{bc} = 0 \qquad\qquad (b, c \neq a, a+1).$$

Of course, the same rules apply on consistently replacing $a \to a-1$, $a \to a-2$, etc. Using these rules on eqn (C.6) gives

$$\nabla_a \cos \theta_a = -(C_{aa}C_{a-1a-1})^{-1/2}\left[(C_{aa-1}/C_{aa})\mathbf{d}_a - \mathbf{d}_{a-1} \right]$$
$$\nabla_{a-1} \cos \theta_a = (C_{aa}C_{a-1a-1})^{-1/2}\left[(C_{aa-1}/C_{aa})\mathbf{d}_a - (C_{aa-1}/C_{a-1a-1})\mathbf{d}_{a-1} + \mathbf{d}_a - \mathbf{d}_{a-1} \right]$$
$$\nabla_{a-2} \cos \theta_a = (C_{aa}C_{a-1a-1})^{-1/2}\left[(C_{aa-1}/C_{a-1a-1})\mathbf{d}_{a-1} - \mathbf{d}_a \right].$$

These equations, together with the derivative $(dv_a^\theta(\cos \theta_a)/d\cos \theta_a)$, are used to compute the contributions of the term $v_a^\theta(\cos \theta_a)$ to the forces $\mathbf{f}_a$, $\mathbf{f}_{a-1}$, and $\mathbf{f}_{a-2}$.

Applying the rules to eqn (C.7) gives

$$\nabla_a \cos\phi_a = -(D_{aa-1}D_{a-1a-2})^{-1/2}\big[C_{a-1a-2}\mathbf{d}_{a-1} - C_{a-1a-1}\mathbf{d}_{a-2}$$
$$- D_{aa-1}^{-1}(C_{aa-1}C_{a-1a-2} - C_{aa-2}C_{a-1a-1})(C_{a-1a-1}\mathbf{d}_a - C_{aa-1}\mathbf{d}_{a-1})\big],$$

$$\nabla_{a-1} \cos\phi_a = -(D_{aa-1}D_{a-1a-2})^{-1/2}\big[C_{a-1a-2}\mathbf{d}_a - C_{a-1a-2}\mathbf{d}_{a-1}$$
$$+ C_{aa-1}\mathbf{d}_{a-2} + C_{a-1a-1}\mathbf{d}_{a-2} - 2C_{aa-2}\mathbf{d}_{a-1}$$
$$- D_{a-1a-2}^{-1}(C_{aa-1}C_{a-1a-2} - C_{aa-2}C_{a-1a-1})(C_{a-2a-2}\mathbf{d}_{a-1} - C_{a-1a-2}\mathbf{d}_{a-2})$$
$$- D_{aa-1}^{-1}(C_{aa-1}C_{a-1a-2} - C_{aa-2}C_{a-1a-1})$$
$$\times (C_{aa}\mathbf{d}_{a-1} - C_{a-1a-1}\mathbf{d}_a - C_{aa-1}\mathbf{d}_a + C_{aa-1}\mathbf{d}_{a-1})\big],$$

$$\nabla_{a-2} \cos\phi_a = -(D_{aa-1}D_{a-1a-2})^{-1/2}\big[-C_{a-1a-2}\mathbf{d}_a + C_{aa-1}\mathbf{d}_{a-1}$$
$$- C_{aa-1}\mathbf{d}_{a-2} - C_{a-1a-1}\mathbf{d}_a + 2C_{aa-2}\mathbf{d}_{a-1}$$
$$- D_{a-1a-2}^{-1}(C_{aa-1}C_{a-1a-2} - C_{aa-2}C_{a-1a-1})$$
$$\times (C_{a-1a-1}\mathbf{d}_{a-2} - C_{a-2a-2}\mathbf{d}_{a-1} - C_{a-1a-2}\mathbf{d}_{a-1} + C_{a-1a-2}\mathbf{d}_{a-2})$$
$$- D_{aa-1}^{-1}(C_{aa-1}C_{a-1a-2} - C_{aa-2}C_{a-1a-1})$$
$$\times (-C_{aa}\mathbf{d}_{a-1} + C_{aa-1}\mathbf{d}_a)\big],$$

$$\nabla_{a-3} \cos\phi_a = -(D_{aa-1}D_{a-1a-2})^{-1/2}\big[-C_{aa-1}\mathbf{d}_{a-1} + C_{a-1a-1}\mathbf{d}_a$$
$$- D_{a-1a-2}^{-1}(C_{aa-1}C_{a-1a-2} - C_{aa-2}C_{a-1a-1})$$
$$\times (-C_{a-1a-1}\mathbf{d}_{a-2} + C_{a-1a-2}\mathbf{d}_{a-1})\big].$$

These equations, together with the derivative $(\mathrm{d}v_a^\phi(\cos\phi_a)/\mathrm{d}\cos\phi_a)$, are used to compute the contributions of the term $v_a^\phi(\cos\phi_a)$ to the forces $\mathbf{f}_a$, $\mathbf{f}_{a-1}$, $\mathbf{f}_{a-2}$, and $\mathbf{f}_{a-3}$. When these expressions are used in a computer program, they may be simplified by identifying various common factors, as we show in the examples of Section C.7.

C.3   The molecular fluid with multipoles

The methods for calculating the force and torque in an interaction site model are described in Chapters 1 and 5. Here, we discuss the forces and torques which arise from the permanent electrostatic interactions within the framework of the multipole expansion. For simplicity, we take the example of linear (i.e. axially symmetric) molecules, as shown in Fig. C.2. The centres of the molecules are separated by a vector $\mathbf{r}_{ij} = \mathbf{r}_i - \mathbf{r}_j$, and we define the unit vector $\hat{\mathbf{r}}_{ij} = \mathbf{r}_{ij}/r_{ij}$. θ_i and θ_j are the angles between $\mathbf{r}_{ij}$ and the unit vectors directed along the molecular axes $\mathbf{e}_i$ and $\mathbf{e}_j$, while ϕ_{ij} is the angle between the plane containing $\mathbf{e}_i$ and $\mathbf{r}_{ij}$ and that containing $\mathbf{e}_j$ and $\mathbf{r}_{ij}$:

$$\cos\theta_i \equiv c_i = \frac{\mathbf{e}_i \cdot \mathbf{r}_{ij}}{r_{ij}} = \mathbf{e}_i \cdot \hat{\mathbf{r}}_{ij}, \qquad \cos\theta_j \equiv c_j = \frac{\mathbf{e}_j \cdot \mathbf{r}_{ij}}{r_{ij}} = \mathbf{e}_j \cdot \hat{\mathbf{r}}_{ij},$$

$$\cos\phi_{ij} = \frac{(\mathbf{e}_i \times \mathbf{r}_{ij}) \cdot (\mathbf{e}_j \times \mathbf{r}_{ij})}{|\mathbf{e}_i \times \mathbf{r}_{ij}| |\mathbf{e}_j \times \mathbf{r}_{ij}|}.$$

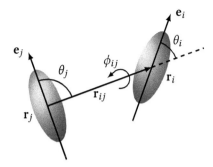

Fig. C.2 The relative orientation of two linear molecules.

It is convenient to define the angle γ_{ij} between the two axis vectors

$$\cos \gamma_{ij} \equiv c_{ij} = \mathbf{e}_i \cdot \mathbf{e}_j = \cos \theta_i \cos \theta_j + \sin \theta_i \sin \theta_j \cos \phi_{ij}.$$

If linear molecules i and j have dipole moments μ_i, μ_j, and quadrupole moments Q_i, Q_j respectively, then the electrostatic energy is (Gray and Gubbins, 1984, Chapter 2):

$$
\begin{aligned}
v_{ij} = {}& (\mu_i \mu_j / r_{ij}^3)\left(c_{ij} - 3c_i c_j\right) \\
& + \tfrac{3}{2}(\mu_i Q_j / r_{ij}^4)\left(c_i(1 - 5c_j^2) + 2c_j c_{ij}\right) \\
& - \tfrac{3}{2}(Q_i \mu_j / r_{ij}^4)\left(c_j(1 - 5c_i^2) + 2c_i c_{ij}\right) \\
& + \tfrac{3}{4}(Q_i Q_j / r_{ij}^5)\left(1 - 5c_i^2 - 5c_j^2 + 2c_{ij}^2 + 35c_i^2 c_j^2 - 20c_i c_j c_{ij}\right).
\end{aligned}
$$

Using the chain rule

$$\nabla_{\mathbf{r}_{ij}} v_{ij} = \left(\frac{\partial v_{ij}}{\partial r_{ij}}\right)\nabla_{\mathbf{r}_{ij}} r_{ij} + \left(\frac{\partial v_{ij}}{\partial c_i}\right)\nabla_{\mathbf{r}_{ij}} c_i + \left(\frac{\partial v_{ij}}{\partial c_j}\right)\nabla_{\mathbf{r}_{ij}} c_j + \left(\frac{\partial v_{ij}}{\partial c_{ij}}\right)\nabla_{\mathbf{r}_{ij}} c_{ij}.$$

The angle γ_{ij} is independent of $\mathbf{r}_{ij}$, so the last term vanishes. From the definition of c_i we obtain

$$\nabla_{\mathbf{r}_{ij}} c_i = \frac{\mathbf{e}_i}{r_{ij}} - \frac{c_i}{r_{ij}^2}\mathbf{r}_{ij} = r_{ij}^{-1}\left(\mathbf{e}_i - c_i \hat{\mathbf{r}}_{ij}\right)$$

and a similar result for $\nabla_{\mathbf{r}_{ij}} c_j$. Combining these expressions gives

$$\mathbf{f}_{ij} = -\left(\frac{\partial v_{ij}}{\partial r_{ij}}\right)\hat{\mathbf{r}}_{ij} - \frac{1}{r_{ij}}\left(\frac{\partial v_{ij}}{\partial c_i}\right)\left(\mathbf{e}_i - c_i \hat{\mathbf{r}}_{ij}\right) - \frac{1}{r_{ij}}\left(\frac{\partial v_{ij}}{\partial c_j}\right)\left(\mathbf{e}_j - c_j \hat{\mathbf{r}}_{ij}\right). \qquad \text{(C.9)}$$

Now we turn to the evaluation of the torque on molecule i due to molecule j, which is defined by

$$\boldsymbol{\tau}_{ij} = -\mathbf{e}_i \times \nabla_{\mathbf{e}_i} v_{ij}.$$

We should only consider the component of the gradient tangential to the vector $\mathbf{e}_i$, but in fact any non-physical radial component will disappear on taking the vector product, and so we can ignore this complication. Again applying the chain rule

$$\nabla_{\mathbf{e}_i} v_{ij} = \left(\frac{\partial v_{ij}}{\partial r_{ij}}\right)\nabla_{\mathbf{e}_i} r_{ij} + \left(\frac{\partial v_{ij}}{\partial c_i}\right)\nabla_{\mathbf{e}_i} c_i + \left(\frac{\partial v_{ij}}{\partial c_j}\right)\nabla_{\mathbf{e}_i} c_j + \left(\frac{\partial v_{ij}}{\partial c_{ij}}\right)\nabla_{\mathbf{e}_i} c_{ij}.$$

The first and third terms vanish, and we obtain finally

$$\tau_{ij} = -\mathbf{e}_i \times \left[\left(\frac{\partial v_{ij}}{\partial c_i} \right) \hat{\mathbf{r}}_{ij} + \left(\frac{\partial v_{ij}}{\partial c_{ij}} \right) \mathbf{e}_j \right]. \tag{C.10a}$$

Note that the force and torque on molecule j due to i can be obtained by interchanging the labels and changing the signs of c_i and c_j. From eqn (C.9) we see that $\mathbf{f}_{ij} = -\mathbf{f}_{ji}$. Applying this prescription to (C.10a) gives

$$\tau_{ji} = -\mathbf{e}_j \times \left[\left(\frac{\partial v_{ij}}{\partial c_j} \right) \hat{\mathbf{r}}_{ij} + \left(\frac{\partial v_{ij}}{\partial c_{ij}} \right) \mathbf{e}_i \right], \tag{C.10b}$$

and we note that $\tau_{ij} \neq -\tau_{ji}$. Instead,

$$\tau_{ij} + \tau_{ji} + \mathbf{r}_{ij} \times \mathbf{f}_{ij} = 0,$$

which satisfies the requirement that angular momentum is locally conserved.

As an example of the use of these equations, the force and torques between a pair of dipoles are

$$\mathbf{f}_{ij} = -\mathbf{f}_{ji} = 3 \frac{\mu_i \mu_j}{r_{ij}^4} \left((c_{ij} - 5c_i c_j) \hat{\mathbf{r}}_{ij} + c_j \mathbf{e}_i + c_i \mathbf{e}_j \right), \tag{C.11a}$$

$$\tau_{ij} = -\frac{\mu_i \mu_j}{r_{ij}^3} \mathbf{e}_i \times \left(\mathbf{e}_j - 3c_j \hat{\mathbf{r}}_{ij} \right), \tag{C.11b}$$

$$\tau_{ji} = -\frac{\mu_i \mu_j}{r_{ij}^3} \mathbf{e}_j \times \left(\mathbf{e}_i - 3c_i \hat{\mathbf{r}}_{ij} \right). \tag{C.11c}$$

The development in this section is based on a paper by Cheung (1976). Price et al. (1984) have given a more formal and thorough treatment, which includes the electrostatic interactions for non-linear molecules. In both these papers, the convention employed is $\mathbf{r}_{ij} = \mathbf{r}_j - \mathbf{r}_i$, which is opposite to that adopted in this book.

C.4   The triple-dipole potential

In this section, we consider the interaction between triplets of atoms through a potential of the Axilrod–Teller form

$$v^{AT}(\mathbf{r}_i, \mathbf{r}_j, \mathbf{r}_k) = v \frac{1 + 3 \cos \theta_i \cos \theta_j \cos \theta_k}{r_{ij}^3 r_{jk}^3 r_{ki}^3} = v \frac{r_{ij}^2 r_{jk}^2 r_{ki}^2 - 3 C_i C_j C_k}{r_{ij}^5 r_{jk}^5 r_{ki}^5} \tag{C.12}$$

where v is a constant, and the geometry is defined in Fig. C.3. Here we have defined

$$C_i = \mathbf{r}_{ki} \cdot \mathbf{r}_{ij}, \quad C_j = \mathbf{r}_{ij} \cdot \mathbf{r}_{jk}, \quad C_k = \mathbf{r}_{jk} \cdot \mathbf{r}_{ki}.$$

For acute-angled triangles, this energy term is positive, but if one of the angles is obtuse it can become negative: thus near-linear configurations are slightly favoured. The net

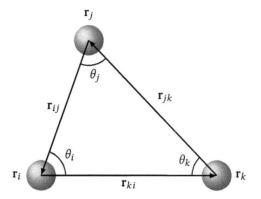

Fig. C.3 A triplet of atoms. The *internal* angles of the triangle are used in calculating the triple-dipole potential. For maximum symmetry in the force expressions, the relative vectors are defined in a cyclic fashion.

contribution in a liquid, however, is typically positive, and may amount to approximately 10 %–15 % of the total energy in, for example, argon.

The forces are readily calculated by differentiation, $\mathbf{f}_i = -\boldsymbol{\nabla}_{\mathbf{r}_i} v^{\text{AT}}(\mathbf{r}_i, \mathbf{r}_j, \mathbf{r}_k)$, etc., and the results are as follows.

$$\mathbf{f}_i = \frac{\nu}{r_{ij}^5 r_{jk}^5 r_{ki}^5}\left[5\left(r_{ij}^2 r_{jk}^2 r_{ki}^2 - 3C_iC_jC_k\right)\left(r_{ij}^{-2}\mathbf{r}_{ij} - r_{ki}^{-2}\mathbf{r}_{ki}\right)\right.$$

$$\left. + 3C_i(C_k - C_j)\mathbf{r}_{jk} + 3C_jC_k(\mathbf{r}_{ki} - \mathbf{r}_{ij}) + 2\left(r_{ij}^2 r_{jk}^2 \mathbf{r}_{ki} - r_{jk}^2 r_{ki}^2 \mathbf{r}_{ij}\right)\right],$$

$$\mathbf{f}_j = \frac{\nu}{r_{ij}^5 r_{jk}^5 r_{ki}^5}\left[5\left(r_{ij}^2 r_{jk}^2 r_{ki}^2 - 3C_iC_jC_k\right)\left(r_{jk}^{-2}\mathbf{r}_{jk} - r_{ij}^{-2}\mathbf{r}_{ij}\right)\right.$$

$$\left. + 3C_j(C_i - C_k)\mathbf{r}_{ki} + 3C_kC_i(\mathbf{r}_{ij} - \mathbf{r}_{jk}) + 2\left(r_{jk}^2 r_{ki}^2 \mathbf{r}_{ij} - r_{ki}^2 r_{ij}^2 \mathbf{r}_{jk}\right)\right],$$

$$\mathbf{f}_k = \frac{\nu}{r_{ij}^5 r_{jk}^5 r_{ki}^5}\left[5\left(r_{ij}^2 r_{jk}^2 r_{ki}^2 - 3C_iC_jC_k\right)\left(r_{ki}^{-2}\mathbf{r}_{ki} - r_{jk}^{-2}\mathbf{r}_{jk}\right)\right.$$

$$\left. + 3C_k(C_j - C_i)\mathbf{r}_{ij} + 3C_iC_j(\mathbf{r}_{jk} - \mathbf{r}_{ki}) + 2\left(r_{ki}^2 r_{ij}^2 \mathbf{r}_{jk} - r_{ij}^2 r_{jk}^2 \mathbf{r}_{ki}\right)\right],$$

where the last two equations may be obtained by cyclic permutation of the indices in the first one. The forces will be evaluated in a triple loop as described in Chapter 1.

C.5   Charged particles using Ewald sum

The interaction between a set of N charges, in a periodic array of cubic boxes, each of volume $V = L^3$, is given by the Ewald sum, see Chapter 6. The potential energy, eqns (6.4) and (6.6), is

$$\mathcal{V}^{qq} = \frac{1}{2}\sum_i \sum_{j \ne i} q_i q_j \frac{\text{erfc}(\kappa r_{ij})}{r_{ij}} + \frac{2\pi}{V}\sum_{\mathbf{k} \ne 0}\hat{G}(\mathbf{k})\hat{\rho}^q(\mathbf{k})\hat{\rho}^q(-\mathbf{k}) + \sum_i \frac{\kappa}{\sqrt{\pi}}q_i^2 + \frac{2\pi}{3V}\left|\sum_i q_i \mathbf{r}_i\right|^2$$

where $\hat{G}(k) = 4\pi \exp(-k^2/4\kappa^2)/k^2$, $\hat{\rho}^q(k) = \sum_i q_i \exp(-ik \cdot r_i)$, $k = 2\pi n/L$ is a reciprocal lattice vector, and $\mathrm{erfc}(x)$ is the complementary error function.

The force on charge i is

$$f_i^{qq} = -\nabla_{r_i} \mathcal{V}^{qq} = q_i \sum_{j \neq i} q_j \left[\frac{2}{\sqrt{\pi}} \kappa r_{ij} \exp(-\kappa^2 r_{ij}^2) + \mathrm{erfc}(\kappa r_{ij}) \right] \frac{r_{ij}}{r_{ij}^3}$$

$$+ \frac{q_i}{2V} \sum_{k \neq 0} \hat{G}(k) ik \left[\hat{\rho}^q(-k) \exp(-ik \cdot r_i) - \hat{\rho}^q(k) \exp(+ik \cdot r_i) \right] - \frac{4\pi q_i}{3V} \sum_j q_j r_j$$

where we have used the result $\mathrm{d}\,\mathrm{erfc}(x)/\mathrm{d}x = -2\exp(-x^2)/\sqrt{\pi}$. The two terms in the k-space part of the force are equivalent and can be combined to give

$$f_i^{qq} = q_i \sum_{j \neq i} q_j \left[\frac{2}{\sqrt{\pi}} \kappa r_{ij} \exp(-\kappa^2 r_{ij}^2) + \mathrm{erfc}(\kappa r_{ij}) \right] \frac{r_{ij}}{r_{ij}^3}$$

$$- \frac{q_i}{V} \sum_{k \neq 0} \hat{G}(k) k \, \mathrm{Im} \left(\hat{\rho}^q(-k) \exp(-ik \cdot r_i) \right) - \frac{4\pi q_i}{3V} \sum_j q_j r_j$$

where $\mathrm{Im}(z)$ is the imaginary part of a complex number.

C.6   The Gay–Berne potential

The potential due to Gay and Berne (1981) was defined by eqns (1.30)–(1.32). Here we discuss the handling of the potential cutoff, and the derivation of forces and torques.

The most common approach to simulations using the Gay–Berne potential is to use a spherical cutoff at $r_{ij} = r_c$, and shift the potential such that it is zero at that distance. It is not usual to apply long-range corrections to energy and pressure, partly because the potential is often used to simulate liquid crystalline phases, the lower symmetry of which would complicate the calculation somewhat. The potential therefore takes the form

$$v_{GB}(r_{ij}, e_i, e_j) - v_{GB}(r_c \hat{r}_{ij}, e_i, e_j), \tag{C.13}$$

where the orientations are represented by unit vectors e_i and e_j, and $\hat{r}_{ij} = r_{ij}/r_{ij}$ is the unit vector pointing along the line of centres. It is important to realize that the extra potential shift term depends on molecular orientations relative to the centre–centre vector, and therefore it acts as an additional source of forces and torques. Also, it is important that the cutoff distance be sufficiently large. For two particles of elongation κ, in the end-to-end arrangement, the effective Lennard-Jones potential crosses the zero axis at $r_{ij} = \kappa \sigma_s$, and the attractive well then extends over a further distance $\sim \sigma_s$ (see Fig. 1.11). Instead of a typical Lennard-Jones value of $r_c = 3.0\sigma_s$, an appropriate value would be $r_c = (\kappa + 2.0)\sigma_s$. Early simulations of the model used a significantly smaller cutoff distance (e.g. Berardi et al. (1993), de Miguel et al. (1996), and Brown et al. (1998) all used $r_c = 4\sigma_s$ for $\kappa = 3$) and this produces noticeably different results.

The procedure for deriving forces and torques from the potential follows closely the approach outlined in the previous sections, especially Section C.3. We can see in eqns (1.30)–(1.32) that v_{GB} depends explicitly on the separation r_{ij}, and through the functions σ and ϵ, on the cosines $c_i = \mathbf{e}_i \cdot \hat{\mathbf{r}}_{ij}$, $c_j = \mathbf{e}_j \cdot \hat{\mathbf{r}}_{ij}$, and $c_{ij} = \mathbf{e}_i \cdot \mathbf{e}_j$. This is the general form of the potential considered by Price et al. (1984). Consequently, we can use eqns (C.9) and (C.10), which simply require the derivatives of the potential with respect to each of these four variables. This is a straightforward, if tedious, exercise using the chain rule (Luckhurst et al., 1990). Writing eqn (1.30) in reduced units with $\sigma_s = 1$,

$$v_{GB}(r_{ij}, c_i, c_j, c_{ij}) = 4\epsilon(c_i, c_j, c_{ij})\left[\rho_{ij}^{-12} - \rho_{ij}^{-6}\right],$$

with $\rho_{ij} = r_{ij} - \sigma(c_i, c_j, c_{ij}) + 1.0$, the derivative with respect to r_{ij} is simply

$$\frac{\partial v_{GB}}{\partial r_{ij}} = -24\epsilon\left[2\rho_{ij}^{-13} - \rho_{ij}^{-7}\right],$$

while the angular derivatives take the form

$$\frac{\partial v_{GB}}{\partial c_i} = 4\left[\rho_{ij}^{-12} - \rho_{ij}^{-6}\right]\frac{\partial \epsilon}{\partial c_i} + 24\epsilon\left[2\rho_{ij}^{-13} - \rho_{ij}^{-7}\right]\frac{\partial \sigma}{\partial c_i}$$

(and similarly for c_j and c_{ij}). For the last term, eqn (1.31) becomes (in reduced units)

$$\sigma = \left[1 - \frac{\chi}{2}\left(\frac{(c_i + c_j)^2}{1 + \chi c_{ij}} + \frac{(c_i - c_j)^2}{1 - \chi c_{ij}}\right)\right]^{-1/2}.$$

Defining coefficients $C_+ = (c_i + c_j)/(1 + \chi c_{ij})$ and $C_- = (c_i - c_j)/(1 - \chi c_{ij})$,

$$\frac{\partial \sigma}{\partial c_i} = \tfrac{1}{2}\chi\sigma^3\left(C_+ + C_-\right), \quad \frac{\partial \sigma}{\partial c_j} = \tfrac{1}{2}\chi\sigma^3\left(C_+ - C_-\right), \quad \frac{\partial \sigma}{\partial c_{ij}} = -\left(\tfrac{1}{2}\chi\right)^2\sigma^3\left(C_+^2 - C_-^2\right).$$

Similar, slightly more complicated, expressions come from differentiating the energy expression ϵ of eqn (1.32).

The forces and torques derived from the cutoff shift term in eqn (C.13) are calculated in the same way, but evaluated at $r_{ij} = r_c$, and omitting the term involving the derivative with respect to r_{ij} (the first term in eqn (C.9)). The full form of all these expressions may be found in the example code (see Section C.7).

The Gay–Berne potential is readily generalized to the case of non-linear molecules, in which case it represents (approximately) a biaxial ellipsoid (Berardi et al., 1995; 1998; Cleaver et al., 1996). The procedure just outlined is straightforwardly extended to this case. If the orientation of molecule i is defined by a mutually orthogonal set of unit vectors $\{\mathbf{e}_i^\alpha\}$, $\alpha = 1, 2, 3$, and that of molecule j by a similar set, the force may be written

$$\mathbf{f}_{ij} = -\frac{\partial v_{GB}}{\partial r_{ij}}\hat{\mathbf{r}}_{ij} - \sum_{\mathbf{e}}\frac{\partial v_{GB}}{\partial(\mathbf{e}\cdot\hat{\mathbf{r}}_{ij})}\frac{\mathbf{e} - (\mathbf{e}\cdot\hat{\mathbf{r}}_{ij})\hat{\mathbf{r}}_{ij}}{r_{ij}},$$

where the sum ranges over all the orientation vectors on both molecules (compare eqn (C.9)). Similarly the torques may be expressed (compare eqns (C.10))

$$\boldsymbol{\tau}_i = -\sum_{\alpha=1}^{3}\mathbf{e}_i^\alpha \times \left(\frac{\partial v_{GB}}{\partial(\mathbf{e}_i^\alpha \cdot \hat{\mathbf{r}}_{ij})}\hat{\mathbf{r}}_{ij} + \sum_{\beta=1}^{3}\frac{\partial v_{GB}}{\partial(\mathbf{e}_i^\alpha \cdot \mathbf{e}_j^\beta)}\mathbf{e}_j^\beta\right),$$

$$\boldsymbol{\tau}_j = -\sum_{\beta=1}^{3}\mathbf{e}_j^\beta \times \left(\frac{\partial v_{GB}}{\partial(\mathbf{e}_j^\beta \cdot \hat{\mathbf{r}}_{ij})}\hat{\mathbf{r}}_{ij} + \sum_{\alpha=1}^{3}\frac{\partial v_{GB}}{\partial(\mathbf{e}_i^\alpha \cdot \mathbf{e}_j^\beta)}\mathbf{e}_i^\alpha\right).$$

Code C.1   Programs to test forces and torques

These files are provided online. test_pot_atom.f90 compares forces for an atomic system with those computed by numerical differentiation of the potential, while test_pot_linear.f90 performs the same task for forces and torques for a system of linear molecules. To go with them, we supply additional files to show how these quantities are calculated for the potentials discussed in this appendix: test_pot_bend.f90 (angle bending), test_pot_twist.f90 (angle torsion), test_pot_dd.f90 (dipole–dipole), test_pot_dq.f90 (dipole–quadrupole), test_pot_qq.f90 (quadrupole–quadrupole), test_pot_at.f90 (Axilrod–Teller), and test_pot_gb.f90 (Gay–Berne). All of these files take the form of a module with name test_pot_module. Utility module routines (Appendix A) are used to generate random positions and orientations, and perform rotations.

```
! test_pot_atom.f90
! Test potential, forces for atoms
PROGRAM test_pot_atom

! test_pot_linear.f90
! Test potential, forces, torques for linear molecule
PROGRAM test_pot_linear
```

Further discussion of the calculation of forces and torques in this kind of model may be found elsewhere (Allen and Germano, 2006).

It is worth noting that the LAMMPS implementation of the Gay–Berne potential uses a different approach to the interactions between two molecules (Everaers and Ejtehadi, 2003), giving a different functional form of the potential, for which analytical expressions for forces and torques have been published (Babadi et al., 2006). Although in many cases this version is, numerically, a close match to the form discussed here, the reader should be aware that the two potentials are not identical.

C.7   Numerically testing forces and torques

In Code C.1 we introduce some simple programs to test the analytical formulae for forces and torques by comparing with quantities obtained directly from the potential by numerical differentiation. We give examples for most of the potentials discussed earlier in this appendix. All these codes have been written with the aim of comparing easily with the analytical expressions given in the earlier sections; some improvements in efficiency would normally be possible, for practical implementations.

Appendix D
Fourier transforms and series

D.1 The Fourier transform

The structural and dynamic results of computer simulations must often be transformed between time and frequency domains or between normal space and reciprocal space. To be compared with experiment a time correlation function $C(t)$ is usually transformed to produce a spectrum $\hat{C}(\omega)$

$$\hat{C}(\omega) = \int_{-\infty}^{\infty} dt\, C(t) \exp(-i\omega t) \equiv \mathfrak{F}\big[C(t)\big], \qquad \text{(D.1a)}$$

and the inverse transform is

$$C(t) = \int_{-\infty}^{\infty} \frac{d\omega}{2\pi}\, \hat{C}(\omega) \exp(i\omega t) \equiv \mathfrak{F}^{-1}\big[\hat{C}(\omega)\big]. \qquad \text{(D.1b)}$$

If $C(t)$ is an even function of time (e.g. a classical autocorrelation function) this may be written

$$\hat{C}(\omega) = 2 \int_{0}^{\infty} dt\, C(t) \cos \omega t, \qquad \text{(D.2a)}$$

and the inverse transform is

$$C(t) = \frac{1}{\pi} \int_{0}^{\infty} d\omega\, \hat{C}(\omega) \cos \omega t. \qquad \text{(D.2b)}$$

It should be noted that true (i.e. quantum-mechanical) autocorrelation functions are not even in time, obeying instead the detailed balance condition (see Section 2.9). The time-reversal symmetry of classical cross-correlation functions is discussed in many places (e.g. Berne and Harp, 1970).

A variety of combinations of numerical prefactors may appear in these definitions. Also, it is sometimes convenient to use, instead of ω the variable $\nu = \omega/2\pi$, when the definitions become symmetrical. We shall stick to ω, reserving ν for use as a discrete frequency index in the following section.

In practice, the time correlation function is known to some finite maximum time, t_{max}, which is determined by the method of analysis and the simulation run time. This value

Computer Simulation of Liquids. Second Edition. M. P. Allen and D. J. Tildesley.
© M. P. Allen and D. J. Tildesley 2017. Published in 2017 by Oxford University Press.

of t_{max} replaces the upper limit in eqn (D.2a). Useful formulae may be derived from the orthogonality relations

$$\int_{-\infty}^{\infty} \frac{d\omega}{2\pi} \exp(i\omega t) \exp(-i\omega t') = \delta(t - t') \tag{D.3a}$$

$$\int_{0}^{\infty} d\omega \cos \omega t \cos \omega t' = \tfrac{1}{2}\pi\delta(t - t'). \tag{D.3b}$$

The convolution/correlation theorem states that if

$$C(t) = \int_{-\infty}^{\infty} dt' \, A(t')B(t - t') \equiv \left(A \star B\right)(t) \tag{D.4}$$

then

$$\hat{C}(\omega) = \hat{A}(\omega)\,\hat{B}(\omega) \tag{D.5}$$

while if

$$C(t) = \int_{-\infty}^{\infty} dt' \, A(t')B(t + t') \tag{D.6}$$

then

$$\hat{C}(\omega) = \hat{A}(-\omega)\,\hat{B}(\omega) = \hat{A}^*(\omega)\,\hat{B}(\omega) \tag{D.7}$$

where * denotes the complex conjugate, and we take $A(t)$ and $B(t)$ to be real. With these definitions Parseval's theorem is

$$\int_{-\infty}^{\infty} dt \, |C(t)|^2 = \int_{-\infty}^{\infty} \frac{d\omega}{2\pi} \, |\hat{C}(\omega)|^2.$$

D.2   Spatial Fourier transforms and series

The spatial analogue of eqns (D.1a) and (D.2b) is

$$\hat{f}(k) = \int_{-\infty}^{\infty} dx \, f(x) \exp(-ikx) \equiv \mathfrak{F}\left[f(x)\right], \tag{D.8a}$$

$$f(x) = \int_{-\infty}^{\infty} \frac{dk}{2\pi} \, \hat{f}(k) \exp(ikx) \equiv \mathfrak{F}^{-1}\left[\hat{f}(k)\right], \tag{D.8b}$$

in one dimension, and

$$\hat{f}(\mathbf{k}) = \iiint d\mathbf{r} \, f(\mathbf{r}) \exp(-i\mathbf{k} \cdot \mathbf{r}) \equiv \mathfrak{F}\left[f(\mathbf{r})\right], \tag{D.9a}$$

$$f(\mathbf{r}) = \iiint \frac{d\mathbf{k}}{(2\pi)^3} \, \hat{f}(\mathbf{k}) \exp(i\mathbf{k} \cdot \mathbf{r}) \equiv \mathfrak{F}^{-1}\left[\hat{f}(\mathbf{k})\right], \tag{D.9b}$$

in three dimensions. Usually we replace the triple integral notation by a single one for brevity. Structural quantities such as $g(r)$ may be related to quantities observed in (say) scattering experiments by a three-dimensional Fourier transform, such as

$$S(k) - 1 = \iiint d\mathbf{r} \, \exp(-i\mathbf{k} \cdot \mathbf{r}) \, \rho \, g(r). \tag{D.10}$$

These may be treated in a manner analogous to the one-dimensional case. In practice, when the functions depend only upon the magnitude of their arguments, it is sensible to integrate over the angular variables to obtain an equation such as

$$S(k) - 1 = 4\pi \int_0^\infty dr\, r^2 \frac{\sin kr}{kr} \rho\, g(r) \tag{D.11}$$

with the inverse transform being

$$\rho\, g(r) = \frac{1}{2\pi^2} \int_0^\infty dk\, k^2 \frac{\sin kr}{kr} (S(k) - 1). \tag{D.12}$$

In a periodic simulation box, the Fourier transform becomes a Fourier series, which we usually write

$$f(x) = \frac{1}{L} \sum_n \hat{f}(k) \exp(ikx), \quad k = n\frac{2\pi}{L} \tag{D.13a}$$

$$\hat{f}(k) = \int_{-L/2}^{L/2} dx\, f(x) \exp(-ikx), \tag{D.13b}$$

in one dimension, and

$$f(\mathbf{r}) = \frac{1}{V} \sum_n \hat{f}(\mathbf{k}) \exp(i\mathbf{k} \cdot \mathbf{r}), \quad \mathbf{k} = \mathbf{n}\frac{2\pi}{L} = (n_x, n_y, n_z)\frac{2\pi}{L} \tag{D.14a}$$

$$\hat{f}(\mathbf{k}) = \iiint_V d\mathbf{r}\, f(\mathbf{r}) \exp(-i\mathbf{k} \cdot \mathbf{r}), \tag{D.14b}$$

in three dimensions, where we have assumed a cubic box with $V = L^3$. The sums are over the discrete wavenumbers or wavevectors that are commensurate with the box; sometimes these are written as $\sum_k$ rather than $\sum_n$, with this understanding. In some contexts, it is convenient to arrange the prefactors differently; the current arrangement means that, as $L \to \infty$, eqn (D.13b) becomes eqn (D.8a), and the sum of eqn (D.13a) goes over smoothly to the integral of eqn (D.8b)

$$\frac{1}{L} \sum_n = \sum_n \frac{\Delta k}{2\pi} \to \int \frac{dk}{2\pi}, \quad \text{with} \quad \Delta k = 2\pi/L$$

and similarly in three dimensions. Parseval's theorem may be written

$$\int_{-L/2}^{L/2} dx\, |f(x)|^2 = \frac{1}{L} \sum_n |\hat{f}(k)|^2, \quad \text{or} \quad \int_V d\mathbf{r}\, |f(\mathbf{r})|^2 = \frac{1}{V} \sum_n |\hat{f}(\mathbf{k})|^2.$$

Quite often in simulations, we deal with the Fourier transforms of single-particle densities, typified by (Hansen and McDonald, 2013)

$$\rho(\mathbf{r}) = \sum_{i=1}^N \delta(\mathbf{r} - \mathbf{r}_i)$$

which is a sum of Dirac delta functions involving the coordinates of all the particles. In a homogeneous system, one can perform a uniform volume average so that

$$\frac{1}{V}\int_V dr\delta(\mathbf{r}-\mathbf{r}_i) = \frac{1}{V} \quad\Rightarrow\quad \langle\rho(\mathbf{r})\rangle = \frac{N}{V}$$

as expected. The instantaneous Fourier-transformed single particle density is then

$$\hat{\rho}(\mathbf{k}) = \int_V d\mathbf{r}\sum_{i=1}^{N}\delta(\mathbf{r}-\mathbf{r}_i)\exp(-i\mathbf{k}\cdot\mathbf{r}) = \sum_{i=1}^{N}\exp(-i\mathbf{k}\cdot\mathbf{r}_i).$$

In a similar way, the charge density and its Fourier transform are

$$\rho^q(\mathbf{r}) = \sum_{i=1}^{N}q_i\delta(\mathbf{r}-\mathbf{r}_i), \qquad \hat{\rho}^q(\mathbf{k}) = \sum_{i=1}^{N}q_i\exp(-i\mathbf{k}\cdot\mathbf{r}_i).$$

Variables of this kind appear in equations for structural and dynamical functions.

D.3   The discrete Fourier transform

A discrete Fourier transform pair is defined

$$\hat{C}(\nu) = \sum_{\tau=0}^{n-1}C(\tau)\exp(-2\pi i\nu\tau/n) \qquad \nu = 0,1,\ldots,n-1, \tag{D.15a}$$

$$C(\tau) = \frac{1}{n}\sum_{\nu=0}^{n-1}\hat{C}(\nu)\exp(2\pi i\nu\tau/n) \qquad \tau = 0,1,\ldots,n-1. \tag{D.15b}$$

This may be interpreted as a relationship between a function $C(t)$ of time, tabulated at n points, δt apart, so that $C(\tau) = C(\tau\delta t)$, and a function $\hat{C}(\omega)$ of frequency ω, also tabulated at n points, so that $\hat{C}(\nu) = \hat{C}(\nu\delta\omega)$. The intervals in time and frequency are related by

$$n\,\delta t\,\delta\omega = 2\pi. \tag{D.16}$$

Note however that neither δt nor $\delta\omega$ actually appear in eqns (D.15), just the integers τ, ν, and n. The orthogonality relation is

$$\frac{1}{n}\sum_{\nu=0}^{n-1}\exp(2\pi i\nu\tau/n)\exp(-2\pi i\nu\tau'/n) = \delta_{\tau\tau'}. \tag{D.17}$$

The analogy between these equations and eqns (D.1)–(D.3) is obvious. There are, however, some subtleties involved in the use of the discrete Fourier transforms (Brigham, 1974; Smith, 1982a,b). In particular, when we use eqns (D.15) and (D.17), we must understand $C(t)$ to be periodic in time with period $n\delta t$ and $\hat{C}(\omega)$ to be periodic in frequency with period $n\delta\omega$. Note that the indices here run from 0 to $n-1$, rather than taking positive and negative values (contrast eqns (D.15) with eqns (D.1)). This is not serious, since a shift of time origin simply implies multiplication of the transform by a complex number. These transforms may be calculated very rapidly on a computer (Cooley and Tukey, 1965).

The discrete convolution/correlation theorem is that if

$$C(\tau) = \sum_{\tau'=0}^{n-1} A(\tau')B(\tau - \tau') \equiv \left(A \star B\right)(\tau) \tag{D.18}$$

then

$$\hat{C}(v) = \hat{A}(v)\hat{B}(v) \tag{D.19}$$

while if

$$C(\tau) = \sum_{\tau'=0}^{n-1} A(\tau')B(\tau + \tau') \tag{D.20}$$

then

$$\hat{C}(v) = \hat{A}^*(v)\hat{B}(v). \tag{D.21}$$

These equations are used in the computation of correlation functions by the FFT method (Section 8.3). In this application, the τ_{run} data items generated in a run are supplemented by τ_{run} zeroes, so that $n = 2\tau_{\text{run}}$ in this case. This avoids the introduction of spurious correlations due to the implied periodicity of the functions mentioned earlier.

In Chapter 6 the discrete spatial Fourier transform is used in the calculation of long-range forces. The usual definitions are modified by the inclusion of a cell length; also we use s_c for the number of mesh points in each dimension. Thus

$$\hat{f}(\mathbf{k}) = \ell^3 \sum_{\mathbf{r}_s} f(\mathbf{r}_s) \exp(-i\mathbf{k} \cdot \mathbf{r}_s)$$

$$f(\mathbf{r}_s) = \frac{1}{s_c^3} \frac{1}{\ell^3} \sum_{\mathbf{k}} \hat{f}(\mathbf{k}) \exp(i\mathbf{k} \cdot \mathbf{r}_s)$$

where (in 3D) the s_c^3 positions, $\mathbf{r}_s$, lie on a mesh of spacing $\ell = L/s_c$ and the s_c^3 wavevectors, $\mathbf{k}$, are from the corresponding Fourier mesh, of spacing $2\pi/L = 2\pi/(s_c\ell)$. Note that these equations have the same form as eqns (D.14), except that the integral of eqn (D.14a) is replaced by a discrete approximation. In these units, a convolution is defined

$$\left(A \star B\right)(\mathbf{r}_s) = \ell^3 \sum_{\mathbf{r}_s'} A(\mathbf{r}_s')B(\mathbf{r}_s - \mathbf{r}_s').$$

D.4 Numerical Fourier transforms

Most functions have to be Fourier transformed numerically. For large values of the frequency, ω, the integrand in the transform oscillates rapidly and methods such as Simpson's rule are inadequate. An accurate method due to Filon (1928) fits a quadratic polynomial between discrete function points and evaluates the resulting integral analytically. For an integral of the form

$$\hat{C}(\omega) = 2 \int_0^{t_{max}} dt \, C(t) \cos \omega t. \tag{D.22}$$

the range is divided into $2n$ equal intervals, so that

$$t_{max} = 2n\delta t. \tag{D.23}$$

If we define

$$\theta = \omega \delta t, \tag{D.24}$$

then

$$\hat{C}(\omega) = 2\delta t \Big(\alpha C(t_{max}) \sin \omega t_{max} + \beta C_e + \gamma C_o \Big), \tag{D.25}$$

where

$$\alpha = (1/\theta^3)\Big(\theta^2 + \theta \sin \theta \cos \theta - 2 \sin^2 \theta\Big),$$
$$\beta = (2/\theta^3)\Big(\theta(1 + \cos^2 \theta) - 2 \sin \theta \cos \theta\Big),$$
$$\gamma = (4/\theta^3)\Big(\sin \theta - \theta \cos \theta\Big). \tag{D.26}$$

C_e is the sum of all the even ordinates of the curve $C(t) \cos \omega t$, less one-half of the first and last ones. C_o is the sum of all the odd ordinates. This algorithm, though accurate, does not preserve the orthogonality of the transform: transformation from t-space to ω-space and back again will not, in general, regenerate the initial correlation function exactly. Lado (1971) has suggested a simple algorithm which preserves the orthogonality of the transform and can be used with the FFT method of Cooley and Tukey (1965). The integral is replaced by a discrete sum

$$\hat{C}(v - \tfrac{1}{2}) = 2\delta t \sum_{\tau=1}^{n} C(\tau - \tfrac{1}{2}) \cos\left[\left(\tau - \tfrac{1}{2}\right)\left(v - \tfrac{1}{2}\right)\pi/\left(n - \tfrac{1}{2}\right)\right]. \tag{D.27}$$

The back transform is

$$C(\tau - \tfrac{1}{2}) = \frac{\delta\omega}{\pi} \sum_{v=1}^{n} \hat{C}(v - \tfrac{1}{2}) \cos\left[\left(\tau - \tfrac{1}{2}\right)\left(v - \tfrac{1}{2}\right)\pi/\left(n - \tfrac{1}{2}\right)\right]. \tag{D.28}$$

The upper limit n of the summations can be replaced by $n - 1$, since the last term vanishes in each case. The points at which the function is evaluated are fixed in this method. $C(t - \tfrac{1}{2})$ means $C((t - \tfrac{1}{2})\delta t)$ where $\delta t = t_{max}/(n - \tfrac{1}{2})$ and $\hat{C}(v - \tfrac{1}{2})$ means $\hat{C}((v - \tfrac{1}{2})\delta\omega)$ with $\delta\omega = \pi/t_{max}$. These 'half-integer' values would usually be calculated by interpolation from the simulation data. Apart from the trivial 'half-integer' phase shifts, these are straightforward discrete Fourier transforms, and they may be computed by the efficient FFT method. This method is less accurate than that of Filon, being essentially a trapezoidal rule, but it can be made more accurate by decreasing δt (i.e. calculating the correlation function at finer intervals). Sine transforms are tackled in a way analogous to cosine

Code D.1   3D Fourier transform example

This file is provided online. `fft3dwrap.f90` performs forward and reverse FFTs on a 3D Gaussian function, as described in the text. It is assumed that the FFTW library is installed; in the SConstruct file, an environment variable may need changing to point to this library. The code also uses Fortran's C-interoperability features. In SConstruct it is assumed that the compiler flags will result in variable KIND declarations that are compatible with the C routines, but this is not guaranteed to be the case on all systems.

```
! fft3dwrap.f90
! 3D fast Fourier transform applied to a Gaussian function
PROGRAM fft3dwrap
```

transforms. The one-dimensional transform of eqn (D.11) may be calculated by Filon's method, if an extra factor of r is incorporated into the function being transformed. Lado (1971) also discusses in detail the calculation of two- and three-dimensional Fourier transforms. Further information about the numerical evaluation of Fourier transforms may be found in Press et al. (2007).

Three-dimensional Fourier transforms, between real and reciprocal space, can be calculated using an FFT, which evaluates the 3D generalization of the transform pair given by eqns (D.15). An example is provided in Code D.1. Consider a function $g(x, y, z)$ which is periodic in a box of side L (for example this could be the charge density discussed in Chapter 6). This function is evaluated on a regular grid

$$g(i, j, k) = g\left(i\ell, j\ell, k\ell\right) \qquad i, j, k = 0, 1, 2, \ldots s_c - 1 \qquad (D.29)$$

where the spacing $\ell = L/s_c$. For simplicity, L and ℓ are assumed to be the same in each of the three coordinate directions.

A 3D array of complex numbers, `fin = CMPLX(g, 0.0)`, is constructed. The FFT is performed using the C subroutine library FFTW (Frigo and Johnson, 2005). In this library the function call

```
plan = fftw_plan_dft_3d (sc, sc, sc, fin, fout, &
                  & FFTW_FORWARD, FFTW_ESTIMATE)
```

establishes a plan for the complex 3D-FFT in the forward direction (from real to reciprocal space). The subroutine `fftw_execute_dft (plan, fin, fout)` performs the planned transform. On exit, the array `fout(i,j,k)` contains the value of the discrete transform, where the Fourier transform of $g(i, j, k)$ is

$$\hat{g}(i, j, k) = \ell^3 f_{\text{out}}(i, j, k). \qquad (D.30)$$

In the output array, `fout(i,j,k)`, the discrete Fourier transform is arranged in the so-called wrap-around order. So for a particular index i, the corresponding wavevector component k_x is

$$k_{x,i} = \begin{cases} 2\pi i/(s_c\ell) & 0 \le i \le s_c/2 - 1 \\ 2\pi(i - s_c)/(s_c\ell) & s_c/2 \le i \le s_c - 1, \end{cases} \tag{D.31}$$

with the same pattern applying to $k_{y,j}$ and $k_{z,k}$. The transform starts at $k_x = 0$ and increases, in increments of $2\pi/L$, to the most positive value $k_x^{\max} = \pi/\ell - 2\pi/L$. The values of the transform for negative k_x follow, starting at $k_x^{\min} = -\pi/\ell$ and ascending towards zero again. The last stored value is $-2\pi/L$. Since the transformed function is periodic in k_x with period $2\pi/\ell$, the values stored in the upper half of the array also correspond to the regular continuation of those in the lower half, $k_x = \pi/\ell \ldots 2\pi/\ell - 2\pi/L$.

To obtain the inverse discrete Fourier transform, fin(i,j,k), a new plan is created using

```
plan = fftw_plan_dft_3d (sc,  sc,  sc,  fout,  fin, &
                    & FFTW_BACKWARD ,  FFTW_ESTIMATE)
```

and the back transform is performed by calling fftw_execute_dft(plan, fout, fin). The original input function can be straightforwardly recovered

$$g(i, j, k) = \frac{1}{s_c^3} f_{\mathrm{in}}(i, j, k). \tag{D.32}$$

In Code D.1 we apply the forward and backward transforms to the Gaussian function, $g(x, y, z) = \exp\left(-\pi(x^2 + y^2 + z^2)\right)$ for which $\mathfrak{F}\left[g(x, y, z)\right] = \exp\left(-(k_x^2 + k_y^2 + k_z^2)/4\pi\right)$.

Appendix E
Random numbers

E.1 Random number generators

Before the develoment of computers, random sequences of numbers had to be generated by physical methods such as rolling dice, tossing coins, picking numbered balls from an urn, or analysing noise generated in an electronic valve. To assist workers in this field large tables of pre-calculated random sequences were published (RAND, 2001).

Many applications, including MC and stochastic dynamics simulation, require random sequences of millions of numbers, which must be generated on the computer. The field of pseudo-random number generation (so-called because the sequences are generated deterministically and repeatably) is fairly well developed, and the desirable (and undesirable) features have been studied in detail.

In the current context of liquid-state computer simulations, the raw speed of random number generation is almost never critical: other parts of the program are almost always more time-consuming. The statistical distribution of the numbers should, of course, conform to the desired one: most commonly, random numbers are generated uniformly over the unit interval $(0, 1)$, and more complicated distributions are obtained from this. However, on top of this, the numbers should be uncorrelated, and to investigate this it is necessary to calculate two-point, three-point, and higher, joint distributions. The sequence will typically repeat itself after a certain period, and the length of this repeat cycle should be as long as possible. A battery of tests for random number generators known as Diehard, due to Marsaglia (1995) has been updated under the name Dieharder by Brown (2015). A very comprehensive test suite, TestU01 (L'Ecuyer and Simard, 2007), is also available online (TestU01, 2009).

The dangers of using a poor random number generator should not be underestimated. Although MC simulations are often (or appear to be) insensitive to the details, occasionally catastrophic results can result from a poor choice (Ferrenberg et al., 1992; Schmid and Wilding, 1995; Bauke and Mertens, 2004; Deng et al., 2008). A serious molecular simulator will want to either research the algorithm used by the built-in generator, or replace it with one whose behaviour is known, selected from a library (Press et al., 2007; Barash and Shchur, 2013). Be aware, however, that software library contents can change from version to version. A collection of random number generators may be found in the Gnu is not Unix (GNU) Scientific Library. A selection is also provided in the Intel Math Kernel library.

Computer Simulation of Liquids. Second Edition. M. P. Allen and D. J. Tildesley.
© M. P. Allen and D. J. Tildesley 2017. Published in 2017 by Oxford University Press.

E.2   Uniformly distributed random numbers

The generation of uniform (pseudo)random numbers is an enormous field and several books are available to give the reader a proper introduction (Knuth, 1997; Gentle, 2003; L'Ecuyer, 2007; Kroese et al., 2011). Jones (2010) has authored a very useful online paper giving advice on good practice.

Typically, a sequence of large positive integers is produced, each obtained from the last by some operation (e.g. multiplication) conducted in finite modulus arithmetic. This is typified by linear congruential generators (LCGs), of the form

$$X_{i+1} = (aX_i + b) \mod M \tag{E.1}$$

where the quantities a, b, and M are large positive integer parameters, and mod is the modulo operation (as for the MOD function in Fortran). The sequence begins by selecting an initial seed, X_0. Thus, each result in the sequence is an integer lying in the range $(0, M-1)$ inclusive. The desired random numbers are returned as

$$\xi_i = X_i/M.$$

Note that this excludes the possibility of generating $X_i = 1$, but allows the possibility of $X_i = 0$; in practice, generators may allow neither, either, or both end values to appear. A special case of this kind of generator is the simple multiplicative congruential generator, for which $b = 0$.

These basic generators are not usually considered good enough for MC simulation work, and may be extended in several ways. Multiple recursive generators (MRGs) make the right-hand side of eqn (E.1) a linear function of several previous values X_k. A linear feedback shift register generator uses the same approach, but calculates the output ξ_i in a different way. A variation of this method is used in the popular and widespread Mersenne twister (Matsumoto and Nishimura, 1998). A drawback of this generator is that it is rather elaborate and complicated to seed properly. Combining MRGs to give a single, better-quality, sequence is a productive approach (L'Ecuyer, 1996; 1999); notable, and widely available, examples of this type are MRG32k3a (L'Ecuyer, 1999) and various versions of KISS from Marsaglia. The WELL generators (Panneton et al., 2006) are also promising. All of these choices pass the overwhelming majority of the statistical tests mentioned in the introductory section.

Increasingly one wishes to generate independent streams of random numbers on parallel computers. Here, a key issue is the seeding: if the different processes use the same seeds, they will generate the same sequences of numbers, which will almost certainly lead to trouble. Calculating a seed from the system clock carries with it the danger that several of the processes will still start with the same value; combining the clock value with the processor i.d. number may avoid the problem. On some operating systems it is possible to use values from the device /dev/random or /dev/urandom.

Fortran provides built in routines RANDOM_SEED and RANDOM_NUMBER for, respectively, initializing the random number generator and producing random numbers, uniformly sampled on the range $(0, 1)$. We use these extensively in the supplied examples. Two points are worthy of comment. First, the RANDOM_SEED routine takes a set of optional arguments which may be used to fine-tune the initialization, for example in order to reliably repeat a given sequence of numbers, or alternatively to generate a 'random' starting point based on (e.g.) the system clock. If called without these arguments, it simply initializes to a default state. The GNU documentation for gfortran provides a way of randomly seeding the built-in generator, using /dev/urandom if available, and combining the clock and processor i.d. values otherwise. We have reproduced this in the function init_random_seed provided in the file maths_module.f90 of Code A.1. Of course, for a different compiler, a different method may be needed, and there is no guarantee that /dev/urandom will be available on a given computer. Second, the Fortran standard does not actually specify that RANDOM_NUMBER uses any particular random number generator: it depends on the particular compiler, and possibly on the computer. In our code, for simplicity, we are ignoring most of the preceding advice, and using this routine instead of including a particular choice of our own. The GNU documentation for gfortran (at the time of writing) indicates that one of Marsaglia's KISS generators is used in RANDOM_NUMBER, which should be satisfactory.

We often wish to generate uniform random integers k within a specified range, (k_{min}, k_{max}) inclusive, for example to choose a particle for an MC move or thermal velocity randomization. This is simply, and reliably, computed from a uniformly distributed variate ξ in the range $(0, 1)$ by

$$k = k_{min} + \left\lfloor \xi(k_{max} - k_{min} + 1) \right\rfloor$$

where $\lfloor x \rfloor$ is the largest integer less than or equal to x, that is, the FLOOR function in Fortran. It is sensible to guard against the (very small) danger of roundoff effects by ensuring, afterwards, that $k_{min} \leq k \leq k_{max}$. This is implemented in the routine random_integer in the file maths_module.f90 of Code A.1.

E.3 Generating non-uniform distributions

Using the random number ξ generated uniformly on $(0, 1)$ it is possible to construct random numbers taken from a variety of distributions. There are many distributions which are of interest to statisticians, but only a limited number which are required in liquid-state simulation. In this section we discuss generating random variables on the normal (Gaussian), and exponential distributions. The interested reader is referred to the standard texts (Knuth, 1997; Gentle, 2003; L'Ecuyer, 2007; Kroese et al., 2011) for a comprehensive discussion of other distributions, and proofs of the results quoted in this section. Many libraries contain routines to generate these distributions automatically.

The normal distribution, with mean μ, and variance σ^2, is defined as

$$\rho(x) = \frac{1}{\sigma\sqrt{2\pi}} \exp\left(-\frac{(x-\mu)^2}{2\sigma^2}\right) \qquad -\infty < x < \infty.$$

The expectation values of the first two moments are given by

$$\langle x \rangle = \mu, \quad \langle x^2 \rangle = \mu^2 + \sigma^2, \quad \text{or} \quad \langle (x-\mu)^2 \rangle = \sigma^2.$$

A random number ζ' chosen from this distribution will be related to a number ζ generated from the normal distribution with zero mean and unit variance by

$$\zeta' = \mu + \sigma\zeta.$$

A typical example is the random selection of velocity components (v_x, v_y, v_z) from the Maxwell–Boltzmann distribution at temperature T in the Andersen thermostat, when $\mu = 0$, $\sigma = \sqrt{k_B T / m}$, for each component, where m is the particle mass (see Section 3.8.1).

The problem is reduced to sampling ζ. One way (of many) to do this involves two steps and the generation of two uniform random numbers (Box and Muller, 1958):

(a) generate independent uniform random numbers ξ_1 and ξ_2 on $(0, 1)$;
(b) calculate $\zeta_1 = (-2 \ln \xi_1)^{-1/2} \cos 2\pi \xi_2$ and $\zeta_2 = (-2 \ln \xi_1)^{-1/2} \sin 2\pi \xi_2$.

This algorithm is provided in the routine `random_normal` in the file `maths_module.f90` of Code A.1. The routine generates ζ_1 and ζ_2 together, typically returning ζ_1 and saving ζ_2 for the next call.

In some applications (e.g. Brownian dynamics, Chapter 12) we need to generate correlated pairs of numbers that are normally distributed. Given two independent normal random numbers ζ_1 and ζ_2, with zero means and unit variances, obtained as before, the variables

$$\zeta_1' = \sigma_1 \zeta_1, \qquad \zeta_2' = \sigma_2 \left(c_{12} \zeta_1 + (1 - c_{12}^2)^{1/2} \zeta_2 \right) \tag{E.2}$$

are sampled from the bivariate Gaussian distribution with zero means, variances σ_1^2 and σ_2^2, and correlation coefficient c_{12}.

In the BD simulations described in Section 12.2, we need to sample a large number, n, of correlated random numbers from a multivariate Gaussian distribution

$$\rho(\mathbf{x}) = \frac{1}{\sqrt{(2\pi)^n |\mathbf{C}|}} \exp\left(-\tfrac{1}{2}\mathbf{x} \cdot \mathbf{C}^{-1} \cdot \mathbf{x}\right)$$

where for simplicity we consider zero means, $\langle x_i \rangle = 0$, $i = 1, \ldots, n$, and the (symmetric) covariance matrix $\mathbf{C}$ is defined such that $\langle x_i x_j \rangle = C_{ij}$. Many libraries include routines to sample directly from this distribution in an efficient way. Otherwise, the correlated set of random numbers ζ_i', $i = 1, \ldots, n$, may be obtained from an independent set of normally distributed random numbers, ζ_i, given the lower-triangular matrix $\mathbf{L}$ which satisfies $\mathbf{C} = \mathbf{L} \cdot \mathbf{L}^T$ (see eqn (12.9)). This may be obtained by a Cholesky decomposition, which is once more a common feature of numerical libraries. The elements of the matrix are

$$L_{jj} = \sqrt{C_{jj} - \sum_{k=1}^{j-1} L_{jk}^2}$$

$$L_{ij} = L_{jj}^{-1}\left(C_{ij} - \sum_{k=1}^{j-1} L_{ik}L_{jk}\right), \qquad i > j$$

and the calculation typically begins with $L_{11} = \sqrt{C_{11}}$, followed by successive rows $\{L_{21}, L_{22}\}$, $\{L_{31}, L_{32}, L_{33}\}$, and so on. The expression inside the square root is guaranteed to be positive provided the matrix $\mathbf{C}$ is positive definite. The desired random variables are

$$\zeta_i' = \sum_{j=1}^{i} L_{ij}\zeta_j.$$

Equation (E.2) is just a special case of this equation, with $C_{11} = \sigma_1^2$, $C_{22} = \sigma_2^2$ and $C_{12} = C_{21} = \sigma_1 \sigma_2 c_{12}$. In the BD application, $n = 3N$, $\mathbf{C} = \mathbf{D}$, the diffusion coefficient, and the random displacements are given by eqn (12.9)

$$R_i = \sqrt{2 \Delta t}\, \zeta_i' = \sqrt{2 \Delta t} \sum_{j=1}^{i} L_{ij} \zeta_j.$$

As mentioned in Section 12.2, it cannot be guaranteed that the Oseen tensor is positive definite, so an alternative form for $\mathbf{D}$ is usually employed.

Now we turn to the problem of sampling from the exponential distribution

$$\rho(x) = \begin{cases} \mu^{-1} \exp(-x/\mu) & 0 < x < \infty \\ 0 & \text{otherwise} \end{cases} \tag{E.3}$$

where μ is a positive parameter defining the mean value $\langle x \rangle = \mu$. A suitable method is

(a) generate a uniform random number ξ on $(0, 1)$;
(b) calculate $\zeta = -\mu \ln \xi$.

One example of the use of such a distribution is the selection of a random angular velocity ω for a linear molecule. The direction of ω may be chosen randomly in a plane perpendicular to the molecular axis (see next section), and then the value of ω^2 selected from the exponential distribution with mean value $\mu = \langle \omega^2 \rangle = 2 k_B T / I$, where I is the moment of inertia.

This is an example of the general technique of *inverse transforms*. It relies on being able to calculate (either analytically or in tabular form) the cumulative probability function

$$P(x) = \int_{-\infty}^{x} dx'\, \rho(x')$$

that is, the integrated probability of occurrence of a value less than or equal to x, given the probability density $\rho(x)$. More specifically, it relies on knowing the *inverse* function $x(P)$. If a random value of P is chosen in the range $0 \leq P \leq 1$, then the variable $x(P)$ will be distributed with probability density $\rho(x)$. In this case, $P(x) = 1 - \exp(-x/\mu)$, the inverse function is $x = -\mu \ln(1 - P)$, and we sample $1 - P$, or equivalently P, uniformly over the range $(0, 1)$.

This method also works for sampling from a discrete distribution. Consider the problem of selecting from a set of m alternatives $k = 1, 2, \ldots, m$ according to their assigned probabilities p_k. Suppose that we have already normalized these, so that $\sum_k p_k = 1$, and that they are stored in an array p. Then the following code snippet is used.

```
CALL RANDOM_NUMBER(ran)
k = 1
p_cumul = p(1)
DO
    IF ( p_cumul >= ran ) EXIT
    k = k+1
    p_cumul = p_cumul + p(k)
END DO
```

The value of k on exit from the loop identifies the 'winner' with the correct probability. A moment's thought reveals that this is the inverse transform method. The first step is to choose a random number uniformly on $(0, 1)$. Then, the loop determines the first value of k for which the cumulative probability exceeds this number. It is possible to add a guard against roundoff errors, to ensure that k never exceeds the dimension of the array. This method is used in CBMC (see Section 9.3.4), where we typically use un-normalized weights w_k, given by the Boltzmann factors associated with trial moves. In this case, a trivial modification of the code is used.

```
w_total = SUM(w)
CALL RANDOM_NUMBER(ran)
ran = ran * w_total
k = 1
w_cumul = w(1)
DO
   IF ( w_cumul >= ran ) EXIT
   k = k+1
   w_cumul = w_cumul + w(k)
END DO
```

E.4   Random vectors on the surface of a sphere

There are a number of suitable methods for generating a vector on the surface of a unit sphere. The simplest of these is an iterative procedure using the acceptance–rejection technique of von Neumann (1951).

(a) Generate three uniform random numbers ξ_1, ξ_2, and ξ_3 on $(0, 1)$.
(b) Calculate $\zeta_i = 2\xi_i - 1$ for $i = 1, \ldots, 3$ so that the vector $\boldsymbol{\zeta} = (\zeta_1, \zeta_2, \zeta_3)$ is distributed uniformly in a cube of side 2 centred at the origin.
(c) Form the sum $\zeta^2 = \zeta_1^2 + \zeta_2^2 + \zeta_3^2$.
(d) If $\zeta^2 < 1$ (i.e. inside the inscribed sphere) take $\hat{\boldsymbol{\zeta}} = \boldsymbol{\zeta}/\zeta$ as the vector.
(e) Otherwise, reject the vector and return to step (a).

This is provided as a routine in the file maths_module.f90 of Code A.1. Marsaglia (1972) has suggested an improvement.

(a) Generate two uniform random numbers ξ_1, ξ_2 on $(0, 1)$.
(b) Calculate $\zeta_i = 2\xi_i - 1$ for $i = 1, 2$.
(c) Form the sum $\zeta^2 = \zeta_1^2 + \zeta_2^2$.
(d) If $\zeta^2 < 1$ take $\hat{\boldsymbol{\zeta}} = \left(2\zeta_1\sqrt{1 - \zeta^2}, 2\zeta_2\sqrt{1 - \zeta^2}, 1 - 2\zeta^2\right)$ as the vector.
(e) Otherwise, reject the vector and return to step (a).

This is also provided as an alternative routine. The method can be readily extended to choosing points on a four-sphere (suitable for quaternion orientations) and Marsaglia gives an appropriate algorithm. To obtain random vectors in a plane normal to a given unit vector $\hat{\mathbf{e}}$, simply proceed as just described and project out the component $\boldsymbol{\zeta}' = \hat{\boldsymbol{\zeta}} - (\hat{\boldsymbol{\zeta}} \cdot \hat{\mathbf{e}})\hat{\mathbf{e}}$, then renormalize $\boldsymbol{\zeta}'$ to unit length.

E.5   Choosing randomly and uniformly from complicated regions

von Neumann (1951) suggested the following algorithm for generating random numbers from an arbitrary distribution $\rho(\mathbf{r})$. The function is split in the following way

$$\rho(\mathbf{r}) = Ca(\mathbf{r})b(\mathbf{r}) \tag{E.4}$$

where $a(\mathbf{r})$ is a simpler (normalized) distribution function, from which it is easy to sample a random number, $b(\mathbf{r})$ is a function which lies between zero and one, and C is a constant, with $C \geq 1$. The following steps generate a random vector sampled from $\rho(\mathbf{r})$.

(a) Sample ζ randomly from the distribution $a(\mathbf{r})$.

(b) Generate a uniform random number ξ on $(0, 1)$.

(c) If $\xi \leq b(\zeta)$ then ζ is sampled from the distribution $\rho(\mathbf{r})$.

(d) Otherwise, reject ζ and return to step (a).

A simple example illustrates the method. Suppose we wish to generate a vector uniformly inside the unit circle $|\mathbf{r}|^2 < 1$, where $\mathbf{r} = (x_1, x_2)$, but we wish to sample uniformly in the square $-1 < x_1, x_2 < 1$. In this case

$$\rho(\mathbf{r}) = \begin{cases} 1/\pi & x_1^2 + x_2^2 < 1 \\ 0 & \text{otherwise} \end{cases} \qquad\qquad a(\mathbf{r}) = \begin{cases} 1/4 & -1 < x_1, x_2 < 1 \\ 0 & \text{otherwise} \end{cases}$$

$$b(\mathbf{r}) = \Theta(1 - x_1^2 - x_2^2) \qquad\qquad\qquad C = 4/\pi$$

The step function Θ ensures that $b(\mathbf{r}) = 1$ inside the region of interest, and zero outside. A little reflection shows that in this case the von Neumann algorithm simplifies considerably:

(a) Sample ζ randomly inside the square.

(b) If ζ lies within the circle then ζ is sampled from the distribution $\rho(\mathbf{r})$.

(c) Otherwise, reject ζ and return to step (a).

It is not necessary to know the value of C. It will be noticed that the basic MC sampling method discussed in Section 4.2 is based on this approach, as well as the rejection techniques for generating random orientations discussed in the previous sections. This approach can be readily extended to higher dimensions and more complicated regions of interest.

Mc simulation of molecules relies on generating small random rotations: a new vector should be sampled uniformly on the surface of the unit sphere, but within a given solid angle of the old vector. A very simple implementation, using the approach just described, is given in a routine in the file maths_module.f90 of Code A.1. This is, however, not the most efficient approach, and several better alternatives are also provided. In general, the secret of success with the rejection method is to choose $a(\mathbf{r})$ so as to be convenient to sample the random variates (a hypersphere, hyperellipsoid, etc.) and which covers the whole of the region that is of interest, but is not too much larger. The closer the match, the smaller the fraction of rejected tries.

E.6   Generating a random permutation

In the Lowe–Andersen thermostat (Lowe, 1999) discussed in Sections 3.8.1 and 12.4, it is necessary to examine pairs of particles and, with a certain probability, randomize their relative velocity. It is essential to do this in a random order; the best way is to prepare the desired list (which typically contains all pairs within a certain interaction range) in a systematic way (looping over particle indices) and then shuffle it. The standard in-place algorithm for doing this is quite simple (Knuth, 1997) and is illustrated by the following code snippet, which shuffles an array a of quantities (numbered from 1 to m).

```
! INTEGER                :: q, p, m
! REAL, DIMENSION(m) :: a
! REAL                :: tmp
DO p = 1, m-1
    q = random_integer ( p, m )
    tmp   = a(p)
    a(p) = a(q)
    a(q) = tmp
END DO
```

In words: a single sweep through the array is undertaken, and each element is swapped with a randomly selected element from further up the array. The function random_integer appears in the file maths_module.f90 of Code A.1 and returns an integer chosen randomly in the specified range $p \ldots m$ inclusive.

Appendix F
Configurational temperature

In this Appendix we discuss the calculation of the configurational temperature, introduced in Chapter 2 and defined in eqn (2.56) for a simple atomic system. Two possible sources of confusion arise: the formula to use for the configurational temperature, and the proper counting of interaction terms when it is evaluated in a simulation program.

F.1 Expression for configurational temperature

In the microcanonical ensemble, the exact expression for the configurational temperature is (Rugh, 1997; Jepps et al., 2000; Rickayzen and Powles, 2001; Powles et al., 2005; Travis and Braga, 2006)

$$\beta = \frac{1}{k_B T} = \left\langle \boldsymbol{\nabla} \cdot \left(\frac{\boldsymbol{\nabla}\mathcal{V}}{\boldsymbol{\nabla}\mathcal{V} \cdot \boldsymbol{\nabla}\mathcal{V}} \right) \right\rangle = \left\langle \frac{\nabla^2\mathcal{V}}{\boldsymbol{\nabla}\mathcal{V} \cdot \boldsymbol{\nabla}\mathcal{V}} - 2\frac{\boldsymbol{\nabla}\mathcal{V}\boldsymbol{\nabla}\mathcal{V} : (\boldsymbol{\nabla}\boldsymbol{\nabla}\mathcal{V})}{(\boldsymbol{\nabla}\mathcal{V} \cdot \boldsymbol{\nabla}\mathcal{V})^2} \right\rangle, \qquad \text{(F.1)}$$

where, as usual, $\mathcal{V}$ stands for the potential energy. The gradient $\boldsymbol{\nabla}$ refers to all coordinates of all atoms, and $(\boldsymbol{\nabla}\boldsymbol{\nabla}\mathcal{V})$ is a $3N \times 3N$ Hessian matrix, which is doubly contracted with $\boldsymbol{\nabla}\mathcal{V}$. The Hessian term arises from the chain rule

$$\boldsymbol{\nabla} \cdot \left(\frac{\boldsymbol{\nabla}\mathcal{V}}{\boldsymbol{\nabla}\mathcal{V} \cdot \boldsymbol{\nabla}\mathcal{V}} \right) = \frac{\nabla^2\mathcal{V}}{\boldsymbol{\nabla}\mathcal{V} \cdot \boldsymbol{\nabla}\mathcal{V}} - \frac{\boldsymbol{\nabla}\mathcal{V} \cdot \boldsymbol{\nabla}(\boldsymbol{\nabla}\mathcal{V} \cdot \boldsymbol{\nabla}\mathcal{V})}{(\boldsymbol{\nabla}\mathcal{V} \cdot \boldsymbol{\nabla}\mathcal{V})^2}$$

and the vector identity (C.8) setting $\mathbf{A} = \mathbf{B} = \boldsymbol{\nabla}\mathcal{V}$, which makes the curl terms vanish. This Hessian term is easily seen to be $O(1/N)$ for an N-atom system, and is often dropped, to give an equation very similar to eqn (2.56) which applies in the canonical ensemble. The remaining discrepancy, also $O(1/N)$, relates to whether the numerator $\nabla^2\mathcal{V}$ and denominator $\boldsymbol{\nabla}\mathcal{V} \cdot \boldsymbol{\nabla}\mathcal{V}$ of the leading term are averaged separately (as in eqn (2.56)) or whether their ratio is computed before averaging (as in eqn (F.1)).

F.2 Implementation details

To illustrate how this works in practice, we assume a central pairwise form $\sum_{i<j} v(r_{ij})$ for the interaction potential, with $r_{ij} = |\mathbf{r}_i - \mathbf{r}_j|$. The gradient, and Laplacian, of the function $v(r)$, with respect to $\mathbf{r} = (x, y, z)$, are

$$\boldsymbol{\nabla}v(r) = \frac{v'}{r}\mathbf{r} \equiv v^{(1)}(r)\,\mathbf{r}, \qquad \nabla^2 v(r) = 2\left(\frac{v'}{r}\right) + v'' \equiv v^{(2)}(r),$$

Computer Simulation of Liquids. Second Edition. M. P. Allen and D. J. Tildesley.
© M. P. Allen and D. J. Tildesley 2017. Published in 2017 by Oxford University Press.

where $v' = dv/dr$, $v'' = d^2v/dr^2$, and we have defined the functions $v^{(1)}(r)$ and $v^{(2)}(r)$ for convenience later. The gradient terms, for each atomic pair, are combined to give the overall gradient, in a way that is familiar from the force calculation in an MD program

$$\nabla_i \mathcal{V} = \sum_{j \neq i} v^{(1)}(r_{ij})\, \mathbf{r}_{ij} = -\sum_{j \neq i} \mathbf{f}_{ij} = -\mathbf{f}_i.$$

Hence

$$\nabla \mathcal{V} \cdot \nabla \mathcal{V} = \sum_i \mathbf{f}_i \cdot \mathbf{f}_i = \sum_i |\mathbf{f}_i|^2,$$

the sum of squared forces on each atom. Each $\mathbf{f}_i$ is calculated inside a double loop over atoms, typically looking at each distinct pair ij just once, but using $\mathbf{f}_{ij} = -\mathbf{f}_{ji}$ to make sure no contributions are missed. The Laplacian term may be calculated in the same double loop, as a sum of terms of the form

$$\nabla_i^2 v(r_{ij}) = \nabla_j^2 v(r_{ij}) = v^{(2)}(r_{ij})$$

but it should be remembered that for a given pair ij, *both* the i and j terms in this equation contribute to the total.

If it is desired to keep the $O(1/N)$ Hessian term, a little more calculation is required. The 3×3 Hessian of the pair potential function is

$$\mathbf{H}(\mathbf{r}) = \nabla\nabla v(r) = \frac{v'}{r}\mathbf{1} + \left(\frac{v''}{r^2} - \frac{v'}{r^3}\right)\mathbf{rr} = v^{(1)}(r)\mathbf{1} + v^{(3)}(r)\mathbf{rr}$$

where $\mathbf{1}$ is the 3×3 unit matrix, $\mathbf{rr}$ is a 3×3 dyadic matrix, and we have defined the function $v^{(3)}(r)$. The full $3N \times 3N$ Hessian matrix can be built up from 3×3 matrices of this kind for each pair of atoms and is then combined twice with the gradients $\nabla \mathcal{V}$ or equivalently the force vectors $\mathbf{f}$. The required components are

$$\nabla_i\nabla_i v(r_{ij}) = \nabla_j\nabla_j v(r_{ij}) = \mathbf{H}(\mathbf{r}_{ij}), \quad \nabla_i\nabla_j v(r_{ij}) = \nabla_j\nabla_i v(r_{ij}) = -\mathbf{H}(\mathbf{r}_{ij}).$$

Hence, having evaluated the total forces $\mathbf{f}_i = -\nabla_i \mathcal{V}$ on each atom, the contribution of pair ij to the Hessian term may be expressed

$$\left[\nabla\mathcal{V}\nabla\mathcal{V} : \left(\nabla\nabla\mathcal{V}\right)\right]_{ij} = (\mathbf{f}_i - \mathbf{f}_j)(\mathbf{f}_i - \mathbf{f}_j) :: \mathbf{H}(\mathbf{r}_{ij})$$

$$= v^{(1)}(r_{ij})\,(\mathbf{f}_i - \mathbf{f}_j) \cdot (\mathbf{f}_i - \mathbf{f}_j) + v^{(3)}(r_{ij})\left[(\mathbf{f}_i - \mathbf{f}_j) \cdot \mathbf{r}_{ij}\right]^2.$$

Let us give an example to show how this translates into computer code. For the Lennard-Jones potential, $v(r) = 4\epsilon\left((\sigma/r)^{12} - (\sigma/r)^6\right)$,

$$v^{(1)}(r) = -24\epsilon\left(2(\sigma/r)^{12} - (\sigma/r)^6\right)r^{-2},$$
$$v^{(2)}(r) = 24\epsilon\left(22(\sigma/r)^{12} - 5(\sigma/r)^6\right)r^{-2},$$
$$v^{(3)}(r) = 96\epsilon\left(7(\sigma/r)^{12} - 2(\sigma/r)^6\right)r^{-4}.$$

The first double loop accumulates contributions to the forces and the Laplacian, and we highlight the appearance of these variables in Code F.1. Notice that the Laplacian term is

Code F.1 Calculating the configurational temperature

Here we give the heart of a force loop, in which additional statements accumulate the quantities needed to compute the configurational temperature. epslj and sigma hold the values of the Lennard-Jones parameters ϵ and σ respectively, and rcut holds the cutoff distance.

```
DIMENSION(3)    :: rij, fij
DIMENSION(3,n) :: r, f
sigma_sq = sigma**2
rcut_sq  = rcut**2
eps4     = epslj * 4
eps24    = epslj * 24

pot = 0.0
lap = 0.0
f   = 0.0

DO i = 1, n-1
   DO j = i+1, n
      rij    = r(:,i) - r(:,j)
      rij    = rij - box * ANINT ( rij / box )
      rij_sq = SUM ( rij**2 )
      IF ( rij_sq < rcut_sq ) THEN
         r2    = 1.0 /  rij_sq
         sr2   = sigma_sq * r2
         sr6   = sr2**3
         sr12  = sr6**2
         v     = eps4  * ( sr12 - sr6 )
         v1    = eps24 * ( sr6 - 2 * sr12 ) * r2
         v2    = eps24 * ( 22 * sr12 - 5 * sr6 ) * r2
         pot   = pot + v
         fij   = -v1 * rij
         f(:,i) = f(:,i) + fij
         f(:,j) = f(:,j) - fij
         lap   = lap + 2 * v2
      END IF
   END DO
END DO

fsq  = SUM ( f**2 )
beta = lap / fsq
```

Code F.2   Correction to configurational temperature

This loop implements the $O(1/N)$ correction to β, from the terms involving the Hessian of the Lennard-Jones pair potential. Notice that fij in this double loop has a different meaning to the same variable in Code F.1.

```
sigma_sq = sigma ** 2
rcut_sq  = rcut**2
eps24    = epslj * 24
eps96    = epslj * 96


hes = 0.0

DO i = 1, n-1
    DO j = i+1, n
        rij     = r(:,i) - r(:,j)
        rij     = rij - box * ANINT ( rij / box )
        rij_sq  = SUM ( rij**2 )
        IF ( rij_sq < rcut_sq ) THEN
            r2    = 1.0 /  rij_sq
            sr2   = sigma_sq * r2
            sr6   = sr2**3
            sr12  = sr6**2
            v1    = eps24 * ( sr6 - 2 * sr12 ) * r2
            v3    = eps96 * ( 7 * sr12 - 2*sr6 ) * r2**2
            fij   = f(:,i) - f(:,j)
            ff    = SUM ( fij**2 )
            rf    = SUM ( rij*fij )
            hes   = hes + v1 * ff + v3 * rf**2
        END IF
    END DO
END DO


fsq  = SUM ( f**2 )
beta = beta - 2.0*hes/(fsq**2)
```

incremented by $2v^{(2)}$ so as to count the contributions of both i and j, as mentioned earlier. The leading term in the expression for β appears at the end, ready for averaging over the course of the simulation. Then a second double loop accumulates the Hessian terms, if they are required, and this is shown in Code F.2.

List of Acronyms

We give the section in which acronyms appear, except for one or two cases, which appear throughout the book.

CG	Coarse-Grained	12.7
CHARMM	Chemistry at HARvard Molecular Mechanics: A class-I force field and simulation package	1.3, 1.5, 3.5, 5.7, 9.3
CMD	Centroid Molecular Dynamics	13.4
CNDO	Complete Neglect of Differential Overlap semi-empirical method	13.3
COMPASS	Condensed-phase Optimized Molecular Potentials for Atomistic Simulation Studies: a class-II force field	1.5
CP	Car–Parrinello	13.2, 13.3
CP2K	Car–Parrinello 2K: an electronic structure code	13.2
CPMD	Car–Parrinello Molecular Dynamics: an electronic structure code	13.2, 13.3
CPMD-QM	CPMD-Quantum Mechanics: molecular mechanics option in the CPMD package	13.3
CPU	Central Processing Unit	3.3, 5.3, 5.7, 7.3, 7.4, A.1
CUDA	Compute Unified Device Architecture	7.1, 7.4
DYNAMO	Dynamics of discrete Objects	3.7
DFS	Damped Force-Shifted	6.10
DFT	Density Functional Theory	13.2, 13.3, 13.6
DL_POLY	Daresbury Laboratory Polymer simulation: a simulation package	1.5
DNA	DeoxyriboNucleic Acid	1.4, 1.5
DOLLS	a shear-flow algorithm	11.3
DPD	Dissipative Particle Dynamics	3.8, 11.5, 12.4, 12.5, B.1
DVR	Discrete Variable Representation	13.2
EB	End Bridging	9.3
FB	Force Bias	9.3
FCC	Face Centred Cubic	1.3, 1.5, 2.2, 5.6, 10.7, 11.5
FCI	Full Configuration Interaction	13.5
FCIQMC	Full Configuration Interaction Quantum Monte Carlo	13.5
FENE	Finitely Extensible Nonlinear Elastic model	1.3, 4.8, 11.5
FFS	Forward Flux Sampling	10.1, 10.6
FFT	Fast Fourier Transform	6.1, 6.3, 6.4, 6.6, 6.10, 7.4, 8.3, 13.4, D.3, D.4
FFTW	Fastest Fourier Transform in the West: C library for discrete Fourier transforms	6.3, D.4
FM	Force Matched	5.2, 6.4, 6.10

FMM	Fast Multipole Method	6.1, 6.6, 6.7, 6.10, 7.4
FS	Force Shifted	6.10
GolP	Gold–Protein: a polarizable force field	1.5
GAFF	General AMBER Force Field	1.5
GCMC	Grand Canonical Monte Carlo	4.6, 9.2, 9.3, 14.5
GGA	Generalized Gradient Approximations	13.2
GNU	GNU is Not Unix	E.1, E.2
GPU	Graphics Processing Unit	3.2, 7.1, 7.4, A.1
GROMACS	GROningen MAchine for Chemical Simulations	3.5, 4.9, 5.7, 7.4, 7.5
GROMOS	GROningen MOlecular Simulation: a class-I force field and simulation package	1.3, 1.5, 3.11, 13.3
HF	Hartree–Fock	13.2, 13.3
HMC	Hybrid Monte Carlo	12.3
HOOMD	Highly Object Oriented Molecular Dynamics	7.1
HPC	High Performance Computing	7.1
IBI	Iterative Boltzmann Inversion	12.7
IBISCO	It is Boltzmann Inversion software for COarse graining simulations	12.7
IDE	Integrated Development Environment	A.3
IDR	Intramolecular Double Rebridging	9.3
IK	Irving–Kirkwood	2.12, 14.1, 14.5
IMC	Inverse Monte Carlo	12.7
IPS	Isotropic Periodic Sum	6.4, 6.10
KISS	Keep It Simple Stupid: a random number generator	E.2
KS	Kohn–Sham	13.2, 13.6
KTHNY	Kosterlitz–Thouless–Halperin–Nelson–Young	8.2
L2L	Local-to-Local	6.6
LAMMPS	Large-scale Atomic/Molecular Massively Parallel Simulator	1.5, 5.7, 7.1, 7.4, C.6
LAPACK	Linear Algebra PACKage	3.4
LB	Lattice-Boltzmann	12.6, 12.7
LCG	Linear Congruential Generator: a class of random number algorithms	E.2
LDA	Local Density Approximation	13.2
LINCS	LINear Constraint Solver	3.4, 7.5
LRC	Long Range Correction	14.1
ls1 mardyn	large systems 1: molecular dynamics	1.1
MagiC	MagiC coarse-grained simulation package	12.7
M2L	Multipole-to-Local	6.6
M2M	Multipole-to-Multipole	6.6

MANIAC	Mathematical Analyzer, Numerical Integrator, And Computer: an early Los Alamos computer	1.1, 4.1
MARTINI	MARrink's Toolkit INItiative: a coarse-grained force field	1.3, 1.5
MBAR	Multi-state Bennett Acceptance Ratio	5.5, 9.2
MC	Monte Carlo	
MCCCS Towhee	Monte Carlo for Complex Chemical Systems: Towhee package	9.3
MCPRO	Monte Carlo simulation for biomolecules package	9.3
MCYna	Matsuoka–Clementi–Yoshimine non-additive model of water	1.3
MD	molecular dynamics	
MEMD	Maxwell Equation Molecular Dynamics	6.1, 6.8, 6.10, 7.4, B.1
MILC-SHAKE	Matrix Inverted Linear Constraints SHAKE: a constraint algorithm	3.4
MM	Molecular Mechanics	1.5, 13.3
MM4	Molecular Mechanics 4: a class-II force field	1.5
MMFF	Merck Molecular Force Field: a class-II force field	1.5
MOP	Method Of Planes	14.1
MP4	Møller–Plesset 4th order	13.2
MPCD	MultiParticle Collision Dynamics	12.5
MPI	Message Passing Interface	5.3, 7.1, 7.3, 7.4, A.3
MRG	Multiple Recursive Generator: a class of random number algorithms	E.2
MSM	Multilevel Summation Method	6.7, 6.10, 7.4
NAMD	NAnoscale Molecular Dynamics	7.4
NEMD	NonEquilibrium Molecular Dynamics	11.1–11.5, 11.7, 11.8
NMR	Nuclear Magnetic Resonance	4.9
NS	Nested Sampling	9.2
openCL	Open Computing Language	7.1, 7.4
openMP	Open Multi-Processing	7.1, 7.2, 7.4, A.3
OPLS	Optimized Potentials for Liquid Simulations: a class-I force field	1.5
PB	Poisson–Boltzmann	6.5
PBE0	Perdew–Burke–Ernzerhof: a hybrid density functional	13.2
PDB	Protein Data Base: a file format	5.7
PIMD	Path-Integral Molecular Dynamics	13.2, 13.4
PME	Particle Mesh Ewald	6.3–6.5, 6.10, 7.4
PPIP	Pairwise Point Interaction Pipeline	3.5

PPPM	Particle–Particle Particle–Mesh	6.1–6.3, 6.6, 6.7, 6.10, 7.4, 14.1, B.1
PUT	Profile Unbiased Thermostat	11.7
PW91	Perdew–Wang '91: a density functional	13.2
QDO	Quantum Drude oscillator model of water	1.3
QM	Quantum Mechanical	13.3
QM/MM	Quantum Mechanics/Molecular Mechanics	1.1, 12.7, 13.1–13.3
ReaxFF	Reactive Force Field: a class-III force field	1.5
RATTLE	a constraint algorithm	3.4, 7.5, 13.2
REMC	Replica Exchange Monte Carlo	4.9, 9.2
REMD	Replica Exchange Molecular Dynamics	4.9
RESP	Restrained ElectroStatic Potential	1.4, 1.5, 13.3
REST	Replica Exchange with Solute Tempering	4.9, 9.2
RMS	Root Mean Square	2.3, 2.5, 3.2, 3.5, 3.6
RNA	RiboNucleic Acid	1.5
RPMD	Ring Polymer Molecular Dynamics	13.4
RUMD	Roskilde University Molecular Dynamics	7.1
SAFT	Statistical Associating Fluid Theory	12.7
SETTLE	a constraint algorithm	3.4
SHAKE	a constraint algorithm	3.4, 7.5, 13.2, 13.4
SI	Système International	3.6, 6.8, B.1
SIESTA	Spanish Initiative for Electronic Simulations with Thousands of Atoms: an electronic structure code	13.2
SLLOD	a shear-flow algorithm	11.3, 11.5, 11.7
SMC	Smart Monte Carlo	9.3, 12.3
SPAM	Smoothed Particle Applied Mechanics	12.4
SPC	Simple Point Charge: a model of water	1.3, 6.3
SPC/E	Simple Point Charge/Extended: a model of water	1.3, 6.4, 6.5, 6.10, 9.4, 13.4, 14.5
SPH	Smoothed Particle Hydrodynamics	12.4
SPMD	Single Program Multiple Data	7.1
SPME	Smooth Particle Mesh Ewald	6.3, 6.10, 7.4, 14.1
SPRES	Stochastic Process Rare Event Sampling	10.6
TA	Test Area	14.1, 14.5
TDDFT	Time-Dependent Density Functional Theory	13.2, 13.6
TIGER	Temperature Intervals with Global Energy Reassignment	4.9
TIP3P	Transferable Intermolecular Potential with 3 Points: a model of water	1.3, 3.5, 4.9, 6.4–6.7, 6.10

TIP4P	Transferable Intermolecular Potential with 4 Points: a model of water	1.3, 7.4
TIP4P/2005	TIP4P/2005: a generalized model of water	1.3, 14.2
TIP5P	Transferable Intermolecular Potential with 5 Points: a model of water	1.3
TIS	Transition Interface Sampling	10.1, 10.6, 10.7
TPS	Transition Path Sampling	10.1, 10.5, 10.7
TST	Transition State Theory	10.1–10.4
UB	Unbonding–Bonding Monte Carlo method	9.3
UFF	Universal Force Field: a class-II force field	1.5
VASP	Vienna *Ab initio* Simulation Package: an electronic structure code	13.2
VMD	Visual Molecular Dynamics	5.7
VOTCA	Versatile Object-oriented Toolkit for Coarse-graining Applications: a coarse-grained simulation package	12.7
WELL	Well-Equidistributed Long-period Linear: a class of random number algorithms	E.2
WHAM	Weighted Histogram Analysis Method	9.2
WIGGLE	a constraint algorithm	3.4
XYZ	XYZ: a file format	5.7

List of Greek Symbols

We give the principal sections in which the symbols are defined or used.

λ	undetermined multiplier	3.4
λ	Hamiltonian perturbation parameter	11.6
λ_T	thermal conductivity	2.7
μ	chemical potential	2.1
μ, ν	Gay–Berne exponents	1.3
μ	fictitious mass	13.2
μ	dipole moment	1.3
μ_0	permeability of free space	6.8, B.1
ν	stoichiometric coefficient	9.5
ξ	fugacity fraction	4.7
ξ	friction coefficient	3.8, 12.2
ξ	random number	E.2
π	transition matrix	4.3
ϱ	density operator	13.1
$\rho(z)$	single particle density profile	14.1
ρ	number density	2.1
ρ	electron density	13.2
$\rho(\Gamma)$	phase space distribution function	2.1
$\rho^{(2)}$	pair density function	14.1
$\sigma(\mathcal{A})$	root-mean-square fluctuation of $\mathcal{A}$	2.3, 8.4
σ	Lennard-Jones range parameter	1.3
τ	torque	1.3
τ	discrete time or trial index	2.1
Υ	wave function	13.5
ϕ	electrostatic potential	1.3, 6.1
ϕ	angle	2.6
φ	angle of rotation	3.3
Φ	total wave function	13.1
χ_{ab}	intramolecular constraint	3.4
χ	nuclear wave function	13.1
ψ_i	orbital	13.2
ψ	angle	1.5
Ψ	electronic wave function	13.1
Ψ	thermodynamic potential	2.1
Ω	molecular orientation	1.3
ω	molecular angular velocity	2.7
ω	frequency	2.7

List of Roman Symbols

We give the principal sections in which the symbols are defined or used.

$\hat{G}$	influence function	6.3
G	Green's function	13.6
$\mathbf{G}$	matrix of Gaussian random variables	12.2
$\mathbf{G}$	metric tensor	2.10
G	symmetry function for neural network	5.2
$\mathbf{g}$	constraint force	3.4
g	number of degrees of freedom	3.8
g	Kirkwood g-factor	6.2
g_{kj}	switching probability between quantum states	13.6
$G_{\ell,m}$	irregular solid harmonic	6.6
$g(r)$	pair distribution function	2.6, 8.2
$g_{ab}(r_{ab})$	site–site pair distribution function	2.6
$g_{\ell\ell'm}(r)$	spherical harmonic coefficients of g	2.6, 8.2
$g(r_{ij}, \Omega_i, \Omega_j)$	molecular pair distribution function	2.6
$\mathcal{H}$	Hamiltonian	1.3
H	magnetic field strength	6.8
H	occupation histogram	9.2
H	enthalpy	2.2
$\hbar$	Planck's constant $h/2\pi$	2.9
h	indicator function	10.2
$H(x, \Delta x)$	top-hat function	9.2, 14.1
I	ionic strength	6.5
$\mathbf{I}$	moment of inertia tensor	3.3
i	atomic or molecular index	1.3
$\mathbf{j}$	electric current	6.8
j	atomic or molecular index	1.3
$\mathcal{K}$	kinetic energy	1.3
$\mathbf{k}$	wavevector	1.6
k	rate constant	10.2
k_B	Boltzmann's constant	1.3
$\mathcal{L}$	Lagrangian	3.1
L	simulation box length	1.6
iL	Liouville operator	2.1
$\boldsymbol{\ell}$	molecular angular momentum	3.3
ℓ	grid spacing	13.2
$L_{\ell,m}$	local expansion coefficient	6.6
$\mathbf{m}$	vector of integers	6.2
m	possible outcome or state label	4.3
m	molecular mass	1.3
M_I	mass of the nucleus I	13.1
$M_{\ell,m}$	multipole expansion coefficient	6.6
N	number of atoms or molecules	1.3
$\mathbf{n}$	axis of rotation	3.3
$\mathbf{n}$	nematic director	2.14, 14.7
n	possible outcome or state label	4.3

O	octopole moment	1.3
P	number of processors in a parallel computer	7.4, 13.4
P	pressure	2.1
$\mathbf{P}$	pressure tensor	14.1
$\mathcal{P}$	instantaneous pressure	2.4
$\mathbf{p}$	molecular momentum	1.3
$\mathbf{P}_I$	nuclear momentum	13.4
P_ℓ	Legendre polynomial	2.14
Q	thermostat inertia	3.8
Q	quadrupole moment	1.3
Q	partition function	2.1
$\mathbf{Q}$	orientational order tensor	2.14
$\mathbf{q}$	generalized coordinate	1.3
q	charge	1.3
$q(\mathbf{r})$	reaction coordinate	10.1
$q^\dagger$	transition state	10.2
$\mathbf{r}$	position	1.3
r_{ij}	i–j separation	1.3
$\mathbf{r}_{ab}$	a–b site–site separation vector $= \mathbf{r}_{ia} - \mathbf{r}_{jb}$	1.3
r_c	potential cutoff distance	1.6
r_e	equimolar dividing surface	14.5
$\mathbf{R}_I$	nuclear position	13.1
$\mathbf{r}_{ij}$	i–j separation vector $= \mathbf{r}_i - \mathbf{r}_j$	1.3
$r_{s,\,m}$	mechanically defined radius of tension	14.5
S	entropy	2.2
$\mathcal{S}$	action	13.6
$\mathbf{s}$	2D position $= (x, y)$	2.12
$\mathbf{s}$	box-scaled position vectors	4.5
s	statistical inefficiency	8.4
s	imaginary time	13.4
s_c	number of cells	6.3
$\mathcal{T}$	instantaneous temperature	2.4
T	temperature	2.1
$\mathbf{T}$	transformation matrix	9.3
t	time	2.1
$t_\mathcal{A}$	correlation time	2.7
U	propagator	3.2
$\mathbf{u}_{ia}$	normal modes	13.4
V	volume	2.1
$\mathcal{V}$	potential energy	1.3
$\mathbf{v}$	molecular velocity	2.7
$v(r_{ij})$	i–j pair potential	1.3
W	weighting function	9.2
W	barostat inertia	3.9
W	charge assignment function	6.3

$\mathcal{W}$	instantaneous total virial	2.4
$\mathcal{W}$	work	11.6
w	Wiener process	12.2
$w(r_{ij})$	pair virial	2.4
$\mathcal{X}$	instantaneous total hypervirial	2.5
$x(r_{ij})$	pair hypervirial	2.5
$Y_{\ell m}$	spherical harmonic function	2.6
Z	configurational integral	2.2
z	valence	6.5
z	activity	4.6

List of Examples

List of Codes

Bibliography

Abascal, J. L. F. and Vega, C. (2005). A general purpose model for the condensed phases of water: TIP4P/2005. *J. Chem. Phys.* **123**, 234505.

Abraham, F. F. (1982). Statistical surface physics: a perspective via computer simulation of microclusters, interfaces and simple films. *Prog. Phys. Rep.* **45**, 1113–1161.

Adams, D. J. (1974). Chemical potential of hard sphere fluids by Monte Carlo methods. *Molec. Phys.* **28**, 1241–1252.

Adams, D. J. (1975). Grand canonical ensemble Monte Carlo for a Lennard-Jones fluid. *Molec. Phys.* **29**, 307–311.

Adams, D. J. (1979). Computer simulation of ionic systems: distorting effects of the boundary conditions. *Chem. Phys. Lett.* **62**, 329–332.

Adams, D. J. (1980). Periodic truncated octahedral boundary conditions. *The problem of long-ranged forces in the computer simulation of condensed media.* NRCC Workshop Proceedings. **9**, 13. (Menlo Park, USA, 1980). Ed. by D. Ceperley.

Adams, D. J. (1983a). Alternatives to the periodic cube in computer simulation. *CCP5 Quarterly* **10**, 30–36.

Adams, D. J. (1983b). On the use of the Ewald summation in computer simulation. *J. Chem. Phys.* **78**, 2585–2590.

Adams, D. J. and McDonald, I. R. (1976). Thermodynamic and dielectric properties of polar lattices. *Molec. Phys.* **32**, 931–947.

Adams, D. J., Adams, E. M., and Hills, G. J. (1979). Computer simulation of polar liquids. *Molec. Phys.* **38**, 387–400.

Adelman, S. A. (1979). Generalized Langevin theory for many-body problems in chemical dynamics: general formulation and the equivalent harmonic chain representation. *J. Chem. Phys.* **71**, 4471–4486.

Adhikari, R., Stratford, K., Cates, M. E., and Wagner, A. J. (2005). Fluctuating lattice Boltzmann. *Europhys. Lett.* **71**, 473–479.

Aguado, A. and Madden, P. A. (2003). Ewald summation of electrostatic multipole interactions up to the quadrupolar level. *J. Chem. Phys.* **119**, 7471–7483.

Ahlrichs, P. and Dünweg, B. (1999). Simulation of a single polymer chain in solution by combining lattice Boltzmann and molecular dynamics. *J. Chem. Phys.* **111**, 8225–8239.

Aidun, C. K. and Clausen, J. R. (2010). Lattice-Boltzmann method for complex flows. *Ann. Rev. Fluid Mech.* **42**, 439–472.

Aimoli, C. G., Maginn, E. J., and Abreu, C. R. A. (2014). Force field comparison and thermodynamic property calculation of supercritical CO_2 and CH_4 using molecular dynamics simulations. *Fluid Phase Equilibria* **368**, 80–90.

Akhmatskaya, E. V. and Reich, S. (2008). GSHMC: an efficient method for molecular simulation. *J. Comput. Phys.* **227**, 4934–4954.

Akhmatskaya, E. V., Bou-Rabee, N., and Reich, S. (2009). A comparison of generalized hybrid Monte Carlo methods with and without momentum flip. *J. Comput. Phys.* **228**, 2256–2265. Erratum: *ibid.* **228**, 7492–7496 (2009).

Akino, N., Schmid, F., and Allen, M. P. (2001). Molecular dynamics study of the nematic–isotropic interface. *Phys. Rev. E* **63**, 041706.

Alder, B. J. (1986). Molecular dynamics simulations. *Molecular dynamics simulation of statistical mechanical systems.* Proc. Int. School Phys. Enrico Fermi. **97**, 66–80. (Varenna, Italy, 1985). Ed. by G. Ciccotti and W. G. Hoover. Bologna: Soc. Italiana di Fisica.

Alder, B. J. and Wainwright, T. E. (1957). Phase transition for a hard sphere system. *J. Chem. Phys.* **27**, 1208–1209.

Alder, B. J. and Wainwright, T. E. (1959). Studies in molecular dynamics. I. General method. *J. Chem. Phys.* **31**, 459–466.

Alder, B. J. and Wainwright, T. E. (1960). Studies in molecular dynamics. II. Behavior of a small number of elastic spheres. *J. Chem. Phys.* **33**, 1439–1451.

Alder, B. J. and Wainwright, T. E. (1969). Enhancement of diffusion by vortex-like motion of classical hard particles. *J. Phys. Soc. Japan Suppl.* **26**, 267–269. Proc. Int. Conf. on Statistical Mechanics, Kyoto, 1968.

Alder, B. J. and Wainwright, T. E. (1970). Decay of the velocity autocorrelation function. *Phys. Rev. A* **1**, 18–21.

Alder, B. J., Frankel, S. P., and Lewinson, V. A. (1955). Radial distribution function calculated by the Monte Carlo method for a hard sphere fluid. *J. Chem. Phys.* **23**, 417–419.

Alder, B. J., Gass, D. M., and Wainwright, T. E. (1970). Studies in molecular dynamics. VIII. The transport coefficients for a hard sphere fluid. *J. Chem. Phys.* **53**, 3813–3826.

Alejandre, J. and Chapela, G. A. (2010). The surface tension of TIP4P/2005 water model using the Ewald sums for the dispersion interactions. *J. Chem. Phys.* **132**, 014701.

Alfe, D., Bartok, A. P., Csanyi, G., and Gillan, M. J. (2014). Analyzing the errors of DFT approximations for compressed water systems. *J. Chem. Phys.* **141**, 014104.

Allen, M. P. (2006a). Computer simulation of liquid crystals. *Computer simulations in condensed matter systems: from materials to chemical biology, vol 2.* Lect. Notes Phys. **704**, 191–210. (Erice, Italy, 2005). Ed. by M. Ferrario, G. Ciccotti, and K. Binder. Heidelberg: Springer.

Allen, M. P., Frenkel, D., and Talbot, J. (1989). Molecular dynamics simulation using hard particles. *Comput. Phys. Rep.* **9**, 301–353.

Allen, M. P. (1980). Brownian dynamics simulation of a chemical reaction in solution. *Molec. Phys.* **40**, 1073–1087.

Allen, M. P. (1982). Algorithms for Brownian dynamics. *Molec. Phys.* **47**, 599–601.

Allen, M. P. (1984). Atomic and molecular representations of molecular hydrodynamic variables. *Molec. Phys.* **52**, 705–716.

Allen, M. P. (1993). Calculating the helical twisting power of dopants in a liquid crystal by computer simulation. *Phys. Rev. E* **47**, 4611–4614.

Allen, M. P. (2006b). Configurational temperature in membrane simulations using dissipative particle dynamics. *J. Phys. Chem. B* **110**, 3823–3830.

Allen, M. P. (2006c). Evaluation of pressure tensor in constant-volume simulations of hard and soft convex bodies. *J. Chem. Phys.* **124**, 214103.

Allen, M. P. and Cunningham, I. C. H. (1986). A computer simulation study of idealized model tetrahedral molecules. *Molec. Phys.* **58**, 615–625.

Allen, M. P. and Germano, G. (2006). Expressions for forces and torques in molecular simulations using rigid bodies. *Molec. Phys.* **104**, 3225–3235.

Allen, M. P. and Imbierski, A. A. (1987). A molecular dynamics study of the hard dumb-bell system. *Molec. Phys.* **60**, 453–473.

Allen, M. P. and Kivelson, D. (1981). Nonequilibrium molecular dynamics simulation and the generalized hydrodynamics of transverse modes in molecular fluids. *Molec. Phys.* **44**, 945–965.

Allen, M. P. and Maréchal, G. (1986). Nonequilibrium molecular dynamics simulation of a molecular fluid subjected to an oscillatory perturbation. *Molec. Phys.* **57**, 7–19.

Allen, M. P. and Masters, A. J. (1993). Computer simulation of a twisted nematic liquid crystal. *Molec. Phys.* **79**, 277–289.

Allen, M. P. and Masters, A. J. (2001). Molecular simulation and theory of liquid crystals: chiral parameters, flexoelectric coefficients, and elastic constants. *J. Mater. Chem.* **11**, 2678–2689.

Allen, M. P. and Quigley, D. (2013). Some comments on Monte Carlo and molecular dynamics methods. *Molec. Phys.* **111**, 3442–3447.

Allen, M. P. and Schmid, F. (2007). A thermostat for molecular dynamics of complex fluids. *Molec. Simul.* **33**, 21–26.

Allen, M. P., Evans, G. T., Frenkel, D., and Mulder, B. (1993). Hard convex body fluids. *Adv. Chem. Phys.* **86**, 1–166.

Allen, M. P., Warren, M. A., Wilson, M. R., Sauron, A., and Smith, W. (1996). Molecular dynamics calculation of elastic constants in Gay–Berne nematic liquid crystals. *J. Chem. Phys.* **105**, 2850–2858.

Allen, R. J., Frenkel, D., and ten Wolde, P. R. (2006a). Forward flux sampling-type schemes for simulating rare events: efficiency analysis. *J. Chem. Phys.* **124**, 194111.

Allen, R. J., Frenkel, D., and ten Wolde, P. R. (2006b). Simulating rare events in equilibrium or nonequilibrium stochastic systems. *J. Chem. Phys.* **124**, 024102.

Allen, R. J., Valeriani, C., and ten Wolde, P. R. (2009). Forward flux sampling for rare event simulations. *J. Phys. Cond. Mat.* **21**, 463102.

Allinger, N. L., Chen, K. S., and Lii, J. H. (1996). An improved force field (MM4) for saturated hydrocarbons. *J. Comput. Chem.* **17**, 642–668.

Almarza, N. G. and Lomba, E. (2007). Cluster algorithm to perform parallel Monte Carlo simulation of atomistic systems. *J. Chem. Phys.* **127**, 084116.

Ambrosiano, J., Greengard, L., and Rokhlin, V. (1988). The fast multipole method for gridless particle simulation. *Comput. Phys. Commun.* **48**, 117–125.

Andersen, H. C. (1980). Molecular dynamics simulations at constant pressure and/or temperature. *J. Chem. Phys.* **72**, 2384–2393.

Andersen, H. C. (1983). RATTLE: a 'velocity' version of the SHAKE algorithm for molecular dynamics calculations. *J. Comput. Phys.* **52**, 24–34.

Anderson, J. B. (1975). A random-walk simulation of the Schrödinger equation: H_3^+. *J. Chem. Phys.* **63**, 1499–1503.

Anderson, J. B. (1976). Quantum chemistry by random walk. H ^2P, H$_3^+$ $D_{3h}^1 A_1'$, H$_2$ $^3\Sigma_u^+$, H$_4$ $^1\Sigma_g^+$, Be ^1S. *J. Chem. Phys.* **65**, 4121–4127.

Anderson, J. B. (1980). Quantum chemistry by random walk: higher accuracy. *J. Chem. Phys.* **73**, 3897–3899.

Ando, T., Chow, E., Saad, Y., and Skolnick, J. (2012). Krylov subspace methods for computing hydrodynamic interactions in Brownian dynamics simulations. *J. Chem. Phys.* **137**, 064106.

Andrea, T. A., Swope, W. C., and Andersen, H. C. (1983). The role of long ranged forces in determining the structure and properties of liquid water. *J. Chem. Phys.* **79**, 4576–4584.

Anisimov, V. M., Lamoureux, G., Vorobyov, I. V., Huang, N., Roux, B., and MacKerell, A. D. (2005). Determination of electrostatic parameters for a polarizable force field based on the classical Drude oscillator. *J. Chem. Theor. Comput.* **1**, 153–168.

Antila, H. S. and Salonen, E. (2013). Polarizable force fields. *Biomolecular simulations, methods and protocols.* Methods Molec. Biol. **924**. Chap. 9, 215–241. Ed. by L. Monticelli and E. Salonen. New York: Humana Press.

Applequist, J., Carl, J. R., and Fung, K. K. (1972). Atom dipole interaction model for molecular polarizability. Application to polyatomic molecules and determination of atom polarizabilities. *J. Am. Chem. Soc.* **94**, 2952–2960.

Apte, P. A. and Zeng, X. C. (2008). Anisotropy of crystal–melt interfacial free energy of silicon by simulation. *Appl. Phys. Lett.* **92**, 221903.

Arias, T. A., Payne, M. C., and Joannopoulos, J. D. (1992). *Ab initio* molecular dynamics: analytically continued energy functionals and insights into iterative solutions. *Phys. Rev. Lett.* **69**, 1077–1080.

Arnold, A., Fahrenberger, F., Holm, C., Lenz, O., Bolten, M., Dachsel, H., Halver, R., Kabadshow, I., Gaehler, F., Heber, F., Iseringhausen, J., Hofmann, M., Pippig, M., Potts, D., and Sutmann, G. (2013). Comparison of scalable fast methods for long-range interactions. *Phys. Rev. E* **88**, 063308.

Asenjo, D., Paillusson, F., and Frenkel, D. (2014). Numerical calculation of granular entropy. *Phys. Rev. Lett.* **112**, 098002.

Ashurst, W. T. and Hoover, W. G. (1972). Nonequilibrium molecular dynamics: shear viscosity and thermal conductivity. *Bull. Amer. Phys. Soc.* **17**, 1196.

Ashurst, W. T. and Hoover, W. G. (1973). Argon shear viscosity via a Lennard-Jones potential with equilibrium and nonequilibrium molecular dynamics. *Phys. Rev. Lett.* **31**, 206–208.

Ashurst, W. T. and Hoover, W. G. (1975). Dense fluid shear viscosity via nonequilibrium molecular dynamics. *Phys. Rev. A* **11**, 658–678.

Assfeld, X. and Rivail, J. L. (1996). Quantum chemical computations on parts of large molecules: the *ab initio* local self consistent field method. *Chem. Phys. Lett.* **263**, 100–106.

Attard, P. (1992). Simulation results for a fluid with the Axilrod–Teller triple dipole potential. *Phys. Rev. A* **45**, 5649–5653.

Attard, P. (1995). On the density of volume states in the isobaric ensemble. *J. Chem. Phys.* **103**, 9884–9885.

Aviram, I., Tildesley, D. J., and Streett, W. B. (1977). Virial pressure in a fluid of hard polyatomic molecules. *Molec. Phys.* **34**, 881–885.

Axilrod, B. M. and Teller, E. (1943). Interaction of the van der Waals type between three atoms. *J. Chem. Phys.* **11**, 299–300.

Babadi, M., Ejtehadi, M., and Everaers, R. (2006). Analytical first derivatives of the RE-squared interaction potential. *J. Comput. Phys.* **219**, 770–779.

Backer, J. A., Lowe, C. P., Hoefsloot, H. C. J., and Iedema, P. D. (2005). Poiseuille flow to measure the viscosity of particle model fluids. *J. Chem. Phys.* **122**, 154503.

Bagchi, K., Balasubramanian, S., Mundy, C. J., and Klein, M. L. (1996). Profile unbiased thermostat with dynamical streaming velocities. *J. Chem. Phys.* **105**, 11183–11189.

Bai, P. and Siepmann, J. I. (2013). Selective adsorption from dilute solutions: Gibbs ensemble Monte Carlo simulations. *Fluid Phase Equilibria* **351**, 1–6.

Bailey, A. G., Lowe, C. P., and Sutton, A. P. (2008). Efficient constraint dynamics using MILCSHAKE. *J. Comput. Phys.* **227**, 8949–8959.

Bailey, A. G., Lowe, C. P., and Sutton, A. P. (2009). REVLD: a coarse-grained model for polymers. *Comput. Phys. Commun.* **180**, 594–599. Special issue based on the Conference on Computational Physics 2008.

Baker, C. M. (2015). Polarizable force fields for molecular dynamics simulations of bio-molecules. *Comp. Mol. Sci. (Wiley Interdisc. Rev.)* **5**, 241–254.

Ballenegger, V., Cerda, J. J., and Holm, C. (2012). How to convert SPME to P3M: influence functions and error estimates. *J. Chem. Theor. Comput.* **8**, 936–947.

Banchio, A. J. and Brady, J. F. (2003). Accelerated Stokesian dynamics: Brownian motion. *J. Chem. Phys.* **118**, 10323–10332.

Bannerman, M. N., Sargant, R., and Lue, L. (2011). DynamO: a free $O(N)$ general event-driven molecular dynamics simulator. *J. Comput. Chem.* **32**, 3329–3338.

Baranyai, A. and Cummings, P. T. (1999). Steady state simulation of planar elongation flow by nonequilibrium molecular dynamics. *J. Chem. Phys.* **110**, 42–45.

Barash, L. Y. and Shchur, L. N. (2013). RNGSSELIB: Program library for random number generation. More generators, parallel streams of random numbers and Fortran compatibility. *Comput. Phys. Commun.* **184**, 2367–2369.

Barducci, A., Bussi, G., and Parrinello, M. (2008). Well-tempered metadynamics: a smoothly converging and tunable free-energy method. *Phys. Rev. Lett.* **100**, 020603.

Barker, A. A. (1965). Monte Carlo calculations of the radial distribution functions for a proton–electron plasma. *Aust. J. Phys.* **18**, 119–133.

Barker, J. A. (1979). A quantum-statistical Monte Carlo method: path integrals with boundary conditions. *J. Chem. Phys.* **70**, 2914–2918.

Barker, J. A. (1980). Reaction field method for polar fluids. *The problem of long-range forces in the computer simulation of condensed matter.* NRCC Workshop Proceedings. **9**, 45–46. (Menlo Park, USA, 1980). Ed. by D. Ceperley.

Barker, J. A. and Henderson, D. (1976). What is 'liquid'? Understanding the states of matter. *Rev. Mod. Phys.* **48**, 587–671.

Barker, J. A. and Watts, R. O. (1969). Structure of water: a Monte Carlo calculation. *Chem. Phys. Lett.* **3**, 144–145.

Barker, J. A. and Watts, R. O. (1973). Monte Carlo studies of dielectric properties of water-like models. *Molec. Phys.* **26**, 789–792.

Barker, J. A., Fisher, R. A., and Watts, R. O. (1971). Liquid argon: Monte Carlo and molecular dynamics calculations. *Molec. Phys.* **21**, 657–673.

Barojas, J., Levesque, D., and Quentrec, B. (1973). Simulation of diatomic homonuclear liquids. *Phys. Rev. A* **7**, 1092–1105.

Baroni, S. and Moroni, S. (1999). Reptation quantum Monte Carlo: a method for unbiased ground-state averages and imaginary-time correlations. *Phys. Rev. Lett.* **82**, 4745–4748.

Barrat, J.-L. and Bocquet, L. (1999a). Influence of wetting properties on hydrodynamic boundary conditions at a fluid/solid interface. *Faraday Disc. Chem. Soc.* **112**, 119–127.

Barrat, J.-L. and Bocquet, L. (1999b). Large slip effect at a non-wetting fluid–solid interface. *Phys. Rev. Lett.* **82**, 4671–4674.

Barrat, J.-L. and Hansen, J.-P. (2003). *Basic concepts for simple and complex liquids.* Cambridge: Cambridge University Press.

Barrett, J. (2006). Some estimates of the surface tension of curved surfaces using density functional theory. *J. Chem. Phys.* **124**, 144705.

Barth, E., Kuczera, K., Leimkuhler, B., and Skeel, R. D. (1995). Algorithms for constrained molecular dynamics. *J. Comput. Chem.* **16**, 1192–1209.

Basconi, J. E. and Shirts, M. R. (2013). Effects of temperature control algorithms on transport properties and kinetics in molecular dynamics simulations. *J. Chem. Theor. Comput.* **9**, 2887–2899.

Bates, M. A. and Frenkel, D. (1998). Influence of polydispersity on the phase behavior of colloidal liquid crystals: a Monte Carlo simulation study. *J. Chem. Phys.* **109**, 6193–6199.

Batôt, G., Dahirel, V., Mériguet, G., Louis, A. A., and Jardat, M. (2013). Dynamics of solutes with hydrodynamic interactions: comparison between Brownian dynamics and stochastic rotation dynamics simulations. *Phys. Rev. E* **88**, 043304.

Bauke, H. and Mertens, S. (2004). Pseudo random coins show more heads than tails. *J. Stat. Phys.* **114**, 1149–1169.

Baxter, R. J. (1970). Ornstein–Zernike relation and Percus–Yevick approximation for fluid mixtures. *J. Chem. Phys.* **52**, 4559–4562.

Baxter, R. J. (1982). *Exactly solved models in statistical mechanics.* London: Academic Press.

Bayly, C. I., Cieplak, P., Cornell, W. D., and Kollman, P. A. (1993). A well-behaved electrostatic potential based method using charge restraints for deriving atomic charges: the RESP model. *J. Phys. Chem.* **97**, 10269–10280.

Bechinger, C., Sciortino, F., and Ziherl, P., eds. (2013). *Physics of complex colloids.* Vol. 184. Proc. Int. School Phys. Enrico Fermi. Italian Physical Society. Amsterdam: IOS Press. (Varenna, Italy, 2012).

Becke, A. D. (1988). Density-functional exchange-energy approximation with correct asymptotic behavior. *Phys. Rev. A* **38**, 3098–3100.

Becker, S., Urbassek, H. M., Horsch, M., and Hasse, H. (2014). Contact angle of sessile drops in Lennard-Jones systems. *Langmuir* **30**, 13606–13614.

Bedrov, D. and Smith, G. D. (2000). Thermal conductivity of molecular fluids from molecular dynamics simulations: application of a new imposed-flux method. *J. Chem. Phys.* **113**, 8080–8084.

Beenakker, C. W. J. (1986). Ewald sum of the Rotne–Prager tensor. *J. Chem. Phys.* **85**, 1581–1582.

Begau, C. and Sutmann, G. (2015). Adaptive dynamic load-balancing with irregular domain decomposition for particle simulations. *Comput. Phys. Commun.* **190**, 51–61.

Behler, J. (2011). Neural network potential-energy surfaces in chemistry: a tool for large-scale simulations. *Phys. Chem. Chem. Phys.* **13**, 17930–17955.

Behler, J. and Parrinello, M. (2007). Generalized neural-network representation of high-dimensional potential-energy surfaces. *Phys. Rev. Lett.* **98**, 146401.

Belardinelli, R. E. and Pereyra, V. D. (2007a). Fast algorithm to calculate density of states. *Phys. Rev. E* **75**, 046701.

Belardinelli, R. E. and Pereyra, V. D. (2007b). Wang–Landau algorithm: a theoretical analysis of the saturation of the error. *J. Chem. Phys.* **127**, 184105.

Bellemans, A., Orban, J., and van Belle, D. (1980). Molecular dynamics of rigid and non-rigid necklaces of hard disks. *Molec. Phys.* **39**, 781–782.

Benet, J., MacDowell, L. G., and Sanz, E. (2014). A study of the ice–water interface using the TIP4P/2005 water model. *Phys. Chem. Chem. Phys.* **16**, 22159–22166.

Benet, J., MacDowell, L. G., and Sanz, E. (2015). Interfacial free energy of the NaCl crystal–melt interface from capillary wave fluctuations. *J. Chem. Phys.* **142**, 134706.

Benjamin, R. and Horbach, J. (2014). Crystal–liquid interfacial free energy via thermodynamic integration. *J. Chem. Phys.* **141**, 044715.

Benjamin, R. and Horbach, J. (2015). Crystal–liquid interfacial free energy of hard spheres via a thermodynamic integration scheme. *Phys. Rev. E* **91**, 032410.

Bennett, C. H. (1975). Mass tensor molecular dynamics. *J. Comput. Phys.* **19**, 267–279.

Bennett, C. H. (1976). Efficient estimation of free-energy differences from Monte-Carlo data. *J. Comput. Phys.* **22**, 245–268.

Bennett, C. H. (1977). Molecular dynamics and transition state theory: the simulation of infrequent events. *Algorithms for chemical computations.* ACS Symp. Ser. **46**. Chap. 4, 63–97. (New York, USA, 1976). Ed. by R. E. Christoffersen. Washington: American Chemical Society.

Benzi, R., Succi, S., and Vergassola, M. (1992). The lattice Boltzmann equation: theory and applications. *Physics Reports* **222**, 145–197.

Berardi, R., Emerson, A. P. J., and Zannoni, C. (1993). Monte Carlo investigations of a Gay–Berne liquid crystal. *J. Chem. Soc. Faraday Trans.* **89**, 4069–4078.

Berardi, R., Fava, C., and Zannoni, C. (1995). A generalized Gay–Berne intermolecular potential for biaxial particles. *Chem. Phys. Lett.* **236**, 462–468.

Berardi, R., Fava, C., and Zannoni, C. (1998). A Gay–Berne potential for dissimilar biaxial particles. *Chem. Phys. Lett.* **297**, 8–14.

Bereau, T. and Deserno, M. (2009). Generic coarse-grained model for protein folding and aggregation. *J. Chem. Phys.* **130**, 235106.

Berendsen, H. J. C. and van Gunsteren, W. F. (1984). Molecular dynamics simulations: techniques and approaches. *Molecular liquids, dynamics and interactions.* NATO ASI series C. **135**, 475–500. (Florence, Italy, 1983). Ed. by A. J. Barnes, W. J. Orville-Thomas, and J. Yarwood. New York: Reidel.

Berendsen, H. J. C. and van Gunsteren, W. F. (1986). Practical algorithms for dynamic simulations. *Molecular dynamics simulation of statistical mechanical systems.* Proc. Int.

School Phys. Enrico Fermi. **97**, 43–65. (Varenna, Italy, 1985). Ed. by G. Ciccotti and W. G. Hoover. Bologna: Soc. Italiana di Fisica.

Berendsen, H. J. C., Postma, J. P. M., van Gunsteren, W. F., and Hermans, J. (1981). Interaction models for water in relation to protein hydration. *Intermolecular forces*. Jerusalem Symp. Quantum Chemistry and Biochemistry. **14**, 331–342. (Jerusalem, Israel, 1981). Ed. by B. Pullman. Dordrecht: Reidel.

Berendsen, H. J. C., Postma, J. P. M., van Gunsteren, W. F., Dinola, A., and Haak, J. R. (1984). Molecular dynamics with coupling to an external bath. *J. Chem. Phys.* **81**, 3684–3690.

Berendsen, H. J. C., Grigera, J. R., and Straatsma, T. P. (1987). The missing term in effective pair potentials. *J. Phys. Chem.* **91**, 6269–6271.

Berens, P. H. and Wilson, K. R. (1981). Molecular dynamics and spectra. 1. Diatomic rotation and vibration. *J. Chem. Phys.* **74**, 4872–4882.

Berens, P. H., Mackay, D. H. J., White, G. M., and Wilson, K. R. (1983). Thermodynamics and quantum corrections from molecular dynamics for liquid water. *J. Chem. Phys.* **79**, 2375–2389.

Berg, B. A., Hansmann, H. E., and Okamoto, Y. (1995). Monte Carlo simulation of a first-order transition for protein folding. *J. Phys. Chem.* **99**, 2236–2237.

Berg, B. and Celik, T. (1992). New approach to spin-glass simulations. *Phys. Rev. Lett.* **69**, 2292–2295.

Bergsma, J. P., Reimers, J. R., Wilson, K. R., and Hynes, J. T. (1986). Molecular dynamics of the A+BC reaction in rare gas solution. *J. Chem. Phys.* **85**, 5625–5643.

Berman, C. L. and Greengard, L. (1994). A renormalization method for the evaluation of lattice sums. *J. Math. Phys.* **35**, 6036–6048.

Bernal, J. D. and King, S. V. (1968). Experimental studies of a simple liquid model. *Physics of Simple Liquids*, 231–252. Ed. by H. N. V. Temperley, J. S. Rowlinson, and G. S. Rushbrooke. Amsterdam: North Holland.

Bernard, E. P. and Krauth, W. (2011). Two-step melting in two dimensions: first-order liquid–hexatic transition. *Phys. Rev. Lett.* **107**, 155704.

Bernardi, S., Todd, B. D., and Searles, D. J. (2010). Thermostating highly confined fluids. *J. Chem. Phys.* **132**, 244706.

Bernaschi, M., Melchionna, S., Succi, S., Fyta, M., Kaxiras, E., and Sircar, J. (2009). MUPHY: a parallel MUlti PHYsics/scale code for high performance bio-fluidic simulations. *Comput. Phys. Commun.* **180**, 1495–1502.

Berne, B. J. (1986). Path integral Monte Carlo methods: static correlation and time correlation functions. *J. Stat. Phys.* **43**, 911–929.

Berne, B. J. and Harp, G. D. (1970). On the calculation of time correlation functions. *Adv. Chem. Phys.* **17**, 63–227.

Berne, B. J. and Pechukas, P. (1972). Gaussian model potentials for molecular interactions. *J. Chem. Phys.* **56**, 4213–4216.

Berne, B. J. and Pecora, R. (1976). *Dynamic light scattering*. New York: Wiley.

Berryman, J. T. and Schilling, T. (2010). Sampling rare events in nonequilibrium and nonstationary systems. *J. Chem. Phys.* **133**, 244101.

Besold, G., Vattulainen, I., Karttunen, M., and Polson, J. M. (2000). Towards better integrators for dissipative particle dynamics simulations. *Phys. Rev. E* **62**, 7611–7614.

Best, R. B. and Hummer, G. (2005). Reaction coordinates and rates from transition paths. *Proc. Nat. Acad. Sci.* **102**, 6732–6737.

Bethkenhagen, M., French, M., and Redmer, R. (2013). Equation of state and phase diagram of ammonia at high pressures from *ab initio* simulations. *J. Chem. Phys.* **138**, 234504.

Bhargava, B. L. and Balasubramanian, S. (2006). Intermolecular structure and dynamics in an ionic liquid: a Car–Parrinello molecular dynamics simulation study of 1,3-dimethylimidazolium chloride. *Chem. Phys. Lett.* **417**, 486–491.

Bhatnagar, P. L., Gross, E. P., and Krook, M. (1954). A model for collision processes in gases. I. Small amplitude processes in charged and neutral one-component systems. *Phys. Rev.* **94**, 511–525.

Bhupathiraju, R., Cummings, P. T., and Cochran, H. D. (1996). An efficient parallel algorithm for non-equilibrium molecular dynamics simulations of very large systems in planar Couette flow. *Molec. Phys.* **88**, 1665–1670.

Binder, K. (1982). Monte Carlo calculation of the surface tension for two- and three-dimensional lattice-gas models. *Phys. Rev. A* **25**, 1699–1709.

Binder, K. (1984). *Applications of the Monte Carlo method in statistical physics.* Topics Curr. Phys. Vol. 36. Berlin: Springer.

Binder, K. (1995). *Monte Carlo and molecular dynamics simulations in polymer science.* New York: Oxford University Press.

Binder, K. and Landau, D. P. (1984). Finite-size scaling at first-order phase transitions. *Phys. Rev. B* **30**, 1477–1485.

Binder, K. (2014). Simulations of interfacial phenomena in soft condensed matter and nanoscience. *J. Phys. Conf. Ser.* **510**, 012002.

Binder, K., Sengupta, S., and Nielaba, P. (2002). The liquid–solid transition of hard discs: first–order transition or Kosterlitz-Thouless-Halperin-Nelson-Young scenario? *J. Phys. Cond. Mat.* **14**, 2323–2333.

Binder, K., Block, B., Das, S. K., Virnau, P., and Winter, D. (2011). Monte Carlo methods for estimating interfacial free energies and line tensions. *J. Stat. Phys.* **144**, 690–729.

Binder, K., Block, B. J., Virnau, P., and Tröster, A. (2012). Beyond the van der Waals loop: what can be learned from simulating Lennard-Jones fluids inside the region of phase coexistence. *Amer. J. Phys.* **80**, 1099–1109.

Biscay, F., Ghoufi, A., Goujon, F., Lachet, V., and Malfreyt, P. (2009). Calculation of the surface tension from Monte Carlo simulations: does the model impact on the finite-size effects? *J. Chem. Phys.* **130**, 184710.

Bishop, M., Ceperley, D., Frisch, H. L., and Kalos, M. H. (1980). Investigations of static properties of model bulk polymer fluids. *J. Chem. Phys.* **72**, 3228–3235.

Bixon, M. and Lifson, S. (1967). Potential functions and conformations in cycloalkanes. *Tetrahedron* **23**, 769–784.

Block, B. J., Das, S. K., Oettel, M., Virnau, P., and Binder, K. (2010). Curvature dependence of surface free energy of liquid drops and bubbles: a simulation study. *J. Chem. Phys.* **133**, 154702.

Blokhuis, E. M. and Bedeaux, D. (1992). Pressure tensor of a spherical interface. *J. Chem. Phys.* **97**, 3576–3586.

Blokhuis, E. M., Bedaux, D., Holcomb, C. D., and Zollweg, J. A. (1995). Tail corrections to the surface tension of a Lennard-Jones liquid–vapour interface. *Molec. Phys.* **85**, 665–669.

Blokhuis, E. M. (2009). On the spectrum of fluctuations of a liquid surface: from the molecular scale to the macroscopic scale. *J. Chem. Phys.* **130**, 014706.

Blokhuis, E. M., Kuipers, J., and Vink, R. L. C. (2008). Description of the fluctuating colloid–polymer interface. *Phys. Rev. Lett.* **101**, 086101.

Bocquet, L. and Barrat, J.-L. (1994). Hydrodynamic boundary conditions, correlation functions, and Kubo relations for confined fluids. *Phys. Rev. E* **49**, 3079–3092.

Bocquet, L. and Barrat, J.-L. (2013). On the Green–Kubo relationship for the liquid–solid friction coefficient. *J. Chem. Phys.* **139**, 044704.

Boinepalli, S. and Attard, P. (2003). Grand canonical molecular dynamics. *J. Chem. Phys.* **119**, 12769–12775.

Bolhuis, P. G. and Dellago, C. (2015). Practical and conceptual path sampling issues. *Euro. Phys. J. Spec. Top.* **224**, 2409–2427.

Bolhuis, P. G. and Kofke, D. A. (1996). Monte Carlo study of freezing of polydisperse hard spheres. *Phys. Rev. E* **54**, 634–643.

Bolhuis, P. G., Chandler, D., Dellago, C., and Geissler, P. L. (2002). Transition path sampling: throwing ropes over rough mountain passes, in the dark. *Ann. Rev. Phys. Chem.* **53**, 291–318.

Bolhuis, P. G. (2008). Rare events via multiple reaction channels sampled by path replica exchange. *J. Chem. Phys.* **129**, 114108.

Bolhuis, P. G. and Dellago, C. (2010). Trajectory-based rare event simulations. *Rev. Comput. Chem.* **27**. Chap. 3, 111–210. Ed. by K. B. Lipkowitz. Wiley.

Bond, S. D., Leimkuhler, B. J., and Laird, B. B. (1999). The Nosé–Poincaré method for constant temperature molecular dynamics. *J. Comput. Phys.* **151**, 114–134.

Bonthuis, D. J., Horinek, D., Bocquet, L., and Netz, R. R. (2009). Electrohydraulic power conversion in planar nanochannels. *Phys. Rev. Lett.* **103**, 144503.

Bonthuis, D. J., Rinne, K. F., Falk, K., Kaplan, C. N., Horinek, D., Berker, A. N., Bocquet, L., and Netz, R. R. (2011). Theory and simulations of water flow through carbon nanotubes: prospects and pitfalls. *J. Phys. Cond. Mat.* **23**, 184110.

Booth, G. H., Thom, A. J. W., and Alavi, A. (2009). Fermion Monte Carlo without fixed nodes: a game of life, death, and annihilation in Slater determinant space. *J. Chem. Phys.* **131**, 054106.

Booth, G. H., Grueneis, A., Kresse, G., and Alavi, A. (2013). Towards an exact description of electronic wavefunctions in real solids. *Nature* **493**, 365–370.

Bordat, P. and Müller-Plathe, F. (2002). The shear viscosity of molecular fluids: a calculation by reverse nonequilibrium molecular dynamics. *J. Chem. Phys.* **116**, 3362–3369.

Borgs, C. and Kotecký, R. (1990). A rigorous theory of finite-size scaling at first-order phase transitions. *J. Stat. Phys.* **61**, 79–119.

Born, M. and von Karman, T. (1912). Über Schwingungen in Raumgittern. *Physik. Z.* **13**, 297–309.

Borrero, E. E. and Dellago, C. (2010). Overcoming barriers in trajectory space: mechanism and kinetics of rare events via Wang–Landau enhanced transition path sampling. *J. Chem. Phys.* **133**, 134112.

Borrero, E. E. and Escobedo, F. A. (2008). Optimizing the sampling and staging for simulations of rare events via forward flux sampling schemes. *J. Chem. Phys.* **129**, 024115.

Borrero, E. E., Weinwurm, M., and Dellago, C. (2011). Optimizing transition interface sampling simulations. *J. Chem. Phys.* **134**, 244118.

Bottaro, S., Boomsma, W., Johansson, K. E., Andreetta, C., Hamelryck, T., and Ferkinghoff-Borg, J. (2012). Subtle Monte Carlo updates in dense molecular systems. *J. Chem. Theor. Comput.* **8**, 695–702.

Boublik, T. (1974). Statistical thermodynamics of convex molecule fluids. *Molec. Phys.* **27**, 1415–1427.

Bowers, K. J., Dror, R. O., and Shaw, D. E. (2005). Overview of neutral territory methods for the parallel evaluation of pairwise particle interactions. *J. Phys. Conf. Ser.* **16**, 300–304.

Bowers, K. J., Dror, R. O., and Shaw, D. E. (2007). Zonal methods for the parallel execution of range-limited N-body simulations. *J. Comput. Phys.* **221**, 303–329.

Box, G. E. P. and Muller, M. E. (1958). A note on the generation of random normal deviates. *Ann. Math. Stat.* **29**, 610–611.

Bradbury, T. C. (1968). *Theoretical mechanics.* New York: Wiley.

Braga, C. and Travis, K. P. (2005). A configurational temperature Nosé–Hoover thermostat. *J. Chem. Phys.* **123**, 134101.

Braga, C. and Travis, K. P. (2006). Configurational constant pressure molecular dynamics. *J. Chem. Phys.* **124**, 104102.

Brandt, E. G., Braun, A. R., Sachs, J. N., Nagle, J. F., and Edholm, O. (2011). Interpretation of fluctuation spectra in lipid bilayer simulations. *Biophys. J.* **100**, 2104–2111.

Brender, C. and Lax, M. (1983). A Monte Carlo off-lattice method: the slithering snake in a continuum. *J. Chem. Phys.* **79**, 2423–2425.

Brennan, J. K. and Madden, W. G. (1998). Efficient volume changes in constant-pressure Monte Carlo simulations. *Molec. Simul.* **20**, 139–157.

Brigham, E. O. (1974). *The fast Fourier transform.* Englewood Cliffs, NJ: Prentice-Hall.

Brooks, B. R., Bruccoleri, R. E., Olafson, B. D., States, D. J., Swaminathan, S., and Karplus, M. (1983). CHARMM: a program for macromolecular energy, minimization, and dynamics calculations. *J. Comput. Chem.* **4**, 187–217.

Brorsen, K. R., Willow, S. Y., Xantheas, S. S., and Gordon, M. S. (2015). The melting temperature of liquid water with the effective fragment potential. *J. Phys. Chem. Lett.* **6**, 3555–3559.

Broughton, J. Q. and Gilmer, G. H. (1986). Molecular dynamics investigation of the crystal–fluid interface. VI. Excess surface free energies of crystal–liquid systems. *J. Chem. Phys.* **84**, 5759–5768.

Broughton, J. Q., Abraham, F., Bernstein, N., and Kaxiras, E. (1999). Concurrent coupling of length scales: methodology and application. *Phys. Rev. B* **60**, 2391–2403.

Brown, D. and Clarke, J. H. R. (1983). The rheological properties of model liquid normal-hexane determined by non-equilibrium molecular dynamics. *Chem. Phys. Lett.* **98**, 579–583.

Brown, D., Clarke, J. H. R., Okuda, M., and Yamazaki, T. (1994). A domain decomposition parallel-processing algorithm for molecular dynamics simulations of polymers. *Comput. Phys. Commun.* **83**, 1–13.

Brown, E. W., Clark, B. K., DuBois, J. L., and Ceperley, D. M. (2013). Path-integral Monte Carlo simulation of the warm dense homogeneous electron gas. *Phys. Rev. Lett.* **110**, 146405.

Brown, J. T., Allen, M. P., Martín del Río, E., and de Miguel, E. (1998). Effects of elongation on the phase behavior of the Gay–Berne fluid. *Phys. Rev. E* **57**, 6685–6699.

Brown, R. G. (2015). *Dieharder: a random number test suite.* http://www.phy.duke.edu/~rgb/General/dieharder.php (Accessed Nov 2015).

Brown, W. M., Kohlmeyer, A., Plimpton, S. J., and Tharrington, A. N. (2012). Implementing molecular dynamics on hybrid high performance computers: particle–particle particle–mesh. *Comput. Phys. Commun.* **183**, 449–459.

Brualla, L., Sakkos, K., Boronat, J., and Casulleras, J. (2004). Higher order and infinite Trotter-number extrapolations in path integral Monte Carlo. *J. Chem. Phys.* **121**, 636–643.

Brumby, P. E., Haslam, A. J., de Miguel, E., and Jackson, G. (2011). Subtleties in the calculation of the pressure and pressure tensor of anisotropic particles from volume-perturbation methods and the apparent asymmetry of the compressive and expansive contributions. *Molec. Phys.* **109**, 169–189.

Brünger, A., Brooks, C. L., and Karplus, M. (1984). Stochastic boundary conditions for molecular dynamics simulations of ST2 water. *Chem. Phys. Lett.* **105**, 495–500.

Bu, W., Kim, D., and Vaknin, D. (2014). Density profiles of liquid/vapor interfaces away from their critical points. *J. Phys. Chem. C* **118**, 12405–12409.

Buff, F. P. (1955). Spherical interface. II. Molecular theory. *J. Chem. Phys.* **23**, 419–427.

Buff, F. P., Lovett, R. A., and Stillinger, F. H. (1965). Interfacial density profile for fluids in the critical region. *Phys. Rev. Lett.* **15**, 621–623.

Buff, F. P. (1952). Some considerations of surface tension. *Zeit. Elecktrochem.* **56**, 311–313.

Buhl, M., Chaumont, A., Schurhammer, R., and Wipff, G. (2005). *Ab initio* molecular dynamics of liquid 1,3-dimethylimidazolium chloride. *J. Phys. Chem. B* **109**, 18591–18599.

Bulo, R. E., Ensing, B., Sikkema, J., and Visscher, L. (2009). Toward a practical method for adaptive QM/MM simulations. *J. Chem. Theor. Comput.* **5**, 2212–2221.

Bunker, A. and Dünweg, B. (2001). Parallel excluded volume tempering for polymer melts. *Phys. Rev. E* **63**, 16701–16710.

Burgos, E., Murthy, C. S., and Righini, R. (1982). Crystal structure and lattice dynamics of chlorine. The role of electrostatic and anisotropic atom–atom potentials. *Molec. Phys.* **47**, 1391–1403.

Bussi, G. and Parrinello, M. (2007). Accurate sampling using Langevin dynamics. *Phys. Rev. E* **75**, 056707.

Bussi, G., Donadio, D., and Parrinello, M. (2007). Canonical sampling through velocity rescaling. *J. Chem. Phys.* **126**, 014101.

Butler, B. D., Ayton, G., Jepps, O. G., and Evans, D. J. (1998). Configurational temperature: verification of Monte Carlo simulations. *J. Chem. Phys.* **109**, 6519–6522.

Caccamo, C., Costa, D., and Pellicane, G. (1998). A comprehensive study of the phase diagram of symmetrical hard-core Yukawa mixtures. *J. Chem. Phys.* **109**, 4498–4507.

Çağin, T. and Pettitt, B. M. (1991a). Grand molecular dynamics: a method for open systems. *Molec. Simul.* **6**, 5–26.

Çağin, T. and Pettitt, B. M. (1991b). Molecular dynamics with a variable number of molecules. *Molec. Phys.* **72**, 169–175.

Çağin, T. and Ray, J. R. (1988). Isothermal molecular-dynamics ensembles. *Phys. Rev. A* **37**, 4510–4513.

Cahn, J. W. and Hilliard, J. E. (1958). Free energy of a nonuniform system. I. Interfacial free energy. *J. Chem. Phys.* **28**, 258–267.

Camley, B. A., Lerner, M. G., Pastor, R. W., and Brown, F. L. H. (2015). Strong influence of periodic boundary conditions on lateral diffusion in lipid bilayer membranes. *J. Chem. Phys.* **143**, 243113.

Cannon, D. A., Ashkenasy, N., and Tuttle, T. (2015). Influence of solvent in controlling peptide–surface interactions. *J. Phys. Chem. Lett.* **6**, 3944–3949.

Cao, J. and Martyna, G. J. (1996). Adiabatic path integral molecular dynamics methods. 2. Algorithms. *J. Chem. Phys.* **104**, 2028–2035.

Cao, J. S. and Voth, G. A. (1994). The formulation of quantum-statistical mechanics based on the Feynman path centroid density. 1. Equilibrium properties. *J. Chem. Phys.* **100**, 5093–5105.

Car, R. and Parrinello, M. (1985). Unified approach for molecular dynamics and density-functional theory. *Phys. Rev. Lett.* **55**, 2471–2474.

Carberry, D. M., Reid, J. C., Wang, G. M., Sevick, E. M., Searles, D. J., and Evans, D. J. (2004). Fluctuations and irreversibility: an experimental demonstration of a second-law-like theorem using a colloidal particle held in an optical trap. *Phys. Rev. Lett.* **92**, 140601.

Care, A., Bergquist, P. L., and Sunna, A. (2015). Solid-binding peptides: smart tools for nanobiotechnology. *Trends Biotech.* **33**, 259–268.

Care, C. M. and Cleaver, D. J. (2005). Computer simulation of liquid crystals. *Rep. Prog. Phys.* **68**, 2665–2700.

Carnahan, N. F. and Starling, K. E. (1969). Equation of state for nonattracting rigid spheres. *J. Chem. Phys.* **51**, 635–636.

Carnie, S. L. and Torrie, G. M. (1984). The statistical mechanics of the electrical double layer. *Adv. Chem. Phys.* **56**, 141–253.

Carrier, J., Greengard, L., and Rokhlin, V. (1988). A fast adaptive multipole algorithm for particle simulations. *SIAM J. Sci. Stat. Comput.* **9**, 669–686.

Carter, E. A., Ciccotti, G., Hynes, J. T., and Kapral, R. (1989). Constrained reaction coordinate dynamics for the simulation of rare events. *Chem. Phys. Lett.* **156**, 472–477.

Celledoni, E., Fassò, F., Säfström, N., and Zanna, A. (2008). The exact computation of the free rigid body motion and its use in splitting methods. *SIAM J. Sci. Comput.* **30**, 2084–2112.

Ceperley, D. M. and Alder, B. J. (1980). Ground state of the electron gas by a stochastic method. *Phys. Rev. Lett.* **45**, 566–569.

Ceperley, D. M., Chester, G. V., and Kalos, M. H. (1977). Monte Carlo simulation of a many-fermion study. *Phys. Rev. B* **16**, 3081–3099.

Ceriotti, M., Tribello, G. A., and Parrinello, M. (2011). Simplifying the representation of complex free energy landscapes using sketch-map. *Proc. Nat. Acad. Sci.* **108**, 13023.

Ceriotti, M. and Markland, T. E. (2013). Efficient methods and practical guidelines for simulating isotope effects. *J. Chem. Phys.* **138**, 014112.

Ceriotti, M., Parrinello, M., Markland, T. E., and Manolopoulos, D. E. (2010). Efficient stochastic thermostatting of path integral molecular dynamics. *J. Chem. Phys.* **133**, 124104.

Chacón, E. and Tarazona, P. (2003). Intrinsic profiles beyond the capillary wave theory: a Monte Carlo study. *Phys. Rev. Lett.* **91**, 166103.

Chandler, D. (1978). Statistical mechanics of isomerization dynamics in liquids and the transition state approximation. *J. Chem. Phys.* **68**, 2959–2970.

Chandler, D. (1982). Equilibrium theory of polyatomic fluids. *The liquid state of matter: fluid simple and complex*. Studies in Statistical Mechanics. **8**, 275–340. Ed. by E. W. Montroll and J. L. Lebowitz. Amsterdam: North Holland.

Chandler, D. (1987). *Introduction to modern statistical mechanics*. New York: Oxford University Press.

Chandler, D. and Berne, B. J. (1979). Comment on the role of constraints on the structure of n-butane in liquid solvents. *J. Chem. Phys.* **71**, 5386–5387.

Chandler, D. and Wolynes, P. G. (1981). Exploiting the isomorphism between quantum theory and classical statistical mechanics of polyatomic fluids. *J. Chem. Phys.* **74**, 4078–4095.

Chandra, R., Kohr, D., Menon, R., Dagum, L., Maydan, D., and McDonald, J. (2000). *Parallel programming in OpenMP*. Morgan Kaufmann.

Chandrasekhar, S. (1943). Stochastic problems in physics and astronomy. *Rev. Mod. Phys.* **15**, 1–89.

Chapela, G. A., Saville, G., and Rowlinson, J. (1975). Computer simulation of the gas/liquid surface. *Faraday Disc. Chem. Soc.* **59**, 22–28.

Chapela, G. A., Saville, G., Thompson, S. M., and Rowlinson, J. S. (1977). Computer simulation of a gas–liquid surface. 1. *J. Chem. Soc. Faraday Trans. II* **73**, 1133–1144.

Chapela, G. A., Martinez-Casas, S. E., and Alejandre, J. (1984). Molecular dynamics for discontinuous potentials. 1. General method and simulation of hard polyatomic molecules. *Molec. Phys.* **53**, 139–159.

Chapman, W. and Quirke, N. (1985). Metropolis Monte Carlo simulation of fluids with multiparticle moves. *Physica B* **131**, 34–40.

Chapman, W. G., Jackson, G., and Gubbins, K. E. (1988). Phase equilibria of associating fluids. *Molec. Phys.* **65**, 1057–1079.

Chapman, W. G., Gubbins, K. E., Jackson, G., and Radosz, M. (1989). SAFT: equation-of-state solution model for associating fluids. *Fluid Phase Equilibria* **52**, 31–38.

Chapman, W. G., Gubbins, K. E., Jackson, G., and Radosz, M. (1990). New reference equation of state for associating liquids. *Ind. Eng. Chem. Res.* **29**, 1709–1721.

Chatfield, C. (1984). *The analysis of time series. An introduction*. 3rd ed. Chapman and Hall.

Chayes, J. T. and Chayes, L. (1984). On the validity of the inverse conjecture in classical density functional theory. *J. Stat. Phys.* **36**, 471–488.

Chayes, J. T., Chayes, L., and Lieb, E. H. (1984). The inverse problem in classical statistical mechanics. *Commun. Math. Phys.* **93**, 57–121.

Chen, B. and Siepmann, J. I. (2000). A novel Monte Carlo algorithm for simulating strongly associating fluids: applications to water, hydrogen fluoride, and acetic acid. *J. Phys. Chem. B* **104**, 8725–8734.

Chen, L. J. (1995). Area dependence of the surface tension of a Lennard-Jones fluid from molecular dynamics simulations. *J. Chem. Phys.* **103**, 10214–10216.

Chen, S. and Doolen, G. D. (1998). Lattice Boltzmann method for fluid flows. *Ann. Rev. Fluid Mech.* **30**, 329–364.

Chen, S., Wang, H., Qian, T., and Sheng, P. (2015). Determining hydrodynamic boundary conditions from equilibrium fluctuations. *Phys. Rev. E* **92**, 043007.

Chen, T., Smit, B., and Bell, A. T. (2009). Are pressure fluctuation-based equilibrium methods really worse than nonequilibrium methods for calculating viscosities? *J. Chem. Phys.* **131**, 246101.

Cheng, H., Greengard, L., and Rokhlin, V. (1999). A fast adaptive multipole algorithm in three dimensions. *J. Comput. Phys.* **155**, 468–498.

Chesnut, D. A. (1963). Monte Carlo calculations for the two-dimensional triangular lattice gas: supercritical region. *J. Chem. Phys.* **39**, 2081–2084.

Cheung, P. S. Y. (1976). On efficient evaluation of torques and forces for anisotropic potentials in computer simulation of liquids composed of linear molecules. *Chem. Phys. Lett.* **40**, 19–22.

Cheung, P. S. Y. (1977). On the calculation of specific heats, thermal pressure coefficients, and compressibilities in molecular dynamics simulations. *Molec. Phys.* **33**, 519–526.

Cheung, P. S. Y. and Powles, J. G. (1975). The properties of liquid nitrogen. IV. A computer simulation. *Molec. Phys.* **30**, 921–949.

Cho, K., Joannopoulos, J. D., and Kleinman, L. (1993). Constant-temperature molecular dynamics with momentum conservation. *Phys. Rev. E* **47**, 3145–3151.

Chodera, J. D., Swope, W. C., Pitera, J. W., Seok, C., and Dill, K. A. (2007). Use of the weighted histogram analysis method for the analysis of simulated and parallel tempering simulations. *J. Chem. Theor. Comput.* **3**, 26–41.

Christiansen, D., Perram, J. W., and Petersen, H. G. (1993). On the fast multipole method for computing the energy of periodic assemblies of charged and dipolar particles. *J. Comput. Phys.* **107**, 403–405.

Chu, J.-W., Izveko, S., and Voth, G. A. (2006). The multiscale challenge for biomolecular systems: coarse-grained modeling. *Molec. Simul.* **32**, 211–218.

Chu, J.-W., Ayton, G. S., Izvekov, S., and Voth, G. A. (2007). Emerging methods for multiscale simulation of biomolecular systems. *Molec. Phys.* **105**, 167–175.

Chung, K. L. (1960). *Markov chains with stationary state probabilities, Vol. 1.* Heidelberg: Springer.

Ciccotti, G. and Jacucci, G. (1975). Direct computation of dynamical response by molecular dynamics: mobility of a charged Lennard-Jones particle. *Phys. Rev. Lett.* **35**, 789–792.

Ciccotti, G. and Ryckaert, J. P. (1980). Computer simulation of the generalized Brownian motion. 1. The scalar case. *Molec. Phys.* **40**, 141–159.

Ciccotti, G. and Ryckaert, J. P. (1986). Molecular dynamics simulation of rigid molecules. *Comput. Phys. Rep.* **4**, 345–392.

Ciccotti, G., Orban, J., and Ryckaert, J. P. (1976a). *Stochastic approach to the dynamics of large molecules in a solvent.* Tech. rep. CECAM. Rapport d'activité scientifique du CECAM Models for Protein Dynamics.

Ciccotti, G., Jacucci, G., and McDonald, I. R. (1976b). Transport properties of molten alkali halides. *Phys. Rev. A* **13**, 426–436.

Ciccotti, G., Jacucci, G., and McDonald, I. R. (1978). Thermal response to a weak external field. *J. Phys. C* **11**, 509–513.

Ciccotti, G., Jacucci, G., and McDonald, I. R. (1979). 'Thought experiments' by molecular dynamics. *J. Stat. Phys.* **21**, 1–22.

Ciccotti, G., Ferrario, M., and Ryckaert, J.-P. (1982). Molecular dynamics of rigid systems in Cartesian coordinates. A general formulation. *Molec. Phys.* **47**, 1253–1264.

Ciccotti, G. and Ferrario, M. (2004). Blue moon approach to rare events. *Molec. Simul.* **30**, 787–793.

Cicero, G., Grossman, J. C., Schwegler, E., Gygi, F., and Galli, G. (2008). Water confined in nanotubes and between graphene sheets: a first principle study. *J. Am. Chem. Soc.* **130**, 1871–1878.

Cieplak, P., Caldwell, J., and Kollman, P. (2001). Molecular mechanical models for organic and biological systems going beyond the atom centered two body additive approximation: aqueous solution free energies of methanol and N-methyl acetamide, nucleic acid base, and amide hydrogen bonding and chloroform/water partition coefficients of the nucleic acid bases. *J. Comput. Chem.* **22**, 1048–1057.

Clausen, J. R., Reasor Jr., D. A., and Aidun, C. K. (2010). Parallel performance of a lattice-Boltzmann/finite element cellular blood flow solver on the IBM Blue Gene/P architecture. *Comput. Phys. Commun.* **181**, 1013–1020.

Cleaver, D. J., Care, C. M., Allen, M. P., and Neal, M. P. (1996). Extension and generalization of the Gay–Berne potential. *Phys. Rev. E* **54**, 559–567.

Coffey, W. T. and Kalmykov, Y. P. (2012). *The Langevin equation. With applications to stochastic problems in physics, chemistry and electrical engineering.* 3rd ed. World Scientific Series in Contemporary Chemical Physics. Vol. 27. Singapore: World Scientific.

Coker, D. F. and Bonella, S. (2006). Linearized path integral methods for quantum time correlation functions. *Computer simulations in condensed matter systems: from materials to chemical biology, vol 1.* Lect. Notes Phys. **703**, 553–590. (Erice, Italy, 2005). Ed. by M. Ferrario, G. Ciccotti, and K. Binder. Heidelberg: Springer.

Cole, D. J., Tirado-Rives, J., and Jorgensen, W. L. (2015). Molecular dynamics and Monte Carlo simulations for protein-ligand binding and inhibitor design. *Biochim. Biophys. Acta General Subjects* **1850**, 966–971.

Cole, D. R., Herwig, K. W., Mamontov, E., and Larese, J. Z. (2006). Neutron scattering and diffraction studies of fluids and fluid–solid interactions. *Neutron scattering in earth sciences.* Rev. Mineral. Geochem. **63**, 313–362. Ed. by H. R. Wenk.

Collin, D., Ritort, F., Jarzynski, C., Smith, S. B., Tinoco, I., and Bustamante, C. (2005). Verification of the Crooks fluctuation theorem and recovery of RNA folding free energies. *Nature* **437**, 231–234.

Coluzza, I. and Frenkel, D. (2005). Virtual-move parallel tempering. *ChemPhysChem* **6**, 1779–1783.

Cooke, B. and Schmidler, S. C. (2008). Preserving the Boltzmann ensemble in replica-exchange molecular dynamics. *J. Chem. Phys.* **129**, 164112.

Cooley, J. N. and Tukey, J. W. (1965). An algorithm for the machine calculation of complex Fourier series. *Math. Comput.* **19**, 297–301.

Cormier, J., Rickman, J. M., and Delph, T. J. (2001). Stress calculation in atomistic simulations of perfect and imperfect solids. *J. Appl. Phys.* **89**, 99–104. Erratum: *ibid.* **89**, 4198 (2001).

Corner, J. (1948). The second virial coefficient of a gas of nonspherical molecules. *Proc. Roy. Soc. Lond. A* **192**, 275–292.

Corni, S., Hnilova, M., Tamerler, C., and Sarikaya, M. (2013). Conformational behavior of genetically-engineered dodecapeptides as a determinant of binding affinity for gold. *J. Phys. Chem. C* **117**, 16990–17003.

Corti, D. S. and Soto-Campos, G. (1998). Deriving the isothermal–isobaric ensemble: the requirement of a 'shell' molecule and applicability to small systems. *J. Chem. Phys.* **108**, 7959–7966.

Costa, D., Sergi, A., and Ferrario, M. (2013). Transient behavior of a model fluid under applied shear. *J. Chem. Phys.* **138**, 184501.

Cotter, C. J. and Reich, S. (2006). Time stepping algorithms for classical molecular dynamics. *Handbook of theoretical and computational nanotechnology.* Ed. by M. Rieth and W. Schommers. American Scientific Publishers.

Couturier, R. and Chipot, C. (2000). Parallel molecular dynamics using openMP on a shared memory machine. *Comput. Phys. Commun.* **124**, 49–59.

Cracknell, R. F., Nicholson, D., and Quirke, N. (1993). A grand canonical Monte Carlo study of Lennard-Jones mixtures in slit shaped pores. *Molec. Phys.* **80**, 885–897.

Craig, C. F., Duncan, W. R., and Prezhdo, O. V. (2005). Trajectory surface hopping in the time-dependent Kohn–Sham approach for electron–nuclear dynamics. *Phys. Rev. Lett.* **95**, 163001.

Craig, I. R. and Manolopoulos, D. E. (2004). Quantum statistics and classical mechanics: real time correlation functions from ring polymer molecular dynamics. *J. Chem. Phys.* **121**, 3368–3373.

Craig, I. R. and Manolopoulos, D. E. (2005a). A refined ring polymer molecular dynamics theory of chemical reaction rates. *J. Chem. Phys.* **123**, 034102.

Craig, I. R. and Manolopoulos, D. E. (2005b). Chemical reaction rates from ring polymer molecular dynamics. *J. Chem. Phys.* **122**, 084106.

CRC (1984). *CRC handbook of chemistry and physics.* Florida: Chemical Rubber Company.

Creutz, M. (1983). Microcanonical Monte Carlo simulation. *Phys. Rev. Lett.* **50**, 1411–1414.

Crooks, G. E. (1999). Entropy production fluctuation theorem and the nonequilibrium work relation for free energy differences. *Phys. Rev. E* **60**, 2721–2726.

Crowell, A. D. (1958). Potential energy functions for graphite. *J. Chem. Phys.* **29**, 446–447.

Croxton, C. A. (1980). *Statistical mechanics of the liquid surface.* New York: Wiley.

Cunningham, G. W. and Meijer, P. H. E. (1976). A comparison of two Monte Carlo methods for computations in statistical mechanics. *J. Comput. Phys.* **20**, 50–63.

Curro, J. (1974). Computer simulation of multiple chain systems: the effect of density on the average chain dimensions. *J. Chem. Phys.* **61**, 1203–1207.

Cygan, R. T., Liang, J. J., and Kalinichev, A. G. (2004). Molecular models of hydroxide, oxyhydroxide, and clay phases and the development of a general force field. *J. Phys. Chem. B* **108**, 1255–1266.

Dahlquist, G. and Björk, A. (1974). *Numerical methods*. Englewood Cliffs, NJ: Prentice-Hall.

Daivis, P. J. (2014). Personal Communication.

Daivis, P. J. and Todd, B. D. (2006). A simple, direct derivation and proof of the validity of the SLLOD equations of motion for generalized homogeneous flows. *J. Chem. Phys.* **124**, 194103.

Daivis, P. J., Dalton, B. A., and Morishita, T. (2012). Effect of kinetic and configurational thermostats on calculations of the first normal stress coefficient in nonequilibrium molecular dynamics simulations. *Phys. Rev. E* **86**, 056707.

D'Alessandro, M., Tenenbaum, A., and Amadei, A. (2002). Dynamical and statistical mechanical characterization of temperature coupling algorithms. *J. Phys. Chem. B* **106**, 5050–5057.

Damasceno, P. F., Engel, M., and Glotzer, S. C. (2012). Predictive self-assembly of polyhedra into complex structures. *Science* **337**, 453–457.

Darden, T., York, D., and Pedersen, L. (1993). Particle mesh Ewald: an $N \log(N)$ method for Ewald sums in large systems. *J. Chem. Phys.* **98**, 10089–10092.

Davidchack, R. L. (2010). Hard spheres revisited: accurate calculation of the solid–liquid interfacial free energy. *J. Chem. Phys.* **133**, 234701.

Davidchack, R. L. and Laird, B. B. (2000). Direct calculation of the hard-sphere crystal / melt interfacial free energy. *Phys. Rev. Lett.* **85**, 4751–4754.

Davidchack, R. L., Morris, J. R., and Laird, B. B. (2006). The anisotropic hard-sphere crystal–melt interfacial free energy from fluctuations. *J. Chem. Phys.* **125**, 094710.

de Gennes, P. G. (1971). Reptation of a polymer chain in the presence of fixed obstacles. *J. Chem. Phys.* **55**, 572–579.

de Leeuw, S. W., Perram, J. W., and Smith, E. R. (1980). Simulation of electrostatic systems in periodic boundary conditions. 1. Lattice sums and dielectric constants. *Proc. Roy. Soc. Lond. A* **373**, 27–56.

de Miguel, E., Martín del Río, E., Brown, J. T., and Allen, M. P. (1996). Effect of the attractive interactions on the phase behavior of the Gay–Berne liquid crystal model. *J. Chem. Phys.* **105**, 4234–4249.

de Miguel, E. and Jackson, G. (2006). Detailed examination of the calculation of the pressure in simulations of systems with discontinuous interactions from the mechanical and thermodynamic perspectives. *Molec. Phys.* **104**, 3717–3734.

de Pablo, J. J., Bonnin, M., and Prausnitz, J. M. (1992). Vapor–liquid equilibria for polyatomic fluids from site–site computer simulations: pure hydrocarbons and binary mixtures containing methane. *Fluid Phase Equilibria* **73**, 187–210.

de Pablo, J. J., Laso, M., Siepmann, J. I., and Suter, U. W. (1993). Continuum-configurational-bias Monte Carlo simulations of long-chain alkanes. *Molec. Phys.* **80**, 55–63.

De Raedt, B., Sprik, M., and Klein, M. L. (1984). Computer simulation of muonium in water. *J. Chem. Phys.* **80**, 5719–5724.

Debolt, S. E. and Kollman, P. A. (1993). AMBERCUBE MD, parallelization of Amber's molecular dynamics module for distributed-memory hypercube computers. *J. Comput. Chem.* **14**, 312–329.

Deem, M. W. and Bader, J. S. (1996). A configurational bias Monte Carlo method for linear and cyclic peptides. *Molec. Phys.* **87**, 1245–1260.

Del Popolo, M. G., Lynden-Bell, R. M., and Kohanoff, J. (2005). *Ab initio* molecular dynamics simulation of a room temperature ionic liquid. *J. Phys. Chem. B* **109**, 5895–5902.

Del Popolo, M. G., Kohanoff, J., Lynden-Bell, R. M., and Pinilla, C. (2007). Clusters, liquids, and crystals of dialkyimidazolium salts. A combined perspective from *ab initio* and classical computer simulations. *Acc. Chem. Res.* **40**, 1156–1164.

Delhommelle, J. and Millié, P. (2001). Inadequacy of the Lorentz–Berthelot combining rules for accurate predictions of equilibrium properties by molecular simulation. *Molec. Phys.* **99**, 619–625.

Dellago, C., Bolhuis, P. G., and Geissler, P. L. (2002). Transition path sampling. *Adv. Chem. Phys.* **123**, 1–78.

Dellago, C. and Hummer, G. (2014). Computing equilibrium free energies using nonequilibrium molecular dynamics. *Entropy* **16**, 41–61.

den Otter, W. K. and Briels, W. J. (2003). The bending rigidity of an amphiphilic bilayer from equilibrium and nonequilibrium molecular dynamics. *J. Chem. Phys.* **118**, 4712–4720.

den Otter, W. K. and Clarke, J. H. R. (2001). A new algorithm for dissipative particle dynamics. *Europhys. Lett.* **53**, 426–431.

Deng, L.-Y., Guo, R., Lin, D. K. J., and Bai, F. (2008). Improving random number generators in the Monte Carlo simulations via twisting and combining. *Comput. Phys. Commun.* **178**, 401–408.

Denniston, C., Orlandini, E., and Yeomans, J. M. (2000). Simulations of liquid crystal hydrodynamics in the isotropic and nematic phases. *Europhys. Lett.* **52**, 481–487.

Denniston, C., Orlandini, E., and Yeomans, J. M. (2001). Lattice Boltzmann simulations of liquid crystal hydrodynamics. *Phys. Rev. E* **63**, 056702.

Deserno, M. and Holm, C. (1998). How to mesh up Ewald sums. I. A theoretical and numerical comparison of various particle mesh routines. *J. Chem. Phys.* **109**, 7678–7693.

Dettmann, C. P. and Morriss, G. P. (1996). Hamiltonian formulation of the Gaussian isokinetic thermostat. *Phys. Rev. E* **54**, 2495–2500.

D'Evelyn, M. P. and Rice, S. A. (1981). Comment on the configuration space diffusion criterion for optimization of the force bias Monte Carlo method. *Chem. Phys. Lett.* **77**, 630–633.

d'Humières, D., Ginzburg, I., Krafczyk, M., Lallemand, P., and Luo, L. S. (2002). Multiple-relaxation-time lattice Boltzmann models in three dimensions. *Phil. Trans. Roy. Soc. A* **360**, 437–451.

Di Felice, R. and Corni, S. (2011). Simulation of peptide–surface recognition. *J. Phys. Chem. Lett.* **2**, 1510–1519.

Dijkstra, M. (1997). Confined thin films of linear and branched alkanes. *J. Chem. Phys.* **107**, 3277–3288.

Ding, W., Palaiokostas, M., Wang, W., and Orsi, M. (2015). Effects of lipid composition on bilayer membranes quantified by all-atom molecular dynamics. *J. Phys. Chem. B* **119**, 15263–15274.

Dinur, U. and Hagler, A. T. (1991). New approaches to empirical force fields. *Rev. Comput. Chem.* **9**. Chap. 4, 99–164. Ed. by K. B. Lipkowitz and D. B. Boyd. VCH Publishers.

Dixon, M. and Hutchinson, P. (1977). Method for extrapolation of pair distribution functions. *Molec. Phys.* **33**, 1663–1670.

Dixon, M. and Sangster, M. J. L. (1976). The structure of molten NaCl from a simulation model which allows for polarization of both ions. *J. Phys. C* **9**, 5–9.

Do, H., Wheatley, R. J., and Hirst, J. D. (2010). Gibbs ensemble Monte Carlo simulations of binary mixtures of methane, difluoromethane, and carbon dioxide. *J. Phys. Chem. B* **114**, 3879–3886.

Do, H. and Wheatley, R. J. (2013). Density of states partitioning method for calculating the free energy of solids. *J. Chem. Theor. Comput.* **9**, 165–171.

Do, H., Hirst, J. D., and Wheatley, R. J. (2011). Rapid calculation of partition functions and free energies of fluids. *J. Chem. Phys.* **135**, 174105.

Dodd, L. R., Boone, T. D., and Theodorou, D. (1993). A concerted rotation algorithm for atomistic Monte Carlo simulation of polymer melts and glasses. *Molec. Phys.* **78**, 961–996.

Doi, M. and Edwards, S. F. (1988). *The theory of polymer dynamics*. International Series of Monographs on Physics. Oxford University Press.

Doll, J. D. and Dion, D. R. (1976). Generalized Langevin equation approach for atom/solid–surface scattering: numerical techniques for Gaussian generalized Langevin dynamics. *J. Chem. Phys.* **65**, 3762–3766.

Domański, J., Stansfeld, P. J., Sansom, M. S. P., and Beckstein, O. (2010). Lipidbook: a public repository for force-field parameters used in membrane simulations. *J. Membr. Biol.* **236**, 255–258.

Donadio, D., Cicero, G., Schwegler, E., Sharma, M., and Galli, G. (2009). Electronic effects in the IR spectrum of water under confinement. *J. Phys. Chem. B* **113**, 4170–4175.

Dong, R.-Y. and Cao, B.-Y. (2012). Application of the uniform source-and-sink scheme to molecular dynamics calculation of the self-diffusion coefficient of fluids. *Int. J. Num. Meth. Eng.* **92**, 229–237.

Doran, M. B. and Zucker, I. J. (1971). Higher order multipole three-body van der Waals interactions and stability of rare gas solids. *J. Phys. C* **4**, 307–312.

Du, W. and Bolhuis, P. G. (2014). Sampling the equilibrium kinetic network of Trp-cage in explicit solvent. *J. Chem. Phys.* **140**, 195102.

Duane, S., Kennedy, A. D., Pendleton, B. J., and Roweth, D. (1987). Hybrid Monte Carlo. *Phys. Lett. B* **195**, 216–222.

Dubbeldam, D., Ford, D. C., Ellis, D. E., and Snurr, R. Q. (2009). A new perspective on the order-n algorithm for computing correlation functions. *Molec. Simul.* **35**, 1084–1097.

Dullweber, A., Leimkuhler, B., and McLachlan, R. (1997). Symplectic splitting methods for rigid body molecular dynamics. *J. Chem. Phys.* **107**, 5840–5851.

Dünweg, B. and Ladd, A. J. C. (2009). Lattice Boltzmann simulations of soft matter systems. *Advanced computer simulation approaches for soft matter sciences III.* Adv. Polym. Sci. **221**, 89–166. Ed. by C. Holm and K. Kremer. Berlin, Heidelberg: Springer.

Duong-Hong, D., Phan-Thien, N., and Fan, X. (2004). An implementation of no-slip boundary conditions in DPD. *Comput. Mech.* **35**, 24–29.

Dykstra, C. E. (1988). Efficient calculation of electrically based intermolecular potentials of weakly bonded clusters. *J. Comput. Chem.* **9**, 476–487.

Dymond, J. H. and Smith, E. B. (1980). *The virial coefficients of pure gases and mixture. A critical compilation.* Oxford Science Research Paper Series. Oxford: Oxford University Press.

Dysthe, D. K., Fuchs, A. H., and Rousseau, B. (2000). Fluid transport properties by equilibrium molecular dynamics. III. Evaluation of united atom interaction potential models for pure alkanes. *J. Chem. Phys.* **112**, 7581–7590.

E, W. and Vanden-Eijnden, E. (2010). Transition-path theory and path-finding algorithms for the study of rare events. *Ann. Rev. Phys. Chem.* **61**, 391–420.

Earl, D. J and Deem, M. W. (2005). Parallel tempering: theory, applications, and new perspectives. *Phys. Chem. Chem. Phys.* **7**, 3910–3916.

Earl, D. J. and Deem, M. W. (2004). Optimal allocation of replicas to processors in parallel tempering simulations. *J. Phys. Chem. B* **108**, 6844–6849.

Eastwood, J. W., Hockney, R. W., and Lawrence, D. N. (1980). P3M3DP: the three-dimensional periodic particle–particle/particle–mesh program. *Comput. Phys. Commun.* **19**, 215–261.

Eastwood, M. P., Stafford, K. A., Lippert, R. A., Jensen, M. Ø., Maragakis, P., Predescu, C., Dror, R. O., and Shaw, D. E. (2010). Equipartition and the calculation of temperature in biomolecular simulations. *J. Chem. Theor. Comput.* **6**, 2045–2058.

Ebert, F., Dillmann, P., Maret, G., and Keim, P. (2009). The experimental realization of a two-dimensional colloidal model system. *Rev. Sci. Instr.* **80**, 083902.

Economou, I. G. (2002). Statistical Associating Fluid Theory: a successful model for the calculation of thermodynamic and phase equilibrium properties of complex fluid mixtures. *Ind. Eng. Chem. Res.* **41**, 953–962.

Edvinsson, T., Rasmark, P. J., and Elvingson, C. (1999). Cluster identification and percolation analysis using a recursive algorithm. *Molec. Simul.* **23**, 169–190.

Edwards, B. J., Baig, C., and Keffer, D. J. (2006). A validation of the p-SLLOD equations of motion for homogeneous steady-state flows. *J. Chem. Phys.* **124**, 194104.

Edwards, S. F. (1990). The flow of powders and of liquids of high viscosity. *J. Phys. Cond. Mat.* **2**, SA63–SA68.

Edwards, S. F. and Oakeshott, R. B. S. (1989). Theory of powders. *Physica A* **157**, 1080–1090.

Egelstaff, P. A. (1961). The theory of the thermal neutron scattering law. *Inelastic scattering of neutrons in solids and liquids*, 25–38. Vienna: IAEA.

Egorov, S. A., Rabani, E., and Berne, B. J. (1999). On the adequacy of mixed quantum–classical dynamics in condensed phase systems. *J. Phys. Chem. B* **103**, 10978–10991.

Eike, D. M., Brennecke, J. F., and Maginn, E. J. (2005). Toward a robust and general molecular simulation method for computing solid–liquid coexistence. *J. Chem. Phys.* **122**, 014115.

Ekdawi-Sever, N. C., Conrad, P. B., and de Pablo, J. J. (2001). Molecular simulation of sucrose solutions near the glass transition temperature. *J. Phys. Chem. A* **105**, 734–742.

Elliott, W. D and Board, J. A. (1996). Fast Fourier transform accelerated fast multipole algorithm. *SIAM J. Sci. Comput.* **17**, 398–415.

Engel, M., Anderson, J. A., Glotzer, S. C., Isobe, M., Bernard, E. P., and Krauth, W. (2013). Hard-disk equation of state: first-order liquid–hexatic transition in two dimensions with three simulation methods. *Phys. Rev. E* **87**, 042134.

Engle, R. D., Skeel, R. D., and Drees, M. (2005). Monitoring energy drift with shadow Hamiltonians. *J. Comput. Phys.* **206**, 432–452.

English, C. A. and Venables, J. A. (1974). Structure of diatomic molecular solids. *Proc. Roy. Soc. Lond. A* **340**, 57–80.

English, N. J. (2013). Massively-parallel molecular dynamics simulation of clathrate hydrates on Blue Gene platforms. *Energies* **6**, 3072–3081.

English, N. J. (2014). Massively parallel molecular dynamics simulation of ice crystallisation and melting: the roles of system size, ensemble, and electrostatics. *J. Chem. Phys.* **141**, 234501.

English, N. J. and Tse, J. S. (2015). Massively parallel molecular dynamics simulation of formation of ice-crystallite precursors in supercooled water: incipient-nucleation behavior and role of system size. *Phys. Rev. E* **92**, 032132.

English, N. J., Johnson, J. K., and Taylor, C. E. (2005). Molecular dynamics simulations of methane hydrate dissociation. *J. Chem. Phys.* **123**, 244503.

Eppenga, R. and Frenkel, D. (1984). Monte Carlo study of the isotropic and nematic phases of infinitely thin hard platelets. *Molec. Phys.* **52**, 1303–1334.

Ercolessi, F. and Adams, J. B. (1994). Interatomic potentials from first-principles calculations: the force-matching method. *Europhys. Lett.* **26**, 583–588.

Ermak, D. L. (1975). A computer simulation of charged particles in solution. 1. Technique and equilibrium properties. *J. Chem. Phys.* **62**, 4189–4196.

Ermak, D. L. and Buckholz, H. (1980). Numerical integration of the Langevin equation: Monte Carlo simulation. *J. Comput. Phys.* **35**, 169–182.

Ermak, D. L. and McCammon, J. A. (1978). Brownian dynamics with hydrodynamic interactions. *J. Chem. Phys.* **69**, 1352–1360.

Ermak, D. L. and Yeh, Y. (1974). Equilibrium electrostatic effects on behavior of polyions in solution: polyion–mobile ion interaction. *Chem. Phys. Lett.* **24**, 243–248.

Erpenbeck, J. J. and Wood, W. W. (1977). Molecular dynamics techniques for hard core systems. *Statistical mechanics B*. Modern Theoretical Chemistry. **6**, 1–40. Ed. by B. J. Berne. New York: Plenum.

Errington, J. R. (2003). Evaluating surface tension using grand-canonical transition-matrix Monte Carlo simulation and finite-size scaling. *Phys. Rev. E* **67**, 012102.

Errington, J. R. (2004). Solid–liquid phase coexistence of the Lennard-Jones system through phase-switch Monte Carlo simulation. *J. Chem. Phys.* **120**, 3130–3141.

Errington, J. R. and Kofke, D. A. (2007). Calculation of surface tension via area sampling. *J. Chem. Phys.* **127**, 174709.

Escobedo, F. A. (2005). A unified methodological framework for the simulation of nonisothermal ensembles. *J. Chem. Phys.* **123**, 044110.

Eslami, H. and Müller-Plathe, F. (2007). Molecular dynamics simulation in the grand canonical ensemble. *J. Comput. Chem.* **28**, 1763–1773.

Español, P. and Warren, P. (1995). Statistical mechanics of dissipative particle dynamics. *Europhys. Lett.* **30**, 191–196.

Espinosa, J. R., Vega, C., and Sanz, E. (2014). The mold integration method for the calculation of the crystal–fluid interfacial free energy from simulations. *J. Chem. Phys.* **141**, 134709.

Essmann, U., Perera, L., Berkowitz, M. L., Darden, T., Lee, H., and Pedersen, L. G. (1995). A smooth particle mesh Ewald method. *J. Chem. Phys.* **103**, 8577–8593.

Evans, D. J. (1977). On the representation of orientation space. *Molec. Phys.* **34**, 317–325.

Evans, D. J. (1979a). Nonlinear viscous flow in the Lennard-Jones fluid. *Phys. Lett. A* **74**, 229–232.

Evans, D. J. (1979b). Nonsymmetric pressure tensor in polyatomic fluids. *J. Stat. Phys.* **20**, 547–555.

Evans, D. J. (1979c). The frequency dependent shear viscosity of methane. *Molec. Phys.* **37**, 1745–1754.

Evans, D. J. (1981a). Equilibrium fluctuation expressions for the wave-vector- and frequency-dependent shear viscosity. *Phys. Rev. A* **23**, 2622–2626.

Evans, D. J. (1981b). Nonequilibrium molecular dynamics study of the rheological properties of diatomic liquids. *Molec. Phys.* **42**, 1355–1365.

Evans, D. J. (1982). Homogeneous NEMD algorithm for thermal conductivity: application of non-canonical linear response theory. *Phys. Lett. A* **91**, 457–460.

Evans, D. J. (1983). Computer 'experiment' for non-linear thermodynamics of Couette flow. *J. Chem. Phys.* **78**, 3297–3302.

Evans, D. J. and Gaylor, K. (1983). NEMD algorithm for calculating the Raman spectra of dense fluids. *Molec. Phys.* **49**, 963–972.

Evans, D. J. and Hanley, H. J. M. (1982). Fluctuation expressions for fast thermal transport processes: vortex viscosity. *Phys. Rev. A* **25**, 1771–1774.

Evans, D. J. and Morriss, G. P. (1983a). Isothermal–isobaric molecular dynamics. *Chem. Phys.* **77**, 63–66.

Evans, D. J. and Morriss, G. P. (1983b). The isothermal/isobaric molecular dynamics ensemble. *Phys. Lett. A* **98**, 433–436.

Evans, D. J. and Morriss, G. P. (1984a). Non-Newtonian molecular dynamics. *Comput. Phys. Rep.* **1**, 297–343.

Evans, D. J. and Morriss, G. P. (1984b). Nonlinear-response theory for steady planar Couette flow. *Phys. Rev. A* **30**, 1528–1530.

Evans, D. J. and Morriss, G. P. (1986). Shear thickening and turbulence in simple fluids. *Phys. Rev. Lett.* **56**, 2172–2175.

Evans, D. J. and Morriss, G. P. (2008). *Statistical mechanics of nonequilibrium liquids.* 2nd ed. Cambridge: Cambridge University Press.

Evans, D. J. and Murad, S. (1977). Singularity-free algorithm for molecular dynamics simulation of rigid polyatomics. *Molec. Phys.* **34**, 327–331.

Evans, D. J. and Searles, D. J. (1994). Equilibrium microstates which generate second law violating steady states. *Phys. Rev. E* **50**, 1645–1648.

Evans, D. J. and Searles, D. J. (2002). The fluctuation theorem. *Adv. Phys.* **51**, 1529–1585.

Evans, D. J. and Streett, W. B. (1978). Transport properties of homonuclear diatomics. 2. Dense fluids. *Molec. Phys.* **36**, 161–176.

Evans, D. J. and Watts, R. O. (1976). Structure of liquid benzene. *Molec. Phys.* **32**, 93–100.

Evans, D. J., Hoover, W. G., Failor, B. H., Moran, B., and Ladd, A. J. C. (1983). Non-equilibrium molecular dynamics via Gauss' principle of least constraint. *Phys. Rev. A* **28**, 1016–1021.

Evans, H., Tildesley, D. J., and Sluckin, T. (1984). Boundary effects in the orientational ordering of adsorbed nitrogen. *J. Phys. C* **17**, 4907–4926.

Evans, R., Stewart, M. C., and Wilding, N. B. (2016). Critical drying of liquids. Preprint, arXiv:1607.03772 [cond-mat.stat-mech].

Evans, R. and Wilding, N. B. (2015). Quantifying density fluctuations in water at a hydrophobic surface: evidence for critical drying. *Phys. Rev. Lett.* **115**, 016103.

Evans, W. A. B. and Ojeda, C. J. (1992). NEMD 'q-kick response' method for the evaluation of the dynamic liquid structure factor. *J. Molec. Liquids* **54**, 297–309.

Evans, W. A. B. and Powles, J. G. (1982). The computer simulation of the dielectric properties of polar liquids. The dielectric constant and relaxation of liquid hydrogen chloride. *Molec. Phys.* **45**, 695–707.

Everaers, R. and Ejtehadi, M. R. (2003). Interaction potentials for soft and hard ellipsoids. *Phys. Rev. E* **67**, 041710.

Ewald, P. P. (1921). Die Berechnung optischer und elektrostatischer Gitterpotentiale. *Ann. Phys.* **369**, 253–287.

Eyring, H. (1932). Steric hindrance and collision diameters. *J. Am. Chem. Soc.* **54**, 3191–3203.

Ezra, G. S. (2006). Reversible measure-preserving integrators for non-Hamiltonian systems. *J. Chem. Phys.* **125**, 034104.

Fahrenberger, F. and Holm, C. (2014). Computing the Coulomb interaction in inhomogeneous dielectric media via a local electrostatics lattice algorithm. *Phys. Rev. E* **9**, 063304.

Fahrenberger, F., Xu, Z., and Holm, C. (2014). Simulation of electric double layers around charged colloids in aqueous solution of variable permittivity. *J. Chem. Phys.* **141**, 064902.

Falcioni, M. and Deem, M. W. (1999). A biased Monte Carlo scheme for zeolite structure solution. *J. Chem. Phys.* **110**, 1754–1766.

Faradjian, A. K. and Elber, R. (2004). Computing time scales from reaction coordinates by milestoning. *J. Chem. Phys.* **120**, 10880–10889.

Fassò, F. (2003). Comparison of splitting algorithms for the rigid body. *J. Comput. Phys.* **189**, 527–538.

Fedosov, D. A., Pan, W., Caswell, B., Gompper, G., and Karniadakis, G. E. (2011). Predicting human blood viscosity in silico. *Proc. Nat. Acad. Sci.* **108**, 11772–11777.

Fedosov, D. A., Caswell, B., and Karniadakis, G. E. (2010). A multiscale red blood cell model with accurate mechanics, rheology, and dynamics. *Biophys. J.* **98**, 2215–2225.

Fedosov, D. A., Noguchi, H., and Gompper, G. (2013). Multiscale modeling of blood flow: from single cells to blood rheology. *Biomech. Model. Mechanobiology* **13**, 239–258.

Feig, M., Onufriev, A., Lee, M. S., Im, W., Case, D. A., and Brooks, C. L. (2004). Performance comparison of generalized Born and Poisson methods in the calculation of electrostatic solvation energies for protein structures. *J. Comput. Chem.* **25**, 265–284.

Felderhof, B. U. (1977). Hydrodynamic interaction between two spheres. *Physica A* **89**, 373–384.

Felderhof, B. U. (1980). Fluctuation theorems for dielectrics with periodic boundary conditions. *Physica A* **101**, 275–282.

Feller, W. (1957). *An introduction to probability theory and its applications.* 2nd ed. Vol. 1. New York: Wiley.

Feng, J., Pandey, R. B., Berry, R. J., Farmer, B. L., Naik, R. R., and Heinz, H. (2011). Adsorption mechanism of single amino acid and surfactant molecules to Au {111} surfaces in aqueous solution: design rules for metal-binding molecules. *Soft Matter* **7**, 2113–2120.

Fennell, C. J. and Gezelter, J. D. (2006). Is the Ewald summation still necessary? Pairwise alternatives to the accepted standard for long-range electrostatics. *J. Chem. Phys.* **124**, 234104.

Ferrenberg, A. M. and Swendsen, R. H. (1989). Optimized Monte Carlo data analysis. *Phys. Rev. Lett.* **63**, 1195–1198.

Ferrenberg, A. M., Landau, D. P., and Wong, Y. J. (1992). Monte Carlo simulations: hidden errors from 'good' random number generators. *Phys. Rev. Lett.* **69**, 3382–3384.

Feynman, R. P. and Hibbs, A. R. (1965). *Quantum mechanics and path integrals.* New York: McGraw-Hill.

Field, M. J., Bash, P. A., and Karplus, M. (1990). A combined quantum mechanical and molecular mechanical potential for molecular dynamics simulations. *J. Comput. Chem.* **11**, 700–733.

Filon, L. N. G. (1928). On a quadrature formula for trigonometric integrals. *Proc. Roy. Soc. Edinburgh A* **49**, 38–49.

Fincham, D. (1981). An algorithm for the rotational motion of rigid molecules. *CCP5 Quarterly* **2**, 6–10.

Fincham, D. (1983). RDF on the DAP: easier than you think. *CCP5 Quarterly* **8**, 45–46.

Fincham, D. (1992). Leapfrog rotational algorithms. *Molec. Simul.* **8**, 165–178.

Fincham, D. and Heyes, D. M. (1982). Integration algorithms in molecular dynamics. *CCP5 Quarterly* **6**, 4–10.

Fincham, D., Quirke, N., and Tildesley, D. J. (1986). Computer simulation of molecular liquid mixtures. 1. A diatomic Lennard-Jones model mixture for CO_2/C_2H_6. *J. Chem. Phys.* **84**, 4535–4546.

Finn, J. E. and Monson, P. A. (1989). Prewetting at a fluid–solid interface via Monte Carlo simulation. *Phys. Rev. A* **39**, 6402–6408.

Fisher, M. E. (1964). The free energy of a macroscopic system. *Arch. Rational Mech. Anal.* **17**, 377–410.

Fixman, M. (1974). Classical statistical mechanics of constraints: a theorem and application to polymers. *Proc. Nat. Acad. Sci.* **71**, 3050–3053.

Fixman, M. (1978a). Simulation of polymer dynamics. I. General theory. *J. Chem. Phys.* **69**, 1527–1537.

Fixman, M. (1978b). Simulation of polymer dynamics. II. Relaxation rates and dynamic viscosity. *J. Chem. Phys.* **69**, 1538–1545.

Fixman, M. (1986). Construction of Langevin forces in the simulation of hydrodynamic interaction. *Macromolecules* **19**, 1204–1207.

Flinn, P. A. and McManus, G. M. (1961). Monte Carlo calculation of the order disorder transformation in the body-centred cubic lattices. *Phys. Rev.* **124**, 54–59.

Flyvbjerg, H. and Petersen, H. G. (1989). Error estimates on averages of correlated data. *J. Chem. Phys.* **91**, 461–466.

Focher, P., Chiarotti, G. L., Bernasconi, M., Tosatti, E., and Parrinello, M. (1994). Structural phase transformations via first-principles simulation. *Europhys. Lett.* **26**, 345–351.

Fosdick, L. D. and Jordan, H. F. (1966). Path-integral calculation of the two-particle Slater sum for He^4. *Phys. Rev.* **143**, 58–66.

Foulkes, W. M. C., Mitas, L., Needs, R. J., and Rajagopal, G. (2001). Quantum Monte Carlo simulations of solids. *Rev. Mod. Phys.* **73**, 33–83.

Fowler, R. F. (1984). The identification of a droplet in equilibrium with its vapour. *CCP5 Quarterly* **13**, 58–62.

Frank, F. C. (1958). On the theory of liquid crystals. *Discuss. Faraday Soc.* **25**, 19–28.

Freasier, B. C. (1980). Equation of state of fused hard spheres revisited. *Molec. Phys.* **39**, 1273–1280.

Freasier, B. C., Jolly, D. L., Hamer, N. D., and Nordholm, S. (1979). Effective vibrational potentials of bromine in argon. Monte Carlo simulation in a mixed ensemble. *Chem. Phys.* **38**, 293–300.

Frenkel, D. (1980). Intermolecular spectroscopy and computer simulations. *Intermolecular spectroscopy and dynamical properties of dense systems*. Proc. Int. School Phys. Enrico Fermi. **75**, 156–201. (Varenna, Italy, 1978). Ed. by J. van Kranendonk. Bologna: Soc. Italiana di Fisica.

Frenkel, D. (1986). Free-energy computation and first-order phase transitions. *Molecular dynamics simulation of statistical mechanical systems*. Proc. Int. School Phys. Enrico Fermi. **97**, 151–188. (Varenna, Italy, 1985). Ed. by G. Ciccotti and W. G. Hoover. Bologna: Soc. Italiana di Fisica.

Frenkel, D. and Ladd, A. J. C. (1984). New Monte Carlo method to compute the free energy of arbitrary solids. Application to the fcc and hcp phases of hard spheres. *J. Chem. Phys.* **81**, 3188–3193.

Frenkel, D. and Maguire, J. F. (1983). Molecular dynamics study of the dynamical properties of an assembly of infinitely thin hard rods. *Molec. Phys.* **49**, 503–541.

Frenkel, D. and McTague, J. P. (1979). Evidence for an orientationally ordered two-dimensional fluid phase from molecular-dynamics calculations. *Phys. Rev. Lett.* **42**, 1632–1635.

Frenkel, D. and Mulder, B. M. (1985). The hard ellipsoid-of-revolution fluid. 1. Monte Carlo simulations. *Molec. Phys.* **55**, 1171–1192.

Frenkel, D. and Smit, B. (2002). *Understanding molecular simulation: from algorithms to applications*. 2nd ed. San Diego: Academic Press.

Frenkel, D. (2014). Why colloidal systems can be described by statistical mechanics: some not very original comments on the Gibbs paradox. *Molec. Phys.* **112**, 2325–2329.

Friedberg, R. and Cameron, J. E. (1970). Test of the Monte Carlo method: fast simulation of a small Ising lattice. *J. Chem. Phys.* **52**, 6049–6058.

Friedman, H. L. (1975). Image approximation to reaction field. *Molec. Phys.* **29**, 1533–1543.

Friedman, H. L. (1985). *A course in statistical mechanics.* Englewood Cliffs, NJ.: Prentice-Hall.

Frigo, M. and Johnson, S. G. (2005). The design and implementation of FFTW3. *Proc. IEEE* **93**, 216–231. available from http://www.fftw.org.

Frisch, M. J., Trucks, G. W., Schlegel, H. B., Scuseria, G. E., Robb, M. A., Cheeseman, J. R., Scalmani, G., Barone, V., Mennucci, B., Petersson, G. A., Nakatsuji, H., Caricato, M., Li, X., Hratchian, H. P., Izmaylov, A. F., Bloino, J., Zheng, G., Sonnenberg, J. L., Hada, M., Ehara, M., Toyota, K., Fukuda, R., Hasegawa, J., Ishida, M., Nakajima, T., Honda, Y., Kitao, O., Nakai, H., Vreven, T., Montgomery Jr., J. A., Peralta, J. E., Ogliaro, F., Bearpark, M., Heyd, J. J., Brothers, E., Kudin, K. N., Staroverov, V. N., Kobayashi, R., Normand, J., Raghavachari, K., Rendell, A., Burant, J. C., Iyengar, S. S., Tomasi, J., Cossi, M., Rega, N., Millam, J. M., Klene, M., Knox, J. E., Cross, J. B., Bakken, V., Adamo, C., Jaramillo, J., Gomperts, R., Stratmann, R. E., Yazyev, O., Austin, A. J., Cammi, R., Pomelli, C., Ochterski, J. W., Martin, R. L., Morokuma, K., Zakrzewski, V. G., Voth, G. A., Salvador, P., Dannenberg, J. J., Dapprich, S., Daniels, A. D., Farkas, Ö., Foresman, J. B., Ortiz, J. V., Cioslowski, J., and Fox, D. J. (2009). *Gaussian 09 Revision D.01.* Gaussian Inc. Wallingford CT 2009.

Fritsch, S., Poblete, S., Junghans, C., Ciccotti, G., Delle Site, L., and Kremer, K. (2012). Adaptive resolution molecular dynamics simulation through coupling to an internal particle reservoir. *Phys. Rev. Lett.* **108**, 170602.

Fröhlich, H. (1949). *Theory of dielectrics; dielectric constant and dielectric loss.* Oxford: Oxford University Press.

Fukuda, I. (2013). Zero-multipole summation method for efficiently estimating electrostatic interactions in molecular system. *J. Chem. Phys.* **139**, 174107.

Fukuda, I. and Nakamura, H. (2012). Non-Ewald methods: theory and application in molecular systems. *Biophys. Rev.* **4**, 161–170.

Fukuda, I., Yonezawa, Y., and Nakamura, H. (2011). Molecular dynamics scheme for precise estimation of electrostatic interaction via zero-dipole summation principle. *J. Chem. Phys.* **134**, 164107.

Fukunishi, H., Watanabe, O., and Takada, S. (2002). On the Hamiltonian replica exchange method for efficient sampling of biomolecular systems: application to protein structure prediction. *J. Chem. Phys.* **116**, 9058–9067.

Futrelle, R. P. and McGinty, D. J. (1971). Calculation of spectra and correlation functions from molecular dynamics data using fast Fourier transform. *Chem. Phys. Lett.* **12**, 285–287.

Gaiduk, A. P., Zhang, C., Gygi, F., and Galli, G. (2014). Structural and electronic properties of aqueous NaCl solutions from *ab initio* molecular dynamics simulations with hybrid density functionals. *Chem. Phys. Lett.* **604**, 89–96.

Galbraith, A. L. and Hall, C. K. (2007). Solid–liquid phase equilibria for mixtures containing diatomic Lennard-Jones molecules. *Fluid Phase Equilibria* **262**, 1–13.

Galli, G. and Parrinello, M. (1991). *Ab-initio* molecular dynamics: principles and practical implementations. *Computer simulation in materials science.* NATO ASI Series E. **205**, 283–304. (Aussois, France, 1991). Ed. by M. Meyer and V. Pontikis. Dordrecht: Kluwer.

Galli, G. and Pasquarello, A. (1993). First-principles molecular dynamics. *Computer simulation in chemical physics.* NATO ASI Series C. **397**, 261–313. (Alghero, Italy, 1992). Ed. by M. P. Allen and D. J. Tildesley. Dordrecht: Kluwer.

Gao, J., Luedtke, W. D., and Landman, U. (1997a). Layering transitions and dynamics of confined liquid films. *Phys. Rev. Lett.* **79**, 705–708.

Gao, J., Luedtke, W. D., and Landman, U. (1997b). Structure and solvation forces in confined films: linear and branched alkanes. *J. Chem. Phys.* **106**, 4309–4318.

Gay, J. G. and Berne, B. J. (1981). Modification of the overlap potential to mimic a linear site–site potential. *J. Chem. Phys.* **74**, 3316–3319.

Gear, C. W. (1966). *The numerical integration of ordinary differential equations of various orders.* Tech. rep. 7126. Argonne National Laboratory.

Gear, C. W. (1971). *Numerical initial value problems in ordinary differential equations.* Englewood Cliffs, NJ: Prentice-Hall.

Gentle, J. E. (2003). *Random number generation and Monte Carlo methods.* 2nd ed. New York: Springer.

Georgescu, I., Brown, S. E., and Mandelshtam, V. A. (2013). Mapping the phase diagram for neon to a quantum Lennard-Jones fluid using Gibbs ensemble simulations. *J. Chem. Phys.* **138**, 134502.

Germano, G., Allen, M. P., and Masters, A. J. (2002). Simultaneous calculation of the helical pitch and the twist elastic constant in chiral liquid crystals from intermolecular torques. *J. Chem. Phys.* **116**, 9422–9430.

Geyer, C. J. (1991). Markov chain Monte Carlo maximum likelihood. *23rd symposium on the interface between computing science and statistics*, 156–163. (Seattle, USA, 1991). Ed. by E. M. Keramidas. New York: American Statistical Association.

Geyer, T. and Winter, U. (2009). An $O(N^2)$ approximation for hydrodynamic interactions in Brownian dynamics simulations. *J. Chem. Phys.* **130**, 114905.

Ghiringhelli, L. M. and Meijer, E. J. (2005). Phosphorus: first principle simulation of a liquid–liquid phase transition. *J. Chem. Phys.* **122**, 184510.

Ghoufi, A. and Malfreyt, P. (2011). Mesoscale modeling of the water liquid–vapor interface: a surface tension calculation. *Phys. Rev. E* **83**, 051601.

Ghoufi, A., Goujon, F., Lachet, V., and Malfreyt, P. (2008). Surface tension of water and acid gases from Monte Carlo simulations. *J. Chem. Phys.* **128**, 154716.

Gibson, W. G. (1974). Quantum corrections to the radial distribution function of a fluid. *Molec. Phys.* **28**, 793–800.

Gibson, W. G. (1975a). Quantum corrections to the properties of a dense fluid with nonanalytic intermolecular potential function. I. The general case. *Molec. Phys.* **30**, 1–11.

Gibson, W. G. (1975b). Quantum corrections to the properties of a dense fluid with nonanalytic intermolecular potential function. II. Hard spheres. *Molec. Phys.* **30**, 13–30.

Gilge, M. (2012). IBM system blue gene solution: Blue Gene/Q applications development. *IBM Redbooks.* Rochester, New York: IBM international technical support organisation.

Gillan, M. J. and Dixon, M. (1983). The calculation of thermal conductivities by perturbed molecular dynamics simulation. *J. Phys. C* **16**, 869–878.

Gillan, M. J. and Towler M.D.and Alfè, D. (2011). Petascale computing opens new vistas for quantum Monte Carlo. *Psi-k Newsletter* **103**, 1–43. http://www.psi-k.org/newsletters.shtml.

Gingold, R. A. and Monaghan, J. J. (1977). Smoothed particle hydrodynamics: theory and application to non-spherical stars. *Mon. Not. Roy. Astr. Soc.* **181**, 375–389.

Giovambattista, N., Almeida, A. B., Alencar, A. M., and Buldyrev, S. V. (2016). Validation of capillarity theory at the nanometer scale by atomistic computer simulations of water droplets and bridges in contact with hydrophobic and hydrophilic surfaces. *J. Phys. Chem. C* **120**, 1597–1608.

Glaser, J., Nguyen, T. D., Anderson, J. A., Lui, P., Spiga, F., Millan, J. A., Morse, D. C., and Glotzer, S. C. (2015). Strong scaling of general-purpose molecular dynamics simulations on GPUs. *Comput. Phys. Commun.* **192**, 97–107.

Gloor, G. J., Jackson, G., Blas, F. J., and de Miguel, E. (2005). Test-area simulation method for the direct determination of the interfacial tension of systems with continuous or discontinuous potentials. *J. Chem. Phys.* **123**, 134703.

Gō, N. and Scheraga, H. A. (1976). On the use of classical statistical mechanics in the treatment of polymer chain conformation. *Macromolecules* **9**, 535–542.

Goedecker, S., Teter, M., and Hutter, J. (1996). Separable dual-space Gaussian pseudopotentials. *Phys. Rev. B* **54**, 1703–1710.

Goetz, R. and Lipowsky, R. (1998). Computer simulations of bilayer membranes: self-assembly and interfacial tension. *J. Chem. Phys.* **108**, 7397–7409.

Goetz, R., Gompper, G., and Lipowsky, R. (1999). Mobility and elasticity of self-assembled membranes. *Phys. Rev. Lett.* **82**, 221–224.

Goldstein, H. (1980). *Classical mechanics.* 2nd ed. Reading, Massachusetts: Addison Wesley.

Golio, M. (2015). Fifty years of Moore's Law. *Proc. IEEE* **103**, 1932–1937.

Gompper, G., Ihle, T., Kroll, D. M., and Winkler, R. G. (2009). Multi-particle collision dynamics: a particle-based mesoscale simulation approach to the hydrodynamics of complex fluids. *Advanced computer simulation approaches for soft matter sciences III.* Adv. Polym. Sci. **221**, 1–87. Ed. by C Holm and K Kremer. Springer.

Gonnet, P. (2007). A simple algorithm to accelerate the computation of non-bonded interactions in cell-based molecular dynamics simulations. *J. Comput. Chem.* **28**, 570–573.

Gonze, X., Stumpf, R., and Scheffler, M. (1991). Analysis of separable potentials. *Phys. Rev. B* **44**, 8503–8513.

Gordon, R. (1968). Correlation functions for molecular motion. *Adv. Magn. Reson.* **3**, 1–42.

Gosling, E. M., McDonald, I. R., and Singer, K. (1973). Calculation by molecular dynamics of shear viscosity of a simple fluid. *Molec. Phys.* **26**, 1475–1484.

Goujon, F., Malfreyt, P., and Tildesley, D. J (2004). Dissipative particle dynamics simulations in the grand canonical ensemble: applications to polymer brushes. *ChemPhysChem* **5**, 457–464.

Goujon, F., Malfreyt, P., and Tildesley, D. J. (2005). The compression of polymer brushes under shear: the friction coefficient as a function of compression, shear rate and the properties of the solvent. *Molec. Phys.* **103**, 2675–2685.

Goujon, F., Malfreyt, P., and Tildesley, D. J. (2014). The gas–liquid surface tension of argon: a reconciliation between experiment and simulation. *J. Chem. Phys.* **140**, 244710.

Goujon, F., Ghoufi, A., Malfreyt, P., and Tildesley, D. J. (2015). Controlling the long-range corrections in atomistic Monte Carlo simulations of two-phase systems. *J. Chem. Theor. Comput.* **11**, 4573–4585.

Graben, H. W. and Fowler, R. (1969). Triple-dipole potentials in classical nonpolar fluids. *Phys. Rev.* **177**, 288–292.

Graben, H. W. and Ray, J. R. (1993). Eight physical systems of thermodynamics, statistical mechanics, and computer simulations. *Molec. Phys.* **80**, 1183–1193.

Grafmüller, A., Shillcock, J., and Lipowsky, R. (2007). Pathway of membrane fusion with two tension-dependent energy barriers. *Phys. Rev. Lett.* **98**, 218101.

Gray, C. G. and Gubbins, K. E. (1984). *Theory of molecular fluids. 1. Fundamentals.* Oxford: Clarendon Press.

Gray, S. K., Noid, D. W., and Sumpter, B. G. (1994). Symplectic integrators for large scale molecular dynamics simulations: a comparison of several explicit methods. *J. Chem. Phys.* **101**, 4062–4072.

Green, H. S. (1951). The quantum mechanics of assemblies of interacting particles. *J. Chem. Phys.* **19**, 955–962.

Greengard, L. and Gropp, W. D. (1990). A parallel version of the fast multipole method. *Comput. Math. Appl.* **20**, 63–71.

Greengard, L. and Rokhlin, V. (1988). The rapid evaluation of potential fields in 3 dimensions. *Lect. Notes Math.* **1360**, 121–141.

Griebel, M., Knapek, S., and Zumbusch, G. (2007). *Numerical simulation in molecular dynamics: numerics, algorithms, parallelization, applications.* Texts in computational science and engineering. Vol. 5. Berlin: Springer.

Grochola, G. (2004). Constrained fluid λ-integration: constructing a reversible thermodynamic path between the solid and liquid state. *J. Chem. Phys.* **120**, 2122–2126.

Grochola, G. (2005). Further application of the constrained fluid λ-integration method. *J. Chem. Phys.* **122**, 046101.

Groenhof, G. (2013). Introduction to QM/MM simulations. *Biomolecular simulations, methods and protocols.* Methods Molec. Biol. **924**. Chap. 3, 43–66. Ed. by L. Monticelli and E. Salonen. New York: Humana Press.

Groot, R. D. and Rabone, K. L. (2001). Mesoscopic simulation of cell membrane damage, morphology change and rupture by nonionic surfactants. *Biophys. J.* **81**, 725–736.

Groot, R. D. and Warren, P. B. (1997). Dissipative particle dynamics: bridging the gap between atomistic and mesoscopic simulation. *J. Chem. Phys.* **107**, 4423–4435.

Groot, R. D., Madden, T. J., and Tildesley, D. J. (1999). On the role of hydrodynamic interactions in block copolymer microphase separation. *J. Chem. Phys.* **110**, 9739–9749.

Grubmüller, H., Heller, H., Windemuth, A., and Schulten, K. (1991). Generalized Verlet algorithm for efficient molecular dynamics simulations with long-range interactions. *Molec. Simul.* **6**, 121–142.

Grubmüller, H. (1995). Predicting slow structural transitions in macromolecular systems: conformational flooding. *Phys. Rev. E* **52**, 2893–2906.

Grzelak, E. M. and Errington, J. R. (2010). Nanoscale limit to the applicability of Wenzel's equation. *Langmuir* **26**, 13297–13304.

Grzybowski, A., Gwóźdź, E., and Bródka, A. (2000). Ewald summation of electrostatic interactions in molecular dynamics of a three-dimensional system with periodicity in two directions. *Phys. Rev. B* **61**, 6706–6712.

Guldbrand, L., Jönsson, B., Wennerström, H., and Linse, P. (1984). Electrical double layer forces. A Monte Carlo study. *J. Chem. Phys.* **80**, 2221–2228.

Guo, M. and Lu, B. C. Y. (1997). Long range corrections to thermodynamic properties of inhomogeneous systems with planar interfaces. *J. Chem. Phys.* **106**, 3688–3695.

Habershon, S., Fanourgakis, G. S., and Manolopoulos, D. E. (2008). Comparison of path integral molecular dynamics methods for the infrared absorption spectrum of liquid water. *J. Chem. Phys.* **129**, 074501.

Habershon, S., Manolopoulos, D. E., Markland, T. E., and Miller, T. F. (2013). Ring-Polymer Molecular Dynamics: quantum effects in chemical dynamics from classical trajectories in an extended phase space. *Ann. Rev. Phys. Chem.* **64**, 387–413.

Hadley, K. R. and McCabe, C. (2012). Coarse-grained molecular models of water: a review. *Molec. Simul.* **38**, 671–681.

Hafskjold, B., Liew, C. C., and Shinoda, W. (2004). Can such long time steps really be used in dissipative particle dynamics simulations? *Molec. Simul.* **30**, 879–885.

Hagler, A. T., Lifson, S., and Dauber, P. (1979). Consistent force field studies of inter-molecular forces in hydrogen-bonded crystals. 2. Benchmark for the objective comparison of alternative force fields. *J. Am. Chem. Soc.* **101**, 5122–5130.

Haile, J. M. and Gray, C. G. (1980). Spherical harmonic expansions of the angular pair correlation function in molecular fluids. *Chem. Phys. Lett.* **76**, 583–588.

Haile, J. M. and Gupta, S. (1983). Extensions of the molecular dynamics simulation method. 2. Isothermal systems. *J. Chem. Phys.* **79**, 3067–3076.

Halgren, T. A. (1992). Representation of van der Waals (vdW) interactions in molecular mechanics force fields: potential form, combination rules, and vdW parameters. *J. Am. Chem. Soc.* **114**, 7827–7843.

Halgren, T. A. and Damm, W. (2001). Polarizable force fields. *Curr. Opin. Struct. Biol.* **11**, 236–242.

Halperin, B. I. and Nelson, D. R. (1978). Theory of two-dimensional melting. *Phys. Rev. Lett.* **41**, 121–124.

Halverson, J. D., Brandes, T., Lenz, O., Arnold, A., Bevc, S., Starchenko, V., Kremer, K., Stuehn, T., and Reith, D. (2013). espresso++: a modern multiscale simulation package for soft matter systems. *Comput. Phys. Commun.* **184**, 1129–1149.

Hammersley, J. M. and Handscomb, D. C. (1964). *Monte Carlo methods*. New York: Wiley.

Han, K.-K. and Son, H. S. (2001). On the isothermal–isobaric ensemble partition function. *J. Chem. Phys.* **115**, 7793–7794.

Hansen, D. P. and Evans, D. J. (1994). A parallel algorithm for nonequilibrium molecular dynamics simulation of shear flow on distributed memory machines. *Molec. Simul.* **13**, 375–393.

Hansen, J.-P. and McDonald, I. R. (2013). *Theory of simple liquids with applications to soft matter*. 4th ed. Amsterdam: Academic Press.

Hansen, J.-P. and Weis, J. J. (1969). Quantum corrections to the coexistence curve of neon near the triple point. *Phys. Rev.* **188**, 314–318.

Hansen, J.-P., Levesque, D., and Weis, J. J. (1979). Self-diffusion in the two-dimensional, classical electron gas. *Phys. Rev. Lett.* **43**, 979–982.

Hansen, J. S., Daivis, P. J., and Todd, B. D. (2009). Viscous properties of isotropic fluids composed of linear molecules: departure from the classical Navier–Stokes theory in nano-confined geometries. *Phys. Rev. E* **80**, 046322.

Hansen, J. S., Bruus, H., Todd, B. D., and Daivis, P. J. (2010). Rotational and spin viscosities of water: application to nanofluidics. *J. Chem. Phys.* **133**, 144906.

Hansen, J. S., Todd, B. D., and Daivis, P. J. (2011). Prediction of fluid velocity slip at solid surfaces. *Phys. Rev. E* **84**, 016313.

Hansmann, U. H. E. (1997). Parallel tempering algorithm for conformational studies of biological molecules. *Chem. Phys. Lett.* **281**, 140–150.

Harasima, A. (1958). Molecular theory of surface tension. *Adv. Chem. Phys.* **1**, 203–237.

Hardacre, C., Holbrey, J. D., McMath, S. E. J., Bowron, D. T., and Soper, A. K. (2003). Structure of molten 1,3-dimethylimidazolium chloride using neutron diffraction. *J. Chem. Phys.* **118**, 273–278.

Hardy, D. J., Wu, Z., Phillips, J. C., Stone, J. E., Skeel, R. D., and Schulten, K. (2015). Multilevel summation method for electrostatic force evaluation. *J. Chem. Theor. Comput.* **11**, 766–779.

Hardy, D. J., Stone, J. E., and Schulten, K. (2009). Multilevel summation of electrostatic potentials using graphics processing units. *Parallel Comput.* **35**, 164–177.

Harmandaris, V. A. and Deserno, M. (2006). A novel method for measuring the bending rigidity of model lipid membranes by simulating tethers. *J. Chem. Phys.* **125**, 204905.

Harp, G. D. and Berne, B. J. (1968). Linear and angular momentum autocorrelation functions in diatomic liquids. *J. Chem. Phys.* **49**, 1249–1254.

Harp, G. D. and Berne, B. J. (1970). Time correlation functions, memory functions, and molecular dynamics. *Phys. Rev. A* **2**, 975–996.

Harrison, N. M. (2003). An introduction to density functional theory. *Computational materials science.* NATO Science Series III. **187**, 45–70. (Il Ciocco, Italy, 2001). Ed. by R. Catlow and E. Kotomin. Amsterdam: IOS Press.

Härtel, A., Oettel, M., Rozas, R. E., Egelhaaf, S. U., Horbach, J., and Löwen, H. (2012). Tension and stiffness of the hard sphere crystal–fluid interface. *Phys. Rev. Lett.* **108**, 226101.

Hartwigsen, C., Goedecker, S., and Hutter, J. (1998). Relativistic separable dual-space Gaussian pseudopotentials from H to Rn. *Phys. Rev. B* **58**, 3641–3662.

Harvey, S. C., Tan, R. K.-Z., and Cheatham, T. E. (1998). The flying ice cube: velocity rescaling in molecular dynamics leads to violation of energy equipartition. *J. Comput. Chem.* **19**, 726–740.

Haslam, A. J., Galindo, A., and Jackson, G. (2008). Prediction of binary intermolecular potential parameters for use in modelling fluid mixtures. *Fluid Phase Equilibria* **266**, 105–128.

Hastings, W. K. (1970). Monte Carlo sampling methods using Markov chains, and their applications. *Biometrika* **57**, 97–109.

Hautman, J. and Klein, M. (1992). An Ewald summation method for planar surfaces and interfaces. *Molec. Phys.* **75**, 379–395.

Hawlicka, E., Pálinkás, G., and Heinzinger, K. (1989). A molecular dynamics study of liquid methanol with a flexible six-site model. *Chem. Phys. Lett.* **154**, 255–259.

Haye, M. J. and Bruin, C. (1994). Molecular dynamics study of the curvature correction to the surface tension. *J. Chem. Phys.* **100**, 556–559.

Hecht, M., Harting, J., Ihle, T., and Herrmann, H. J. (2005). Simulation of claylike colloids. *Phys. Rev. E* **72**, 011408.

Heffelfinger, G. S. and van Swol, F. (1994). Diffusion in Lennard-Jones fluids using dual control volume grand canonical molecular dynamics simulation (DCV-GCMD). *J. Chem. Phys.* **100**, 7548–7552.

Heinz, H. and Ramezani-Dakhel, H. (2016). Simulations of inorganic–bioorganic interfaces to discover new materials: insights, comparisons to experiment, challenges, and opportunities. *Chem. Soc. Rev.* **45**, 412–448.

Heinz, H., Vaia, R. A., Farmer, B. L., and Naik, R. R. (2008). Accurate simulation of surfaces and interfaces of face-centered cubic metals using 12–6 and 9–6 Lennard-Jones potentials. *J. Phys. Chem. C* **112**, 17281–17290.

Heinz, H., Lin, T.-J., Mishra, R. K., and Emami, F. S. (2013). Thermodynamically consistent force fields for the assembly of inorganic, organic, and biological nanostructures: the INTERFACE force field. *Langmuir* **29**, 1754–1765.

Heinz, T. N. and Hünenberger, P. H. (2004). A fast pairlist-construction algorithm for molecular simulations under periodic boundary conditions. *J. Comput. Chem.* **25**, 1474–1486.

Helfand, E. (1979). Flexible vs rigid constraints in statistical mechanics. *J. Chem. Phys.* **71**, 5000–5007.

Helfand, E. and Rice, S. A. (1960). Principle of corresponding states for transport properties. *J. Chem. Phys.* **32**, 1642–1644.

Helfrich, W. (1973). Elastic properties of lipid bilayers: theory and possible experiments. *Z. Naturf. C. Biosciences* **28**, 693–703.

Hemmer, P. C. (1968). The hard core quantum gas at high temperatures. *Phys. Lett. A* **27**, 377–378.

Henderson, J. R. (1983). Statistical mechanics of fluids at spherical structureless walls. *Molec. Phys.* **50**, 741–761.

Henderson, R. L. (1974). A uniqueness theorem for fluid pair correlation functions. *Phys. Lett. A* **49**, 197–198.

Henrich, O., Marenduzzo, D., Stratford, K., and Cates, M. E. (2010). Domain growth in cholesteric blue phases: hybrid lattice Boltzmann simulations. *Comput. Math. Appl.* **59**, 2360–2369.

Herman, M. F., Bruskin, E. J., and Berne, B. J. (1982). On path integral Monte Carlo simulations. *J. Chem. Phys.* **76**, 5150–5155.

Hernandez, R., Cao, J. S., and Voth, G. A. (1995). On the Feynman path centroid density as a phase space distribution in quantum statistical mechanics. *J. Chem. Phys.* **103**, 5018–5026.

Hess, B., Bekker, H., Berendsen, H. J. C., and Fraaije, J. G. E. M. (1997). LINCS: a linear constraint solver for molecular simulations. *J. Comput. Chem.* **18**, 1463–1472.

Hess, B. (2008). P-LINCS: a parallel linear constraint solver for molecular simulation. *J. Chem. Theor. Comput.* **4**, 116–122.

Hess, B., Kutzner, C., van der Spoel, D., and Lindahl, E. (2008). GROMACS4: algorithms for highly efficient, load-balanced, and scalable molecular simulation. *J. Chem. Theor. Comput.* **4**, 435–447.

Heyes, D. M. (1981). Electrostatic potentials and fields in infinite point charge lattices. *J. Chem. Phys.* **74**, 1924–1929.

Heyes, D. M. (1983a). MD incorporating Ewald summations on partial charge polyatomic systems. *CCP5 Quarterly* **8**, 29–36.

Heyes, D. M. (1983b). Molecular dynamics at constant pressure and temperature. *Chem. Phys.* **82**, 285–301.

Heyes, D. M. and Clarke, J. H. R. (1981). Computer simulation of molten-salt interphases. Effect of a rigid boundary and an applied electric field. *J. Chem. Soc. Faraday Trans. II* **77**, 1089–1100.

Heyes, D. M. and Singer, K. (1982). A very accurate molecular dynamics algorithm. *CCP5 Quarterly* **6**, 11–23.

Heyes, D. M., Smith, E. R., Dini, D., and Zaki, T. A. (2011). The equivalence between volume averaging and method of planes definitions of the pressure tensor at a plane. *J. Chem. Phys.* **135**, 024512.

Hill, T. L. (1956). *Statistical mechanics.* New York: McGraw-Hill.

Hinsen, K. (2000). The molecular modeling toolkit: a new approach to molecular simulations. *J. Comput. Chem.* **21**, 79–85.

Hinsen, K. (2014). MOSAIC: a data model and file formats for molecular simulations. *J. Chem. Inf. Model.* **54**, 131–137.

Hirschfelder, J. O. (1960). Classical and quantum mechanical hypervirial theorems. *J. Chem. Phys.* **33**, 1462–1466.

Hirshfeld, F. L. (1977). Bonded-atom fragments for describing molecular charge densities. *Theor. Chim. Acta* **44**, 129–138.

Hiyama, M., Kinjo, T., and Hyodo, S. (2008). Angular momentum form of Verlet algorithm for rigid molecules. *J. Phys. Soc. Japan* **77**, 064001.

Hockney, R. W. (1970). The potential calculation and some applications. *Methods Comput. Phys.* **9**, 136–211.

Hockney, R. W. and Eastwood, J. W. (1988). *Computer simulation using particles.* Bristol: Adam Hilger.

Hohenberg, P. and Kohn, W. (1964). Inhomogeneous electron gas. *Phys. Rev. B* **136**, 864–871.

Homman, A.-A., Bourasseau, E., Stoltz, G., Malfreyt, P., Strafella, L., and Ghoufi, A. (2014). Surface tension of spherical drops from surface of tension. *J. Chem. Phys.* **140**, 034110.

Honeycutt, J. D. and Andersen, H. C. (1984). The effect of periodic boundary conditions on homogeneous nucleation observed in computer simulations. *Chem. Phys. Lett.* **108**, 535–538.

Hoogerbrugge, P. J. and Koelman, J. M. V. A. (1992). Simulating microscopic hydrodynamic phenomena with dissipative particle dynamics. *Europhys. Lett.* **19**, 155–160.

Hoover, W. G. (1983a). Atomistic nonequilibrium computer simulations. *Physica A* **118**, 111–122.

Hoover, W. G. (1983b). Nonequilibrium molecular dynamics. *Ann. Rev. Phys. Chem.* **34**, 103–127.

Hoover, W. G. and Ashurst, W. T. (1975). Nonequilibrium molecular dynamics. *Theoretical chemistry: advances and perspectives.* **1**, 1–51. Ed. by H. Eyring and D. Henderson. New York: Academic Press.

Hoover, W. G. and Hoover, C. G. (2001). SPAM-based recipes for continuum simulations. *Comput. Sci. Eng.* **3**, 78–85.

Hoover, W. G. (1985). Canonical dynamics: equilibrium phase-space distributions. *Phys. Rev. A* **31**, 1695–1697.

Hoover, W. G. (1986). Constant-pressure equations of motion. *Phys. Rev. A* **34**, 2499–2500.

Hoover, W. G. and Alder, B. J. (1967). Studies in molecular dynamics. IV. The pressure, collision rate, and their number dependence for hard disks. *J. Chem. Phys.* **46**, 686–691.

Hoover, W. G. and Ree, F. H. (1968). Melting transition and communal entropy for hard spheres. *J. Chem. Phys.* **49**, 3609–3617.

Hoover, W. G., Ross, M., Johnson, K. W., Henderson, D., Barker, J. A., and Brown, B. C. (1970). Soft-sphere equation of state. *J. Chem. Phys.* **52**, 4931–4941.

Hoover, W. G., Gray, S. G., and Johnson, K. W. (1971). Thermodynamic properties of the fluid and solid phases for inverse power potentials. *J. Chem. Phys.* **55**, 1128–1136.

Hoover, W. G., Ladd, A. J. C., Hickman, R. B., and Holian, B. L. (1980a). Bulk viscosity via nonequilibrium and equilibrium molecular dynamics. *Phys. Rev. A* **21**, 1756–1760.

Hoover, W. G., Evans, D. J., Hickman, R. B., Ladd, A. J. C., Ashurst, W. T., and Moran, B. (1980b). Lennard-Jones triple-point bulk and shear viscosities. Green–Kubo theory, Hamiltonian mechanics, and nonequilibrium molecular dynamics. *Phys. Rev. A* **22**, 1690–1697.

Hoover, W. G., Ladd, A. J. C., and Moran, B. (1982). High-strain-rate plastic flow studied via nonequilibrium molecular dynamics. *Phys. Rev. Lett.* **48**, 1818–1820.

Horowitz, A. M. (1991). A generalized guided Monte Carlo algorithm. *Phys. Lett. B* **268**, 247–252.

Hoyt, J. J., Asta, M., and Karma, A. (2001). Method for computing the anisotropy of the solid–liquid interfacial free energy. *Phys. Rev. Lett.* **86**, 5530–5533.

Hu, J., Ma, A., and Dinner, A. R. (2006). Monte Carlo simulations of biomolecules: the MC module in CHARMM. *J. Comput. Chem.* **27**, 203–216.

Hu, M., Briguglio, J. J., and Deserno, M. (2012). Determining the Gaussian curvature modulus of lipid membranes in simulations. *Biophys. J.* **102**, 1403–1410.

Hub, J. S., de Groot, B. L., Grubmüller, H., and Groenhof, G. (2014). Quantifying artifacts in Ewald simulations of inhomogeneous systems with a net charge. *J. Chem. Theor. Comput.* **10**, 381–390.

Huitema, H. E. A., van Hengstum, B., and van der Eerden, J. (1999). Simulation of crystal growth from Lennard-Jones solutions. *J. Chem. Phys.* **111**, 10248–10260.

Hummer, G. and Szabo, A. (2001). Free energy reconstruction from nonequilibrium single-molecule pulling experiments. *Proc. Nat. Acad. Sci.* **98**, 3658–3661.

Hummer, G. and Szabo, A. (2005). Free energy surfaces from single-molecule force spectroscopy. *Acc. Chem. Res.* **38**, 504–513.

Hummer, G. (2004). From transition paths to transition states and rate coefficients. *J. Chem. Phys.* **120**, 516–523.

Humpert, A. and Allen, M. P. (2015a). Elastic constants and dynamics in nematic liquid crystals. *Molec. Phys.* **113**, 2680–2692.

Humpert, A. and Allen, M. P. (2015b). Propagating director bend fluctuations in nematic liquid crystals. *Phys. Rev. Lett.* **114**, 028301.

Humphrey, W., Dalke, A., and Schulten, K. (1996). VMD: visual molecular dynamics. *J. Molec. Graphics* **14**, 33–38. http://www.ks.uiuc.edu/Research/vmd/.

Hünenberger, P. H. and van Gunsteren, W. F. (1998). Alternative schemes for the inclusion of a reaction-field correction into molecular dynamics simulations: influence on the simulated energetic, structural, and dielectric properties of liquid water. *J. Chem. Phys.* **108**, 6117–6134.

Hünenberger, P. H. (2005). Thermostat algorithms for molecular dynamics simulations. *Advanced computer simulation.* Adv. Polym. Sci. **173**, 105–149. Ed. by C. Holm and K. Kremer. Berlin Heidelberg: Springer.

Hunt, T. A. (2015). Periodic boundary conditions for the simulation of uniaxial extensional flow of arbitrary duration. *Molec. Simul.* **42**, 347–352.

Hunt, T. A. and Todd, B. D. (2003). On the Arnold cat map and periodic boundary conditions for planar elongational flow. *Molec. Phys.* **101**, 3445–3454.

Hunt, T. A., Bernardi, S., and Todd, B. D. (2010). A new algorithm for extended nonequilibrium molecular dynamics simulations of mixed flow. *J. Chem. Phys.* **133**, 154116.

Hunter, J. E. I. and Reinhardt, W. P. (1995). Finite-size scaling behavior of the free energy barrier between coexisting phases: determination of the critical temperature and interfacial tension of the Lennard-Jones fluid. *J. Chem. Phys.* **103**, 8627–8637.

Huwald, J., Richter, S., Ibrahim, B., and Dittrich, P. (2016). Compressing molecular dynamics trajectories: breaking the one-bit-per-sample barrier. *J. Comput. Chem.* **37**, 1897–1906.

Ihle, T. and Kroll, D. M. (2001). Stochastic rotation dynamics: a Galilean-invariant mesoscopic model for fluid flow. *Phys. Rev. E* **63**, 020201.

Ihle, T. and Kroll, D. M. (2003a). Stochastic rotation dynamics. I. Formalism, Galilean invariance, and Green–Kubo relations. *Phys. Rev. E* **67**, 066705.

Ihle, T. and Kroll, D. M. (2003b). Stochastic rotation dynamics. II. Transport coefficients, numerics, and long-time tails. *Phys. Rev. E* **67**, 066706.

Ikeguchi, M. (2004). Partial rigid-body dynamics in NPT, $NPAT$ and $NP\gamma T$ ensembles for proteins and membranes. *J. Comput. Chem.* **25**, 529–541.

Impey, R. W., Madden, P. A., and Tildesley, D. J. (1981). On the calculation of the orientational correlation parameter g_2. *Molec. Phys.* **44**, 1319–1334.

Ingebrigtsen, T. and Toxvaerd, S. (2007). Contact angles of Lennard-Jones liquids and droplets on planar surfaces. *J. Phys. Chem. C* **111**, 8518–8523.

in't Veld, P. J., Ismail, A. E., and Grest, G. S. (2007). Application of Ewald summations to long-range dispersion forces. *J. Chem. Phys.* **127**, 144711.

Iori, F., Di Felice, R., Molinari, E., and Corni, S. (2009). GOLP: an atomistic force-field to describe the interaction of proteins with Au(111) surfaces in water. *J. Comput. Chem.* **30**, 1465–1476.

Ippoliti, E., Dreyer, J., Carloni, P., and Röthlisberger, U. (2012). *Hierarchical methods for dynamics in complex molecular systems.* http://juser.fz-juelich.de/record/127337/files/IAS-Series-10.pdf (Accessed Dec 2015).

Irving, J. H. and Kirkwood, J. G. (1950). The statistical mechanical theory of transport processes. IV. The equations of hydrodynamics. *J. Chem. Phys.* **18**, 817–829.

Isele-Holder, R. E., Mitchell, W., and Ismail, A. E. (2012). Development and application of a particle–particle particle–mesh Ewald method for dispersion interactions. *J. Chem. Phys.* **137**, 174107.

Isele-Holder, R. E., Mitchell, W., Hammond, J. R., Kohlmeyer, A., and Ismail, A. E. (2013). Reconsidering dispersion potentials: reduced cutoffs in mesh-based Ewald solvers can be faster than truncation. *J. Chem. Theor. Comput.* **9**, 5412–5420.

Ishida, H. and Kidera, A. (1998). Constant temperature molecular dynamics of a protein in water by high-order decomposition of the Liouville operator. *J. Chem. Phys.* **109**, 3276–3284.

Ishida, H., Nagai, Y., and Kidera, A. (1998). Symplectic integrator for molecular dynamics of a protein in water. *Chem. Phys. Lett.* **282**, 115–120.

Ismail, A. E., Grest, G. S., and Stevens, M. J. (2006). Capillary waves at the liquid–vapor interface and the surface tension of water. *J. Chem. Phys.* **125**, 014702.

Itoh, S. G., Morishita, T., and Okumura, H. (2013). Decomposition-order effects of time integrator on ensemble averages for the Nosé–Hoover thermostat. *J. Chem. Phys.* **139**, 064103.

Izvekov, S., Swanson, J. M. J., and Voth, G. A. (2008). Coarse-graining in interaction space: a systematic approach for replacing long-range electrostatics with short-range potentials. *J. Phys. Chem. B* **112**, 4711–4724.

Izvekov, S. and Voth, G. A. (2005a). A multiscale coarse-graining method for biomolecular systems. *J. Phys. Chem. B* **109**, 2469–2473.

Izvekov, S. and Voth, G. A. (2005b). Multiscale coarse graining of liquid-state systems. *J. Chem. Phys.* **123**, 134105.

Jackson, G., Chapman, W. G., and Gubbins, K. E. (1988). Phase equilibria of associating fluids. *Molec. Phys.* **65**, 1–31.

Jackson, J. L. and Mazur, P. (1964). On the statistical mechanical derivation of the correlation formula for the viscosity. *Physica* **30**, 2295–2304.

Jacucci, G. (1984). Path integral Monte Carlo. *Monte Carlo methods in quantum problems.* NATO ASI Series C. **125**, 117–144. Ed. by M. H. Kalos. New York: Reidel.

Jacucci, G. and McDonald, I. R. (1975). Structure and diffusion in mixtures of rare-gas liquids. *Physica A* **80**, 607–625.

Jacucci, G. and Omerti, E. (1983). Monte Carlo calculation of the radial distribution function of quantum hard spheres at finite temperatures using path integrals with boundary conditions. *J. Chem. Phys.* **79**, 3051–3054.

Jacucci, G. and Quirke, N. (1980a). Monte Carlo calculation of the free energy difference between hard and soft core diatomic liquids. *Molec. Phys.* **40**, 1005–1009.

Jacucci, G. and Quirke, N. (1980b). Structural correlation functions in the coexistence region from molecular dynamics. *Nuovo Cimento B* **58**, 317–322.

Jacucci, G. and Rahman, A. (1984). Comparing the efficiency of Metropolis Monte Carlo and molecular dynamics methods for configuration space sampling. *Nuovo Cimento D* **4**, 341–356.

Jain, A., Sunthar, P., Dünweg, B., and Prakash, J. R. (2012). Optimization of a Brownian dynamics algorithm for semidilute polymer solutions. *Phys. Rev. E* **85**, 066703.

Jakobsen, A. F., Mouritsen, O. G., and Besold, G. (2005). Artifacts in dynamical simulations of coarse-grained model lipid bilayers. *J. Chem. Phys.* **122**, 204901.

Jakobsen, A. F., Besold, G., and Mouritsen, O. G. (2006). Multiple time step update schemes for dissipative particle dynamics. *J. Chem. Phys.* **124**, 094104.

Jancovici, B. (1969). Quantum mechanical equation of state of a hard sphere gas at high temperature. *Phys. Rev.* **178**, 295–297.

Janeček, J. (2006). Long range corrections in inhomogeneous simulations. *J. Phys. Chem. B* **131**, 6264–6269.

Jang, S., Sinitskiy, A. V., and Voth, G. A. (2014). Can the ring polymer molecular dynamics method be interpreted as real time quantum dynamics? *J. Chem. Phys.* **140**, 154103.

Janke, W. and Paul, W. (2016). Thermodynamics and structure of macromolecules from flat-histogram Monte Carlo simulations. *Soft Matter* **12**, 642–657.

Jansoone, V. M. (1974). Dielectric properties of a model fluid with the Monte Carlo method. *Chem. Phys.* **3**, 78–86.

Jarzynski, C. (1997). Nonequilibrium equality for free energy differences. *Phys. Rev. Lett.* **78**, 2690–2693.

Jepps, O. G., Ayton, G., and Evans, D. J. (2000). Microscopic expressions for the thermodynamic temperature. *Phys. Rev. E* **62**, 4757–4763.

Jiménez-Serratos, G., Vega, C., and Gil-Villegas, A. (2012). Evaluation of the pressure tensor and surface tension for molecular fluids with discontinuous potentials using the volume perturbation method. *J. Chem. Phys.* **137**, 204104.

Jmol (2016). *Jmol: an open-source Java viewer for chemical structures in 3D.* http://www.jmol.org/ (Accessed 8/2/2016).

Johner, N., Harries, D., and Khelashvili, G. (2016). Implementation of a methodology for determining elastic properties of lipid assemblies from molecular dynamics simulations. *BMC Bioinformatics* **17**, 161. Erratum: *ibid.* **17**, 236 (2016).

Johnson, J. K., Zollweg, J. A., and Gubbins, K. E. (1993). The Lennard-Jones equation of state revisited. *Molec. Phys.* **78**, 591–618.

Johnson, J. K., Panagiotopoulos, A. Z., and Gubbins, K. E. (1994). Reactive canonical Monte Carlo. A new simulation technique for reacting or associating fluids. *Molec. Phys.* **81**, 717–733.

Johnson, M. E., Head-Gordon, T., and Louis, A. A. (2007). Representability problems for coarse-grained water potentials. *J. Chem. Phys.* **126**, 144509.

Jolly, D. L., Freasier, B. C., and Bearman, R. J. (1976). The extension of simulation radial distribution functions to an arbitrary range by Baxter's factorisation technique. *Chem. Phys.* **15**, 237–242.

Jones, A., Cipcigan, F., Sokhan, V. P., Crain, J., and Martyna, G. J. (2013). Electronically coarse-grained model for water. *Phys. Rev. Lett.* **110**, 227801.

Jones, D. (2010). *Good practice in (pseudo) random number generation for bioinformatics applications.* http://www0.cs.ucl.ac.uk/staff/d.jones/GoodPracticeRNG.pdf (Accessed Nov 2015).

Jones, R. A. L. (2002). *Soft condensed matter.* Oxford Master Series in Physics. Oxford: Oxford University Press.

Jordan, H. F. and Fosdick, L. D. (1968). Three-particle effects in the pair distribution function for He^4 gas. *Phys. Rev.* **171**, 128–149.

Jorgensen, W. L. and Ravimohan, C. (1985). Monte Carlo simulation of differences in free energies of hydration. *J. Chem. Phys.* **83**, 3050–3054.

Jorgensen, W. L. and Thomas, L. L. (2008). Perspective on free-energy perturbation calculations for chemical equilibria. *J. Chem. Theor. Comput.* **4**, 869–876.

Jorgensen, W. L. and Tirado-Rives, J. (2005). Molecular modeling of organic and biomolecular systems using BOSS and MCPRO. *J. Comput. Chem.* **26**, 1689–1700.

Jorgensen, W. L., Chandrasekhar, J., Madura, J. D., Impey, R. W., and Klein, M. L. (1983). Comparison of simple potential functions for simulating liquid water. *J. Chem. Phys.* **79**, 926–935.

Jorgensen, W. L., Maxwell, D. S., and Tirado-Rives, J. (1996). Development and testing of the OPLS all-atom force field on conformational energetics and properties of organic liquids. *J. Am. Chem. Soc.* **118**, 11225–11236.

Kaestner, J. (2011). Umbrella sampling. *Comp. Mol. Sci. (Wiley Interdisc. Rev.)* **1**, 932–942.

Kalia, R. K., de Leeuw, S., Nakano, A., and Vashishta, P. (1993). Molecular dynamics simulations of coulombic systems on distributed-memory MIMD machines. *Comput. Phys. Commun.* **74**, 316–326.

Kalos, M. H. and Whitlock, P. A. (2008). *Monte Carlo methods.* 2nd ed. Weinheim, Germany: Wiley-VCH.

Kalos, M. H., Levesque, D., and Verlet, L. (1974). Helium at zero temperature with hard-sphere and other forces. *Phys. Rev. A* **9**, 2178–2195.

Kaminski, G. and Jorgensen, W. L. (1996). Performance of the AMBER94, MMFF94, and OPLS-AA force fields for modeling organic liquids. *J. Phys. Chem.* **100**, 18010–18013.

Kannam, S. K., Todd, B. D., Hansen, J. S., and Daivis, P. J. (2011). Slip flow in graphene nanochannels. *J. Chem. Phys.* **135**, 144701.

Kannam, S. K., Todd, B. D., Hansen, J. S., and Daivis, P. J. (2012a). Interfacial slip friction at a fluid–solid cylindrical boundary. *J. Chem. Phys.* **136**, 244704.

Kannam, S. K., Todd, B. D., Hansen, J. S., and Daivis, P. J. (2012b). Slip length of water on graphene: limitations of non-equilibrium molecular dynamics simulations. *J. Chem. Phys.* **136**, 024705.

Kannam, S. K., Todd, B. D., Hansen, J. S., and Daivis, P. J. (2013). How fast does water flow in carbon nanotubes? *J. Chem. Phys.* **138**, 094701.

Kapfer, S. C. and Krauth, W. (2015). Two-dimensional melting: from liquid–hexatic coexistence to continuous transitions. *Phys. Rev. Lett.* **114**, 035702.

Kapral, R. and Ciccotti, G. (1999). Mixed quantum–classical dynamics. *J. Chem. Phys.* **110**, 8919–8929.

Karasawa, N. and Goddard, W. A. (1989). Acceleration of convergence for lattice sums. *J. Phys. Chem.* **93**, 7320–7327.

Karayiannis, N. C., Mavrantzas, V. G., and Theodorou, D. N. (2002). A novel Monte Carlo scheme for the rapid equilibration of atomistic model polymer systems of precisely defined molecular architecture. *Phys. Rev. Lett.* **88**, 105503.

Karimi-Varzaneh, H. A., Qian, H.-J., Chen, X., Carbone, P., and Müller-Plathe, F. (2011). IBISCO: a molecular dynamics simulation package for coarse-grained simulation. *J. Comput. Chem.* **32**, 1475–1487.

Karma, A. (1993). Fluctuations in solidification. *Phys. Rev. E* **48**, 3441–3458.

Karney, C. F. F. (2007). Quaternions in molecular modeling. *J. Molec. Graphics Model.* **25**, 595–604.

Kawamoto, S., Nakamura, T., Nielsen, S. O., and Shinoda, W. (2013). A guiding potential method for evaluating the bending rigidity of tensionless lipid membranes from molecular simulation. *J. Chem. Phys.* **139**, 034108.

Kestemont, E. and Van Craen, J. (1976). On the computation of correlation functions in molecular dynamics experiments. *J. Comput. Phys.* **22**, 451–458.

Kikuchi, N., Pooley, C. M., Ryder, J. F., and Yeomans, J. M. (2003). Transport coefficients of a mesoscopic fluid dynamics model. *J. Chem. Phys.* **119**, 6388–6395.

Kikuchi, N., Ryder, J. F., Pooley, C. M., and Yeomans, J. M. (2005). Kinetics of the polymer collapse transition: the role of hydrodynamics. *Phys. Rev. E* **71**, 061804.

Kim, W. K. and Falk, M. L. (2014). A practical perspective on the implementation of hyperdynamics for accelerated simulation. *J. Chem. Phys.* **140**, 044107.

Kirk, D. B. and Hu, W. W. (2012). *Programming massively parallel processors: a hands-on approach.* 2nd ed. Morgan Kaufmann.

Kirkpatrick, S., Gelatt, C. D., and Vecchi, M. P. (1983). Optimization by simulated annealing. *Science* **220**, 671–680.

Kirkwood, J. G. (1933). Quantum statistics of almost classical assemblies. *Phys. Rev.* **44**, 31–37.

Klein, M. L. and Weis, J. J. (1977). Dynamical structure factor $S(q, \omega)$ of solid β-N_2. *J. Chem. Phys.* **67**, 217–224.

Kleinman, L. and Bylander, D. M. (1982). Efficacious form for model pseudopotentials. *Phys. Rev. Lett.* **48**, 1425–1428.

Knuth, D. (1973). *The art of computer programming.* 2nd ed. Reading MA: Addison-Wesley.

Knuth, D. E. (1997). *The art of computer programming: seminumerical algorithms.* 3rd ed. Vol. 2. Addison-Wesley.

Koehl, P. (2006). Electrostatics calculations: latest methodological advances. *Curr. Opin. Struct. Biol.* **16**, 142–151.

Kofke, D. A. and Bolhuis, P. G. (1999). Freezing of polydisperse hard spheres. *Phys. Rev. E* **59**, 618–622.

Kofke, D. A. and Cummings, P. T. (1998). Precision and accuracy of staged free-energy perturbation methods for computing the chemical potential by molecular simulation. *Fluid Phase Equilibria* **151**, 41–49.

Kofke, D. A. and Glandt, E. D. (1988). Monte Carlo simulation of multicomponent equilibria in a semigrand canonical ensemble. *Molec. Phys.* **64**, 1105–1131.

Kofke, D. A. (1993a). Direct evaluation of phase coexistence by molecular simulation via integration along the saturation line. *J. Chem. Phys.* **98**, 4149–4162.

Kofke, D. A. (1993b). Gibbs–Duhem integration: a new method for direct evaluation of phase coexistence by molecular simulation. *Molec. Phys.* **78**, 1331–1336.

Kofke, D. A. (2002). On the acceptance probability of replica-exchange Monte Carlo trials. *J. Chem. Phys.* **117**, 6911–6914. Erratum: *ibid.* **120,** 10852 (2004).

Kofke, D. A. (2004). Getting the most from molecular simulation. *Molec. Phys.* **102**, 405–420.

Kofke, D. A. (2005). Free energy methods in molecular simulation. *Fluid Phase Equilibria* **228-229**, 41–48.

Kofke, D. A. (2006). On the sampling requirements for exponential-work free-energy calculations. *Molec. Phys.* **104**, 3701–3708.

Kohn, W. and Sham, L. J. (1965). Self-consistent equations including exchange and correlation effects. *Phys. Rev. A* **140**, 1133–1138.

Kolafa, J. and Perram, J. W. (1992). Cutoff errors in the Ewald summation formulas for point-charge systems. *Molec. Simul.* **9**, 351–368.

Kolafa, J., Labík, S., and Malíjevsky, A. (2002). The bridge function of hard spheres by direct inversion of computer simulation data. *Molec. Phys.* **100**, 2629–2640.

Kone, A. and Kofke, D. A. (2005). Selection of temperature intervals for parallel-tempering simulations. *J. Chem. Phys.* **122**, 206101.

Kong, A., McCullagh, P., Meng, X.-L., Nicolae, D., and Tan, Z. (2003). A theory of statistical models for Monte Carlo integration. *J. Roy. Stat. Soc. Ser. B* **65**, 585–604.

Kony, D., Damm, W., Stoll, S., and van Gunsteren, W. F. (2002). An improved OPLS-AA force field for carbohydrates. *J. Comput. Chem.* **23**, 1416–1429.

Koper, G. J. M. and Reiss, H. (1996). Length scale for the constant pressure ensemble: application to small systems and relation to Einstein fluctuation theory. *J. Phys. Chem.* **100**, 422–432.

Kornfeld, H. (1924). Die Berechnung elektrostatischer Potentiale und der Energie von Dipol- und Quadrupolgittern. *Z. Phys.* **22**, 27–43.

Kosterlitz, J. M. and Thouless, D. J. (1973). Ordering, metastability and phase transitions in two-dimensional systems. *J. Phys. C* **6**, 1181–1203.

Kosztin, I., Faber, B., and Schulten, K. (1996). Introduction to the diffusion Monte Carlo method. *Amer. J. Phys.* **64**, 633–644.

Kouza, M. and Hansmann, U. H. E. (2011). Velocity scaling for optimizing replica exchange molecular dynamics. *J. Chem. Phys.* **134**, 044124.

Kranenburg, M. and Smit, B. (2004). Simulating the effect of alcohol on the structure of a membrane. *FEBS Lett.* **568**, 15–18.

Kranenburg, M., Venturoli, M., and Smit, B. (2003). Phase behavior and induced interdigitation in bilayers studied with dissipative particle dynamics. *J. Phys. Chem. B* **107**, 11491–11501.

Kratky, K. W. (1980). New boundary conditions for computer experiments of thermodynamic systems. *J. Comput. Phys.* **37**, 205–217.

Kratky, K. W. and Schreiner, W. (1982). Computational techniques for spherical boundary conditions. *J. Comput. Phys.* **47**, 313–320.

Kratzer, K., Arnold, A., and Allen, R. J. (2013). Automatic, optimized interface placement in forward flux sampling simulations. *J. Chem. Phys.* **138**, 164112.

Kratzer, K., Berryman, J. T., Taudt, A., Zeman, J., and Arnold, A. (2014). The flexible rare event sampling harness system (FRESHS). *Comput. Phys. Commun.* **185**, 1875–1885.

Kräutler, V. and Hünenberger, P. H. (2006). A multiple time step algorithm compatible with a large number of distance classes and an arbitrary distance dependence of the time step size for the fast evaluation of nonbonded interactions in molecular simulations. *J. Comput. Chem.* **27**, 1163–1176.

Kraynik, A. M. and Reinelt, D. A. (1992). Extensional motions of spatially periodic lattices. *Int. J. Multiphase Flow* **18**, 1045–1059.

Kremer, K. and Grest, G. S. (1990). Dynamics of entangled linear polymer melts: a molecular dynamics simulation. *J. Chem. Phys.* **92**, 5057–5086.

Kroese, D. P., Taimre, T., and Botev, Z. I. (2011). *Handbook of Monte Carlo methods.* Wiley series in probability and statistics. New York: Wiley.

Kubo, R. (1957). Statistical-mechanical theory of irreversible processes. I. General theory and simple applications to magnetic and conduction problems. *J. Phys. Soc. Japan* **12**, 570–586.

Kubo, R. (1966). The fluctuation–dissipation theorem. *Rep. Prog. Phys.* **29**, 255–284.

Kuharski, R. A. and Rossky, P. J. (1984). Quantum mechanical contributions to the structure of liquid water. *Chem. Phys. Lett.* **103**, 357–362.

Kulik, H. J., Schwegler, E., and Galli, G. (2012). Probing the structure of salt water under confinement with first-principles molecular dynamics and theoretical x-ray absorption spectroscopy. *J. Phys. Chem. Lett.* **3**, 2653–2658.

Kumar, S., Bouzida, D., Swendsen, R. H., Kollman, P. A., and Rosenberg, J. M. (1992). The weighted histogram analysis method for free-energy calculations on biomolecules. 1. The method. *J. Comput. Chem.* **13**, 1011–1021.

Kumar, V. and Errington, J. R. (2013). Wetting behavior of water near nonpolar surfaces. *J. Phys. Chem. C* **117**, 23017–23026.

Kunz, A.-P. E., Liu, H., and van Gunsteren, W. F. (2011). Enhanced sampling of particular degrees of freedom in molecular systems based on adiabatic decoupling and temperature or force scaling. *J. Chem. Phys.* **135**, 104106.

Kuo, I. F. W., Mundy, C. J., McGrath, M. J., and Siepmann, J. I. (2006). Time-dependent properties of liquid water: a comparison of Car–Parrinello and Born–Oppenheimer molecular dynamics simulations. *J. Chem. Theor. Comput.* **2**, 1274–1281.

Kurzak, J. and Pettitt, B. M. (2005). Massively parallel implementation of a fast multipole method for distributed memory machines. *J. Parallel Distr. Com.* **65**, 870–881.

Kurzak, J. and Pettitt, B. M. (2006). Fast multipole methods for particle dynamics. *Molec. Simul.* **32**, 775–790.

Kutzelnigg, W. (1997). The adiabatic approximation. 1. The physical background of the Born–Handy ansatz. *Molec. Phys.* **90**, 909–916.

Laasonen, K., Sprik, M., Parrinello, M., and Car, R. (1993). *Ab-initio* liquid water. *J. Chem. Phys.* **99**, 9080–9089.

Lachet, V., Boutin, A., Tavitian, B., and Fuchs, A. H. (1997). Grand canonical Monte Carlo simulations of adsorption of mixtures of xylene molecules in faujasite zeolites. *Faraday Disc. Chem. Soc.* **106**, 307–323.

Ladd, A. J. C. and Verberg, R. (2001). Lattice-Boltzmann simulations of particle–fluid suspensions. *J. Stat. Phys.* **104**, 1191–1251.

Ladd, A. J. C. (1984). Equations of motion for non-equilibrium molecular dynamics simulations of viscous flow in molecular fluids. *Molec. Phys.* **53**, 459–463.

Ladd, A. J. C. (1994a). Numerical simulations of particulate suspensions via a discretized Boltzmann equation. Part 1. Theoretical foundation. *J. Fluid Mech.* **271**, 285–309.

Ladd, A. J. C. (1994b). Numerical simulations of particulate suspensions via a discretized Boltzmann equation. Part 2. Numerical results. *J. Fluid Mech.* **271**, 311–339.

Lado, F. (1971). Numerical Fourier transforms in one, two, and three dimensions for liquid state calculations. *J. Comput. Phys.* **8**, 417–433.

Laio, A. and Parrinello, M. (2002). Escaping free-energy minima. *Proc. Nat. Acad. Sci.* **99**, 12562–12566.

Laio, A., VandeVondele, J., and Röthlisberger, U. (2002a). A Hamiltonian electrostatic coupling scheme for hybrid Car–Parrinello molecular dynamics simulations. *J. Chem. Phys.* **116**, 6941–6947.

Laio, A., VandeVondele, J., and Röthlisberger, U. (2002b). D-RESP: Dynamically generated electrostatic potential derived charges from quantum mechanics/molecular mechanics simulations. *J. Phys. Chem. B* **106**, 7300–7307.

Laio, A., Gervasio, F. L., VandeVondele, J., Sulpizi, M., and Röthlisberger, U. (2004). A variational definition of electrostatic potential derived charges. *J. Phys. Chem. B* **108**, 7963–7968.

Laio, A. and Gervasio, F. L. (2008). Metadynamics: a method to simulate rare events and reconstruct the free energy in biophysics, chemistry and material science. *Rep. Prog. Phys.* **71**, 126601.

Laird, B. B. and Davidchack, R. L. (2005). Direct calculation of the crystal–melt interfacial free energy via molecular dynamics computer simulation. *J. Phys. Chem. B* **109**, 17802–17812.

Lal, M. and Spencer, D. (1971). Monte Carlo computer simulation of chain molecules. III. Simulation of n-alkane molecules. *Molec. Phys.* **22**, 649–659.

Lamoureux, G. and Roux, B. (2003). Modeling induced polarization with classical Drude oscillators: theory and molecular dynamics simulation algorithm. *J. Chem. Phys.* **119**, 3025–3039.

Landau, D. P. and Binder, K. (2009). *A guide to Monte Carlo simulations in statistical physics.* 3rd ed. Cambridge: Cambridge Press.

Landau, L. D. and Lifshitz, E. M. (1958). *Statistical physics.* 1st ed. Course of theoretical physics. Vol. 5. Translated by E. Peierls and R. F. Peierls. London: Pergamon Press.

Landau, L. D. and Lifshitz, E. M. (1980). *Statistical physics.* 3rd ed. Course of theoretical physics. Vol. 5. Oxford: Pergamon Press. Revised E. M. Lifshitz and L. P. Pitaevskii.

Lane, T. J., Shukla, D., Beauchamp, K. A., and Pande, V. S. (2013). To milliseconds and beyond: challenges in the simulation of protein folding. *Curr. Opin. Struct. Biol.* **23**, 58–65.

Larini, L., Lu, L., and Voth, G. A. (2010). The multiscale coarse-graining method. VI. Implementation of three-body coarse-grained potentials. *J. Chem. Phys.* **132**, 164107.

Larsson, P., Hess, B., and Lindahl, E. (2011). Algorithm improvements for molecular dynamics simulations. *Comp. Mol. Sci. (Wiley Interdisc. Rev.)* **1**, 93–108.

Lau, G. V., Ford, I. J., Hunt, P. A., Mueller, E. A., and Jackson, G. (2015). Surface thermodynamics of planar, cylindrical, and spherical vapour–liquid interfaces of water. *J. Chem. Phys.* **142**, 114701.

Lebowitz, J. L., Percus, J. K., and Verlet, L. (1967). Ensemble dependence of fluctuations with application to machine computations. *Phys. Rev.* **153**, 250–254.

Lechner, W. and Dellago, C. (2008). Accurate determination of crystal structures based on averaged local bond order parameters. *J. Chem. Phys.* **129**, 114707.

Lechner, W., Dellago, C., and Bolhuis, P. G. (2011). Reaction coordinates for the crystal nucleation of colloidal suspensions extracted from the reweighted path ensemble. *J. Chem. Phys.* **135**, 154110.

L'Ecuyer, P. (1996). Combined multiple recursive random number generators. *Operations Res.* **44**, 816–822.

L'Ecuyer, P. (1999). Good parameters and implementations for combined multiple recursive random number generators. *Operations Res.* **47**, 159–164.

L'Ecuyer, P. (2007). Random number generation. *Handbook of simulation.* Chap. 4, 93–137. Ed. by J. Banks. Wiley.

L'Ecuyer, P. and Simard, R. (2007). TestU01: a C library for empirical testing of random number generators. ACM *Trans. Math. Softw.* **33**, 22.

Lee, C. T., Yang, W. T., and Parr, R. G. (1988). Development of the Colle–Salvetti correlation-energy formula into a functional of the electron density. *Phys. Rev. B* **37**, 785–789.

Lee, C. Y. and Scott, H. L. (1980). The surface tension of water: a Monte Carlo calculation using an umbrella sampling algorithm. *J. Chem. Phys.* **73**, 4591–4596.

Lee, H. S. and Tuckerman, M. E. (2006). *Ab initio* molecular dynamics with discrete variable representation basis sets: techniques and application to liquid water. *J. Phys. Chem. A* **110**, 5549–5560.

Lee, H. S. and Tuckerman, M. E. (2007). Dynamical properties of liquid water from *ab initio* molecular dynamics performed in the complete basis set limit. *J. Chem. Phys.* **126**, 164501.

Lee, J. (1993). New Monte Carlo algorithm: entropic sampling. *Phys. Rev. Lett.* **71**, 211–214.

Lee, J. K., Barker, J. A., and Pound, G. M. (1974). Surface structure and surface tension: perturbation theory and Monte Carlo calculation. *J. Chem. Phys.* **60**, 1976–1980.

Lee, S.-H. (2012). Molecular dynamics simulation of a small drop of liquid argon. *Bull. Korean Chem. Soc.* **33**, 3805–3809.

Lee, S.-H., Palmo, K., and Krimm, S. (2005). WIGGLE: a new constrained molecular dynamics algorithm in Cartesian coordinates. *J. Comput. Phys.* **210**, 171–182.

Lee, W.-G., Gallardo-Macias, R., Frey, K. M., Spasov, K. A., Bollini, M., Anderson, K. S., and Jorgensen, W. L. (2013). Picomolar inhibitors of HIV reverse transcriptase featuring bicyclic replacement of a cyanovinylphenyl group. *J. Am. Chem. Soc.* **135**, 16705–16713.

Lees, A. W. and Edwards, S. F. (1972). The computer study of transport processes under extreme conditions. *J. Phys. C* **5**, 1921–1929.

Legoll, F. and Monneau, R. (2002). Designing reversible measure invariant algorithms with applications to molecular dynamics. *J. Chem. Phys.* **117**, 10452–10464.

Leimkuhler, B. J. and Reich, S. (2004). *Simulating Hamiltonian dynamics.* Cambridge: Cambridge University Press.

Leimkuhler, B. J. and Skeel, R. D. (1994). Symplectic numerical integrators in constrained Hamiltonian systems. *J. Comput. Phys.* **112**, 117–125.

Leimkuhler, B. and Shang, X. (2015). On the numerical treatment of dissipative particle dynamics and related systems. *J. Comput. Phys.* **280**, 72–95.

Leimkuhler, B. J. and Matthews, C. (2013a). Rational construction of stochastic numerical methods for molecular sampling. *Appl. Math. Res. eXpress* **2013**, 34–56.

Leimkuhler, B. J. and Matthews, C. (2013b). Robust and efficient configurational molecular sampling via Langevin dynamics. *J. Chem. Phys.* **138**, 174102.

Leimkuhler, B. J. and Sweet, C. R. (2004). The canonical ensemble via symplectic integrators using Nosé and Nosé–Poincaré chains. *J. Chem. Phys.* **121**, 108–116.

Leimkuhler, B. J. and Sweet, C. R. (2005). A Hamiltonian formulation for recursive multiple thermostats in a common timescale. *SIAM J. Appl. Dyn. Syst.* **4**, 187–216.

Lekkerkerker, H. N. W and Tuinier, R. (2011). *Colloids and the depletion interaction.* Lect. Notes Phys. Vol. 833. Berlin: Springer-Verlag.

Lekner, J. (1989). Summation of dipolar fields in simulated liquid–vapour interfaces. *Physica A* **157**, 826–838.

Lekner, J. (1991). Summation of Coulomb fields in computer-simulated disordered systems. *Physica A* **176**, 485–498.

Lennard-Jones, J. E. and Devonshire, A. F. (1939a). Critical and cooperative phenomena. III. A theory of melting and the structure of liquids. *Proc. Roy. Soc. Lond. A* **169**, 317–338.

Lennard-Jones, J. E. and Devonshire, A. F. (1939b). Critical and cooperative phenomena. IV. A theory of disorder in solids and liquids and the process of melting. *Proc. Roy. Soc. Lond. A* **170**, 464–484.

LeSar, R. (1984). Improved electron-gas model calculations of solid N_2 to 10 GPa. *J. Chem. Phys.* **81**, 5104–5108.

LeSar, R. and Gordon, R. G. (1982). Density-functional theory for the solid alkali cyanides. *J. Chem. Phys.* **77**, 3682–3692.

LeSar, R. and Gordon, R. G. (1983). Density-functional theory for solid nitrogen and carbon dioxide at high pressure. *J. Chem. Phys.* **78**, 4991–4996.

Levesque, D., Verlet, L., and Kürkijarvi, J. (1973). Computer 'experiments' on classical fluids. IV. Transport properties and time-correlation functions of the Lennard-Jones liquid near its triple point. *Phys. Rev. A* **7**, 1690–1700.

Levine, Z. A., Venable, R. M., Watson, M. C., Lerner, M. G., Shea, J.-E., Pastor, R. W., and Brown, F. L. H. (2014). Determination of biomembrane bending moduli in fully atomistic simulations. *J. Am. Chem. Soc.* **136**, 13582–13585.

Li, D., Liu, L., Yu, H., Zhai, Z., Zhang, Y., Guo, B., Yang, C., and Liu, B. (2014). A molecular simulation study of the protection of insulin bioactive structure by trehalose. *J. Molec. Model.* **20**, 2496.

Li, J., Zhou, Z., and Sadus, R. J. (2007a). Role of nonadditive forces on the structure and properties of liquid water. *J. Chem. Phys.* **127**, 154509.

Li, X., Latour, R. A., and Stuart, S. J. (2009). TIGER2: an improved algorithm for temperature intervals with global exchange of replicas. *J. Chem. Phys.* **130**, 174106.

Li, X., O'Brien, C. P., Collier, G., Vellore, N. A., Wang, F., Latour, R. A., Bruce, D. A., and Stuart, S. J. (2007b). An improved replica-exchange sampling method: temperature intervals with global energy reassignment. *J. Chem. Phys.* **127**, 164116.

Liem, S. Y., Brown, D., and Clarke, J. H. R. (1991). Molecular dynamics simulations on distributed memory machines. *Comput. Phys. Commun.* **67**, 261–267.

Lifson, S. and Warshel, A. (1968). Consistent force field for calculations of conformations vibrational spectra and enthalpies of cycloalkane and n-alkane molecules. *J. Chem. Phys.* **49**, 5116–5129.

Lin, I.-C. and Tuckerman, M. E. (2010). Enhanced conformational sampling of peptides via reduced side-chain and solvent masses. *J. Phys. Chem. B* **114**, 15935–15940.

Lin, I.-C., Seitsonen, A. P., Tavernelli, I., and Röthlisberger, U. (2012). Structure and dynamics of liquid water from *ab initio* molecular dynamics. Comparison of BLYP, PBE, and revPBE density functionals with and without van der Waals corrections. *J. Chem. Theor. Comput.* **8**, 3902–3910.

Lin, Y., Baumketner, A., Deng, S., Xu, Z., Jacobs, D., and Cai, W. (2009). An image-based reaction field method for electrostatic interactions in molecular dynamics simulations of aqueous solutions. *J. Chem. Phys.* **131**, 154103.

Lin, Z. and van Gunsteren, W. F. (2015). On the use of a weak-coupling thermostat in replica-exchange molecular dynamics simulations. *J. Chem. Phys.* **143**, 034110.

Lindahl, E. and Edholm, O. (2000). Mesoscopic undulations and thickness fluctuations in lipid bilayers from molecular dynamics simulations. *Biophys. J.* **79**, 426–433.

Lindan, P. J. D. (1995). Dynamics with the shell model. *Molec. Simul.* **14**, 303–312.

Lindbo, D. and Tornberg, A.-K. (2012). Fast and spectrally accurate Ewald summation for 2-periodic electrostatic systems. *J. Chem. Phys.* **136**, 164111.

Lindorff-Larsen, K., Piana, S., Dror, R. O., and Shaw, D. E. (2011). How fast-folding proteins fold. *Science* **334**, 517–520.

Liphardt, J., Dumont, S., Smith, S. B., Tinoco, I., and Bustamante, C. (2002). Equilibrium information from nonequilibrium measurements in an experimental test of Jarzynski's equality. *Science* **296**, 1832–1835.

Lisal, M., Brennan, J. K., and Smith, W. R. (2006). Mesoscale simulation of polymer reaction equilibrium: combining dissipative particle dynamics with reaction ensemble Monte Carlo. I. Polydispersed polymer systems. *J. Chem. Phys.* **125**, 164905.

Lisal, M., Brennan, J. K., and Avalos, J. B. (2011). Dissipative particle dynamics at isothermal, isobaric, isoenergetic, and isoenthalpic conditions using Shardlow-like splitting algorithms. *J. Chem. Phys.* **135**, 204105.

Liu, J., Davidchack, R. L., and Dong, H. B. (2013). Molecular dynamics calculation of solid–liquid interfacial free energy and its anisotropy during iron solidification. *Comput. Mater. Sci.* **74**, 92–100.

Liu, M. B., Liu, G. R., Zhou, L. W., and Chang, J. Z. (2014). Dissipative particle dynamics (DPD): an overview and recent developments. *Arch. Comp. Meth. Eng.* **22**, 529–556.

Liu, P., Kim, B., Friesner, R. A., and Berne, B. J. (2005). Replica exchange with solute tempering: a method for sampling biological systems in explicit water. *Proc. Nat. Acad. Sci.* **102**, 13749–13754.

Lo, C. M. and Palmer, B. (1995). Alternative Hamiltonian for molecular dynamics simulations in the grand canonical ensemble. *J. Chem. Phys.* **102**, 925–931.

Lobanova, O., Avendano, C., Lafitte, T., Müller, E. A., and Jackson, G. (2015). SAFT-γ force field for the simulation of molecular fluids. 4. A single-site coarse-grained model of water applicable over a wide temperature range. *Molec. Phys.* **113**, 1228–1249.

Lobanova, O., Mejía, A., Jackson, G., and Müller, E. A. (2016). SAFT-γ force field for the simulation of molecular fluids. 6. Binary and ternary mixtures comprising water, carbon dioxide, and n-alkanes. *J. Chem. Thermo.* **93**, 320–336.

Loeffler, T. D., Sepehri, A., and Chen, B. (2015). Improved Monte Carlo scheme for efficient particle transfer in heterogeneous systems in the grand canonical ensemble: application to vapor–liquid nucleation. *J. Chem. Theor. Comput.* **11**, 4023–4032.

Lopes, J. N. C. and Tildesley, D. J. (1997). Multiphase equilibria using the Gibbs ensemble Monte Carlo method. *Molec. Phys.* **92**, 187–195.

Lopes, J. N. C, Deschamps, J., and Padua, A. A. H. (2004). Modeling ionic liquids using a systematic all-atom force field. *J. Phys. Chem. B* **108**, 2038–2047.

López-Lemus, J. and Alejandre, J. (2002). Thermodynamic and transport properties of simple fluids using lattice sums: bulk phases and liquid–vapour interface. *Molec. Phys.* **100**, 2983–2992.

Louis, A. A. (2002). Beware of density dependent pair potentials. *J. Phys. Cond. Mat.* **14**, 9187–9206.

Lowden, L. J. and Chandler, D. (1974). Theory of intermolecular pair correlations for molecular liquids. Applications to liquid carbon disulfide, carbon diselenide and benzene. *J. Chem. Phys.* **61**, 5228–5241.

Lowe, C. P. (1999). An alternative approach to dissipative particle dynamics. *Europhys. Lett.* **47**, 145–151.

Lu, N. D. and Kofke, D. A. (2001a). Accuracy of free-energy perturbation calculations in molecular simulation. I. Modeling. *J. Chem. Phys.* **114**, 7303–7311.

Lu, N. D. and Kofke, D. A. (2001b). Accuracy of free-energy perturbation calculations in molecular simulation. II. Heuristics. *J. Chem. Phys.* **115**, 6866–6875.

Lu, N. D., Singh, J. K., and Kofke, D. A. (2003). Appropriate methods to combine forward and reverse free-energy perturbation averages. *J. Chem. Phys.* **118**, 2977–2984.

Luckhurst, G. R. (2006). The Gay–Berne mesogen: a paradigm shift? *Liq. Cryst.* **33**, 1389–1395.

Luckhurst, G. R. and Simpson, P. (1982). Computer simulation studies of anisotropic systems. VIII. The Lebwohl–Lasher model of nematogens revisited. *Molec. Phys.* **47**, 251–265.

Luckhurst, G. R., Stephens, R. A., and Phippen, R. W. (1990). Computer simulation studies of anisotropic systems. 19. Mesophases formed by the Gay–Berne model mesogen. *Liq. Cryst.* **8**, 451–464.

Luckhurst, G. R. and Satoh, K. (2010). The director and molecular dynamics of the field-induced alignment of a Gay–Berne nematic phase: an isothermal–isobaric nonequilibrium molecular dynamics simulation study. *J. Chem. Phys.* **132**, 184903.

Lucy, L. B. (1977). Numerical approach to testing of fission hypothesis. *Astron. J.* **82**, 1013–1024.

Luehr, N., Markland, T. E., and Martínez, T. J. (2014). Multiple time step integrators in *ab initio* molecular dynamics. *J. Chem. Phys.* **140**, 084116.

Lundborg, M., Apostolov, R., Spångberg, D., Gärdenäs, A., van der Spoel, D., and Lindahl, E. (2013). An efficient and extensible format, library, and API for binary trajectory data from molecular simulations. *J. Comput. Chem.* **35**, 260–269.

Lustig, R. (1994a). Statistical thermodynamics in the classical molecular dynamics ensemble. 1. Fundamentals. *J. Chem. Phys.* **100**, 3048–3059.

Lustig, R. (1994b). Statistical thermodynamics in the classical molecular dynamics ensemble. 2. Application to computer simulation. *J. Chem. Phys.* **100**, 3060–3067.

Lustig, R. (1998). Microcanonical Monte Carlo simulation of thermodynamic properties. *J. Chem. Phys.* **109**, 8816–8828.

Lustig, R. (2012). Statistical analogues for fundamental equation of state derivatives. *Molec. Phys.* **110**, 3041–3052.

Luttinger, J. M. (1964). Theory of thermal transport coefficients. *Phys. Rev.* **135**, 1505–1514.

Lyklema, J. W. (1979a). Computer simulations of a rough sphere fluid I. *Physica A* **96**, 573–593.

Lyklema, J. W. (1979b). Computer simulations of a rough sphere fluid II. Comparison with stochastic models. *Physica A* **96**, 594–605.

Lynch, G. C. and Pettitt, B. M. (1997). Grand canonical ensemble molecular dynamics simulations: reformulation of extended system dynamics approaches. *J. Chem. Phys.* **107**, 8594–8610.

Lynden-Bell, R. M. (1995). Landau free energy, Landau entropy, phase transitions and limits of metastability in an analytical model with a variable number of degrees of freedom. *Molec. Phys.* **86**, 1353–1373.

Lynden-Bell, R. M. (2010). Towards understanding water: simulation of modified water models. *J. Phys. Cond. Mat.* **22**, 284107.

Lyubartsev, A. P. and Laaksonen, A. (1995). Calculation of effective interaction potentials from radial distribution functions: a reverse Monte-Carlo approach. *Phys. Rev. E* **52**, 3730–3737.

Lyubartsev, A., Mirzoev, A., Chen, L., and Laaksonen, A. (2010). Systematic coarse-graining of molecular models by the Newton inversion method. *Faraday Disc. Chem. Soc.* **144**, 43–56.

MacDowell, L. G. and Blas, F. J. (2009). Surface tension of fully flexible Lennard-Jones chains: role of long-range corrections. *J. Chem. Phys.* **131**, 074705.

MacDowell, L. G., Shen, V. K., and Errington, J. R. (2006). Nucleation and cavitation of spherical, cylindrical, and slablike droplets and bubbles in small systems. *J. Chem. Phys.* **125**, 034705.

Mackerell, A. D., Bashford, D., Bellott, M., Dunbrack, R. L., Evanseck, J. D., Field, M. J., Fischer, S., Gao, J., Guo, H., Ha, S., Joseph-McCarthy, D., Kuchnir, L., Kuczera, K., Lau, F. T. K., Mattos, C., Michnick, S., Ngo, T., Nguyen, D. T., Prodhom, B., Reiher, W. E., Roux, B., Schlenkrich, M., Smith, J. C., Stote, R., Straub, J., Watanabe, M., Wiorkiewicz-Kuczera, J., Yin, D., and Karplus, M. (1998). All-atom empirical potential for molecular modeling and dynamics studies of proteins. *J. Phys. Chem. B* **102**, 3586–3616.

Madden, P. A. and Kivelson, D. (1984). A consistent molecular treatment of dielectric phenomena. *Adv. Chem. Phys.* **56**, 467–566.

Madelung, E. (1918). Das elektrische Feld in Systemen von regelmässig angeordneten Punktladungen. *Physik. Z.* **19**, 524–532.

Madurga, S., Rey-Castro, C., Pastor, I., Vilaseca, E., David, C., Lluis Garces, J., Puy, J., and Mas, F. (2011). A semi-grand canonical Monte Carlo simulation model for ion binding to ionizable surfaces: proton binding of carboxylated latex particles as a case study. *J. Chem. Phys.* **135**, 184103.

Maerzke, K., Gai, L., Cummings, P. T., and McCabe, C. (2012). Incorporating configurational-bias Monte Carlo into the Wang–Landau algorithm for continuous molecular systems. *J. Chem. Phys.* **137**, 204105.

Maggs, A. C. and Rossetto, V. (2002). Local simulation algorithms for Coulomb interactions. *Phys. Rev. Lett.* **88**, 196402.

Mahoney, M. W. and Jorgensen, W. L. (2000). A five-site model for liquid water and the reproduction of the density anomaly by rigid, nonpolarizable potential functions. *J. Chem. Phys.* **112**, 8910–8922.

Maitland, G. C., Rigby, M., Smith, E. B., and Wakeham, W. A. (1981). *Intermolecular forces: their origin and determination.* Oxford: Clarendon Press.

Malani, A., Auerbach, S. M., and Monson, P. A. (2010). Probing the mechanism of silica polymerization at ambient temperatures using Monte Carlo simulations. *J. Phys. Chem. Lett.* **1**, 3219–3224.

Malevanets, A. and Kapral, R. (1999). Mesoscopic model for solvent dynamics. *J. Chem. Phys.* **110**, 8605–8613.

Malevanets, A. and Kapral, R. (2000). Solute molecular dynamics in a mesoscale solvent. *J. Chem. Phys.* **112**, 7260–7269.

Malijevský, A. and Jackson, G. (2012). A perspective on the interfacial properties of nanoscopic liquid drops. *J. Phys. Cond. Mat.* **24**, 464121.

Malins, A., Williams, S. R., Eggers, J., and Royall, C. P. (2013). Identification of structure in condensed matter with the topological cluster classification. *J. Chem. Phys.* **139**, 234506.

Mandell, M. J. (1976). Properties of a periodic fluid. *J. Stat. Phys.* **15**, 299–305.

Marais, P., Kenwood, J., Smith, K. C., Kuttel, M. M., and Gain, J. (2012). Efficient compression of molecular dynamics trajectory files. *J. Comput. Chem.* **33**, 2131–2141.

Marcelli, G. and Sadus, R. J. (2012). Molecular simulation of the phase behavior of noble gases using accurate two-body and three-body intermolecular potentials. *J. Chem. Phys.* **111**, 1533–1540.

Marconi, U. M. B., Puglisi, A., Rondoni, L., and Vulpiani, A. (2008). Fluctuation–dissipation: response theory in statistical physics. *Physics Reports* **461**, 111–195.

Maréchal, G. and Ryckaert, J. P. (1983). Atomic vs molecular description of transport properties in polyatomic fluids: n-butane as an example. *Chem. Phys. Lett.* **101**, 548–554.

Markland, T. E. and Manolopoulos, D. E. (2008a). A refined ring polymer contraction scheme for systems with electrostatic interactions. *Chem. Phys. Lett.* **464**, 256–261.

Markland, T. E. and Manolopoulos, D. E. (2008b). An efficient ring polymer contraction scheme for imaginary time path integral simulations. *J. Chem. Phys.* **129**, 024105.

Marques, M. A. L. and Gross, E. K. U. (2004). Time-dependent density functional theory. *Ann. Rev. Phys. Chem.* **55**, 427–455.

Marrink, S. J., de Vries, A. H., and Mark, A. E. (2004). Coarse grained model for semiquantitative lipid simulations. *J. Phys. Chem. B* **108**, 750–760.

Marrink, S. J., Risselada, H. J., Yefimov, S., Tieleman, D. P., and de Vries, A. H. (2007). The MARTINI force field: coarse grained model for biomolecular simulations. *J. Phys. Chem. B* **111**, 7812–7824.

Marsaglia, G. (1972). Choosing a point from the surface of a sphere. *Ann. Math. Stat.* **43**, 645–646.

Marsaglia, G. (1995). *The Marsaglia random number CDROM including the diehard battery of tests of randomness.* http://stat.fsu.edu/pub/diehard/ (Accessed Nov 2015).

Marsh, C. A. and Yeomans, J. M. (1997). Dissipative particle dynamics: the equilibrium for finite time steps. *Europhys. Lett.* **37**, 511–516.

Marsh, D. (2007). Lateral pressure profile, spontaneous curvature frustration, and the incorporation and conformation of proteins in membranes. *Biophys. J.* **93**, 3884–3899.

Martin, M. G. (2013). MCCCS Towhee: a tool for Monte Carlo molecular simulation. *Molec. Simul.* **39**, 1184–1194.

Martin, M. G. and Siepmann, J. I. (1999). Novel configurational-bias Monte Carlo method for branched molecules. Transferable potentials for phase equilibria. 2. United-atom description of branched alkanes. *J. Phys. Chem. B* **103**, 4508–4517.

Martin, R. (2008). *Electronic structure.* New York: Cambridge University Press.

Martínez-Ruiz, F. J., Moreno-Ventas Bravo, A. I., and Blas, F. J. (2015). Liquid–liquid interfacial properties of a symmetrical Lennard-Jones binary mixture. *J. Chem. Phys.* **143**, 104706.

Martyna, G. J., Klein, M. L., and Tuckerman, M. E. (1992). Nosé–Hoover chains: the canonical ensemble via continuous dynamics. *J. Chem. Phys.* **97**, 2635–2643.

Martyna, G. J., Tobias, D. J., and Klein, M. L. (1994). Constant-pressure molecular dynamics algorithms. *J. Chem. Phys.* **101**, 4177–4189.

Martyna, G. J., Tuckerman, M. E., Tobias, D. J., and Klein, M. L. (1996). Explicit reversible integrators for extended systems dynamics. *Molec. Phys.* **87**, 1117–1157.

Martyna, G. J., Hughes, A., and Tuckerman, M. E. (1999). Molecular dynamics algorithms for path integrals at constant pressure. *J. Chem. Phys.* **110**, 3275–3290.

Martyna, G. J. (1994). Remarks on 'Constant-temperature molecular dynamics with momentum conservation'. *Phys. Rev. E* **50**, 3234–3236.

Marx, D. and Hutter, J. (2012). Ab initio *molecular dynamics, basic theory and advanced methods.* New York: Cambridge University Press.

Marx, D., Tuckerman, M. E., Hutter, J., and Parrinello, M. (1999). The nature of the hydrated excess proton in water. *Nature* **397**, 601–604.

Maseras, F. and Morokuma, K. (1995). IMOMM: a new integrated *ab-initio* plus molecular mechanics geometry optimization scheme of equilibrium structures and transition states. *J. Comput. Chem.* **16**, 1170–1179.

Massopust, P. (2010). *Interpolation and approximation with splines and fractals.* New York: Oxford University Press.

Matin, M. L., Daivis, P. J., and Todd, B. D. (2003). Cell neighbor list method for planar elongational flow: rheology of a diatomic fluid. *Comput. Phys. Commun.* **151**, 35–46.

Matsumoto, M. and Nishimura, T. (1998). Mersenne twister: a 623-dimensionally equidistributed uniform pseudo-random number generator. *ACM Trans. Model. Comp. Simul.* **8**, 3–30.

Mattson, W. and Rice, B. M. (1999). Near-neighbor calculations using a modified cell-linked list method. *Comput. Phys. Commun.* **119**, 135–148.

Matubayasi, N. and Nakahara, M. (1999). Reversible molecular dynamics for rigid bodies and hybrid Monte Carlo. *J. Chem. Phys.* **110**, 3291–3301.

Mavrantzas, V. G., Boone, T. D., Zervopoulou, E., and Theodorou, D. N. (1999). End-bridging Monte Carlo: a fast algorithm for atomistic simulation of condensed phases of long polymer chains. *Macromolecules* **32**, 5072–5096.

Mazars, M. (2005). Lekner summations and Ewald summations for quasi-two-dimensional systems. *Molec. Phys.* **103**, 1241–1260.

Mazur, P. (1982). On the motion and Brownian motion of N spheres in a viscous fluid. *Physica A* **110**, 128–146.

Mazur, P. and van Saarloos, W. (1982). Many-sphere hydrodynamic interactions and mobilities in a suspension. *Physica A* **115**, 21–57.

Mazzeo, M. D., Ricci, M., and Zannoni, C. (2010). The linked neighbour list (LNL) method for fast off-lattice Monte Carlo simulations of fluids. *Comput. Phys. Commun.* **181**, 569–581.

McCabe, C. and Galindo, A. (2010). SAFT Associating Fluids and Fluid Mixtures. *Applied Thermodynamics of Fluids*. Chap. 8, 215–279. Ed. by A. R. Goodwin, J. Sengers, and C. J. Peters. Royal Society of Chemistry.

McCoustra, M., Bain, C., Buch, V., Finney, J., Hansen, J.-P., Held, G., Russell, A., and Wheatley, R., eds. (2009). *Water: from interfaces to the bulk*. Faraday Disc. Chem. Soc. Vol. 141. Royal Society of Chemistry, 1–488.

McDaniel, J. G. and Schmidt, J. R. (2013). Physically-motivated force fields from symmetry-adapted perturbation theory. *J. Phys. Chem. A* **117**, 2053–2066.

McDonald, I. R. (1969). Monte Carlo calculations for one- and two-component fluids in the isothermal–isobaric ensemble. *Chem. Phys. Lett.* **3**, 241–243.

McDonald, I. R. (1972). NpT-ensemble Monte Carlo calculations for binary liquid mixtures. *Molec. Phys.* **23**, 41–58.

McDonald, I. R. (1986). Molecular liquids: orientational order and dielectric properties. *Molecular dynamics simulation of statistical mechanical systems*. Proc. Int. School Phys. Enrico Fermi. **97**, 341–370. (Varenna, Italy, 1985). Ed. by G. Ciccotti and W. G. Hoover. Bologna: Soc. Italiana di Fisica.

McGrath, M. J., Ghogomu, J. N., Mundy, C. J., Kuo, I.-F. W., and Siepmann, J. I. (2010). First principles Monte Carlo simulations of aggregation in the vapor phase of hydrogen fluoride. *Phys. Chem. Chem. Phys.* **12**, 7678–7687.

McMillan, W. L. (1965). Ground state of liquid He^4. *Phys. Rev.* **138**, 442–451.

McNeil, W. J. and Madden, W. G. (1982). A new method for the molecular dynamics simulation of hard core molecules. *J. Chem. Phys.* **76**, 6221–6226.

McNeil-Watson, G. C. and Wilding, N. B. (2006). Freezing line of the Lennard-Jones fluid: a phase switch Monte Carlo study. *J. Chem. Phys.* **124**, 064504.

McQuarrie, D. A. (1976). *Statistical mechanics*. New York: Harper and Row.

McTague, J. P., Frenkel, D., and Allen, M. P. (1980). Simulation studies of the 2D melting mechanism. *Ordering in two dimensions*, 147–153. Ed. by S. K. Sinha. Amsterdam: North Holland.

Mecke, K. R. and Dietrich, S. (1999). Effective Hamiltonian for liquid–vapor interfaces. *Phys. Rev. E* **59**, 6766–6784.

Mehrotra, P. K., Mezei, M., and Beveridge, D. L. (1983). Convergence acceleration in Monte Carlo computer simulation on water and aqueous solutions. *J. Chem. Phys.* **78**, 3156–3166.

Mehta, M. and Kofke, D. A. (1994). Coexistence diagrams of mixtures by molecular simulation. *Chem. Eng. Sci.* **49**, 2633–2645.

Melchionna, S. (2007). Design of quasisymplectic propagators for Langevin dynamics. *J. Chem. Phys.* **127**, 044108.

Melchionna, S., Ciccotti, G., and Holian, B. L. (1993). Hoover NPT dynamics for systems varying in shape and size. *Molec. Phys.* **78**, 533–544.

Mentch, F. and Anderson, J. B. (1981). Quantum chemistry by random walk: importance sampling for H_3^+. *J. Chem. Phys.* **74**, 6307–6311.

Menzeleev, A. R., Ananth, N., and Miller, T. F. I. (2011). Direct simulation of electron transfer using ring polymer molecular dynamics: comparison with semiclassical instanton theory and exact quantum methods. *J. Chem. Phys.* **135**, 074106.

Mertz, J. E., Tobias, D. J., Brooks, C. L., and Singh, U. C. (1991). Vector and parallel algorithms for the molecular dynamics simulation of macromolecules on shared-memory computers. *J. Comput. Chem.* **12**, 1270–1277.

Mester, Z. and Panagiotopoulos, A. Z. (2015). Mean ionic activity coefficients in aqueous NaCl solutions from molecular dynamics simulations. *J. Chem. Phys.* **142**, 044507.

Metropolis, N. and Ulam, S. (1949). The Monte Carlo method. *J. Amer. Stat. Ass.* **44**, 335–341.

Metropolis, N., Rosenbluth, A. W., Rosenbluth, M. N., Teller, A. H., and Teller, E. (1953). Equation of state calculations by fast computing machines. *J. Chem. Phys.* **21**, 1087–1092.

Mezei, M. (1980). A cavity-biased $(TV\mu)$ Monte Carlo method for the computer simulation of fluids. *Molec. Phys.* **40**, 901–906.

Mezei, M. (1983). Virial-bias Monte Carlo methods. *Molec. Phys.* **48**, 1075–1082.

Miao, Y. L., Feixas, F., Eun, C., and McCammon, J. A. (2015). Accelerated molecular dynamics simulations of protein folding. *J. Comput. Chem.* **36**, 1536–1549.

Michielssens, S., van Erp, T. S., Kutzner, C., Ceulemans, A., and de Groot, B. L. (2012). Molecular dynamics in principal component space. *J. Phys. Chem. B* **116**, 8350–8354.

Míguez, J. M., González-Salgado, D., Legido, J. L., and Piñeiro, M. M. (2010). Calculation of interfacial properties using molecular simulation with the reaction field method: results for different water models. *J. Chem. Phys.* **132**, 184102.

Míguez, J. M., Piñeiro, M. M., and Blas, F. J. (2013). Influence of the long-range corrections on the interfacial properties of molecular models using Monte Carlo simulation. *J. Chem. Phys.* **138**, 034707.

Miller, T. F. (2008). Isomorphic classical molecular dynamics model for an excess electron in a supercritical fluid. *J. Chem. Phys.* **129**, 194502.

Miller, T. F. and Manolopoulos, D. E. (2005). Quantum diffusion in liquid water from ring polymer molecular dynamics. *J. Chem. Phys.* **123**, 154504.

Miller, T. F., Eleftheriou, M., Pattnaik, P., Ndirango, A., Newns, D., and Martyna, G. J. (2002). Symplectic quaternion scheme for biophysical molecular dynamics. *J. Chem. Phys.* **116**, 8649–8659.

Minary, P., Martyna, G. J., and Tuckerman, M. E. (2003). Algorithms and novel applications based on the isokinetic ensemble. I. Biophysical and path integral molecular dynamics. *J. Chem. Phys.* **118**, 2510–2526.

Mirzoev, A. and Lyubartsev, A. P. (2013). MAGIC: software package for multiscale modeling. *J. Chem. Theor. Comput.* **9**, 1512–1520.

Misquitta, A. J. and Stone, A. J. (2013). *CAMCASP 5.6, a user manual.* http://www-stone.ch.cam.ac.uk/documentation/camcasp/users_guide.pdf.

Mitchell, P. J. and Fincham, D. (1993). Shell model simulations by adiabatic dynamics. *J. Phys. Cond. Mat.* **5**, 1031–1038.

Miyamoto, S. and Kollman, P. A. (1992). SETTLE: an analytical version of the SHAKE and RATTLE algorithm for rigid water models. *J. Comput. Chem.* **13**, 952–962.

Miyazaki, J., Barker, J. A., and Pound, G. M. (1976). A new Monte Carlo method for calculating surface tension. *J. Chem. Phys.* **64**, 3364–3369.

Molinero, V. and Moore, E. B. (2009). Water modeled as an intermediate element between carbon and silicon. *J. Phys. Chem. B* **113**, 4008–4016.

Mon, K. K. and Griffiths, R. B. (1985). Chemical potential by gradual insertion of a particle in Monte Carlo simulation. *Phys. Rev. A* **31**, 956–959.

Monticelli, L. and Salonen, E. (2013). *Biomolecular simulations, methods and protocols.* Methods Molec. Biol. Vol. 924. New York: Humana Press.

Monticelli, L., Kandasamy, S. K., Periole, X., Larson, R. G., Tieleman, D. P., and Marrink, S. J. (2008). The MARTINI coarse-grained force field: extension to proteins. *J. Chem. Theor. Comput.* **4**, 819–834.

Moore, E. B. and Molinero, V. (2011). Structural transformation in supercooled water controls the crystallization rate of ice. *Nature* **479**, 506–508.

Moore, G. E. (1965). Cramming more components onto integrated circuits. *Electronics* **38**, 114–117.

Moore, S. G. and Crozier, P. S. (2014). Extension and evaluation of the multilevel summation method for fast long-range electrostatics calculations. *J. Chem. Phys.* **140**, 234112.

Mor, A., Ziv, G., and Levy, Y. (2008). Simulations of proteins with inhomogeneous degrees of freedom: the effect of thermostats. *J. Comput. Chem.* **29**, 1992–1998.

Morales, M. A., Clay, R., Pierleoni, C., and Ceperley, D. M. (2014). First principles methods: a perspective from quantum Monte Carlo. *Entropy* **16**, 287–321.

Morawietz, T., Singraber, A., Dellago, C., and Behler, J. (2016). How van der Waals interactions determine the unique properties of water. *Proc. Nat. Acad. Sci.* **113**, 8368–8373.

Mori, H. (1965a). A continued-fraction representation of the time-correlation functions. *Prog. Theor. Phys.* **34**, 399–416.

Mori, H. (1965b). Transport, collective motion, and Brownian motion. *Prog. Theor. Phys.* **33**, 423–455.

Mori, Y. and Okamoto, Y. (2010). Replica-exchange molecular dynamics simulations for various constant temperature algorithms. *J. Phys. Soc. Japan* **79**, 074001.

Morishita, T. (2000). Fluctuation formulas in molecular dynamics simulations with the weak coupling heat bath. *J. Chem. Phys.* **113**, 2976–2982.

Moroni, D., van Erp, T. S., and Bolhuis, P. G. (2004a). Investigating rare events by transition interface sampling. *Physica A* **340**, 395–401.

Moroni, D., van Erp, T. S., and Bolhuis, P. G. (2005a). Simultaneous computation of free energies and kinetics of rare events. *Phys. Rev. E* **71**, 056709.

Moroni, D. (2005). Efficient sampling of rare event pathways: from simple models to nucleation. PhD thesis. University of Amsterdam.

Moroni, D., Bolhuis, P. G., and van Erp, T. S. (2004b). Rate constants for diffusive processes by partial path sampling. *J. Chem. Phys.* **120**, 4055–4065.

Moroni, D., ten Wolde, P. R., and Bolhuis, P. G. (2005b). Interplay between structure and size in a critical crystal nucleus. *Phys. Rev. Lett.* **94**, 235703.

Morrell, W. E. and Hildebrand, J. H. (1936). The distribution of molecules in a model liquid. *J. Chem. Phys.* **4**, 224–227.

Morris, J. R. (2002). Complete mapping of the anisotropic free energy of the crystal–melt interface in Al. *Phys. Rev. B* **66**, 144104.

Morris, J. R. and Song, X. (2002). The melting lines of model systems calculated from coexistence simulations. *J. Chem. Phys.* **116**, 9352–9358.

Morris, J. R. and Song, X. (2003). The anisotropic free energy of the Lennard-Jones crystal–melt interface. *J. Chem. Phys.* **119**, 3920–3925.

Morse, M. D. and Rice, S. A. (1982). Tests of effective pair potentials for water: predicted ice structures. *J. Chem. Phys.* **76**, 650–660.

Moucka, F., Nezbeda, I., and Smith, W. R. (2013). Molecular simulation of aqueous electrolytes: water chemical potential results and Gibbs–Duhem equation consistency tests. *J. Chem. Phys.* **139**, 124505.

Mouritsen, O. G. and Berlinsky, A. J. (1982). Fluctuation-induced first-order phase transition in an anisotropic planar model of N_2 on graphite. *Phys. Rev. Lett.* **48**, 181–184.

Mu, Y. G. and Stock, G. (2002). Conformational dynamics of trialanine in water: a molecular dynamics study. *J. Phys. Chem. B* **106**, 5294–5301.

Mullen, R. G., Shea, J.-E., and Peters, B. (2015). Easy transition path sampling methods: flexible-length aimless shooting and permutation shooting. *J. Chem. Theor. Comput.* **11**, 2421–2428.

Mullen-Schultz, G. L. (2005). IBM system blue gene solution: Blue Gene/L applications development. *IBM Redbooks*. Rochester, New York: IBM international technical support organisation.

Müller, E. A. and Jackson, G. (2014). Force-field parameters from the SAFT-γ equation of state for use in coarse-grained molecular simulations. *Ann. Rev. Chem. Bio. Eng.* **5**, 405–437. Ed. by J. M. Prausnitz, M. F. Doherty, and R. Segalman.

Müller, E. A. and Gubbins, K. E. (2001). Molecular-based equations of state for associating fluids: a review of SAFT and related approaches. *Ind. Eng. Chem. Res.* **40**, 2193–2211.

Müller, M. and Daoulas, K. C. (2008). Single-chain dynamics in a homogeneous melt and a lamellar microphase: a comparison between Smart Monte Carlo dynamics, slithering-snake dynamics, and slip-link dynamics. *J. Chem. Phys.* **129**, 164906.

Müller, M. and Pastorino, C. (2008). Cyclic motion and inversion of surface flow direction in a dense polymer brush under shear. *Europhys. Lett.* **81**, 28002.

Müller, U. and Stock, G. (1997). Surface-hopping modeling of photoinduced relaxation dynamics on coupled potential energy surfaces. *J. Chem. Phys.* **107**, 6230–6245.

Müller-Krumbhaar, H. and Binder, K. (1973). Dynamic properties of the Monte Carlo method in statistical mechanics. *J. Stat. Phys.* **8**, 1–24.

Müller-Plathe, F. (1999). Reversing the perturbation in nonequilibrium molecular dynamics: an easy way to calculate the shear viscosity of fluids. *Phys. Rev. E* **59**, 4894–4898.

Müller-Plathe, F. and Reith, D. (1999). Cause and effect reversed in non-equilibrium molecular dynamics: an easy route to transport coefficients. *Comput. Theor. Polym. Sci.* **9**, 203–209.

Müller-Plathe, F. (1997). A simple nonequilibrium molecular dynamics method for calculating the thermal conductivity. *J. Chem. Phys.* **106**, 6082–6085.

Mundy, C. J., Balasubramanian, S., and Klein, M. L. (1996). Hydrodynamic boundary conditions for confined fluids via a nonequilibrium molecular dynamics simulation. *J. Chem. Phys.* **105**, 3211–3214.

Mundy, C. J., Balasubramanian, S., and Klein, M. L. (1997). Computation of the hydrodynamic boundary parameters of a confined fluid via non-equilibrium molecular dynamics. *Physica A* **240**, 305–314.

Münster, A. (1969). *Statistical thermodynamics.* Berlin/New York: Springer/Academic Press.

Murad, S. (1978). LINEAR and NONLINEAR. QCPE **12**, 357. Indiana University Quantum Chemistry Program Exchange.

Murad, S. and Gubbins, K. E. (1978). Molecular dynamics simulation of methane using a singularity free algorithm. *Computer modelling of matter.* ACS Symp. Ser. **86**, 62–71. (Anaheim, USA, 1978). Ed. by P. Lykos. Washington: American Chemical Society.

Murthy, C. S., Singer, K., Klein, M. L., and McDonald, I. R. (1980). Pairwise additive effective potentials for nitrogen. *Molec. Phys.* **41**, 1387–1399.

Murthy, C. S., O'Shea, S. F., and McDonald, I. R. (1983). Electrostatic interactions in molecular crystals. Lattice dynamics of solid nitrogen and carbon dioxide. *Molec. Phys.* **50**, 531–541.

Nadler, R. and Sanz, J. F. (2012). Effect of dispersion correction on the Au(111)–H_2O interface: a first-principles study. *J. Chem. Phys.* **137**, 114709.

Naitoh, T. and Ono, S. (1976). The shear viscosity of hard-sphere fluid via non-equilibrium molecular dynamics. *Phys. Lett. A* **57**, 448–450.

Naitoh, T. and Ono, S. (1979). The shear viscosity of a hard-sphere fluid via nonequilibrium molecular dynamics. *J. Chem. Phys.* **70**, 4515–4523.

Nam, K. (2014). Acceleration of *ab initio* QM/MM calculations under periodic boundary conditions by multiscale and multiple time step approaches. *J. Chem. Theor. Comput.* **10**, 4175–4183.

Nangia, S. and Garrison, B. J. (2010). Theoretical advances in the dissolution studies of mineral–water interfaces. *Theor. Chem. Acc.* **127**, 271–284.

Neale, C., Bennett, W. F. D., Tieleman, D. P., and Pomès, R. (2011). Statistical convergence of equilibrium properties in simulations of molecular solutes embedded in lipid bilayers. *J. Chem. Theor. Comput.* **7**, 4175–4188.

Nelson, D. R. and Halperin, B. I. (1979). Dislocation-mediated melting in two dimensions. *Phys. Rev. B* **19**, 2457–2484.

Neto, N., Righini, R., Califano, S., and Walmsley, S. H. (1978). Lattice dynamics of molecular crystals using atom–atom and multipole–multipole potentials. *Chem. Phys.* **29**, 167–179.

Neumann, M. and Steinhauser, O. (1980). The influence of boundary conditions used in machine simulations on the structure of polar systems. *Molec. Phys.* **39**, 437–454.

Neumann, M. and Steinhauser, O. (1983a). On the calculation of the dielectric constant using the Ewald–Kornfeld tensor. *Chem. Phys. Lett.* **95**, 417–422.

Neumann, M. and Steinhauser, O. (1983b). On the calculation of the frequency-dependent dielectric constant in computer simulations. *Chem. Phys. Lett.* **102**, 508–513.

Neumann, M. (1983). Dipole moment fluctuation formulas in computer simulations of polar systems. *Molec. Phys.* **50**, 841–858.

Neumann, M., Steinhauser, O., and Pawley, G. S. (1984). Consistent calculation of the static and frequency-dependent dielectric constant in computer simulations. *Molec. Phys.* **52**, 97–113.

Neyt, J.-C., Wender, A., Lachet, V., Ghoufi, A., and Malfreyt, P. (2014). Quantitative predictions of the interfacial tensions of liquid–liquid interfaces through atomistic and coarse grained models. *J. Chem. Theor. Comput.* **10**, 1887–1899.

Nezbeda, I. (1977). Statistical thermodynamics of interaction-site molecules. *Molec. Phys.* **33**, 1287–1299.

Nicholson, D. (1984). Grand ensemble Monte Carlo. *CCP5 Quarterly* **11**, 19–24.

Nicholson, D. and Parsonage, N. (1982). *Computer simulation and the statistical mechanics of adsorption.* New York: Academic Press.

Nicolas, J. J., Gubbins, K. E., Streett, W. B., and Tildesley, D. J. (1979). Equation of state for the Lennard-Jones fluid. *Molec. Phys.* **37**, 1429–1454.

Nielsen, S. O. (2013). Nested sampling in the canonical ensemble: direct calculation of the partition function from *NVT* trajectories. *J. Chem. Phys.* **139**, 124104.

Niethammer, C., Becker, S., Bernreuther, M., Buchholz, M., Eckhardt, W., Heinecke, A., Werth, S., Bungartz, H.-J., Glass, C. W., Hasse, H., Vrabec, J., and Horsch, M. (2014). ls1 mardyn: the massively parallel molecular dynamics code for large systems. *J. Chem. Theor. Comput.* **10**, 4455–4464.

Nikunen, P., Karttunen, M., and Vattulainen, I. (2003). How would you integrate the equations of motion in dissipative particle dynamics simulations? *Comput. Phys. Commun.* **153**, 407–423.

Noguchi, H. and Gompper, G. (2005). Shape transitions of fluid vesicles and red blood cells in capillary flows. *Proc. Nat. Acad. Sci.* **102**, 14159–14164.

Noid, W. G., Chu, J.-W., Ayton, G. S., Krishna, V., Izvekov, S., Voth, G. A., Das, A., and Andersen, H. C. (2008a). The multiscale coarse-graining method. I. A rigorous bridge between atomistic and coarse-grained models. *J. Chem. Phys.* **128**, 244114.

Noid, W. G., Liu, P., Wang, Y., Chu, J.-W., Ayton, G. S., Izvekov, S., Andersen, H. C., and Voth, G. A. (2008b). The multiscale coarse-graining method. II. Numerical implementation for coarse-grained molecular models. *J. Chem. Phys.* **128**, 244115.

Nordholm, S. and Zwanzig, R. (1975). A systematic derivation of exact generalized Brownian motion theory. *J. Stat. Phys.* **13**, 347–371.

Norman, G. E. and Filinov, V. S. (1969). Investigations of phase transitions by a Monte Carlo method. *High Temp. Res. USSR* **7**, 216–222.

Northrup, S. H. and McCammon, J. A. (1980). Simulation methods for protein structure fluctuations. *Biopolymers* **19**, 1001–1016.

Nosé, S. (1984). A molecular dynamics method for simulations in the canonical ensemble. *Molec. Phys.* **52**, 255–268.

Nosé, S. (1991). Constant temperature molecular dynamics methods. *Prog. Theor. Phys. Suppl.* **103**, 1–46.

Nosé, S. and Klein, M. L. (1983). Constant pressure molecular dynamics for molecular systems. *Molec. Phys.* **50**, 1055–1076.

Noya, E. G., Vega, C., and McBride, C. (2011). A quantum propagator for path-integral simulations of rigid molecules. *J. Chem. Phys.* **134**, 054117.

Oberhofer, H., Dellago, C., and Geissler, P. L. (2005). Biased sampling of nonequilibrium trajectories: can fast switching simulations outperform conventional free energy calculation methods? *J. Phys. Chem. B* **109**, 6902–6915.

Oberhofer, H. and Dellago, C. (2009). Efficient extraction of free energy profiles from nonequilibrium experiments. *J. Comput. Chem.* **30**, 1726–1736.

O'Dell, J. and Berne, B. J. (1975). Molecular dynamics of the rough sphere fluid. I. Rotational relaxation. *J. Chem. Phys.* **63**, 2376–2394.

Ohno, Y., Yokota, R., Koyama, H., Morimoto, G., Hasegawa, A., Masumoto, G., Okimoto, N., Hirano, Y., Ibeid, H., Narumi, T., and Taiji, M. (2014). Petascale molecular dynamics simulation using the fast multipole method on K computer. *Comput. Phys. Commun.* **185**, 2575–2585.

O'Keeffe, C. J. and Orkoulas, G. (2009). Parallel canonical Monte Carlo simulations through sequential updating of particles. *J. Chem. Phys.* **130**, 134109.

Okumura, H., Itoh, S. G., and Okamoto, Y. (2007). Explicit symplectic integrators of molecular dynamics algorithms for rigid-body molecules in the canonical, isobaric–isothermal, and related ensembles. *J. Chem. Phys.* **126**, 084103.

Okumura, H., Itoh, S. G., Ito, A. M., Nakamura, H., and Fukushima, T. (2014). Manifold correction method for the Nosé–Hoover and Nosé–Poincaré molecular dynamics simulations. *J. Phys. Soc. Japan* **83**, 024003.

Olafsen, J. S. and Urbach, J. S. (2005). Two-dimensional melting far from equilibrium in a granular monolayer. *Phys. Rev. Lett.* **95**, 098002.

Ollila, O. H. S. and Vattulainen, I. (2010). Lateral pressure profiles in lipid membranes: dependence on molecular composition. *Molecular simulations and biomembranes.* Chap. 2, 26–55. Ed. by M. S. P. Sansom and P. C. Biggin. Royal Society of Chemistry (RSC).

Omelyan, I. P. (1998). Algorithm for numerical integration of the rigid-body equations of motion. *Phys. Rev. E* **58**, 1169–1172.

Omelyan, I. P. (1999). A new leapfrog integrator of rotational motion. The revised angular-momentum approach. *Molec. Simul.* **22**, 213–236.

Onsager, L. (1936). Electric moments of molecules in liquids. *J. Am. Chem. Soc.* **58**, 1486–1493.

Oppenheim, I. and Ross, J. (1957). Temperature dependence of distribution functions in quantum statistical mechanics. *Phys. Rev.* **107**, 28–32.

Orban, J. and Ryckaert, J. P. (1974). *Methods in molecular dynamics.* Tech. rep. Rapport d'activité scientifique du CECAM.

Orea, P., López-Lemus, J., and Alejandre, J. (2005). Oscillatory surface tension due to finite-size effects. *J. Chem. Phys.* **123**, 114702.

Orkoulas, G. and Panagiotopoulos, A. Z. (1993). Chemical potentials in ionic systems from Monte Carlo simulations with distance-biased test particle insertions. *Fluid Phase Equilibria* **83**, 223–231.

Orlandini, E., Swift, M. R., and Yeomans, J. M. (1995). A lattice Boltzmann model of binary fluid mixtures. *Europhys. Lett.* **32**, 463–468.

Orsi, M., Haubertin, D. Y., Sanderson, W. E., and Essex, J. W. (2008). A quantitative coarse-grain model for lipid bilayers. *J. Phys. Chem. B* **112**, 802–815.

Oseen, C. (1933). Theory of liquid crystals. *Trans. Faraday Soc.* **29**, 883–899.

O'Shea, S. F. (1978). Monte Carlo study of the classical octopolar solid. *J. Chem. Phys.* **68**, 5435–5441.

O'Shea, S. F. (1983). Neighbour lists again. *CCP5 Quarterly* **9**, 41–46.

Owicki, J. C. (1978). Optimizing of sampling algorithms in Monte Carlo calculations on fluids. *Computer modelling of matter.* ACS Symp. Ser. **86**, 159–171. (Anaheim, USA, 1978). Ed. by P. Lykos. Washington: American Chemical Society.

Owicki, J. C. and Scheraga, H. A. (1977a). Monte Carlo calculations in the isothermal-isobaric ensemble. 2. Dilute aqueous solutions of methane. *J. Am. Chem. Soc.* **99**, 7413–7418.

Owicki, J. C. and Scheraga, H. A. (1977b). Preferential sampling near solutes in Monte Carlo calculations on dilute solutions. *Chem. Phys. Lett.* **47**, 600–602.

Pacheco, P. (1996). *Parallel programming with MPI.* Morgan Kaufmann.

Padding, J. T. and Louis, A. A. (2006). Hydrodynamic interactions and Brownian forces in colloidal suspensions: coarse-graining over time and length scales. *Phys. Rev. E* **74**, 031402.

Padilla, P. and Toxvaerd, S. (1991). Self-diffusion in n-alkane fluid models. *J. Chem. Phys.* **94**, 5650–5654.

Padilla, P., Toxvaerd, S., and Stecki, J. (1995). Shear flow at liquid–liquid interfaces. *J. Chem. Phys.* **103**, 716–724.

Pagonabarraga, I., Hagen, M. H. J., and Frenkel, D. (1998). Self-consistent dissipative particle dynamics algorithm. *Europhys. Lett.* **42**, 377–382.

Paliwal, H. and Shirts, M. R. (2013). Using multistate reweighting to rapidly and efficiently explore molecular simulation parameters space for nonbonded interactions. *J. Chem. Theor. Comput.* **9**, 4700–4717.

Palmer, B. J. and Lo, C. M. (1994). Molecular dynamics implementation of the Gibbs ensemble calculation. *J. Chem. Phys.* **101**, 10899–10907.

Pan, A. C., Weinreich, T. M., Piana, S., and Shaw, D. E. (2016). Demonstrating an order-of-magnitude sampling enhancement in molecular dynamics simulations of complex protein systems. *J. Chem. Theor. Comput.* **12**, 1360–1367.

Pan, D., Spanu, L., Harrison, B., Sverjensky, D. A., and Galli, G. (2013). Dielectric properties of water under extreme conditions and transport of carbonates in the deep earth. *Proc. Nat. Acad. Sci.* **110**, 6646–6650.

Pan, G., Ely, J. F., McCabe, C., and Isbister, D. J. (2005). Operator splitting algorithm for isokinetic SLLOD molecular dynamics. *J. Chem. Phys.* **122**, 094114.

Panagiotopoulos, A. Z., Quirke, N., Stapleton, M., and Tildesley, D. J. (1988). Phase equilibria by simulation in the Gibbs ensemble. *Molec. Phys.* **63**, 527–545.

Panagiotopoulos, A. Z. (1987). Direct determination of phase coexistence properties of fluids by Monte Carlo simulation in a new ensemble. *Molec. Phys.* **61**, 813–826.

Pangali, C., Rao, M., and Berne, B. J. (1978). On a novel Monte Carlo scheme for simulating water and aqueous solutions. *Chem. Phys. Lett.* **55**, 413–417.

Panneton, F., L'Ecuyer, P., and Matsumoto, M. (2006). Improved long-period generators based on linear recurrences modulo 2. *ACM Trans. Math. Softw.* **32**, 1–16.

Pant, P. V. K. and Theodorou, D. N. (1995). Variable connectivity method for the atomistic Monte Carlo simulation of polydisperse polymer melts. *Macromolecules* **28**, 7224–7234.

Papadopoulou, A., Becker, E. D., Lupkowski, M., and van Swol, F. (1993). Molecular dynamics and Monte Carlo simulations in the grand canonical ensemble: local versus global control. *J. Chem. Phys.* **98**, 4897–4908.

Papoulis, A. (1965). *Probability, random variables, and stochastic processes*. New York: McGraw-Hill.

Park, K., Goetz, A. W., Walker, R. C., and Paesani, F. (2012). Application of adaptive QM/MM methods to molecular dynamics simulations of aqueous systems. *J. Chem. Theor. Comput.* **8**, 2868–2877.

Parrinello, M. and Rahman, A. (1980). Crystal structure and pair potentials: a molecular-dynamics study. *Phys. Rev. Lett.* **45**, 1196–1199.

Parrinello, M. and Rahman, A. (1981). Polymorphic transitions in single crystals: a new molecular dynamics method. *J. Appl. Phys.* **52**, 7182–7190.

Parrinello, M. and Rahman, A. (1982). Strain fluctuations and elastic constants. *J. Chem. Phys.* **76**, 2662–2666.

Parrinello, M. and Rahman, A. (1984). Study of an F center in molten KCl. *J. Chem. Phys.* **80**, 860–867.

Parry, A. O., Rascón, C., and Evans, R. (2015). Liquid–gas asymmetry and the wave-vector-dependent surface tension. *Phys. Rev. E* **91**, 030401.

Pártay, L. B., Bartók, A. P., and Csányi, G. (2010). Efficient sampling of atomic configurational spaces. *J. Phys. Chem. B* **114**, 10502–10512.

Pártay, L. B., Bartók, A. P., and Csányi, G. (2014). Nested sampling for materials: the case of hard spheres. *Phys. Rev. E* **89**, 022302.

Pasichnyk, I. and Dünweg, B. (2004). Coulomb interactions via local dynamics: a molecular-dynamics algorithm. *J. Phys. Cond. Mat.* **16**, S3999–S4020.

Pastorino, C., Binder, K., Kreer, T., and Müller, M. (2006). Static and dynamic properties of the interface between a polymer brush and a melt of identical chains. *J. Chem. Phys.* **124**, 064902.

Patey, G. N., Levesque, D., and Weis, J. J. (1982). On the theory and computer simulation of dipolar fluids. *Molec. Phys.* **45**, 733–746.

Patkowski, K., Murdachaew, G., Fou, C. M., and Szalewicz, K. (2005). Accurate *ab initio* potential for argon dimer including highly repulsive region. *Molec. Phys.* **103**, 2031–2045.

Patkowski, K. and Szalewicz, K. (2010). Argon pair potential at basis set and excitation limits. *J. Chem. Phys.* **133**, 094304.

Pavese, M., Jang, S., and Voth, G. A. (2000). Centroid molecular dynamics: a quantum dynamics method suitable for the parallel computer. *Parallel Comput.* **26**, 1025–1041.

Pawley, G. S., Bowler, K. C., Kenway, R. D., and Wallace, D. J. (1985). Concurrency and parallelism in MC and MD simulations in physics. *Comput. Phys. Commun.* **37**, 251–260.

Payne, M. C., Tarnow, E., Bristowe, P. D., and Joannopoulos, J. D. (1989). *Ab initio* materials science and engineering using the molecular dynamics method for total energy pseudopotential calculations. *Molec. Simul.* **4**, 79–94.

Payne, M. C., Teter, M. P., Allan, D. C., Arias, T. A., and Joannopoulos, J. D. (1992). Iterative minimization techniques for *ab initio* total-energy calculations: molecular dynamics and conjugate gradients. *Rev. Mod. Phys.* **64**, 1045–1097.

Pear, M. R. and Weiner, J. H. (1979). Brownian dynamics study of a polymer chain of linked rigid bodies. *J. Chem. Phys.* **71**, 212–224.

Peng, Z. W., Ewig, C. S., Hwang, M. J., Waldman, M., and Hagler, A. T. (1997). Derivation of class II force fields. 4. van der Waals parameters of alkali metal cations and halide anions. *J. Phys. Chem. A* **101**, 7243–7252.

Perdew, J. P. and Zunger, A. (1981). Self-interaction correction to density-functional approximations for many-electron systems. *Phys. Rev. B* **23**, 5048–5079.

Perdew, J. P., Chevary, J. A., Vosko, S. H., Jackson, K. A., Pederson, M. R., Singh, D. J., and Fiolhais, C. (1992). Atoms, molecules, solids, and surfaces: applications of the generalized gradient approximation for exchange and correlation. *Phys. Rev. B* **46**, 6671–6687.

Perego, C., Salvalaglio, M., and Parrinello, M. (2015). Molecular dynamics simulations of solutions at constant chemical potential. *J. Chem. Phys.* **142**, 144113.

Perez, A., Tuckerman, M. E., and Muser, M. H. (2009a). A comparative study of the centroid and ring-polymer molecular dynamics methods for approximating quantum time correlation functions from path integrals. *J. Chem. Phys.* **130**, 184105.

Perez, D., Uberuaga, B. P., Shim, Y., Amar, J. G., and Voter, A. F. (2009b). Accelerated molecular dynamics methods: introduction and recent developments. *Ann. Rep. Comp. Chem.* **5**, 79–98. Ed. by R. A. Wheeler. Elsevier.

Peristeras, L. D., Rissanou, A. N., Economou, I. G., and Theodorou, D. N. (2007). Novel Monte Carlo molecular simulation scheme using identity-altering elementary moves for the calculation of structure and thermodynamic properties of polyolefin blends. *Macromolecules* **40**, 2904–2914.

Perram, J. W., Peteresen, H. G., and de Leeuw, S. W. (1988). An algorithm for the simulation of condensed matter which grows as the 3/2 power of the number of particles. *Molec. Phys.* **65**, 875–893.

Pershan, P. S. (2009). X-ray scattering from liquid surfaces: effect of resolution. *J. Phys. Chem. B* **113**, 3639–3646.

Peskun, P. H. (1973). Optimum Monte Carlo sampling using Markov chains. *Biometrika* **60**, 607–612.

Peters, B. and Trout, B. L. (2006). Obtaining reaction coordinates by likelihood maximization. *J. Chem. Phys.* **125**, 054108.

Peters, E. A. J. F. (2004). Elimination of time step effects in DPD. *Europhys. Lett.* **66**, 311–317.

Petersen, H. G. (1995). Accuracy and efficiency of the particle mesh Ewald method. *J. Chem. Phys.* **103**, 3668–3679.

Petravic, J. (2007). Equivalence of nonequilibrium algorithms for simulations of planar Couette flow in confined fluids. *J. Chem. Phys.* **127**, 204702.

Petravic, J. and Harrowell, P. (2007). On the equilibrium calculation of the friction coefficient for liquid slip against a wall. *J. Chem. Phys.* **127**, 174706. Erratum: *ibid.* **128**, 209901 (2007).

Phillips, J. C., Braun, R., Wang, W., Gumbart, J., Tajkhorshid, E., Villa, E., Chipot, C., Skeel, R. D., Kale, L., and Schulten, K. (2005). Scalable molecular dynamics with NAMD. *J. Comput. Chem.* **26**, 1781–1802.

Piana, S., Klepeis, J. L., and Shaw, D. E. (2014). Assessing the accuracy of physical models used in protein-folding simulations: quantitative evidence from long molecular dynamics simulations. *Curr. Opin. Struct. Biol.* **24**, 98–105.

Pieprzyk, S., Heyes, D. M., Maćkowiak, S., and Brańka, A. C. (2015). Galilean-invariant Nosé–Hoover-type thermostats. *Phys. Rev. E* **91**, 033312.

Pieranski, P., Malecki, J., Kuczynski, W., and Wojciechowski, K. (1978). A hard disc system, an experimental model. *Phil. Mag. A* **37**, 107–115.

Pierleoni, C. and Ceperley, D. M. (2006). The coupled electron–ion Monte Carlo method. *Computer simulations in condensed matter systems: from materials to chemical biology, vol 1.* Lect. Notes Phys. **703**, 641–683. (Erice, Italy, 2005). Ed. by M. Ferrario, G. Ciccotti, and K. Binder. Heidelberg: Springer.

Pinches, M. R. S., Tildesley, D. J., and Smith, W. (1991). Large scale molecular dynamics on parallel computers using the link-cell algorithm. *Molec. Simul.* **6**, 51–87.

Pitzer, K. S. and Gwinn, W. D. (1942). Energy levels and thermodynamic functions for molecules with internal rotation. 1. Rigid frame with attached tops. *J. Chem. Phys.* **10**, 428–440.

Plimpton, S. (1995). Fast parallel algorithms for short-range molecular dynamics. *J. Comput. Phys.* **117**, 1–19. LAMMPS website http://lammps.sandia.gov.

Pluhackova, K. and Böckmann, R. A. (2015). Biomembranes in atomistic and coarse-grained simulations. *J. Phys. Cond. Mat.* **27**, 323103.

Poblete, S., Praprotnik, M., Kremer, K., and Delle Site, L. (2010). Coupling different levels of resolution in molecular simulations. *J. Chem. Phys.* **132**, 114101.

Pohorille, A., Jarzynski, C., and Chipot, C. (2010). Good practices in free-energy calculations. *J. Phys. Chem. B* **114**, 10235–10253.

Pollock, E. L. and Glosli, J. (1996). Comments on P3M, FMM, and the Ewald method for large periodic Coulombic systems. *Comput. Phys. Commun.* **95**, 93–110.

Polyakov, P., Müller-Plathe, F., and Wiegand, S. (2008). Reverse nonequilibrium molecular dynamics calculation of the Soret coefficient in liquid heptane/benzene mixtures. *J. Phys. Chem. B* **112**, 14999–15004.

Ponder, J. W. and Case, D. A. (2003). Force fields for protein simulations. *Protein simulations.* Adv. Protein Chem. **66**, 27–85. Ed. by V. Daggett. San Diego: Academic Press.

Ponder, J. W., Wu, C., Ren, P., Pande, V. S., Chodera, J. D., Schnieders, M. J., Haque, I., Mobley, D. L., Lambrecht, D. S., DiStasio Jr., R. A., Head-Gordon, M., Clark, G. N. I., Johnson, M. E., and Head-Gordon, T. (2010). Current status of the AMOEBA polarizable force field. *J. Phys. Chem. B* **114**, 2549–2564.

Pooley, C. M. and Yeomans, J. M. (2005). Kinetic theory derivation of the transport coefficients of stochastic rotation dynamics. *J. Phys. Chem. B* **109**, 6505–6513.

Potestio, R., Fritsch, S., Español, P., Delgado-Buscalioni, R., Kremer, K., Everaers, R., and Donadio, D. (2013a). Hamiltonian adaptive resolution simulation for molecular liquids. *Phys. Rev. Lett.* **110**, 108301.

Potestio, R., Español, P., Delgado-Buscalioni, R., Everaers, R., Kremer, K., and Donadio, D. (2013b). Monte Carlo adaptive resolution simulation of multicomponent molecular liquids. *Phys. Rev. Lett.* **111**, 060601.

Potoff, J. J. and Panagiotopoulos, A. Z. (2000). Surface tension of the three-dimensional Lennard-Jones fluid from histogram reweighting Monte Carlo simulations. *J. Chem. Phys.* **112**, 6411–6415.

Potter, D. (1972). *Computational physics.* New York: Wiley.

Poulsen, J. A., Nyman, G., and Rossky, P. J. (2004). Quantum diffusion in liquid para-hydrogen: an application of the Feynman–Kleinert linearized path integral approximation. *J. Phys. Chem. B* **108**, 19799–19808.

Powles, J. G. (1984). The liquid–vapour coexistence line for Lennard-Jones-type fluids. *Physica A* **126**, 289–299.

Powles, J. G. and Rickayzen, G. (1979). Quantum corrections and the computer simulation of molecular fluids. *Molec. Phys.* **38**, 1875–1892.

Powles, J. G., Evans, W. A. B., and Quirke, N. (1982). Non-destructive molecular dynamics simulation of the chemical potential of a fluid. *Molec. Phys.* **46**, 1347–1370.

Powles, J. G., Fowler, R. F., and Evans, W. A. B. (1983a). A new method for computing surface tension using a drop of liquid. *Chem. Phys. Lett.* **96**, 289–292.

Powles, J. G., Fowler, R. F., and Evans, W. A. B. (1983b). The surface thickness of simulated microscopic liquid drops. *Phys. Lett. A* **98**, 421–425.

Powles, J. G., Mallett, M. J. D., and Evans, W. A. B. (1994). Measuring the properties of liquids by counting. *Proc. Roy. Soc. Lond. A* **446**, 429–439.

Powles, J. G., Rickayzen, G., and Heyes, D. M. (2005). Temperatures: old, new and middle aged. *Molec. Phys.* **103**, 1361–1373.

Prabhu, N. V., Zhu, P. J., and Sharp, K. A. (2004). Implementation and testing of stable, fast implicit solvation in molecular dynamics using the smooth-permittivity finite difference Poisson–Boltzmann method. *J. Comput. Chem.* **25**, 2049–2064.

Praprotnik, M., Delle Site, L., and Kremer, K. (2005). Adaptive resolution molecular-dynamics simulation: changing the degrees of freedom on the fly. *J. Chem. Phys.* **123**, 224106.

Praprotnik, M., Delle Site, L., and Kremer, K. (2008). Multiscale simulation of soft matter: from scale bridging to adaptive resolution. *Ann. Rev. Phys. Chem.* **59**, 545–571.

Pratt, L. R. and Haan, S. W. (1981). Effects of periodic boundary conditions on equilibrium properties of computer simulated fluids. I. Theory. *J. Chem. Phys.* **74**, 1864–1872.

Press, W. H., Teukolsky, S. A., Vetterling, W. T., and Flannery, B. P. (2007). *Numerical recipes: the art of scientific computing.* 3rd ed. New York: Cambridge University Press.

Preto, J. and Clementi, C. (2014). Fast recovery of free energy landscapes via diffusion-map-directed molecular dynamics. *Phys. Chem. Chem. Phys.* **16**, 19181–19191.

Price, S. L. (2008). Computational prediction of organic crystal structures and polymorphism. *Int. Rev. Phys. Chem.* **27**, 541–568.

Price, S. L., Stone, A. J., and Alderton, M. (1984). Explicit formulae for the electrostatic energy, forces and torques between a pair of molecules of arbitrary symmetry. *Molec. Phys.* **52**, 987–1001.

Priezjev, N. V. and Troian, S. M. (2004). Molecular origin and dynamic behavior of slip in sheared polymer films. *Phys. Rev. Lett.* **92**, 018302.

Priezjev, N. V., Darhuber, A. A., and Troian, S. M. (2005). Slip behavior in liquid films on surfaces of patterned wettability: comparison between continuum and molecular dynamics simulations. *Phys. Rev. E* **71**, 041608.

Procacci, P. and Marchi, M. (1996). Taming the Ewald sum in molecular dynamics simulations of solvated proteins via a multiple time step algorithm. *J. Chem. Phys.* **104**, 3003–3012.

Pusey, P. N. and van Megen, W. (1986). Phase behaviour of concentrated suspensions of nearly hard colloidal spheres. *Nature* **320**, 340–342.

Qian, Y. H., D'Humières, D., and Lallemand, P. (1992). Lattice BGK models for Navier–Stokes equation. *Europhys. Lett.* **17**, 479.

Quentrec, B. and Brot, C. (1973). New method for searching for neighbors in molecular dynamics computations. *J. Comput. Phys.* **13**, 430–432.

Quigley, D. and Probert, M. I. J. (2004). Langevin dynamics in constant pressure extended systems. *J. Chem. Phys.* **120**, 11432–11441.

Quinn, M. J. (2003). *Parallel programming in C with MPI and OpenMP.* McGraw-Hill Education.

Raabe, D. (2004). Overview of the lattice Boltzmann method for nano- and microscale fluid dynamics in materials science and engineering. *Model. Sim. Mat. Sci. Eng.* **12**, 13–46.

Rahman, A. (1964). Correlations in the motion of atoms in liquid argon. *Phys. Rev. A* **136**, 405–411.

Rahman, A. and Stillinger, F. H. (1971). Molecular dynamics study of liquid water. *J. Chem. Phys.* **55**, 3336–3359.

Ramírez, J., Sukumaran, S. K., Vorselaars, B., and Likhtman, A. E. (2010). Efficient on the fly calculation of time correlation functions in computer simulations. *J. Chem. Phys.* **133**, 154103.

RAND (2001). *A million random digits with 100,000 normal deviates.* RAND.

Rao, M. and Levesque, D. (1976). Surface structure of a liquid film. *J. Chem. Phys.* **65**, 3233–3236.

Rao, M., Pangali, C., and Berne, B. J. (1979). On the force bias Monte Carlo simulation of water: methodology, optimization and comparison with molecular dynamics. *Molec. Phys.* **37**, 1773–1798.

Rapaport, D. C. (1978). Molecular dynamics simulation of polymer chains with excluded volume. *J. Phys. A Math. Gen.* **11**, 213–217.

Rapaport, D. C. (1979). Molecular dynamics study of a polymer chain in solution. *J. Chem. Phys.* **71**, 3299–3303.

Rapaport, D. C. (1980). The event scheduling problem in molecular dynamic simulation. *J. Comput. Phys.* **34**, 184–201.

Rapaport, D. C. (2004). *The art of molecular dynamics simulation.* 2nd ed. Cambridge University Press.

Rappe, A. K., Casewit, C. J., Colwell, K. S., Goddard, W. A., and Skiff, W. M. (1992). UFF, a full periodic table force field for molecular mechanics and molecular dynamics simulations. *J. Am. Chem. Soc.* **114**, 10024–10035.

Rathore, N., Chopra, M., and de Pablo, J. J. (2005). Optimal allocation of replicas in parallel tempering simulation. *J. Chem. Phys.* **122**, 024111.

Ray, J. R. (1991). Microcanonical ensemble Monte Carlo method. *Phys. Rev. A* **44**, 4061–4064.

Ray, J. R. and Graben, H. W. (1981). Direct calculation of fluctuation formulae in the microcanonical ensemble. *Molec. Phys.* **43**, 1293–1297.

Ray, J. R. and Graben, H. W. (1991). Small systems have non-Maxwellian momentum distributions in the microcanonical ensemble. *Phys. Rev. A* **44**, 6905–6908.

Rebertus, D. W. and Sando, K. M. (1977). Molecular dynamics simulation of a fluid of hard spherocylinders. *J. Chem. Phys.* **67**, 2585–2590.

Ree, F. H. (1970). Statistical mechanics of single-occupancy systems of spheres, disks and rods. *J. Chem. Phys.* **53**, 920–931.

Reith, D., Putz, M., and Müller-Plathe, F. (2003). Deriving effective mesoscale potentials from atomistic simulations. *J. Comput. Chem.* **24**, 1624–1636.

Remler, D. K. and Madden, P. A. (1990). Molecular dynamics without effective potentials via the Car–Parrinello approach. *Molec. Phys.* **70**, 921–966.

Reynolds, P. J., Ceperley, D. M., Alder, B. J., and Lester, W. A. (1982). Fixed-node quantum Monte Carlo for molecules. *J. Chem. Phys.* **77**, 5593–5603.

Richardson, J. O. and Althorpe, S. C. (2009). Ring-polymer molecular dynamics rate-theory in the deep-tunneling regime: connection with semiclassical instanton theory. *J. Chem. Phys.* **131**, 214106.

Rick, S. W., Stuart, S. J., and Berne, B. J. (1994). Dynamical fluctuating charge force fields: application to liquid water. *J. Chem. Phys.* **101**, 6141–6156.

Rick, S. W. and Stuart, S. J. (2003). Potentials and algorithms for incorporating polarizability in computer simulations. *Rev. Comput. Chem.* **18**, 89–146. Wiley.

Rickayzen, G. and Powles, J. G. (2001). Temperature in the classical microcanonical ensemble. *J. Chem. Phys.* **114**, 4333–4334.

Righini, R., Maki, K., and Klein, M. L. (1981). An intermolecular potential for methane. *Chem. Phys. Lett.* **80**, 301–305.

Rodger, P. M., Stone, A. J., and Tildesley, D. J. (1988a). Intermolecular interactions in halogens: bromine and iodine. *Chem. Phys. Lett.* **145**, 365–370.

Rodger, P. M., Stone, A. J., and Tildesley, D. J. (1988b). The intermolecular potential of chlorine. A three phase study. *Molec. Phys.* **63**, 173–188.

Rodgers, J. M. and Smit, B. (2012). On the equivalence of schemes for simulating bilayers at constant surface tension. *J. Chem. Theor. Comput.* **8**, 404–417.

Rohrdanz, M. A., Zheng, W., Maggioni, M., and Clementi, C. (2011). Determination of reaction coordinates via locally scaled diffusion map. *J. Chem. Phys.* **134**, 124116.

Rohrdanz, M. A., Zheng, W., and Clementi, C. (2013). Discovering mountain passes via torchlight: methods for the definition of reaction coordinates and pathways in complex macromolecular reactions. *Ann. Rev. Phys. Chem.* **64**, 295–316.

Romano, S. and Singer, K. (1979). Calculation of the entropy of liquid chlorine and bromine by computer simulation. *Molec. Phys.* **37**, 1765–1772.

Romero-Bastida, M. and López-Rendón, R. (2007). Anisotropic pressure molecular dynamics for atomic fluid systems. *J. Phys. A Math. Gen.* **40**, 8585–8598.

Rosenbluth, M. N. and Rosenbluth, A. W. (1955). Monte-Carlo calculation of the average extension of molecular chains. *J. Chem. Phys.* **23**, 356–359.

Rosenbluth, M. N. and Rosenbluth, A. W. (1954). Further results on Monte Carlo equations of state. *J. Chem. Phys.* **22**, 881–884.

Rossky, P. J., Doll, J. D., and Friedman, H. L. (1978). Brownian dynamics as smart Monte Carlo simulation. *J. Chem. Phys.* **69**, 4628–4633.

Rosta, E., Buchete, N.-V., and Hummer, G. (2009). Thermostat artifacts in replica exchange molecular dynamics simulations. *J. Chem. Theor. Comput.* **5**, 1393–1399.

Röthlisberger, U. and Carloni, P. (2006). Drug-target binding investigated by quantum mechanical/molecular mechanical (QMMM) methods. *Computer simulations in condensed matter systems: from materials to chemical biology, vol 2.* Lect. Notes Phys. **704**, 449–479. (Erice, Italy, 2005). Ed. by M. Ferrario, G. Ciccotti, and K. Binder. Heidelberg: Springer.

Rotne, J. and Prager, S. (1969). Variational treatment of hydrodynamic interaction in polymers. *J. Chem. Phys.* **50**, 4831–4837.

Rottler, J. and Maggs, A. C. (2004). Local molecular dynamics with Coulombic interactions. *Phys. Rev. Lett.* **93**, 170201.

Rowley, L. A., Nicholson, D., and Parsonage, N. G. (1975). Monte Carlo grand canonical ensemble calculation in a gas–liquid transition region of 12–6 argon. *J. Comput. Phys.* **17**, 401–414.

Rowley, L. A., Nicholson, D., and Parsonage, N. G. (1978). Long range corrections to grand canonical ensemble Monte Carlo calculations for adsorption systems. *J. Comput. Phys.* **26**, 66–79.

Rowlinson, J. S. (1963). *The perfect gas.* Oxford: Pergamon Press.

Rowlinson, J. S. (1969). *Liquids and liquid mixtures.* 2nd ed. London: Butterworth.

Rowlinson, J. S. and Swinton, F. L. (1982). *Liquids and liquid mixtures.* 3rd ed. London: Butterworth.

Rowlinson, J. S. and Widom, B. (1982). *Molecular theory of capillarity.* Oxford: Clarendon Press.

Royall, C. P. and Williams, S. R. (2015). The role of local structure in dynamical arrest. *Physics Reports* **560**, 1–75.

Rubinstein, R. Y. (1981). *Simulation and Monte Carlo methods*. New York: Wiley.

Rudzinski, J. F. and Noid, W. G. (2011). Coarse-graining entropy, forces, and structures. *J. Chem. Phys.* **135**, 214101.

Rugh, H. H. (1997). Dynamical approach to temperature. *Phys. Rev. Lett.* **78**, 772–774.

Rühle, V., Junghans, C., Lukyanov, A., Kremer, K., and Andrienko, D. (2009). Versatile object-oriented toolkit for coarse-graining applications. *J. Chem. Theor. Comput.* **5**, 3211–3223.

Rusanov, A. I. and Brodskaya, E. N. (1977). The molecular dynamics simulation of a small drop. *J. Coll. Int. Sci.* **62**, 542–555.

Ryckaert, J.-P. (1985). Special geometrical constraints in the molecular dynamics of chain molecules. *Molec. Phys.* **55**, 549–556.

Ryckaert, J.-P. and Bellemans, A. (1975). Molecular dynamics of liquid n-butane near its boiling point. *Chem. Phys. Lett.* **30**, 123–125.

Ryckaert, J.-P. and Bellemans, A. (1978). Molecular dynamics of liquid alkanes. *Faraday Disc. Chem. Soc.* **66**, 95–106.

Ryckaert, J.-P. and Ciccotti, G. (1983). Introduction of Andersen's demon in the molecular dynamics of systems with constraints. *J. Chem. Phys.* **78**, 7368–7374.

Ryckaert, J.-P. and Ciccotti, G. (1986). Andersen's canonical-ensemble molecular dynamics for molecules with constraints. *Molec. Phys.* **58**, 1125–1136.

Ryckaert, J.-P., Ciccotti, G., and Berendsen, H. J. C. (1977). Numerical integration of the Cartesian equations of motion of a system with constraints: molecular dynamics of n-alkanes. *J. Comput. Phys.* **23**, 327–341.

Sadus, R. J. and Prausnitz, J. M. (1996). Three-body interactions in fluids from molecular simulation: vapor–liquid phase coexistence of argon. *J. Chem. Phys.* **104**, 4784–4787.

Saffman, P. G. (1976). Brownian motion in thin sheets of viscous fluid. *J. Fluid Mech.* **73**, 593–602.

Saffman, P. G. and Delbrück, M. (1975). Brownian motion in biological membranes. *Proc. Nat. Acad. Sci.* **72**, 3111–3113.

Safran, S. A. (1994). *Statistical thermodynamics of surfaces, interfaces, and membranes*. Reading, Massachusetts: Addison Wesley.

Salanne, M., Rotenberg, B., Jahn, S., Vuilleumier, R., Simon, C., and Madden, P. A. (2012). Including many-body effects in models for ionic liquids. *Theor. Chem. Acc.* **131**, 1143.

Salsburg, Z. W., Jacobson, J. D., Fickett, W., and Wood, W. W. (1959). Application of the Monte Carlo method to the lattice gas model. 1. Two dimensional triangular lattice. *J. Chem. Phys.* **30**, 65–72.

Sansom, M. S. P. and Biggin, P. C. (2010). *Molecular simulations and biomembranes: from biophysics to function*. RSC Biomolecular Sciences. The Royal Society of Chemistry.

Sarman, S. and Laaksonen, A. (2009a). Evaluation of the viscosities of a liquid crystal model system by shear flow simulation. *Chem. Phys. Lett.* **479**, 47–51.

Sarman, S. and Laaksonen, A. (2009b). Flow alignment phenomena in liquid crystals studied by molecular dynamics simulation. *J. Chem. Phys.* **131**, 144904.

Sarman, S. and Laaksonen, A. (2011). Molecular dynamics simulation of viscous flow and heat conduction in liquid crystals. *J. Comput. Theo. Nanosci* **8**, 1081–1100.

Sarman, S. and Laaksonen, A. (2015). Molecular dynamics simulation of planar elongational flow in a nematic liquid crystal based on the Gay–Berne potential. *Phys. Chem. Chem. Phys.* **17**, 3332–3342.

Saunders, M. G. and Voth, G. A. (2013). Coarse-graining methods for computational biology. *Ann. Rev. Biophys.* **42**, 73–93.

Schaefer, H. (1972). *The electronic structure of atoms and molecules; a survey of rigorous quantum mechanical results.* Reading, Mass: Addison-Wesley Pub. Co.

Schäfer, L. V., de Jong, D. H., Holt, A., Rzepiela, A. J., de Vries, A. H., Poolman, B., Killian, J. A., and Marrink, S. J. (2011). Lipid packing drives the segregation of transmembrane helices into disordered lipid domains in model membranes. *Proc. Nat. Acad. Sci.* **108**, 1343–1348.

Schlichting, H. (1979). *Boundary layer theory.* 7th ed. New York: McGraw-Hill.

Schmid, F. and Wilding, N. B. (1995). Errors in Monte Carlo simulations using shift register random number generators. *Int. J. Mod. Phys. C* **6**, 781–787.

Schmidt, K. E. and Kalos, M. H. (1984). Few- and many-fermion problems. *Applications of the Monte Carlo method in statistical physics.* Topics Curr. Phys. **36**, 125–143. Ed. by K. Binder. Berlin: Springer.

Schmidt, K. E., Niyaz, P., Vaught, A., and Lee, M. A. (2005). Green's function Monte Carlo method with exact imaginary-time propagation. *Phys. Rev. E* **71**, 016707.

Schmitz, F. and Virnau, P. (2015). The ensemble switch method for computing interfacial tensions. *J. Chem. Phys.* **142**, 144108.

Schmitz, F., Virnau, P., and Binder, K. (2014). Logarithmic finite-size effects on interfacial free energies: phenomenological theory and Monte Carlo studies. *Phys. Rev. E* **90**, 012128.

Schmitz, R. and Felderhof, B. U. (1982). Mobility matrix for two spherical particles with hydrodynamic interaction. *Physica A* **116**, 163–177.

Schoen, M., Vogelsang, R., and Hoheisel, C. (1984). The recurrence time in molecular dynamics ensembles. *CCP5 Quarterly* **13**, 27–37.

Schofield, P. (1960). Space time correlation function formalism for slow neutron scattering. *Phys. Rev. Lett.* **4**, 239–240.

Schofield, P. (1973). Computer simulation studies of the liquid state. *Comput. Phys. Commun.* **5**, 17–23.

Schofield, P. and Henderson, J. R. (1982). Statistical mechanics of inhomogeneous fluids. *Proc. Roy. Soc. Lond. A* **379**, 231–246.

Schommers, W. (1983). Pair potentials in disordered many-particle systems: a study for liquid gallium. *Phys. Rev. A* **28**, 3599–3605.

Schrader, M., Virnau, P., Winter, D., Zykova-Timan, T., and Binder, K. (2009). Methods to extract interfacial free energies of flat and curved interfaces from computer simulations. *Euro. Phys. J. Spec. Top.* **177**, 103–127.

Schuler, L. D., Daura, X., and van Gunsteren, W. F. (2001). An improved GROMOS96 force field for aliphatic hydrocarbons in the condensed phase. *J. Comput. Chem.* **22**, 1205–1218.

Schultz, A. J. and Kofke, D. A. (2011). Algorithm for constant-pressure Monte Carlo simulation of crystalline solids. *Phys. Rev. E* **84**, 046712.

Schultz, A. J. and Kofke, D. A. (2015). Etomica: an object-oriented framework for molecular simulation. *J. Comput. Chem.* **36**, 573–583.

Schweizer, K. S., Stratt, R. M., Chandler, D., and Wolynes, P. G. (1981). Convenient and accurate discretized path integral methods for equilibrium quantum mechanical calculations. *J. Chem. Phys.* **75**, 1347–1364.

Semenov, A. N. (1993). Theory of block copolymer interfaces in the strong segregation limit. *Macro. Chem. Phys.* **26**, 6617–6621.

Semenov, A. N. (1994). Scattering of statistical structure of polymer/polymer interfaces. *Macromolecules* **27**, 2732–2735.

Sergi, A. and Ferrario, M. (2001). Non-Hamiltonian equations of motion with a conserved energy. *Phys. Rev. E* **64**, 056125.

Severin, E. S. and Tildesley, D. J. (1980). A methane molecule adsorbed on a graphite surface. *Molec. Phys.* **41**, 1401–1418.

Severin, E. S., Freasier, B. C., Hamer, N. D., Jolly, D. L., and Nordholm, S. (1978). An efficient microcanonical sampling method. *Chem. Phys. Lett.* **57**, 117–120.

Shalf, J. M. and Leland, R. (2015). Computing beyond Moore's Law. *Computer* **48**, 14–23.

Shan, Y. B., Klepeis, J. L., Eastwood, M. P., Dror, R. O., and Shaw, D. E. (2005). Gaussian split Ewald: a fast Ewald mesh method for molecular simulation. *J. Chem. Phys.* **122**, 054101.

Shardlow, T. (2003). Splitting for dissipative particle dynamics. *SIAM J. Sci. Comput.* **24**, 1267–1282.

Shaw, D. E., Dror, R. O., Salmon, J. K., Grossman, J. P., Mackenzie, K. M., Bank, J. A., Young, C., Deneroff, M. M., Batson, B., Bowers, K. J., Chow, E., Eastwood, M. P., Ierardi, D. J., Klepeis, J. L., Kuskin, J. S., Larson, R. H., Lindorff-Larsen, K., Maragakis, P., Moraes, M. A., Piana, S., Shan, Y., and Towles, B. (2009). Millisecond-scale molecular dynamics simulations on Anton. *SC09: Proceedings of the conference on high performance computing, networking, storage and analysis.* (Portland, USA, 2009). New York: ACM.

Shaw, D. E., Maragakis, P., Lindorff-Larsen, K., Piana, S., Dror, R. O., Eastwood, M. P., Bank, J. A., Jumper, J. M., Salmon, J. K., Shan, Y., and Wriggers, W. (2010). Atomic-level characterization of the structural dynamics of proteins. *Science* **330**, 341–346.

Shaw, D. E., Grossman, J. P., Bank, J. A., Batson, B., Butts, J. A., Chao, J. C., Deneroff, M. M., Dror, R. O., Even, A., Fenton, C. H., Forte, A., Gagliardo, J., Gill, G., Greskamp, B., Ho, C. R., Ierardi, D. J., Iserovich, L., Kuskin, J. S., Larson, R. H., Layman, T., Lee, L.-S., Lerer, A. K., Li, C., Killebrew, D., Mackenzie, K. M., Mok, S. Y.-H., Moraes, M. A., Mueller, R., Nociolo, L. J., Peticolas, J. L., Quan, T., Ramot, D., Salmon, J. K., Scarpazza, D. P., Schafer, U. B., Siddique, N., Snyder, C. W., Spengler, J., Tang, P. T. P., Theobald, M., Toma, H., Towles, B., Vitale, B., Wang, S. C., and Young, C. (2014). Anton 2: raising the bar for performance and programmability in a special-purpose molecular dynamics supercomputer. *SC14: Proceedings of the conference on high performance computing, networking, storage and analysis.* (New Orleans, USA, 2014). Piscataway: IEEE press.

Shell, M. S. (2008). The relative entropy is fundamental to multiscale and inverse thermodynamic problems. *J. Chem. Phys.* **129**, 144108.

Shepherd, J. J., Booth, G., Grueneis, A., and Alavi, A. (2012). Full configuration interaction perspective on the homogeneous electron gas. *Phys. Rev. B* **85**, 081103.

Shi, Q. and Geva, E. (2004). A derivation of the mixed quantum–classical Liouville equation from the influence functional formalism. *J. Chem. Phys.* **121**, 3393–3404.

Shi, Q., Izvekov, S., and Voth, G. A. (2006). Mixed atomistic and coarse-grained molecular dynamics: simulation of a membrane-bound ion channel. *J. Phys. Chem. B* **110**, 15045–15048.

Shillcock, J. C. and Lipowsky, R. (2006). The computational route from bilayer membranes to vesicle fusion. *J. Phys. Cond. Mat.* **18**, 1191–1219.

Shing, K. S. and Chung, S. T. (1987). Computer simulation methods for the calculation of solubility in supercritical extraction systems. *J. Phys. Chem.* **91**, 1674–1681.

Shing, K. S. and Gubbins, K. E. (1981). The chemical potential from computer simulation. *Molec. Phys.* **43**, 717–721.

Shing, K. S. and Gubbins, K. E. (1982). The chemical potential in dense fluids and fluid mixtures via computer simulation. *Molec. Phys.* **46**, 1109–1128.

Shirts, M. R. and Chodera, J. D. (2008). Statistically optimal analysis of samples from multiple equilibrium states. *J. Chem. Phys.* **129**, 124105. A Python implementation of MBAR is available from https://github.com/choderalab/pymbar.

Shirts, R. B., Burt, S. R., and Johnson, A. M. (2006). Periodic boundary condition induced breakdown of the equipartition principle and other kinetic effects of finite sample size in classical hard-sphere molecular dynamics simulation. *J. Chem. Phys.* **125**, 164102.

Shuttleworth, R. (1950). The surface tension of solids. *Proc. Phys. Soc. A* **63**, 444–457.

Shvab, I. and Sadus, R. J. (2013). Intermolecular potentials and the accurate prediction of the thermodynamic properties of water. *J. Chem. Phys.* **139**, 194505.

Siegbahn, P. E. M. and Himo, F. (2009). Recent developments of the quantum chemical cluster approach for modeling enzyme reactions. *J. Bio. Inorg. Chem* **14**, 643–651.

Siepmann, J. I. and Frenkel, D. (1992). Configurational bias Monte Carlo: a new sampling scheme for flexible chains. *Molec. Phys.* **75**, 59–70.

Sikkenk, J. H., Indekeu, J. O., van Leeuwen, J. M. J., Vossnack, E. O., and Bakker, A. F. (1988). Simulation of wetting and drying at solid–fluid interfaces on the Delft molecular dynamics processor. *J. Stat. Phys.* **52**, 23–44.

Silva, C. C. and Martins, R. D. (2002). Polar and axial vectors versus quaternions. *Amer. J. Phys.* **70**, 958–963.

Sindhikara, D. J., Emerson, D. J., and Roitberg, A. E. (2010). Exchange often and properly in replica exchange molecular dynamics. *J. Chem. Theor. Comput.* **6**, 2804–2808.

Singer, J. V. L. and Singer, K. (1984). Molecular dynamics based on the first order correction in the Wigner–Kirkwood expansion. *CCP5 Quarterly* **14**, 24–26.

Singer, K. and Smith, W. (1990). Quantum dynamics and the Wigner–Liouville equation. *Chem. Phys. Lett.* **167**, 298–304.

Singer, K., Taylor, A., and Singer, J. V. L. (1977). Thermodynamic and structural properties of liquids modelled by two-Lennard-Jones centres pair potentials. *Molec. Phys.* **33**, 1757–1795.

Singh, S., Chopra, M., and de Pablo, J. J. (2012). Density of states-based molecular simulations. *Ann. Rev. Chem. Bio. Eng.* **3**, 369–394.

Skeel, R. D., Tezcan, I., and Hardy, D. J. (2002). Multiple grid methods for classical molecular dynamics. *J. Comput. Chem.* **2**, 673–684.

Skilling, J. (2006). Nested sampling for general Bayesian computation. *Bayesian Anal.* **1**, 833–859.

Slater, J. C. and Kirkwood, J. G. (1931). The van der Waals forces in gases. *Phys. Rev.* **37**, 682–697.

Slingsby, J. G., Vyas, S., and Maupin, C. M. (2015). A charge-modified general amber force field for phospholipids: improved structural properties in the tensionless ensemble. *Molec. Simul.* **41**, 1449–1458.

Smirnova, Y. G. and Müller, M. (2015). Calculation of membrane bending rigidity using field-theoretic umbrella sampling. *J. Chem. Phys.* **143**, 243155.

Smit, B. (1992). Phase diagrams of Lennard-Jones fluids. *J. Chem. Phys.* **96**, 8639–8640.

Smit, B. (1995). Grand canonical Monte Carlo simulations of chain molecules: adsorption isotherms of alkanes in zeolites. *Molec. Phys.* **85**, 153–172.

Smit, B., De Smedt, P., and Frenkel, D. (1989). Computer simulations in the Gibbs ensemble. *Molec. Phys.* **68**, 931–950.

Smith, E. B. and Wells, B. H. (1984). Estimating errors in molecular simulation calculations. *Molec. Phys.* **52**, 701–704.

Smith, E. R. (1981). Electrostatic energy in ionic crystals. *Proc. Roy. Soc. Lond. A* **375**, 475–505.

Smith, E. R. (1987). Boundary conditions on hydrodynamics in simulations of dense suspensions. *Faraday Disc. Chem. Soc.* **83**, 193–198.

Smith, E. R., Snook, I. K., and van Megen, W. (1987). Hydrodynamic interactions in Brownian dynamics: I. Periodic boundary conditions for computer simulations. *Physica A* **143**, 441–467.

Smith, S. and Morin, P. A. (2005). Optimal storage conditions for highly dilute DNA samples: a role for trehalose as a preserving agent. *J. Forensic Sci.* **50**, 1101–1108.

Smith, W. (1982a). An introduction to the discrete Fourier transform. *CCP5 Quarterly* **5**, 34–41.

Smith, W. (1982b). Correlation functions and the fast Fourier transform. *CCP5 Quarterly* **7**, 12–24.

Smith, W. (1983). The periodic boundary condition in non-cubic MD cells. Wigner–Seitz cells with reflection symmetry. *CCP5 Quarterly* **10**, 37–42.

Smith, W. R. and Triska, B. (1994). The reaction ensemble method for the computer simulation of chemical and phase equilibria. 1. Theory and basic examples. *J. Chem. Phys.* **100**, 3019–3027.

Smith, W. R., Henderson, D. J., Leonard, P. J., Barker, J. A., and Grundke, E. W. (2008). Fortran codes for the correlation functions of hard sphere fluids. *Molec. Phys.* **106**, 3–7.

Snook, I. K. (2007). *The Langevin and generalised Langevin approach to the dynamics of atomic, polymeric and colloidal systems*. Amsterdam: Elsevier.

Sokhan, V. P., Jones, A. P., Cipcigan, F. S., Crain, J., and Martyna, G. J. (2015). Signature properties of water: their molecular electronic origins. *Proc. Nat. Acad. Sci.* **112**, 6341–6346.

Sølvason, D., Kolafa, J., Petersen, H. G., and Perram, J. W. (1995). A rigorous comparison of the Ewald method and the fast multipole method in 2 dimensions. *Comput. Phys. Commun.* **87**, 307–318.

Son, C. Y., McDaniel, J. G., Schmidt, J. R., Cui, Q., and Yethiraj, A. (2016). First-principles united atom force field for the ionic liquid $BMIM^+BF_4^-$: an alternative to charge scaling. *J. Phys. Chem. B* **120**, 3560–3568.

Soper, A. K. (1996). Empirical potential Monte Carlo simulation of fluid structure. *Chem. Phys.* **202**, 295–306.

Sorensen, M. R. and Voter, A. F. (2000). Temperature-accelerated dynamics for simulation of infrequent events. *J. Chem. Phys.* **112**, 9599–9606.

Souaille, M., Loirat, H., Borgis, D., and Gaigeot, M. P. (2009). MDVRY: a polarizable classical molecular dynamics package for biomolecules. *Comput. Phys. Commun.* **180**, 276–301.

Spångberg, D., Larsson, D. S. D., and van der Spoel, D. (2011). Trajectory NG: portable, compressed, general molecular dynamics trajectories. *J. Molec. Model.* **17**, 2669–2685.

Spill, Y. G., Bouvier, G., and Nilges, M. (2013). A convective replica-exchange method for sampling new energy basins. *J. Comput. Chem.* **34**, 132–140.

Spohr, E. (1997). Effect of electrostatic boundary conditions and system size on the interfacial properties of water and aqueous solutions. *J. Chem. Phys.* **107**, 6342–6348.

Sprik, M. (1991). Computer simulation of the dynamics of induced polarization fluctuations in water. *J. Phys. Chem.* **95**, 2283–2291.

Sprik, M. and Klein, M. L. (1988). A polarizable model for water using distributed charge sites. *J. Chem. Phys.* **89**, 7556–7560.

Sprik, M., Klein, M. L., and Chandler, D. (1985). Staging: a sampling technique for the Monte Carlo evaluation of path integrals. *Phys. Rev. B* **31**, 4234–4244.

St. Pierre, A. G. and Steele, W. A. (1969). The rotational Wigner function. *Ann. Phys.* **52**, 251–292.

Stapleton, M. R., Tildesley, D. J., Panagiotopoulos, A. Z., and Quirke, N. (1989). Phase equilibria of quadrupolar fluids by simulation in the Gibbs ensemble. *Molec. Simul.* **2**, 147–162.

Stecher, T. and Althorpe, S. C. (2012). Improved free-energy interpolation schemes for obtaining gas-phase reaction rates from ring-polymer dynamics. *Molec. Phys.* **110**, 875–883.

Steele, W. A. (1969). Time correlation functions. *Transport phenomena in fluids*, 209–312. Ed. by H. J. M. Hanley. New York: Dekker.

Steele, W. A. (1980). Molecular reorientation in dense systems. *Intermolecular spectroscopy and dynamical properties of dense systems*. Proc. Int. School Phys. Enrico Fermi. **75**, 325–374. (Varenna, Italy, 1978). Ed. by J. van Kranendonk. Bologna: Soc. Italiana di Fisica.

Steinhardt, P. J., Nelson, D. R., and Ronchetti, M. (1983). Bond-orientational order in liquids and glasses. *Phys. Rev. B* **28**, 784–805.

Stern, H. A. and Calkins, K. G. (2008). On mesh-based Ewald methods: optimal parameters for two differentiation schemes. *J. Chem. Phys.* **128**, 214106.

Stillinger, F. H. and Weber, T. A. (1985). Computer simulation of local order in condensed phases of silicon. *Phys. Rev. B* **31**, 5262–5271. Erratum: *ibid.* **31**, 5262 (1985).

Stoddard, S. D. and Ford, J. (1973). Numerical experiments on the stochastic behavior of a Lennard-Jones gas system. *Phys. Rev. A* **8**, 1504–1512.

Stoddard, S. D. (1978). Identifying clusters in computer experiments on systems of particles. *J. Comput. Phys.* **27**, 291–293.

Stone, A. J. (1981). Distributed multipole analysis, or how to describe a molecular charge distribution. *Chem. Phys. Lett.* **83**, 233–239.

Stone, A. J. (2013). *The theory of intermolecular forces.* 2nd ed. Oxford: Clarendon Press.

Stone, J. E., Phillips, J. C., Freddolino, P. L., Hardy, D. J., Trabuco, L. G., and Schulten, K. (2007). Accelerating molecular modeling applications with graphics processors. *J. Comput. Chem.* **28**, 2618–2640.

Straatsma, T. P. and McCammon, J. A. (1990). Molecular dynamics simulations with interaction potentials including polarization development of a noniterative method and application to water. *Molec. Simul.* **5**, 181–192.

Strandburg, K. J. (1988). Two-dimensional melting. *Rev. Mod. Phys.* **60**, 161–207.

Stratford, K. and Pagonabarraga, I. (2008). Parallel simulation of particle suspensions with the lattice Boltzmann method. *Comput. Math. Appl.* **55**, 1585–1593.

Stratt, R. M., Holmgren, S. L., and Chandler, D. (1981). Constrained impulsive molecular dynamics. *Molec. Phys.* **42**, 1233–1143.

Streett, W. B. and Tildesley, D. J. (1976). Computer simulations of polyatomic molecules. I. Monte Carlo studies of hard diatomics. *Proc. Roy. Soc. Lond. A* **348**, 485–510.

Streett, W. B. and Tildesley, D. J. (1977). Computer simulations of polyatomic molecules. II. Molecular dynamics studies of diatomic liquids with atom–atom and quadrupole–quadrupole potentials. *Proc. Roy. Soc. Lond. A* **355**, 239–266.

Streett, W. B. and Tildesley, D. J. (1978). Computer simulations of polyatomic molecules. III. Monte Carlo simulation studies of heteronuclear and homonuclear hard diatomics. *J. Chem. Phys.* **68**, 1275–1284.

Streett, W. B., Tildesley, D. J., and Saville, G. (1978). Multiple time step methods and an improved potential function for molecular dynamics simulations of molecular liquids. *Computer modelling of matter.* ACS Symp. Ser. **86**, 144–158. (Anaheim, USA, 1978). Ed. by P. Lykos. Washington: American Chemical Society.

Sturgeon, J. B. and Laird, B. B. (2000). Symplectic algorithm for constant-pressure molecular dynamics using a Nosé–Poincaré thermostat. *J. Chem. Phys.* **112**, 3474–3482.

Subramanian, G. and Davis, H. T. (1975). Molecular dynamical studies of rough-sphere fluids. *Phys. Rev. A* **11**, 1430–1439.

Subramanian, G. and Davis, H. T. (1979). Molecular dynamics of a hard sphere fluid in small pores. *Molec. Phys.* **38**, 1061–1066.

Succi, S. (2013). *The lattice Boltzmann equation: for fluid dynamics and beyond.* Oxford: Oxford University Press.

Sugita, Y. and Okamoto, Y. (1999). Replica-exchange molecular dynamics methods for protein folding. *Chem. Phys. Lett.* **314**, 141–151.

Sun, H. (1998). COMPASS: an *ab initio* force-field optimized for condensed-phase applications. Overview with details on alkane and benzene compounds. *J. Phys. Chem. B* **102**, 7338–7364.

Suzuki, K. and Nakata, Y. (1970). The three-dimensional structure of macromolecules. 1. The conformation of ethylene polymers by the Monte Carlo method. *Bull. Chem. Soc. Japan* **43**, 1006–1010.

Svoboda, M., Malijevský, A., and Lísal, M. (2015). Wetting properties of molecularly rough surfaces. *J. Chem. Phys.* **143**, 104701.

Swendsen, R. H. and Wang, J.-S. (1986). Replica Monte Carlo simulation of spin-glasses. *Phys. Rev. Lett.* **57**, 2607–2609.

Swendsen, R. H. (2002). Statistical mechanics of classical systems with distinguishable particles. *J. Stat. Phys.* **107**, 1143–1166.

Swendsen, R. H. (2006). Statistical mechanics of colloids and Boltzmann's definition of the entropy. *Amer. J. Phys.* **74**, 187–190.

Swendsen, R. H. (2012). *An introduction to statistical mechanics and thermodynamics.* Oxford: Oxford University Press.

Swift, M. R., Osborn, W. R., and Yeomans, J. M. (1995). Lattice Boltzmann simulation of non-ideal fluids. *Phys. Rev. Lett.* **75**, 830–833.

Swift, M. R., Orlandini, E., Osborn, W. R., and Yeomans, J. M. (1996). Lattice Boltzmann simulations of liquid–gas and binary fluid systems. *Phys. Rev. E* **54**, 5041–5052.

Swope, W. C. and Andersen, H. C. (1990). 10^6-particle molecular dynamics study of homogeneous nucleation of crystals in a supercooled atomic liquid. *Phys. Rev. B* **41**, 7042–7054.

Swope, W. C., Andersen, H. C., Berens, P. H., and Wilson, K. R. (1982). A computer simulation method for the calculation of equilibrium constants for the formation of physical clusters of molecules: application to small water clusters. *J. Chem. Phys.* **76**, 637–649.

Swope, W. C. and Andersen, H. C. (1984). A molecular dynamics method for calculating the solubility of gases in liquids and the hydrophobic hydration of inert-gas atoms in aqueous solution. *J. Phys. Chem.* **88**, 6548–6556.

Szleifer, I., Kramer, D., Ben-Shaul, A., Gelbart, W. M., and Safran, S. A. (1990). Molecular theory of curvature elasticity in surfactant films. *J. Chem. Phys.* **92**, 6800–6817.

Takahashi, K., Narumi, T., and Yasuoka, K. (2010). Cutoff radius effect of the isotropic periodic sum method in homogeneous system. II. Water. *J. Chem. Phys.* **133**, 014109.

Takahashi, M. and Imada, M. (1984). Monte Carlo calculation of quantum-systems. II. Higher order correction. *J. Phys. Soc. Japan* **53**, 3765–3769.

Tang, Z., Palafox-Hernandez, J. P., Law, W.-C., Hughes, Z. E., Swihart, M. T., Prasad, P. N., Knecht, M. R., and Walsh, T. R. (2013). Biomolecular recognition principles for bionanocombinatorics: an integrated approach to elucidate enthalpic and entropic factors. *ACS Nano* **7**, 9632–9646.

Tarazona, P. and Chacón, E. (2004). Monte Carlo intrinsic surfaces and density profiles for liquid surfaces. *Phys. Rev. B* **70**, 235407.

Tarazona, P., Chacón, E., and Bresme, F. (2012). Intrinsic profiles and the structure of liquid surfaces. *J. Phys. Cond. Mat.* **24**, 284123.

Tarmyshov, K. B. and Müller-Plathe, F. (2005). Parallelizing a molecular dynamics algorithm on a multiprocessor workstation using openMP. *J. Chem. Inf. Model.* **45**, 1943–1952.

ten Wolde, P. R. and Frenkel, D. (1998). Computer simulation study of gas–liquid nucleation in a Lennard-Jones system. *J. Chem. Phys.* **109**, 9901–9918.

ten Wolde, P. R., Ruiz-Montero, M. J., and Frenkel, D. (1996). Numerical calculation of the rate of crystal nucleation in a Lennard-Jones system at moderate undercooling. *J. Chem. Phys.* **104**, 9932–9947.

ten Wolde, P. R., Ruiz-Montero, M. J., and Frenkel, D. (1995). Numerical evidence for bcc ordering at the surface of a critical fcc nucleus. *Phys. Rev. Lett.* **75**, 2714–2717.

Tenenbaum, A., Ciccotti, G., and Gallico, R. (1982). Stationary nonequilibrium states by molecular dynamics. Fourier's law. *Phys. Rev. A* **25**, 2778–2787.

Tenney, C. M. and Maginn, E. J. (2010). Limitations and recommendations for the calculation of shear viscosity using reverse nonequilibrium molecular dynamics. *J. Chem. Phys.* **132**, 014103.

Terada, T. and Kidera, A. (2002). Generalized form of the conserved quantity in constant-temperature molecular dynamics. *J. Chem. Phys.* **116**, 33–41.

TestU01 (2009). *TestU01: a C library for empirical testing of random number generators.* http://simul.iro.umontreal.ca/testu01/tu01.html (Accessed Nov 2015).

Teter, M. P., Payne, M. C., and Allan, D. C. (1989). Solution of Schrödinger's equation for large systems. *Phys. Rev. B* **40**, 12255–12263.

Theodorou, D. N. (2010). Progress and outlook in Monte Carlo simulations. *Ind. Eng. Chem. Res.* **49**, 3047–3058.

Thirumalai, D. and Berne, B. J. (1983). On the calculation of time correlation functions in quantum systems: path integral techniques. *J. Chem. Phys.* **79**, 5029–5033.

Thirumalai, D. and Berne, B. J. (1984). Time correlation functions in quantum systems. *J. Chem. Phys.* **81**, 2512–2513.

Thole, B. T. (1981). Molecular polarizabilities calculated with a modified dipole interaction. *Chem. Phys.* **59**, 341–350.

Thomas, J. C. and Rowley, R. L. (2007). Transient molecular dynamics simulations of viscosity for simple fluids. *J. Chem. Phys.* **127**, 174510.

Thompson, S. M. (1983). Use of neighbour lists in molecular dynamics. *CCP5 Quarterly* **8**, 20–28.

Thompson, S. M., Gubbins, K. E., Walton, J. P. R. B., Chantry, R. A. R., and Rowlinson, J. S. (1984). A molecular dynamics study of liquid drops. *J. Chem. Phys.* **81**, 530–542.

Thota, A., Luckow, A., and Jha, S. (2011). Efficient large-scale replica-exchange simulations on production infrastructure. *Phil. Trans. Roy. Soc.* **369**, 3318–3335.

Tieleman, D. P. (2010). Methods and parameters for membrane simulations. *Molecular simulations and biomembranes.* Chap. 1, 1–25. Ed. by M. S. P. Sansom and P. C. Biggin. Royal Society of Chemistry.

Tieleman, D. P. (2012). Computer simulation of membrane dynamics. *Comprehensive Biophysics.* **5**, 312–336. Elsevier BV.

Tildesley, D. J. and Madden, P. A. (1981). An effective pair potential for liquid carbon disulphide. *Molec. Phys.* **42**, 1137–1156.

Tiller, W. A. (1991). *The science of crystallization: microscopic interfacial phenomena.* Cambridge: Cambridge University Press.

Tironi, I. G., Sperb, R., Smith, P. E., and van Gunsteren, W. F. (1995). A generalized reaction field method for molecular dynamics simulations. *J. Chem. Phys.* **102**, 5451–5459.

Tobochnik, J. and Chester, G. V. (1980). The melting of two-dimensional solids. *Ordering in two dimensions*, 339–340. Ed. by S. K. Sinha. Amsterdam: North Holland.

Tobochnik, J. and Chester, G. V. (1982). Monte Carlo study of melting in two dimensions. *Phys. Rev. B* **25**, 6778–6798.

Todd, B. D. and Daivis, P. J. (1998). Nonequilibrium molecular dynamics simulations of planar elongational flow with spatially and temporally periodic boundary conditions. *Phys. Rev. Lett.* **81**, 1118–1121.

Todd, B. D. and Daivis, P. J. (1999). A new algorithm for unrestricted duration nonequilibrium molecular dynamics simulations of planar elongational flow. *Comput. Phys. Commun.* **117**, 191–199.

Todd, B. D. and Daivis, P. J. (2007). Homogeneous non-equilibrium molecular dynamics simulations of viscous flow: techniques and applications. *Molec. Simul.* **33**, 189–229.

Todd, B. D., Evans, D. J., and Daivis, P. J. (1995). Pressure tensor for inhomogeneous fluids. *Phys. Rev. E* **52**, 1627–1638.

Todorov, I. T. and Smith, W. (2011). *The DLPOLY user manual version 4.02.0.* available from http://www.ccp5.ac.uk/dl_poly/.

Tolman, R. C. (1938). *The principles of statistical mechanics.* Oxford: Clarendon Press.

Torrie, G. M. and Valleau, J. P. (1974). Monte Carlo free energy estimates using non-Boltzmann sampling: application to the sub-critical Lennard-Jones fluid. *Chem. Phys. Lett.* **28**, 578–581.

Torrie, G. M. and Valleau, J. P. (1977a). Monte Carlo study of a phase-separating liquid mixture by umbrella sampling. *J. Chem. Phys.* **66**, 1402–1408.

Torrie, G. M. and Valleau, J. P. (1977b). Nonphysical sampling distributions in Monte Carlo free-energy estimation: umbrella sampling. *J. Comput. Phys.* **23**, 187–199.

Torrie, G. M. and Valleau, J. P. (1979). A Monte Carlo study of an electrical double layer. *Chem. Phys. Lett.* **65**, 343–346.

Toukan, K. and Rahman, A. (1985). Molecular-dynamics study of atomic motions in water. *Phys. Rev. B* **31**, 2643–2648.

Towse, C.-L. and Daggett, V. (2015). Modeling protein folding pathways. *Rev. Comput. Chem.* **28**, 87–135. Ed. by A. L. Parrill and K. B. Lipkowitz. Wiley-Blackwell.

Toxvaerd, S. (1972). Surface structure of simple fluids. *Prog. Surf. Sci.* **3**, 189–220.

Toxvaerd, S. (1994). Hamiltonians for discrete dynamics. *Phys. Rev. E* **50**, 2271–2274.

Toxvaerd, S., Heilmann, O. J., and Dyre, J. C. (2012). Energy conservation in molecular dynamics simulations of classical systems. *J. Chem. Phys.* **136**, 224106.

Travis, K. P. and Braga, C. (2006). Configurational temperature and pressure molecular dynamics: review of current methodology and applications to the shear flow of a simple fluid. *Molec. Phys.* **104**, 3735–3749.

Travis, K. P. and Braga, C. (2008). Configurational temperature control for atomic and molecular systems. *J. Chem. Phys.* **128**, 014111.

Travis, K. P., Daivis, P. J., and Evans, D. J. (1995). Thermostats for molecular fluids undergoing shear flow: application to liquid chlorine. *J. Chem. Phys.* **103**, 10638–10651. Erratum: *ibid.* **105**, 3893–3894 (1996).

Tribello, G. A., Ceriotti, M., and Parrinello, M. (2012). Using sketch-map coordinates to analyze and bias molecular dynamics simulations. *Proc. Nat. Acad. Sci.* **109**, 5196.

Tribello, G. A., Ceriotti, M., and Parrinello, M. (2010). A self-learning algorithm for biased molecular dynamics. *Proc. Nat. Acad. Sci.* **107**, 17509–17514.

Trokhymchuk, A. and Alejandre, J. (1999). Computer simulations of liquid/vapor interface in Lennard-Jones fluids: some questions and answers. *J. Chem. Phys.* **111**, 8510–8523.

Troullier, N. and Martins, J. L. (1991). Efficient pseudopotentials for plane-wave calculations. *Phys. Rev. B* **43**, 1993–2006.

Trudu, F., Donadio, D., and Parrinello, M. (2006). Freezing of a Lennard-Jones fluid: from nucleation to spinodal regime. *Phys. Rev. Lett.* **97**, 105701.

Truhlar, D. G., Garrett, B. C., and Klippenstein, S. J. (1996). Current status of transition-state theory. *J. Phys. Chem.* **100**, 12771–12800.

Tschöp, W., Kremer, K., Batoulis, J., Burger, T., and Hahn, O. (1998). Simulation of polymer melts. I. Coarse-graining procedure for polycarbonates. *Acta Polymerica* **49**, 61–74.

Tuckerman, M. E. (2010). *Statistical mechanics: theory and molecular simulation.* Oxford: Oxford University Press.

Tuckerman, M. E. and Parrinello, M. (1994). Integrating the Car–Parrinello equations. 1. Basic integration techniques. *J. Chem. Phys.* **101**, 1302–1315.

Tuckerman, M. E., Berne, B. J., and Martyna, G. J. (1992). Reversible multiple time scale molecular dynamics. *J. Chem. Phys.* **97**, 1990–2001.

Tuckerman, M. E., Berne, B. J., Martyna, G. J., and Klein, M. L. (1993). Efficient molecular dynamics and hybrid Monte Carlo algorithms for path integrals. *J. Chem. Phys.* **99**, 2796–2808.

Tuckerman, M. E., Mundy, C. J., Balasubramanian, S., and Klein, M. L. (1997). Modified nonequilibrium molecular dynamics for fluid flows with energy conservation. *J. Chem. Phys.* **106**, 5615–5621.

Tuckerman, M. E., Mundy, C. J., and Martyna, G. J. (1999). On the classical statistical mechanics of non-Hamiltonian systems. *Europhys. Lett.* **45**, 149–155.

Tuckerman, M. E., Liu, Y., Ciccotti, G., and Martyna, G. J. (2001). Non-Hamiltonian molecular dynamics: generalizing Hamiltonian phase space principles to non-Hamiltonian systems. *J. Chem. Phys.* **115**, 1678–1702.

Tuckerman, M. E., Marx, D., and Parrinello, M. (2002). The nature and transport mechanism of hydrated hydroxide ions in aqueous solution. *Nature* **417**, 925–929.

Tuckerman, M. E., Alejandre, J., López-Rendón, R., Jochim, A. L., and Martyna, G. J. (2006). A Liouville-operator derived measure-preserving integrator for molecular dynamics simulations in the isothermal–isobaric ensemble. *J. Phys. A Math. Gen.* **39**, 5629–5651.

Tully, J. C. (1990). Molecular dynamics with electronic transitions. *J. Chem. Phys.* **93**, 1061–1071.

Tully, J. C. (1998). Mixed quantum–classical dynamics. *Faraday Disc. Chem. Soc.* **110**, 407–419.

Turci, F., Schilling, T., Yamani, M. H., and Oettel, M. (2014). Solid phase properties and crystallization in simple model systems. *Euro. Phys. J. Spec. Top.* **223**, 421–438.

Turci, F. and Schilling, T. (2014). Crystal growth from a supersaturated melt: relaxation of the solid–liquid dynamic stiffness. *J. Chem. Phys.* **141**, 054706.

Turner, C. H., Brennan, J. K., Lisal, M., Smith, W. R., Johnson, J. K., and Gubbins, K. E. (2008). Simulation of chemical reaction equilibria by the reaction ensemble Monte Carlo method: a review. *Molec. Simul.* **34**, 119–146.

Tüzel, E., Strauss, M., Ihle, T., and Kroll, D. M. (2003). Transport coefficients for stochastic rotation dynamics in three dimensions. *Phys. Rev. E* **68**, 036701.

Tuzun, R. E., Noid, D. W., and Sumpter, B. G. (1997). Efficient treatment of out-of-plane bend and improper torsion interactions in MM2, MM3, and MM4 molecular mechanics calculations. *J. Comput. Chem.* **18**, 1804–1811.

Uhlherr, A., Leak, S. J., Adam, N. E., Nyberg, P. E., Doxastakis, M., Mavrantzas, V. G., and Theodorou, D. N. (2002). Large scale atomistic polymer simulations using Monte Carlo methods for parallel vector processors. *Comput. Phys. Commun.* **144**, 1–22.

Uline, M. J., Siderius, D. W., and Corti, D. S. (2008). On the generalized equipartition theorem in molecular dynamics ensembles and the microcanonical thermodynamics of small systems. *J. Chem. Phys.* **128**, 124301.

Ullrich, C. A. (2011). *Time-dependent density-functional theory: concepts and applications*. Oxford Graduate Texts. Oxford University Press.

Ulmschneider, J. P. and Jorgensen, W. L. (2003). Monte Carlo backbone sampling for polypeptides with variable bond angles and dihedral angles using concerted rotations and a Gaussian bias. *J. Chem. Phys.* **118**, 4261–4271.

Ulmschneider, J. P. and Jorgensen, W. L. (2004). Monte Carlo backbone sampling for nucleic acids using concerted rotations including variable bond angles. *J. Phys. Chem. B* **108**, 16883–16892.

Vaikuntanathan, S. and Jarzynski, C. (2011). Escorted free energy simulations. *J. Chem. Phys.* **134**, 054107.

Valleau, J. P. (1980). The problem of Coulombic forces in computer simulation. *The problem of long range forces in the computer simulation of condensed matter.* NRCC Workshop Proceedings. **9**, 3–8. (Menlo Park, USA, 1980). Ed. by D. Ceperley.

Valleau, J. P. and Card, D. N. (1972). Monte Carlo estimation of the free energy by multistage sampling. *J. Chem. Phys.* **57**, 5457–5462.

Valleau, J. P. and Torrie, G. M. (1977). A guide to Monte Carlo for statistical mechanics. 2. Byways. *Statistical mechanics B. Modern theoretical chemistry.* **6**, 169–194. Ed. by B. J. Berne. New York: Plenum.

Valleau, J. P. and Whittington, S. G. (1977a). A guide to Monte Carlo for statistical mechanics. 1. Highways. *Statistical mechanics A. Modern theoretical chemistry.* **5**, 137–168. Ed. by B. J. Berne. New York: Plenum Press.

Valleau, J. P. and Whittington, S. G. (1977b). Monte Carlo in statistical mechanics: choosing between alternative transition matrices. *J. Comput. Phys.* **24**, 150–157.

Valleau, J. P. (1999). Thermodynamic-scaling methods in Monte Carlo and their application to phase equilibria. *Adv. Chem. Phys.* **105**, 369–404. Ed. by I. Prigogine and S. A. Rice. Wiley-Blackwell.

van Anders, G., Klotsa, D., Ahmed, N. K., Engel, M., and Glotzer, S. C. (2014). Understanding shape entropy through local dense packing. *Proc. Nat. Acad. Sci.* **111**, E4812–E4821.

van den Bogaart, G., Meyenberg, K., Risselada, H. J., Amin, H., Willig, K. I., Hubrich, B. E., Dier, M., Hell, S. W., Grubmüller, H., Diederichsen, U., and Jahn, R. (2011). Membrane protein sequestering by ionic protein–lipid interactions. *Nature* **479**, 552–555.

van Duin, A. C. T., Dasgupta, S., Lorant, F., and Goddard, W. A. (2001). ReaxFF: a reactive force field for hydrocarbons. *J. Phys. Chem. A* **105**, 9396–9409.

van Erp, T. S. (2007). Reaction rate calculation by parallel path swapping. *Phys. Rev. Lett.* **98**, 268301.

van Erp, T. S. (2012). Dynamical rare event simulation techniques for equilibrium and nonequilibrium systems. *Adv. Chem. Phys.* **151**, 27–60.

van Erp, T. S. and Bolhuis, P. G. (2005). Elaborating transition interface sampling methods. *J. Comput. Phys.* **205**, 157–181.

van Erp, T. S., Moroni, D., and Bolhuis, P. G. (2003). A novel path sampling method for the calculation of rate constants. *J. Chem. Phys.* **118**, 7762–7774.

van Giessen, A. E. and Blokhuis, E. M. (2009). Direct determination of the Tolman length from the bulk pressures of liquid drops via molecular dynamics simulations. *J. Chem. Phys.* **131**, 164705.

van Gunsteren, W. F. (1980). Constrained dynamics of flexible molecules. *Molec. Phys.* **40**, 1015–1019.

van Gunsteren, W. F. and Berendsen, H. J. C. (1977). Algorithms for macromolecular dynamics and constraint dynamics. *Molec. Phys.* **34**, 1311–1327.

van Gunsteren, W. F. and Berendsen, H. J. C. (1982). Algorithms for Brownian dynamics. *Molec. Phys.* **45**, 637–647.

van Gunsteren, W. F. and Berendsen, H. J. C. (1988). A leap-frog algorithm for stochastic dynamics. *Molec. Simul.* **1**, 173–185.

van Gunsteren, W. F. and Karplus, M. (1982). Effect of constraints on the dynamics of macromolecules. *Macromolecules* **15**, 1528–1544.

van Saarloos, W. and Mazur, P. (1983). Many-sphere hydrodynamic interactions. II. Mobilities at finite frequencies. *Physica A* **120**, 77–102.

van Swol, F. and Henderson, J. R. (1989). Wetting and drying transitions at a fluid–wall interface: density-functional theory versus computer simulation. *Phys. Rev. A* **40**, 2567–2578.

van Swol, F. and Henderson, J. R. (1991). Wetting and drying transitions at a fluid–wall interface: density-functional theory versus computer simulation. II. *Phys. Rev. A* **43**, 2932–2942.

van Zon, R. and Schofield, J. (2007a). Numerical implementation of the exact dynamics of free rigid bodies. *J. Comput. Phys.* **225**, 145–164.

van Zon, R. and Schofield, J. (2007b). Symplectic algorithms for simulations of rigid-body systems using the exact solution of free motion. *Phys. Rev. E* **75**, 056701.

Vanden-Eijnden, E. and Tal, F. A. (2005). Transition state theory: variational formulation, dynamical corrections, and error estimates. *J. Chem. Phys.* **123**, 184103.

Vanderbilt, D. (1985). Optimally smooth norm-conserving pseudopotentials. *Phys. Rev. B* **32**, 8412–8415.

VandeVondele, J., Krack, M., Mohamed, F., Parrinello, M., Chassaing, T., and Hutter, J. (2005). QUICKSTEP: fast and accurate density functional calculations using a mixed Gaussian and plane waves approach. *Comput. Phys. Commun.* **167**, 103–128.

Varnik, F. and Binder, K. (2002). Shear viscosity of a supercooled polymer melt via nonequilibrium molecular dynamics simulations. *J. Chem. Phys.* **117**, 6336–6349.

Vattulainen, I., Karttunen, M., Besold, G., and Polson, J. M. (2002). Integration schemes for dissipative particle dynamics simulations: from softly interacting systems towards hybrid models. *J. Chem. Phys.* **116**, 3967–3979.

Vega, C., Sanz, E., Abascal, J. L. F., and Noya, E. G. (2008). Determination of phase diagrams via computer simulation: methodology and applications to water, electrolytes and proteins. *J. Phys. Cond. Mat.* **20**, 153101.

Venable, R. M., Chen, L. E., and Pastor, R. W. (2009). Comparison of the extended isotropic periodic sum and particle mesh Ewald methods for simulations of lipid bilayers and monolayers. *J. Phys. Chem. B* **113**, 5855–5862.

Venneri, G. D. and Hoover, W. G. (1987). Simple exact test for well-known molecular dynamics algorithms. *J. Comput. Phys.* **73**, 468–475.

Venturoli, M. and Smit, B. (1999). Simulating the self-assembly of model membranes. *Phys. Chem. Comm.* **2**, 45–49.

Venturoli, M., Smit, B., and Sperotto, M. M. (2005). Simulation studies of protein-induced bilayer deformations, and lipid-induced protein tilting, on a mesoscopic model for lipid bilayers with embedded proteins. *Biophys. J.* **88**, 1778–1798.

Venturoli, M., Sperotto, M. M., Kranenburg, M., and Smit, B. (2006). Mesoscopic models of biological membranes. *Physics Reports* **437**, 1–54.

Verlet, L. (1967). Computer experiments on classical fluids. I. Thermodynamical properties of Lennard-Jones molecules. *Phys. Rev.* **159**, 98–103.

Verlet, L. (1968). Computer experiments on classical fluids. II. Equilibrium correlation functions. *Phys. Rev.* **165**, 201–214.

Vesely, F. J. (1982). Angular Monte Carlo integration using quaternion parameters: a spherical reference potential for carbon tetrachloride. *J. Comput. Phys.* **47**, 291–296.

Vieillard-Baron, J. (1972). Phase transitions of the classical hard-ellipse system. *J. Chem. Phys.* **56**, 4729–4744.

Vila Verde, A., Acres, J. M., and Maranas, J. K. (2009). Investigating the specificity of peptide adsorption on gold using molecular dynamics simulations. *Biomacromolecules* **10**, 2118–2128.

Vila Verde, A., Beltramo, P. J., and Maranas, J. K. (2011). Adsorption of homopolypeptides on gold investigated using atomistic molecular dynamics. *Langmuir* **27**, 5918–5926.

Villarreal, M. A. and Montich, G. G. (2005). On the Ewald artifacts in computer simulations. The test case of the octaalanine peptide with charged termini. *J. Biomolec. Struc. Dynam.* **23**, 135–142.

Vink, R. L. C., Horbach, J., and Binder, K. (2005). Capillary waves in a colloid–polymer interface. *J. Chem. Phys.* **122**, 134905.

Virnau, P. and Müller, M. (2004). Calculation of free energy through successive umbrella sampling. *J. Chem. Phys.* **120**, 10925–10930.

Virnau, P., Müller, M., MacDowell, L. G., and Binder, K. (2004). Phase behavior of n-alkanes in supercritical solution: a Monte Carlo study. *J. Chem. Phys.* **121**, 2169.

Vlugt, T. J. H. and Smit, B. (2001). On the efficient sampling of pathways in the transition path ensemble. *Phys. Chem. Comm.* **4**, 11–17.

von Lilienfeld, O. A., Tavernelli, I., Röthlisberger, U., and Sebastiani, D. (2005). Variational optimization of effective atom centered potentials for molecular properties. *J. Chem. Phys.* **122**, 014113.

von Neumann, J. (1951). Various techniques used in connection with random digits. *U. S. Nat. Bur. Stand. Appl. Math. Ser.* **12**, 36–38.

von Neumann, J. and Ulam, S. (1945). Random ergodic theorems. *Bull. Amer. Math. Soc.* **51(9)**. No. 165, 660.

Vorholz, J., Harismiadis, V. I., Panagiotopoulos, A. Z., Rumpf, B., and Maurer, G. (2004). Molecular simulation of the solubility of carbon dioxide in aqueous solutions of sodium chloride. *Fluid Phase Equilibria* **226**, 237–250.

Voronstov-Vel'Yaminov, P. N., El'y-Ashevich, A. M., Morgenshtern, L. A., and Chakovskikh, V. P. (1970). Investigations of phase transitions in argon and Coulomb gas by the Monte Carlo method using an isothermically isobaric ensemble. *High Temp. Res. USSR* **8**, 261–268.

Voth, G. A. (1996). Path-integral centroid methods in quantum statistical mechanics and dynamics. *Adv. Chem. Phys.* **93**, 135–218.

Wagner, A. J. and Pagonabarraga, I. (2002). Lees–Edwards boundary conditions for lattice Boltzmann. *J. Stat. Phys.* **107**, 521–537.

Wagner, A. J. and Yeomans, J. M. (1999). Phase separation under shear in two-dimensional binary fluids. *Phys. Rev. E* **59**, 4366–4373.

Waheed, Q. and Edholm, O. (2011). Quantum corrections to classical molecular dynamics simulations of water and ice. *J. Chem. Theor. Comput.* **7**, 2903–2909.

Wales, D. (2004). *Energy landscapes.* Cambridge: Cambridge University Press.

Wall, F. T. and Mandel, F. (1975). Macromolecular dimensions obtained by an efficient Monte Carlo method without sample attrition. *J. Chem. Phys.* **63**, 4592–4595.

Wall, F. T., Chin, J. C., and Mandel, F. (1977). Configurations of macromolecular chains confined to strips or tubes. *J. Chem. Phys.* **66**, 3066–3069.

Walton, J. P. R. B., Tildesley, D. J., Rowlinson, J. S., and Henderson, J. R. (1983). The pressure tensor at the planar surface of a liquid. *Molec. Phys.* **48**, 1357–1368. Erratum: *ibid.* **50**, 1381 (1983).

Wang, F. G. and Landau, D. P. (2001a). Determining the density of states for classical statistical models: a random walk algorithm to produce a flat histogram. *Phys. Rev. E* **64**, 056101.

Wang, F. G. and Landau, D. P. (2001b). Efficient multiple-range random walk algorithm to calculate the density of states. *Phys. Rev. Lett.* **86**, 2050–2053.

Wang, G. M., Sevick, E. M., Mittag, E., Searles, D. J., and Evans, D. J. (2002). Experimental demonstration of violations of the second law of thermodynamics for small systems and short time scales. *Phys. Rev. Lett.* **89**, 050601.

Wang, H. Y. and LeSar, R. (1996). An efficient fast-multipole algorithm based on an expansion in the solid harmonics. *J. Chem. Phys.* **104**, 4173–4179.

Wang, H., Hartmann, C., Schütte, C., and Delle Site, L. (2013). Grand-canonical-like molecular dynamics simulations by using an adaptive-resolution technique. *Phys. Rev. X* **3**, 011018.

Wang, J., Yoo, S., Bai, J., Morris, J. R., and Zeng, X. C. (2005). Melting temperature of ice Ih calculated from coexisting solid–liquid phases. *J. Chem. Phys.* **123**, 036101.

Wang, J. M., Cieplak, P., and Kollman, P. A. (2000). How well does a restrained electrostatic potential (RESP) model perform in calculating conformational energies of organic and biological molecules? *J. Comput. Chem.* **21**, 1049–1074.

Wang, J. M., Wolf, R. M., Caldwell, J. W., Kollman, P. A., and Case, D. A. (2004). Development and testing of a general AMBER force field. *J. Comput. Chem.* **25**, 1157–1174.

Wang, L., Friesner, R. A., and Berne, B. J. (2011). Replica exchange with solute scaling: a more efficient version of replica exchange with solute tempering (REST2). *J. Phys. Chem. B* **115**, 9431–9438.

Wang, S. S. and Krumhansl, J. A. (1972). Superposition assumption. 2. High-density fluid argon. *J. Chem. Phys.* **56**, 4287–4290.

Wang, Y., Izvekov, S., Yan, T., and Voth, G. A. (2006). Multiscale coarse-graining of ionic liquids. *J. Phys. Chem. B* **110**, 3564–3575.

Warren, P. B. (1998). Combinatorial entropy and the statistical mechanics of polydispersity. *Phys. Rev. Lett.* **80**, 1369–1372. Erratum: *ibid.* **80**, 3671 (1998).

Warshel, A. and Levitt, M. (1976). Theoretical studies of enzymic reactions: dielectric, electrostatic and steric stabilization of the carbonium ion in the reaction of lysozyme. *J. Molec. Biol.* **103**, 227–249.

Warshel, A., Kato, M., and Pisliakov, A. V. (2007). Polarizable force fields: history, test cases, and prospects. *J. Chem. Theor. Comput.* **3**, 2034–2045.

Watanabe, H., Ito, N., and Hu, C.-K. (2012). Phase diagram and universality of the Lennard-Jones gas–liquid system. *J. Chem. Phys.* **136**, 204102.

Watanabe, K. and Klein, M. L. (1989). Effective pair potentials and the properties of water. *Chem. Phys.* **131**, 157–167.

Watanabe, N. and Tsukada, M. (2002). Efficient method for simulating quantum electron dynamics under the time-dependent Kohn–Sham equation. *Phys. Rev. E* **65**, 036705.

Watson, M. C., Brandt, E. G., Welch, P. M., and Brown, F. L. H. (2012). Determining biomembrane bending rigidities from simulations of modest size. *Phys. Rev. Lett.* **109**, 028102.

Weber, H., Marx, D., and Binder, K. (1995). Melting transition in 2 dimensions: a finite-size scaling analysis of bond-orientational order in hard disks. *Phys. Rev. B* **51**, 14636–14651.

Weber, T. A. and Stillinger, F. H. (1981). Gaussian core model in two dimensions. II. Solid and fluid phase topological distribution functions. *J. Chem. Phys.* **74**, 4020–4028.

Weeks, J. D., Chandler, D., and Andersen, H. C. (1971). Role of repulsive forces in determining the equilibrium structure of simple liquids. *J. Chem. Phys.* **54**, 5237–5247.

Weerasinghe, S. and Pettitt, B. M. (1994). Ideal chemical potential contribution in molecular dynamics simulations of the grand canonical ensemble. *Molec. Phys.* **82**, 897–912.

Weinbach, Y. and Elber, R. (2005). Revisiting and parallelizing SHAKE. *J. Comput. Phys.* **209**, 193–206.

Welling, U. and Germano, G. (2011). Efficiency of linked cell algorithms. *Comput. Phys. Commun.* **182**, 611–615.

Werner, A., Schmid, F., Müller, M., and Binder, K. (1999). 'Intrinsic' profiles and capillary waves at homopolymer interfaces: a Monte Carlo study. *Phys. Rev. E* **59**, 728–738.

Wertheim, M. S. (1984a). Fluids with highly directional attractive forces. I. Statistical thermodynamics. *J. Stat. Phys.* **35**, 19–34.

Wertheim, M. S. (1984b). Fluids with highly directional attractive forces. II. Thermodynamic perturbation theory and integral equations. *J. Stat. Phys.* **35**, 35–47.

Wertheim, M. S. (1986a). Fluids with highly directional attractive forces. III. Multiple attraction sites. *J. Stat. Phys.* **42**, 459–476.

Wertheim, M. S. (1986b). Fluids with highly directional attractive forces. IV. Equilibrium polymerization. *J. Stat. Phys.* **42**, 477–492.

Wheatley, R. J. (2013). Calculation of high-order virial coefficients with applications to hard and soft spheres. *Phys. Rev. Lett.* **110**, 200601.

White, G. W. N., Goldman, S., and Gray, C. G. (2000). Test of rate theory transmission coefficient algorithms. An application to ion channels. *Molec. Phys.* **98**, 1871–1885.

Whitehouse, J. S., Nicholson, D., and Parsonage, N. G. (1983). A grand ensemble Monte Carlo study of krypton adsorbed on graphite. *Molec. Phys.* **49**, 829–847.

Widom, B. (1963). Some topics in the theory of fluids. *J. Chem. Phys.* **39**, 2808–2812.

Widom, B. (1982). Potential distribution theory and the statistical mechanics of fluids. *J. Phys. Chem.* **86**, 869–872.

Wierzchowski, S. and Kofke, D. A. (2001). A general-purpose biasing scheme for Monte Carlo simulation of associating fluids. *J. Chem. Phys.* **114**, 8752–8762.

Wigner, E. (1932). On the quantum correction for thermodynamic equilibrium. *Phys. Rev.* **40**, 749–759.

Wilding, N. B. (2009). Solid–liquid coexistence of polydisperse fluids via simulation. *J. Chem. Phys.* **130**, 104103.

Wilding, N. B. and Bruce, A. D. (2000). Freezing by Monte Carlo phase switch. *Phys. Rev. Lett.* **85**, 5138–5141.

Wilding, N. B. and Sollich, P. (2004). Phase equilibria and fractionation in a polydisperse fluid. *Europhys. Lett.* **67**, 219–225.

Wilding, N. B. (2016). Improved grand canonical sampling of vapour–liquid transitions. *J. Phys. Cond. Mat.* **28**, 414016.

Willemsen, S. M., Vlugt, T. J. H., Hoefsloot, H. C. J., and Smit, B. (1998). Combining dissipative particle dynamics and Monte Carlo techniques. *J. Comput. Phys.* **147**, 507–517.

Williams, D. E. (1971). Accelerated convergence of crystal-lattice potential sums. *Acta Cryst. A* **27**, 452–455.

Wilson, B. A., Gelb, L. D., and Nielsen, S. O. (2015). Nested sampling of isobaric phase space for the direct evaluation of the isothermal–isobaric partition function of atomic systems. *J. Chem. Phys.* **143**, 154108.

Wilson, M. and Madden, P. A. (1993). Polarization effects in ionic systems from 1st principles. *J. Phys. Cond. Mat.* **5**, 2687–2706.

Wilson, M. R. (2005). Progress in computer simulations of liquid crystals. *Int. Rev. Phys. Chem.* **24**, 421–455.

Winkler, R. G., Singh, S. P., Huang, C.-C., Fedosov, D. A., Mussawisade, K., Chatterji, A., Ripoll, M., and Gompper, G. (2013). Mesoscale hydrodynamics simulations of particle suspensions under shear flow: from hard to ultrasoft colloids. *Euro. Phys. J. Spec. Top.* **222**, 2773–2786.

Witt, A., Ivanov, S. D., Shiga, M., Forbert, H., and Marx, D. (2009). On the applicability of centroid and ring polymer path integral molecular dynamics for vibrational spectroscopy. *J. Chem. Phys.* **130**, 194510.

Wojcik, M. and Gubbins, K. E. (1983). Thermodynamics of hard dumbbell mixtures. *Molec. Phys.* **49**, 1401–1415.

Wolf, D., Keblinski, P., Phillpot, S. R., and Eggebrecht, J. (1999). Exact method for the simulation of Coulombic systems by spherically truncated, pairwise r^{-1} summation. *J. Chem. Phys.* **110**, 8254–8282.

Wolf-Gladrow, D. A. (2000). *Lattice-gas cellular automata and lattice Boltzmann models: an introduction*. Lect. Notes Math. Vol. 1725. Springer.

Wolfsheimer, S., Tanase, C., Shundyak, K., van Roij, R., and Schilling, T. (2006). Isotropic–nematic interface in suspensions of hard rods: mean-field properties and capillary waves. *Phys. Rev. E* **73**, 061703.

Wood, W. W. (1968a). Monte Carlo calculations for hard disks in the isothermal–isobaric ensemble. *J. Chem. Phys.* **48**, 415–434.

Wood, W. W. (1968b). Monte Carlo studies of simple liquid models. *Physics of simple liquids*. Chap. 5, 115–230. Ed. by H. N. V. Temperley, J. S. Rowlinson, and G. S. Rushbrooke. Amsterdam: North Holland.

Wood, W. W. (1970). NpT-ensemble Monte Carlo calculations for the hard disk fluid. *J. Chem. Phys.* **52**, 729–741.

Wood, W. W. (1986). Early history of computer simulation in statistical mechanics. *Molecular dynamics simulation of statistical mechanical systems*. Proc. Int. School Phys. Enrico Fermi. **97**, 2–14. (Varenna, Italy, 1985). Ed. by G. Ciccotti and W. G. Hoover. Bologna: Soc. Italiana di Fisica.

Wood, W. W. and Jacobson, J. D. (1957). Preliminary results from a recalculation of the Monte Carlo equation of state of hard spheres. *J. Chem. Phys.* **27**, 1207–1208.

Wood, W. W. and Jacobson, J. D. (1959). Monte Carlo calculations in statistical mechanics. *Proceedings of the western joint computer conference*, 261–269. (San Francisco, USA, 1959).

Wood, W. W. and Parker, F. R. (1957). Monte Carlo equation of state of molecules interacting with the Lennard-Jones potential. I. A supercritical isotherm at about twice the critical temperature. *J. Chem. Phys.* **27**, 720–733.

Woodcock, L. V. (1971). Isothermal molecular dynamics calculations for liquid salts. *Chem. Phys. Lett.* **10**, 257–261.

Wright, L. B. and Walsh, T. R. (2013). Efficient conformational sampling of peptides adsorbed onto inorganic surfaces: insights from a quartz binding peptide. *Phys. Chem. Chem. Phys.* **15**, 4715–4726.

Wright, L. B., Rodger, P. M., Walsh, T. R., and Corni, S. (2013a). First-principles-based force field for the interaction of proteins with Au(100)(5×1): an extension of GolP-CHARMM. *J. Phys. Chem. C* **117**, 24292–24306.

Wright, L. B., Rodger, P. M., Corni, S., and Walsh, T. R. (2013b). GolP–CHARMM: first-principles based force fields for the interaction of proteins with Au(111) and Au(100). *J. Chem. Theor. Comput.* **9**, 1616–1630.

Wright, L. B., Palafox-Hernandez, J. P., Rodger, P. M., Corni, S., and Walsh, T. R. (2015). Facet selectivity in gold binding peptides: exploiting interfacial water structure. *Chem. Sci.* **6**, 5204–5214.

Wu, D. and Kofke, D. A. (2004). Asymmetric bias in free-energy perturbation measurements using two Hamiltonian-based models. *Phys. Rev. E* **70**, 066702.

Wu, D. and Kofke, D. A. (2005). Phase-space overlap measures. I. Fail-safe bias detection in free energies calculated by molecular simulation. *J. Chem. Phys.* **123**, 054103.

Wu, X. W. and Brooks, B. R. (2005). Isotropic periodic sum: a method for the calculation of long-range interactions. *J. Chem. Phys.* **122**, 044107.

Wu, X. W. and Brooks, B. R. (2008). Using the isotropic periodic sum method to calculate long-range interactions of heterogeneous systems. *J. Chem. Phys.* **129**, 154115.

Xia, J., Flynn, W. F., Gallicchio, E., Zhang, B. W., He, P., Tan, Z., and Levy, R. M. (2015). Large-scale asynchronous and distributed multidimensional replica exchange molecular simulations and efficiency analysis. *J. Comput. Chem.* **36**, 1772–1785.

Yamakawa, H. (1970). Transport properties of polymer chains in dilute solution: hydrodynamic interaction. *J. Chem. Phys.* **53**, 436–443.

Yan, Q. and de Pablo, J. J. (1999). Hyper-parallel tempering Monte Carlo: Application to the Lennard-Jones fluid and the restricted primitive model. *J. Chem. Phys.* **111**, 9509–9516.

Yao, J., Greenkorn, R. A., and Chao, K. C. (1982). Monte Carlo simulation of the grand canonical ensemble. *Molec. Phys.* **46**, 587–594.

Yao, Z. H., Wang, H. S., Liu, G. R., and Cheng, M. (2004). Improved neighbor list algorithm in molecular simulations using cell decomposition and data sorting method. *Comput. Phys. Commun.* **161**, 27–35.

Yee, K. S. (1966). Numerical solution of initial boundary value problems involving Maxwell's equations in isotropic media. *IEEE Trans. Ant. Prop.* **14**, 302–307.

Yeh, I. C. and Berkowitz, M. L. (1999). Ewald summation for systems with slab geometry. *J. Chem. Phys.* **111**, 3155–3162.

Yi, P. and Rutledge, G. C. (2012). Molecular origins of homogeneous crystal nucleation. *Ann. Rev. Chem. Bio. Eng.* **3**, 157–182.

Yoo, S., Zeng, X. C., and Morris, J. R. (2004). The melting lines of model silicon calculated from coexisting solid–liquid phases. *J. Chem. Phys.* **120**, 1654–1656.

Young, A. P. (1979). Melting and the vector Coulomb gas in two dimensions. *Phys. Rev. B* **19**, 1855–1866.

Yu, H. and van Gunsteren, W. F. (2005). Accounting for polarization in molecular simulation. *Comput. Phys. Commun.* **172**, 69–85.

Yu, T.-Q., Alejandre, J., López-Rendón, R., Martyna, G. J., and Tuckerman, M. E. (2010). Measure-preserving integrators for molecular dynamics in the isothermal–isobaric ensemble derived from the Liouville operator. *Chem. Phys.* **370**, 294–305.

Zamuner, S., Rodriguez, A., Seno, F., and Trovato, A. (2015). An efficient algorithm to perform local concerted movements of a chain molecule. *PLOS ONE* **10**, e0118342.

Zannoni, C. (1979). Computer simulation. *The molecular physics of liquid crystals*. Chap. 9, 191–220. Ed. by G. R. Luckhurst and G. W. Gray. London: Academic Press.

Zannoni, C. (2001). Molecular design and computer simulations of novel mesophases. *J. Mater. Chem.* **11**, 2637–2646.

Zannoni, C. (1994). An introduction to the molecular dynamics method and to orientational dynamics in liquid crystals. *The molecular dynamics of liquid crystals.* NATO ASI Series C. **431**, 139–164. (Il Ciocco, Italy, 1989). Ed. by G. R. Luckhurst and C. A. Veracini. Dordrecht: Kluwer.

Zannoni, C. (2000). Liquid crystal observables: static and dynamic properties. *Advances in the computer simulations of liquid crystals.* NATO Science Series C. **545**, 17–50. (Erice, Italy, 1998). Ed. by P. Pasini and C. Zannoni. Dordrecht: Kluwer.

Zeidler, A., Salmon, P. S., Fischer, H. E., Neuefeind, J. C., Simonson, J. M., and Markland, T. E. (2012). Isotope effects in water as investigated by neutron diffraction and path integral molecular dynamics. *J. Phys. Cond. Mat.* **24**, 284126.

Zhang, C., Donadio, D., and Galli, G. (2010). First-principle analysis of the IR stretching band of liquid water. *J. Phys. Chem. Lett.* **1**, 1398–1402.

Zhang, C., Donadio, D., Gygi, F., and Galli, G. (2011a). First principles simulations of the infrared spectrum of liquid water using hybrid density functionals. *J. Chem. Theor. Comput.* **7**, 1443–1449.

Zhang, C., Wu, J., Galli, G., and Gygi, F. (2011b). Structural and vibrational properties of liquid water from van der Waals density functionals. *J. Chem. Theor. Comput.* **7**, 3054–3061.

Zhang, F. (1997). Operator-splitting integrators for constant-temperature molecular dynamics. *J. Chem. Phys.* **106**, 6102–6106.

Zhang, M., Lussetti, E., de Souza, L. E. S., and Müller-Plathe, F. (2005). Thermal conductivities of molecular liquids by reverse nonequilibrium molecular dynamics. *J. Phys. Chem. B* **109**, 15060–15067.

Zhang, Y. H., Feller, S. E., Brooks, B. R., and Pastor, R. W. (1995). Computer simulation of liquid/liquid interfaces. 1. Theory and application to octane/water. *J. Chem. Phys.* **103**, 10252–10266.

Zhang, Y. and Maginn, E. J. (2012). A comparison of methods for melting point calculation using molecular dynamics simulations. *J. Chem. Phys.* **136**, 144116.

Zheng, W., Rohrdanz, M. A., Maggioni, M., and Clementi, C. (2011). Polymer reversal rate calculated via locally scaled diffusion map. *J. Chem. Phys.* **134**, 144109.

Zillich, R. E., Mayrhofer, J. M., and Chin, S. A. (2010). Extrapolated high-order propagators for path integral Monte Carlo simulations. *J. Chem. Phys.* **132**, 044103.

Zwanzig, R. (1960). Ensemble method in the theory of irreversibility. *J. Chem. Phys.* **33**, 1338–1341.

Zwanzig, R. (1961a). Memory effects in irreversible thermodynamics. *Phys. Rev.* **124**, 983–992.

Zwanzig, R. (1961b). Statistical mechanics of irreversibility. *Lectures in theoretical physics.* **3**, 106–141. Ed. by W. E. Brittin, B. W. Downs, and J. Downs. New York: Interscience.

Zwanzig, R. (1965). Time correlation functions and transport coefficients in statistical mechanics. *Ann. Rev. Phys. Chem.* **16**, 67–102.

Zwanzig, R. (1969). Langevin theory of polymer dynamics in dilute solution. *Adv. Chem. Phys.* **15**, 325–331.

Zwanzig, R. and Ailawadi, N. K. (1969). Statistical error due to finite time averaging in computer experiments. *Phys. Rev.* **182**, 280–283.

Zykova-Timan, T., Rozas, R. E., Horbach, J., and Binder, K. (2009). Computer simulation studies of finite-size broadening of solid–liquid interfaces: from hard spheres to nickel. *J. Phys. Cond. Mat.* **21**, 464102.

Zykova-Timan, T., Horbach, J., and Binder, K. (2010). Monte Carlo simulations of the solid–liquid transition in hard spheres and colloid–polymer mixtures. *J. Chem. Phys.* **133**, 014705.

Index